TAXONOMY OF ANGIOSPERMS

BP PANDEY
MSc, PhD, FPSI
Former Head
Department of Botany
J.V. College, Baraut
Uttar Pradesh

S. CHAND
PUBLISHING

S Chand And Company Limited
(ISO 9001 Certified Company)

S Chand And Company Limited

(ISO 9001 Certified Company)

Head Office: D-92, Sector–2, Noida – 201301, U.P. (India), Ph. 91-120-4682700

Registered Office: A-27, 2nd Floor, Mohan Co-operative Industrial Estate, New Delhi – 110 044, Phone: 011-49731800

www.**schandpublishing.com;** e-mail: **info@schandpublishing.com**

Marketing Offices:

Chennai : Ph: 23632120; chennai@schandpublishing.com
Guwahati : Ph: 2738811, 2735640; guwahati@schandpublishing.com
Hyderabad : Ph: 40186018; hyderabad@schandpublishing.com
Jalandhar : Ph: 4645630; jalandhar@schandpublishing.com
Kolkata : Ph: 23357458, 23353914; kolkata@schandpublishing.com
Lucknow : Ph: 4003633; lucknow@schandpublishing.com
Mumbai : Ph: 25000297; mumbai@schandpublishing.com
Patna : Ph: 2260011; patna@schandpublishing.com

© S Chand And Company Limited, 1969

All rights reserved. No part of this publication may be reproduced or copied in any material form (including photocopying or storing it in any medium in form of graphics, electronic or mechanical means and whether or not transient or incidental to some other use of this publication) without written permission of the copyright owner. Any breach of this will entail legal action and prosecution without further notice.

Jurisdiction: *All disputes with respect to this publication shall be subject to the jurisdiction of the Courts, Tribunals and Forums of New Delhi, India only.*

First Edition 1969

Subsequent Editions and Reprints 1971, 76, 79, 81, 82, 89, 92, 93, 95, 97, 99, 2001, 2003, 2005, 2007, 2009, 2010, 2012, 2013, 2014, 2015, 2016, 2017, 2019, 2020 (Twice), 2021, 2022 (Twice), 2023

Reprint 2025

ISBN: 978-81-219-0932-7 **Product Code:** H6TAN68BOTN10ENAF0XO

PRINTED IN INDIA

By Vikas Publishing House Private Limited, Plot 20/4, Site-IV, Industrial Area Sahibabad, Ghaziabad – 201 010 and Published by S Chand And Company Limited, A-27, 2nd Floor, Mohan Co-operative Industrial Estate, New Delhi – 110 044.

PREFACE TO THE SIXTH EDITION

This sixth edition of 'Taxonomy of Angiosperms' has been thoroughly revised. In this edition, the portion of the general account of the book has been revised, enlarged and supplemented with new topics, such as alternation of generations, chemotaxonomy, floral anatomy, biosytematics, anatomy of leaf, embryology *vs* taxonomy, epidermal emergences, ontogenetical, palaeontological and palynological parameters, serotaxonomy, and life-cycle of angiosperms. Several new genera, comparative account of the systems of classification, principles of phylogeny, characteristics of the orders and an elaborate account of botanic gardens have been added. The orders Boraginales, Loganiales, Lythrales, Sarraceniales, Podostemales, and the families Boraginaceae, Droseraceae, Lythraceae, Onagraceae, Oleaceae, Lentibulariaceae and Podostemaceae have been added to cope with the study of taxonomy of post-graduate and honours students, and therefore, the utility of the book has been increased a lot to the students of taxonomy.

I am greatly indebted to the works of numerous authors from which the substance of present text has been obtained. My sincere thanks are due to my colleagues, friends, botanists and educationists who have accorded their full co-operation and well wishes during the preparation of the manuscript of this edition.

I am also thankful to my publisher Sri Ravindra Kumar Gupta and Sri T.N. Goel, who inspired me for the task, and brought the book in this fine get-up.

B.P. Pandey

'Neelkanth Kuteer'
Subhash Nagar
BARAUT-250 611

PREFACE TO THE FIRST EDITION

The present text, 'Taxonomy of Angiosperms' has been written according to the syllabi of degree classes of all Indian Universities. In all, sixty-five families have been described in detail, of which fortynine are from Dicots and sixteen from Monocots. The families have been arranged according to the phylogenetic system of classification propounded by the eminent taxonomist John Hutchinson. He has divided the phylum Angiospermae into two sub phyla, *i.e.*, 1. Dicotyledones and 2. Monocotyledones. The sub-phylum Dicotyledones has been further divided into two divisions, *i.e.*, 1. Lignosae and 2. Herbaceae groups. The sub-phylum Monocotyledones has been divided into three divisions, *i.e.*, Calyciferae, 2. Corolliferae and 3. Glumiflorae. The flowers of group Calyciferae possess distinct calyx and corolla; the flowers of group Corolliferae possess more or less similar calyx and corolla; and the flowers of group Glumiflorae possess much reduced perianth or lodicules. The first seven chapters of the book deal with the fundamentals of the study of Systematic Botany. The important topics such as, Evolution of Classification; Principles of Taxonomy and Phylogeny of Angiosperms; Binomial system of Nomenclature; Plant collection and Preparation of Herbarium; How to describe an Angiospermic Plant, etc., have been treated in detail. I hope, the present text will meet all the requirements of the study of Systematic Botany both in theory and practicals. Though the book has been written for the students of degree classes of all Indian Universities, yet the students of B.Sc. (Hons.), postgraduate diploma, and postgraduate classes, may derive equal benefit from it.

Several plants found around us have been described in semitechnical language. The descriptions of the various families and genera have been supported with more than two hundred illustrations. The subject matter and many illustrations have been taken from many journals, floras and other authentic sources. Most of the diagrams have been drawn by the author himself from natural specimens and dissected flowers.

In conclusion, the author desires to express his gratitude to Dr. M.P. Kaushik, Professor and Head of the Department of Botany, D.A.V. College, Muzaffarnagar, who has given every kind of help, inspiration and valuable suggestions from time to time during the preparation of the manuscript. My sincere thanks are also due to Prof. S.D. Sharma, D.A.V. College, Dehradun, for the help he sponsored during the preparation of the book.

The criticisms and suggestions for further improvement of the book will be most welcomed and thankfully accepted.

Department of Botany,
Janta Vedic College,
Baraut.

B.P. Pandey

CONTENTS

THE MONOCOTYLEDONES

1

INTRODUCTION

What is Systematic Botany? Systematic botany is the science which deals with the classification and naming of the plants. The science of classifying the plants is said to be *plant taxonomy* and lays emphasis upon phylogenetic relationships. The naming of the plants is known as *nomenclature* and provides each plant with a name. This way, the systematic botany consists of classifying and naming of the plants. The important task included in the field of systematic botany is the collection of the plant specimens from all over the world and building the great herbaria. The plants are being classified chiefly on the basis of the study of their comparative morphology. In modern days the study of systematic botany needs a good background in general botany, cytology, genetics, ecology, plant geography and palaeobotany; otherwise, the origin of the plants cannot be traced out.

Nomenclature makes the important part of systematic botany. It deals with names which may or may not indicate relationships. The science of plant taxonomy classify the plants on the basis of similarities and differences, which are now known as "phylogenetic relationships". Without naming the plants, they cannot be classified and therefore, nomenclature makes a very important part of systematic botany.

Ecology and systematic botany are interrelated to each other. Certain species are definitely hydrophytic, xerophytic, mesophytic or halophytic. The soil characters are being indicated by the vegetation growing upon. The character of the soil may be indicated by its vegetation.

Scope of Systematic Botany As already mentioned, the study of systematic botany consists of two important steps–(1) plant taxonomy–which establishes the phylogenetic relationships that exist naturally between many groups of plants, (2) nomenclature, *i.e.*, giving of names to all kinds of plants. There are millions of plants all over the world, and it is very clear, that no two individuals are *exactly* alike. Most of the individuals are so different from each other that hardly they show any bond of relationship. On the basis of the comparative study of their morphology, the taxonomists have arranged the individual plants in a fairly orderly system. Thousands of specimens have been provided with their names and descriptions. The group relationships, distribution, properties, etc., of the plants have been studied. This way, systematic botany prepares foundation for all sciences concerning plants.

The systematic botany is of great value in forestry. Every type of tree of the forest must be named and classified. At the same time, the characteristic features, distribution and abundance of the plants must be learned.

The study of systematic botany is equally essential in agriculture and horticulture. To get resistant varieties etc., the crossing is needed, which is furnished by foreign seeds and plants. The foreign seed and plant are introduced by the men who have sufficient knowledge of systematic botany.

The work of soil conservation needs the good study of systematic botany. The ecologists who are well trained in systematic botany choose the native grasses and plants which act as sand soil binders.

To study the plant ecology, the knowledge of systematic botany becomes essential. The plant ecologist must be knowing the names of the plants and their relationships to habitat and environment. The character of the soil of a land area may easily be judged by the vegetation growing there.

The study of systematic botany is of great use in tracing the origin of plants. The fossils, scattered here and there in fragments are found to be very useful in tracing out the characters of the ancestral forms of the plants. The study of these fossils requires the sound knowledge of systematic botany.

Objectives of Systematic Botany– The study of systematic botany fulfils four main objectives. They are as follows :-

1. The first objective of systematic botany is to know the various kinds of plants on the surface of earth with their names, affinities, geographical distribution, habit characteristics and their economic importance.

2. The second objective of this science is to have a reference system for plants, where the scientists can work with named entities such as species, genus and family.

3. The third objective is to demonstrate the manifold diversities of plant kingdom and their relation to evolution of plants. A systematic reconstruction of plant kingdom can be made only after the complete knowledge of the individual plants. After making an inventory of the components of plant kingdom, the different facts of evolutionary knowledge with an accurate phylogenetic scheme can be obtained.

4. The fourth objective is to ascertain nomenclature. To every plant a binomial name is given, *e.g., Solanum tuberosum* Linn. The first name refers to its genus, the second to its species and Linn., or L. for Linnaeus, who first observed and reported the plant.

ANGIOSPERMS OR FLOWERING PLANTS

Angiosperms or flowering plants form the largest group of plant kingdom, including about 300 families (411 families, Hutchinson), 8,000 genera and 300,000 species. They are considered to be highest evolved plants on the surface of the earth. From Cretaceous age, the angiosperms eclipsed all other vegetation and now they are dominant. They are found almost every where in each possible type of habitat and climate. They occur in deep lakes, deserts, in beds of seas and even on high peaks of mountains. The species of *Opuntia* (Cactaceae) can survive without water in acute desert conditions, whereas on the other hand the species of *Hydrilla* (aquatic plant) are extremely sensitive to drought conditions. Some species are found on rocks, some in waterfalls and also some are marine. The species of *Rhizophora*, popularly known as 'mangrove vegetation' are found near the water of the sea. The epiphytes, parasites, saprophytes, symbionts and even insectivorous plants are also not uncommon.

They may be annual, biennial or perennial herbs, shrubs, trees, climbers, twiners and lianes. On one hand the angiosperms may be as minute in size as a pin head, *e.g., Wolffia microscopica*, on the other extremity like eucleptiles of Australia may reach upto 300 feet in height.

ALTERNATION OF GENERATIONS (LIFE CYCLE OF ANGIOSPERMS)

In the life-cycle of angiosperms, there is alternation of nutritionally independent and more complex sporophyte with the inconspicuous, reduced and parasitic gametophytes. The sporophyte, which may be a *herb, shrub* or a *tree* is differentiated into roots, stem and leaves each with a vascular tissue with the highest degree of perfection. In addition, it bears floral leaves or sporophylls organised into a structre called the *flower.* The sporophylls are of two kinds : *microsporophylls* and *megasporophylls*. Each microsporophyll bears four *microsporangia*. The latter contains *microspores* which are differentiated by meiosis from the diploid *microspore mother cells.* The megasporophyll in a flower part, which is called the *ovary*, contains one or more *ovules*. The ovule consists of one or two integuments enclosing the *nucellus* or *megasporangium.* Towards the micropylar end of the megasporangium is differentiated, by meiosis, a single functional *megaspore* which is haploid. The *microspores* and *megaspores* are the first structures of the gametophyte generation. The *sporophyte generation* in the angiosperms, thus consists of the sporophyte plant, flowers, microsporophylls and megasporophylls, microsporangia, megasporangia, microspore mother cells and the megaspore mother cells.

With meiosis the sporophyte phase switches on the gametophyte phase. The first structures of the gametophyte phase are the *microspores* and the *megaspores.* On germination, they produce the alternate structures in the life-cycle which are the male and the female gametophytes and not the sporophyte plant. The male and the female gametophytes are extremely reduced and parasitic. The female gametophyte consists only of six cells and two nuclei. Three cells at the micropylar end form the *egg apparatus.*The other three at the chalazal end form the *antipodal group.* There are two *polar nuclei* in the centre. The male gametophyte consists of a pollen tube containing the *tube nucleus* and

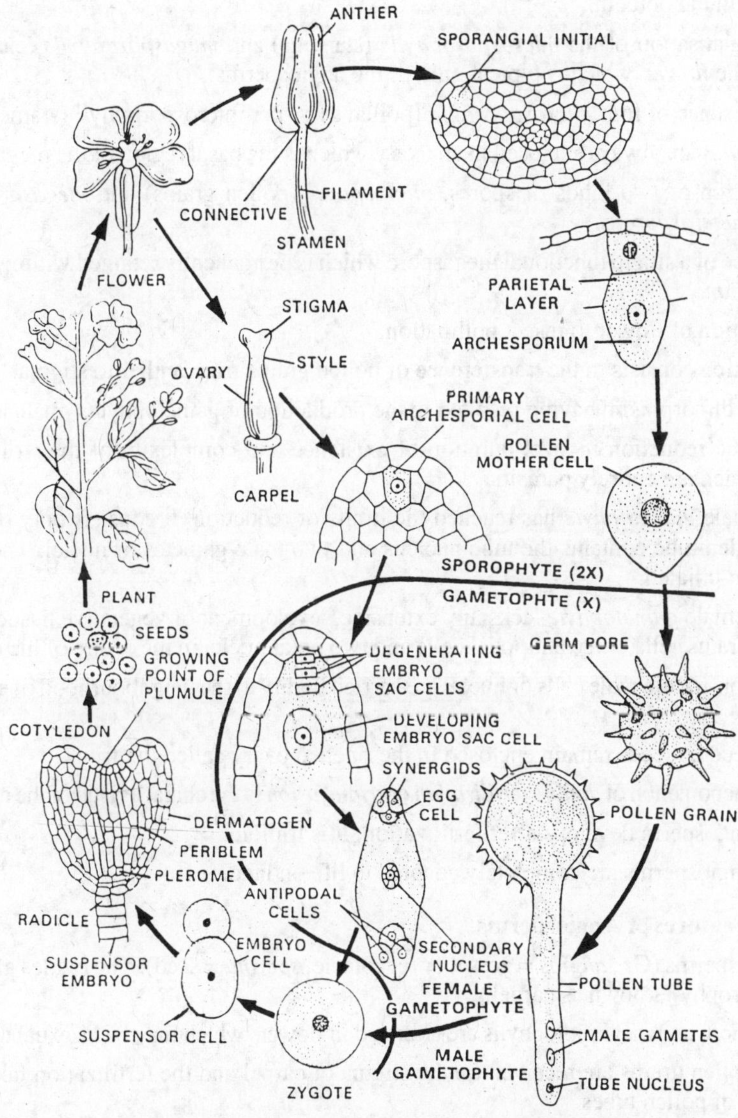

Fig 1.1. Life-history of a typical angiosperm.

two *male nuclei.* The eggs and the male nuclei represent the gametes which are the last structures of the gametophyte phase. With the fusion of one of the male nuclei with the egg, the gametophyte phase ends. *Fertilization* thus is the second critical point in the life cycle. With it the gametophyte generation switches on to the sporophyte generation. The fertilized egg or *oospore* is the first structure of the next sporophyte. By segmentation it produces the embryo (*i.e.*, the baby plant in the seed); the ovule to the seed and the ovary as a whole to the fruit. The embryo lies dormant in the seed and the latter lies embedded in the fruit. The seed and the fruit give adequate protection to the embryo, store up food material for it, and are often well adapted for dispersal. Sooner or later as the seed germinates the embryo grows into a *seedling* which gradually grows into a mature plant.

Salient Features of Angiosperms

1. The sporophyte which is the dominant plant in the life-cycle is differentiated into roots, stem and leaves.

2. The highest degree of perfection of the vascular system with true vessels in the xylem and companion cells in the phloem.

3. The organisation of the *microsporophylls* (stamens) and *megasporophylls* (carpels) into a structure called the *flower,* which is typical only of the angiosperms.

4. The presence of four *microsporangia* (pollen sacs) per microsporophyll (stamen).

5. The *ovules* are always enclosed in an ovary which is the basal region of the megasporophyll.

6. Production of two kinds of spores, *microspores* (pollen grains) and *megaspores.* Angiosperms thus are *heterosporous.*

7. Presence of a single functional megaspore which is permanently retained within the *nucellus* or *megasporangium.*

8. Adaptation of flower to insect pollination.

9. Pollination consists in the transference of pollen grains from anther to stigma.

10. Spore dimorphisim having resulted in the production of gametophytes, male and female.

11. Extreme reduction in size, duration of existence and complexity of the structure of the gametophytes which are entirely parasitic.

12. The male *gametophyte* has reached the limits of reduction. It consists only of the pollen grain and the pollen tube contains the tube nucleus and two male gametes or nuclei. The male cells (gametes) are non-ciliated.

13. The female *gametophyte* lacks any extensive development of vegetative tissue. It consists of three egg apparatus cells, three antipodal cells and two polar nuclei in the centre of the embryo sac.

14. The non-motile male cells or nuclei are carried bodily to the neighbourhood of egg appratus by the pollen tube.

15. The seed or seeds remain enclosed in the ripened ovary called the fruit.

16. The phenomenon of *double fertilization* or *triple fusion* is the characteristic of the angiosperms.

17. The endosperm develops after fertilization. It is triploid.

18. The angiosperms are completely adapted to life on land.

Characteristic Features of Angiosperms

(1) Angiosperms (Gr., *angios* = vessel or receptacle; *sperma* = seed). The ovules always borne in the closed sporophyll known as carpel.

(2) The micro and megasporophylls are arranged in flowers which may be bisexual or unisexual.

(3) The pollen grains are received on the stigma of carpel and the fertilization takes place by the development of pollen tubes.

(4) The two male gametes are without flagella, one unites with egg and the other with secondary nucleus.

Taxa or Taxon – Taxonomic group of any rank or unit, *viz.*, family, class, order, genus, species, etc., is called *taxon* (plural : *taxa*). For example, the genus *Solanum* (the name of the genus begins always with a capital letter) consists of many species (the name of the species begins always with a small letter) such as *tuberosum, melongena, nigrum*, etc., but certain common characters of the genus *Solanum* can separate this genus from all other genera of the family Solanaceae.

Higher to genus is the family, where one or many genera are correlated between greater number of morphological characters are grouped together. The name of the family ends in *aceae, e.g.*, Brassicaeae, Poaceae, etc.

The family represents a more natural *taxon* or unit than any of the higher categories, e.g., Brassicaceae, Palmaceae, Solanaceae, Rosaceae are identified as natural ***taxa*** by certain characters peculiar to them. The angiospermic families may be separated from one another by floral characters, inflorescence, type, nature of perianth, ovary position, number of carpels, placentation, etc.

The families which are allied to one another are grouped into an order which makes a natural taxon. For example, the order Ranales comprises of Ranunculaceae, Nymphaeaceae, etc. Here the families are related to each other in general floral constitution.

What is a Species

According to new definition the species are biological units that have evolved from a series of often identifiable ancestors, by which the smaller taxa of plants (genera, species, and their subdivisions) may be circumscribed by a combination of genetical, morphological and ecological criteria. According to this view, a species is considered as an objective definitive unit.

TAXONOMY AND ITS SIGNIFICANCE

The study of taxonomy has among its objectives the learning of the kinds of plants on earth and their names, of their distinctions and their affinities, their distributions and characteristics of habitats, and the correlation of these facets of knowledge with pertinent scientific data. A secondary objective of taxonomy is the assemblage of knowledge gained. Floras are published to account for the plants of a given area.

All the products of taxonomic research add to the resources available to the scientist. They are essential to any study of the natural resources of an area, to studies of land potentials, to evaluation of resources of raw materials possible suited to man's needs (for example, forest products, medicines food, agricultural crops, ornamentals and industry). A third important and scientific objective is the demonstration of the tremendous diversity of the plants and their relation to man's understanding of evolution.

Interrelationships with allied sciences. Taxonomy is dependent on many other sciences and they in turn are equally dependent on it. One who studies taxonomy must have a knowledge of morphology, embryology, floral anatomy, ontogenetical development and teratological variations of the plants with which he works. Modern taxonomists place considerable value on the importance of cytogenetic findings as criteria in delimiting the species and its elements. In addition to an appreciation and understanding of the contributory value of morphological, anatomical and cytogenetical findings, modern taxonomic studies reflect the significance of distributional patterns and of more detailed data of the extent of normal variation and its causes. All these viewpoints demonstrate the increasing dependence of taxonomy on the findings of related sciences; the product of modern taxonomic research is rapidly becoming one of synthesis rather than of individual conclusions.

Problems in taxonomy: During the eighteenth and early nineteenth centuries the study of taxonomy dominated the field of botanical activity. Today the taxonomist is interested in the problems associated with the distribution of the plants. Knowledge of plant distribution is relevant to the determination of gegraphic areas of origins of species, of genera, and often of families. These studies in distribution and geography bring taxonomy into the field of phytogeography, the inquiry into why a group occupies that area, how long it has been there, how rapidly it is migrating, and what evolutionary trends it is showing. Studies with this wider viewpoint represent a synthesis of ecologic, genetic and taxonomic aspects leading to a better understanding of a series of common problems.

ORIGIN OF ANGIOSPERMS

The angiosperms appeared suddenly in Cretaceous age about 65 million years back. **Charles Darwin** described this sudden appearance of angiosperms in lower or upper Cretaceous as an '*abominable mystery*'. When angiosperms appeared for the first time in lower or upper Cretaceous, they were full fledged like the trees and the herbs of today. In support of this view *Prof. Knowlton* advocates in his '*Plant of the past*', "*from the time of their apperance they did not progress at all due to their full-fledged appearance in the Cretaceous*".

The fossil records of the angiosperms also support their appearance full-fledged in lower or upper Cretaceous. The fossils of that age are so characteristic and modern in appearance that most of them can be referred unmistakably to living families, general and even to some species. *Prof. Knowlton* stated in his book, '*Plant of the past*' "*if a student of present day trees and shrubs, could have wandered over the hills and vales in those days, he would have found himself quite at home among the trees and shrubs growing there*". The forms of cycads and conifers, which long dominated the universe were already pushed background and the earth had become infact the earth of flowering plants. Charles Darwin has called this sudden appearance of angiosperms as an "*abominable mystery*".

However, some workers do not agree with the doctrine of 'abominable mystery'. According to **H.H. Thomas** (1936), the angiosperms of the past replaced many of older gymnosperms in asturine and marshy waters. **Graud Eury** (1906) believes that the angiosperms came into existence through mutation. **Guppy** (1919) however, supported the view of mutation. **Prof. Bertrand** is of opinion that all the great groups of vascular plants (Pteridophyta, Gymnospermae and Angiosperms) not only arose quite independently of each other but also they originated simultaneously as far back in the Archian period (2000 million years old–oldest).

There is a very considerable but scattered literature on the origin and phylogeny of angiosperms. The palaeobotanical evidence shows that there seems three possibilities as regards the origin of angiosperms. These possibilities are–1. that the angiosperms are monophyletic in their origin but have had a very much longer history than at present known, perhaps stretching back into Palaeozoic times and with a whole series of missing links; 2. that the angiosperms are monophyletic but that the first and at present unknown group diverged quickly in terms of geological time, into a considerable number of different groups; 3. that the angiosperms are polyphyletic.

According to Campbell, "both comparative morphology and the geological record indicate that the existing angiosperms represent a number of distinct phyla which cannot be traced back to a single ancestral type". This statement shows that he does not believe in monophyletic origin of angiosperms.

According to Thomas, the evolutionary tendencies detected in the three groups, *i.e.*, Caytoniales, Bennettitales and Pteriodosperms furnish reasonable grounds for the idea that the angiosperms were derived from various pteriodsperms early in the Mesozoic period.

Parkin also argues for the monophyletic origin of the angiosperms.

In conclusion we can say, that the history of the angiosperms is still almost as great a mystery as it was in the time of Darwin. We do not know, when, where or from what the presumably most recent and now dominant large group of existing terrestrial plants arose.

Lotsy says, that hybridization is the key to evolution of angiosperms. This view has been supported by Anderson. He has suggested on the basis of cytological investigations that the angiosperms may have arisen as a result of hybridization between two gymnosperms.

According to Hagerup, the origin of some angiosperms took place from the Coniferales through Gnetales.

According to Eames, Sinnott and Bailey the more primitive angiosperms were arboreal in habit and the herbaceous angiosperms have been evolved from them.

According to Hutchinson, in certain groups, trees and shrubs are probably more primitive than herbs.

Evidently views in this respect are divergent and speculative, the available data being meagre, fragmentary and isolated. Some of the theories proposed from time to time in this connection, are as follows:

1. **Bennettitales-Ranales Theory.** (a) Hallier's view (1906) regarding the origin of angiosperms is that Ranales, (*e.g., Magnolia*) seems to be related to Bennettitales and may have been derived from *Cycadeoidea,* and that the monocotyledons are an offshoot of the dicotyledons. The elongated floral axis of Ranales with spirally arranged male and female sporophylls and the cone of *Cycadeoidea* are definitely alike. Both the groups were also abundant in the Cretaceous. Ranales is regarded as the earliest stock from which the polycarpic families of dicotyledons have arisen, and also the monocotyledons as an offshoot. But the differences in the anatomical structure of the wood, types of sporophylls, nature and position of ovules, etc., in the two groups, *i.e.,* Bennettitales and Ranales, have led to difficulties in accepting this view. It is more probable that both the groups have evolved from a common ancestry and developed in unrelated parallel lines.

(b) **Arber and Parkin** (1907), proposed the existence of an imaginary group (Hemiangiospermae) having cycadeoid type of flowers linking the above two groups. From this imaginary group the Ranalian type of flowers might have originated, and from this Ranalian type all other angiosperms had sprung up. They were of the opinion that the earliest monocotyledons were more primitive than the dicotyledons. They also supported the Hallier's view.

(c) **Hutchinson** (1925) considered the origin of angiosperms as *monophyletic,* and supported the views of Hallier (1906) and Arber and Parkin (1907). He believed in the Bennettitalean origin of angiosperms and stressed on two parallel evolutionary lines for the primitive dicotyledons: a woody (arborescent) line, called *Lignosae,* starting from Magnoliales, and a herbaceous line, called *Herbaceae* starting from Ranales. He further held the view that the monocotyledons were derived from a primitive dicotyledonous order, The Ranales.

2. **Coniferae–Amentiferae Theory.** Engler and Prantl (1924) rejected the Cycadeoidean origin of angiosperms, as proposed by Hallier (1906) earlier. They held the view that dicotyledons and monocotyledons had arisen independently from a hypothetical group of extinct gymnosperms (allied coniferae) with unisexual strobilus which developed in the Mesozoic. Thus, according to them, the angiosperms had a *polyphyletic* origin, and evolution took place on several parallel lines from the beginning. They considered the monocotyledons to be more primitive than the dicotyledons. The unisexual naked (without perianth) condition of the angiospermic flowers, such as Pandanales (monocotyledons) and Amentiferae (catkin-bearing dicotyledons, *e.g., Casuarina, Salix, Betula,* etc.) was most primitive. But according to the modern classification these orders are regarded more advanced.

3. **Gnetales-Casuarinales Theory.** Wettstein (1935) held the view that angiosperms of *Casuarina* type evolved from Gnetales, (particularly *Ephedra*), a highly advanced group of gymnosperms, which branched off from the main gymnospermic line. However, there are some angiospermic features in Gnetales but this group has meagre fossil records, and not gone earlier than Tertiary. Fagerland (1947) was of the opinion that both Gnetales and Proangiosperms had a common ancestor and the modern angiosperms evolved from the Proangiosperms in Polyphyletic lines.

4. **Caytoniales-Angiosperm Theory.** Thomas (1925) suggested that Caytoniales, a small group of angiosperm-like plants, discovered in the Jurassic rocks of Yorkshire (England), might be the ancestor of angiosperms. However, Harris (1932-33) opposed this view. Knowlton (1927) expressed the view in his book *Plants of the Past* that both Caytoniales and angiosperms evolved from the large extinct Palaeozoic group of Pteridosperms. Arnold (1948), expressed the view that Caytoniales were allied to the Pteridosperms rather than to the angiosperms.

5. **Pteridosperm-Angiosperm Theory.** Andrews (1947) was of opinion that seed ferns or Pteridosperms, an ancient group of the Palaeozoic, might be the starting point for the angiospermic plants.

Origin of Dicotyledons

There are two large groups of angiosperms (1) Dicotyledons and (2) Monocotyledons. The dicots are more important and they are supposed to have originated before the monocotyledons. It is thought that for the first time dicotyledons appeared in the early Mesozoic era, or perhaps the late Palaeozoic. The oldest known fossils of dicots are from the lower Cretaceous rocks.

It is generally believed and agreed that the dicotyledons have been originated from gymnosperms of a type somewhat different from present day forms of gymnosperms. According to some similarities noted in these two groups, the development seems to be parallel, or in some respects convergent development of the two groups.

Origin of Monocotyledons

This is the subject of great controversy. For some time this was thought that the monocoytledons were more primitive than the dicotyledons and probably they have given rise to the dicotyledons. Now, this belief has been totally given up. It is now generally thought that the dicotyledons are more primitive and they have given rise to the monocotyledons. This is also believed, that the monocotyledons were an offshoot of the primitive dicotyledons back in the early Mesozoic era. They are thought to be monophyletic (*i.e.*, of one origin) in their origin. According to another conception the monocotyledons are not monophyletic but polyphyletic in their origin.

2

EVOLUTION OF CLASSIFICATION

In India, botanical science had its origin in a very remote age with the development of agriculture, when Indians of the Vedic Period (2000 B.C. - 800 B.C.) cultivated various food crops. In the Vedic literature several technical terms are used in the description of plants and plant parts. Both the external features and the anatomy were studied. One of the most ancient works dealing with plant life in scientific manner is the *Vrikshayurveda* (science of plants and plant life), compiled by Parasara even before the beginning of the Christian era, and which formed the basis of botanical teaching and the medical studies in ancient India. This ancient work deals with the morphology and anatomy of the plant organs, nature and properties of soil, description and distribution of forests in the country, etc. According to Parasara there are innumerable cells (*rasakosa*) which serve as storehouse of sap derived from the soil, through the roots by transporting system (*syandani*) and that the sap is digested with the help of chlorophyll (*ranjakena pacyamanat*) into nutritive substances and by-products. A system of classification was formulated, and the plants were divided into *eight* groups : 1. *Vriksha*; 2. *Vanaspati*; 3. *Guccha*; 4. *Gulma*; 5. *Trina*; 6. *Aushadhi;* 7. *Valli* and 8. Charaka was devoted for medicinal plants, and he wrote the famous work *Charaka Samhita.* The ancient system of classification was formulated based on the study of the comparative morphology of plants, which according to Majumdar was more advanced than any system developed in Europe before the eighteenth century. Many families (*ganas*) were clearly distinguished as to be recognizable today. The present day family Cruciferae was called *Swastikaganiyam* because the calyx resembles a *Swastika*, and was characterized by having superior ovary, four free sepals, four free petals, six stamens two of which are shorter and two carpels fused and forming a two-locular fruit. The present day Cucurbitaceae was known as *Tripusaganiyam* and was characterized by having flowers which were epigynous, sometimes bisexual with five sepals, five fused petals, three stamens and a trilocular ovary with three rows of ovules.

In middle ages there was an extraordinary rise of Islam among the Arabians who spread their culture from Spain to Bokhara. Islamic scholars contributed a lot to the botanical science. A Persian writer Abu Mansur wrote a book on medicinal plants, also referred Indian plants. Ibn Sina (Avicenna) (980 - 1037A.D.) also a Persian, compiled a book *Canon of Medicine* containing numerous medicinal plants, which was much valued and had more than twenty editions in the sixteenth century.

The problem of classifing the heterogenous groups of angiosperms is very interesting and as well as too tough for taxonomists. The angiosperms are widely distributed with so many variations that sometimes it seems almost impossible to arrange them in systematic order. Since the prehistoric times the people were interested in the solution of the problem, and for the first time a few plants were classified according to their medicinal and food value and thus the taxonomy of flowering plants originated. After the discovery of some useful plants, the mind of the people of the past diverted towards the cultivation of food producing and other benefical plants. They prepared fields, sown the seeds and harvested the crops which were assumed to be God given gifts.

There can be three different possible types of systems of classification, *the aritifical, the natural* and *the phylogenetic* classifications:-

1. **Artifical System**– In artifical system of classification only a few characters of the plants are being considered, for example, the grouping of plants into *herbs, shrubs* and *trees* or the sexual system of Linnaeus based on number of stamens and styles. There are so many drawbacks in this system the most important one is that the plants closely resembling each other are very often placed in widely separated groups, while those quite different from each other are being placed in the same group. This system can be compared to the alphabetical arrangement of words in a dictionary, where two words arranged one after the other have nothing to do with each other. The artificial systems are not commonly in current use. The best known artificial system is of Linnaeus published in 1735.

2. **Natural System**–In the natural system all the important characters of the plant are being considered, and the plants are classified according to their related affinities. This system reflects the situation as it is thought to exist under natural conditions, *e.g.,* Bentham and Hooker's system (1862-83).

3. **Phylogenetic System**–In phylogenetic system, the plants are classified according to their evolutionary and genetic affinities. But it seems a bit difficult to clasify the plants perfectly on the basis of evolutionary tendencies due to imperfection of fossil records, and therefore, at present the plants are classified partly according to natural and partly according to phylogenetic basis. The systems of classification proposed by Engler in 1886, by Hutchinson in 1926, and by Tippo in 1942 are phylogenetic. The plant taxonomy begins with the Greeks.

ANCIENT SYSTEMS OF CLASSIFICATION

From Theophrastus to Linnaeus

Theophrastus (370-285 B.C.), the father of Botany, was a pupil of Plato (426-347 B.C.) and friend of Aristotle (384-322 B.C.) classified plants on the basis of form and texture. In his *Historia Plantarum* he had classified and described about 450 cultivated plants. He had classified them as herbs, under-shrubs, shrubs and trees. Theophrastus considered the trees to be highest evolved while herbs very primitive. He also distinguished between annual, biennial and perennial plants. He also noted the difference between centripetal and centrifugal inflorescences. His classification was strictly artificial.

Dioscorides (64 A.D.) of Christian era prepared a '*Materia Medica*' and described many plants of medicinal value, but his descriptions were imperfect and classification of little importance.

Pliny (23-79 A.D) also a botanist of Christian era had written seventeen volumes on the subject and gave many details of the cultivation of gum and spices yielding plants. He had described medicinal plants in three volumes of his great work, *'Historia Naturalis'* (77 A.D.), largely compiled from Greek authors.

Albertus Magnus (1193-1280), for the first time distinguished between monocotyledons and dicotyledons on the anatomical basis with the help of crude lenses available at that time. In his *De vegetalis* he had described many plants and methods of gardening and orcharding.

Otto Brunfels (1464-1534), could recognise for the first time between perfecti and imperfecti groups of plant kingdom on the basis of the presence and absence of flowers. Sometimes it is also said that modern study of systematic botany has started from the time of Brunfels.

Jerome Bock (1498-1554), classified the plants as trees, shrubs, and herbs but he tried to follow the natural system of classification and brought many plants close to each other according to their related affinities.

William Turner (1512), is the acknowledged 'Father of English Botany' and was the author of several botanical works. He wrote the first *English Herbal.* In this work the arrangement is alphabetical and the woodcuts are inferior to those of Brunfels. The work was published in three parts.

The Herbal of Valerius Cordus (1561) of Wittenberg showed a great advance and is a landmark in descriptive botany.

Andrea Caesalpino(1519-1603), an Italian botanist and physician, wrote a set of sixteen volumes on plants. The first volume was philosophical describing some minor differences of cultivated and wild plants, trees and shrubs, shrubs and herbs and some medicinal plants. In the other fifteen volumes he divided the plant kingdom into fifteen heterogeneous groups. In his work *De plqntis* published in 1583, he classified about 1500 plants chiefly on the basis of the characters of the seed and embryo. He classified the plants first on the basis of their habit (trees and herbs) and then subdivided them on the basis of their characters of the seed and embryo. He classified the plants first on the basis of their habit (trees and herbs) and then subdivided them on the basis of their characters of the seeds and fruits. He also used the characters like the presence or absence of bulbs, presence of latex and number of locules, etc. He prepared a herbarium of 768 plants which still exists as one of the oldest herbaria. Caesalpiono's work was remarkable in that time to clssiffy the whole plant kingdom but it also lacked in many ways.

The most important of the British herbals was written by John Gerard (1545-1612) entitled *Herbal or General Historie of Plante.* This was published in London in 1597. The groups of plants recognised are based on well marked characters of general form, manner of growth and economic uses. They started with supposed simpler forms, grass-like plants with narrow leaves, and advanced through the broader-leaved bulbous and rhizomatous monocotyledons to dicotyledonous herbs, culminating in shrubs and trees, the latter being regarded as the most perfect.

Jean Bauhin(1541-1613), a swiss physician, wrote an illustrated *Historia Plantarum Universalis* which was published after his death in 1650-51 in three volumes, which contained the descriptions of about 5,000 plants with 3,500 illustrations.

G.Bauhin (1560-1624), a Swiss botanist and brother of Jean Bauhin wrote '*An Introduction to Botany*' in which he described about 6,000 species of plants. He also made considerable advance in the frequent use of binomial nomenclature. He described many plants with generic and specific names. His great work *'phytopinax'* of 1596 is great contribution of Botany. Many of the plant descriptions of Bauhin's work are very much alike modern description of species.

M. de Lobelius (1538-1616), he differentiated between monocotyledons and dicotyledons chiefly on the basis of leaves. He clearly recognized certain natural affinities in plants. His great work *'Kruydtboeck'* was published in 1581, which contains many excellent illustrations.

Robert Morison (1620-1683), published his *Plantarum Historia Universalis* at Oxford in 1680. His system much resembles that of Caesalpino. Woody plants are kept distinct from herbaceous, and the latter are divided into fifteen sections based partly on habit and partly on characters derived from the fruit and seed. Morison had the distinction of publishing the first systematic monograph of a limited group. This was his *Plantarum Umbelliferarum Distributio Nova*, issued in Oxford in 1672. The arrangement is based on the form of the fruit, which is still employed for the classification of Umbelliferae.

John Ray (1628 1705), an English botanist contributed much and the foundation of modern Botany was laid. His most important work was the *'Historia Plantarum'* which was published in three volumes between 1686 and 1704. He clearly showed the importance of the monocotyledonous and dicotyledonous nature of the embryo from the point of view of related affinities and classification. He divided the plant kingdom in two main groups, A. Herbae, B. Arbores. For the first time he used the terms *Dicotyledons* and *Monocotyledons.* In descriptions he used many binomial names. His classification was as follows –

Herbae : (*i*) Imperfecti (flowerless)

(*ii*) Perfecti (flowering)

(*a*) Dicotyledons (with two cotyledons),

(*b*) Monocotyledons (with one cotyledon).

Arbores: (*i*) Dicotyledons (with two cotyledons).

(*ii*) Monocotyledons (with one cotyledon).

J.P. de Tournefort (1656-1708), is said to be the founder of the concept of modern genera. Though generic names were used before, yet Tournefort was the first to provide genera with descriptions and placed them definitely apart from species. His classification was inferior in many ways to John Ray's classification. He never distinguished between Cryptogams and Phanerogams whereas monocots and dicots. He classified the plants on the basis of presence and absence of petals. He called the plants with petals as *Petalodes* and without petals as *Apetali.* His famous work *Elements de Botanique* was published in 1684 which was later translated into *Latin.*

Camerarius (1665-1721), a German botanist could demonstrate for the first time the sexuality in the flowering plants on the experimental basis. He proved experimentally that pollen is absolutely necessary for fertilization and seed formation in the life history of plants.

MODERN SYSTEMS OF CLASSIFICATION

From Linnaeus to Hutchinson

Carolus Linnaeus (1707-1778), a great Swedish botanist and said to be the father of modern botany was born in 1707. Only due to this great man, botany and other sciences got inspiration and expanded day by day. From the time of Linnaeus actually the concept of modern botany began. He followed the binomial system of nomenclature and described hundreds of plants from different parts of the world. He also revised many older genera and gave great stability to species. He published a number of important works, *e.g., Genera Plantarum, Species Plantarum, Flora Lapponica,* etc., in which he described hundreds of plants known in that time. Linnaeus proposed an *artificial sexual system* of classification containing twentyfour classes. The plants were classified on the basis of number, cohesion, length and certain other characters of stamens. The outline of the classes of his system was published in his *Systema Nature* in 1735, and again in *General Plantarum* in 1737 along with all the known plants of that time arranged systematically. The outline of the classification followed by him is as follows :

1.	**Monandria**	one stamen	*e.g., Canna, Lemna, Salicornia, Scirpus., etc.*
2.	**Diandria**	two stamens	*Veronica, Lepidium.*
3.	**Triandria**	three stamens	*Iris, Galium,* Grasses.
4.	**Tetrandria**	four stamens	*Protea, Ulmus.*
5.	**Pentandria**	five stamens	*Ipomoea, Primula.*
6.	**Hexandria**	six stamens	*Lilium, Alisma.*
7.	**Heptandria**	seven stamens	*Trientalis, Aesculus.*
8.	**Octandria**	eight stamens	*Vaccinium, Dirca.*
9.	**Enneandria**	nine stamens	*Rheum, Ranunculus.*
10.	**Decandria**	ten stamens	*Linum, Acer, Oxalis.*
11.	**Dodecandria**	11-19 stamens	*Reseda, Asarum, Calla.*
12.	**Icosandria**	20 or more stamens attached to the calyx	*Rosa, Rubus, Spiraea.*
13.	**Polyandria**	20 or more stamens attached to the receptacle	*Tilia, Papaver.*
14.	**Didynamia**	Didynamous stamens	Labiatae
15.	**Tetradynamia**	Tetradynamous stamens	Cruciferae
16.	**Monadelphia**	Stamens in one bundle	*Linum,* Malvaceae.

17.	**Diadelphia**	Stamens in two bundles	Papilionatae.
18.	**Polyadelphia**	Stamens in many bundles	*Tilia, Hypericum.*
19.	**Syngenesia**	Syngenesious stamens	Compositae, *Solanum.*
20.	**Gynandria**	Stamens grown to the pistil	Orchidaceae
21.	**Monoecia**	Plants monoecious	*Morus, Typha, Betula.*
22.	**Dioecia**	Plants dioecious	*Salix, Juniperus.*
23.	**Polygamia**	Plants polygamous	Compositae, *Acer.*
24.	**Cryptogamia**	Hidden flowers	Ferns, Mosses, Algae, Fungi, etc.

Though system was artificial, yet due to its simplicity and easy way of identification of plants, the system continued for more than one hundred years after the death of Linnaeus. He recognised the weakness of his system and was fully aware of the shortcomings. He attempted once again a more natural system but anyhow he could not complete it. On the whole, whatever Linnaeus presented to the scientific world is undoubtedly remarkable and of great value.

B.de Jussieu (1699-1777) and **A.L.de Jussieu** (1748-1836). The two French botanists Bernard de Jussieu and his nephew A.L. de Jussieu at the end of eighteenth century proposed a scheme of classification which was based upon the sexual system of Linnaeus but in an improved form. B. de Jussieu made many changes in the classification of Linnaeus and more or less gave a new form to it, but this was never published. His nephew, A.L. de Jussieu modified his uncle's plan with many changes and published it in his own way, in his famous work "*Genera Plantarum*" in 1789. He included fifteen classes and more than one hundred orders (now known as families) in his system. His system was chiefly based on the number and position of cotyledons and adhesion of petals. His system of classification can be claimed to be a natural one. The outline of his plan is as follows :

Acotyledones	Class 1
Monocotyledones	
Stamina hypogyna	Class 2
Stamina perigyna	Class 3
Stamina epigyna	Class 4
Dicotyledones	
Apetalae	
Stamina epigyna	Class 5
Stamina perigyna	Class 6
Stamina hypogyna	Class 7
Monopetalae	
Corolla hypogyna	Class 8
Corolla perigyna	Class 9
Corolla epigyna	
Antheris connatus	Class 10
Antheris distinctis	Class 11
Polypetalae	
Stamina epigyna	Class 12
Stamina hypogyna	Class 13
Stamina perigyna	Class 14
Diclines Irregulares	Class 15

Michel Adanson (1727-1806), published his *Families des plantes* in Paris in 1763. This contains an exhaustive account of previous systems, and an arrangement of genera in fifty-eight families, which are named and characterized.

A.P. de Candolle (1778-1841), a great French botanist proposed a very reasonable natural system of classification. In his younger days, he spent much of his time at Paris but after 1816 he shifted to Geneva, Switzerland, where he carried out several researches and made the city centre of botanical research. He wrote many monographs concerning classification and plant families, he was also interested in Physiology and Plant Geography. His famous work "*Theorie elementaire de la botanique*" was published in 1813, in which he proposed a classification completely based on natural system. He declared certain principles which later provided the base of origin of Bentham and Hooker's classification. His system contains 161 orders or families. The last edition of his *"Theorie elementaire"* was published in 1844 by his son Alphonse de Candolle in which he gave 213 orders or families.

The outline of de Candollean classification is as follows –

Vasculares – Plants with vascular bundles or plants with cotyledons.

Class 1. Exogenae – Dicotyledoneae, embryo with two cotyledons and vascular bundles arranged in a ring.

A. Diplochlamydeae – Flowers with two whorls of perianth, *i.e.*, calyx and corolla both present.

(a) Thalamiflorae – Flowers polypetalous and hypogynous.

Cohort 1. Carpels many and stamens opposite the petals, orders 1–8 Ranunculaceae, Berberidaceae, etc.

Cohort 2. Carpels solitary or joined; placenta parietal, order 9-20. Cruciferae, Violaceae, etc.

Cohort 3. Ovary solitary; placenta central, orders 21-44. Caryophyllaceae, Malvaceae, etc.

Cohort 4. Fruit gynobasic, orders 45-46.

(b) Calyciflorae –Flowers perigynous or epigynous, stamens inserted on the calyx tube, orders 47-84.

(c) Corolliflorae –Flowers gamopetalous and hypogynous, stamens epipetalous, orders 85-108.

B. Monochlamydeae –Flowers with single whorl of perianth, orders 109-128.

Class 2. Endogenae –Monocotyledoneae, embryo with single cotyledon; vascular bundles scattered.

A. Phanerogamae –Flowers present, *i.e.*, visible and regular, orders 129-150.

B. Cryptogamae –Flowers absent, orders 151-155.

Cellulares –Plants without vascular bundles and cots.

Class 1. Foliaceae –Leafy; sexuality known, orders 156-157 (mosses and liverworts).

Class 2. Aphyllae –Not leafy; sexuality unknown, orders 158-161 (algae, fungi, lichens).

A.P. de Candolle's system was definitely superior to Jussieu's system of classification including about 161 orders (families) of plants as compared with 167 families of Jussieu and Linnaeus respectively. The last edition of de Candolle's *"Theorie elementaire"* was published by his son Alphonse de Candolle in 1844, which contained 213 orders or families.

Robert Brown (1773-1858)–The name of Robert Brown seems fairly large in the chronicles of systematic botany, because he has often been referred to as *botanicorum facile princeps*. Brown was the first keeper of the Botanical Department of the British Museum. In 1810 Brown published his *Prodromus Florae Noval Hollandiae,* containing descriptions of Australian plants collected by Banks and Solander. In 1827 Brown published a paper entitled, *'The female flower in Cycadaceae and Coniferae'* and in this he announced the important discovery of the distinction between Angiosperms and Gymnosperms.

John Lindley (1799-1865), an English botanist, who was Professor of Botany at the University College, proposed a system of classification and published *"Introduction to the Natural order of Plants"*. His system was based on de-Candollean system with few remarkable improvements.

Stephen Endlicher (1804-1849), was a prominent systematist of the first half of the nineteenth century. His scheme of classification was published in *"Genera Plantarum Secundum Ordines Disposita"* in 1836-40 which was widely used in the continental countries. He divided the plant kingdom into two main groups 1. *Thallophyta*, 2. *Cormophyta*. In *Thallophyta* he included algae, fungi and lichens while in *Cormophyta* mosses, ferns and seed plants were included. In his published work *"Genera plantarum"* he described 6,235 vascular plants.

Henri Baillon (1827-1895), Henri Baillons' *Histoire des plantes* (1867-1895), a work of thirteen volumes, contain useful descriptions of the families and genera of flowering plants. Its chief merits, however, are the splendid illustrations with accurate dissections. He also published monographs on Euphorbiaceae and Buxaceae, etc, accompanied by beautifully reproduced plates.

In 1851, **Hofmeister** carried out many brilliant researches and established the existence of alternation of generations in lower as well as in higher plants. Following the researches of Hofmeister in 1859, Charles Darwin published his famous doctrine of evolution *"Origin of Species"*. The doctrine of evolution was known to every educated man in the world, which brought revolutionary changes in solving the problem of classification of plants and animals.

Bentham and Hooker, the well known English systematists, established an important system of classification which appeared in the last half of the nineteenth century. George Bentham (1800-1884) and Sir Joseph Dalton Hooker (1817-1911), two English botanists, whose researches were carried over at the great herbarium of the Royal botanical garden at Kew. After the death of his father Sir W.J. Hooker, Sir Joseph Dalton Hooker was director of the Kew gardens for twenty years. Bentham and Hooker published a joint system of classification, containing 202 orders (families) in a monumental work *"Genera Plantarum"* (1862-1883) which became most popular classification in European countries except France for many years. The system of classification by Bentham and Hooker is based on that of De Candolle's and Jussieu's systems, but it is a great improvement over those and other previous schemes. An outline of Bentham and Hooker's classification is given below:

System of Classification by Bentham and Hooker (1862-1883)

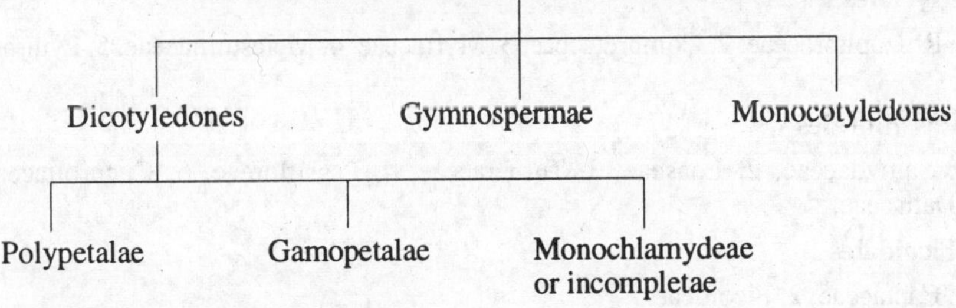

1– Polypetalae – Petals separate.

Series (A) Thalamiflorae–Hypogynous, stamens and pistils many, indefinite, mostly free, distinct sepals free from ovary.

Order 1. **Ranales**

Families –1. Ranunculaceae, 2. Dilleniaceae, 3. Calycanthaceae, 4. Magnoliaceae, 5. Anonaceae, 6. Menispermaceae, 7. Berberidaceae, 8. Nymphaeaceae.

Order 2. **Parietales.**

Families –1. Sarraceniaceae, 2. Papaveraceae, 3. Cruciferae, 4. Capparidaceae, 5. Resedaceae, 6. Cistineae. 7. Violaceae, 8. Canellaceae, 9. Bixineae.

Order 3. **Polygalineae**

Families–1. Pittosporeae, 2. Tremandreae, 3. Polygaleae, 4. Vochysiaceae.

Order 4. **Caryophyllineae**

Families–1. Frankeniaceae, 2. Caryophyllaceae, 3. Portulacaceae, 4. Tamariscineae.

Order 5. **Guttiferales**

Families–1. Elatineae, 2. Hypericineae, 3. Guttiferae, 4. Ternstroemiaceae, 5. Dipterocarpeae, 6. Chlaenaceae.

Order 6. **Malvales**

Families–1. Malvaceae, 2. Sterculiaceae, 3. Tiliaceae.

Series (B) Disciflorae. Hypogynous, calyx consists of free or united sepals which are usually free from ovary. A prominent cushion shaped disc present below the ovary. Ovary free or embedded in the receptacular disc. Stamens usually definite in number.

Order 7. **Geraniales**

Families–1. Lineae, 2. Humiriaceae, 3. Malpighiaceae, 4. Zygophylleae, 5. Geraniaceae, 6. Rutaceae, 7. Simarubeae, 8. Ochnaceae, 9. Burseraceae, 10. Meliaceae, 11. Chailletiaceae.

Order 8. **Olacales**

Families–1. Olacineae, 2. Ilicineae, 3. Cyrilleae.

Order 9. **Celastrales**

Families-1. Celastrineae, 2. Stackhousieae, 3. Rhamneae, 4. Ampelideae.

Order 10. **Sapindales**

Families–1. Sapindaceae, 2. Sabiaceae, 3. Anacardiaceae, 4. Coriarieae, 5. Moringeae.

Series (C) Calyicflorae–Calyx consists of united sepals rarely free and adnate to ovary. Perigynous or epigynous. Ovary enclosed by the axis sometimes inferior.

Order 11. **Rosales**

Families–1. Connaraceae, 2. Leguminosae, 3. Rosaceae, 4. Saxifrageae, 5. Crassulaceae, 6. Droseraceae, 7. Hamamelideae, 8. Bruniaceae, 9. Halorageae.

Order 12. **Myrtales**

Families–1. Rhizophoraceae, 2. Combretaceae, 3. Myrtaceae, 4. Melastomaceae, 5. Lythrarieae, 6. Onagrarieae.

Order 13. **Passiflorales**

Families–1. Samydaceae, 2. Loaseae, 3. Turneraceae, 4. Passifloreae, 5. Cucurbitaceae, 6. Begoniaceae, 7. Datisceae.

Order 14. **Ficoidales**

Families–1. Cactaceae, 2. Ficoideae.

Order 15. **Umbellales**

Families–1. Umbelliferae, 2. Araliaccae, 3. Cornaceae.

2. Gamopetalae : Petals united.

Series(A) Inferae–Ovary inferior, stamens as many as corolla lobes or fewer.

Order 1. **Rubiales**

Families–1. Caprifoliaceae, 2. Rubiaceae.

Order 2. **Asterales**

Families–1. Valerianeae, 2. Dipsaceae, 3. Calycereae, 4. Compositae.

Series(B) Heteromerae–Ovary usually superior, carpels more than two, stamens as many as corolla lobes or fewer, epipetalous or free.

Order 4. **Ericales**

Families–1. Ericaceae, 2. Vaccinieae, 3. Monotropeae, 4. Epacrideae, 5. Diapensiaceae, 6. Lennoaceae.

Order 5. **Primulales**

Families–1. Plumbagineae, 2. Primulaceae, 3. Myrsineae

Order 6. **Ebenales**

Families–1. Sapotaceae, 2. Ebenaceae, 3. Styraceae.

Series (C) Bicarpellatae–Ovary usually superior, carpels usually two, rarely 1 to 3, stamens alternate with the corolla lobes and equal in number or fewer.

Order 7. **Gentianales**

Families–1. Oleaceae, 2. Salvadoraceae, 3. Apocynaceae, 4. Asclepiadaceae, 5. Loganiaceae, 6. Gentianaceae.

Order 8. **Polemoniales**

Families–1. Polemoniaceae, 2. Hydrophyllaceae, 3. Poragineae, 4. Convolvulaceae, 5. Solanaceae.

Order 9. **Personales**

Families–1. Scrophularineae, 2. Orobanchaceae, 3. Lentibularieae, 4. Columelliaceae, 5. Gesneraceae, 6. Bignoniaceae, 7. Pedalineae, 8. Acanthaceae.

Order 10. **Lamiales**

Families–1. Myoporineae, 2. Selagineae, 3. Verbenaceae, 4. Labiatae, 5. Plantagineae.

3. Monochlamydeae or Incompletae–Flowers with simple perianth which is usually sepaloid and sometimes altogether absent.

Series 1. Curvembryae

Families–1. Nyctagineae. 2. Illecebraceae, 3. Amarantaceae, 4. Chenopodiaceae. 5. Phytolaccaceae, 6. Batideae, 7. Polygonaceae.

Series 2. Multivulatae Aquaticae

Families–1. Podostemaceae

Series 3. Multivulatae Terrestres

Families–1. Nepenthaceae, 2. Cytinaceae, 3. Aristolochieae.

Series 4. Micrembryae

Families–1. Piperaceae, 2. Chloranthaceae, 3. Myristiceae, 4. Moniaceae.

Series 5. Daphnales

Families–1. Laurineae, 2. Proteaceae, 3. Thymelaceae, 4. Penaeaceae, 5. Eleagnaceae.

Series 6. Achlamydosporeae

Families–1. Loranthaceae, 2. Santalaceae, 3. Balanophoreae.

Series 7. Unisexuales

Families–1. Euphorbiaceae, 2. Balanopseae, 3. Urticaceae. 4. Plantanaceae, 5. Leitnerieae, 6. Juglandeae, 7. Myricaceae, 8. Casuarineae, 9. Cupuliferae.

Series 8. Ordines anomali (*Anomalous Families*)

Families–1. Salicaceae, 2. Lacistemaceae, 3, Empetraceae, 4. Ceratophylleae.

Gymnospermae :

Families–1. Gnetaceae, 2. Coniferae, 3. Cycadaceae.

Monocotyledones :

Series 1. Microspermae : Ovary superior tricarpellary. Seeds very minute and many. Inner perianth leaves petaloid.

Families–1. Hydrocharideae, 2. Burmanniaceae, 3. Orchideae.

Series 2. Epigynae : Ovary inferior.

Families–1. Scitamineae, 2. Bromellaceae, 3. Haemodoraceae, 4. Irideae, 5. Amaryllideae, 6. Taccaceae, 7. Dioscoreaceae.

Series 3. Coronarieae : Ovary superior, albuminous.

Families–1. Roxburghiaceae, 2. Liliaceae, 3. Pontederiaceae, 4. Philydraceae, 5. Xyrideae, 6. Mayacaceae, 7. Commelinaceae, 8. Rapateaceae.

Series 4. Calycineae : Ovary superior ; albumen copious; perianth inconspicuous, sepaloid, stifff or herbaceous.

Families–1. Flagellarieae, 2. Juncaceae, 3. Palmae.

Series 5. Nudiflorae : Ovary superior; perianth absent or reduced to scales.

Families–1. Pandaneae, 2. Cyclanthaceae, 3. Typhaceae, 4. Aroideae, 5. Lemnaceae.

Series 6. Apocarpae : Ovary superior, carpels separate or single.

Families–1. Triurideae, 2. Alismaceae, 3. Najadaceae.

Series 7. Glumaceae : Ovary single celled, single ovuled. Flowers in spikelets or heads. Perianth scale–like.

Families–1. Eriocauleae, 2. Centrolepideae, 3. Restiaceae, 4. Cyperaceae. 5. Gramineae.

Merits and Demerits of Bentham and Hooker's System of Classification

Merits: Though this system is not very natural, yet it is very easily workable, and is important from the point of view of its field applications. The system has been worked out as a result of very careful comparative examination of all the known genera of phanerogams. There are 202 orders (now known as families) beginning with Ranunculaceae and ending in Gramineae. This system was accepted by entire British Empire, U.S.A. and other European countries. Even today many botanists follow this classification because of its simplicity. This classification makes the basis for the arrangement of plants in Kew Herbarium and other important herbaria of Commonwealth Countries.

A special feature of this system is the addition of the *Disciflorae* and a curious arrangement of dividing certain groups on the basis of aquatic or terrestrial characteristics.

In this system, the greater emphasis has been given on the contrast between *free* and *fused* petals, *e.g.*, the Dicots are divided into the three groups –*Polypetalae, Gamopetalae* and *Monochlamydeae.*

In Monocots, the stress is being given to the relative position of ovary and perianth characteristics.

Demerits : The greatest disadvantage of this system is the retention in the group Monochlamydeae, a number of orders which show affinities with those in which a biseriate perianth is the rule.

The position of Gymnosperms between Dicots and Monocots is only for convenience rather than an indication of affinities.

This system of classification is based mainly on single and mostly artificial characters, with the result the closely related families are widely separated from each other.

The extreme simple nature of the parts of certain orders such as Salicineae, Cupuliferae, etc. indicate an affinity with an early group or groups which are now extinct. Such orders would therefore have no near allies in existing orders with an elaborate arrangement of the flower parts.

In Polypetalae, the orders (families) with inferior ovary are placed afterwards, whereas in Gamopetalae they are treated before than those with superior ovary.

In Monocots, much stress is being given to the relative position of the ovary and the perianth characters in determining the affinities, than seems justified by the comparative study of the orders. For example, the families Iridaceae and Amaryllidaceae exhibit greater affinities to Liliaceae than to Scitamineae and Bromeliaceae with which they are allied in this system because of common characters of epigyny. Thus the treatment of Monocots in general, except for position of Glumaceae is not proper.

Eichler's System of Classification

In 1883, a system of classifying the flowering plants was proposed by A.W. Eichler (1839–1887) is also remarkable because of its resemblance with widely accepted Englerian system. An outline of Eichler's system is given below:

A. Cryptogamae.

B. Phanerogamae.

1. Gymnospermae.

2. Angiospermae.

1. *Class:* **Monocotyleae**

Series 1. Liliflorae
Series 2. Enantioblastae
Series 3. Spadiciflorae
Series 4. Glumiflorae
Series 5. Scitamineae
Series 6. Gynandrae
Series 7. Helobiae

2. *Class:* **Dicotyleae**

Sub–class : **Choripetalae**

Series 1 Amentaceae
Series 2. Urticineae
Series 3. Polygoninae
Series 4. Centrospermae
Series 5. Polycarpicae
Series 6. Rhoeadinae
Series 7. Cistiflorae
Series 8. Columniferae
Series 9. Gruinales
Series 10. Terebinthinae
Series 11. Aesculinae
Series 12. Frangulinae
Series 13. Tricoccae
Series 14. Umbelliflorae
Series 15. Saxifraginae
Series 16. Opuntinae
Series 17. Passiflorinae
Series 18. Myrtiflorae
Series 19. Thymelaeinae
Series 20 Rosiflorae
Series 21. Leguminosae

Anhang :	**Hysterophyta**
Sub-class:	**Sympetalae**
Series 1.	Bicornes
Series 2.	Primulinae
Series 3.	Diospyrinae
Series 4.	Contortae
Series 5.	Tubiflorae
Series 6.	Labiatiflorae
Series 7.	Campanulinae
Series 8.	Rubiinae
Series 9.	Aggregatae

The "series' in Eichler's system can easily be compared with modern 'order' and whereas the 'orders' of Eichler's system may be referred to modern families. The inclusion of the Hysterophyta as an "anhang" in the sub-class Choripetalae is the indication that in those days the position of angiospermic parasites was not at all clear. This group includes Loranthaceae, Rafflesiaceae and the Balanophoraceae. Eichler also included one non-parasitic family Aristolochiaceae in this group. Eichler's system also separates the Gymnosperms from their usual position between the Dicotyledons and Monocotyledons (as proposed by Bentham and Hooker) and raised the group to equivalent group with the angiosperms and lower than the monocotyledons.

Engler and Prantl's System of Classification.

Adolf Engler (1844-1938), a German botanist who served as Professor of botany in the University of Berlin for thirty years and director of Botanical Gardens from 1889 to 1921. His phylogenetic system of classification was first published as a guide to the botanical garden of Breslau in 1892. Later on the system expanded in a monumental work called "*Die Naturlichen Pflanzebfamilien*"! with means for the identifiction of the genera of whole plant kingdom. This publication continued with numerous volumes, many supplements, syllabi and revisions from 1895 to the present day. This system has been the dominant one of the plant classification in most of the scientific world since 1900. Most of the prominent herbaria of the world are arranged according to this system. This great work was completed in the collaboration of his associate worker Eugen Prantl (1849–1893).

According to this system the families were arranged in accord to the increasing complexity of the flower, fruit and seed development. The general outline of the system proposed by Engler and Prantl is given below—

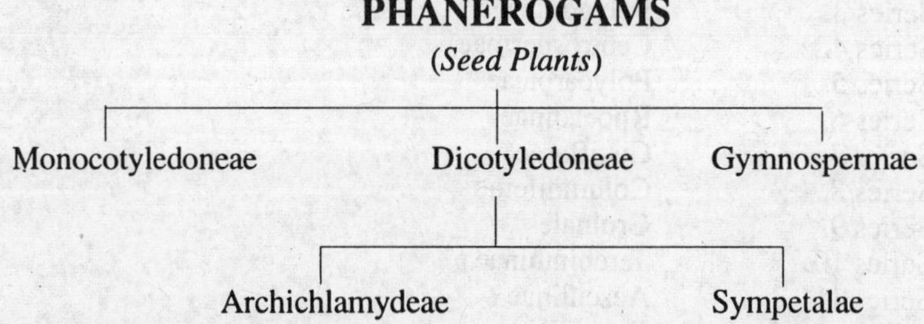

Class 1. MONOCOTYLEDONEAE

Order 1. **Pandanales**

Families–1. Typhaceae, 2. Pandanaceae, 3. Sparganiaceae.

Order 2. **Helobiae**

Families–1. Potamogetonaceae, 2. Najadaceae, 3. Aponogetonaceae, 4. Scheuchzeriaceae, 5. Alismaceae, 6. Butomaceae, 7. Hydrocharitaceae.

Order 3. **Triuridales**

Family–1. Triuridaceae.

Order 4. **Glumiflorae**

Families–1. Gramineae, 2. Cyperaceae.

Order 5. **Principes**

Family–1. Palmae.

Order 6. **Synanthae**

Family–1. Cyclanthaceae.

Order 7. **Spathiflorae**

Families–1. Araceae, 2. Lemnaceae.

Order 8. **Farinosae**

Families–1. Flagellariaceae, 2. Restionaceae, 3. Centrolepidaceae, 4. Mayacaceae, 5. Xyridaceae, 6. Eriocaulaceae, 7. Thurniaceae, 8. Rapateaceae, 9. Bromeliaceae, 10. Commelinaceae, 11. Pontederiaceae, 12. Cyanastraceae, 13. Philydraceae.

Order 9. **Liliflorae**

Families–1. Juncaceae, 2. Stemonaceae, 3. Liliaceae, 4. Haemodoraceae, 5. Amaryllidaceae, 6. Velloziaceae, 7. Taccaceae, 8. Dioscoriaceae, 9. Iridaceae.

Order 10. **Scitamineae**

Families–1. Musaceae, 2. Zingiberaceae, 3. Cannaceae, 4. Marantaceae.

Order 11. **Microspermae**

Families–1. Burmanniaceae, 2. Orchidaceae.

Class 2. DICOTYLEDONEAE

Sub-class A. Archichlamydeae

Order 1. **Verticillatae**

Family–1. Casuarinaceae.

Order 2. **Piperales**

Families–1. Saururaceae, 2. Piperaceae, 3. Chloranthaceae, 4. Lacistemaceae.

Order 3. **Salicales**

Family–1. Salicaceae.

Order 4. **Garryales**

Family–1. Garryaceae.

Order 5. **Myricales**

Family–1. Myricaceae.

Order 6. **Balanopsidales**

Family–1. Balanopsidaceae.

Order 7. **Leitneriales.**

Family–1. Leitneriaceae.

Order 8. **Juglandales**

Family–1. Juglandaceae.

Order 9. **Batidales**

Family–1. Batidaceae.

Order 10. **Julianales**

Family–1. Julianaceae.

Order 11. **Fagales**

Families–1. Betulaceae, 2. Fagaceae.

Order 12. **Urticales**

Families–1. Ulmaceae, 2. Moraceae, 3. Urticaceae.

Order 13. **Proteales**

Family–1. Proteaceae.

Order 14. **Santalales**

Families–1. Myzodendraceae, 2. Santalaceae, 3. Opiliaceae, 4. Grubbiaceae, 5. Olacaceae, 6. Octoknemataceae, 7. Loranthaceae, 8. Balanophoraceae.

Order 15. **Aristolochiales**

Families–1. Aristolochiaceae, 2. Raflessiaceae, 3. Hydnoraceae.

Order 16. **Polygonales**

Family–1. Polygonaceae.

Order 17. **Centrospermeae**

Families–1. Chenopodiaceae, 2. Amarantaceae, 3. Nyctaginaceae, 4. Cynocrambaceae, 5. Phytolaccaceae, 6. Aizoaceae, 7. Portulacaceae, 8. Basellaceae, 9. Caryophyllaceae.

Order 18. **Ranales**

Families–1. Nymphaeaceae, 2. Ceratophyllaceae, 3. Trochodendraceae, 4. Cercidiphyllaceae, 5. Ranunculaceae, 6. Lardizabalaceae, 7. Berberidaceae, 8. Menispermaceae, 9. Magnoliaceae, 10. Calycanthaceae, 11. Lactoridaceae, 12. Anonaceae, 13. Eupomatiaceae, 14. Myristicaceae, 15. Gomortegaceae, 16. Monimiaceae, 17. Lauraceae, 18. Hernandiaceae.

Order 19. **Rhoeadales**

Families–1. Papaveraceae. 2. Capparidaceae, 3. Cruciferae, 4. Tovariaceae, 5. Resedaceae, 6. Moringaceae.

Order 20. **Sarraceniales**

Families–1. Sarraceniaceae, 2. Nepenthaceae, 3. Droseraceae.

Order 21. **Rosales**

Families–1. Podostemaceae, 2. Tristichaceae, 3. Hydrostachnaceae, 4. Crassulaceae, 5. Cephalotaceae, 6. Saxifragaceae, 7. Pittosporaceae, 8. Brunelliaceae, 9. Cunoniaceae, 10. Myrothamnaceae, 11. Bruniaceae, 12. Hamamelidaceae, 13. Eucommiaceae, 14. Platanaceae, 15. Crossosomataceae, 16. Rosaceae, 18. Leguminosae.

Order 22. **Pandales**

Family–Pandaceae.

Order 23. **Geraniales**

Families–1. Geraniaceae, 2. Oxalidaceae, 3. Tropaeolaceae, 4. Linaceae, 5. Humiriaceae, 6. Erythroxylaceae 7. Zygophyllaceae 8. Cneoraceae, 9. Rutaceae, 10. Simarubaceae, 11. Burseraceae, 12. Meliaceae, 13. Malpighiaceae, 14. Trigoniaceae, 15. Vochysiaceae. 16. Tremandraceae, 17. Polygalaceae, 18. Dichapetalaceae, 19. Euphorbiaceae, 20. Callitrichaceae.

Order 24. **Sapindales**

Families–1. Empetraceae, 2. Coriariaceae, 3. Limnanthaceae. 4. Anacardiaceae, 5. Cyrillaceae, 6. Buxaceae, 7. Pentaphylacaceae, 8. Corynocarpaceae, 9. Aquifoliaceae, 10. Celastraceae, 11. Hippocrateaceae, 12. Salvadoraceae, 13. Stackhousiaceae, 14. Staphyleaceae, 15. Icacinaceae, 16. Aceraceae, 17. Hippocastanaceae, 18. Sapindaceae, 19. Sabiaceae, 20. Melianthaceae, 21. Balsaminaceae.

Order 25. **Rhamnales**

Families–1. Rhamnaceae, 2. Vitaceae.

Order 26. **Malvales**

Families–1. Elaeocarpaceae, 2. Chlaenaceae, 3. Gonystilaceae, 4. Tiliaceae, 5. Malvaceae, 6. Bombacaceae, 7. Sterculiaceae, 8. Scytopetalaceae.

Order 27. **Parietales**

Families–1. Dilleniaceae, 2. Eucryphiaceae, 3. Ochnaceae, 4. Caryocaraceae, 5. Marcgraviaceae, 6. Quinaceae, 7. Theaceae, 8. Guttiferae, 9. Dipterocarpaceae, 10. Elatinaceae, 11. Frankeniaceae, 12. Tamaricaceae, 13. Fouquieriaceae, 14 Cistaceae, 15. Bixaceae, 16. Cochlospermaceae, 17. Winteraceae, 18. Violaceae, 19. Flacourtiaceae, 20. Stachyuraceae, 21. Turneraceae, 22. Malesherbiaceae, 23. Passifloraceae, 24. Achariaceae, 25. Caricaceae, 26. Loasaceae, 27. Datiscaceae, 28. Begoniaceae, 29. Ancistrocladaceae.

Order 28. **Opuntiales**

Family–1. Cactaceae.

Order 29. **Myrtiflorae**

Families–1. Geissolomataceae, 2. Penaeaceae, 3. Oliniaceae, 4. Thymelaceae, 5. Elaeagnaceae, 6. Lythraceae, 7. Sonneratiaceae 8. Punicaceae, 9. Lecythidaceae, 10. Rhizophoraceae, 11. Nyssaceae, 12. Alangiaceae, 13. Combretaceae, 14. Myrtaceae, 15. Melastomaceae, 16. Onagraceae, 17. Haloragidaceae, 18. Hippuridaceae, 19. Cynomoriaceae.

Order 30. **Umbelliflorae**

Families–1. Araliaceae, 2. Umbelliferae, 3. Cornaceae.

Sub class B. Sympetalae

Order 1. **Ericales**

Families–1. Clethraceae, 2. Pyrolaceae, 3. Lennoaceae, 4. Ericaceae, 5. Epacridaceae, 6. Diapensiaceae.

Order 2. **Primulales**

Families–1. Theophrastaceae, 2. Myrsinaceae, 3. Primulaceae.

Order 3. **Plumbaginales**

Family–1. Plumbaginaceae.

Order 4. **Ebenales**

Families–1. Sapotaceae, 2. Ebenaceae, 3. Symplocaceae, 4. Styracaceae.

Order 5. **Contortae**

Families–1. Oleaceae, 2. Loganiaceae, 3. Gentianaceae, 4. Apocynaceae, 5. Asclepiadaceae.

Order 6. **Tubiflorae.**

Families–1. Convolvulaceae, 2. Polemoniaceae, 3. Hydrophyllaceae, 4. Boraginaceae, 5. Verbenaceae, 6. Labiatae, 7. Nolanaceae, 8. Solanaceae, 9. Scrophulariaceae, 10. Bignoniaceae, 11. Pedaliaceae, 12. Martyniaceae, 13. Orobanchaceae, 14. Gesneriaceae, 15. Columelliaceae, 16. Lentibulariaceae, 17. Globulariaceae, 18. Acanthaceae, 19. Myoporaceae, 20. Phrymaceae.

Order 7. **Plantaginales**

Family–1. Plantaginaceae.

Order 8. **Rubiales**

Families–1. Rubiaceae, 2. Caprifoliaceae, 3. Adoxaceae 4. Valerianaceae. 5. Dipsaceae.

Order 9. **Cucurbitales**

Family–1. Cucurbitaceae

Order10. Campanulatae

Families–1. Campanulaceae, 2. Goodeniaceae, 3. Brunoniaceae, 4. Stylidiaceae, 5. Calyceraceae, 6. Compositae.

Merits and Demerits of Engler and Prantl's System of Classification

Merits–According to this system, the large artificial group of Bentham and Hooker's system, the *Monochlamydeae* has been completely abolished, and its families have been distributed among the related forms with free petals in the large series of this (Engler's) system, the *Archichlamydeae.*

The Sympetalae of this system corresponds to the Gamopetalae of Bentham and Hooker's system.

In this system the Gymnosperms are treated separately.

The families with inferior ovary have been treated in the last, both in Archichlamydeae and Sympetalae. The advancement is marked from the hypogyny to complete epigyny.

Engler considered the orchids to be more highly evolved than the grasses.

Demerits–In this system the Amentiferae or catkin bearers, (*e.g.,* Salicaceae, Juglandaceae, Betulaceae, etc.) have been regarded as most primitive and precede petaliferous families, (*e.g.* Ranunculaceae and Magnoliaceae). The Amentiferae are a reduced rather than a primitive group. According to Bessey and others the polypetaly was earlier, and apetaly was derived from it through modification.

The acceptance of the derivation of dichlamydeous flowers (perianth in two series) from monochlamydeous ones (perianth in single series) is objectionable.

The other demerits of this system (Engler's) are :

Derivation of parietal placenatation from axile placentation.

Derivation of free–central placentation from parietal placentation.

Derivation of bisexual flowers from unisexual flowers.

Derivation of entomophily from anemophily.

In this system Monocots have been considered to be more primitive than Dicots, which does not correspond to the present day knowledge.

C.E. Bessey, an American botanist (1845–1915) proposed a system of classification based on Bentham and Hooker's system. According to Bessey the Ranales were the primitive angiosperms from which two branches were given out which gave birth to Monocots and Dicots. The system proposed by C.E. Bessey has never been accepted and adopted to any great degree in the world.

Richard von Vettstein (1862–1931), an Austrian botanist also proposed a system of classification parallel to Engler and Prantl's system with minute differences. In this system the unisexual and naked flowers are considered more primitive than bisexual and flowers with perianth. He considered dicots to be more primitive than monocots.

Hans Hallier (1868–1932), a German botanist proposed a system based on phylogenetic relationships parallel to Besseyan system with minor alternations. He considers dicots to be older and more primitive than monocots.

A.B.Rendle (1865–1938), an English botanist has also classified the flowering plants in his two–volumes "*The classification of flowering plants*", Vol. I contains Monocots and Vol. II contains Dicots. His system resembles Englerian system with very minor differences. According to him the Dicots are arranged in three grades: 1. Monochlamydeae, 2. Dialypetalae, and 3. Sympetalae. He considers the family Compositae (Dicot) to be highest evolved.

Oswald Tippo, of the University of Illinois published an outline (1942) of a projected phylogenetic classification. This system, however, was not his original, but was a compilation, a synthesis of several principles of Smith and Eames. The Classification of non–vascular plants was that followed by G.M. Smith (1938) and the basic arrangement of vascular plants was that proposed by A.J. Eames (1936). Tippo studied and synthesised the palaeobotanical data and compared these data with the data obtained through the study of living plants. He emphasized the dependence on the vegatative characters such as plant form, branching and vascular anatomy in arriving at the natural grouping of major groups. This system emphasises that the Pteridophyta is not a homogeneous group and that there is no distinct line of differentiation between Pteridophyta and Spermatophyta. The outline of this classification is as follows :

Kingdom : Plantae

Sub–kingdom: Thallophyta

Phyla. 1.Cyanophyta; 2. Chlorophyta; 3. Chrysophyta; 4. Pyrrophyta; 5. Phaeophyta; 6. Rhodophyta; 7. Schizomycophyta; 8. Myxomycophyta; 9. Eumycophyta.

Sub–kingdom: Embryophyta

Phylum : Bryophyta

Class. 1. Musci; 2. Hepaticae; 3. Anthocerotae

Phylum : Tracheophyta

Sub-phylum: Lycopsida

Class: Lycopodineae

Orders: 1. Lycopodiales; 2. Selaginellales; 3. Lepidodendrales; 4. Pleuromeiales; 5. Isoetales.

Sub–phylum : Sphenopsida.

Class: Equisetineae

Orders: 1. Hyeniales; 2. Sphenophyllales; 3. Equisetales.

Sub–phylum: Pteropsida

Class: Filicineae

Orders: 1. Coenopteridales; 2. Ophioglossales; 3. Marattiales; 4. Filicales

Class : Gymnospermae

Sub–class: Cycadophytae

Orders: 1. Cycadofilicales; 2. Bennettitales; 3. Cycadales.

Sub–class : Coniferophyta

Orders : 1. Cordaitales; 2. Ginkgoales; 3. Coniferales; 4. Gnetales.

Class : Angiospermae

Sub–class: 1. Dicotyledoneae; 2. Monocotyledoneae

John Hutchinson (1884–1972) an English botanist has given the latest information about the phylogenetic classification of angiosperms which has been published in his famous work, *"Families of Flowering Plants"* recently in 1959. Formerly this work was published twice in 1926 and 1934. Hutchinson's classification is more closely related to that of Bentham and Hooker's and Bessey's systems of classification than that of Engler and Prantl's system. According to Hutchinson, the primitive polypetalous forms have been diverged from the very beginning along two separate lines. One of the lines still retains the arboreal habit whereas the other one has adopted the herbaceous habit. The first order of his arboreal line is Magnoliales and the first order of herbaceous line is Ranales. Hutchinson has divided the angiosperms into two large groups 1. Herbaceae and 2. Lignosae. The group Herbaceae contains the herbaceous representatives and the group Lignosae contains the woody representatives. He has proposed that the families having apetalous flowers have been derived partly through the Magnoliales and partly through the Ranales.

Hutchinson has divided the seed plants into two phyla–1. *Gymnospermae* and 2. *Angiospermae.* The phylum Angiospermae has been further divided into two sub-phyla–1. *Dicotyledones* and 2. *Monocotyledones.* The sub–phylum Dicotyledones has been divided into two divisions–1. *Lignosae* and 2. *Herbaceae.* The Lignosae, a woody group and Herbaceae, a herbaceous group. The sub–phylum Monocotyledones has been divided into three groups-1. *Calyciferae,* 2. *Corolliferae* and 3. *Glumiflorae.* The flowers of group Calyciferae possess distinct calyx and corolla; the flowers of group Corolliferae possess more or less similar calyx and corolla; the flowers of group Glumiflorae possess much reduced perianth or lodicules.

Phylum Angiospermae–Ovules enclosed in an ovary usually crowned by a style and stigma, the latter receiving the pollen grains mainly through the agency of insects becoming wind–pollinated when much reduced. Wood when present consisting of true vessels. A more recently evolved phylum than the *Gymnospermae,* and constituting the bulk of the present world vegetation, yielding valuable timbers and practically all food, forage and medicinal plants.

Sub–Phylum **I. Dicotyledones**–Embryonic plant with two cotyledons. Vascular bundles of the stem usually arranged in a circle (except in a few genera of the lower herbaceous families which have scattered bundles). Leaves typically net veined, opposite or alternate. Flowers usually pentamerous or tetramerous.

Division **I. Lignosae**–Trees and shrubs and some herbs clearly derived from and related to other woody plants–a fundamentally woody group.

Division **II. Herbaceae**–Herbs, or rarely shrubby plants related to and derived from herbaceous stocks–a fundamentally herbaceous group.

Sub–Phylum **II. Monocotyledones**–Embryonic plant with only one cotyledon; vascular bundles of the stem closed and scattered. Leaves typically parallel nerved, alternate, often sheathing at the base. Flowers usually trimerous.

Primitive and advanced characters which formed the basis of his classification are : polypetalous condition is more primitive than gamopetalous condition; unisexual flowers are more advanced than bisexual flowers; apetalous forms have originated from petaliferous stock, and represent reduction and advance; free parts on the whole are considered more primitive than connate or adnate parts; thus numerous free stamens are earlier than few or connate stamens; similarly apocarpous pistil is more primitive than syncarpous pistil; spiral arrangement of parts is more primitive than cyclic.

The outline of Hutchinson's system of classification is as follows :

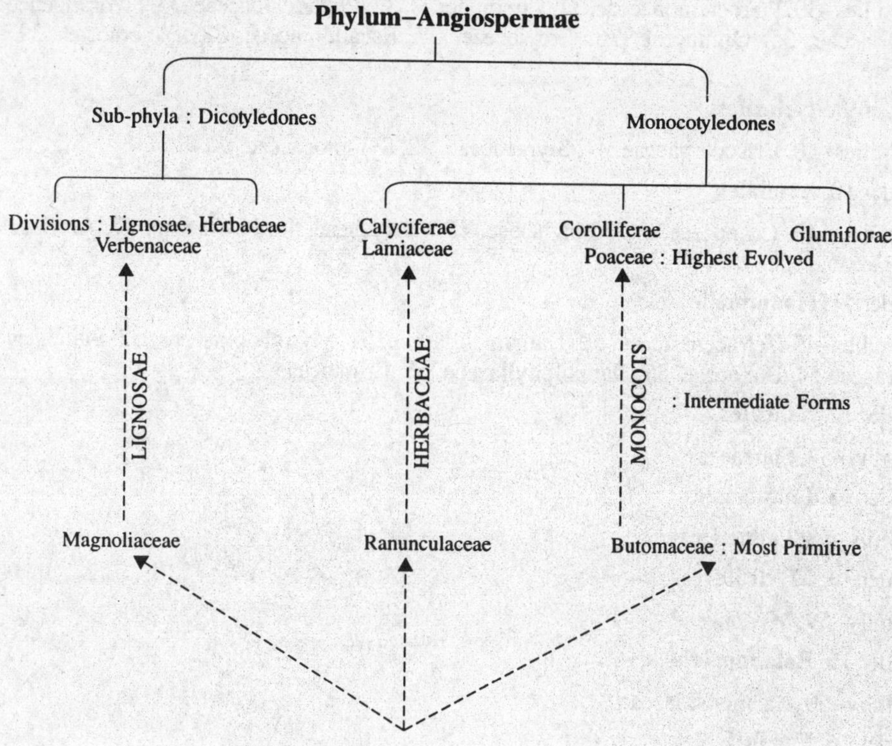

HUTCHINSON'S SYSTEM OF CLASSIFICTION, 1959

Phylum – *Angiospermae*

Sub–Phylum – *Dicotyledones*

Division I – *Lignosae*

Order 1. **Magnoliales**

Families–1. Magnoliaceae, 2. Illiciaceae, 3. Winteraceae, 4. Canellaceae, 5. Schisandraceae, 6. Himantandraceae, 7. Lactoridaceae, 8. Trochodendraceae, 9. Cercidiphyllaceace.

Order 2. **Annonales**

Families–10. Annonaceae, 11. Eupomatiaceae.

Order 3. **Laurales**

Families–12. Monimiaceae, 13. Austrobaileyaceae, 14. Trimeniaceae, 15. Lauraceae, 16. Gomortegaceae, 17. Hernandiaceae, 18. Myristicaceae.

Order 4. **Dilleniales**

Families–19. Dilleniaceae, 20. Connaraceae, 21. Crossosomataceae. 22. Brunelliaceae.

Order 5. **Coriariales**

Family–23. Coriariaceae.

Order 6. **Rosales**

Families–24. Rosaceae, 25. Dichapetalaceae, 26. Calycanthaceae.

Order 7. **Leguminales**

Families–27. Caesalpiniaceae, 28. Mimosaceae, 29. Papilionaceae.

Order 8. **Cunoniales**

Families–30. Pterostemonaceae, 31. Cunoniaceae; 32. Philadelphaceae, 33. Hydrangeaceae, 34. Grossulariaceae; 35. Oliniaceae, 36. Greyiaceae, 37. Escalloniaceae, 38. Baueraceae. 39. Crypteroniaceae.

Order 9. **Styracales**

Families–40. Lissocarpaceae. 41. Styracaceae, 42. Symplocaceae.

Order 10. **Araliales**

Families–43. Cornaceae, 44. Alangiaceae, 45. Garryaceae, 46. Nyssaceae, 47. Araliaceae, 48. Caprifoliaceae.

Order 11. **Hamamelidales**

Families–49. Tetracentraceae, 50. Hamamelidaceae; 51. Myrothamnaceae, 52. Platanaceae, 53. Stachyuraceae 54. Buxaceae, 55. Daphniphyllaceae, 56. Bruniaceae.

Order 12. **Salicales**

Family–57. Salicaceae.

Order 13. **Leitneriales**

Family–58. Leitneriaceae.

Order 14. **Myricales**

Family–59. Myricaceae.

Order 15. **Balanopsidales**

Family–60. Balanopsidaceae.

Order 16. **Fagales**

Families–61. Betulaceae, 62. Fagaceae, 63. Corylaceae.

Order 17. **Juglandales**

Family–64. Rhoipteleaceae, 65. Juglandaceae, 66. Pictrodendraceae.

Order 18. **Casuarinales**

Family–67. Casuarinaceae.

Order19. **Urticales**

Families–68. Ulmaceae, 69. Cannabiaceae, 70. Moraceae, 71. Urticaceae, 72. Barbeyaceae, 73. Eucommiaceae.

Order 20. **Bixales**

Families–74. Bixaceae, 75. Cistaceae, 76. Flacourtiaceae, 77. Cochlospermaceae, 78. Hoplestigmataceae, 79. Achatocarpaceae, 80. Lacistemaceae.

Order 21. **Thymelaeales**

Families–81. Gonystylaceae, 82. Aquilariaceae, 83. Geissolomataceae, 84. Penaeaceae, 85. Thymelaeaceae, 86. Nyctaginaceae.

Order 22. **Proteales**

Family–87. Proteaceae.

Order 23. **Pittosporales**

Families–88. Pittosoporaceae, 89. Byblidaceae, 90. Stegnospermaceae, 91. Vivianiaceae, 92. Tremandraceae.

Order 24. **Capparidales**

Families–93. Capparidaceae, 94. Moringaceae, 95. Tovariaceae.

Order 25. **Tamaricales**

Families–96. Frankeniaceae, 97. Tamaricaceae, 98. Fouquieriaceae.

Order 26. **Violales**

Family–99. Violaceae.

Order 27. **Polygalales**

Families–100. Polygalaceae, 101. Krameriaceae, 102. Trigoniaceae, 103. Vochysiaceae.

Order 28. **Loasales**

Families–104. Turneraceae, 105. Loasaceae.

Order 29. **Passiflorales**

Families–106. Malcsherbiaceae, 107. Passifloraceae, 108. Achariaceae.

Order 30. **Cucurbitales**

Families–109. Cucurbitaceae, 110. Begoniaceae, 111. Datiscaceae, 112. Caricaceae.

Order 31. **Cactales**

Family–113. Cactaceae.

Order 32. **Tiliales**

Families–114. Dirachmaceae, 115. Scytopetalaceae, 116. Tiliaceae, 117. Sterculiaceae, 118. Peridiscaceae, 119. Bombacaceae.

Order 33. **Malvales**

Family–120. Malvaceae.

Order 34. **Malpighiales**

Families–121. Ixonanthaceae, 122. Malpighiaceae. 123. Humiriaceae, 124. Linaceae, 125. Irvingiaceae, 126. Huaceae, 127. Ledocarpaceae, 128. Erythroxylaceae, 129. Ctenolophonaceae, 130. Lepidobotryaceae, 131. Balanitaceae, 132. Zygophyllaceae.

Order 35. **Euphorbiales**

Family–133. Euphorbiaceae.

Order 36. **Theales**

Families–134. Bonnetiaceae, 135. Theaceae, 136. Sauraniaceae, 137. Actinidiaceae, 138. Pelicieraceae, 139. Pentaphylacaceae, 140. Tetrameristaceae, 141. Marcgraviaceae, 142. Caryocaraceae, 143. Medusagynaceae.

Order 37. **Ochnales**

Families–144. Strasburgeriaceae, 145. Ochnaceae, 146. Rhodolaenaceae, 147. Sphaerosepalaceae, 148. Dipterocarpaceae, 149. Ancistrocladaceae.

Order 38. **Ericales**

Families–150. Clethraceae, 151. Pyrolaceae, 152. Ericaceae, 153. Epacridaceae, 154. Diapensiaceae, 155. Monatropaceae, 156. Lennoaceae, 157. Vacciniaceae.

Order 39. **Guttiferales**

Families–158. Hypericaceae, 159. Clusiaceae; 160. Eucryphiaceae, 161. Quiinaceae.

Order 40. **Myrtales**

Families–162. Myrtaceae, 163. Lecythidaceae, 164. Rhizophoraceae, 165. Sonneratiaceae, 166. Punicaceae, 167. Combretaceae, 168. Melastomataceae.

Order 41. **Celastrales**

Families–169. Pandaceae, 170. Aquifoliaceae, 171. Salvadoraceae, 172. Koeberliniaceae, 173. Cneoraceae, 174. Cardiopteridaceae, 175. Cyrillaceae, 176. Icacinaceae, 177. Empetraceae, 178. Aextoxicaceae, 179. Pentadiplandraceae, 180. Celastraceae, 181. Corynocarpaceae, 182. Stackhousiaceae, 183. Goupiaceae, 184. Hippocrateaceae, 185. Erythropalaceae, 186. Capusiaceae, 187. Scyphostegiaceae.

Order 42. **Olacales**

Family–188. Olacaceae, 189. Opiliaceae, 190. Octoknemaceae, 191. Aptandraceae, 192. Dipentodontaceae, 193. Medusandraceae.

Order 43. **Santalales**

Families–194. Loranthaceae, 195. Grubbiaceae, 196. Santalaceae, 197. Myzodendraceae, 198. Balanophoraceae.

Order 44. **Rhamnales**

Families–199. Heteropyxidaceae, 200. Elaeagnaceae, 201. Rhamnaceae, 202. Vitaceae.

Order 45. **Myrsinales**

Families–203. Myrsinaceae, 204. Theophrastaceae, 203. Aegicerataceae.

Order 46. **Ebenales**

Families–206. Ebenaceae, 207. Sapotaceae, 208. Sarcospermaceae.

Order 47. **Rutales**

Families–209. Rutaceae, 210. Simaroubaceae, 211. Burseraceae, 212. Averhoaceae.

Order 48. **Meliales**

Family–213. Meliaceae.

Order 49. **Sapindales**

Families–214. Melianthaceae, 215. Sapindaceae, 216. Podoaceae, 217. Sabiaceae, 218. Anacardiaceae, 219. Aceraceae, 220. Hippocastanaceae, 221. Staphyleaceae, 222. Akaniaceae, 223. Julianiaceae, 224. Didiereaceae.

Order 50. **Loganiales**

Families–225. Potaliaceae, 226. Loganiaceae, 227. Buddleiaceae, 228. Antoniaceae, 229. Spigeliaceae, 230. Strychnaceae, 231. Oleaceae.

Order 51.**Apocynales**

Families–232. Plocospermaceae, 233. Apocynaceae, 234. Periplocaceae, 235. Asclepiadaceae.

Order 52. **Rubiales**

Families–236. Dialypetalanthaceae, 237. Rubiaceae.

Order 53. **Bignoniales**

Families–238. Cobaeaceae, 239. Bignoniaceae, 240. Pedaliaceae, 241. Martyniaceae.

Order 54. **Verbenales**

Families–242. Ehretiaceae, 243. Verbenaceae, 244. Stillbeaceae, 245. Chloanthaceae, 246. Phrymaceae.

Subphylum ... *Dicotyledones*

Division II ... *Herbaceae*

Order 55. **Ranales**

Families-247. Paeoniaceae, 248. Helleboraceae, 249. Ranunculaceae, 250. Nymphaeaceae, 251. Podophyllaceae, 252. Ceratophyllaceae, 253. Cabombaceae.

Order 56. **Berberidales**

Families–254. Sargentodoxaceae, 255. Lardizabalaceae, 256. Menispermaceae, 257. Nandinaceae, 258. Circaeasteraceae, 259. Berberidaceae.

Order 57. **Aristolochiales**

Families–260. Aristolochiaceae, 261. Hydnoraceae, 262. Cytinaceae, 263. Nepenthaceae.

Order 58. **Piperales**

Families–264. Piperaceae, 265. Saururaceae, 266. Chloranthaceae.

Order 59. **Rhoeadales**

Families–267. Papaveraceae, 268. Fumariaceae,

Order 60. **Cruciales** (Brassicales).

Family–269. Cruciferae (Brassicaceae).

Order 61. **Resedales**

Family–270. Resedaceae.

Order 62. **Caryophyllales**

Families–271. Elatinaceae, 272. Molluginaceae, 273. Caryophyllaceae, 274. Ficoidaceae, 275. Portulacaceae.

Order 63. **Polygonales**

Families–276. Polygonaceae, 277. Illecebraceae.

Order 64. **Chenopodiales.**

Families–278. Barbeniaceae. 279. Phytolaccaceae, 280. Gyrostemonaceae, 281. Agdestidaceae, 282. Petiveriaceae, 283. Chenopodiaceae, 284. Amaranthaceae, 285. Cynocrambaceae, 286. Batidaceae, 287. Basellaceae.

Order 65. **Lythrales**

Families–288. Lythraceae, 289. Onagraceae, 290. Trapaceae, 291. Haloragidaceae, 292. Callitrichaceae.

Order 66. **Gentianales**

Family–293. Gentianaceae, 294. Menyanthaceae.

Order 67. **Primulales**

Families–295. Primulaceae, 296. Plumbaginaceae.

Order 68. **Plantaginales**

Family–297. Plantaginaceae.

Order 69. **Saxifragales**

Families–298. Crassulaceae, 299. Cephalotaceae, 300. Saxifragaceae, 301. Eremosynaceae, 302. Vahliaceae, 303. Francoaceae, 304. Donatiaceae, 305. Parnassiaceae, 306. Adoxaceae,

Order 70. **Sarraceniales**

Families–307. Droseraceae, 308. Sarraceniaceae.

Order 71. **Podostemales**

Families–309. Podostemaceae, 310. Hydrostachyaceae.

Order 72. **Umbellales**

Family–311. Umbelliferae (Apiaceae).

Order 73. **Valerianales**

Families–312. Valerianaceae, 313. Dipsacaceae, 314. Calyceraceae.

Order 74. **Campanales**

Families–315. Campanulaceae, 316. Lobeliaceae,

Order 75. **Goodeniales**

Families–317. Goodeniaceae, 318. Brunoniaceae,319. Stylidiaceae.

Order 76. **Asterales**

Family–320. Compositae (Asteraceae).

Order 77. **Solanales**

Families–321. Solanaceae, 322. Convolvulaceae, 323. Nolanaceae.

Order 78. **Personales**

Families–324. Scrophulariaceae, 325. Acanthaceae, 326. Gesneriaceae, 327. Orobanchaceae, 328. Lentibulariaceae, 329. Columelliaceae.

Order 79. **Geraniales**

Families–330. Geraniaceae, 331. Limnanthaceae, 332. Oxalidaceae, 33. Trapaeolaceae, 324. Balsaminaceae.

Order 80. **Polemoniales**

Families–335. Polemoniaceae, 336. Hydrophyllaceae, 337. Cuscutaceae.

Order 81. **Boraginales**

Family–338. Boraginaceae.

Order 82. **Lamiales**

Families–339. Myoporaceae, 340. Selaginaceae, 341. Globuriaceae, 342. Labiatae (Lamiaceae).

Subphylum – *Monocotyledones*

Division I – *Calyciferae*

Order 83. **Butomales**

Families–343. Butomaceae, 344. Hydrocharitaceae.

Order 84. **Alismatales**

Family–345. Alismataceae, 46. Scheuchzeriaceae, 347. Petrosaviaceae.

Order 85. **Triuridales**

Family–348. Triuridaceae.

Order 86. **Juncaginales**

Families–349. Juncaginaceae, 350. Lilaeaceae (Heterostylaceae), 351. Posidoniaceae.

Order 87. **Aponogetonales**

Families–352. Aponogetonaceae, 353. Zosteraceae.

Order 88. **Potamogetonales**

Families–354. Potamogetonaceae, 355. Ruppiaceae.

Order 89. **Najadales**

Families–356. Zannichelliaceae, 357. Najadaceae.

Order 90. **Commelinales**

Families-358. Commelinaceae, 359. Cartonemataceae, 360. Flagellariaceae, 361. Mayacaceae.

Order 91. **Xyridales**

Families–362. Xyridaceae, 363. Rapateaceae

Order 92. **Eriocaulales**

Family–364. Eriocaulaceae.

Order 93. **Bromeliales**

Family–365. Bromeliaceae.

Order 94. **Zingiberales**

Families–366, Musaceae, 367 Sterlitziaceae, 368. Lowiaceae, 369. Zingiberaceae, 370. Cannaceae, 371. Marantaceae.

Division II – *Corolliferae*

Order 95. **Liliales**

Families–372, Liliaceae, 373. Tecophilaceae, 374. Triliaceae, 375. Pontederiaceae, 376. Smilacaceae, 377. Ruscaceae.

Order 96. **Alstroemeriales**

Families–378. Alstroemeriaceae, 379. Petermanniaceae, 380. Philesiaceae.

Order 97. **Arales**

Families–381. Araceae, 382. Lemnaceae.

Order 98. **Typhales**

Family–383. Sparganiaceae, 384. Typhaceae.

Order 99. **Amaryllidales**

Families–385. Amaryllidaceae.

Order 100. **Iridales**

Family–386. Iridaceae.

Order 101. **Dioscoreales**

Families–387. Stenomeridaceae, 388. Trichopodaceae, 389. Roxburghiaceae, 390. Dioscoreaceae.

Order 102. **Agavales**

Family–391. Xanthorrhoeaceae, 392. Agavaceae.

Order 103. **Palmales**

Family–393. Palmae (Palmaceae Arecaceae).

Order 104. **Pandanales**

Family–394. Pandanaceae.

Order 105. **Cyclanthales**

Family—395. Cyclanthaceae.

Order 106. **Haemodorales**

Families–396. Haemodoraceae, 397. Hypoxidaceae, 398. Velloziaceae, 399. Apostasiaceae, 400. Taccaceae, 401. Philydraceae.

Order 107. **Burmanniales**

Families–402. Burmanniaceae, 403. Thismiaceae, 404. Corsiaceae.

Order108. **Orchidales**

Family–405. Orchidaceae.

Division III – *Glumiflorae*

Order 109. **Juncales**

Families–406. Juncaceae, 407. Thurniaceae, 408. Centrolepidaceae, 409. Restionaceae.

Order 110. **Cyperales**

Family–410. Cyperaceae.

Order 111. **Graminales**

Family–411. Gramineae (Poaceae)

Merits and Demerits of Hutchinson's System.

Merits. 1. Most taxonomists are of opinion that this system has given a much better idea of phylogenetic conception and has stimulated phyletic rethinking to a greater extent. 2. Primitive or basic orders are Magnoliales representing arborescent families, and Ranales representing herbaceous families, giving rise to woody and herbaceous forms respectively on parallel lines; 3. Bisexual and polypetalous flowers precede unisexual and gamopetalous flowers; 4. Amentiferae(catkin–bearing families with unisexual, apetalous flowers), *e.g.*, Fagaceae–oak, Betulaceae–birch, Juglandaceae–walnut, Salicaceae–willow and poplar, etc., is regarded as advanced (and not primitive as considered by Engler) and thus transferred (including Urticales) to a new phyletic position close to Rosales and Leguminales; their apparent simplicity represents reduction and specialization (and not primitiveness); 5. Casuarinaceae has been assigned an advanced position and placed at the top of the Amentiferae; 6. Several big orders have been split up into distinct small families, thus simplifying matters, *e.g.*, Rosales Perietales, Malvales, etc. 7. Many families have been raised to the rank of orders, *e.g.*, Leguminosae to Leguminales, Saxifragaceae to Saxifragales, Podostemaceae to Podostemales, etc. 8. Reshuffling of several orders and families, *e.g.* Cactales (Opuntiales of Engler) placed close to Cucurbitales and Passiflorales; 9. Origin of monocotyledons from dicotyledons at an early stage of evolution, the point of origin being the Ranales. 10. Splitting up of Helobiae (of Engler) into separate orders and considering the order Butomales (Butomaceae and Hydrocharitaceae) as the starting point of monocotyledons. 11. Reshuffling of genera of Liliaceae and Amaryllidaceae on the basis of inflorescence characters. 12. Rearrangement of several orders, finally ending in Cyperales and Graminales.

Demerits. 1. Many taxonomists do not agree with his rigid bifurcation of dicotyledons into Lignosae (woody plants) and Herbaceae (herbaceous types). 2. Many taxonomists hold different views regarding the relationship between various orders and families. 3. Monophyletic view regarding the origin of angiosperms is not universally accepted. 4. Monophyletic origin of Monocotyledons from the Ranales, as against the polyphyletic (diphyletic) views of Lotsty (1911) and Hallier (1912). 5. Urticales, Umbellales, Euphorbiales, etc., as originating from different ancestors.

COMPARISON OF THREE SYSTEMS

Bentham & Hooker (1862) *Genera Plantarum*	**Engler and Prantl (1931)** *Die naturlichen Pflanzenfamilien*	**Hutchinson (1959)** *The Families of Flowering Plants*
Flowering plants divided into a few large groups based mainly on single and mostly artificial characters. Closely allied families often widely separated in consequence.	A phylogenetic system proposed subsequent to the acceptance of the theory of descent. Based on Eichler's system, and on the assumption that in fiowering plants, the *absence of a perianth is a primitive* feature.	A phylogenetic system based on the assumption that plants with sepals and petals, associate with other floral and anatomical characters regarded as also primitive are more ancient phylogenetically than plants without sepals or petals.

This was never claimed to be a phylogenetic system but for practical purposes has served a very useful term.	Thus families such as Casuarinaccae, Piperaceae, Salicaceae, Betulacec, Fagaceae, Urticaceae, etc., are regarded as primitive and are placed before families such as Magnoliaceae and Ranunculaceae.	This seems a logical interpretation of the theory that the parts of an angiospermous flower are modi-fied leaves. Free parts are regarded as primitive, and connate or adnate parts as more recent. The spiral arrangement of parts is more primitive than the cyclic (whorled) and numerous free stamens are earlier than the few or connate. Also the bisexual precedes the unisexual flower.
Based on the earlier de Candoile's System (1818), itself an elaboration and slight, modification of Jussieu's arrangement proposed as early as 1779. This system was therefore cenceived under the influence of the old dogma of the fixity of species, and long before the theory of descent was proposed.	Disregards the fact that many of these families without a perianth are characterized by having a gynoecium composed of the union of two or more carpels, a condition at variance with the universally accepted theory that the carpel is a modified leaf. With such an accepted view as to the origin of the carpel a syncarpous ovary could hardly have preceded an apocarpous one. Moreover a syncarpous ovary is characteristic of nearly all the highly placed of Engler's Archichlamydeae and of Meta–chlamydeae (Gamopeta-lae), the latter group accepted by all authors as being the most highly evolved of the Dicotyledons.	In this system the groups (orders of cohorts) are smaller and families are associated which show close general relationship Resemblan-ces emphasized rather than differences. The monochlamdeae and Gamopetalae of Bentham and Hooker are distributed amongst the Polypetalae according to their relationships.
As a *Genera Plantarum* will ever remain a classic.	Monocotyledons precede the Dicotyledons. Recent work on the fossil records gives little support to this system.	The monocotyledons are placed after the Dicotyledons, from which they were derived at an early stage, the point of origin being the Ranales.

List of Families of Dicotyledones in the Present Text with certain more or less constant characters.

Leaves : Opposite (or verticillate) leaves constant

Families with constantly opposite (or verticillate) leaves.

Caryophyllaceae; Casuarinaceae; Rubiaceae.

Leaves : Opposite (or verticillate) leaves predominant.

Families in which opposite (or verticillate) leaves are predominant.

Acanthaceae ; Apocynaceae ; Asclepiadaceae ; Loranthaceae ; Myrtaceae; Nyctaginaceae ; Pedaliaceae ; Rutaceae ; Scrophulariaceae.

Leaves : Compound : leaves predominant

Families in which compound leaves are predominant.

Umbelliferae (Apiaceae); Caesalpiniaceae; Papilionaceae (Fabaceae); Mimosaceae ; Rosaceae; Rutaceae.

Leaves : Sometimes Compound

Families in which compound leaves sometimes occur.

Convolvulaceae ; Cucurbitaceae; Euphorbiaceae ; Ranunculaceae; Sterculiaceae; Verbenaceae.

Stipules : Leaves always stipulate.

Stipules are constant in the following families;

Caesalpiniaceae; Papilionaceae (Fabaceae); Magnoliaceae; Malvaceae; Mimosaceae; Polygonaceae; Rubiaceae.

Stipules : Leaves mostly stipulate

Families in which the leaves are mostly stipulate.

Linaceae; Rosaceae; Sterculiaceae.

Glandular Leaves

Families in which glandular or pellicid dots occur or sometimes occur in the leaves.

Acanthaceae; Anacardiaceae; Annonaceae; Asteraceae (Compositae); Combretaceae; Euphorbiaceae; Magnoliaceae; Meliaceae; Myrtaceae; Nymphaeaceae; Piperaceae; Polygonaceae; Rubiaceae; Rutaceae; Urticaceae; Verbenaceae; Violaceae.

Anther opening by terminal pores

Loranthaceae (some); Malvaceae; Solanaceae.

Inferior Ovary

Apiaceae (Umbelliferae); Asteraceae (Compositae); Cactaceae; Combretaceae; Cucurbitaceae; Loranthaceae; Myrtaceae; Nymphaeaceae (some); Rosaceae (some); Rubiaceae; Saxifragaceae (some).

CURRENT SYSTEMS

Cronquist's System

Arthur Cronquist (1919—) of U.S.A. presented a comprehensive system of classification in 1968 and thereafter its revised version in 1981. His system is parallel to that of Takhtajan's system in so many respects.

He divided the angiosperms in two classes–Magnoliopsida (=Dicotyledons) and Liliopsida (=Monocotyledons). Magnoliopsida is further divided into six sub–classes, *viz.,* Magnoliidae, Hamamelidae, Caryophyllidae, Dilleniidae, Rosidae and Asteridae; Liliopsida has five sub–classes, *viz.,* Alismatidae, Arecidae, Commelinidae, Zingiberidae and Liliidae.

The division Magnoliophyta has two classes, *viz.,* Magnoliopsida (= Dicotyledons) and Liliopsida (= Monocotyledons). There are six sub–classes, sixty four orders, 318 families, and about 165,000 species in Magnoliopsida (=Dicotyledons) whereas in Liliopsida (= Monocotyledons) there are five sub–classes, nineteen orders, 65 families and about 50,000 species. The outline of Cronquist's (1981) system of classification is as follows :

CLASS A. MAGNOLIOPSIDA (= DICOTYLEDONS)

Sub–class I Magnoliidae

Order	1.	Magnoliales
" "	2.	Laurales
" "	3.	Piperales
" "	4.	Aristolochiales
" "	5.	Illiciales
" "	6.	Nymphaeales
" "	7.	Ranunculales
" "	8.	Papaverales

Sub–Class II. Hamamelidae

Order	1.	Trochodendrales
" "	2.	Hamamelidales
" "	3.	Daphnophyllales
" "	4.	Dedymelales
" "	5.	Eucommiales
" "	6.	Urticales
" "	7.	Leitneriales
" "	8.	Juglandales
" "	9.	Myricales
" "	10.	Fagales
" "	11.	Casuarinales.

Sub–class III Caryophyllidae

Order	1.	Caryophyllales
" "	2.	Polygonales
" "	3.	Plumbaginales

Sub–class IV. Dilleniidae

Order	1.	Dilleniales
" "	2.	Theales
" "	3.	Malvales
" "	4.	Lecythidales
" "	5.	Nepenthales
" "	6.	Violales
" "	7.	Salicales
" "	8.	Capparales
" "	9.	Batales
" "	10.	Ericales
" "	11.	Diapensiales
Order	12.	Ebenales
" "	13.	Primulales

Sub–Class V. Rosidae

Order	1.	Rosales
" "	2.	Fabales
" "	3.	Proteales
" "	4.	Podostemales
" "	5.	Haloragales
" "	6.	Myrtales
" "	7.	Rhizophorales
" "	8.	Cornales
" "	9.	Santalales
" "	10.	Rafflesiales
" "	11.	Celastrales
" "	12.	Euphorbiales
" "	13.	Rhamnales
" "	14.	Linales
" "	15.	Polygalales
" "	16.	Sapindales
" "	17.	Apiales

Sub–class VI. Asteridae

Order	1.	Gentianales
" "	2.	Solanales
" "	3.	Lamiales
" "	4.	Callitrichales
" "	5.	Plantaginales
" "	6.	Scrophulariales
" "	7.	Campanulales
" "	8.	Rubiales
" "	9.	Dipsacales
" "	10.	Calycerales
" "	11.	Asterales

CLASS B. LILIOPSIDA (=MONOCOTYLEDONS)

Sub–class I. Alismatidae

Order	1.	Alismatales
" "	2.	Hydrocharitales
Order	3.	Najadales
" "	4.	Triuridales

Sub–class II. Arecidae

Order	1.	Arecales
" "	2.	Cyclanthales
" "	3.	Pandanales

" "	4.	Arales	" "	7.	Typhales
Sub–class III.	Commelinidae		Sub–class IV.	Zingiberidae	
Order	1.	Commelinales	Order	1.	Bromeliales
" "	2.	Eriocaulales	" "	2.	Zingiberales
" "	3.	Restionales	Sub–Class V.	Liliidae	
Order	4.	Juncales	Order	1.	Liliales
" "	5.	Cyperales	" "	2.	Orchidales
" "	6.	Hydatallales			

Takhtajan's System

Armen Takhtajan, an academician of Leningrad, Russia, presented a system of classification of angiosperms for the first time in 1942, and thereafter its revised versions in 1954, 1966, 1969 and 1980 respectively. Takhtajan's approach is quite similar to that of Cronquist. According to him the angiosperms are of monophyletic origin and evolved from some ancient group of gymnosperms.

The evolutionary tendencies followed by Takhtajan are as follows:

Woody plants precede herbaceous plants.

Sparingly branced trees gave rise to the trees with numerous slender branches.

Evergreen trees precede deciduous woody plants.

Simple pinnately–nerved leaves precede pinnately lobed, pinnatified and pinnatisect leaves with pinnate venation.

Simple leaves precede compound leaves and *vice–versa.*

Pinnate venation precedes palmate venation. Reticulate venation precedes parallel venation.

Alternate leaf arrangement precedes opposite and verticillaster arrangements.

Stomata with subsidiary cells are primitive and without subsidiary cells are advanced.

Tripentalacunar type of nodal structure is primitive, and unilacunar type is advanced.

The vessel members with scalariform perforations precede the vessel members with simple perforati ns.

Tracheids gave rise to fibre tracheids, and then to libriform fibres.

Cymose inflorescence is primitive, and racemose is derived one.

Primitive flowers are spiral whereas advanced are cyclic in arrangement of their floral organs.

The primitive stamens were leaf–like, and with marginally situated microsporangia.

The pollen is derived from monocolpate to tricolpate and from tricolpate to triporate. The exine of primitive pollen was without external sculpturing whereas in advanced pollen there was sculpturing of various types.

Apocarpy precedes syncarpy.

Bitegmic ovules are primitive, whereas unitegmic are advanced.

Crassinucellate ovules precede tenuinucellate ovules. Anatropous ovule is primitive; the other types are derived from it.

Entomophily gave rise to anemophily and *vice versa.*

Polygonum–type of embryo sac is primitive, the other types are derived from it.

Endospermic seeds are primitive, whereas non–endospermic seeds are advanced. The monocot embryo is evolved from dicot embryo.

Many seeded follicle is most primitive type of fruit, from this other types are evolved.

Takhtajan has divided his Division–Magnoliophyta (=Angiospermae) into two classes–Magnoliopsida (= Dicotyledones) and Liliopsida (=Monocotyledones). He has further divided–Class Magnoliopsida into seven sub–classes and twenty superorders, and Liliopsida into three sub–classes and eight superorders.

According to him the order Magnoliales is most primitive among the flowering plants. All other angiosperms have been derived from the Magnoliales.

The order Alismatales has been supposed to be most primitive among monocotyledons and the members of the other monocotyledonous orders have been derived from them.

According to him the highest evolved order among dicotyledons is Asterales; and Arales among monocotyledons.

The outline of Taktajan's (1980) system is as follows :

DIVISION MAGNOLIOPHYTA (=ANGIOSPERMAE)
Class MAGNOLIOPSIDA (=DICOTYLEDONES)

Sub–class A. Magnoliidae
Superorder I. Magnolianae
Order 1. Magnoliales(Annonales)
" " 2. Illiciales
" " 3. Laurales
" " 4. Piperales
" " 5. Aristolochiales
Superorder II. Rafflesianae
Order 6. Rafflesiales
Superorder III. Nymphaeanae
Order 7. Nymphaeales
" " 8. Nelumbonales
Sub–class B. Ranunculidae
Superorder IV. Ranunculanae
Order 9. Ranunculales
" " 10. Papaverales
" " 11. Sarraceniales
Sub–class C. Hamamelididae
Superorder V. Hamamelidanae
Order 12. Trochodendrales
" " 13. Cercidiphyllales
" " 14. Eupteleales
" " 15. Didymelales
" " 16. Hamamelidales
" " 17. Eucommiales
" " 18. Urticales
Order 19. Barbeyales
" " 20 Casuarinales
" " 21. Fagales
" " 22. Balanopales
" " 23. Leitneriales
Superorder VI. Juglandanae
Order 24. Myricales
" " 25. Juglandales
Sub–class D. Caryophyllidae
Superorder VII. Caryophyllanae
Order 26. Caryophyllales
" " 27. Polygonales
Super order VIII. Plumbaginanae
Order 28. Plumbaginales
Sub–class E. Dilleniidae
Order 29. Dilleniales
" " 30. Paeoniales
" " 31. Theales
" " 32. Violales
" " 33. Bigoniales(Datiscales)
" " 34. Capparales
" " 35. Tamaricales
" " 36. Salicales
Superorder X. Ericanae
Order 37. Ericales
" " 38. Ebenales
Order 39. Primulales
Superorder XI. Malvanae
Order 40 Malvales
" " 41. Euphorbiales

" " 42. Thymelaeales

Sub–class F. Rosidae

Superorder XII Rosanae

Order 43. Saxifragales

" " 44. Rosales

" " 45. Fabales

" " 46. Connarales

" " 47. Podostemales

" " 48. Nepenthales

Superorder XIII. Myrtanae

Order 49. Myrtales

Superorder XIV. Rutanae

Order 50. Rutales

" " 51. Sapindales(Acerales)

" " 52. Geraniales

" " 53. Polygalales

Superorder XV. Aralianae

Order 54. Cornales

" " 55. Araliales (Apiales)

Superorder XVI. Celastranae

Order 56. Celastrales

" " 57. Santalales

" " 58. Balanophorales

" " 59. Rhamnales

" " 60. Elaeagnales

Superorder XVII. Proteanae

Order 61. Proteales

Sub–class G. Asteridae

Superorder XVIII. Gentiananae

Order 62. Gentianales

" " 63. Oleales

" " 64. Dipsacales

" " 65. Loasales

Superorder XIX. Lamianae

Order 66. Polemoniales

" " 67. Lamiales

" " 68. Scrophulariales

Superorder XX. Asteranae

Order 69. Campanulales (including Goodeniales)

" " 70 Calycerales

" " 71. Asterales

CLASS–LILIOPSIDA (=MONOCOTYLEDONES)

Sub–class A. Alismatidae

Superorder I. Alismatanae

Order 1. Alismatales

" " 2. Najadales (including Aponogetonales, Potomogetonales and Zosterales)

Sub–class B. Lilidae (including Commelinidae and Zingiberidae)

Superorder II. Triuridanae

Order 3. Triuridales

Superorder III. Lilianae

Order 4. Liliales

" " 5. Smilacales

Order 6. Burmanniales

" " 7. Orchidales

" " 8. Bromeliales

Superorder IV. Juncanae

Order 9. Juncales

" " 10. Cyperales

Superorder V. Commelinanae

Order 11. Commelinales

" " 12. Eriocaulales

" " 13. Restionales

" " 14. Hydatellales

" " 15. Poales

Superorder VI. Zingiberanae

Order 16. Zingiberales

Sub–class C. Arecidae

Superorder VII. Arecanae

Order 17. Arecales

" " 18. Cyclanthales

" " 19. Pandanales

" " 20. Typhales

Superorder VII. Aranae

Order 21. Arales

Dahlgren's System

Rolf Dahlgren, a botanist of Copenhagen, Denmark published a system of classification of angiosperms in 1975, and thereafter a revised and improved version in 1980, 1981 and 1983 respectively. He has taken into consideration the utmost possible information. He has taken into consideration the following : apocarpy, syncarpy, and monocarpy; choripetaly and sympetaly; monoaperturate pollen grains; sclereid idioblasts; microsporogenesis types; bi–and unitegmic ovules, tenuinucellate, pseudocrassinucellate and crassinucellate ovules; bi–and trinucleate pollens; oxalate raphides, silica bodies; iridoid compounds; ellagic acid and ellagitannins; benzylisoquinoline alkaloids, tropane alkaloids, pyrrolizidine alkaloids; polyacetylene and several flavonoids. This shows that Dahlgren has used chemical characters extensively in his system of classification.

According to him the angiosperms have evolved in one particular line of gymnosperms and thus treated of monophyletic origin. He is not of opinion that Alimatiflorae are closely connected with the monocotyledonous ancestors. According to him "the appearance of the monocotyledons may have its counterpart in the evolution of the mainly herbaceous Aristolochiaceae, which also evolved out of early Magnoliales with trimerous flowers and two perianth whorls Aristolochiaceae share with the monocotyledons in possession of cuneate crystalloids in the sieve element plastids". According to him the "Amentiferae" have been interpreted as derived from trimerous ancestral type.

Dahlgren is in agreement that the discontinuity between monocotyledons and dicotyledons is not particularly marked. He is of opinion that the monocotyledons can be clearly circumscribed on the combination of a single–cotyledoned embryo and sieve element plastids. This system has also been advocated by serological studies. In Dahlgren's classification the angiosperms are made up of a class, Magnoliopsida. It is as follows:

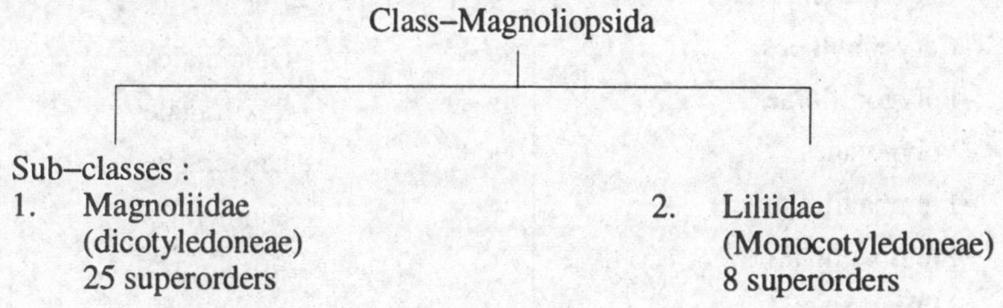

The superorders end with the suffix–florae.

According to Dahlgren the superorder magnoliiflorae have been treated to be most primitive among dicotyledons. They possess the following characters: woody habit, primitive wood; P–type plastids in the sieve elements; alternate, exstipulate leaves, cells with essential oils in leaves; benzylisoquinoline alkaloid present, trimerous tepals; flat and leaf–like stamens with sporangia below the apex, pollengrains sulcate, 2–nucleate; gynoecium apocarpous, or monocarpellary; endosperm formation cellular. The order Nelumbonales is included in superorder Magnoliiflorae and not in Ranunculiflorae.

The superorder Alismatiflorae has the following characteristics which are generally unusual for monocotyledons : intervaginal squamules present; vessels absent in the stem and often in the roots; root hairs attached to short epidermal hairs, amoeboid tapetum; 3–nucleate pollen grains; apocarpous or monocarpellary condition of gynoecia; embryo formation of caryophyllad type; oxalate raphides and silica crystals absent.

The outline classification is as follows:

CLASS–MAGNOLIOPSIDA (ANGIOSPERMAE)

Sub–class MAGNOLIIDAE (=DICOTYLEDONEAE)

Superorder –Magnoliiflorae
Order –Annonales
" " –Aristolochiales
" " –Rafflesiales
" " –Magnoliales
" " –Lactoridales
" " –Chloranthales
" " –Illiciales
" " –laurales
" " –Nelumbonales
Superorder –Nymphaeiflorae
Order –Piperales
" " –Nymphaeales
Superorder –Ranunculiflorae
Order –Ranunculales
" " –Papaverales
Superorder –Caryophylliflorae
Order –Caryophyllales
Superorder –Polygoniflorae
Order –Polygonales
Superorder –Plumbaginiflorae
Order –Plumbaginales
Superorder –Malviflorae
Order –Malvales
" " –Urticales
" " –Euphorbiales
" " –Thymelaeales
" " –Rhamnales
" " –Elaegnales
Superorder –Violiflorae
Order –Violales
" " –Cucurbitales
" " –Salicales
" " –Tamaricales
" " –Capparales
Order –Salvadorales
Superorder –Theiflorae
Order –Dilleniales
" " –Paeoniales
" " –Theales
Superorder –Primuliflorae
Order –Primulales
" " –Ebenales
Superorder –Rosiflorae
Order –Trochodendrales
" " –Cercidiphyllales
" " –Hamamelidales
" " –Geissolomomatales
" " –Balanopales
" " –Fagales
" " –Juglandales
" " –Myricales
" " –Casuarinales
" " –Buxales
" " –Cunoniales
" " –Saxifragales
" " –Droserales
" " –Gunnerales
" " –Rosales
Superorder –Podostemiflorae
Order –Podostemales
Superorder –Proteiflorae
Order –Proteales
Superorder –Myrtiflorae
Order –Haloragales
" " –Rhizophorales
" " –Myrtales
" " –Chrysobalanales
Superorder –Fabiflorae
Order –Fabales
Superorder –Rutiflorae
Order –Sapindales

Order –Rutales
" " –Polygalales
" " –Geraniales
" " –Balsaminales
" " –Tropaeolales
Superorder –Santaliflorae
Order –Celastrales
" " –Vitales
" " –Santalales
Superorder –Balanophoriflorae
Order –Balanophorales
Superorder –Araliflorae
Order –Pittosporales
" " –Araliales
Superorder –Asteriflorae
Order –Campanulales
" " –Asterales
Superorder –Solaniflorae
Order –Solanales
" " –Boraginales
Superorder –Corniflorae
Order –Fouquieriales
" " –Ericales
" " –Eucommiales
" " –Sarraceniales
" " –Cornales
" " –Dipsacales
Superorder –Losaliflorae
Order –Losales
Superorder –Gentianiflorae
Order –Goodeniales
" " –Oleales
" " –Gentianales
Superorder –Lamiiflorae
Order –Scrophulariales
" " –Hippuridales
" " –Lamiales

Order –Hydrostachyales
Sub–class –LILIIDAE
(=MONOCOTYLEDONEAE)
Superorder –Alismatiflorae
Order –Hydrocharitales
" " –Alismatales
" " –Zosterales
Superorder –Triuridiflorae
Order –Triuridales
Superorder –Ariflorae
Order –Arales
Superorder –Liliiflorae
Order –Dioscoreales
" " –Asparagales
" " –Liliales
" " –Burmanniales
" " –Orchidales
Superorder –Bromeliiflorae
Order –Velloziales
" " –Bromeliales
" " –Haemodorales
" " –Philydrales
" " –Pontederiales
" " –Typhales
Superorder –Zingiberiiflorae
Order –Zingiberales
Superorder –Commeliniflorae
Order –Commelinales
" " –Hydatellales
" " –Juncales
" " –Cyperales
" " –Poales
Superorder –Areciflorae
Order –Arecales
" " –Cyclanthales
" " –Pandanales

Thorne's system

Robert F. Thorne of Calfornia U.S.A., published a system of classification in 1968. Thereafter a revised outline of this system was published in 1981 and 1983 respectivelty. This system has been used to express the phylogenetic relationships among the higher taxa of flowering plants. The modern trends of taxonomy have been taken into consideration in the depiction of more accurate relationships of different taxa. He has laid much emphasis on phytochemical approach. The other modern approaches such as palynology, seed morphology, anatomy, embryology, host–parasite relationships, plant geography, palaeobotany, microstructure and cytology have been uitlized in the differentiation of taxa. According to Thorne the Angiospermae are monophyletic in origin. He has divided the flowering plants into two sub–classes (1) Dicotyledoneae (Annonidae) and (2) Monocotyledoneae (Liliidae). There are 19 superorders in Dicotyledoneae and 9 superorders in Monocotyledoneae. The outline of Thorne's system is as follows:

CLASS–ANGIOSPERMAE (ANNONOPSIDA)

Sub–classes : (1) DICOTYLEDONEAE (ANNONIDAE)
19 Superorders

(2) MONOCOTYLEDONEAE
9 Superorders

Sub–CLASS–DICOTYLEDONEAE (ANNONIDAE)

Superorder	–Annoniflorae
Order	–Annonales
" "	–Nelumbonales
" "	–Paeoniales
" "	–Berberidales
Superorder	–Nymphaeiflorae
Order	–Nymphaeales
Superorder	–Rafflesiiflorae
Order	–Rafflesiales
Superorder	–Theiflorae
Order	–Theales
" "	–Ericales
" "	–Ebenales
" "	–Primulales
" "	–Polygonales
Superorder	–Chenopodiiflorae (Centrospermae)
Order	–Chenopodiales
Superorder	–Geraniiflorae
Order	–Geraniales
Superorder	–Santaliflorae
Order	–Celastrales
" "	–Santalales
" "	–Balanophorales
Superorder	–Violiflorae
Order	–Violales
" "	–Caparales
Superorder	–Malviflorae
Order	–Malvales
" "	–Urticales
" "	–Rhamnales
" "	–Euphorbiales
Superorder	–Rutiflorae
Order	–Rutales
Superorder	-Proteiflorae
Order	–Proteales
Superorder	–Hamamelidiflorae
Order	–Hamamelidales
" "	–Casuarinales
" "	–Fagales
Superorder	–Rosiflorae
Order	–Rosales
" "	–Pittosporales
Superorder	–Loasiflorae
Order	–Loasales
Superorder	–Myrtiflorae
Order	–Myrtales
Superorder	–Gentianiflorae
Order	–Oleales
" "	–Gentianales
" "	–Bignoniales
" "	–Lamiales

Superorder –Solaniflorae
Order –Solanales
" " –Campanulales
Superorder –Corniflorae
Order –Cornales
" " –Araliales
" " –Dipsacales
Superorder –Asteriflorae
Order –Asterales

SUB–CLASS MONOCOTYLEDONEAE (LILIIDAE)

Superorder –Liliiflorae
Order –Liliales
Superorder –Triuridiflorae
Order –Triuridales
Superorder –Alismatiflorae
Order –Alismatales
Order –Zosterales
" " –Najadales
Superorder –Ariflorae
Order –Arales
Superorder –Cyclanthiflorae
Order –Cyclanthales
Superorder –Pandaniflorae
Order –Pandanales
Superorder –Areciflorae
Order –Arecales
Superorder –Typhiflorae
Order –Typhales
Superorder –Commeliniflorae
Order –Commelinales
" " –Pontederiineae
" " –Juncineae
" " –Zingiberales

The European botanists have contributed a lot in the field of systematic botany. The modern systematic botany begins from the time of Linnaeus (1707–1778), which has given birth to various systems till today. Some important systems have been discussed in brief in the preceding pages.

3

PRINCIPLES OF TAXONOMY AND PHYLOGENY OF ANGIOSPERMS

The problem of classifying the angiospermic plants in a systematic way may either be termed as, *'taxonomy of angiosperms'* or *'systematic botany'*. This is a functional science and deals with identification, nomenclature and classification of the plants found all over the world. The angiosperms are widely distributed with so many morphological variations that sometimes it seems almost impossible to arrange them in systematic order. Since the prehistoric times, people were interested in placing the plants in a systematic way and for the first time a few plants were classified according to their medicinal and food value and thus the study of taxonomy of flowering plants began. In nineteenth century, the formation of the principles of taxonomy began and several principles were formed, which are still very helpful in arranging the plants in a systematic order.

Swingle has proposed the principles, which have been uniformly accepted by the botanists. They are as follows:

1. Plant relationships are up and down genetic lines and these must constitute the framework of phylogenetic taxonomy. This will naturally form a branching but not reticulate structure.

2. Some evolutionary processess are progressive while others are regressive.

3. Evolution does not necessarily involve all organs of the plant at one time or in the same direction. One organ may be advancing while another is stationary or retrogressing

4. Evolution has generally been consistent and when a particular progression or regression has set in, it is persisted into the end of the phylum.

5. In any phylum the chlorophyll bearing plants precede the chlorophyll less ones. Saprophytes are derived from independent forms and parasites usually from the saprophytes among the lower plants, and from independent forms among the flowering plants.

6. Usually structures with many similar parts are more primitive, and those with fewer and dissimilar parts are more advanced.

7. Among seed plants the stem structure with collateral bundles arranged in a cylinder is more primitive than that with scattered bundles.

8. In most groups of seed plants woody members have preceded the herbaceous ones.

9. In most groups of seed plants erect members have preceded the vines.

10. Perennials are more primitive than biennials and biennials are usually more primitive than annuals.

11. Scalariform vessels are more primitive than vessels with round pits.

12. The spiral arrangement of the leaves on the stem and of the floral leaves precedes that of opposite and whorled types.

13. Simple leaves are more primitive than compound leaves.

14. Historically leaves were first persistent (evergreen) and later deciduous. This way, the deciduous leaves show advanced character.

15. Among the seed plants the netted venation of leaves is more primitive than the parallel venation.

16. The many-parted flower is the more primitive, the type with few parts being derived from it, and the change is accompanied by a progressive sterilization of sporophylls.

17. A condition in which the perianth consists of like segments is more primitive than one in which sepals and petals are unlike each other.

18. Flowers with petals preceded apetalous ones, the latter being derived by reduction.

19. Polypetalous flowers are more primitive than gamopetaolus ones, the latter being derived from the former by symphysis.

20. Regular flowers are more primitive than irregular ones.

21. Spirally imbricate floral parts are more primitive than those that are whorled and valvate.

22. Hypogyny is the primitive structure and from it perigyny and epigyny have been derived.

23. Numerous carpels represent a more primitive condition (than united carpels).

24. Separate carpels represent a more primitive condition than united carpels.

25. Axile placentation preceded parietal and central placentation of ovaries.

26. The presence of numerous stamens indicates a more primitive condition than that of a few stamens.

27. Separate stamens preceded united stamens.

28. Evolution in angiosperms is believed to have proceeded from seeds with two seed coats to those with only one.

29. The primitive seed contains endosperm and a small embryo, the advanced type has little or no endosperm, with the food stored in a large embryo.

30. A straight embryo is usually more primitive than a curved one.

31. The solitary flower is more primitive than the inflorescence.

32. Bisexual flowers preceded unisexual flowers.

33. The monoecious condition is primitive than dioecious.

34. Simple and aggregate fruits preceded multiple fruits.

35. The same evolutionary phenomena have often been repeated as separate occurrences in different parts of the plant kingdom, *e.g.*, loss of chlorophyll, loss of petals, stamens and carpels, acquisition of fleshy texture in fruits and of various types of thorns, change from simple to compound leaves, from erect to prostrate habit and from hypogynous to perigynous or epigynous insertion of floral parts, and lateral union (symphysis) of petals, stamens and carpels.

36. In determining the closeness of relationship between two families or other groups, it is usually best to compare with each other the most primitive, or basal members of each group, rather than those that are simplified by reduction or those that are most highly specialized.

Besides the above mentioned principles proposed by Swingle, the following parallel principles of taxonomy proposed by C.E. Bessey (1915) are of equal importance and are used by all modern taxonomists of twentieth century. These are known as Besseyan principles. They are as follows:

1. There is progressive evolution, *i.e.*, life advances from simplicity to complexity.

2. The existing simpler forms resemble more to their simple ancestors than to the present complex ones.

3. The evolved forms never become like ancestors.

4. The herbaceous habit of plants is more advanced than arboreal habit.

5. Annuals are more advanced than biennials and biennials are more advanced than perennials.

6. Hydrophytes, epiphytes, saprophytes and parasites are more advanced than ordinary terrestrial forms.

7. Compound leaves are more advanced than simple leaves.

8. Bisexual flowers are more primitive than unisexual ones.

9. Dioecious plants are more advanced than monoecious ones.

10. Inflorescence is more advanced than solitary flowers.

11. Gamopetalous flowers are more advanced than polypetalous flowers. The gamopetaly has been derived from polypetaly by the fusion of the petals.

12. Flowers lacking petals (apetaly) are more advanced.

13. Actinomorphic flowers are more primitive than zygomorphic ones.

14. Hypogyny precedes epigyny.

15. Apocarpous condition is more primitive than syncarpous condition.

16. Exalbuminous seeds are more advanced than albuminous seeds.

17. Polyandrous stamens precede jointed condition of stamens.

18. Simple fruits are more primitive than compound fruits.

Turril states, that "the angiosperms were the last of the great groups (Thallophyta, Bryophyta, Pteridophyta, Gymnosperms, Angiosperms) to appear. Hence, changes towards characters peculiar to the angiosperms may with some justification be read as progressive, the more so that they are now the dominant group ecologically over much of the land surface of the globe".

The 'conservative' characters have been supposed by many taxonomists to be most useful and valuable in phylogenetic studies. For plants, the reproductive organs are supposed to be more conservative than the vegetative organs. For example, the morphogenetic evolutionary changes in carpels have been far greater than in any vegetative organ considered throughout Spermophyta.

According to Tuzson, the monocots form an older group than the dicots, because the smaller groups of monocots are separable into series and families which show greater gaps on the whole than those in dicots.

Asa Gray is of opinion that a natural system of classification of plants aim to arrange all the known plants of the plant kingdom, in a series of grades according to their resemblances, in all respects, so that each species, genus, tribe, family, order, etc., shall stand next to those which it most resembles in all respects, or rather in the whole plan of structure. For example, two plants may be very much alike in external appearance, yet very different in their principal structure.

According to Sprague the monophyletic groups are regarded as the only natural ones at and above the family level in existing angiosperms. On the other hand, Hutchinson in his phylogenetic scheme and classification definitely retains groups he considers polyphyletic, *e.g.*, Asterales and Euphorbiales.

Hutchinson's principles of taxonomy and phylogeny of Angiosperms are most acceptable to the present taxonomists. They are as follows —

Hutchinson's General Principles Adopted for the Classification of Flowering Plants:

Other things being equal, it may be stated that:

1. Evolution is both (1) upwards and (2) downwards. The latter involving degradation and degeneration; examples : (1) towards the sympetalous condition; epigyny; (2) towards the apetalous state of many flowers; unisexuality in flowering plants.

2. Evolution does not necessarily involve all organs of the plant at the same time; and one organ or set of organs may be advancing while another set is stationary or retrograding. For example in the family Crassulaceae the carpels have remained free while some genera are sympetalous.

3. Evolution has generally been consistent, and when a particular progression or retrogression has set in, it is persisted into the end of the phylum; examples: the strong tendency to zygomorphy of the corolla coupled with the reduction in the number of stamens in Engler's hypogynous *Metachlamydeae :* the great tendency to perigyny and epigyny in the *Archichlamydeae* and *Metachlamydeae* as exhibited in the families *Apiaceae* and *Rubiaceae* respectively.

Relating to the General Habit of Plants:

4. In certain groups, trees and shrubs are probably more primitive than herbs; examples:Mimosaceae and Caesalpiniaceae (trees and shrubs) as compared with the family Fabaceae (Papilionaceae) (becoming herbaceous).

5. Trees and shrubs are older than climbers, the latter habit having been acquired through particular environment.

6. Perennials are older than biennials, and from them annuals have been derived; note the extraordinarily few annuals in the primitive family Ranunculaceae and in Rosaceae; the gret number in the more advanced and natural family Brassicaceae.

7. Aquatic phanerogams are as a rule more recent than terrestrial (at any rate in the members of the same family or genus), and the same may be said of epiphytes and parasites.

Relating to the General Structure of Flowering Plants:

8. Plants with collateral vascular bundles arranged in a cylinder (Dicotyledons) are more primitive in origin than those with scattered bundles (Monocotyledons), though it does not necessarily follows that the latter have been directly derived from the former.

9. The spiral arrangement of leaves on the stem and of the floral leaves precedes that of the opposite and whorled type.

10. As a rule simple leaves precede compound leaves.

Relating to the Flowers and Fruits of Plants:

11. Bisexual (hermaphrodite) precede unisexual flowers, and the dioecious is probably more recent than the monoecious condition.

12. The solitary flower is more primitive than the inflorescence, the highest forms of the latter being the umbel and capitulum; examples of the latter: Apiaceae and Asteraceae respectively.

13. Spirally imbricate floral parts are more primitive than whorled and valvate; examples : *Magnolia* and *Clematis.*

14. Many-parted flowers (polymerous) precede, and the type with few parts (oligomerous) follows from it, being accompanied by a progressive sterilisation of reproductive parts (sporophylls); examples : *Magnolia* and *Cherianthus.*

15. Petaliferous flowers precede apetalous ones, the latter being the result of reduction.

16. Free petals (polypetaly) are more primitive than connate petals (sympetaly).

17. Actinomorphy (regularity) of the flower is an earlier type than zygomorphy (irregularity); examples: *Caltha* and *Delphinium.*

18. Hypogyny is the primitive structure, and from it perigyny and epigyny were derived later.

19. Free carpels (apocarpy) are more primitive and from them connate carpels resulted; sometimes however, when the carpels have remained loosely united during evolution they may again become quite free, especially in fruit; example : Asclepiadaceae.

20. Many carpels (polycarpy) precede few carpels (oligocarpy); example: *Ranunculus* and *Nigella.*

21. The endospermic seed with small embryo is primitive and the non-endospermic seed more recent; examples: Ranunculaceae and Rosaceae.

22. In primitive flowers there are many stamens, in more advanced flowers few stamens; examples: *Ranunculus* and *Cherianthus.* This condition may, however, be reversed within the confined of a single family like Papaveraceae, where bees feed on the pollen.

23. Separate stamens precede connate stamens; Examples : Tiliaceae and Malvaceae; Campanulaceae and Lobeliaceae.

24. Aggregate fruits (fruits formed several separate flowers) are more recent than single fruits, and as a rule the capsule.

4

MODERN TRENDS IN TAXONOMY

It has now been generally accepted by taxonomists that morphological characters alone should not be regarded in systematical classification of plants. Other disciplines like cytology, genetics, physiology, geographical distribution, anatomy, embryology, ecology, palynology, phenology, bio-chemistry, numerical taxonomy and transplant experiments have been found to be especially useful in solving some of the taxonomical problems by providing additional characters. The cytogenetical data may also indicate the origin and relationship of different taxa. The impact of above mentioned disciplines on the present day taxonomy has changed the face of it from alpha (classical) to omega (modern kind). New systematics (taxonomy) thus has become a very interesting subject, claiming to have solved many problems of classification and to have evolved a better taxonomy in which relationships of all kinds are knit together. However, it has also created confusion to the discomfort of classical taxonomists (Chennaveeraiah, 1970). There are amongst the taxonomists lumpers as well as the splitters. In such considerations morphology must still occupy a prime place although amongst the other disciplines cytology and genetics have played a very significant role (P.N. Mehra, 1970).

1. GROSS MORPHOLOGICAL

Turril says, the morphology, either of plants or animals, now a days includes a good deal more than the mere study of shape. The term 'morphology' is very wide and includes anatomy, and much of cytology, ontogeny, embryology, palaeontology and genetics. He says, "even within morphology in a very strict sense the taxonomist is forced to consider matters which involve physiology".

Turril says, that some one or other morphological classification provides the best basis for a general classification of the widest possible use, because morphological characters are, in general easier to use than others. According to him the great value of a classification based essentially on gross morphology is that it enables correlations of the characters to be easily determined.

2. ANATOMICAL

The study of morphology with a compound microscope is known as anatomy (histology). According to some anatomists the anatomical features are more conservative than those of gross morphology and are, therefore, of greater use in tracing phylogeny or organogenesis. Vesque has given much emphasis upon the value of anatomy in taxonomic and phylogenetic research. Tippo has proposed the lines of evolution of stem structure, more particularly for the secondary xylem of angiosperms. The view has been accepted by most plant anatomists. Stem anatomy shows that the Magnoliales are a relatively primitive of the angiosperms.

Chalk advocates that the structure is probably more conservative than that of the flower and the minor floral differences by which many species are distinguished from one another are not reflected in the wood; thus the wood of different species of a genus are often indistinguishable. According to

Chalk there is one danger in using wood anatomy in classification is that wood usually means secondary wood, and the stressing of criteria derived, atleast, mainly from the secondary wood of trees and large shrubs may give results statistically biased against herbaceous parts of woody groups.

The following anatomical characters are of taxonomic significance:

1. Epidermal appendages, trichomes and emergences of all kinds.
2. Structure of stomata.

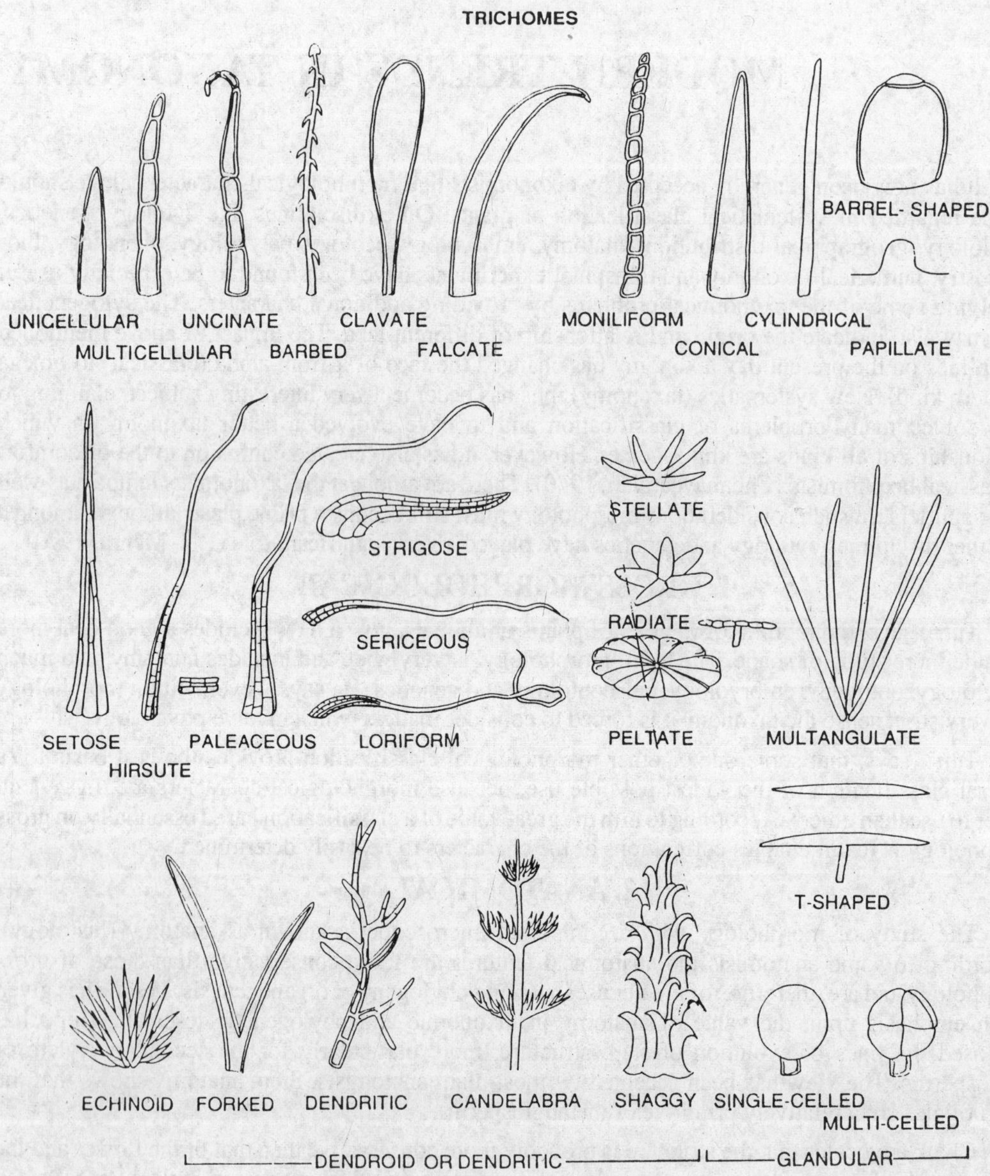

Fig. 4.1. Trichomes of different types

3. The wood anatomy including the elements of the secondary xylem; this is very useful in modern taxonomic studies, particularly in establishing the phylogenetic relationships among taxa, and in many cases serves as an aid in the identification of orders.

4. The nodal anatomy of the axis, the leaf trace nature.

5. Floral anatomy.

Epidermal emergences of all kinds: All unicellular and multicellular appendages are known as **trichomes;** more massive structures, *e.g.,* warts, spinous bodies both epidermal and subepidermal are called **emergences.** The use of trichomes in taxonomic studies are of much value, and has been known for many years. In identification of some families of angiosperms, the presence of a particular type of trichomes helps a lot, *e.g.,* simple unicellular or multicellular uniseriate trichomes occur in many taxa; similarly the occurrence of glandular trichomes of various kinds forms an aid to taxonomy in some respects.

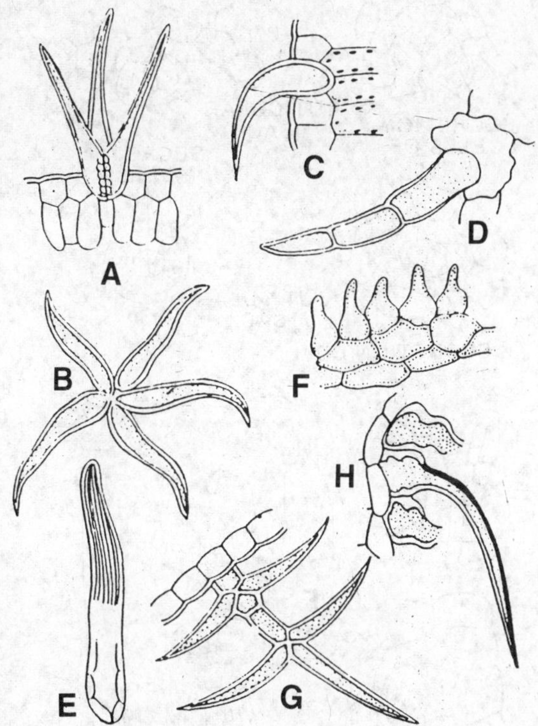

Fig. 4.2. Various kinds of trichomes. A and B, leaf of *Hamamelis;* C, leaf of *Cassia;* D, leaf of *Datura*; E, testa of *Strychnos nuxvomica;* F, petal of *Arnica;* G, leaf of *Verbascum*; H, testa of *Strophanthus.*

Structure of stomata: Stomata which also belong to the epidermal tissue system have been the subject matter of studies, which is very useful in the fields of taxonomy. Stomata have been classified according to the position of subsidiary cells, guard cells in relation to the aperture. They are of the following types :

1. Ranunculaceous or anomocytic. Type A.(*Anomocytic* = irregular celled). This type is characterised by having a limited number of subsidiary cells which are quite alike the remaining epidermal cells; the accessory or subsidiary cells may be four or five in number. In most of the cases, these subsidiary cells are just like the other epidermal cells. This type of stomata occur in Ranunculaceae, Capparidaceae, Malvaceae and some other families.

2. Cruciferous or anisocytic. Type B. (*Anisocytic* = unequal celled). This type of stomata occur in Cruciferae (Brassicaceae) and many genera of Solanaceae. In this type, each stoma remains surrounded by three subsidiary cells of which one is distinctly smaller than the other two.

3. Rubiaceous or paracytic. Type C. (*Paracytic* =parallel celled). In this type, the stoma remains surrounded by two subsidiary or accessory cells which are parallel to the long axis of the pore and guard cells. This type of stomata occur in Rubiaceae and allied families.

4. Caryophyllaceous or diacytic. Type D. (*Diacytic* = cross celled). In this type each stoma remains surrounded by a pair of subsidiary or accessory cells and whose common wall is at right angles to the guard cells. This type of stomata occur in Caryophyllaceae and allied families.

5. Gramineous type. The gramineous stoma possesses guard cells of which the middle portions are much narrower than the ends so that the cells appear in surface view like dump-bells. They are commonly found in Gramineae (Poaceae) and Cyperaceae of monocotyledons.

Wood anatomy: The plant anatomists have established a number of lines of phylogenetic specialisation in the stelar structure on the evolutionary basis and have applied the same in establishing

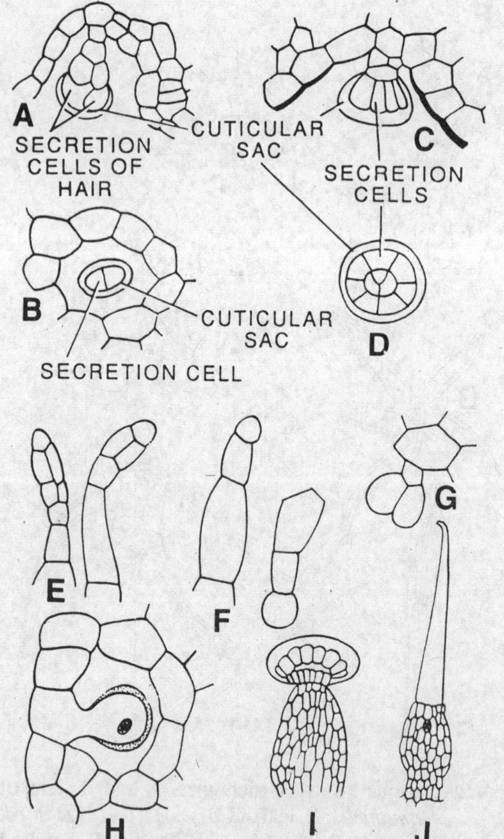

Fig. 4.3. Types of glandular hairs. A and B, section and surface view, *Artemisia*; C and D section and surface view, *Thymus;* E, *Hyascyamus*; F, *Atropa;* G, *Digitalis;* H, *Dryopteris filix-mas* (rhizome); I, *Cannabis*; J, *Urtica dioica.*

data regarding the primitiveness of taxa and relationships between them and ultimate determination of taxon position.

These lines of anatomical specialisation are as follows:

1. The protostele is more primitive than siphonostele or dictyostele.

2. In angiosperms the woody axis of trees and shrubs is more primitive than the stem of herbs.

3. Vessel elements with scalariform perforation plates are more primitive than vessels with single opening in the perforation plate (*i.e.*, simple perforation).

4. The type with numerous bars and narrow openings is more primitive than the type with few bars and wide openings.

5. Vessel elements, long, small in diameter, angular in cross section are more primitive than those with short ones having broad circular cross sections; also the vascular elements with long sloping end walls are more primitive than those having non-sloping transverse walls.

6. Scalariform type of pitting on the lateral walls of the vessel is more primitive; opposite and alternate pitting are advanced.

7. Solitary or single pores occurring in vessel arrangement throughout the wood (*i.e.*, solitary pores) is more primitive than aggregate groupings of various types, such as pore multiples, pore fascicles, pore chains, etc.

8. Diffuse porous conditions of wood is more primitive than ring porous condition.

9. Evolutionary line of origin runs as follows:

Tracheids-to fibre-tracheids- to libriform wood fibres-to scalariform tracheids-to circular bordered pitted tracheids.

10. Diffuse wood parenchyma is more primitive than various aggregate arrangement. Apotracheal wood parenchyma is more primitive than paratracheal wood parenchyama.

heterogeneous rays, *i.e.*, rays with both the vertically as well as radially elongated cells are less specialised and primitive than homogeneous rays in which cells are only radially elongated.

Schemes based on floral morphology and together with anatomical characters as mentioned above must go together in determintion, of taxonomic position, generic identification, etc.

For example, Amentiferae according to Engler is most primitive in the Dicotyledons, but the wood anatomy shows that this taxon is very advanced in comparison to Magnoliales-Ranales.

Amentiferae : Simple perforation plates, *i.e.*, an advanced character; vessels are ring porous, short, round in cross section, large in diameter with alternate or opposite pitting, *i.e.*, an advanced character; libriform wood fibres prevail, an advanced character.

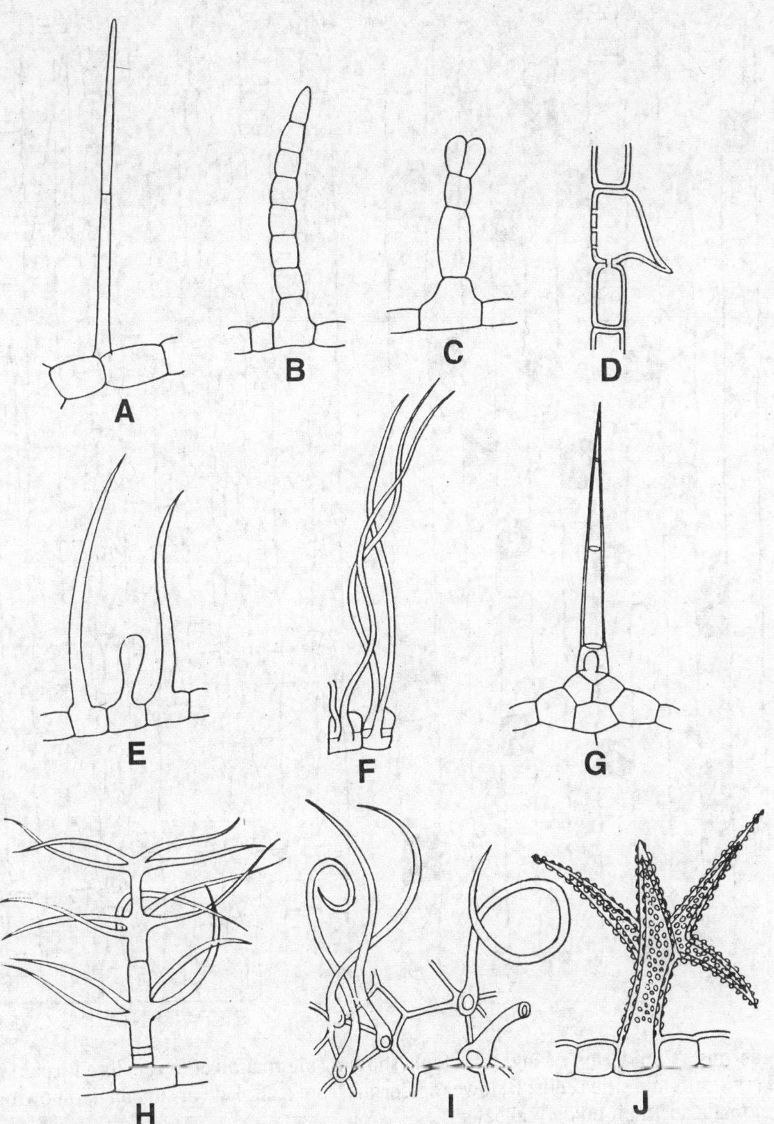

Fig. 4.4. Various types of trichomes (hairs)—A, from petal of *Epigaea;* B, from leaf of *Coreopsis;* C, from petal of *Phryma;* D, from leaf of *Avena;* E, from speal of *Helotropium;* F, from stem of *Onopordum;* G, from *Cucumis* leaf; H, from *Platanus* leaf; I, from *Rubus* fruit; J, from *Aubrietia* stem. (After E. & M.)

Phylogenetic specialisation of tracheary and fibre cells: The specialisation of tracheary elements (*i.e.*, tracheids and vessels) was concomitant of the separation of functions of conduction and imparting mechanical rigidity in the vascular plants which took place during the evolution of terrestrial plants (Bailey, 1953). In the less specialised state, conduction of water and mechanical support are both undertaken by tracheids; with the increasing specialisation woods evolved with better conducting elements, the vessels or tracheae which are more efficient in conduction than in support. On the other hand fibres took their origin as mainly strengthening elements. Thus, from primitive tracheids two lines of specialisation evolved in two directions, one towards the vessel, another towards the fibres.

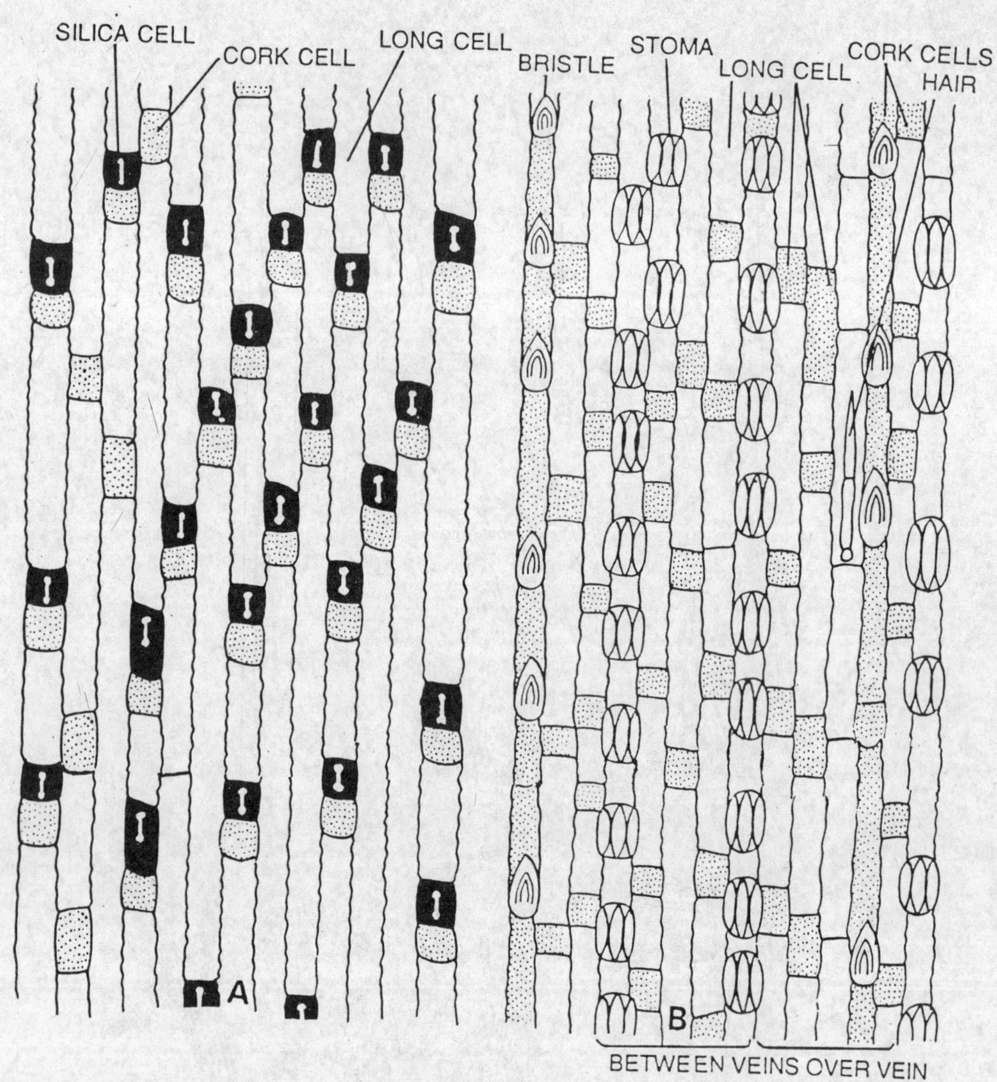

Fig. 4.5. Epidermis. A, epidermis of sugarcane stem showing alternation of long cells with pairs of short cells, the cork cells and silica cells; B, lower epidermis from leaf blade of sugarcane, showing distribution of stomata and their epidermal cells.

Vessels arose separately in several groups (taxa) among Tracheophyta, *i.e.,* vascular plants. The common accepted idea is that the vessels evolved independently in angiosperms. Evidence supports that in the Dicotyledons vessels originated and undergone specialisation first in the secondary xylem, then in metaxylem and were lost in the early primary xylem, *i.e.*, protoxylem.

The evolutionary sequence of the vessel members of the secondary xylem of the Dicotyledons began with long scalariform pitted tracheids (primitive character) as found in the very primitive dicotyledons, such as Winteraceae, Trochodendraceae, Tetracentraceae, Chloranthaceae, etc. These tracheids were succeeded by vessel members of long and narrow form with tapering ends. The cells became short progressively became wider and less inclined and finally transverse. In very primitive state the perforation plate was scalariform with numerous bars resembling a wall with scalariformly arranged pits without pit memberanes; further specialistion went on, as a result number of bars decreased and finally total elimination of bars took place resulting in the simple perforation (advanced). The pitting on vessel walls also underwent change in course of evolution. In intervessel pitting bordered pit pairs in opposite arrangement were changed into smaller bordered pit pairs in opposite arrangement then to alternate arrangement. The evolutionary sequence was from tracheids to fibre tracheids to libriform fibre.

Wood parenchyma: In the secondary wood (xylem), there are two types of parenchyma, *i.e.*, (*a*) axial parenchyma, (*b*) ray parenchyma, *i.e.*, radial parenchyma, these two types of parenchyma are almost similar in their wall structure and contents. They store starch, oils and many other ergastic

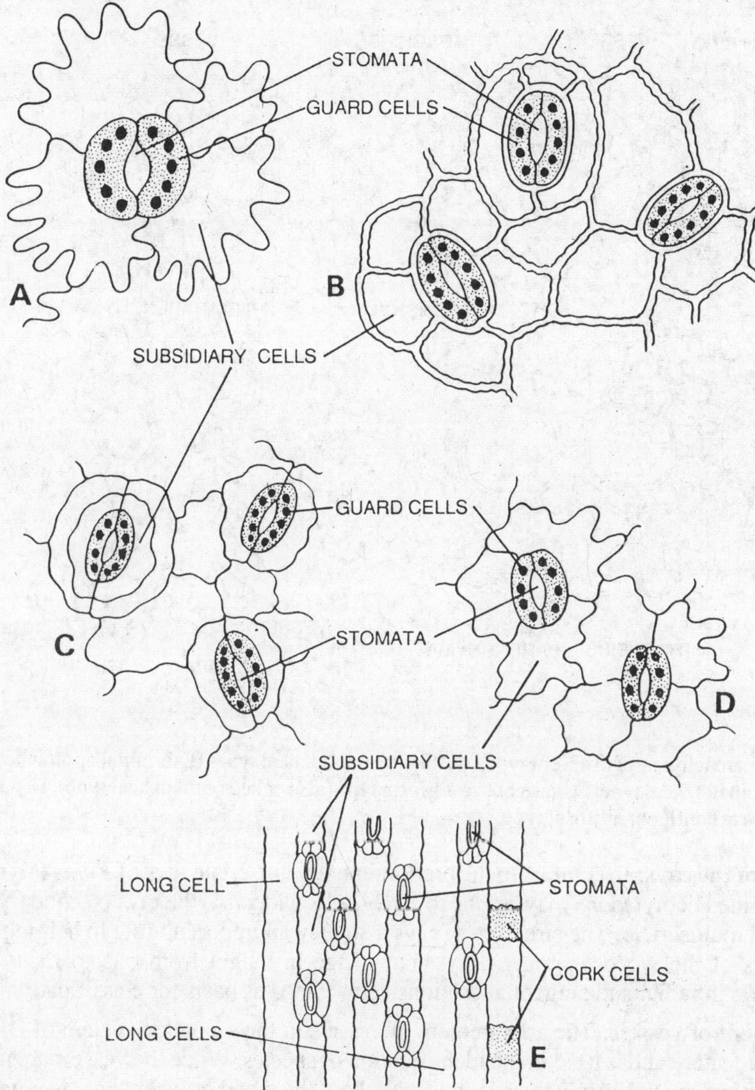

Fig. 4.6. Stomata—types of stomata. A anomocytic or irregular celled type (ranunculaceous type); B, anisocytic or unequal celled type (cruciferous type); C, paracytic or parallel celled type (rubiaceous type); D, diacytic or cross celled type (caryophyllaceous type); E, gramineous type.

substances; tannins and crystals are common, the types of crystals and their arrangements serve much in the identification of wood. The distribution of axial parenchyma in the secondary wood exhibits many intergrading patterns. The touch relation of the parenchyma with regard to the vessels as observed in transverse sections under the microscope shows two principal patterns, *e.g.*, (*a*) **apotra-**

cheal and (*b*) **paratracheal**; The former is primitive condition, the latter is advanced. Apotracheal parenchyma is further classified into (*a*) **diffuse,** in which single parenchyma cell or parenchyma strand scattered among fibres and (*b*) **banded** apotracheal in which parenchyma cells in a line form band (not in touch) over and under the vessels. The paratracheal parenchyma presents two types also, *e.g.*, (*a*) **vasicentric** (very advanced), it forms a complete sheath in touch and round the vessel and (*b*) confluent in which coalesced parenchyma form irregular or tangential or diagonal bands.

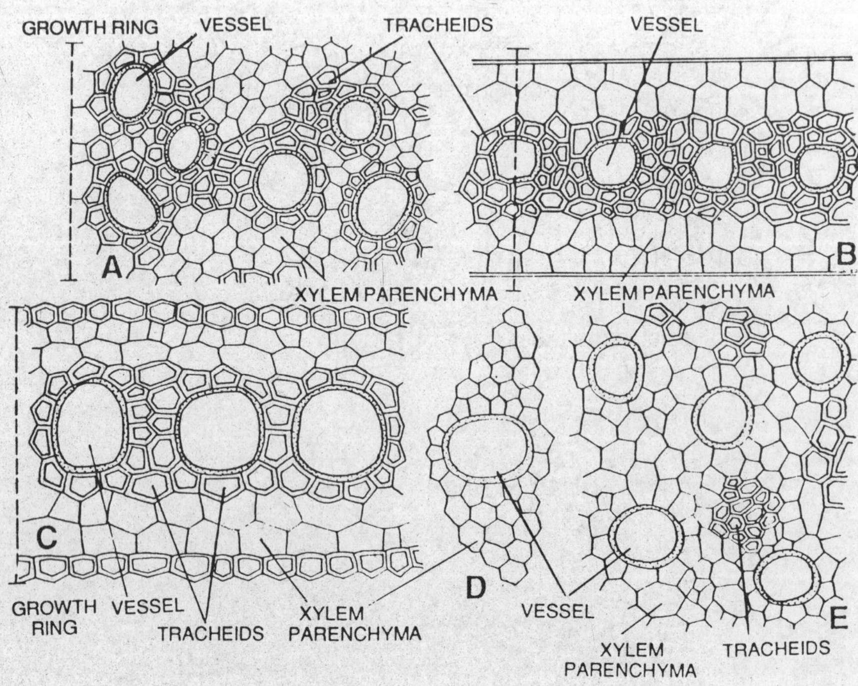

Fig. 4.7. Xylem parenchyma of different types. A, diffuse apotracheal type; B, terminal apotracheal type; C, banded apotracheal type; D, vasicentric ailiform type (also called paratracheal type); E, paratracheal vasicentric confluent ailiform type.

Structure of the rays: In contrast to the predomimantly uniseriate rays of *Pinus* (Gymnosperm-conifer) those of the Dicotyledons may be one to many cells wide, *i.e.*, the rays (secondary rays) may be uniseriate and multiseriate. The structure of rays also play an important role in helping to identify the primitiveness of the wood anatomy. Rays may range in height from one to many cells. The appearance of rays in radial and tangential sections can be used as basis for classification.

Arrangement of vessels: The arrangement of vessels in the secondary xylem of Dicotyledons is a characteristic feature and is used in the identification of species. When the vessels are more or less equal in diameter and are uniformly distributed throughout the wood or when there is gradual change in size and distribution throughout the growth ring the wood is known as "**diffuse porous**", as found in *Salix, Populus, Acacia,* etc. When the wood consists of vessels of different diameters and in which those produced at the beginning of the season are distinctly larger than the late wood, the wood is called "**ring porous**", as found in *Fraxinus, Quercus,* etc. From the phylogenetic point of view diffuse porous wood is primitive than ring porous wood.

Wall thickenings of the vessels: In many vascular plants (Tracheophyta) the secondary wall thickenings of the protoxylem are annular or spiral. From an ontogenetic viewpoint the annular thickenings develop prior to the spiral. In later formed tracheary elements the spiral bands became joined in certain areas giving rise to scalariform thickenings. In tracheary elements developed at later ontogenetic stage, the thickening of wall becomes reticulate; when the openings in the secondary wall of such a network are elongated in a direction perpendicular to the longitudinal axis of the element, the thickening is known as scalariform-reticulate. Ontogenetically the most advanced type of thickening is pitted which occurs in the late primary and secondary xylem (wood). The pittings according to their arrangements are known as (*a*) scalariform, (*b*) opposite and (*c*) alternate. When the pits are transversely elongated and are consequently arranged in ladder-like manner, it is termed scalariform; when circular pits are arranged in horizantal or transverse rows they are known as opposite; when in diagonal series, known as alternate. The different types of pitting are of taxonomic importance.

Vessels and the structure of perforation plates: Vessels have limited length and in those vessel members which end in a tube bearing perforation at the terminal end, sometimes perforations also are present on the side wall. The end or side wall bearing perforation is known as perforation plate. The perforation plate consisting of a single large pore is known as simple perforation. In case of end wall containing numerous perforations they may be scalariform perforation plate in which perforations are arranged in parallel series in ladder-like fashion, reticulate perforation plate is distinguished by the arrangement of perforations in reticulate or net-like fashion; when end wall has many circular perforations arranged in scattered fashion, it is known as **foraminate** type characteristic of many gymnosperms and a few angiosperms.

The following structural features of angiosperm tracheary elements are used as the basis for the study of evolution:

1. The length of the element, *e.g.*, tracheids are long cells, the average length in *Trochodendron* is 4.35 mm., in the monocats the same is 5.07 mm.

2. Diameter of tracheary elements (vessels).

3. The perforation plates; the scalariform oblique perforation plates with numerous perforations are most primitive; simple horizontal one is the most advanced.

4. Shape of the cross-section of the vessel; angular primitive; circular recent and advanced.

Origin and specialisation of vessels: In the Dicotyledons the evolution of vessels took place first in the woody plants, apparently they arose independently in taxa with vessels which occur in most existing dicotyledons. As a result of the facts collected, it has been assumed that the vessels arose first in the secondary xylem, and later in metaxylem of the primary xylem, the specialisation and consequent evolutionary steps gradually advanced from secondary to primary xylem. It could also be assumed that herbaceous plants have evolved from woody plants as a result of reduction of cambial activity only after obvious development of vessels had taken place in the woody ancestral plants. In some specialised Dicotyledons, *e.g.*, in certain members of Cactaceae the secondary xylem lacks vessels which are replaced by so-called vascular tracheids, here the lack of vessels is the result of secondary reduction (Bailey, 1957).

From the evolutionary point of view, vessels in the Monocotyledons first appeared in roots and later in stems and leaves. The specialization of the vessels followed in the same pattern. Phylogenetically the vessels first appeared and became specialised in late formed metaxylem and advanced gradually in early formed metaxylem and finally into protoxylem (Cheadle, 1944). In the existing Monocotyledons of present day the last formed metaxylem roots contain vessels having scalariform perforation plates. A few aquatic Monocotyledons lack vessels in their xylem of roots, stems and leaves; their character is probably secondary one, *i.e.*, reduction stage and advanced.

Floral anatomy — Floral anatomy for delimiting taxa has been utilized by B. Tiagi and G. Dixit (1965) in Asclepiadaceae; Y.D. Tiagi (1964), B. Tiagi (1965, 1966) in *Opuntia, Cuscuta* and *Evolvulus*. V. Puri (1970) has reviewed the opinions concerning the nature of ovule in angiosperms and suggested that the angiospermic ovule has possibly arisen independently of the gymnosperms and the two, due to lack of any direct link, should not be homologised. Y.D. Tiagi (1961, 1963, 1968) on the basis of evidence derived from floral morphology, anatomy, embryology and cytology in Cactaceae considers the family as closely related to Calycanthaceae. The floral anatomy of two species of *Cuscuta* has been examined by B. Tyagi (1966) and its retention in the family Convolvulaceae has been suggested by C.M. Govil (1968). V. Singh (1977) has studied the primary pattern of floral development in all the taxa of the Alismatales as trimerous. Establishment of Hydrangeaceae and Escalloniaceae of sub-families Hydrangeoideae and Escallonioideae of Saxifragaceae, as separate families has been supported by N.P. Saxena (1966) on anatomical and embryological grounds. Eames (1953) has suggested the removal of *Paeonia* from Ranunculaceae and elevation of a new family Paeoniaceae on the basis of floral anatomy.

Anatomy of Leaf — B.C. Kundu and B.Gupta (1963, 1964 a, b) have suggested the use of quantitative microscopy of leaves in distinguishing taxa which is found useful in Solanaceae.

Anatomy of Axis — This has widely been used for comparing taxa, in *Corchorus* (R.M. Datta and K.Roy, 1963), in *Ipomoea* (C.M. Govil, 1971), *Phaseolus* (P.C. Datta and A.Saha, 1970).

Epidermal and Cuticular Anatomy — This has also been used for taxonomic delimitation of medicinal plants (P. Singh and B.C. Kundu, 1962, Y.N. Pandey, 1970; K.J. Ahmad, 1964 *a*, *b*).

Seed Coat Anatomy — Anatomy of seed coats of pulses has been utilised for delimitation of taxa by P.C. Datta and R.K. Maiti (1968). Chuang and Lawrence (1972) studied the anatomy of seed coat in the genus *Cordylanthus* of Scrophulariaceae.

3. POLLEN MORPHOLOGY OR PALYNOLOGICAL

The science of palynology or study of pollen morphology deals with the detailed study of pollen grains, *i.e.*, the microspores of Spermatophyta, particularly of Angiosperms. It mainly deals with the structure, walls, etc., of pollen grains. Palynological studies help in the confirmation of relationship and affinities between the related taxa. The pollen grains of related families and genera are usually of more or less of same type. The sculpture and pattern of the outer wall exine, number of apertures on the wall, size and the shape of the pollen grains, etc., help in taxonomy.

Mature pollen grains usually have two walls, the outer *exine*, and the inner *intine*. The intine is thinner and more tender than exine. The exine at first appears within the special wall as a thin membrane. In further development the exine thickens considerably and two layers can be distinguished on it, the outermost of which is the *sexine* and the inner is *nexine*. The sexine is thin and has high refractive index, and therefore is not easily visible; the surface of sexime is at first smooth, but later on after the formation of nexine many types of projections may develop on it. In the aperture regions of characteristic shapes from which pollen tube comes out, the exine may be completely wanting or it may be represented by nexine (Erdtman, 1952). The nexine is relatively thick and cutinized. The specific cutin is called *sporopollenin* (Frey-Wyssling, 1959). This substance is more stable than cutin. The intine is not of constant thickness, it is thicker at apertures. Numerous characteristic projections and sculpturings are formed on sexine, in addition to various types of sculpturings of the sexine there are other morphological characters which are used in the classification of the pollen and also imparts great value in angiosperm taxonomy.

Pollen grains are aperturate, *i.e.*, provided with pores or apertures. The position, shape, structure and number of apertures are of taxonomic significance. As a result of research on palynology based on the variability of the grains, a detailed nomenclature of the structural characteristics of pollen has been developed by which exact morphological descriptions of various pollen grains can be done accurately. The categories of the structural morphology of pollen that help in plant taxonomy are as follows:

(*i*) Size of pollen grain to be measured in μ (μ = .001 mm); (*ii*) structure of pollen grain, *i.e.*, oval, ellipitical, triangular, etc.; (*iii*) whether grains are free or in pollinia, or in tetrads; (*iv*) The nature of sculpturing of the pollen grains, the nature of exine, *e.g.*, smooth, spiny, warty, pitted, etc.; (*v*) sulcus-an elongated furrow perpendicular to the long axis at the pole of the grain; (*vi*) colpa-this is an elongated furrow at right angle to the equatorial plane, the ends of the furrow are directed towards the poles of the grain; (*vii*) ruga-an elongated furrow, the direction of which differs from both of the above types; (*viii*) pore- a pore is a circular aperture, when the number of pores is small, they are restricted in the equatorial region; when large they occur on all over the surface.

The comparative study of the structure of pollen grains of living and fossil plants plays as important role in the determination of classification of the plants. According to Wodehouse the pollen grain characters of gymnosperms are primitive to those of angiosperms. Wodehouse says, that the one furrowed grain is a primitive character in Saururaceae, Piperaceae, Chloranthaceae and Magnoliaceae all of which families have been considered by various authors as primitive.

According to P.K.K. Nair, pollen morphology of primitive angiosperms has indicated that there are two primary dicot stocks namely the "Magnolian" stock (Magnoliaceae and its allies), and the "Ranalian" stock (Ranunculaceae and its allies), each stock being characterized by monocolpate and tricolpate pollen respectively.

Cronquist, A. (1968) has proposed that the two subclasses Ranunculideae and Magnoliideae, should be considered to represent the two palynological stocks, and should be raised to higher taxa levels:

4. CHEMOTAXONOMY

The science of chemical taxonomy (chemotaxonomy) is based on the classification of plants on the basis of their chemical constituents which are deeply concerned with the molecular characteristics. The method of chemical taxonomy is simple in principle and is based on the investigation of the distribution of chemical compounds, or groups of biosynthetically related compounds in series of related plants. Different plants sometimes contain substances which although belong to different chemical compounds appear to be biosynthetically analogous. Such plants may contain similar enzyme system, and the compounds produced by such enzymes infer that relationships exist between related plants. However, the chemotaxonomic studies include the investigations of the patterns of the compounds existing in plants, and in all the individual parts of the plants, such as bark, wood, leaves, roots, etc. The chemical constituents usually differ much in different organs. Such investigations are needed for obtaining real evidence for alliance and non-alliance of plants.

According to Turril, the physiological aspects, to some extent, lead the taxonomist towards the determination of phylogeny. It has been biochemical rather than directly functional criteria that have been used in taxonomy. The presence or absence of essential oils, resins, latex, glucosides, alkaloids, etc., has been used in classification. According to Molisch, a definite chemical substance may appear in a single species, in several species of the same genus, in a single genus, in several genera of one family, in an entire family, in two to many related families, or in large divisions of the plant kingdom.

Mc Nair showed the distribution of certain chemical substances in the families of angiosperms and its relation to climate. Climatic conditions have a major influence on the distribution of plants containing certain substances, *e.g.*, fats, volatile oils, alkaloids, etc. He concluded, that for tropics, and perhaps for all climates, the chemical products are highly organized. According to Reichert, it is possible to identify many plants by their starch grains.

Alston and Turner (1963) in their work "**Biochemical Systematics**"; Hegnauer (1965) in his "**Chemotaxonomic der Pflanzen**"; Leone (1964) in his "**Taxonomic Biochemistry and Serology**"; Swain (1963) in his "**Chemical Plant Taxonomy**" and Harborne (1967) have given a review of phytotaxonomical chemistry and emphasized on the role of various chemicals in plant taxonomy. Molery (1967) has stressed on the importance of β-cyanins and β-xanthins in plant taxonomy. β-cyanins are commonly met with in the families of Order Centrospermae. The other chemicals are also found specifically in particular orders or families of flowering plants. For example, **isoquinoline** (alkaloids) is found in the families of Ranales; **retanone** in the families of Leguminales; **biflavonyls** in *Casuarina equisetifolia* (Casuarinaceae). The presence of such chemicals in different groups of plants is of great taxonomic significance.

The choice of host plants by parasites is a part determined by the chemical constitution of the host. Thus, many animal parasites choose plants with similar chemical characters–mustard oil, glucosides, HCN glucosides, alkaloids, organic acids (especially oxalic), salts, etc. In correlation insect larvae which, when free to choose, limit themselves to one plant family or to such families as have similar chemical substances.

Applications and methods. There are a few angiosperamic taxa which are characterized by specific compounds of general occurrence. For example, leaving aside the family Caryophyllaceae, the rest of the families, such as Chenopodiaceae, Amaranthaceae, Aizoaceac, *etc.*, of the taxon Caryophyllales (Centrospermae) contain *Betacyanin*, a colured substance, but differs from anthocyanin. It appears that except Caryophyllaceae, the remaining families are closely related, and therefore, the Caryophyllaceae may be isolated which may form a separate monotypic taxon. Betacyanin also occurs in Cactaceae, and therefore, the members of Caryophyllales are phylogenetically related. There are certain other chemical connections between Cactaceae and members of Caryophyllales, *e.g.*, common presence of isoquinole alkaloid in *Salsola* of Cenopdiaceae and Cactaceae. Another example may be cited, *i.e.*, in the family Cruciferae unsaturated acid, erucic acid is prominent; in *Tropaeolum* erucic acid is also present, it indicates the relationship between Geraniales and Rhoeadales. In Umbelliferac and Araliaceae a structural isomer of oleic acid, *i.e.*, petroselinic acid occurs, these two families are related and belong to the same order. The other examples are from Magnoliales–Ranales taxa, where it shows that in Magnoliaceae Lauraceae, Ranunculaceae, Annonaceae, the alkaloid *isoquinoline* is present; this supports that these families are closely related. On the other hand, Asclepiadaceae and Gentianaceae are allied due to the common occurrence of *pyridine.* The Liliaceae and Amaryllidaceae are closely associated, and this is supported by the presence of *isoquinoline* in both.

5. SEROTAXONOMY

Serotaxonomic test and consequent phylogenetic relationships between the taxa of angiosperms were established in Germany by Professor K.C. Mez (1866-1944) at the University of Koenigsberg in 1926, and this was modified in 1936. He established that relationship between larger group of angiospermic plants could be determined by serological test; and the closely related taxa and plants could be arranged accordingly. Serotaxonomy consists of the study and analysis of protein reaction of plants of different families with the blood serum of either rabbit or guineapig.

The serum test is done as follows:

For example, two plants *A* and *B* of two different families, of controversial relationship are taken for study. The protein extract of plant *A*, *i.e.*, extracted from tender shoot bud, is injected into the body of a rabbit, then after sometime, blood serum is collected from the injected rabbit in a sterilized test tube; protein extract of plant *B* is also collected from the tender shoot bud and is added to the rabbit serum haiving *A* protein; if there is precipitation then *A* and *B* plants are closely related and may be placed taxonomically close to each other; if on the other hand, no precipitation taken place, the plants are not closely related. The precipitation result is known as **positive reaction,** and non-precipitation result is known as **negative reaction.** For example, protein extract of *Passiflora* spp., (plant *A*) injected into the body of a rabbit, the serum then collected in a sterilized test tube and to it the protein extract of *Cucurbita* spp., (plant B) is added; the result is non-precipitation, *i.e.*, negative reaction, and therefore, it is considered that the families Passifloraceae and Cucurbitaceae are not related at all. If degree of precipitation is much, the families are closely related, if less the families are distantly related.

6. PALAEONTOLOGICAL

Darrah has proposed the following palaeontological evidences in support of the phylogenetic evolution of the plants.

1. The invasion of land by an undifferentiated thalloid plant took place not later than Silurian.

2. The simple upright undifferentiated and protostelic axis formed in Silurian.

3. Enlargement of the plant body with specialization towards a division of labour-sterile and fertile, took place apparently in late Silurian.

4. Origin of the photosynthetic leaf took place in Devonian.

5. Specialization in the service of support, with the resultant secondary body took place in Devonian.

6. Development of the strobilus also took place in Devonian.

7. The development of heterospory took place in upper Devonian if not earlier.

8. Retention of the gametophyte within megaspore, within the megasporangium, where fertilization takes place with the ultimate attainment of the seed took place probably in upper Devonian.

9. Evolution of the pollen grain took place in Carboniferous if not earlier.

10. Evolution of the bisporangiate flower took place in Triassic period.

11. Attainment of the condition of angiospermy took place in Jurassic and Triassic periods.

12. Evolution of the herbaceous habit of the angiosperms took place in Cretaceous (chiefly Cenozoic).

7. ONTOGENETICAL

According to Gunderson the evolution of the flowering plants is based on floral structure and presumed **organogenesis.** He points out that the flowering tendencies are widely accepted as progressive in organogenesis and agree with the findings in development.

Petals — from separate to united.

Petals — from actinomorphy to zygomorphy.

Sepals — from separate to united.

Ovary — from superior to inferior.

Carpels — from separate to united.

Placentation — from parietal to axile.

8. CYTOGENETICS AND BIOSYSTEMATICS

In modern days cytology has played an important role in taxonomy to achieve a classification which represents mutual relationships and is useful in indexing plants. Modern taxonomy takes up cytological evidence based on chromosome numbers, chromosome morphology, chromosome behaviour in meiosis and in aberrant forms of reproduction. The branch of taxonomy principally based on cytology is known as '**cytotaxonomy**', it is a part of experimental taxonomy. It includes cytological aspects, study of cytogenetics and phenomena together with consideration of classical aspects of taxonomy. Application of data from the fields of cytology and cytogenetics have become very much helpful in taxonomic studies, particularly in determining categories of genus, species and sub-species in controversial cases. In this approach emphasis has been laid on cytogenetics and other aspects of cytotaxonomy supplemented by classical aspects of morphology, and therefore, the cytotaxonomy is the integration of cytology and taxonomy in the attempt of better understanding to solve problems of plant affinities.

The first step in cytotaxonomy research is the thorough sampling of the taxon (it may or may not be species) and its populations and the consequent cytological studies of chromosomes, *i.e.*, the number of chromosomes, chromosome morphology, chromosome behaviour, etc., of many populations within geographic races, species, genera and so on. Difference in chromosome number, chromosome morphology and chromosome behaviour at meiosis usually indicate generic or specific differences of taxonomic significance.

The second step consists in the capacity of different populations to hybridize and a study of vigour and fertility of hybrids. This enables to know the presence or absence of breeding barriers between groups and is of taxonomic value as indicating the natural limits of the taxa of various categories.

The third and last step includes the study of homologies of the chromosomes in the hybrids as determined in the meiosis, this is significant indicator in the degree of genetic relationship. Informations obtained from the above mentioned three steps are compared with the facts obtained from comparative morphology and geographical distribution. The resultant classification of the taxa, *i.e.*, the category of genus and taxa of lower rank, to which it is applied has an increased objective over one obtained from the consideration of morphology.

For example, Gundersen's (1950) system of classification is based on basic chromosome numbers together with anatomical and morphological characters. He has shown in primitive taxon, such as Magniliflorae the basic number of chromosomes are much less than advanced taxon Rubiflorae and others.

The relationship between genera and the relative primitiveness of the constituent genera in the family can be determined on cytological studies, *e.g.,* cytological studies on Dipterocarpaceae (Roy and Jha, 1965) established such relationship, the taxonomic position and the probable evolution of the taxon. Roy and Jha, 1965, estimated the chromosome numbers in different genera of the different tribes of Dipterocarpaceae. The number of chromosomes estimated as such $2n = 22$, $n=11$; $2n = 14$, $n = 7$; $2n = 12$, $n = 6$ according to different genera, *e.g.,* the tribe Vaticae and Dipterocarpae have some basic chromosome numbers as $2n = 22$, $n = 11$; *Shorea* has $2n = 14$, $n = 7$, so Dipterocarpac and Vaticae are more primitive and more closely related than those of others in the taxon.

Synthetic Biosystematic Approach:Taxonomy is an unending synthesis whose basis is becoming broadened decade after decade. It is therefore highly essential to bring about a coordinated synthesis between the "taxonomist" and "biosystematist"–the former posing the various ambiguities wherever located in the understanding of taxa while revising the families and preparing basic frame structure, and the latter working out such problems with the sophisticated technique utilising the relevant data and material supplied by the taxonomists. Such co-ordination between the Government Scientific Institutes and the university departments would go a long way in solving several taxonomic problems of the Indian Flora (Rolla Seshagiri Rao and R. Sundra Raghavan, 1970).

Cytogenetics and Biosystematic in Synthetic Taxonomy–During modern times, there has been spectacular extension of synthetic taxonomy. Taxonomists, unlike in the past, are not only busy collecting data but basic problems are discussed at the outset as to why should data be collected; how should it be collected and what information to get from them. In all scientific experiments, howsoever, comprehensive and sophisticated it may be, we need some check to maintain balance between technique and concepts derived therefrom. In this context cytogenetics, genecology, population genetics, plant breeding and cytotaxonomy have a special role to play. Phylogenetic relationship at the specific and generic level and origin of species in many cases have been convincingly worked out by cytogeneticists. This approach has brought out new realizations and opened up new lines to study taxonomy (R.P. Roy and R.P. Sinha, 1970).

The cytogenetical data also indicate the origin and relationship of different taxa. Such studies include crossability between taxa and cytological analysis of artifical hybrids along with their parental forms. Similar studies can be carried on natural hybrids. Variation in chromosome number is of special interest to the evolutionist as the new group may be biologically isolated by incompatibility or hybrid sterility and may have been formed as a result of sudden doubling of chromosome number. Types formed by duplication of the same genome are difficult to distinguish from their progenitors as the differences are often in degree of expression of same characters. As a general rule, alloploids are distinct in their morphological characteristics. The detection of more than one chromosome number within any one species may therefore demand its detailed study as the species may be a collective one and needs a taxonomic revision (S.L. Tandon and G.R. Rao, 1970).

9. EMBRYOLOGY AND TAXONOMY

The comparative or phylogenetic embryology deals with the embryological data, which are used as a tool for ascertaining inter-relationships and taxonomic positions. In a symposium on comparative Embryology of Angiosperms held at Delhi in 1967 the systematic positions of various families were discussed in the light of embryological data. This symposium enabled to bring together in a comprehensive manner the scattered embryological literature on various families.

P. Maheshwari, B.M. Johri and S.N. Dixit (1957) pointed out that Loranthoideae and Viscoideae should be raised to the status of families. R.N. Kapil (1970) has pointed out several similarities between Crossosomataceae and Dilleniaceae like numerous centrifugal stamens, 2-3 middle layers, secretory tapetum, bitegmic crassinucellate ovules, micropyle formed by both the integuments. *Polygonum* type of embryo sac, nuclear endosperm showing aggregation of nuclei at chalazal end and reniform arillate seeds. These show that the two families are related.

After a comparative study on the embryology of Magnoliaceae and *Schizandra* R.N. Kapil and S. Jalan (1964) supported the view that *Schizandra* should be raised to a family rank. According to K. Subramanyam (1970) Crassulaceae shows close affinity with Saxifragaceae. For example *Crassula aquatica* has a mode of habit similar to Podostemaceae and it has the most reduced endosperm in Crassulaceae. This taxon according to Subramanyam (1962), forms a connecting link between Crassulaceae and Podostemaceae thus supporting Maheshwari (1945) that Podostemaceae are much reduced apetalous derivatives of the Crassulaceae.

V.K. Sharma (1968) has compared the embryological feature of Coriariaceae and allied families and suggested that it may represent an isolated stock distantly related to the Sapindales.

Dnyansagar (1970) stated that the sub-family Mimosoideae, Caesalpinoideae and Papilionatae should be raised to family levels.

10. NUMERICAL TAXONOMY

Numerical plant taxonomy may be defined as the science in which for the purpose of classifying the plants, mathematical methods are used. In other words, the application of simple mathematical principles or techniques or methods in taxonomical studies of plants may be defined as *numerical taxonomy.* Heywood (1967) defines numerical taxonomy as "the numerical evaluation of the similarity between groups of organisms and the ordering of these groups into higher-ranking taxa on the basis of these similarities." According to Sneath and Sokal (1973) "numerical taxonomy aims to develop methods that are objective, explicit and repeatable, both in evaluation of taxonomic relationships and in the erection of taxa." Thus the establishment of this branch of taxonomy is mainly based on the fact that the taxonomists are becoming now a days more susceptible to more clear, more closure and more reasonable criteria and principles.

Taxometrics or numerical taxonomy is mainly concerned with procedural and operational problems, *i.e.*, with the actual methods we take up in classifications. This science is based on evidence of similarities exhibited by observed and recorded characters of taxa, not on phylogenetic probabilities. It is not concerned with producing new data but methods of handling them by means of electronic computers so as to reduce subjective element involved in comparing sets of data. The necessity of mathematical applications in taxonomy has been realized by J.P.M. Brenon (1968) in Royal Botanic Garden, Kew (England). He has given the example of Kew herbarium with nearly four to five million specimens. The proper arrangement of such a huge number of plants is only possible by computerization of the arranging system.

J. Cullen (1968) has precisely revised the botanical problems of numerical taxonomy and has compared the numerical taxonomical methods with the orthodox taxonomical methods. In the orthodox taxonomy, the meaning of "*character*" is quite different from the meaning of "*character*" in numerical taxonomy. In the former the *character* stands for any attribute referring to form, structure and behaviour which a taxonomist separates from the whole organism for a particular purpose such as comparison or interpretation, while in latter the *character* means a taxonomic character of two or more states, which within this study cannot be sub-divided any further. This is known as *unit character* (operational taxonomic unit, OTU), which makes a basic unit of numerical plant taxonomy.The character concept is combined with the use of average individuals or some other forms of selecting samples to represent the basic taxa (OTU) of the group under consideration.

Another botanical problem of numerical taxonomy is the use of weighing technique. On the basis of numerical taxonomy, the new reasonable phylogenetic classification may be worked out.

The logical steps involved in numerical taxonmy are as follows:

1. Choice of units to be studied. The first task is to decide what kinds of units are to be studied These units may be comprised of individuals, strains, species, etc. The units should be as representative as possible of the organisms being considered. The entities of the lowest rank in particular study used are known as *operational taxonomic units,* (OTU's).

2. Character selection. A wide selection of the characters of OTU's is required. Atleast 50 and preferably 100 or more characters are needed to produce fairly stable classification.

3. Measurement of similarity. Overall similarity(s) is calculated by comparing OTU with every other and expressed as percentage, 100% similarity(s) for identity and 0% similarity(s) for non-resemblance. A similarity table may be worked out tabulating similarity coefficients and one for each operational taxonomic unit (OTU).

4. Cluster analysis. The similarity in matrix is now arranged, so as to bring together into clusters OTU'S whose members have highest similarity.

5. Discrimination. This is being done in the last, according to the degree of similarity and tabulated thereon.

In numerical taxonomy, the high-speed electronic computers and the powerful techniques of mathematics are employed to bring out a method of taxonomic analysis. These methods provide results that agree with establised classification of plants. These computerized techniques have highly relevant and increasing repercussions on many of the procedures and concepts of various fields of taxonomy.

PHYLOGENY AND TAXONOMY

Phylogeny is the evolutionary history of a taxon, and attempts to account for its origin and development. The term phylogeny is the autonym of ontogeny. Ontogeny differs from phylogeny in that it accounts for the life history of the individual plant from its development from the zygote to the production of its own gametes. A primary objective of phylogenetic studies in botany is the determination of origins and reletionships of all taxa of both extinct and living plants.

Significance of Phylogeny to Taxonomy–Phylogeny deals with the evolutionary history of all taxa. It is a function of taxonomic research at all levels of classification. The ultimate goal of phylogenetic research is the production of a phylogenetic system of classification. The phylogenetic system shows the genetic and time relationship of any one taxon to another. According to turrill (1942) "taxonomy is based on characters, phylogeny on changes of characters." A phylogenetic system of classification for plants would provide the answer to questions of (*i*) their origin (*ii*) to their modes of evolution, (*iii*) to problems of monophyleticism vs. polyphyleticism, (*iv*) the identity of primitive and advanced characters, etc. It would conclude in a single stable classification of relationships.

Diversity of Phyletic Concepts–The construction of the ultimate phylogenetic classification must be based on established facts regarding the characteristic features of the ancestors of every taxon level. These ancestors existed in remote prehistoric geological time, and because of their relative simplicity their characteristics are known as *primitive* which those of their present day descendants are known as *advanced.* For example, Eichler, Wettstein, Rendle and others considered the unisexual apetalous cyclic flower to be primitive, whereas Bessey, Hutchinson and others treated it as advanced and considered the bisexual polypetalous flower with spiral arrangement of parts to be primitive.

The formulation of phylogenetic classifications requires the teamwork and collaboration of botanists of all disciplines and the considered evaluation of data.

"Opinions have differed as to whether the monocotyledonous plants are more advanced or more primitive than the dicotyledonous plants. If the view is accepted that woody habit preceded the herbaceous, then, since the vast majority of ancient and modern monocots are herbaceous, it would point to the monocots been derived from one or more dicot ancestors. Botanists accepting this view treat the dicots as the more primitive. Ms. Arber has held that the monocots preceded the dicots. Current criteria from morphological and anatomical studies of both vegetative and reproductive structures would place the monocots as probably having been derived from the dicots. There is evidence for the belief that the monocots are more likely to be monophyletic in origin than is true for the dicots".

5

EMBRYOLOGY IN RELATION TO TAXONOMY

The existing systems of classification of angiosperms are based mainly on the external morphology of the flowers, fruits and seeds. However, in modern days it has been accepted that cytology, anatomy, embryology, genetics, palynology, and chemotaxonomy should also be taken into consideration in a number of doubtful cases where external morphology alone is not sufficient for the purpose.

The possibility of using embryological features in taxonomy was suggested for the first time by Hofmeister and Strasburger, but until the invention of the microtome such work was confined to a few skilled workers. Now it has been established that the embryological studies are quite useful in the systematic assignments of several disputed taxa and families.

Some more important embryological characters have been considered to be of major importance in the delimitation of the larger plant groups. They are as follows:

1. **Nature of anther tapetum.** Whether it is of the amoeboid or glandular type.

2. **Nature of anther lobes.** Whether they are fibrous or non-fibrous.

3. **Division of the microspore mother cells.** Whether it takes place by furrowing or by the formation of cell plates and whether the mode of division is simultaneous or successive.

4. **Development and organization of microspores.** Position and number of the germ pores and furrows; exine ornamentation; place of formation of the generative cell; shape and number of the nuclei in the pollen grain at the time of its discharge from the anther.

5. **Development and structure of the ovule.** Number of integuments and the alterations in structure which they undergo during the formation of the seed; presence or absence of vascular bundles in the integuments; shape of the micropyle, whether it is formed by the inner integument, or the outer, or both; presence or absence of an obturator.

6. **Form and extent of the nucellus.** Whether it is broad and massive or thin and ephemeral; whether hypostase is present or absent; place of origin of the integument, or integuments. Whether close to the apex of the nucellus or near its base; persistence or gradual disintegration of the nucellus during seed formation.

7. **Origin and extent of the sporogenous tissue in the ovule.** Nature of archesporium, whether it is one-celled, or many-celled; presence or absence of wall layers; presence or absence of periclinal division in the cells of the nucellar epidermis.

8. **Megasporogenesis and development of the embryo sac.** Arrangement of megaspores; position of functioning megaspore; whether the embryo sac is monosporic, bisporic or tetrasporic;

number of nuclear divisions occurring between the megaspore mother cell stage and the differentiation of the egg.

9. **Organization of mature embryo sac.** Shape of the embryo sac and the number and distribution of its nuclei; persistence, or early disintegration of the synergids and antipodal cells; increase in number of antipodal cells, if any; formation of embryo sac caeca or haustoria.

10. **Fertilization.** Path of the entry of the pollen tube; time between pollination and fertilization; any tendency towards a branching of the pollen tubes during their course to the ovule.

11. **Endosperm.** Whether it is of nuclear, or helobial type; orientation of the first wall in those cases in which it is cellular; presence or absence of endosperm haustoria and the manner in which they formed, if present; nature of food reserves in endosperm cells; persistence or gradual disintegration of endosperm in the mature seed.

12. **Embryo.** Relation of the proembryonal cells to the body regions of the embryo; form and organization of the mature embryo; presence or absence of endosperm hustoria.

13. **Certain abnormalities of development.** Parthenogenesis; apogamy; adventive embryony; polyembryony.

The above mentioned characters are of much value in the determination of proper position of several difficult groups of plants. Some important taxa, families and orders whose affinities have been confirmed on the basis of embryology are described here.

DICOTYLEDONS

Proteaceae. Some members of this family received meticulous attention by Kaushik (1942). A remarkable feature, missed by earlier workers, is the development in post-fertilization stages of a tubular, coiled and coenocytic structure, called **"vermiform appendage"** at the lower end of the embryo sac. According to Venkata Rao (1962), in *Lomatia,* in addition to the main haustorium, numerous 1-celled finger-shaped outgrowths develop all over the surface of the endosperm and thus increases its absorbing area.

Olacaceae. According to Shamanna (1954), in *Olax wightiana,* the ovules are unitegminal when young but the distinction between the integument and the nucellus is lost during subsequent development. The embryo sac is of the *Allium* type and shows prominent haustoria. The micropylar end of the embryo sac begins to grow out even at the fournucleate stage and becomes extraovular. At the eight-nucleate stage of the embryo sac, a lateral chalazal caecum appears just above the level of the antipodals. This too grows beyond the ovule and meets the vascular strand at the tip of the placenta. A comparative study was made by Agarwal (1963) of *Olax scandens, O.stricta* and *O. wightiana.* All the three species show an integument, however, in *O.imbricata,* this was absent (Fagerlind, 1947). *O.scandens, O. imbricata* and *O. stricta* show a *Polygonum* type of embryo sac whereas *O. wightiana* shows an *Allium* type. In *O. scandens,* as in *O. wightiana,* the tip of the 4-nucleate embryo sac extends beyond the ovule and grows towards stylar canal. In *O.stricta,* it remains restricted to the ovule. The endosperm is helobial.

Lauraceae. In 1853 Lindley proposed that the parasite *Cassytha,* should be removed from the predominantly woody family Lauraceae and placed in a separate unigeneric family, Cassythaceae. Sastri's (1962) investigation has shown that *Cassytha* has several features in common with many members of the Lauraceae such as ethereal oil cells; thick walled unicellular trichomes; successive divisions of the microspore mother cells; 2-celled and acolpate pollen grains; pendulous, anatropous, bitegminal and crassinucellar ovules; *Polygonum* type of embryo sac; disappearance of the inner integument in the mature seed; helical thickenings in the inner epidermis of the outer integument and transformation of the inner epidermis of the fruit wall into a stony layer. On the other hand, there are several characters distinctive to *Cassytha* which are not found in other members of the Lauraceae, such as secretory type of anther tapetum as against the plasmodial type in the Lauraceae; formation of multiple embryo sacs; extention of some embryo sacs beyond the nucellus and cellular endosperm. However, Sastri has proposed that they are not so significant as to warrant the removal of *Cassytha* from the Lauraceae.

Cercidiphyllum. The taxonomic status of *Cercidiphyllum* is far from clear. Botanists have discussed its relationships with the Hamamelidaceae, Cunoniaceae and Magnoliaceae. However, on the basis of embryological studies Swamy and Bailey (1949), assigned *Cercidiphyllum* to a separate family Cercidiphyllaceae. They pointed out that in *Cercidiphyllum* the furrows on the pollen grains are considerably broader and less well defined than in Hamamelidaceae and the structure of exine is different. In *Cercidiphyllum,* the nucellar epidermis divides periclinally, the fertilized embryo sac elongates conspicuously, and the endosperm is cellular. No such features occur in Hamamelidaceae. Whether this family should be placed under the order Ranales or placed in a new order required further investigation.

Nelumboniales. While working on *Nelumbo nucifera* of the Nymphaeaceae, Khanna (1965) observed that the antipodal cells multiply, become multinucleate and function as haustorium. Further, the hypostase and perisperm are absent and the endosperm is nuclear. She therefore, supports the creation of the order Nelumboniales for this genus as proposed earlier by Li (1955) and Takhtajan (1959).

Paeonia. This genus has often been included in the Ranunculaceae; some botanists consider it to be near the Magnoliaceae; others suggest a closer affinity with the Berberidaceae; and still others assign it to a separate family Paeoniaceae. Hutchinson regards *Paeonia* as a link between the Magnoliaceae and Helleboraceae. There are so many features, such as the absence of a distinct perianth; and presence of a concave receptacle, broad disc, centrifugal stamens, large ovules with a massive outer integument, frequent occurrence of multiple embryo sacs, a chalazal endosperm haustorium and arillate seeds in which *Paeonia* shows a sharp contrast with the Ranunculaceae. The removal of *Paeonia* to a separate family Paeoniaceae has been fully justified by the embryoloical investigation of Murgai (1962), as shown in the table:

Table 5.1

Characters	*Paeonia*	Ranunculaceae
1. Flower	It is spiral.	It is spirocyclic
2. Stamen	It is centrifugal	It is centripetal
3. Anther tapetum	It is glandular	It is glandular (occasionally plasmodial as in *Actaea*)
4. Ovulae	It is bitegmic, outer integument massive crassinucellate, nucellus get crushed and the embryo sac remains surrounded by the inner integument, hypostase invariably present.	It is uni-or bitegmic, crassinucellate, nucellar epidermis persists and encloses the embryo sac, hypostase rarely present.
5. Embryo sac	It is *Polygonum* type, antipodal cells ephemeral	It is *Polygonum* type (*Allium* type in *Adonis*), antipodal cells persistent, generally become massive and polyploid.
6. Endosperm	It is nuclear; chalazal haustorium is absent	It is nuclear type; chalazal haustorium is present.
7. Embryogeny	Peculiar, one of the cells of the 2-celled proembryo undergoes free nuclear division followed by wall formation and differentiation of several peripheral groups of cells of which one takes the lead and develops into a dicotyledonous embryo.	Irregular or Onagard type.
8. Seed germination	It is epigeal type	It is hypogeal type.

Actinidia. Endliecher (1836-1840), Bentham and Hooker (1862-1867), Baillon (1867-1895) assigned *Actinidia* in the Dilleniaceae, while Soleredar assigned it to the Theaceae. However, Van

Tieghem (1899) did not agree with this view and created a new family, the Actinidiaceae. The embryological studies of Vijayaraghavan (1965) reveal that in *Actinidia polygama* the ovules are anatropous, unitegminal, tenuinucellar and show a prominent hypostase and an incipient integumentary tapetum. The nucellus on the sides and at the tip is ephemeral. The embryo sac is of the *Polygonum* type, the endosperm is cellular, the embryogeny of Solanad type, and the seeds are without an aril. Except for the *Polygonum* type of embryo sac, non of these features are found in the Dilleniaceae. Thus justifying the creation of a new family Actinidiaceae. However, the family Actinidiaceae should be placed in the order Theales rather than the Dilleniales (Johri, 1963). Hutchinson (1959) has also assigned the Actinidiaceae in the order Theales.

Rosales. In many systems of classification the Crassulaceae are placed in the Rosales adjacent to the family Saxifragaceae. Hutchinson (1959) includes the Crassulaceae and Saxifragaceae in the Saxifragales. Subramanyam (1962) compared the embryological features of this family with allied taxa. The Crassulaceae are characterized by the presence of 3-colporate pollen grains, crassinucellar and bitegminal ovules, *Polygonum* type of embryo sac, chalazal endosperm haustorium and an undivided basal cell of the suspensor. On the other hand, the Podostemaceae have single or double pollen grains, a reduced *Allium* type of embryo sac, a pseudoembryo sac and a binucleate or coenocytic suspensor haustorium; and the endosperm is absent. However, the Crassulaceae resemble the Saxifragaceae in the structure of the pollen grains, ovule and development of female gametophyte and embryo, thus indicating that the two families are closely inter-related.

Alangiaceae. Bentham and Hooker (1862-1867) placed *Alangium* in the family Cornaceae, while Engler and Prantl (1931) kept it in a separate family Alangiaceae. This view has also been supported by many embryologists (Schnarf, 1922; Gopinath, 1945; Mitra and Datta, 1949). According to these workers *Alangium* differs from the family Cornaceae in showing the characters : ovules with a single-celled archesporium; *Polygonum* type of embryo sac and nuclear endosperm. However, the Alangiaceae shows closer affinities with the families included in the Araliales than with those in the Myrtiflorae. Thus the Alangiaceae should be placed in the order Araliales.

Celastrales. Bentham and Hooker (1862-1867) included the family Stackhousiaceae in the Celastrales. Engler and Diels placed it in the Sapindales between the Salvadoraceae and Staphyleaceae and Hutchinson has given it a place next to the Salvadoraceae. Stant (1951) stated that *Stackhousia* most closely resembles certain genera of the Scrophulariaceae and Selaginaceae. Narang's (1953) study of *Stackhousia linariaefolia* has shown that the ovule is bitegminal and tenuinucellar, the embryo sac is of *Polygonum* type with three multinucleate antipodal cells, and the endosperm is nuclear. No such characters are found in the Rhamnaceae, Scrophulariaceae or Selaginaceae. The family Stackhousiaceae is, therefore, assigned to the Celastrales.

Turneraceae. Hutchinson (1959) includes the family Turneraceae in the order Loasales along with the Loasaceae. However, Engler (1964) assigns it to the Violales, close to the Passifloraceae. In *Turnera ulmifolia* of Turneraceae the ovules are anatropous, bitegminal and crassinucellar, the pollen tube is persistent, the endosperm is nuclear, the embryogeny of Onagrad type, and the seeds are arillate. The pollen grains are shed at the 2-celled stage. All these features are found in Passifloraceae but not in Loasaceae. The characteristics of Loasaceae are : hemianatropous, unitegminal, tenuinucellar ovules, occasionally persistent antipodal cells, cellular endosperm with micropylar and chalazal haustoria and the Solanad type of embryogeny. An aril is absent and the pollen grains are shed at the 3-celled stage. Thus the Turneraceae do not have any close relationship with the Loasaceae.

Myrtiflorae. Bentham and Hooker (1962) placed the genus *Haloragis* in the family Haloragaceae along with other genera *Callitriche, Gunnera, Hippuris, Myriophyllum,* etc. Engler and Prantl (1889-1897) removed *Callitriche* and *Hippuris* from the Haloragaceae, and placed them in the Callitrichaceae and Hippuridaceae respectively, and the remaining genera were retained in the

Haloragaceae. Wettstein (1935) placed *Hippuris* and *Gunnera* into two separate families, Hippuridaceae and Gunneraceae. However, *Haloragis, Myriophyllum,* etc., were included in Haloragaceae. According to Hutchinson (1959) *Haloragis, Hippuris* and *Gunnera* were retained in the Haloragaceae, and *Callitriche* was removed to a separate family, the Callitrichaceae. *Haloragis* differs from *Hippuris* in possessing the following features: bitegminal, crassinucellar ovules; porogamy and a suspensor haustorium. It differs from *Gunnera* in the absence of a *Peperomia* type of embryo sac, and the presence of a prominent suspensor haustorium. The assignment of these genera to separate families is justified by Kapil (1962), as shown in the table:

Table : 5.2

Characters	*Gunnera*	*Haloragis*	*Hippuris*
1. Flower	It is ebracteate, hermaphrodite, bimerous and crowded in large spike or panicles.	It is bracteate, hermaphrodite, tetramerous and arranged in glomerules in leaf axil.	It is bracteate, hermaphrodite, occasionally unisexual and solitary; rim of perianth is found around the tip of ovary.
2. Androecium	2, antepetalous	4+4, obdiplostemonous	1, placed medianly
3. Pollen	3-5, colpate	4 or 5, rupoidate	4-6, colpate
4. Gynoecium	2 carpels, unilocular, single ovule	4 carpels, tetralocular, single pendulous ovule in each locule.	1 carpel, unilocular, single pendulous ovule.
5. Ovule	Anatropous, bitegmic, crassinucellate; obtura-tor absent.	Anatropous, bitegmic, cra-ssinucellate; feeble funicu-lar obturator present.	Anatropous, ategmic or unitegmic, tenuinucellate; nucellar cap and a feeble funicular obturator present.
6. Embryo sac	*Peperomia* type	*Polygonum* type	*Polygonum* type
7. Embryo	Minute; obcordate suspensor; haustorium absent.	Large; cylindrical suspensor forming a haustorium.	Large; cylindrical suspensor; haustorium absent.

Gentianaceae. Bentham and Hooker and Rendle divided the family Gentianaceae into two sub-families, *i.e.*, the Menyanthoideae including the plants of marshy habitat, and the Gentianoideae comprising the terrestrial genera. However, on the basis of vegetative and floral anatomy Wettstein (1935) raised Menyanthoideae to a rank of family which was supported by Don (1938) and Lindsey (1938). The view is supported by embryogeny also. Further support came from the embryological studies of Maheshwari Devi (1962). The members of the Menyanthoideae are characterized by a secretory anther tapetum with bi-or multinucleate cells, suppressed parietal placentae, well developed endothelium, cellular endosperm and Asterad type of embryogeny. On the other hand, the Gentianoideae have an amoeboid or secretory anther tapetum with uninucleate cells, well developed parietal placentae, nuclear endosperm and Solanad type of embryogeny; an integumentary tapetum is absent. Hence the raising of these sub-families to two independent families is fully justified.

The comparative table is as follows:

Table: 5.3

Characters	Menyanthoideae	Gentianoideae
1. Anther tapetum	Secretory, with bi-or multinucleate cells.	Amoeboid or secretory, with uninucleate cells.
1. Placenate	Suppressed parietal placentae.	Developed parietal placentae.
3. Endothelium	Present	Absent
4. Endosperm	Cellular	Nuclear
5. Embryogeny	Asterad type	Solanad type

Acanthaceae. By earlier taxonomists the Acanthaceae has been divided into four sub-families, *viz.*, Nelsonioideae, Mendoncioideae, Thunbergeoideae and Acanthoideae. Serveral years ago, a Dutch morphologist Bremekarnp (1953) suggested the separation of Nelsonioideae from Acanthaceae and its assignment to the Rhinantheae (Scrophulariaceae). However, an embryological study of two members of the Nelsonioideae, *Elytraria* and *Nelsonia* carried out by Johri and Singh (1959) and Mohan Ram and Masand (1963), at Delhi shows that these genera have an asymmetrical growth of the endosperm and the occurrence of a rudimentary jaculator which are Acanthaceae like characters and are absent in the Rhinantheae (Scrophulariaceae). Moreover, the persistent and ruminate endosperm present in *Elytraria* and *Nelsonia*, has also been noticed in *Andrographis* and *Haplanthus* of the Acanthoideae which further supports the inclusion of the Nelsonioideae in the Acanthaceae.

Orobanchaceae. Some taxonomists consider the family Orobanchaceae as related to the Gesneriaceae while others consider closer to the Scrophulariaceae. According to Tiagi (1963), the similarities in the development of the gametophytes, endosperm, embryo and seed suggest for an allinace between the Orobanchaceae and the Scrophulariaceae. However, Tiagi (1963) considers that the Orobanchaceae can easily be derived from the Scrophulariaceae through forms like *Striga orobanchoides.*

Campanulaceae. The earlier taxonomists (Bentham and Hooker, 1873- 1876; Engler and Prantl, 1889) assigned the genus *Pentaphragma* to the family Campanulaceae. This genus came into prominence when Airy Shaw (1954) removed it from the Campanulaceae and raised it to the status of a new family Pentaphragmataceae. He compared its characters with those of the Begoniaceae, Boraginaceae and even Cucurbitaceae and Solanaceae. Kapil and Vijayaraghavan (1965) pointed out that *Pentaphragma* has a number of distinctive embryological characters, such as the presence of unitegminal, tenuicellular ovules, an extramicrophylar embryo sac, cellular endosperm and Solanad type of embryogeny, which are not known in the Begoniaceae. On the other hand, the Boraginaceae have a superior ovary, multinucleate tapetal cells, crassinucellar ovules, *Polygonum* or *Allium* type of embryo sac with ephemeral antipodal cells, nuclear, cellular or helobial endosperm and Chenopodiad type of embryo. Hence there is no affinity between *Pentaphragma* and the Boraginaceae. Contrary to it, *Pentaphragma* resembles the Campanulaceae in almost all morphological and embryological characters excepting the presence of an extramicropylar protrusion of the embryo sac, a single-celled uninucleate micropylar haustorium, a short suspensor, and the absence of a chalazal haustorium. Hence Kapil and Vijayaraghavan concluded that the genus *Pentaphragma* belongs to the Campanulaceae.

Another genus *Sphenoclea* has been placed in the family Campanulaceae by Engler (1897) and Hutchinson (1959). On the other hand, Airy Shaw (1948) raised it to a rank of family and suggested that the newly formed family Sphenocleaceae has no relationship with the family Campanulaceae. According to him *Sphenoclea* resembles in habit and anatomy with *Phytolacca,* while in certain features it shows similarities with the family Primulaceae. Hence Airy Shaw (1948) considered the family Sphenocleaceae in between the Phytolaccaceae and the Primulaceae. However, on the basis of embryological studies of Subramanyam (1950) it has been suggested that *Sphenoclea* has no close connections with either the Phytolaccaceae or the Primulaceae. Since *Sphenoclea* also shows certain peculiarities like the *Scutellaria* type of endosperm and 4-celled micropylar and chalazal haustoria, it would be better to assign it to a separate family Sphenocleaceae adjacent to the Campanulaceae.

Onagraceae. The family Onagraceae affords one of the best examples of the utility of embryological characters in taxonomic considerations. For example, Bentham and Hooker (1883) placed the genus *Trapa* in family Onagraceae, whereas Engler and Diels (1936) and Hutchinson (1959) assigned it in the family Hydrocaryaceae (Trapaceae). Many embryologists have supported this view. In addition to its eight-nucleate embryo sac, *Trapa* has a well developed suspensor haustorium, both these features are not known in any member of the Onagraceae. Further in the Onagraceae the ovary is inferior and tetralocular, with axile placentae bearing numerous ovules, and the fruit is a loculicidal capsule. On the other hand, in *Trapa,* the ovary is semi-inferior and bilocular, with only one ovule in each locule, and the fruit is a large one-seeded drupe whose fleshy layer soon disappears, leaving only the stony endocarp. Thus the removal of *Trapa* from family Onagraceae and its assignment to the family Trapaceae is justified. Eames (1951) has also supported this view on the basis of anatomical studies.

Balanitaceae. The genus *Balanites* is placed in the family Simarubaceae (Bentham and Hooker, 1862-1867); Engler and Prantl (1931) included this genus in the Zygophyllaceae of the order Geraniales. However, Hutchinson (1959) raised a monogeneric family, the Balanitaceae under the order Malpighiales. The embryological studies show that *Balanites aegyptiaca* has the following features with Zygophyllaceae and Simarubaceae (*i*) secretory anther tapetum, (*ii*) *Polygonum* type of embryo sac and (*iii*) nuclear endosperm. In *Balanites* and the Zygophyllaceae the embryological features are more common. In both, the gynoecium is syncarpous, pollen is shed at 2 or 3-celled stage, the ovule is bitegminal and crassinucellate. Endothelium is present. On the other hand, in the Simarubaceae the gynoecium is apo-or syncarpous, the pollen is 2-celled, the ovule may be anatropous or amphitropous, uni-or bitegmic, carassinucellate. The endothelium is absent. The genus *Balanites* is more associated to the Zygophyllaceae than the Simarubaceae. Though to some extent the position of Balanitaceae has been justified, yet it requires more investigation to be done in future.

Canellaceae. Acoording to Wettstein (1935) and Vestal (1937) the family Canellaceae is adjacent to the family Myristicaceae. Hutchinson (1959) has placed this family in the order Magnoliales in between the families Winteraceae and Schisandraceae. Wilson (1960) has also suggested its relationship to the families Eupeteleaceae, Dilleniaceae, Eupomatiaceae, Illiciaceae and Schisandraceae of the woody Ranales, than to the families in the order Parietales. Parameswaran (1962) suggested on the basis of embryological studies that the Myristicaceae and the Canellaceae are more closely related to each other than to the remaining families of the woody Ranales.

Cneoraceae. The family Cneoraceae is represented by the genus *Cneorum.* Bentham and Hooker placed the genus *Cneorum* in the family Simarubaceae. However, Engler and Prantl (1897) accepted the family status of the Cneoraceae and transferred it to the Geraniales. Van Tieghem (1906) placed this family in the order Rhamnales, and Hutchinson (1959) placed it near the family Celastraceae in the Celastrales. However, on the basis of embryological studies, Soueges (1955) showed that the family Cneoraceae have affinities with the Rhamnaceae and should be placed in the order Rhamnales.

Aizoaceae. The genus *Limeum* has been placed in the family Aizoaceae (Bentham and Hooker, 1862 - 1867). Lindley (1863) and Gunderson (1959) considered this genus to be a member of family Phytolaccaceae. Hutchinson (1959) stated that it should be kept in the family Molluginaceae. On the basis of the embryological studies by Narayana and Jain (1962) have shown resemblances between the genus *Limeum* and the family Aizoaceae, and between *Limeum* and family Phytolaccaceae. Moreover, these workers are of opinion that *Limeum* has more common features with family Phytolaccaceae than Aizoaceae. However, the dispute is yet to be solved by further studies.

Parnassiaceae. Bentham and Hooker (1862-1867), Engler and Prantl (1897) and Diels (1906) placed the genus Parnassia in the family Saxifragaceae. According to Pace (1906) this genus is more closely related to the family Droseraceae. However, Lebegue (1953) stated on the basis of embryological studies that the genus resembles to the family Portulacaceae.

On the other hand, Hutchinson (1959) raised a monogeneric family, the Parnassiaceae, and kept it under the order Saxifragales. Much work has yet to be done for final results.

MONOCOTYLEDONS

Butomaceae. According to Johri (1933 - 1938), the genus *Butomus* shows a *Polygonum* type of embryo sac, the other genera of the Alismaceae are characterized by an *Allium* type of development. The genus *Butomus* and three other genera, *viz., Butomopsis, Limnocharis* and *Hydrocleis* belong to the family Butomaceae. In pollen morphology *Butomus* also differs from the rest of the genera included in the family Butomaceae. Pichon (1946) showed that the family Alismaceae and all other genera of the family Butomaceae, with the exception of the genus *Butomus,* are characterized by the presence of laticiferous ducts, petiolate leaves with expanded lamina, caducous petals, campylotropous ovules and curved embryos. On the contrary, the genus *Butomus* is characterized by absence of laticiferous ducts, sessile and linear leaves, persistent petals, anatropous ovules and straight embryos. This justifies the transfer of *Butomopsis, Hydrocleis* and *Limnocharis* to the family Alismaceae, and the retention of *Butomus* in the family Butomaceae. Additional support for transferring *Butomopsis, Hydrocleis, Limnocharis* and *Ostenia* to the Alismaceae and retaining only *Butomus* in the Butomaceae has been derived from the cytological studies of Sundar Rao (1953). On the other hand, Takhtajan (1959) has splitted the Butomaceae into two families, *i.e.,* Butomaceae (having single genus *Butomus*) and Limnocharitaceae to include the remaining genera.

Lemnaceae. Hutchinson (1959) has included the Lemnaceae and Araceae in the order Arales, while Engler (1889, 1920) has placed these families in the order Spathiflorae.

Most of the taxonomists accepted this view. However, Eichler (1875) placed this family in the order Helobiales. A detailed embryological study of *Lemna* and *Wolffia* was carried out by Maheshwari (1954, 1956, 1958). According to him the development of the endosperm in the Lemnaceae is not of the Helobial type. Further, he reported that the families Lemnaceae and Araceae exhibit similarities. Such similarities are : the presence of cellular endosperm with a basal caecum, a nucellar cap and an operculum formed by the inner integument. Generally in the Helobiales the endosperm is of helobial or nuclear type without a chalazal caecum. Here the basal cell of the 2-celled proembryo becomes hypertrophied. The nucellar cap and operculum are not found. Thus it is thought that the family Lemnaceae is closely related to the family Araceae.

Graminales. Several criteria have been used from time to time to separate the two sub families, *i.e.,* Poaideae and Panicoideae, and the tribes of the Gramineae. These criteria include cytology, anatomy and embryology. According to Chandra (1963) the structure of the ovule makes an important criterion in the systematics of the Gramineae.

Genus *Lilaea*. The genus *Lilaea* is included in the family Juncaginaceae (Scheuchzeriaceae) along with other genera *Scheuchzeria, Tetranicum* and *Triglochin* by Bentham and Hooker (1883), Engler and Prantl (1889) and Wettstein (1935). On the contrary, Hieronymus (1892), Schumann (1892) and Hutchinson (1959) placed *Lilaea* in a separate family, the Lilaeaceae (Heterostylaceae). Agarwal (1952) has described the most important embryological characters of *Lilaea, Scheuchzeria* and *Triglochin.*

It is observed that *Lilaea* is embryologically similar to *Triglochin* but shows differences from *Scheuchzeria* in two respects (*i*) the endosperm is helobial and (*ii*) the pollen are three-celled. Hence Agrawal (1952) also supported the view for including *Lilaea* in the family Scheuchzeriaceae.

Table : 5.4

Characters	*Lilaea*	*Scheuchzeria*	*Triglochin*
1. Pollen	Two-celled	Three-celled	Two-celled
2. Ovule	Basal, anatropous, bitegminal, crassinucellar	Basal, anatropous, bitegminal, crassinucellar	Basal, anatropous, bitegminal, crassinucellar
3. Embryo sac	*Polygonum* type	*Polygonum* type	*Polygonum* type
4. Endosperm	Nuclear	Helobial	Nuclear
5. Embryo	*Sagittaria* type	*Sagittaria* type	*Sagittaria* type
6. Seed	Exalbuminous	Exalbuminous	Exalbuminous

Scitamineae. The endosperm in the Zingiberaceae is of Helobial type but there is considerable variation in the organization of the basal apparatus. In *Elettaria, Hetchenia* and *Zingiber* the basal apparatus is binucleate, in *Costus* it is multinucleate, and in *Hedychium* it may be uni- or binucleate, or partitioned into two cells of which the lower cell is uninucleate and the upper one is multinucleate.

However, the embryologists do not claim these data as tools to erect a phylogenetic system of their own, but it is worth for the workers of embryology along with information from other sources such as morphology, anatomy, cytology, etc, in order to approach a natural system of classification, and thus embryology in relation to taxonomy has its own importance.

6

NOMENCLATURE OF PLANTS

Historical : The present day binomial system of nomenclature has a long evolutionary development. In near about 200 B.C. Cato in his monumental work 'De Re Rustica' used two names for plants, although he lacked the modern concept of genera and species. Later on the descriptive Greek nouns used for genera were translated into Latin, where it took the form of two words, *i.e.,* binary generic names. According to another tendency a descriptive phrase was given for the specific name. These methods resulted in the so called polynomials. In the middle of sixteenth century, Brunfels changed many binary generic names to single ones. A few years later Dodonaeus followed the binomial usage similar in principal to that of today's binomial system of nomenclature. Gaspard Bauhin wrote his monumental work 'Pinax' in 1623, in which he has given a list of about 6,000 plants. He used the binomial system extensively to give names to the plants. None of these workers could make a permanent impression upon the botanical world of that time and works of the writers of those days continued to appear having mixed usage—monomial, binomial, trinomial and polynomial. In those times the names of the plants were composed of several words in a series, constituting more or less a description of the plants. In 1753 Linnaeus published his great work '*Species Plantarum*' where he employed the binomial system to name the plants in true sense. Augustin de Candolle's *Theorie elementaire de la botanique* published in 1813 provided the first significant work since the publications of Linnaeus on nomenclature and morphology of the plants. In this work for the first time, a complete and detailed set of rules on plant nomenclature was given. In 1821, Stendel's *Nomenclator botanicus* was published, which comprised a list of the Latin names of all flowering plants known by then. The second edition of this work appeared in 1840. *Nomenclator* was used universally by the botanists, and formed the basis of the *Index Kewensis.*

Common Names of the Plants : The plants have been provided with names by the different people of the different regions of the world in their own languages. These common or local names of the plants have their own weaknesses. There are three main defects in common names–(*i*) They may be quite indefinite, *e.g.,* Pansy has been provided with fifty English names. (*ii*) They are restricted to the people of one language or even one section of a country. (*iii*) They are not regulated by any constituted authority. These common names give great trouble to a plant collector in a foreign country. There he feels the local names to be more difficult than the botanical names.

Botanical or Binomial Names of the Plants: At the present time each kind of plant is given a generic name followed by a specific name or epithet. The generic name is a noun, and the specific epithet is an adjective indicating which of the several members of the genus has been considered. The genus name is always capitalized. The modern tendency is not to capitalize the specific name regardless of its derivation. Binomials are frequently descriptive of the plants and are usually derived from Greek or Latin, since these languages are internationally known by scholars. For example *Coffea arabica*

Linn., stands for coffee; *Ficus elastica* Roxb., for India rubber, *Helianthus annuus* Linn., for sunflower, etc.

Binomials when written should always be underlined and when printed they must be in italicized type. The name or the abbreviated name of the scientist who first described the species follows the binomial–such as *Coffea arabica* Linn. ; *Mangifera Indica* Linn. ; *Rosa alba* Linn. ; *Santalum album* Linn., etc. Linn., stands for the name of the great taxonomist Linnaeus; *Albizia procera* Benth. ; *Aloe barbadensis* Mill. ; *Kochia indica* Wight; *Mussaenda glabrata* Hutch.; *Solanum grandiflorum* Ruiz and Pav., here each binomial is followed by the abbreviated name of the scientist who first described the species. Some binomials are followed by two names, the first of which is within the brackets, thus *Albizia lebbeck* (Linn.) Benth. The name within brackets is that of the person who first described the species. The name which follows the parenthesis is that of the person who is responsible for the currently recognized binomial. Accordingly, Linnaeus described the plant for the first time and Bentham corrected the name later on. The other such examples are – *Albizia chinensis* (Osbeck) Merr.; *Alocacia indica* (Roxb.) Schott. ; *Amorphophallus campanulatus* (Roxb.) Blume ex Dcne ; *Brassica juncea* (Linn.) Czern. and Cors. var. *Cuneifolia* Roxb. ; *Butea monosperma* (Lamk.) Taubert; *Feronia limonia* (Linn.) Swingle, etc. The reason for such author citations aids the taxonomist in finding the original and subsequent descriptions of a plant when necessary, and to avoid confusion when different workers use the same binomial accidentally to name different species. To aid the taxonomist further, the year in which the plant was described is sometimes written after the author's name following the binomial.

International Rules of Botanical Nomenclature : In 1950, the seventh international botanical congress was held at Stockholm and the rules of nomenclature were the subject of lengthy and considered discussions. Many minor changes and refinements were legislated.

Thereafter in 1956 the rules have been propounded by Lanjouw as the result of thorough study of many years.

They are as follows : 1. Long generic names with difficult pronunciation must be avoided; 2. A generic and a specific name is given to a perfectly named plant; 3. The specific name must differentiate a plant from all allied species; 4. The species is not distinguished by its size; 5. The specific differences of a plant have no concern with the original place of the plant; 6. colour should not be treated as a specific difference; 7. Each plant species must possess a generic name; 8. The generic name must be followed by a specific name; 9. The shorter specific name is always well and good.

The above Linnean principles are quite simple, the role of codifying such rules has been taken up by various International Congresses, which have clarified some rules, modified others, and when necessary have established new ones. In all these changes the main aim has been to produce an uniform system.

The 1956 edition of the code gives detailed instructions on the steps to be taken when any change appears necessary. Such a change is first proposed to the various nomenclature committees elected by the International Botanical Congress and these committees make their recommendations on the subject in preliminary way. The recommendations are then submitted to a vote by post from all the members of the International Association for Plant Taxonomy, from the authors themselves, and from the members of the committees. During the International Botanical Congress all officially enrolled members of the Section cast their final vote. After the inception of the journal '*Taxon*', every change that has been proposed is fully discussed in the journal before any decision is taken on the subject. Any proposal for a change receives full consideration before it is put up for confirmation or rejection during the session of the congress. It is to be noted that during the last three or four International Botanical Congresses there have been but few changes, except perhaps in the rearrangement of the code and in some less important points of nomenclature.

Main Points of the International Code

The code is divided into three parts, (*i*) **principles,** (*ii*) **rules** and (*iii*) **recommendations.** In addition there are few interesting appendices. The **principles** are basic points on which the code is based. They do not give detailed rules about nomenclature, but show the main ideas that have guided the compilers of the Code, and should be kept in view by any botanist attempting to publish a new *taxon.* According to this section of the code, every *taxon* or group of plants can bear only one correct name and *vice versa* a name could be applied to one group of plants. The *rules* give detailed prescriptions on all the points connected with the naming of plants. The *recommendations* are practical applications of the rules.

Some More Important Rules

1. **The rank of taxa.** In the 1956 edition of the code, the word '*taxon*' has been introduced for the first time, to signify "*taxonomic groups of any rank*" (Article 1). The rank of *species* is basic; one or more species make up a *genus*; one or more genera make up a *family*; one or more families form an *order*; one or more orders make up a *class*; one or more classess make up a *division;* several divisions make up the *Vegetable Kingdom.* It is to be noted that the Code does not recognize "*Natural Order*", a term often used in older books for the modern equivalent of "*family*" of the code.

The code has special provisions for the naming of the various *taxa.* The names of families must ordinarily end in – *aceae,* those of orders in – *ales,* and both should be treated as feminine plural nouns. Accordingly it is wrong to say, *e.g.,* "The Amaranthaceae consists of ten genera". The correct way is to say, "The Amaranthaceae consist of ten genera" or the Papilionaceae are a family of the Leguminosae, which include two other families, Caesalpiniaceae and Mimosaceae." Names of sub-families, *e.g.,* Amaranthoideae; of tribes, *e.g.,* Amarantheae; of sub-tribes, *e.g.,* Amaranthinae, are all plural names and should be treated like the name of the family in this respect.

2. The type method. This is a modification introduced in the code. According to Principle II, "The application of name of taxonomic groups is determined by means of nomenclatural types". This means that when a species is described as new, the author must indicate which is the type of specimen on which the new species is based. In the case of species, the type is an individual specimen, which is the *type specimen* on which the new species is based. In the small herbaceous plants all mounted on one herbarium sheet, the whole sheet may be marked as the type. When a specimen cannot be preserved, an illustration or figure or a description may be a type. The type of a genus is a species; that of a family is a genus.

In the *nomenclature of types,*several terms are used which are as follows :

The *type* or *holotype* is the one element chosen by the author as the nomenclatural type. An *isotype* is a duplicate of the type or holotype, *e.g.,* if several branches of a tree are collected at the same time, one specimen may be chosen as the *type,* the rest are *isotypes*. Similarly, if a number of small herbaceous plants are collected at the same time, all belonging to the same new species, one may be selected as the *holotype,* the rest are *isotypes.* If they are mounted on the same herbarium sheet, they may be all part of the holotype (Article, 10).

A '*paratype*' is a "specimen cited with the original description other than the holotype or isotype(s)". A paratype usually is a specimen from a collection other than the type on which the description of the new *taxon* has been based. Thus, there can be only one sheet of holotype, whereas a number of sheets of isotypes and paratypes. When an author describes a new species and cites two or more specimens, they become the *syntypes,* if none of them individually has been designated the type or if more than one has been so designated.

"A *lectotype* is a specimen or other element selected from the original material to serve as nomenclatural type when the holotype was not designated at the time of publication or for so long as it is missing." When two or more specimens have been designated as types by the author, they become syntypes, and one must be chosen as a *lectotype.* However, when all the material, on which the description of a new species was based, is missing, then a *neotype* must be selected, and the selection once made must be followed by subsequent taxonomists. This seems to be the case with many of the new taxa described by Roxburgh (1814), for which no actual specimen was either designated or preserved.

The fixing of the type specimen for a species or of the type in general for *taxa* below the rank of order is considered so important in the Code that Article 35 prescribes that "publication on or after January 1, 1958, of the name of a new *taxon* of recent plants of the rank of order or below is valid only when the nomenclatural type is indicated" (Santapau, 1959)

3. **Priority.** With a few exceptions, where alternative names are permitted, each *taxon* of plants can bear only one correct name; "the correct name is the earliest legitimate one validly published; except in cases of limitation of priority by conservation."

4. **Tautonyms.** "The name of a species is a binary combination consisting of the name of the *genus* followed by a single *specific epithet*"(Article, 23). The Code gives to authors very great latitude in the selection of names, always provided that the specific name has not been used previously in the genus. There is one case, however, which is specifically condemned as illegitimate and that is the use of *tautonyms, i.e.,* when the specific name repeats unaltered the generic name, *e.g., Malus malus* Britt. This method is accepted in zoological nomenclature, where such names are coined, *e.g., Turdus turdus* or *Corvus corvus,* etc. In botanical nomenclature, however, this is condemned, Art., 70(4). Thus, doubts have arisen about the validity of such names as *Sesbania sesban* or *Samanea saman,* etc. Some authors have expressed strongly against such names, while on the other hand, Merrill (1912, 1916) accepts them as valid on the ground that a tautonym is only a name that repeats *unaltered* the generic name, and the names mentioned above do not repeat the generic name in full. Thus, such names can be taken as valid until a further, International Botanical Congress decides the matter.

5. **Names of families.** The name of a family is a plural adjective and is formed by adding the suffix-*aceae* to the stem of a legitimate name of an included genus, *e.g.,* Rosaceae from *Rosa,* Linaceae from *Linum* and Cucurbitaceae from *Cucurbita.* The new names of the following eight families are :

Palmae	=	Arecaceae
Gramineae	=	Poaceae
Cruciferae	=	Brassicaceae
Leguminosae	=	Fabaceae
Guttiferac	=	Clusiaceae
Umbelliferae	=	Apiaceae
Labiatae	=	Lamiaceae
Compositae	=	Asteraceae

6. **Names of infraspecific taxa.** There is one case where repetition of the names is not only allowed, but ordered. Article 26 prescribes that the name of an infraspecific taxon (*i.e.,* sub-species, variety or form) must be the same as that of the higher taxon, if the lower taxon includes the type of the higher one. Thus, if a species is divided into varieties, the one that contains the type of the species must be called by the same name as the species and not var., *genuina* or var., *typica* (Art., 71), as often used to be called before. For example, *Carissa congesta* wt., of Khandala on the Western Ghats has two varieties, (*i*) the typical plant, var., *congesta* (without name of author after the variety) and (*ii*) var., *albida* Sant. If the species has been divided into sub-species, varieties, forms, etc., then a plant is called *Carissa congesta* Wt., subsp., *congesta,* var., *congesta,* f., *congesta*; if the type of each higher rank is included in the lower division, the name *Andropogon sorghum* subsp., *halepensis* (L) Hackel var., *halepensis* is legitimate since the subspecies and the variety have the same type and the epithet must be repeated under (Art., 26; Art. 27, Santapau).

7. **Effective and valid publication of new taxa.** Publication is effective when one can expect that the general botanical public may come to know of the publication (Santapau). According to Art., 29, publication is effective only when printed matter is distributed to the general public or atleast to botanical institutions with libraries accessible to botanists. It is not effective by communication of a new names at a public meeting such as during the annual session of the Indian Science Congress, nor by placing of names in collections or gardens open to the public as for instance by labelling one's

specimens and placing them in public herbaria, where visitors may see the names. The date of the effective publication is the date on which the printed matter became available.

The following are the conditions attached to the publication of new *taxa* for validity.

1. Publication must be effective.

2. There must be a description of the new *taxon,* or a reference to a previously and effectively published description.

3. A combination is not validly published unless the author actually makes it (Art., 32).

4. Publication must be done in Latin (Art., 34) or with reference to a previously published Latin description.

5. When all the conditions for valid publication are not fulfilled at one and the same time, publication becomes valid when the last condition has been fulfilled. Thus, a publication is invalid if a botanist forgets to indicate the type of his new taxa, but it becomes valid when the type of the taxon has been indicated. The date of such publication will be when the type was indicated.

8. **Changes in name when taxa are divided, fused or altered.** When a genus is divided into two or more genera, or a species is split into two or more species, the original generic or specific name must be retained for the new *taxon* containing the type; this applies also to infra-specific *taxa.*

When a section of a genus or species is transferred to another genus or species *without alteration in rank,* the original name must be retained whenever possible.

When the rank of a genus or infra-generic *taxon* is changed, the correct name or epithet is the earliest legitimate one available in the new rank.

When *taxa* of the same rank are united into one, the oldest legitimate name must be used for the new combined *taxon*; if the names are of the same date, the author who first unites them has the right to choose one of the names and his choice must be followed by subsequent botanists.

9. **Form and spelling of names.** As per Code, authors have full liberty to coin their names, Authors are recommended not to make such names too long, nor too difficult to pronounce. In any case such names should be treated as Latin names. When the name is actually taken from the Latin language, it retains the original gender and spelling. When the name is taken from the vernaculars, the author assigns a gender to it, and this must be followed by subsequent botanists. Names of genera and higher *taxa* are written with a capital initial letter. Specific names should be written with a small initial letter.

Advantages of Binomial System : The binomials are more definite and precise than the common names. They are usually derived from Greek or Latin, since these languages are internationally known by scholars. They are frequently descriptive of the organism or plants. They aid the taxonomists in finding the original and subsequent descriptions which indicate the systematic relationships.

7

PLANT COLLECTION AND PREPARATION OF HERBARIUM

What is Herbarium: A herbarium is a collection of plants, which have been dried, pressed, mounted on herbarium sheets, identified and classified according to some approved system of classification. The succulent plants are usually preserved as such in some suitable preservative, such as 4% formaline solution or F.A.A. Many types of dry fruits; cones of gymnosperms and inflorescences of palms, etc., are dried and kept in large containers.

About 1550, Cesalpini and his co-workers began quite definitely to preserve the materials they studied, and since then herbarium making became a great and interesting feature of botanical work. All civilized countries possess their own plant collections (herbaria). The greatest herbarium of the world is at the Royal Botanic Gardens, Kew, England, possessing about six million specimens. A few good herbaria are there in our country. The biggest herbarium of our country is at the Indian Botanic Garden, Calcutta, possessing about one million specimens. The herbarium of the Forest Research Institute, Dehradun has about 3,00,000 specimens. The herbaria of Agricultural College-and Research Institute, Coimbatore and National Botanical Gardens, Lucknow, have about 200,000 and 40,000 specimens respectively. There are about 25,000 specimens in the herbarium of the Divisions of Mycology and Plant Pathyolog at Indian Agricultural Research Institute, New Delhi. The hsebarium of the Division of Botany at I.A.R.I. New Delhi, contains about 3000 specimens.

Some Important Indian Herbaria

1. **Forest Research Institute (FRI), Dehradun.** The botany branch, one of the first to be created today comprises of three sections, *viz.,* systematic botany, wood anatomy and plant physiology. The division today maintains a botanical garden, an arboretum, having one of the richest live collections of both indigenous and exotic tree species, and a bambusetum, containing germ plasm of forty species of indigenous and exotic bamboos. All this provides a rich source of material for research, teaching and training.

Systematic Botany and Herbarium. This section has been largely responsible for the collection and indentification of a large portion of the floristic diversity of the Indian sub-continent. Started by Gamble (1890), the herbarium of the FRI has grown to become one of the largest of its kind in Asia. It today houses 3,25,000 authenticated plant specimens, including 1300 type specimens, as well as a carpological collection. Some of the notable publications of the division, based primarily upon the herbarium collection, are 'Manual of Indian Forest Botany, 'World Monograph on the genus *Toona'*, 'Forest Flora of Punjab', Kumaon, Chakrata, Andamans, Orchids of north-western India and food from Forests.

2. **Herbarium of the Indian Botanic Gardens, Calcutta.** Foundation, 1787; Status, directed by the State of West Bengal, Department of Agriculture, Animal Husbandry and Forests; Number of specimens, more than one million. Herbarium. World-wide, mainly phanerogams and ferns of India and neighbouring countries of South and South East Asia. In addition to these there are some authentic collections of Cryptogams, Associated Botanic Gardens: Indian Botanic Garden, Calcutta, and Lloyd Botanic Garden, Darjeeling.

3. **Herbarium of the National Botanic Gardens, Lucknow.** Foundation, 1948; in 1953 taken over by the Council of Scientific and Industrial Research (CSIR), New Delhi, Government of India. Status, C.S.I.R., Govt. of India. Number of specimens, about half million. The garden has been established by C.S.I.R., as a Central Garden for India.

4. **Madras Herbarium, Agricultural College and Research Institute, Coimbatore.** Status, Department of Agriculture, Government of Tamil Nadu. Foundation, 1874. Number of specimens, about half million. Herbarium, world-wide, mainly Phanerogams. Important collections, collections of the flora of Madras. Associated Botanic garden, Botanic garden and Research Institute, coimbatore.

5. **Herbarium of the Division of Botany, Indian Agricultural Research Institute, (I.A.R.I.), New Delhi.** Status, maintained by Government of India. Number of specimens about 10,000. Herbarium, North India, introduced plants of economic value, wild relatives of crop plants and associated botanic garden.

Purposes of Herbaria : The drawings, photographs and descriptions of the plants cannot show, what is desired to show by them about individual variation and details of structure and development. In most cases it becomes necessary to compare descriptions with actual specimens which are found in widely separated regions. Most comparative studies for taxonomic purposes are made in the herbarium and laboratory. The plant collections are much used for comparison with new material. The newly collected specimens are identified by the comparison of their morphology and the morphology of the plants collected in the herbarium.

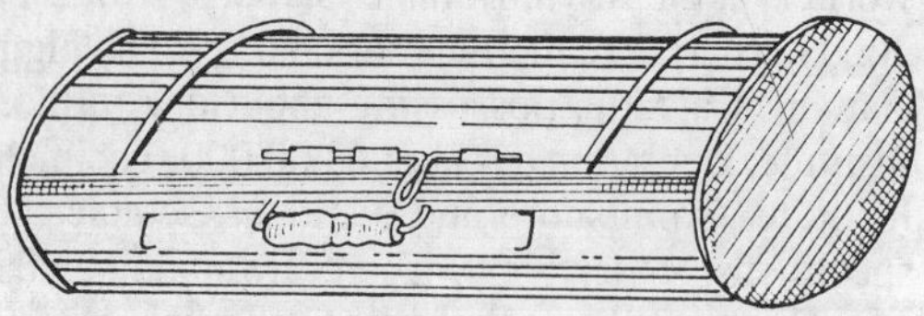

Fig. 7.1. A typical vasculum for botanical collections.

Methods of Plant Collection: The specimens must be collected in every stage of their growth and reproduction and from different localities and habitats. The single specimen may be collected in the late flowering stage having both flowers and young fruits. A complete specimen possesses all parts of the plant including the root system. In certain cases, the roots and other underground parts of the plant are necessary for identification. For collection of the plants one should go out on excursion several times in a season. A list of instruments required by a plant collector, while on excursion is given here :

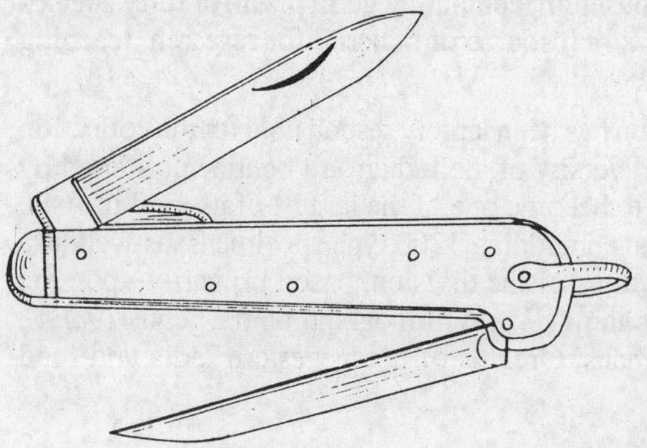

Fig. 7.2. Gardener's knife–one should carry such knife in field trips.

1. A pair of secateurs for cutting woody twigs; 2. A khurpi for digging up roots and underground stems; 3. A knife; 4. A pair of forceps; 5. A vasculum for keeping the collected plants and their twigs; 6. A wooden plant press; 7. Blotting papers and newspapers; 8. A field book; 9. A pocket lens; 10. A small diary to write notes about plants and their habitat.

Pressing of the Plant Materials: Generally the plants are pressed flat. To get good specimens, the plants must be pressed before their wilting. Generally the plants are collected in a cork lined tin vasculum, and then they are pressed after the day's trip. To get the finest specimens the plants must be pressed directly in the plant press, then and there, as soon as they are collected in the field.

Methods of Pressing the Plant materials: The plants are pressed in between the sheets of newspaper. These sheets are alternated between sheets of blotting paper. The wooden presses are too heavy to carry them in the fields, and therefore, while going on a collection trip the presses with aluminium covers are preferred. The field press remains tied with the leather straps. The plants become dried by transferring their moisture into the blotting sheets. The weights are kept upon the press or the press is tied tightly with leather straps, so that, the wrinkles of the plant materials are without crushing the plant tissues. The standard size of the press is 12" × 18".

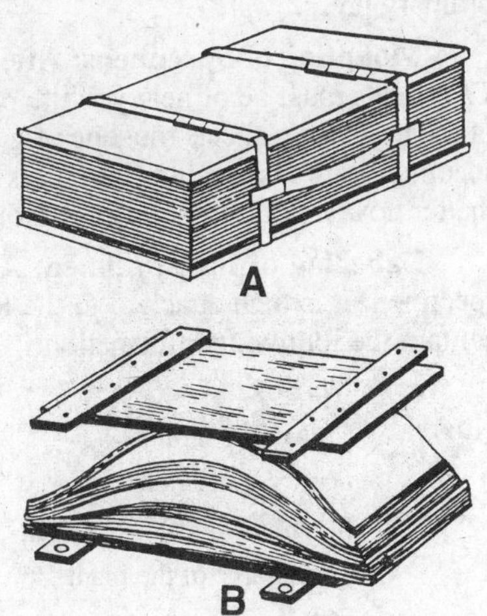

Fig. 7.3. Pressing of specimens in press boards; A, the right way of packing the press; B, the wrong way of packing the press.

Drying of Plant Materials : The blotting sheets or newspaper sheets which have been used for pressing the plant materials must frequently be changed after every 24, 48 or 72 hours. The changing depends very much upon the weather conditions. Changing every 24 hours for two or three days and then two or three more times at longer intervals is a good rule, but this is too much for the plants of arid region. The succulent plants require changing more frequently, and they require a long time to become dry. The drying of some plants such as saprophytes (*Monotropa* sp.), succulent hydrophytes, and plants wet with rain or dew is very difficult.

Such plants become black during their drying process. This becomes necessary that the driers (blotting sheets, newspapers, etc.) must have changed twice or thrice in a day continuously for several days. Even the great care does not prevent the discolouration of some species.

Drying of Plant Materials with Artificial Heat: Now a days, this method is generally used for drying of plants by plant collectors. Usually this method is employed in the damp climates especially in hilly regions during rainy season. The method is as follows–The specimens are pressed in the field press for 24 hours in usual way. The plants are then rearranged within the finely corrugated aluminium sheets. The press is then strapped and thereafter, the heated dry air is passed over through the corrugations of the aluminium sheets. These presses with plant specimens are placed in specially constructed ventilated boxes fitted with electrical incandescent light bulbs. The strapped press may also be placed in a hot air oven at desired temperature. A small electric fan also be placed in the oven, so that the warm air may be forced through the corrugations of the aluminium sheets. Instead of electrical appliances, kerosene lanterns and stoves may also be used for the purpose.

Small and rare specimens may be dried quickly with an electric iron by keeping them in between layers of thick cloth or paper. The time required for drying the plant specimens by artificial heat ranges from 6 to 24 hours.

Fig. 7.4. The wrong method of pressing plants.Here the parts of the plant are folded.

Generally the plants are taken out from the press after 24 hours, and then they are transferred to an ordinary press.

Mounting of Specimens: After drying, the plant specimens are mounted on herbarium sheets. The sheets must be of heavy white paper to give support to the plants in handling. The standard size of the herbarium sheets must be of good quality, so that it will not turn yellow or brittle with age. The specimens are fastened to the sheets by glue or gummed cellophane tape. Usually one specimen, no matter how small it is, should be mounted on one herbarium sheet.

Labelling of the Specimen: A label must contain as much information as possible about the specimen. It is then attached to the lower right hand corner on the herbarium sheet. The label must contain the following information.

Flora of ... ...
Collector's number ... ...
Date of collection ... ...
Botanical name of the plant ... ...
Local name of the plant ... ...
Family... ... ...
Locality... ... ...
Attitude... ... ...
Slope... ... ...
Soil... ... ...
Habit... ... ...
Use... ... ...
Distribution... ... ...
Abundance... ... ...
Remarks... ... ...
Collector's name ... ...

Now-a-days the ecologists show the effects of habitat on morphology and also the distribution of species according to habitat, and such data are only possible when the plant specimens are labelled very properly.

The Field Data : There are two important methods for recording field data. They are as follows–1. While collecting the plants a plant collector keeps a field book. The field book contains the numbers, which correspond to the numbers of specimens collected. The ecological and other data about the collected plants must also be given in the same field book. 2. According to this method the printed forms with headings indicating the required information are put in convenient sized pads. While making collections each printed form duly filled is attached with a specimen, and then the specimen is pressed in the field press.

Storage of Herbarium Sheets: After mounting the plant specimens on herbarium sheets they are usually kept in specially constructed herbarium cases. The wooden or steel cabinets with shelves slightly bigger than the size of the herbarium sheets are used for keeping the mounted plant specimens. The keeping of mounted plant specimen for further record is known as

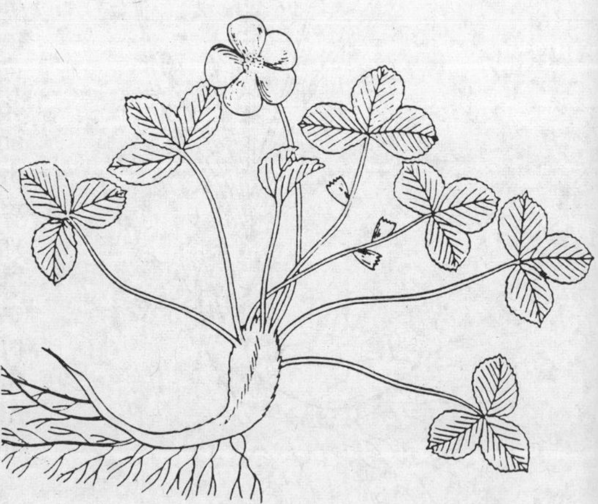

Fig. 7.5. The right method of pressing plants. Here every part of the plant is unfolded.

'filling the specimen'. In comparison to wooden cabinets, the steel almirahs are being preferred as they protect the plants from being damaged by the insects and fire. The specimens are arranged in their cases according to any well-known taxonomic system of classification. In our country and most of Commonwealth countries and British colonies the Bentham and Hooker's system of classification is being used for this purpose. Engler and Prantl's system of classification is also used to classify the plants in many countries. The plants are being arranged familywise and an index card of the families is being prepared. Within the families, the genera are being arranged alphabetically. The specimens of one genus are placed in a folded genus cover. The species are also arranged alphabetically within the genus.

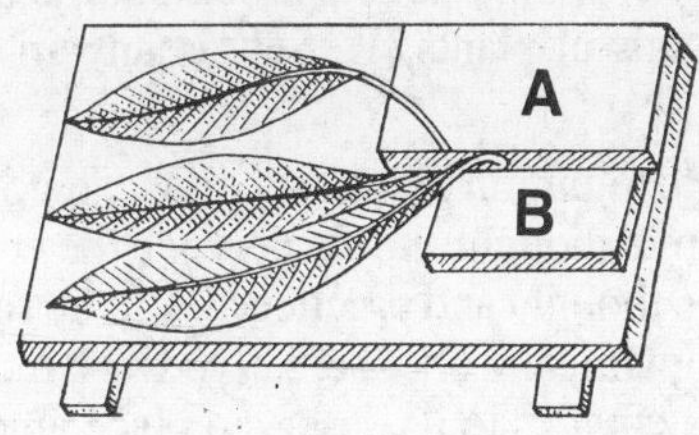

Fig. 7.6. Pressing of specimens with stout branches; A and B-newspaper cushions supporting the stout branch.

Protection of Herbarium Sheets against Molds and Insects: The mold fungi do sufficient harm to the plant specimens, if they are not well dressed by some effective chemical. For this purpose the plant specimens are sprayed with 2% solution of mercuric chloride which is highly poisonous. The insects do much harm to the specimens. To overcome this difficulty, various chemicals have been applied to the specimens. The moth balls, and naphthalene flakes are being placed in the shelves of herbarium to check the insects. These acts as repellents to the insects. The mixture of paradichlorobenzene act as an effective repellent to the insects.

The other chemicals which are used as insect repellents are D.D.T. (*i.e.*, dichlorodiphenyltrichloroethane), carbon disulphite gas, (insecticidal fumigant) etc. The heat is also being employed to kill the insects. If the mounted sheets are being placed in suitably designed steel almirahs for four to six hours at 60°C, all the insects and their eggs are being destroyed.

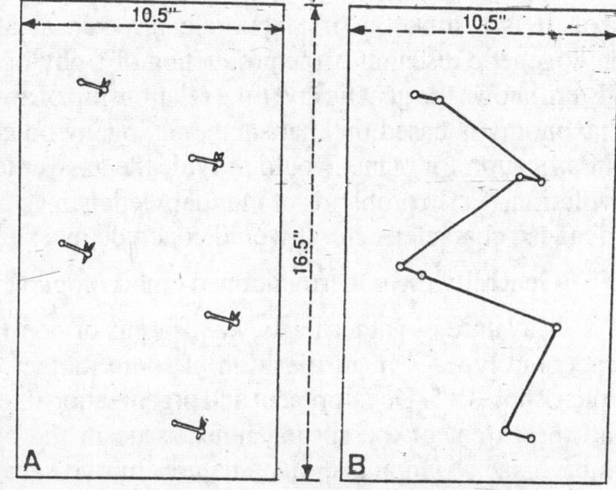

Fig. 7.7. Sticking of specimen on herbarium sheet; A, the right way (back view); B, the wrong way (back view).

Preservation of Type specimens: The type specimens are those on which the name of some taxon has been based. These specimens are those which are used by the author of a name as a basis for the name and description of a species and designed by him as a type specimen, and therefore, they are supposed to be so valuable that they are always preserved with special care. The type specimens are handled with care and are kept away from regular specimens in separate steel cases by the curators. Usually such specimens are kept in separate cellophane envelopes.

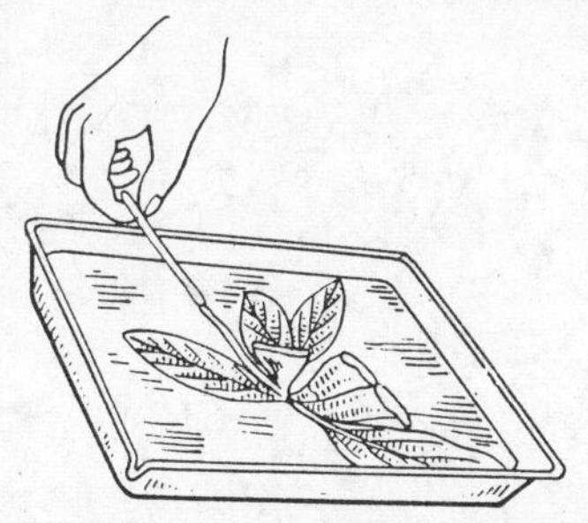

Fig. 7.8. Holding of specimen for poisoning.

Role of Herbaria in Modern Taxonomic Research: Herbaria and herbarium taxonomists have contributed so significantly to our understanding of Natural History that there is no need to defend their existence. There is an urgent need to undertake extensive and intensive explorations of all parts of the country. This is possible and could be carried out very effectively if the Universities start taking an interest in developing herbaria.

Frequently the herbaria are called upon to perform certain duties. While public supported herbaria may be approached for identification of plant specimens and supply of information not easily available from public libraries. Without active basic research identification of specimens may be often difficult or impossible. Hence the herbarium based botanists (taxonomists) should devote at least some time to carry out revisions and monographic studies of difficult groups of plants (K. Subramanyam and C.P. Sreemadhavan, 1970).

Designing of Modern Flora–The species concept and the recognition of intraspecific categories need particular examination; in fact a rational process of taxonomic thought should pervade the entire work. It should reflect adequately the current trends in taxonomic thought and practice and at the same time, keep itself within the limits of a well balanced, integrated botanical work–a creative work meant to inspire botanists of all disciplines and thus help to elevate systematics to its deserved place among the more sophisticated branches of Botany (M.A. Rau, 1970).

Phylogeny is the evolutionary history of a taxon, and attempts to account for its origin and development. The term phylogeny is the autonym of ontogeny. Ontogeny differs from phylogeny in that it accounts for the life history of the individual plant from its development from the zygote to the production of its own gametes. A primary objective of phylogenetic studies in botany is the determination of origins and relationships of all taxa of both extinct and living plants.

Significance of Phylogeny to Taxonomy – Phylogeny deals with the evolutionary history of all taxa. It is a function of taxonomic research at all levels of classification. The ultimate goal of phylogenetic research is the production of a phylogenetic system of classification. The phylogenetic system shows the genetic and time relationship of any one taxon to another. According to turrill (1942), "taxonomy is based on characters, phylogeny on changes of characters." A phylogenetic system of classification for plants would provide the answer to questions of (*i*) their orign (*ii*) to their modes of evolution, (*iii*) to problems of monophyleticism vs., polyphyleticism, (*iv*) the identity of primitive and advanced characters, etc. It would conclude in a single stable classification of relationships.

Generally the undermentioned embryological features are found to be of taxonomic value:

1. Nature of anther lobes, *i.e.,* fibrous or non-fibrous; 2. Nature of Anther tapetum-glandular or amoeboid type; 3. quadriposition of spore mother cell; 4. Development and organization of pollen (microspores); 5. Development and organization of ovule; 6. Form and extent of the nucellus; 7. Origin and the extent of the sporogenous tissue in the ovule; 8. Megasporogenesis and development of embryo-sac; 9. Organization of mature embryo-sac; 10. Fertilization; 11. Nature of the endosperm and 11. Type of the embryo.

8

BOTANICAL SURVEY OF INDIA

The Botanical Survey of India started working in 1890, and conducted exploration in different parts of the country, which led ultimately to the preparation of the Flora of British India and the Regional Floras, *e.g.*, Flora of Madras, etc. One of the important achievements during this period was the establishment of the Herbarium in the Royal Botanic Garden, Sibpur, Calcutta (now named as Indian Botanic Garden). After sometime, there has been considerable reduction in the activities of the survey and the post of the Director was abolished, the work of which was continued by the Superintendent of the Indian Botanic Garden.

The survey has been reorganized in 1954, and some progress has been made during the second and third five year plan periods, with the establishment of the Regional Circles and Herbaria at Dehradun, Coimbatore, Allahabad, Poona and Shillong, in addition to the organization of Central National Herbarium which was originally located in the Indian Botanic Garden, and was under the charge of the Bengal Government.

Further progress has been made in establishment of the Experimental Gardens for working out various aspects of botany and usefulness of plants at Shillong, Poona and Yercaud (Shevroy Hills). National Orchidarium has been added for the collection of the orchids from the eastern part of the country at Shillong and the southern part of the country at Yercaud. The present organization of the Botanical Survey of India consists of :

1. Plant exploration and writing up of the regional floras.

2. Preparation of the Flora of India.

3. Establishment of the National Orchidarium for collection of orchids of India and their improvement.

4. Assessment of the plant products and their display at the Indian Museum, Calcutta.

5. Establishment of Botanical Laboratory to work out the botany of plant in different aspects like cytology, anatomy, phytochemistry, ecology, pharmacognosy, palynology and plant physiology.

Organization and Set-up of the Botanical Survey of India. By the end of the nineteenth century, there had been a decline in efficiency and standard in respect of the botanical work done in the gardens set up by the local Governments of Bengal, Bombay, North Western Provinces, Madras; Calcutta, Pune and Saharanpur, and on account of this deterioration, it was felt that the gardens should be placed under an eminent botanist who in independent charge could correspond with his counterparts in Bengal, North Western Provinces, Madras and Kew. It was thus decided by the Government of India that a central organization for the botanical work be set up and that J.F. Duthie who was then the Superintendent of the Gardens at Saharanpur be dissociated from the gardens then under the North

Western Provinces, to enable him to undertake explorations in then North Western Provinces. These proposals were accepted by the Government of North Western Provinces. As the co-ordination of the Botanical work of the country through a central organization proved useful, and the scheme for the organization of a survey having met with the approval of the Secretary of State for India in July, 1887, the survey was formally constituted on 13th February, 1890, under the designation "**Botanical Survey of India".** Its main functions were laid down, as follows:

1. To explore the vegetable resources of the then Indian Empire.
2. To co-ordinate the botanical work of others in different parts of India.

The country was divided for the purpose of the survey into *four* botanical regions, *viz.,* the East, North, West and South, with centres at Calcutta (Sibpur), Saharanpur, Bombay, Pune and Madras respectively exactly corresponding to the centres where botanical activities were evident in early years. The Calcutta unit was acknowledged as the centre of the Botanical Survey of India. Sir George King, who was then the Superintendent of Sibpur Gardens was made the Director of the Botanical Survey of India. J.F. Duthie, Superintendent of Saharanpur Gardens, J.Cook, the Principal of Science College, Pune and M.A. Lawson, the Government Botanist to Madras and Director of Government Cinchona Plantations on the Nilgiris, became also the first directors of the respective surveys under their jurisdiction under this scheme.

The survey had no separate staff of its own. Co-ordination only appeared to have been attempted initially, each of the existing institutions at Calcutta, Saharanpur, Bombay and Madras being asked to submit reports of its activities to the Government of India through the Director, Botanical survey of India. This co-ordination, however, did not work satisfactorily and the Madras Survey ultimately severed its connection with the Botanical Survey by about 1907. Later on, Bombay also severed its connection, and the Governments of then Madras and Bombay entrusted their potential work to their own Economic Botanists. The Eastern and Northern Surveys were engaged in explorations for some years but in due course there was a decline in the enthusiasm for exploration even in these regions. The link with the Saharanpur survey was also disconnected on the retirement of J.F. Duthie On 31st December, 1902, when the post held by him was also abolished.

Botanical Survey of India (1937 - 1952)

C.C. Calder was the last incumbent to the post of the Director, Botanical Survey of India (BSI). Till that time the Superintendent, Royal Botanic Gardens, Calcutta used to be the ex-officio Director of the Botanical Survey of India. In 1936 the Government of India sanctioned the separation of the post of the Superintendent, Royal Botanic Gardens, Calcutta and the Director, Botanical Survey of India (BSI), Calcutta, from the time of the retirement of C.C. Calder.

After the Second World War (1945), and on the achievement of independence by India, there arose social and economic aspirations of Free India which provided the opportunity for scientific advancement and economic development. In 1952, the Government of India decided to reorganize and expand the Botanical Survey. In 1952-53 when the Botanical survey was to be reorganized, it comprised of the Industrial Section, Indian Museum and Systematic Division, Indian Botanic Gardens with an overall budget of about 1.5 lac rupees.

Reorganization of the Botanical Survey of India

Dr. E.K. Janaki Ammal was appointed on 14th October, 1952 as Officer on Special Duty, a post temporarily created by the Government of India to help them in drafting a suitable plan for the Botanical Survey of India. A small office was given to her in the wing of the Museum House on the premises of the Indian Museum at Calcutta. The plan drafted by her was approved by the Government of India and the President of India was pleased to give sanction for the expansion and reorganization of the Botanical Survey of India (BSI) on 29th, March, 1954. The sanction permitted the formation of the following organisations :

1. A Headquarters establishment under a chief botanist at Calcutta. This was meant for controlling

and co-ordinating the activities of the various units of the Botanical Survey of India for implementing the policy laid down by the Government of India.

2. Four regional circles based on phytogeographical affinities namely, (*i*) Eastern; (*ii*) Western; (*iii*) Northern and (*iv*) Southern Circles with their headquarters at Shillong, Pune, Dehradun and Coimbatore respectively under a Regional Botanist. Each Regional Botanist will be responsible for the survey of his area and will utilise the facilities provided by the states and universities.

3. A Central Botanical Laboratory under a Director at a suitable place in Uttar Pradesh where the living plant will be studied in relation to its botany and utility to the nation.

4. A National Herbarium which will house the type specimens and a fully representative collection of the plants comprising the flora of the country.

5. Maintenance of a Botanical Museum on modern lines at Calcutta.

In the reorganized survey it was decided that the Botanical Survey of India (BSI) will collaborate with universities and research centres in the country and will encourage research students in the universities and other research centres to take up problems relating to the Indian Flora. The Survey will also collaborate with various State Governments and will take advantage of all the facilities made available by them.

The Government of India also decided that the above organisations included in the scheme will be set up in stages, but the preference will be given to headquarters establishment, the Eastern and Western Circles, and the Central Botanical Laboratory in the first year. The expansion of the survey was intended to be under a phase programme, consistent with the Five-Year plans launched in respect of various projects in the country.

Fr. H. Santapau, was appointed on 1st December, 1954 as the Chief Botanist and Head of the Department of the Botanical Survey of India. The powers previously delgated to the Director were withdrawn, on the post having been abolished and the Chief Botanist was declared as Head of the Department by the Ministry of Finance in January 1955. Fr. Santapau, did not continue long as Chief Botanist, and he relinquished charge of 'The Chief Botanist, on 1st, October, 1955, and on the same date Dr. J.C. Sengupta, Principal, presidency College, Calcutta assumed charge as the Chief Botanist.

Setting up of Four Regional Circles

1. **Southern Circle.** It was the first to be set up, on 10th October, 1955, at Coimbatore with its jurisdiction as Madras (Tamil Nadu), Andhra Pradesh, major parts of Madhya Pradesh, Old Vindhya Pradesh, Andaman and Nicobar islands.

2. Eastern Circle. This was set up on December 10, 1955 at Shillong with its jurisdiction as Orissa, Bihar, West Bengal, Assam, Bhutan, Sikkim, then NEFA, Tripura and Manipur.

3. Western Circle. This was set up on December 17, 1955 at Pune with its jurisdiction as States of Rajasthan, Gujarat, Maharashtra, Karnatak, Kerala and Tamil Nadu not falling within Southern Circle and Laccadive and Minicoy Islands.

4. Northern Circle. This was set up at Calcutta on March 21, 1956 and by August 1, 1956 this was shifted to its headquarters at Dehra Dun with its jurisdiction as States of Jammu and Kashmir, Punjab, Himachal Pradesh, Haryana, Uttar Pradesh and Delhi.

Transfer of Principal Local Herbaria to BSI.

The important herbaria hitherto with the State Governments were transferred to the Botanical Survey of India (BSI). The herbarium of the Forest Department of Assam Government was handed over to the Botanical Survey to form a part of the Eastern Circle. The Herbarium of the Economic Botanist to the Government of Bombay (now Maharashtra) was transferred to the Western Circle at Pune. The herbarium at the Sibpur (Calcutta) Botanic Gardens hitherto with the Government of West Bengal was transferred to the Botanical Survey to form the National Herbarium of the BSI. The Madras

Herbarium at the Agriculture College and Research Institute, Government of Madras (now Tamil Nadu), was also transferred to the BSI to form a part of the Southern Circle. There was, however, no transfer of any herbarium of the Northern Circle, since the necessary facilities of the herbarium existed at the herbarium of the Forest Research Institute (FRI), Dehra Dun.

Reprinting of old Standard Floras

Revision and compilation of Floras become the most important activity of the Circles in the reorganised BSI. As the work on taxonomy and teaching of Systematic Botany in universities and colleges was adversely affected on account of non-availability of standard floras and books which had become out of print, it was decided by the BSI to get them reprinted. A number of floras, such as Flora of Madras Presidency by Gamble and Fischer, Flora of Bombay Presidency by Cooke, Flora of Upper Gangetic Plains by Duthie, Botany of Bihar and Orissa by Haines, and Bengal Plants by Prain have been reprinted.

The new set-up of BSI

For scientific and executive functions of the survey, a Central Botanical Laboratory under a Director of the Laboratory, a Botanical Museum under a curator, a Central National Herbarium under a keeper and four Regional Circles each under a Regional Botanist have been established.

Apart from the normal activities of the Botanical Survey of India, the department also works in collaboration with the following schemes and organisations in carrying out various research projects and programmes.

1. Central Indian Medicinal Plants Organisation. The Counsil of Scientific and Industrial Research (CSIR) set up an organisation for a detailed work on the medicinal plants in the country and the Botanical Survey of India (BSI) was recognised as a centre of co-ordinated researches on medicinal plants.

2. Monographs and Research Schemes. Monographs on the gerera of Angiosperms in India and the Chromosome Atlas of Indian flowering plants are being worked out at the Botanical Survey of India under the Council of Scientific and Industrial Research (CSIR).

3. Indian Council of Ecological Research (ICER). The Government of India, Ministry of Food and Agriculture, established the Indian Council of Ecological Research in 1957 and it was decided that the scheme will be controlled by the Botanical Survey of India at Calcutta and will include studies on climate diagrams and vegetation mapping.

4. National Orchidarium. An orchidarium has been set up at Shillong under the Botanical Survey of India (BSI) for development and collection of orchids of India.

5. Economic and Medicinal Plants. The Botanical Survey of India (BSI) has decided apart frόm the study of Indigenous flora of the country, that a considerable amount of experimental work will also be provided in respect of the studies on plants of economic and medicinal value.

The Indian Botanic Garden at Calcutta is concerned with plant introduction and horticulture, and development of the arboretum and garden. Attached to the Headquarters Office at Calcutta, there is a section of Cryptogamic Botany dealing mainly with the mosses and ferns. The Central National Herbarium which has at present a collection of about 6.5 lac specimens, is now engaged in making monographic studies of the families included in the first volume of the Hooker's Flora of British India with the object of ultimately bringing out the new Flora of India on the basis of the collections already made by the survey during the last fifteen years, in addition to those that were available at the Royal Botanic Garden at Calcutta.

There is also a section of publication, which deals with the publication of various workers of this survey and also on systematic botany by workers in the universities and other institutes in India, which has led to the publication of the Bulletin of Botanical Survey of India, Records, printing of Roxburgh's Icones, Chapeters on History of Botany, reprinting of the Regional Floras, etc.

It has been proposed to further strengthen the activities of this organization and establish a circle in the Andamans and a Botanic Garden at Delhi and for the high altitude areas in the Himalayas, somewhere in U.P. The various sections of this organization have also been strengthened so that the work on the preparation of the Flora of India and the regional Floras is expedited. The Central Botanical Laboratory is also going to be housed in the new building which will be fitted up and its work organized on the lines of the Jordell Laboratory at Kew. The work of the survey of the country has been expedited, and a map has been prepared showing the lacuna in our knowledge regarding areas unexplored or underexplored so that the exploration work can be taken up on a priority basis in such areas to improve our knowledge on plant wealth of the country.

The Botanical Survey of India (BSI) has been serving the nation through inventorisation of the plant resources and advising on all matters related to plants. About 65% of the total area of the country has been surveyed and the herbaria of the BSI hold about three million plant specimens including about 15,000 type specimens. The BSI hold about three million plant specimens including about 15,000 type specimens. The research wing has **flora, ecology, cryptogams, plant chemistry,** and **pharmacognosy** (study of medicinal plants) units.

The schemes of setting up of many **botanic gardens** and **field centres** have also been taken up during 1991-92, to augment the activities for conservation and propagation of plant genetic resources in different regions of the country through a network of these centres. Under this scheme, a one time non-recurring financial assistance is provided to botanic gardens in different phytogeographic regions of the country for strengthening their existing facilities for conservation and propagation of threatened and endangered endemic plant species of that region, and education and public awareness of endemic plant species.

The Indian Botanic Garden in Howrah is the largest Botanic Garden in the country and is used for study, introduction and conservation of flora, educational and secretarial purposes. It is a living repository of about 15,000 trees and shrubs distributed over 2,000 species. The greatest attraction of this Garden are the Great Banyan Tree, Giant Water Lilies and the large Palm House in addition to a number of interesting botanical specimens from different parts of India and other countries.

9

INDIAN WORK IN TAXONOMY

During the nineteenth century, taxonmy was the main theme of botany in India. In this period the production of comprehensive works on the study of flora and vegetation was pre-eminent.

Sir Joseph D. Hooker in 1897 on completion of the seventh and last volume of his Flora of British India (1872-1897) clearly indicated that this pioneer work would facilitate the compilation of local floras and monographs of genera of this "richest and most varied botanical area on the surface of the globe" where ample work is still to be done. Simultaneously, the establishment of Botanical Survey of India in 1890 and the organization of extensive botanical explorations in the then British-India, were closely followed by the publication of regional floras of Bombay Presidency (Cooke, 1901-08; Talbot, 1909-11), Bengal (Prain, 1903), Upper Gangetic plain (Duthie, 1903-22), Hyderabad State (Partridge, 1911), Madras Presidency (Gamble and Fischer, 1915-36), Nilgiris and Pulney hill-tops, (Fyson, 1915), Bihar and Orissa (Haines, 1921-25), Rajasthan (Blatter and Hallberg, 1918-21, besides several other smaller floristic accounts on limited regions, monographic studies on plant groups and genera such as Bambuseae (Gamble, 1896), Orchids (King and Pantling, 1898), Asiatic palms (Beccari, 1908-31), *Dalbergia* (Prain, 1904), *Impatiens* (Hooker, 1904-11), *Dioscorea* (Prain and Burkill, 1936-38).

With such background of extensive botanical exploration work and floristic studies and with the successful introduction of Tea, Cinchona, Rubber, Coffee and several other economic and medicinal plants in different parts of India, Burma and Malaya, most of such scientific pursuits were kept in abeyance during World War I (1914-18) and immediately after, in 1920, a dark curtain was drawn over the activities of Botanical Survey of India and a few Indian Universities Botany Departments. With advances in other modern fields, taxonomic studies have come to occupy a place of decreasing importace in the curriculum of the University Biologists as though the Indian Flora has been completely known. However, the forest departments of various states and a few research workers attached to Universities have been keeping up the interest in taxonomy and morphology, publishing general floristic accounts on regions, sometimes limited.

All such contributions have no doubt enrich our knowledge of the Indian Flora considerably; but the range of work covered on a sub-continent like this, is so meagre that the main task of preparation of modern flora for India is considered to be far beyond, unfulfilled. Further, the omission of serious taxonomic studies in the University curricula and stopping of recruitment of posts connected with systematic botany even in important herbaria of the country had the unfortunate effect of depressing still further the descending spiral of the training of taxonomists.

With the advent of new thinking in the very concept of biological species and natural variation as illustrated in several essays in New Systematic (Huxley, 1940) followed by careful analytical presentation by Edgar Anderson (1949), Heslop-Harrison (1953), Turrill *et al.* (1964), Davis and

Heywood (1965) and others, giving importance to field studies of populations, combined with the analysis of the data on their cytology, anatomy, embrylogy and palynology as worked out in close collaboration with the basic structure of herbarium taxonomy. Such fresh appraisal in the fields of morphology and taxonomy has brought about increasing activities particularly in the Universities in the study of embryology, cytology palynology and anatomy in relation to taxonomy. These studies have substantially enhanced the proper understanding of the status of many genera and families and their affinities and systematic position.

Utilizing the data from various branches of botany, it is rather surprising how acutely some important problems in various taxa and families were picked out and analysed by earlier workers like Hooker and others before the advent of cytology and other branches of botany and even by Hutchinson of the modern era, following purely the external morphological characters coupled with acute understanding of their phytogeography, comparative analysis of regional floras and critical study of the natural variation among the species and other allied aspects. In recent years extensive and intensive botanical explorations of different parts of the country have been renewed and useful floristic accounts at various regions such as Khandala (Santapau. 1953), Jammu and Kashmir (T.A. Rao, 1961), Lahul (M.A.Rau, 1960), North Garhwal (M.A. Rau, 1961), Delhi, (J.K. Maheshwari, 1963), Jodhpur division (Rolla et al, 1962, 1963), Rajasthan (Puri et al., 1964), Saurashtra (Santapau, 1962), Sikkim (Rolla, 1963), Siang district, NEFA (Rolla *et al.*, 1965), Manipur (Deb, 1961), Andaman and Nicobar (Thothathri, 1962), Rampa and Gudemghatas (Rolla, 1958, 1964), Tirap district, NEFA (Panigrahi *et al.*, 1966) have been published in the Bulletin of the Botanical Survey of India and other journals besides revision work on *Viscum* (Rolla, 1957), *Derris* (Thothathri, 1961), *Cynodon* (Jain, 1967) *Rubia* (Deb *et al.*, 1968), Bhutan Orchids (Balakrishnan *et al.*, 1966, 1967) and publications on vegetation in India (Gaussen *et al.*, 1961, 1968), Poaceae (Bor, 1960), aquatic angiosperms (Subramanyam, 1963), and several other contributions during Seminars and Symposia. These contributions in addition to the reprinting of the old floras of different states by the Botanical Survey of India have created a considerable enthusiasm among various young workers in the Universities and Colleges to undertake floristic work of districts or regions in different states of India and the Botanical Survey of India is now keen on formulating various projects for the revision of families for the Flora of India and also the regional and State Floras so as to present the data with modern concepts which is considered as extremely essential for better exploitation of vegetable wealth of the country.

10

BOTANIC GARDENS

What is a Botanic Garden?

A botanic garden is essentially a collection of living plants maintained for both pure and applied studies. The earliest botanical gardens were intended mainly for the training of medical students and for the study of drug yielding plants. During the 18th century some of the physic or drug gardens cultivated not only medicinal herbs but also many other types of flowering plants. The original concept of the Royal Botanic Gardens, Kew, however, was the cultivation of flowers for beauty and delight. During the 19th century, botanical gardens in the tropics became the centre of extensive botanical research with the result that quite a sizable number of useful plants become known or better known for their usefulness to the mankind. Just to mention para rubber (*Hevea brasiliensis),* quinine (*Cinchona* spp.), coffee (*Coffea arabica),* and tea (*Camellia sinensis)* today once their detailed understanding and economic exploitation to the comprehensive studies in their botanical gardens. It is now universally appreciated that the merit of a true botanical garden lies as much in its use in botanical education and research as in the display of plants of ornamental and horticultural interest. Botanical and horticultural research is, at present, an accepted and well established sphere of activity of most of the botanical gardens of the world. Several botanical gardens also organize courses in horticultural education for the benefit of botanists, gardeners and the laymen.

Origin of Botanic Gardens

The development of the idea of a botanical garden has been gradual. As the knowledge about plants became progressively scientific, many pleasure gardens were converted into botanical gardens. These not only became the centre of advanced learning but also began to play a vital role in the economic and aesthetic needs of various nations. For instance, the credit goes to the Bogor (Buiten-zorg) Botanical Garden, Java, for providing impetus to the cultivation of cinchona, rubber and coffee; to the economic gardens of the Dutch for the cultivation of spices; and to the Oxford Botanical Garden, England, for the introduction of some of the parents of modern asters, chrysanthemums, dahlias, fuchsias, irises, narcissi and roses. In the field of fundamental research, it may be recalled, the laws of genetics were worked out by Mendel in a monastic garden in Austria. The research problems being studied by the present day botanical gardens vary from the development of the drought resistant varieties of plant to weed control. However, behind the apparent diversity in various lines of work lies the common function which all botanical gardens share, namely, the collection and display of, and research on, plants of medicinal, horticultural and other economic interests.

Royal Botanic Gardens, Kew

This garden was started as private garden of Sir Henery Capel. The history of Kew as a true botanic garden begins in 1759 with appointment of William Aiton as superintendent. In 1841 the gardens were transferred from crown to the state and W.J. Hooker was appointed as Director and under his directorship the gardens were increased from 20 to 250 acres. At present it is, 'under the Ministry of Agriculture, Fisheries and Food and has increased to 300 acres. The herbarium contains 5,000,000 specimens; the arboretum 700 sp. and varieties; about 13,000 sp. and varieties under glass, while 8,000 herbaceous sp. are grown outdoor. The important publications are Kew Bulletin and Index Kewensis.

National Botanic Gardens (NBG), Lucknow

Historically the present garden dates back to the time of Nawab Saadat Ali Khan (1789-1814), who laid out a royal garden in the picturesque surroundings along the southern bank of the river Gomti. Later on, it was improved by Nawab Wajid Ali Shah, who named it Sikanderbagh after his wife, Sikander Mahal Begum. Even today this garden is popularly known as Sikanderbagh. For a long time the garden existed as Government Horticultural gardens and continued to be a centre of horticultural activities in Uttar Pradesh. In 1948, the State government approved the re-organization of the Government Horticultural Gardens into the National Botanic Gardens which was ultimately taken over by the Council of Scientific and Industrial Research (CSIR) in 1953. At present, the main garden occupies an area of 27 hectares.

Since its take-over by the CSIR, the initial activities of NBG have mostly been of a formative nature, which of late have been re-organized, with a number of research laboratories having facilities for both fundamental and applied research. The gardens have a conservatory for shade- and moisture-loving plants, Fern House, Palm House, Cactus House, Orchid House, Vinetum, Citrus Orchard, Arboretum, and a large collection of different ornamental plants. The laboratories for floriculture and Plant Propagation, Plant Breeding, Cytogenetics, Radiation Biology, Plant Morphology, Entomology, Viticulture, Mycology and Utilization Research, are located in the garden, while the new building, situated at the eastern end of the garden, houses Tissue culture, Virology, Plant Physiology, Floristic Botany, Pteridology, Algae, Pharmaco-botany, and Phytochemistry laboratories, besides a large Herbarium with 85,000 specimens, a well stocked library having books on all branches of botany, horticulture and allied subjects, and a modern auditorium.

In brief the main functions of NBG are—

1. Collection, introduction and propagation of plants of ornamental, industrial, medicinal and other economic importance.

2. Maintenance of the botanic gardens at Lucknow and organizational of its field stations and sub-centres.

3. To conduct studies in plant improvement, pharmacobotany, aromatic and oil yielding plants.

4. Study of growth, flowering and pre-and post-harvest physiology.

5. To develop control measures for diseases and parts of ornamental and economic plants.

6. Setting up of a herbarium, especially of economic plants.

7. Organization of a botanical museum to display the achievements of NBG and other interesting facts of plant life.

8. To exchange plants and plant material with scientific institutions and Universities in India and abroad.

9. To conserve rare and interesting plant species.

10. To serve as a centre of advance training in gardening and systematic botany.

11. Dissemination of technical information; and to advise and assist in identification, cultivation; and economic utility of plants, garden lay-out, land scaping, etc.

12. Publication of scientific and semi-popular literature for the benefit of research workers, industrialists and the laymen.

Lloyd Botanic Garden, Darjeeling: In 1878 William Lloyd donated a piece of beautiful land with an area of 40 acres which was developed as a branch establishment of the Royal Botanic Garden, Calcutta. There is good collection of exotic species from China and Japan and representative plants from North Western India, Burma and Nilgiris.

Indian Botanic Garden, Calcutta

Introduction: The Calcutta garden is in the category of gardens started for economic reasons and then growing into scientific institution. It was Roxburgh who made the rest of the world interested in the botany of India's immensely rich and varied flora, at that time new to science and to horticulture. It was also Roxburgh who brought the Botanic Gardens to life and made them world famous, bringing both live plants and dried material for the Herbarium from all over India so that students could study them in one place.

The site of the garden was perfectly flat; it had none of the topographical features which help to make the beauty of a garden. George King sezied the opportunity presented by its ruinous condition to set about creating them artificially. For example, he got rid of the new swamp by excavating a number of artificial lakes, and he made use of the large quantity of soil from these excavations to give the gardens several different levels, to endow it with small hills, slopes, little valley. If any garden designer ever created his land scape from the ground up, King did so at Calcutta.

Although King is remembered in Indian Botanical Garden as its chief horticultural designer, he was also an excellent botanist, it was he who, in 1887 started publishing the **Annals of the Royal Botanic Gardens, Calcutta;** still more important, in 1890 he initiated the Botanical Survey of India.

Dr. Mitra, a botanist but also the first landscape gardener to become Director since King, used the garden in his teaching course in landscape gardening design as a superlative example of its kind.

The Great Banyan Tree: The outstanding feature of interest of Indian Botanical Garden is the great Banyan Tree. This specimen of *Ficus bengalensis* is certainly one of the largest living creatures in the world and it may well be the largest in terms of sheer bulk of tissue. It is about two centuries old. This species of *Ficus,* drops aerial roots from its branches, which, rooting themselves when they reach the ground, in time form subsidiary trunks. This great Banyan Tree possesses about one thousand trunks and the walk round the tree is about quarter of a mile. It is still growing vigorously. It is about 100 feet high. The canopy of tree covers four acres of ground.

Palm Houses : The palm houses in the garden are interesting. In the climate of Calcutta such houses are not, of course, necessary to exclude cold but to exclude excessive sunshine; glass would be quite inappropriate. They are elegantly shaped metal structures prefabricated in England in the nineteenth century, covered with wire netting, and over grown with a number of interesting and pretty climbing plants whose foliage provides the shade for the collection of rare palms, ferns, and moisture or shade-loving plants inside.

About forty species of palms are grown and this collection includes some fine specimens of an interesting rarity, the branching palm *Hyphaene thebaica;* branching monocotyledons are an exception to the rule for their kind. One of the Garden's best vistas is through the Palmetum.

Ornamentals : There is a beautiful collection of ornamental leguminous trees. The Garden's specimens of *Saraca declinata* and *S.indica,* the first a mass of rich orange flowers, the other scarlet; and of its avenue of *Brownia* and *Amherstia.*

A special flower garden is devoted to cultivation during the Bengal winter. of European summer garden flowers, chiefly for the guidance of private, and institutional gardeners. Remarkabiy good results are obtained with many of these exotics in the very difficult climate of Calcutta, but some for example roses are far from easy to grow, and with these the Botanical Garden does useful work in selecting the varieties best able to tolerate the local conditions.

Nursery : The big propagation nursery is one of the best in Asia, and although it is maintained to supply the Botanical Garden's need for replacement plants, the surplus of young plants, is available for sale to the general public and other institutions.

Succulent Plants: The pleasantly planted garden of succulent plants, Euphorbiaceae and Cactaceae, has one especially interesting and attractive plant belonging to the genus *Pereschia*, a cactus which has tree leaves, it is a tall, woody shrub with the exceptionally beautiful flowers.

List of Great Botanic Gardens of the World

Europe

Padua, Pisa, Palermo and Villatanto botanic gardens of Italy; Glasnevin botanic garden of Ireland; Jarden Das Plantes and Les Cedres botanic gardens of France; Leyden botanic garden of Netherland; Meise botanic garden of Belgium; Royal botanic garden of Edinburgh, Scotland; Munich, Berlin and Dahlem botanic gardens of Germany, Oxford and Kew botanic gardens of England; Belvedere and Schenbrunn botanic gardens of Austria; Uppsala and Gateborz botanic of Scandinavia.

North-America

Arnold Arboretum, Brooklyn, Long-wood, Missouri, Strybing Arboretum, Huntington and Fairchild botanic gardens of U.S.A.; Montreal botanic garden of Canada.

Russia

Moscow, Kiev, Tashkent, Minsk and Yalta botanic gardens of Russia and other allied countries.

Tropics

Bogor botanic garden of Java; Singapore; botanic garden of Singapore; Entebbe botanic garden of Uganda; Peradeniya botanic garden of Ceylon; Indian botanic gardens, Calcutta, India; Rio-de-janeiro botanic garden of Brazil.

Southern Hemisphere

Christ Church botanic garden of Newzealand; Kirstenbosch and Stellenbosch botanic gardens of South Africa, Japan botanic gardens of Japan; Melbourne and Sydeny botanic gardens of Australia; Buenos aires botanic garden of Argentina.

Importance of Botanic Gardens

In 1889 Torrey Botanical Club in New York passed a resolution which itself reveals the importance of botanic gardens. The resolution was "The uses of a botanic garden may be reckoned as of four sorts. First and foremost is the purely scientific and educational use. Subsidiary to this but still of marked degree of importance are pharmaceutical and horticultural uses and lastly the general use as a place of agreeable resort for the public at large".

The botanical gardens are very much helpful in science of taxonomy. They are also acting as research stations in exploring the flora of native and other regions. The botanic gardens can also be used as conservatories.

They act as acclimatization station for the introduction of economically valuable plants.

The botanic gardens also play an important role in the field of horticulture.

The botanical gardens act as 'outdoor laboratories' for the students of Botany and give practical information about plants.

THE DICOTYLEDONES

11

ORDER–MAGNOLIALES

According to Hutchinson this is the first order of the phylum Angiospermae, sub-phylum Dicotyledones and division Lignosae. This order comprises of nine families. They are–1. Magnoliaceae, 2. Illiciaceae, 3. Winteraceae, 4. Canellaceae. 5. Schisandraceae, 6. Himantandraceae, 7. Lactoridaceae, 8. Trochodendraceae and 9.Cercidiphyllaceae. The Magnoliaceae has been supposed to be most primitive within the order. This family has been described in the present text in detail. Bentham and Hooker have placed the family Magnoliaceae along with other seven families in the first order—the Ranales of class Dicotyledones, division Polypetalae and series Thalamiflorae. Engler and Prantl have included the family Magnoliaceae in eighteenth order–the Ranales of class Dicotyledoneae and sub-class Archichlamydeae together with fourteen other families.

Hypogynous, bisexual, becoming unisexual with reduction of perianth; apocarpous; petals rarely absent; stamens numerous, usually free; endosperm copious, not ruminate; embryo minute; stipules present or absent.

Family–Magnoliaceae

There are about 18 genera and 300 species in this family (Rendle).

Distribution: The members of this family are distributed mainly in tropical and sub-tropical regions of the northern hemisphere. Some of them are also found in the temperate regions.

Habit : The plants are mostly trees or shrubs. Few plants are climbers. Oil passages occur in parenchyma of the stem and leaves. The wood shows very primitive characters anatomically.

Leaves : Usually the leaves are simple, entire alternate and stipulate. In *Magnolia,* large stipules are formed, which cover the young leaves and fall as the leaves expand. In most of the cases the stipules are quite large in size which unite to form a hood like structure protecting the young leaves.

Inflorescence : The inflorescence is of cymose type. Usually the flowers are terminal or solitary axillary and often of large size.

Flowers: The flowers are mostly hermaphrodite and rarely, unisexual, *e.g.*, in *Kmeria.* They are actinomorphic (regular), hypogynous and bracteate. The bracts are very often large and variously coloured. The floral axis is convex or elongated.

Perianth: Usually the perianth is whorled and consists of three whorls, sometimes it is somewhat distinguished into calyx and corolla. Each whorl of perianth usually consists of three perianth leaves. In *Magnolia,* the arrangement of perianth leaves is cyclic, whereas in other genera it is spiral.

Androecium: It consists of a large number of stamens which are arranged spirally in all genera. The stamens are distinct and hypogynous. The anthers are bi-lobed and dehisce longitudinally.

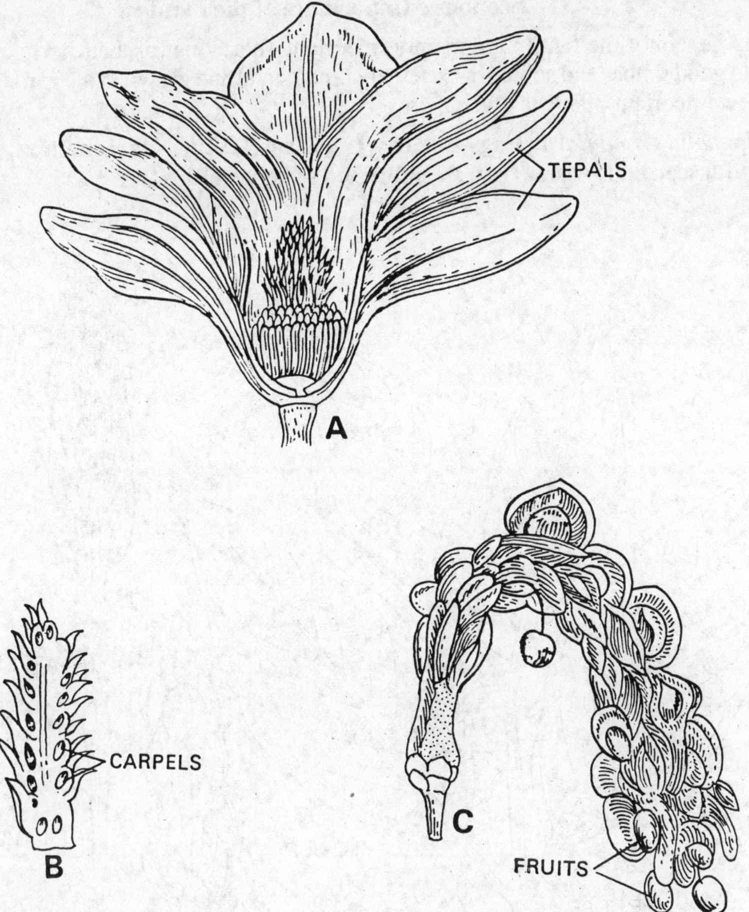

Fig. 11.1. Magnoliaceae–*Magnolia* sp.–A, flower cut lengthwise to show androecium and gynoecium; B, gynoecium cut longitudinally showing carpels; C, fruit.

Gynoecium : It consists of indefinite number of carpels which are arranged spirally in all genera. The carpels are free (apocarpous). The gynoecium may be sessile or elongated. The number of ovules ranges from one to many. The ovules are anatropous and being arranged on parietal placenta.

Fruit : The fruit may be follicle, a winged nut (samara) or a berry.

Seeds : The seed are endospermic. Each seed bears a small embryo. The seeds of *Magnolia* possess bright red or orange coloured testas and are therefore highly decorative.

Pollination : The pollination takes place through the agency of insects.

Floral Formula : Br $\oplus$ ⚥ P3 + 3 + 3 or K3, C3, +3, A ∞, $\underline{G\infty}$

Bentham & Hooker (1862)	**Engler & Prantl (1931)**	**Hutchinson (1959)**
Phanerogams	Phanerogams	Angiospermae
Dicotyledones	Dicotyledoneae	Dicotyledones
Polypetalae	Archichlamydeae	Lignosae
Thalamiflorae	Ranales	Magnoliales
Ranales	Magnoliaceae	Magnoliaceae,
Magnoliaceae		

Economic Importance of the Family

Many genera of the family are mainly important on account of their horticultural value. Some plants yield good timber and some are extremely ornamental and grown in the gardens for this purpose. A list of few important plants is given below.

1. *Magnolia grandiflora* : Eng.–Bull bay, Great laurel; Verna–**Bari champa**–This is a small tree with beautiful scented flowers. They are cultivated as hedge plants.

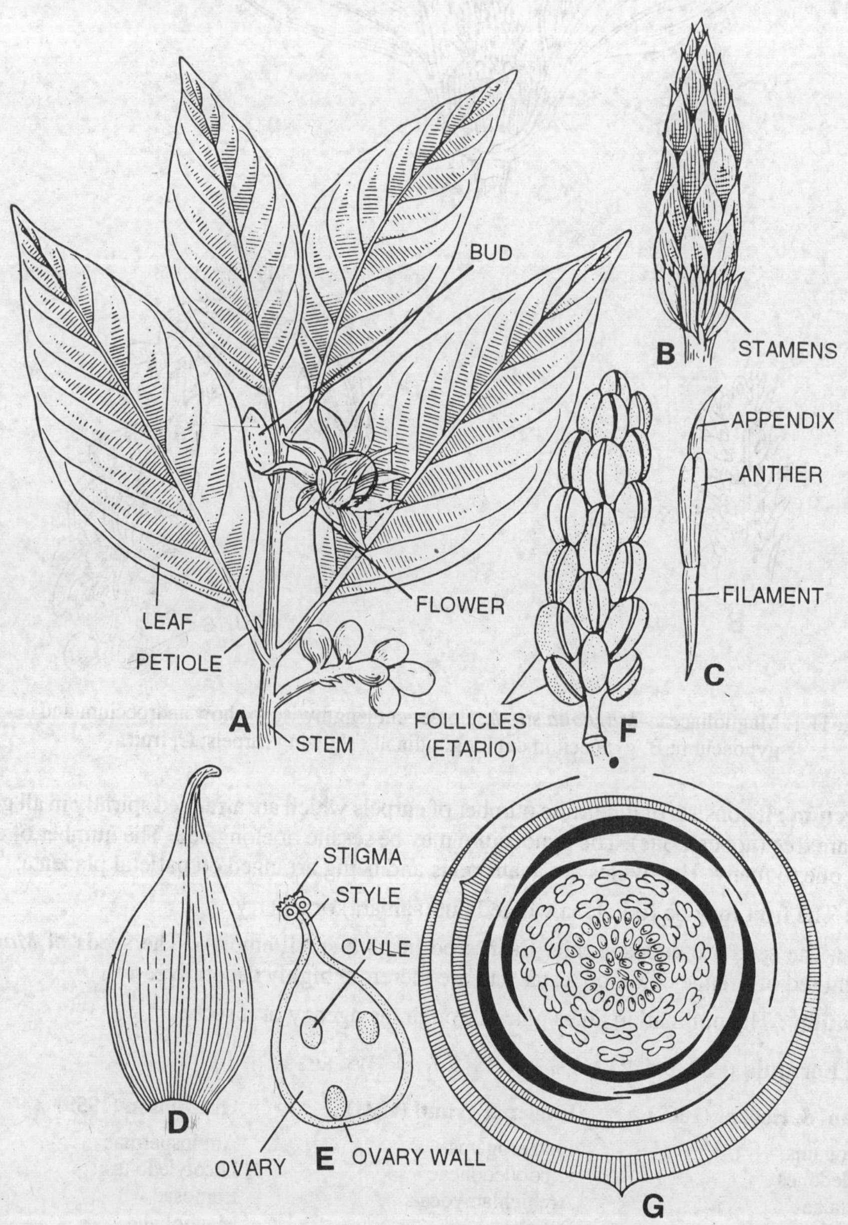

Fig. 11.2. Magnoliaceae–*Michelia champaca.* A, flowering twig; B, flower without perianth; C, stamen; D, tepal; E, L.S. of gynoecium; F, fruit-etaerio of follicles; G, floral diagram.

2. *Michelia champaca;* Eng.–Champac; Verna–**Champa**–This is a large tree, cultivated mainly in South India, the Western Ghats, Assam and Bengal. The oil obtained from the flowers is used in perfumery. Several species yield timber. Root and root bark are used as purgative. The flowers and fruits are used in renal diseases. The oil is used in gout. Flowers yield a dye.

3. *Michelia fuscata;* Eng.–Banana shrub.–This is a small tree. The flowers are used for scenting hair oil.

4. *Liriodendron tulipifera;* Eng.–Tulip tree.–The tree is highly ornamental and grown in avenues.

5. *Talauma*–This is alo an ornamental tree.

Systematic Relationships : Hutchinson has divided the order Ranales into two–1. The Ranales include herbaceous types and 2. The Magnoliales include the arborescent (trees,etc.) forms. He has treated the Magnoliales to be more primitive. Hallier, Hutchinson and many others have considered the Magnoliaceae to be most primitive taxon of living angiosperms.

12

ORDER–ANNONALES

According to Hutchinson this is the second order of the phylum Angiospermae, sub-phylum Dicotyledones and division Lignosae. The order consists of only two families, *i.e.*, Annonaceae and Eupomatiaceae. The family Annonaceae has been described in the present text in detail. Bentham and Hooker put the family Annonaceae together with seven others in their first order–the Ranales of class Dicotyledones, division Polypetalae and series Thalamiflorae. Engler and Prantl have placed this family in eighteenth order–the Ranales of class Dicotyledoneae and sub-class Archichlamydeae, together with fourteen other families.

Hypogynous to rarely perigynous; mostly bisexual; apocarpous to very rarely syncarpous with parietal placentation; petals present, rarely united; stamens numerous, free; endosperm copious, markedly ruminate; embryo minute; leaves without stipules.

Family–Annonaceae

There are about 80 genera and 850 species in this family.

Distribution: The members of this family are widely distributed in the tropical regions of both worlds. Mostly they occur in the old world. In our country the family is represented by several important genera such as *Annona, Artabotrys, Polyalthia* ,etc.

Habit : The family consists of trees or shrubs. Certain plants are climbing, *e.g., Artabotrys, Uvaria,* etc. Oil passages are present in the stems and sometimes even in the flowers.

Leaves: The leaves are simple, petiolate, alternate, entire and exstipulate.

Inflorescence : Cymose, usually the flowers are solitary but sometimes they are found to be arranged in inflorescences of various types. In certain cases the flowers are said to be cauliflorous, *i.e.,* they are borne on the old wood.

Flowers : The flowers are large, regular hermaphrodite; bisexual, rarely unisexual, *e.g., Stelechocarpus* and hypogynous. However, the flowers of Australian geneus *Eupomatia* are perigynous.

Perianth: The perianth consists of three whorls,

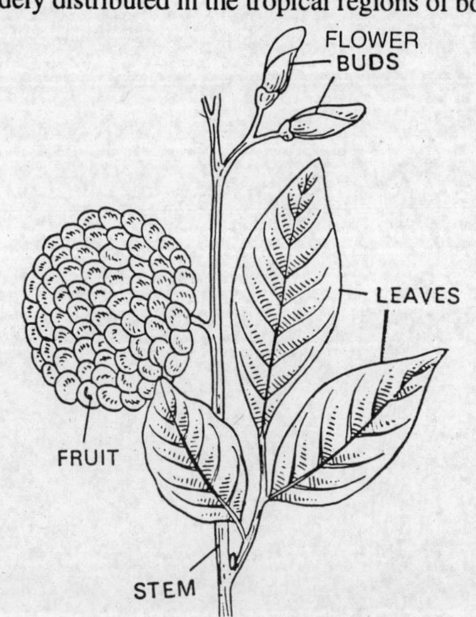

Fig. 12.1. Annonaceae—*Annona squamosa*—A twig with leaves, buds and fruit.

each consisting of three and rarely of two segments. The members of outer whorl form a calyx and have valvate aestivation, the segments are free or somewhat connate at the base. The two inner whorls possess petaloid (coloured) segments, they have valvate or slightly imbricate aestivation. The numbers of the inner whorls are similar but that of *Monodora* are quite different in shape and size. The perianth is not found in the Australian genus *Eupomatia.*

Androecium: The androecium consists of many (indefinite) stamens arranged spirally. Rarely the stamens are few in number. The stamens are found to be arranged on a large convex receptacle

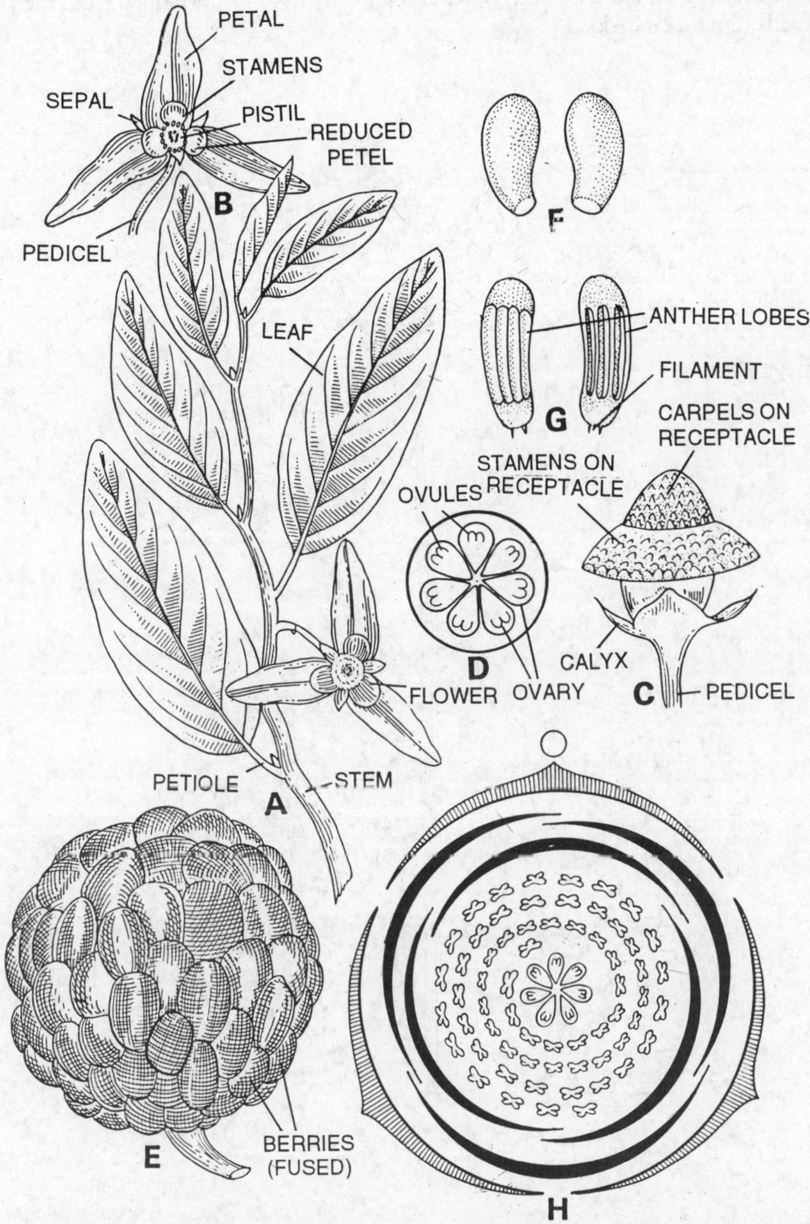

Fig. 12.2. Annonaceae—*Annona squamosa.* A, flowering twig; B, flower; C, flower without perianth; D, T. S. of ovary; E, fruit; F, seeds; G, stamens; H, floral diagram.

above the perinath. The filaments are thick and short, anthers bicelled, introrse and the connective is somewhat elongated outside the anther.

Gynoecium: It consists of many free carpels (apocarpous). The carpels are spirally arranged on a large convex receptacle situated above the perianth axis (the stamens are also arranged on this axis). Rarely they are fewer in number. In *Monodora* sp., the carpels are somewhat connate at the base. The ovary is superior, unilocular with numerous anatropous ovules arranged in a double row on the ventral suture of the carpel.

Fruits: Generally they are etaerio of berries. They are fleshy and form separate berries. When the fruits are many-seeded the constrictions develop between individual seeds. In *Annona squamosa* (**Sharifa**), the ovaries are united with each other and a fleshy mass is developed from the floral axis in which the seeds remain embedded.

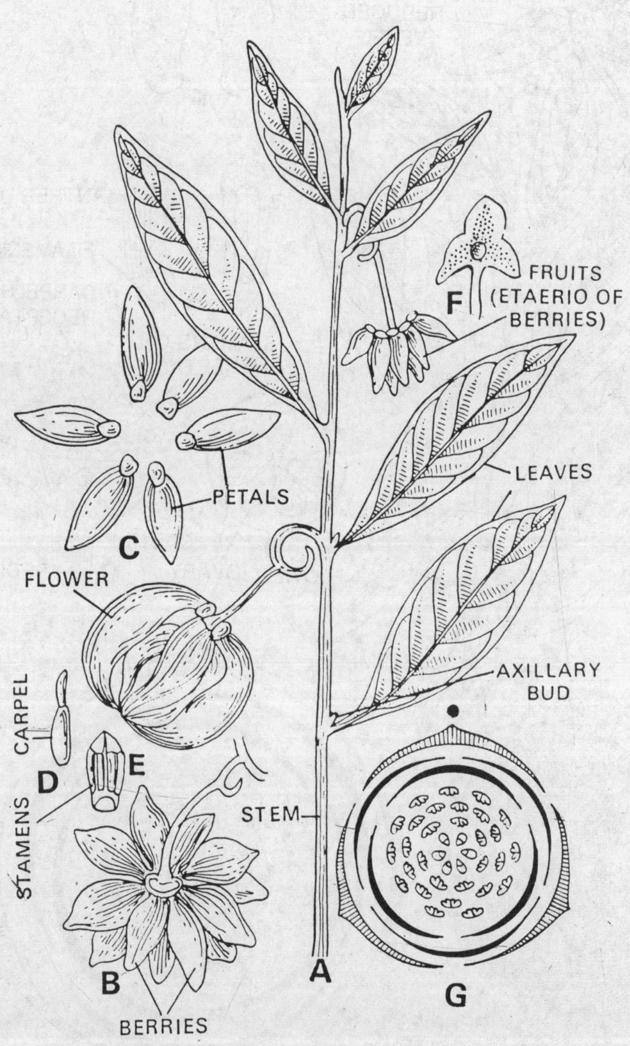

Fig. 12.3. Annonaceae–*Artabotrys odoratissimus*–A, a twig with leaves, flower and fruit; B, fruit-etaerio of berries; C, petals (corolla); D, carpel; E, stamen; F, sepals (calyx); G, floral diagram.

Seeds : The seeds are large and endospermic. The embryo is small.

Pollination : The pollination takes place through the agency of insects.

Flora formula — ⊕ ⚥ P3 + 3 + 3, A∝, G∝

Bentham & Hooker(1862)	**Engler & Prantl(1931)**	**Hutchinson(1959)**
Phanerogams	Phanerogams	Angiospermae
Dicotyledones	Dicotyledoneae	Dicotyledones
Polypetalae	Archichlamydeae	Lignosae
Thalamiflorae	Ranales	Annonales
Ranales	Annonaceae	Annonaceae
Annonaceae		

Economic Importance of the Family

Economically the family is fairly important. Some plants produce edible fruits while others are grown as ornamentals in the gardens. A list of few important plants is given below-

1. Annona squamosa: Eng–Custard apple, Sugar apple; Verna–**Sharifa, Sitafal**–This is native of South America, and the West Indies. This is now cultivated in our country in Assam, Bengal, Uttar Pradesh and Maharashtra for its edible fruits. The powdered seeds and leaves are used as insecticides.

2. Annona muricata : Verna–**Mamphal**–This is native of the West Indies but now grown in Assam, Maharashtra and South India for its edible fruits.

3. Annona reticulata; Eng-Common custard apple ; Verna. **Ramphal**-This is also native of the West Indies, but now cultivated in Bengal and South India for its edible fruits. An oil is extracted from its seeds which could be used as an insecticide and for soap-making.

4. Annona cherimola; Eng–Cherimoya ; Verna–**Hanumanphal**–This is native of South America but now grown in the Nilgiris and Pulni hills for its edible fruits.

5. Annona diversifolia: Eng–Ilama–This is native of Mexico but now grown in South India for its edible fruits.

6. Artabotrys uncinatus; Syn. A. *odoratissimus :* Eng. – Climbing ylang–ylang; Verna–**Harichampa**–This is a large, scandent shrub with greenish-yellow, fragrant flowers. Very often grown in gardens as an ornamental. The flowers yield an essential oil which is used in making perfumes.

7. Polyalthia longifolia; Verna–**Devdaru, Ashok**–They are grown as avenue trees on road sides, the fruits are edible.

8. Polyalthia cerasoides; Verna.–**Kudumi**–This is a small tree, generally found in Bihar, the Peninsulas and the Western Ghats. The fruits are edible.

9. Polyalthia fragrans; Verna–**Gauri**–This is a large tree found in the Western Ghats. The wood is used for making cricket stumps and Billiard cues.

10. Xylopia parvifolia –This is a tree, mainly found in Travancore. The fruits are used as masticatory and the wood is used for making plywood, light furniture and other fittings.

11. Cananga odorata; Syn–*Canangium odoratum;* Eng–Ylang–ylang, Macassar oil–This is a cultivated tree. An oil obtained from the petals, is used in perfumery. The plant is grown especially for this purpose in the Philippines, Indonesia and Madagaskar.

Systematic Relationships : Hutchinson has placed the Annonaceae together with the Eupomatiaceae in the order Annonales. He has treated the Annonales to be more advanced than the Magnoliales. In general it is agreed by most of the taxonomists that the Annonaceae has been derived from Magnoliaceous stocks.

13

ORDER–ROSALES

According to Hutchinson this is the sixth order of the phylum Angiospermae, subphylum Dicotyledones and division Lignosae. The order comprises of three families, *i.e.*, Rosaceae, Dichapetalaceae and Calycanthaceae, of which the Rosaceae has been described here in detail. Bentham and Hooker have placed this family together with other five families in their eleventh order–the Rosales of class Dicotyledones, division Polypetalae and series Calyciflorae. Engler and Prantl have also included the family Roscaceae with other seventeen families in their twenty first order–the Rosales of class Dicotyledoneae and sub-class Archichlamydeae.

Perigynous to epigynous, mostly apocarpous to syncarpous with central placentation; no endosperm; leaves simple to compound, stipulate.

Family-Rosaceae

This family consists of about 115 genera and 3,200 species (Lawrence).

Distribution : Though the members belonging to this family are cosmopolitan, yet they are mainly distributed in temperate and colder regions. In India most of the genera have been reported from hilly regions. The family is represented in our country by the following few important genera, *e.g., Prunus, Pyrus, Rubus, Rosa, Eriobotrya, Fragaria, Potentilla,* etc.

Habit : There are great variations in the habit of the plants included in this family. They may be trees, shrubs or herbs. Very often they are thorny and sometimes climbing. Most of the species are herbaceous and wild, while on the other hand some species are grown for their delicious fruits, and as ornamentals.

Root : Tap and branched.

Stem : The stem is erect or creeping, herbaceous or hard and woody, cylindrical, branched; many shrubs are spiny (*e.g., Rosa);* in *Rosa* and *Rubus* these spiny outgrowths are cortical and arise from both nodes and internodes. Vegetative propagation takes place by means of runners, suckers and cuttings. The bud grafting is very common in this family.

Leaves : The leaves are alternate (rarely opposite, *e.g.*, in seedling stage of *Prunus*), simple or pinnately compound, petiolate, usually stipulate, the stipules are adnate to the petiole; the leaf base is conspicuous, small spines are often present on the rachis; the margins of the leaves are either serrated or entire.

Inflorescence : There are great variations in different species. It may be corymb, corymbose, umbellate, racemose, or the flowers may be solitary or in the small groups in the twos or threes. Sometimes the inflorescence is compound raceme (panicled flowers).

Flower: Usually hermaphrodite (bisexual) rarely unisexual (*e.g., Spiraea aruncus,* where the

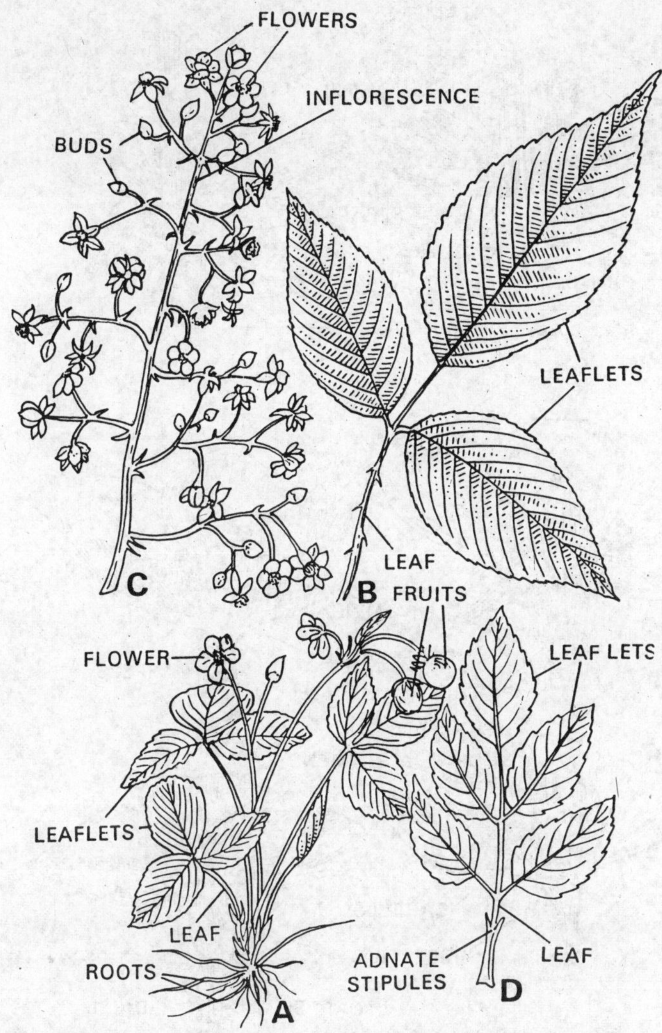

Fig. 13.1. Rosaceae–A, a plant of *Fragaria vesca*: B. a leaf of *Rubus* sp.; C. inflorescence of *Rubus* sp.; D. leaf of *Rosa* sp. showing adnate stipules.

plants are dioecious), actinomorphic, regular (irregular in *Parinarium* and *Parastemon*), bracteolate.

It may be hypogynous (*e.g.*, in *Fragaria*), perigynous (*e.g.*, in *Prunus*) or epigynous (*e.g.*, in *Pyrus*).

Calyx : It consists of 5 sepals, which are connate at the base (gamosepalous); basal portions are usually adnate forming a hypanthium; the lobes of the calyx are free, valvate or imbricate and green; sometimes the epicalyx is formed by bracteoles; the lobes of epicalyx remain alternate to the sepals.

Corolla: It consists of 5 petals (rarely four petals) which are polypetalous; the petals usually arise from the rim of hypanthium; imbricate, very often the hypanthium bears a nectariferous glandular disc. The number of petals increases sufficiently in cultivated varieties because of the conversion of the stamens into petals. They are variously colured. Sometimes the corolla is altogether absent (*i.e.*, in *Alchemilla, Poterium,* etc.).

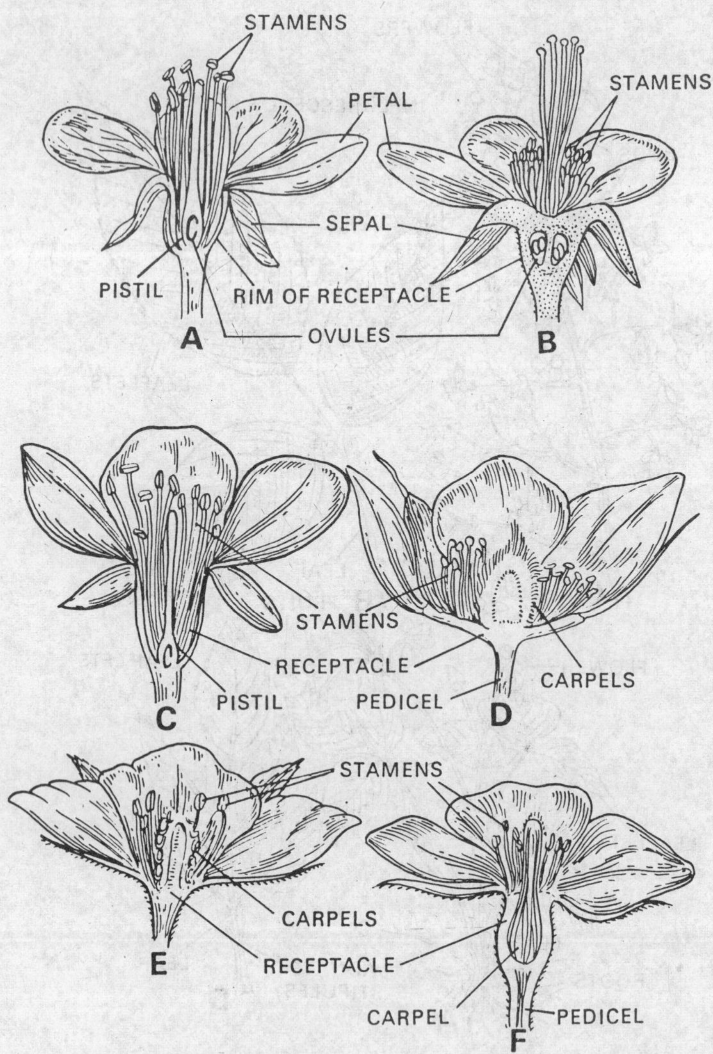

Fig. 13.2 : Rosaceae–flowers cut lengthwise; A. *Prunus* sp.; B. *Malus sylvestris*; C. *Prunus avium* (sweet cherry); D. *Rubus* sp. (raspberry); E. *Fragaria vesca;* F. *Rosa* sp.

Androecium : Usually the number of stamens is indefinite (15-60), sometimes they are 5-10, the stamens are generally arranged in one to many whorls of five each; they are perigynous around the gynoecium and arise from the hypanthium, free; the anthers are small, dithecous (two-celled), introrse and dorsifixed; the dehiscence takes place by means of longitudinal splits; the filaments are usually incurved in bud.

Gynoecium : The number of carpels is one to many; the gynoecium consists of either one compound carpel (syncarpous) or many simple carpels (apocarpous) arranged in cyclic or spiral way. The carpels are usually situated within the hypanthium or the hypanthium remains adnate to the compound ovary; the ovary is either superior or inferior or half superior half inferior (*i.e.,* perigynous condition), when syncarpous, 2-5 locules are found, the placentation is axile, and the stigmatic lobes are as many as the number of carpels. The placentation is basal when one carpel is present (apocarpous); the ovules are one to many in each carpel. The style is free or connate, the stigma is simple, lobed or capitate.

Fruit: They may be dry or fleshy. It may be a pome (*e.g.,* in *Pyrus*), pyriform berry (*e.g.,* in *Eriobotrya*), an etaerio of achenes (*e.g.,* in *Rubus*), or one seeded drupe (*e.g.,* in *Prunus*), etc.

Seeds: the seeds are exalbuminous, generally with small embryo .

Pollination: The pollination takes place usually through the agency of insects (*i.e.,* entomophily), but sometimes there is wind pollination (*i.e.,* anemophily)

Floral Formula : Br ⊕ ⚥ K(5), C5, A ∝, G ∝ or 1 or (2 – 5)

Bentham & Hooker(1862)	**Engler & Prantl(1931)**	**Hutchinson (1959)**
Phanerogams	Phanerogams	Angiospermae
Dicotyledones	Dicotyledoneae	Dicotyledones
Polypetalae	Archichlamydeae	Lignosae
Calyciflorae	Rosales	Rosales
Rosales	Rosaceae	Rosaceae
Rosaceae		

Description of some important plants in semitechnical language:-

Prunus persica ; Verna. **Aru.** (See fig. 13.3 on page 110).

Habit : Either shrubs or small trees, usually grown in the gardens for their delicious fruits.

Root : Tap and branched.

Stem : Erect, branched, woody, cylindrical and glabrous.

Leaf : Alternate, simple, stipulate, petiolate, entire or toothed, very frequently two glands present on the petiole. In *Prunus persica* the leaves are oblong–lanceolate or oblong oblanceolate, 4 to 6 inches long, finely serrate, short petioles, fimbriate stipules.

Inflorescence: Flowers are solitary or in fascicles, corymbose or racemose.

Flower : Bisexual (hermaphrodite), pedicellate (small pedicels), actinomorphic, regular, perigynous, complete, pink.

Calyx : Five sepals, gamosepalous, sepaloid, imbricate, deciduous in fruit, calyx-tube bell shaped, 5-lobed, inferior.

Corolla: 5 polypetalous (petals free), petals large, arranged on hypanthium, conspicuous, pink, inferior.

Androecium : Stamens indefinite (15-60), perigynous, inserted on the hypanthium, anthers bi-celled, dorsifixed, reddish or brown, dehiscence by longitudinal split, introrse, filaments free.

Gynoecium : One carpel (monocarpellary), ovary situated within the hypanthium (calyx tube), ovary unilocular, with two pendulous ovules, style terminal, stigma broad.

Fruit: Single seeded drupe, fleshy, enclosing an indehiscent or two valved furrowed stone.

Seeds : Exalbuminous, pendulous.

Generic Characters : Ripe carpels not enclosed within the calyx tube (hypanthium), carpels free; carpel one, fruit a drupe with a hard furrowed stone; shrubs or trees with simple leaves; calyx 5 lobed, petals large.

Floral Formula : ⊕ ⚥ K (5), C5, A ∝, G1–

Eriobotrya japonica : Verna-**Lokat.**

Habit : Small trees, grown in the gardens for their edible fruits.

Root : Tap and branched.

Stem : Erect, woody branched with stout branches, cylindrical.

Leaves : Simple, alternate, entire, serrate or crenate, coriaceous; in *Eriobotrya japonica,* the leaves are large, 6-12 inches long, subsessile, lanceolate, or oblanceolate, acuminate, dentate or serrate, net veined, and wooly below.

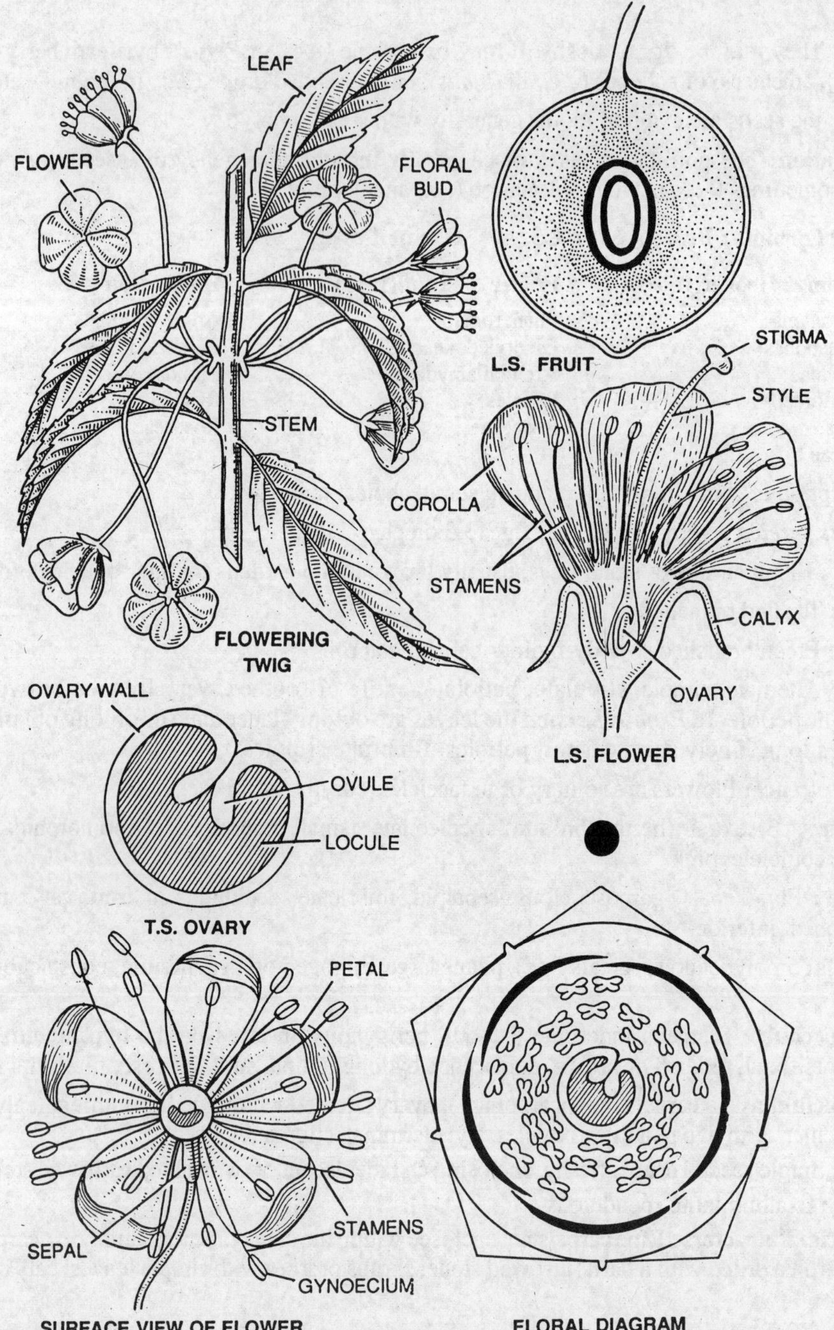

Fig. 13.3. Rosaceae–*Prunus persica* (L.) Stokes; (Verna. **Aru**).

Inflorescence: Flowers arranged in dense panicles, the inflorescence is wooly.

Flower : Pedicellate (short pedicels), actinomorphic, regular, hermaphrodite (bisexual), epigynous, yellowish white in colour, bracteate and bracteolate.

Calyx: 5 sepals, gamosepalous, small persistent, arranged on the turbinate or clavate hypanthium, sepals also wooly.

Corolla: 5 petals, free, polypetalous contorted or imbricate in bud, obovate or orbicular, usually oblique and notched at margins, claw glabrous and wooly, yellowish white in colour.

Androecium : Indefinite in number, stamens curved inward in the bud, filaments free, anthers 2-celled, dorsifixed, dehiscing longitudinally, introrse.

Gynoecium: Carpels enclosed in the hypanthium, 2-5 celled ovary, with connate and wooly below styles, each cell contains two ovules, axile placentation.

Fruit: Succulent or dry; in *E. japonica,* the fruit is ovoid, yellow, succulent and pyriform berry, 1–1.5 inches long. 1–5 seeded.

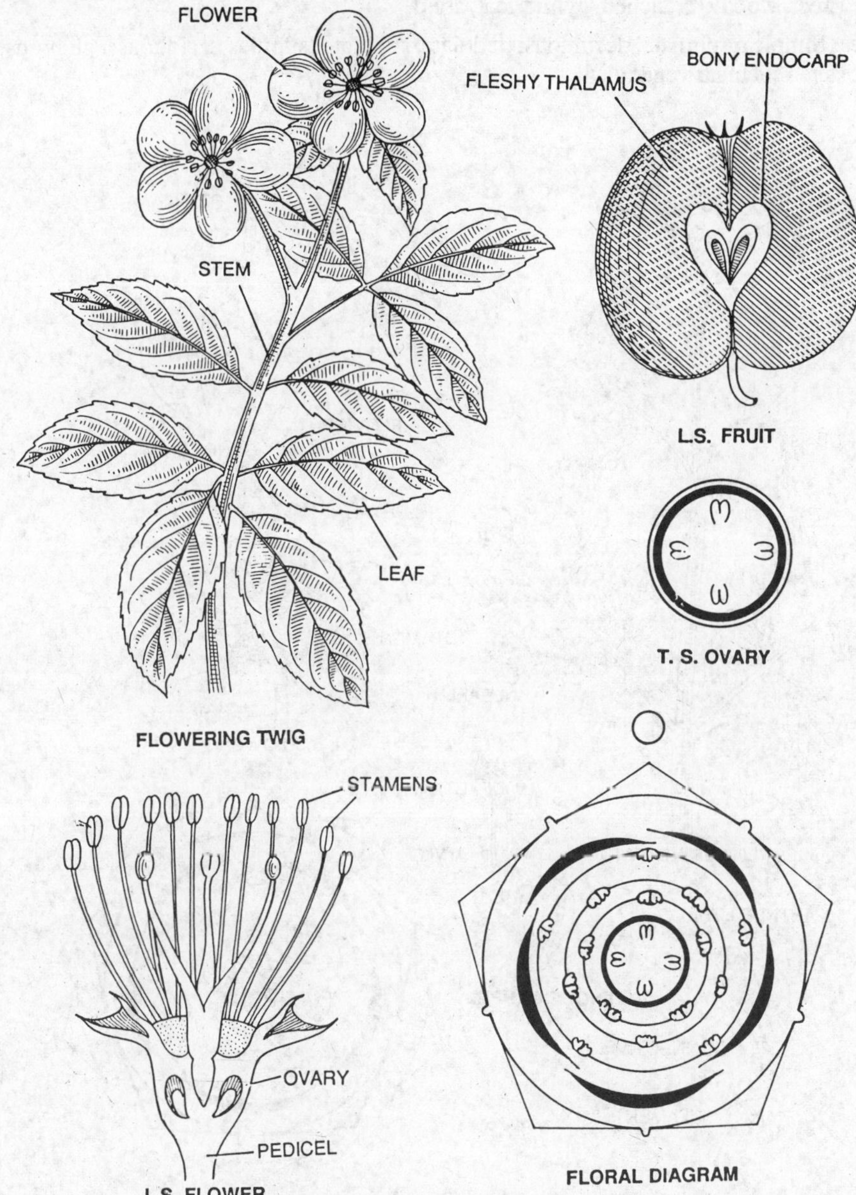

Fig. 13.4. Rosaceae–*Malus sylvestris* (L.) Mill.; syn. *Pyrus malus* Linn. (Verna, *Seb;* Eng. Apple).

Seed: Exalbuminous, large and angled with thick cotyledons.

Generic Characters: Carpels definite, 2-ovuled, enclosed in the hypanthium, and adnate to its walls, more or less connate with one another to form a syncarpous ovary; fruit a berry or pome; flowers arranged in panicles.

Floral Formula : Br Brl $\oplus$ ⚥ K(5), C5, A∞, $G_{\overline{(2-5)}}$.

Malus pumila; Verna. **Seb.**

Habit: Small tree or large shrub.

Stem: Erect, woody, branched, cylindrical, solid.

Leaves: Simple or pinnate, deciduous, petiolate, stipulate (stipules deciduous), glabrous, acute, ovate, unicostate reticulate venation.

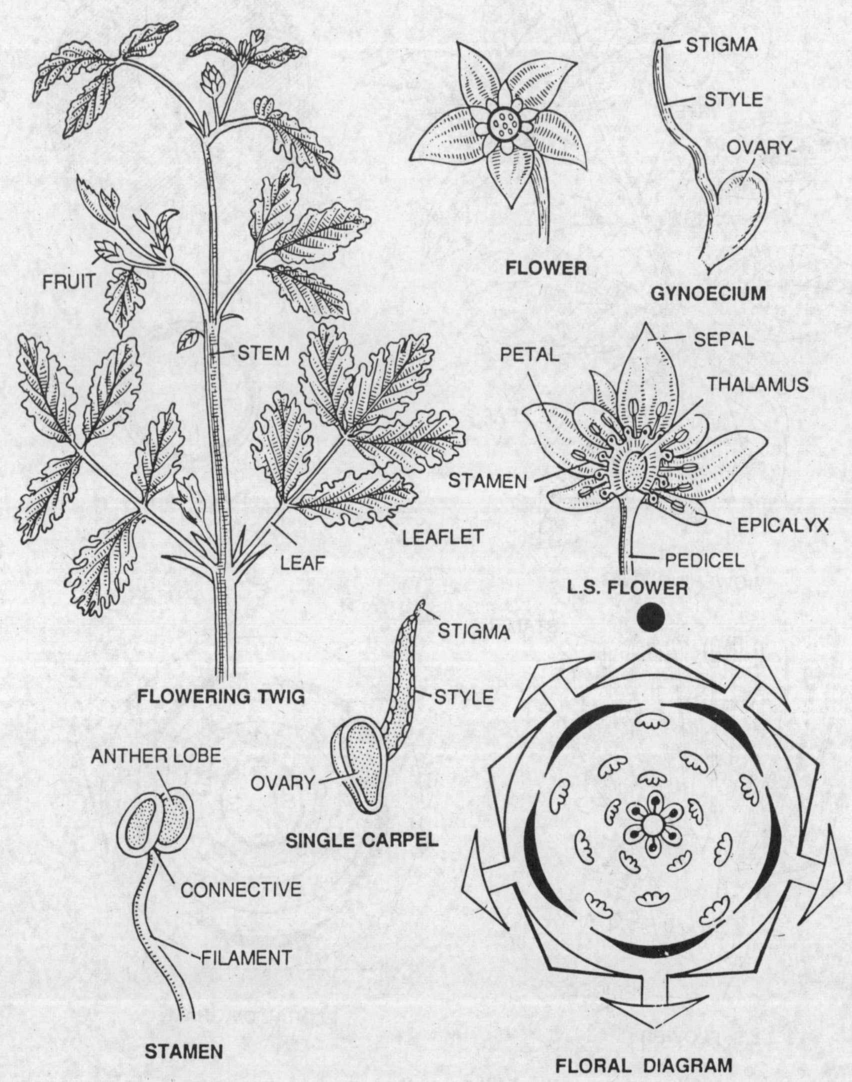

Fig. 13.5. Rosaceae--*Potentilla supina* Linn.

Inflorescence: Cymose, terminal cymes or corymbs.

Flower: Pedicellate, bracteate (bracts linear), bracteolate (two bracteoles), hermaphrodite, actinomorphic; complete, epigynous.

Calyx : Five sepals, gamosepalous, calyx-tube obconical lobes erect or reflexed, persistent or deciduous.

Corolla: Five petals, polypetalous, quincuncially imbricate in bud.

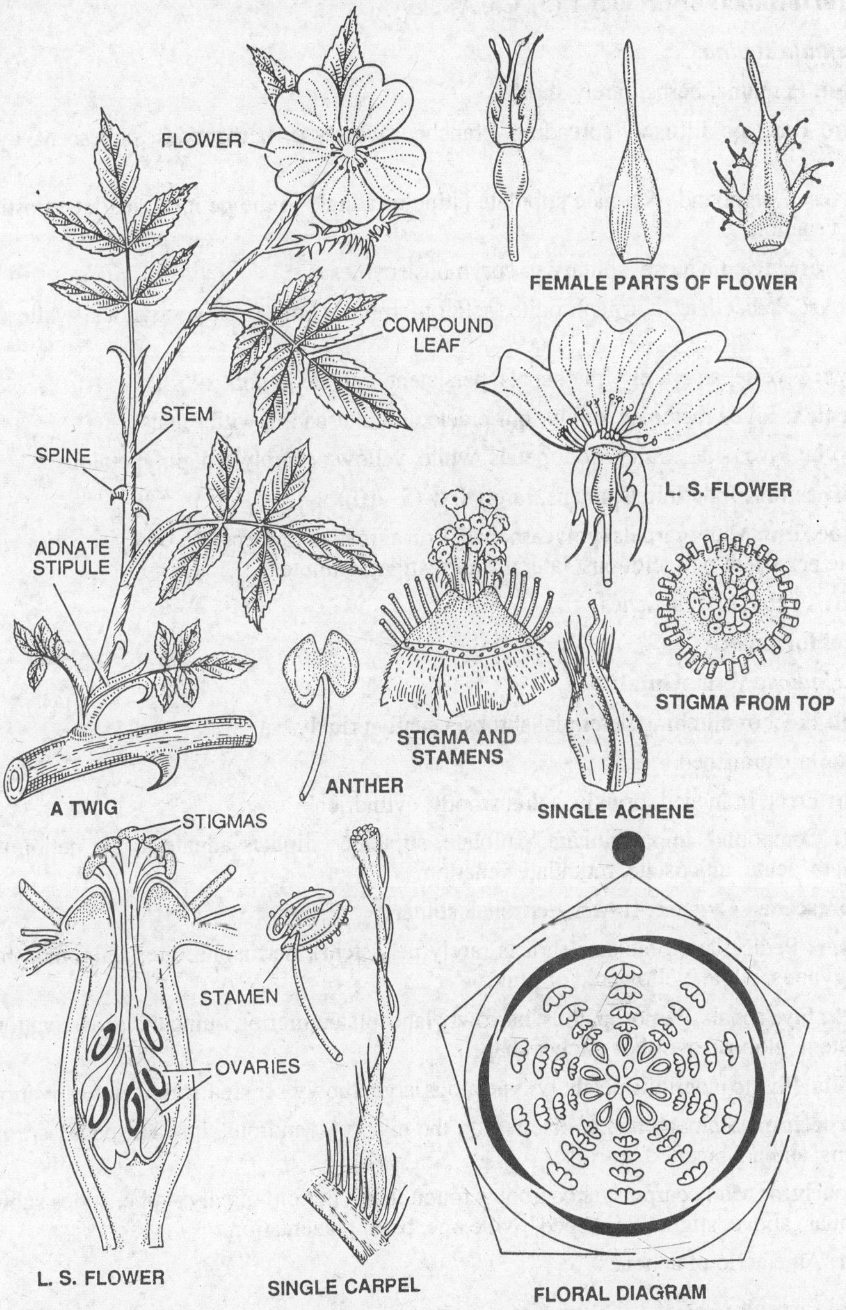

Fig. 13.6. Rosaceae–*Rosa indica,* Eng., rose; Verna. **Gulab, Jangligulab.**

Androecium: Indefinite stamens (20 or more), filaments long, connate at the base, anthers dorsifixed, introrse; dehisce longitudinally.

Gynoecium: Two to five carpels, connate and adnate to calyx-tube, syncarpous, ovary inferior, 2-5 locular, axile placentation, styles 2-5, free or connate below, stigmas truncate; ovules 2 in each loculus.

Fruit: A pome, fleshy.

Floral formula: Br Brl ⊕ ⚥ K(5), C5, A∝, $\overline{G_{(2-5)}}$

Potentilla supina.

Habit: Perennial herbs, rarely shrubs.

Stem: Erect or diffusely spreading, branched, woody or herbaceous, pubescent, cylindrical, solid.

Leaves: Compound, petiolate stipulate (stipules adnate to the petiole), leaflets mostly serrate, reticulate venation.

Inflorescence: Cymose, solitary or corymbose cymes.

Flower: Pedicellate, hermaphrodite, actinomorphic, complete, hypogynous, white or yellow rarely red.

Calyx: Five sepals, rarely four sepals, persistent, valvate in bud.

Epicalyx: Five bracteoles, rarely four bracteoles, alternating with sepals.

Corolla: Five petals, rarely four petals, white, yellow or rarely red, polypetalous.

Androecium: Indefinite stamens, rarely few (5 - 10).

Gynoecium: Many carpels (polycarpellary), on a small dry receptacle, ovary superior with single ovule, style persistent or deciduous, lateral, short, stigma minute.

Fruit: An etaerio of achenes.

Floral formula:

Rosa indica; Verna-**Gulab.**

Habit: Erect or climbing perennial shrubs, usually prickly.

Root: Tap, branched.

Stem: Erect, branched, prickly, solid, woody, cylindrical.

Leaf: Compound, imparipinnate, petiolate, stipulate (stipules adnate to the petiole), leaflets serrate, ovate, acute, unicostate, reticulate venation.

Inflorescene : Cymose, flowers terminal, solitary.

Flower: Pedicellate, bracteate (bracts rarely persistent), hermaphrodite, actinomorphic, complete, perigynous, white, yellow, red or pink.

Calyx: Five sepals, gamosepalous, narrowly lanceolate, inferior, quincuncial aestivation, calyx-tube persistent, globose ovoid or pitcher shaped.

Corolla: Five to indefinite petals, polypetalous large, showy, scented, imbricate aestivation in bud.

Androecium: Stamens many, inserted on the disk, polyandrous, inferior, petals are modified into stamens, anthers bicelled, introrse.

Gynoecium: Many carpels, apocarpous, found in the bottom of calyx-tube, styles subterminal, free or connate above, stigma thickened, ovule one, basal placentation.

Fruit: An etaerio of achenes.

Floral formula: ⊕ ⚥ K5, C5 A∝ $\overline{G}$∝

Economic Importance of the Family

Economically this family is of much importance, containing various types of valuable fruit plants. Several other plants are ornamental and are planted in the gardens, public places and houses for their beautiful and fragrant flowers. A list of some important plants is given here–

1. *Prunus persica;* Syn. *Amygdalus persica; Persica vulgaris;* Eng.–Peach; Verna–**Aru**–This is native of China, but now cultivated in Kashmir, the hills of the Punjab, Himachal Pradesh, Kulu and Kumaon.The fruits are edible and rich in vitamins; the oil is obtained from the seeds which is used for cooking and other purposes.

2. *Prunus domestica* subs. *insititia;* Syn. *Prunus communis* var. *insititia;* Eng.–Plum; Verna-**Alucha, Alu-bokhara**–This is cultivated in the hills of Kashmir, Punjab, Himachal Pradesh, Kulu and Kumaon for the edible vitaminized fruits.

3. *Prunus amygdalus;* Syn. *P. communis; Amygdalus communis;* Eng.–Almond; Verna.–**Badam**–This is a tree. This is native of the East mediterranean region but now chiefly cultivated in the Punjab and Kashmir for the edible seeds. There are two varieties of almonds, *i.e.,* sweet and bitter. The oil obtained from the seeds of both the varieties is used in perfumery and medicinal purposes.

4. *Prunus armeniaca;* Syn. *Armeniaca vulgaris;* Eng.–Apricot; Verna.–**Khubani, Zardalu**–This is native of Asia, now cultivted in Kashmir, Kulu, Kumaon and Himachal Pradesh. The fruits are edible and rich in vitamins.

5. *Prunus avium;* Eng.–Sweet cherry; Verna.–**Gilas**-This is native of Eurasia but now cultivated in Kashmir, Kulu, Kumaon and Himachal Pradesh for their delicious edible fruits.

6. *Prunus cerasus;* Eng.–Sweet cherry; Verna.–**Alubalu, Gilas.**–This is native of Eurasia, but now cultivated in the Punjab, Kashmir, Himachal Pradesh and Kumaon hills. The fruits are edible.

7. *Prunus cerasoides;* Syn. *P. puddum;* Eng.–Wild Himalayan cherry ; Verna.–**Paddam**–They are commonly found in the temperate Himalayas from Garhwal to Sikkim and also in Nilgiris. The fruits are edible. The stones (*i.e.,* endocarp) are made into rosaries and necklaces, and the wood is used for walking sticks.

8. *Prunus salicina ;* Syn. *P. triflora ;* Eng.–Japanese plum–This is native of China but now grown in the Nilgiris. The fruits are edible.

9. *Prunus padus:* Syn. *Padus avium;* Eng.–European birdcherry; Verna.–**Jamana.**–This is found in the temperate Himalayas. The fruits are edible.

10. *Prunus jenkinsii;* Syn.–*Cerasus jenkinsii.*–This is found in Assam and the Khasia hills. The fruits are edible.

11. *Eriobotrya japonica;* Eng.–Loquat; Verna.–**Lokat**–This is a small tree. It is native of China, now chiefly cultivated in Uttar Pradesh and the Punjab for the edible fruits.

12. *Fragaria vesca;* Eng.–Perpetual strawberry.–This is native of Europe but now cultivated in Himachal Pradesh, Assam, Kulu, Kumaon and Maharashtra. The fruits are edible.

13. *Fragaria nilgerrensis;* Eng.–Houtbois strawberry.–This is found in the Khasia hills and Nilgiris. The fruits are edible .

14. *Fragaria chiloensis;* Eng.–Garden strawberry.–This is a herb, cultivated in Kumaon, Kulu and Himachal Pradesh for the edible fruits.

15. *Duchesnea indica;* Syn. *Fragaria indica;* Eng.–Indian strawberry; Verna–**Kiphaliya.**–This is a perennial herb, commonly found in the temperate Himalayas, the Punjab, Assam, the Khasia hills and the Nilgiris. The fruits are edible.

16. *Pyrus communis;* Eng.–Pear; Verna.–**Nakh,**–This is a small tree. It is native of Eurasia but now cultivated in Kashmir, Kulu, Kumaon and Himachal Pradesh. The fruits are edible.

17. *Malus sylvestris;* Syn. *Pyrus malus; Malus pumila;* Eng.–Apple; Verna–**Seb.**–This is a small

tree; native of Eastern Europe and Western Asia but now cultivated in Kashmir, Himachal Pradesh, Kulu, Kumaon, Assam and in the Nilgiris. The fruits are edible, delicious, rich in iron and vitamins.

18. *Malus baccata;* Syn. *Pyrus baccata;* Eng.–Siberian crab apple; Verma **Jangli Seb**–This is a small tree, cultivated in Kashmir and the Khasia hills. It is used as a stock for apple. The fruits are eaten as vegetable.

19. *Docynia indica;* Syn. *Pyrus indica;* Eng.–Indian crab apple.–This is a tree found in the Eastern Himalayas and the Khasia hills. The fruits are edible.

20. *Pyrus pashia ;* Verna.–**Mehal**–This is commonly found in the Western Himalayas. It is used as a root stock for pear. The wood is used in making walking sticks and tobacco pipes.

21. *Pyrus pyrifolia* var. *culta;* Eng.–Chinese pear; Verna.–**Nashpati**–This is native of China but now cultivated in Kashmir, Assam, Kumaon, Darjeeling, the Nilgiris, Himachal Pradesh and Delhi for the edible fruits.

22. *Pyrus vestita;* Verna.**Mauli**–This is found in the Eastern Himalayas and the Khsia hills. The fruits are edible.

23. *Sorbus aucuparia;* Syn. *Pyrus aucuparia;* Eng. Europen mountain ash; Verna.–**Battal**–This is a tree. It is native of Eurasia but now found in the temperate Himalayas from Kashmir to Kumaon. The fruits are edible.

24. *Spiraea cantoniensis ;* This is a native of China and Japan. This is a small herb grown as an ornamental.

25.*Spiraea corymbosa*–This is a small shrub, grown in the gardens as an ornamental.

26. *Rubus fruticosus;* Eng. European black berry; Verna.–**Vilayati–anchu**–This is cultivated in Kashmir, Assam and Conoor for the edible fruits.

27. *Rubus ellipticus;* Eng. Himalayan yellow raspberry; Verna.–**Lal anchu, Hisalu**–This is a shrub found in the Western Himalayas, South India, the Western ghats and the Khasia hills. The fruits are edible.

28. *Rubus rosaefolius;* Eng. Mauritius raspberry; Verna.–**Yeshul**–This is cultivated in Kumaon, the Khasia hills and Manipur. The fruits are edible.

29. *Rubus lanatus;* Verna.–**Hisalu**–This is found in Kumaon. The fruits are edible.

30. *Rubus lasiocarpus;* Eng. Ceylon Raspberry; Verna.–**Kala anchu, Kala hisalu**–This is found in the Western Himalayas, Sikkim, the Khasia hills, the Western Ghats, Karnataka and the Pulney hills. The fruits are edible.

31. *Rubus moluccanus;* Eng. Black cherry; Verna–**Katsol.**–This is found in Assam, the Khasia hills the Nilgiris and Travancore. The fruits are edible.

32. *Rosa alba;* Eng. Indian white rose ; Verna–**Gulab, Gulchini**–This is a herb, grown as an ornamental in the gardens.

33. *Rosa chinensis ;* Syn. *Rosa indica ;* Eng. Edward rose; Verna.–**Cheenia gulab, Desi Gulab**–This is chiefly cultivated in the Kanauj, Kanpur and Hathras regions of Uttar Pradesh. The petals are the sources of rose water, essential oil, pankhuri and gulkand.

34. *Rosa centifolia;* Eng. Cabbage rose; Verna–**Gulab**–This is a herb or small shrub. This is cultivated mainly in the Uttar Pradesh and Bihar. The petals yield an essential oil which is used in perfumery and cosmetics.

35. *Rosa damascena;* Eng. Damask rose; Verna–**Gulab**–They are cultivated chiefly in Aligarh, Ghazipur and Kanauj of Uttar Pradesh. An essential oil known as 'otto of roses' or 'oil of roses' obtained from the petals is used in perfumery. The other important rose products are–rose water, attar, gulkand, pankhuri. Gulkand is used as a tonic and laxative. The dried rose petals are known as pankhuri which are used for preparing cold drinks.

36. *Rosa banksiae;* Eng. Banksia rose–This is climbing herb. This is native of China. It is grown as an ornamental in the gardens.

37. *Crataegus oxycantha;* It is found in the Himalayas from 6,000 feet to 9,000 feet elevation a.s.1. The wood is hard and, therefore, is used in making tool handles.

38. *Potentilla fruticosa*–This is a small herb. The leaves possess medicinal properties and are used as substitute for tea.

39. *Potentilla nepalensis*–This is medicinal plant; used as a remedy of burns.

40. *Potentilla reptans*–The infusion of the herb is used as a remedy of diarrhoea.

41. *Cotoneaster acuminata;* Verna–**Riu**–This is a shrub found in Himalyas. The wood is used for making walking sticks.

42. *Chrysobalanus icaco;* Eng. Coco plum–This is a shrub or small tree, introduced into Malabar. The fruits are edible.

43. *Cydonia oblonga;* Syn. *Cydonia vulgaris;* Verna–**Bihi**–This is a small tree. It is native of West Asia, now grown in Punjab, Kashmir and the Nilgiris for the edible fruits. It is used as a root--stock for pear.

Systematic Relationships : According to Engler the order Rosales consists of about twenty families. Among these families there are very few points of similarity. Hutchinson has splitted the order into a number of distinct orders.

14

ORDER–LEGUMINALES

According to Hutchinson this is the seventh order of the phylum Angiospermae, sub-phylum Dicotyledones and division Lignosae. This order consists of only three families, *i.e.,* Caesalpiniaceae, Mimosaceae and Papilionaceae. The Papilionaceae is the highest evolved family within the order. All the three families have been described in detail in this text. Bentham and Hooker have treated these families as sub-families of the family Leguminosae. They have placed the Leguminosae in the eleventh order — the Rosales, of class Dicotyledones, division Polypetalae and series Calyciflorae. Engler and Prantl have also adopted Bentham and Hooker. They have placed the family Leguminosae with its three sub-families, *i.e.,* Caesalpinioideae, Mimosoideae and Papilionatae in the twentieth order the Rosales of class Dicotyledonae and sub-class Archichlamydeae.

Progressive group with increasing zygomorphy of the flower; free to variously united stamens; carpel one; no endosperm, the embryo filling the seed.

Family–Caesalpiniaceae

There are about 135 genera in the family Caesalpiniaceae.

Distribution: Most of the members belonging to this family are tropical and sub-tropical in distribution. In India the family is represented by many genera, *e.g., Cassia, Bauhinia, Tamarindus, Saraca, Poinciana, Parkinsonia, Caesalpinia* etc.

Habit : The plants show great variations in their habit, they may be trees or shrubs and very rarely herbs. Mostly they are mesophytes, but xerophytes (*e.g., Parkinsonia*) are also reported. Most of them are wild, but many are cultivated for their beautiful flowers and timber. Some plants are woody climbers (*e.g., Bauhinia* sp.).

Root : Tap and branched.

Stem : Erect, woody, cylindrical, solid, branched, sometimes herbaceous or climbing. In many species tannin sacs and gum passages are found.

Leaves: The leaves may be simple or compound. If compound they may be pinnate or bi-pinnate. Usually the pinnate leaves are arranged in pairs; petiolate, pulvinus present at the base of the petiole; the pinna ovate or obovate, glabrous, net veined, entire. Usually exstipulate, sometimes minute caducous stipules present.

Inflorescence: Usually racemose, raceme, sometimes pendulous.

Flower: Pedicellate, zygomorphic rarely actionomorphic, hermaphrodite, hypogynous, complete, variously coloured, showy, large or small.

Calyx : Five sepals, free or fused, often petaloid, aestivation imbricate or valvate.

Corolla: Five petals, polypetalous(free); ascending imbricate aestivation, *i.e.*, the posterior petal innermost in bud; inferrior, spathulate, showy.

Androecium: Stamens usually ten, free, rarely connate, but, sometimes reduced to staminodes or altogether abortive. In *Cassia* 3 to 5 stamens reduced to staminodes, in *Tamarindus* sp. 3 stamens are well developed, while rest are reduced into staminodes.

Gynoecium: Carpel one (monocarpellary); ovary superior, unilocular; marginal placentation; style long; stigma simple.

Fruit: Pod or lomentum, sessile or stalked, dehiscent or indehiscent.

Seed : Exalbuminous.

Pollination : Entomophilous.

Floral Formula: ⊕ or ·|· ⚥ K5, or (5) C5, A 10 $\underline{G.1.}$

Bentham & Hooker (1862)	**Engler & Prantl (1931)**	**Hutchinson(1959)**
Phanerogams	Phanerogams	Angiospermae
Dicotyledones	Dicotyledones	Dicotyledones
Polypetalae	Archichlamydeae	Lignosae
Calyciflorae	Rosales	Leguminales
Rosales	Leguminosae	Caesalpiniaceae
Caesalpinioideae		

1. Cassia tora Linn. Verna. **Pamaar, Chakunda;** Eng. Sickle senna.

Habit : A medium size tree or shrub.

Root : Tap, branched.

Stem : Erect, aerial, woody, branched, cylinderical, glabrous, solid.

Leaves: Cauline and ramal, compound, alternate, paripinnate, pulvinus at the base, petiolate, stipulate (stipules caducous), leaflets 4-8 pairs, ovate, entire, acute, glabrous, venation unicostate reticulate.

Inflorescence : Racemose, typical raceme.

Flower : Pedicellate (long pedicels), bracteate (bracts minute and caducous) or ebracteate, hermaphrodite, actinomorphic, complete, hypogynous, yellow, pentamerous, cyclic.

Calyx: Five sepals, polysepalous, more or less petaloid (yellowish green), inferior, quincunicial aestivation, odd sepals anterior.

Corolla : Five petals, polypetalous, yellow, clawed, ascending imbricate aestivation, inferior.

Androecium: Ten stamens, polyandrous, unequal in length, three posterior stamens reduced to staminodes, basifixed anthers with abortive and indehiscent lobes, dithecous, introrse.

Gynoecium: One carpel (monocarpellary), ovary superior, unilocular, marginal placentation, style short, stigma terminal, hairy.

Fruit: A legume, black when ripe.

Floral formula: Br ·|· ⚥ K5, C5, A7 + 3, G $\underline{1}$.

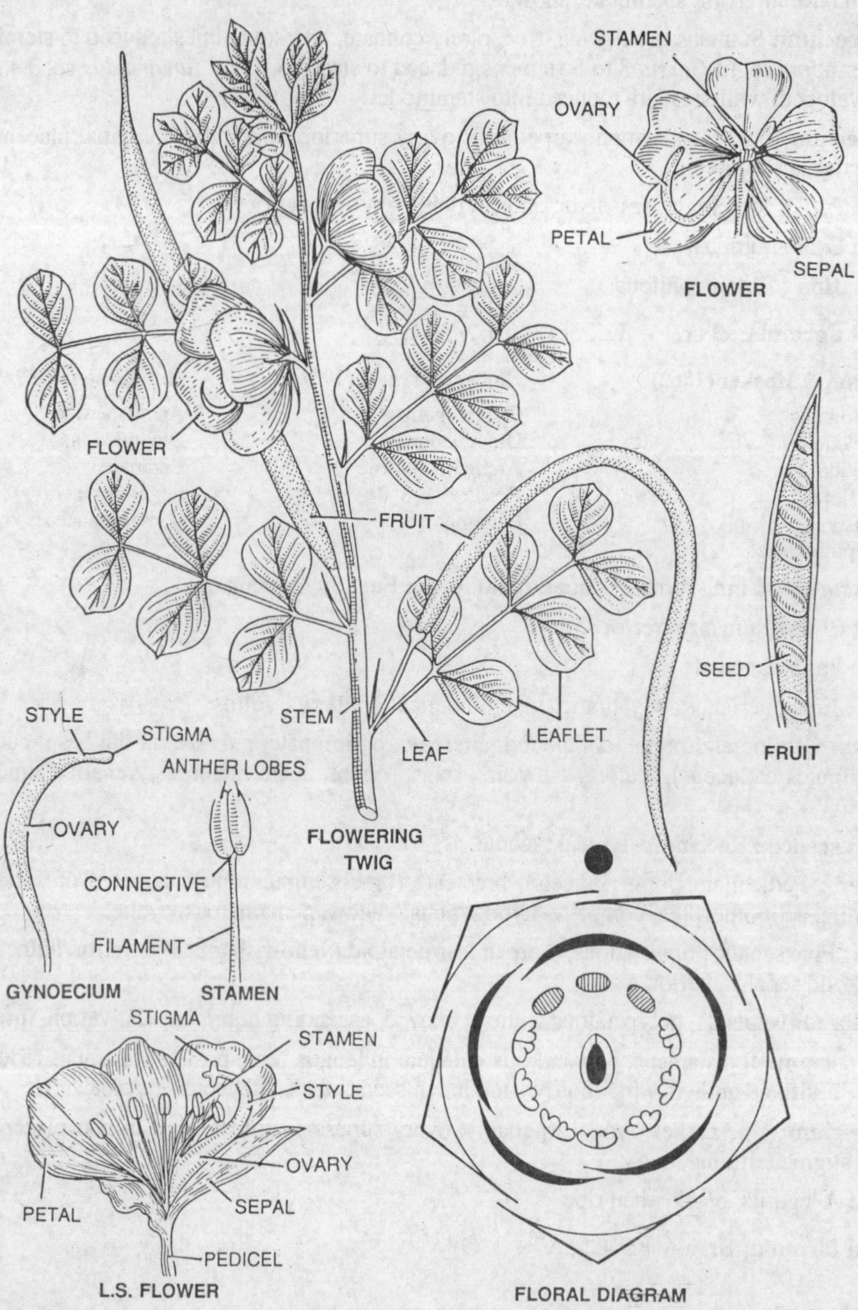

Fig.14.1: Caesalpiniaceae (Leguminosae). *Cassia tora* Linn.; Eng. Sickle senna; Verna. **Pamaar, chakunda.**

Economic value. The seeds are used in the preparation of sweets and the tender young leaves are used as vegetable. Decoction of leaves is laxative. Leaves and seeds are used in skin diseases, for ringworm and itch. Root is used in snake-bite.

Cassia fistula; Verna–**Amaltas.**

Habit: A medium sized tree.

Root: Tap branched,

Stem: Erect, woody, branched, cylindrical, glabrous, solid.

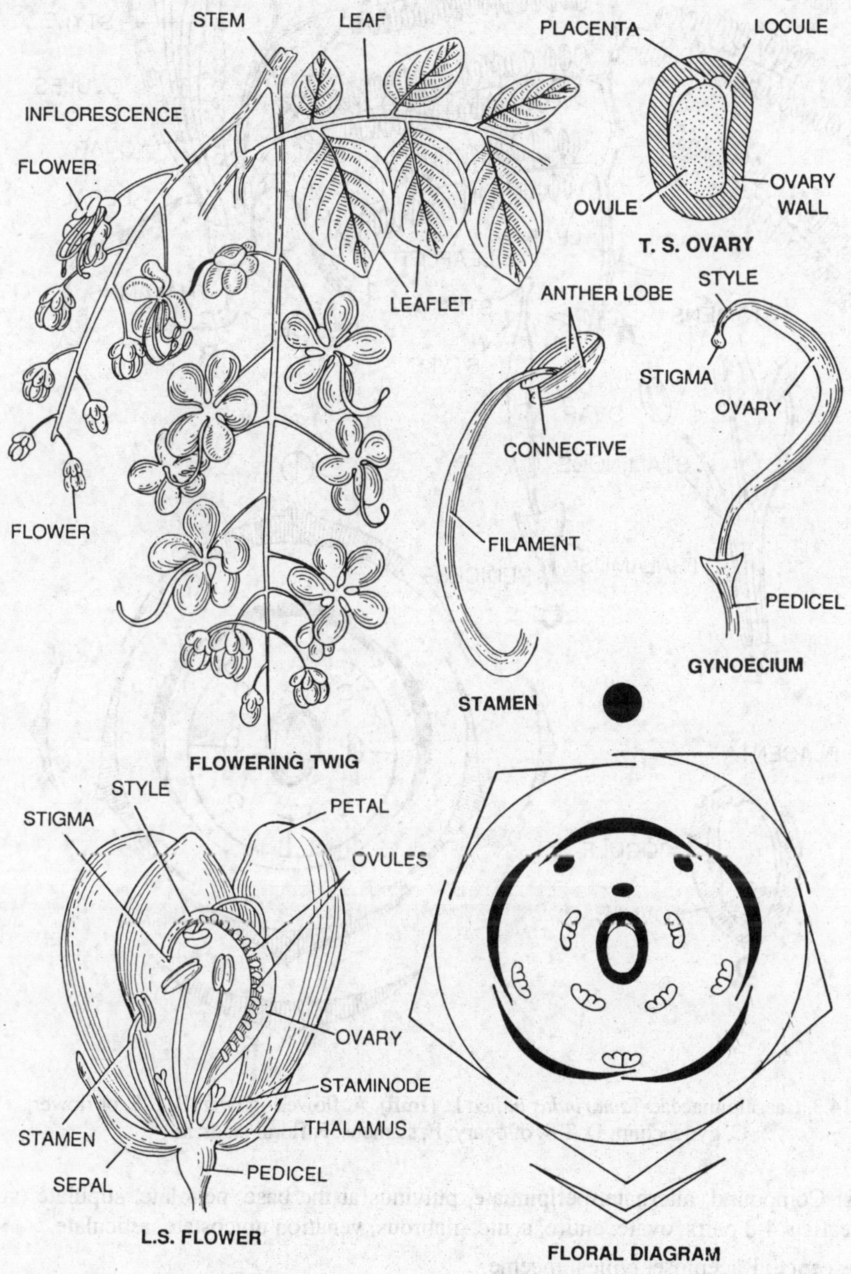

Fig. 14.2. Caesalpiniaceae—*Cassia fistula* Linn; Eng., golden shower; Verna., **amaltas.**

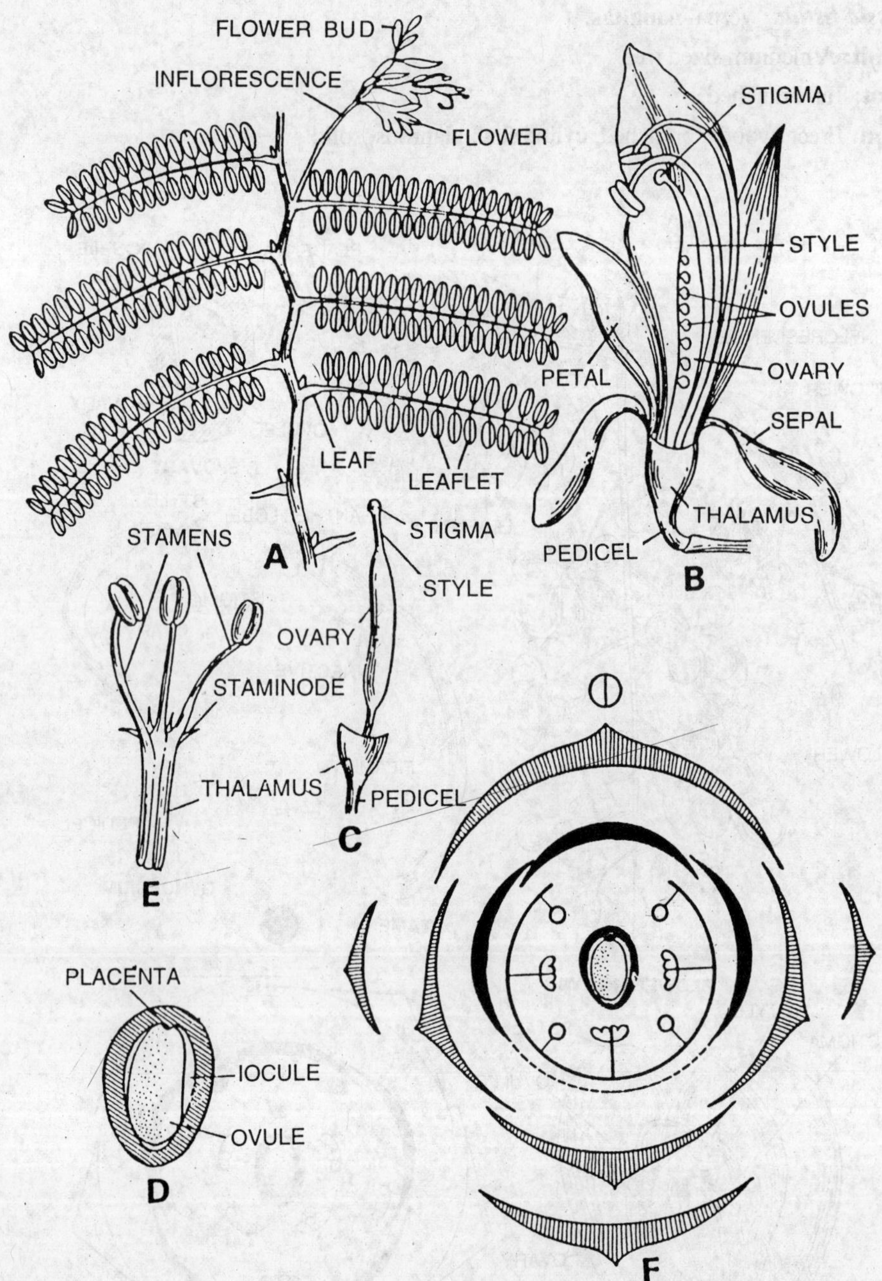

Fig. 14.3. Caesalpiniaceae-*Tamarindus indica* L. (**Imli**). A, flowering twig; B, L.S. of flower; C, gynoecium; D, T.S. of ovary; E, stamens; F, floral diagram.

Leaves: Compound, alternate, peripinnate, pulvinus at the base, petiolate, stipulate (stipules caducous), leaflets 4-8 pairs, ovate, entire, acute, glabrous, venation unicostate, reticulate.

Inflorescence: Racemose, typical raceme.

Flower: Pedicellate (long pedicels), bracteate (bracts minute and caducous) hermaphrodite, actinomorphic, complete, hypogynous, yellow.

Calyx: Five sepals, polysepalous, more or less petaloid (yellowish green), inferior, quincuncial aestivation, odd sepals anterior.

Corolla: Five petals, polypetalous, yellow, clawed, ascending imbricate aestivation, inferior.

Androecium: Ten stamens, polyandrous, unequal in length, three posterior stamens reduced to staminodes, basifixed anthers with abortive and indehiscent lobes.

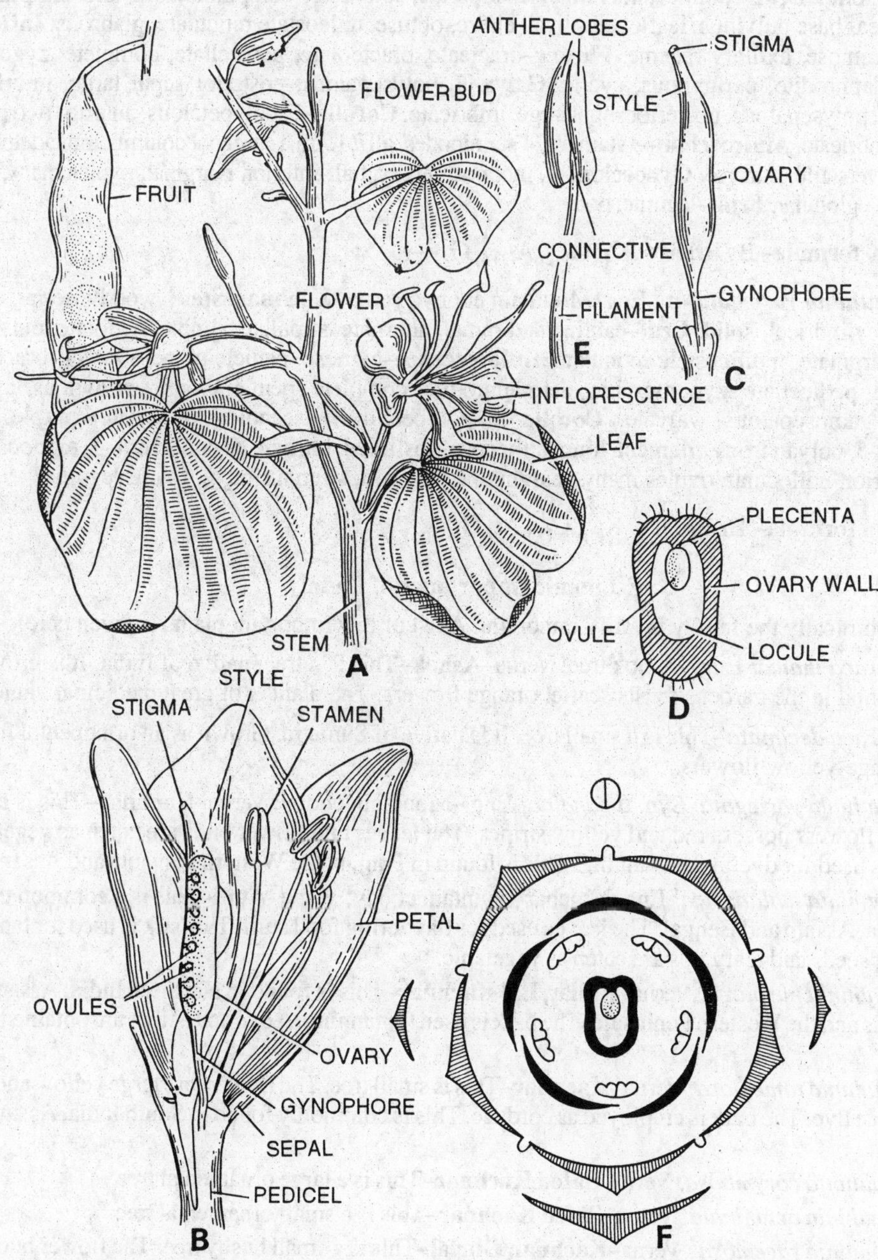

Fig. 14.4. Caesalpiniaceae—*Bauhinia variegata* (**Kachnar**). A, flowering twig and fruit; B, L.S. of flower, C, gynoecium; D, T.S. of ovary; E, stamen, F, floral diagram

Gynoecium: One carpel (monocarpellary), ovary superior, unilocular, marginal placentation, style short, stigma terminal, hairy.

Fruit: A legume, 1-2 feet long, cylindrical, black when ripe.

Floral Formula : Br ·|· ⚥ K 5, C 5, A 7 + 3, G $\underline{1}$.

3. ***Tamarindus indica*** L., Eng. Tamarind; Verna. **Imli. Stem**–woody, aerial, erect, cylindrical, branched, solid. **Leaf**–cauline and ramal, exstipulate, alternate, compound, uni–and paripinnate, petiolate, leaf base pulvinus, leaflet elliptical, entire, obtuse, unicostate reticulate, glabrous. **Inflorescence.** Racemose, axillary raceme. **Flower**–bracteate, bracteolate, pedicellate, complete, zygomorphic, hermaphrodite, perigynous, cyclic. **Calyx–5,** polysepalous, posterior sepal large, imbricate. **Corolla–5,** polysepalous, posterior sepal large, imbricate. **Corolla–5**, polypetalous, anterior two petals reduced, imbricate. **Androecium–3** stamens, 4 staminodes, all 7 form a staminal column, monadelphous, dithecous, versatile, introrse. **Gynoecium–1,** monocarpellary, half-inferior, marginal, ovules many, style long, stigma globular. **Fruit**–legume, pod.

Floral formula– Br Brl ·|· ⚥ K_4, C_{3+2}, $A_{(3+4)}$ G1—

2. ***Bauhinia variegata.*** L., Eng. Mountain ebony; Verna. **Kachnar. Stem**–woody, aerial, erect, branched, cylindrical, solid. **Leaf**–cauline and ramal, alternate, stipulate, simple, petiolate, pulvinus, entire, emarginate, multicostate reticulate. **Inflorescence**–cymose, panicle cyme. **Flower**–bracteate, bracteolate, pedicellate, zygomorphic, hermaphrodite, complete, pentamerous, perigynous, cyclic. **Calyx**–5, gamosepalous, valvate. **Corolla**–5, polypetalous, ascending imbricate, purple. **Androecium**–5, polyandrous, filament, long, dithecous, basifixed, introrse. **Gynoecium**–1, monocarpellary, superior, unilocular, ovules many, marginal, style short, stigma simple, **Fruit.** legume.

Floral formula– Br Brl ·|· ⚥ $K_{(5)}$, $C_{(5)}$, A_5, G_1—.

Economic Importance of Family

Economically the family is fairly important. A list of few important plants is given here.

1. *Saraca indica;* Eng.–Ashok tree; Verna–**Ashok**–This is a tree; native of India. It is grown as an ornamental in the gardens for its scarlet orange flowers. The plant is of great medicinal value.

2. *Saraca declinata*–This is a small tree. It is native of Sumatra. Grown as an ornamental for the bright orange-yellow flowers.

3. *Bauhinia variegata;* Syn. *B candida;* Eng.–Mountain ebony; Verna–**Kachnar**–This is a tree. The white flowers possess red and yellow stripes. The leaves and flower buds are eaten as vegetable. The bark is used for dyeing and tanning. This is found in Punjab, the Western Penisula and Assam.

4. *Bauhinia malabarica;* Eng–Malabar mountain ebony; Verna–**Amli–**This is a common tree of South India, Assam and Bengal. The leaves used for flavouring food stuff. The bark is used for tanning. The tender seeds and leafy tips are eaten as vegetable.

5. *Bauhinia purpurea;* Verna–**Kaliar, Lal Kachnar–**This is found in Northern India. Assam, the Khasia hills and the Western Peninsula. The bark is used for tanning. A dye and a fibre are obtained from its bark.

6. *Bauhinia tomentosa;* Verna–**Kachnar**–This is small tree. The flowers are large yellow and possess a purple dye. The bark is employed as cordage. This is commonly found in South India, Assam and Bihar.

7. *Bauhinia corymbosa;* Verna–**Safed Kachnar**–This is a large ornamental tree.

8. *Bauhinia acuminata;* Verna–**Safed Kachnar**–This is a small ornamental tree.

9. *Bauhinia racemosa ;* Verna–**Kachnar, Gurial**–This is a small bushy tree. The flower buds and fruits are eaten as vegetable. The bark yields a fibre.

10. *Bauhinia galpinii*–This is a straggling or prostrate shrub. This is native of South and tropical Africa. This is grown in the gardens as an ornamental for its bright scarlet flowers.

11. *Bauhinia retusa;* Verna–**Kandla**–This is found in Bihar, Madhya Pradesh and Punjab. It yields a gum which is used in paper and textile industry.

12. *Bauhinia vahlii;* Verna–**Malijhan, Jallur**–This is a large climber. It is found in Assam, Madhya Pradesh, Bihar and the Punjab. The bark yields a fibre which is used for making ropes. The bark is also employed for tanning.

13. *Bauhinia monandra;* Eng.–Butterfly flowers–This is a small tree grown as a hedge plant.

14. *Haematoxylon campechianum;* Eng–Logwood; Verna–**Patang**–This is an ornamental thorny tree. It is native of Central America and West Indies. The heartwood contains haeamotoxylin, which is used in dyeing and as nuclear stain.

15. *Cassia fistula;* Verna–**Amaltas**–This is a tree, grown as an ornamental. The pulp of the fruits is used as a purgative and laxative.

16. *Cassia alata;* Verna–**Dadmurdan–** This is a small shrub. It is found in Tamil Nadu, Bengal. The leaves are used for ringworms and skin diseases. Also grown as an ornamental.

17. *Cassia grandis*–This is an ornamental and grown as a hedge plant in the gardens.

18. *Cassia angustifolia;* Verna–**Sena**–This is native of Arabia but now cultivated in Tamil Nadu and Karnataka. The leaves and fruits are used as a laxative and purgative.

19. *Cassia artemisioides*–This is a shrub grown as an ornamental.

20. *Cassia javanica;* Verna–**Java–ki-Rani**–This is small tree, grown as an ornamental for its pink flowers. It is native of Malaya.

21. *Cassia auriculata;* Verna–**Tarvar**–The bark is employed for tanning. It is found in Madhya Pradesh, the Western Peninsula, Tamil Nadu and Rajasthan.

22. *Cassia nodosa*–This is grown as an ornamental. The flowers are pink.

23. *Cassia siamea*–This is an ornamental tree. The flowers are used as vegetable.

24. *Cassia tora;* Verna–**Pamaar, Chukunda**–This is a herb or undershrub, found throughout our country. The seeds are employed in the preparation of sweets. The young leaves are eaten as vegetable.

26. *Cassia sophera;* Verna.–**Kasaundi**–This is herb grown as an ornamental in the gardens.

26. *Cassia renigera;* Eng–pink cassia–this is an ornamental grown as a hedge plant.

27. *Caesalpinia bonducalla ;* Verna–**Katkaranj, Karanju**–This is a prickly shrub. The seeds are mixed with black pepper, and used as a tonic. The bark is also used as a tonic. The seeds are used as beads of necklaces, bracelets and rosaries.

28. *Caesalpinia digyna;* Verna–**Vakeri–mul**–This is a climbing shrub. The roots and fruits are employed for tanning. It is found in Bengal, Assam and Andamn islands.

29. *Caesalpinia bonduc;* Eng.–Fever nut–This is a scandent prickly shrub, cultivated as a hedge plant. The seeds are used medicinally to cure diarrhoea and rheumatism.

30. *Caesalpinia crista;* Syn.–*C. nuga*; Verna,– **Karanju, Naktamala**–This is a woody climbing shrub, commonly found in South India and Bengal. A kind of fat is obtained from the seeds which is used for making cosmetics.

31. *Caesalpinia sappan*; Verna–**Patang**–This is native of India and Malaya. It is now cultivated in Bengal and South India. A red dye is obtained from its heart wood. This dye is used for colouring cotton and wooleen cloth. The red ink is also prepared from it.

32. *Caesalpinia coriaria*; Verna–**Libi dibi**–This is a small tree. It is native of South America and West Indies. It is found abundantly in South India. It is also cultivated in North–Western India and Bengal. The dried fruits are used for tanning.

33. *Caesalpinia giliiesii*; Eng–Bird of paradise–This is a shrub. It is native of South America. It is grown as an ornamental.

34. *Poinciana pulcherrima*; Syn.–*Caesalpinia pulcherrima;* Eng–Peacock flower; Verna.–**Gulmohur**–This is a shrub or small tree, cultivated as an ornamental.

35. *Delonix rigia*; Syn–*Poinciana regia*; Eng.–Peacock flower; Verna.–**Gulmohur**–This is a tree. It is native of Madagascar but now grown in our country as an ornamental for the large red flowers.

36. *Delonix elata;* Syn–*Poinciana elata*; Eng–white gulmohur; Verna.–**Sankesar**–This is a small tree; native of Abyssinia. It is grown as an ornamental for its yellow–white flowers.

37. *Parkinsonia aculeata;* Verna–**Vilayati Kikar**–This is a small tree. It is native of Panama. It is grown as a hedge plant. The flowers are yellow.

38. *Tamarindus indica*; Eng–Tamarind; Verna–**Imli**–This is a tree found throughout our country. It is grown as an avenue. The fruits are edible and are also used as carminative, and laxative. The bark and leaves are employed in tanning. The seeds yield jellose which is used for sizing jute and cotton. The polyose obtained from the seeds, is good substitute for food pectins. The unripe fruits are rich source of tartaric acid. Tartaric acid and its salts are extensively used in various foods, chemical and pharmaceutical industries.

Family—Mimosaceae

The family Mimosaceae includes about 40 genera, and at the same time has been considered to be the smallest one of all the three families of the order Leguminales.

Distribution: The plants of this family are almost exclusively tropical or sub–tropical in distribution. In our country this family is represented by many important genera, *e.g., Acacia, Mimosa, Neptunia, Prosopis, Senegel,* etc.

Habit : They are either shrubs or trees very rarely herbs, sometimes climbers. Many of them are thorny and xeromorphic, *e.g., Acacia, Prosopis;* hydrophytes are also found, *e.g., Neptunia oleracea.* Mostly wild, but some are grown as sand binders to check soil erosion and desert extension.

Root : Tap and branched, deep rooted in the soil.

Stem: Erect, branched, terete, woody, solid. The stem tissue is often rich in tannin–sacs and gum passages.

Leaves: Alternate, petiolate, usually the base of petiole is provided with pulvinus; stipulate, usually stipules are modified into thorns; compound, pinnate and generally bipinnate. Leaf movements are common, reaching a climax in the ultra–sensitivity of *Mimosa pudica* and *Neptunia oleracea.* In some *Acacias* from Australia the leaves are bipinnate and fall earlier. The petioles are being flattended and developing into phyllodes.

Inflorescence: Racemose, head or spike.

Flower: Usually sessile, actinomorphic, regular, hermaphrodite, hypogynous, complete, small.

Calyx: 5 or 4, gamosepalous, the sepals are more or less connate, green (sepaloid), small, inferior, aestivation generally valvate.

Corolla: 5 or 4, free, polypetalous, valvate, inferior, slightly united towards the base, usually pentamerous.

Androecium: Usually indefinite stamens, sometimes the number is reduced to ten or even four (*e.g.,* in *Mimosa pudica*). Stamens are conspicuous, attractive, bright coloured somewhat scented. Anthers 2–celled, dorsifixed, dehiscence by longitudinal slit. The filaments are long and slender.

Gynoecium: Carpel one (monocarpellary); ovary superior, unilocular; marginal placentation; style long filiform; stigma terminal and simple.

Fruit; Legume or lomentum.

Seeds: Exalbuminous.

Pollination: Usually through the agency of insects (entomophily).

Floral formula: $\oplus$ ⚥ K (5), C(5), A ∞, $\underline{G1}$.

Bentham and Hooker (1862)	Engler and Prantl (1931)	Hutchinson (1959)
Phanerogams	Phanerogams	Angiospermae
Dicotyledones	Dicotyledones	Dicotyledones
Polypetalae	Archichlamydeae	Lignosae
Calyciflorae	Rosales	Leguminales
Rosales	Leguminosae	Mimosaceae
Leguminosae	Mimosoideae	
Mimosoideae		

Two important plants naming *Acacia nilotica* and *Mimosa pudica* have been described in this text in semi–technical language:

Acacia nilotica; Verna—**Babul, Kikar.**

Habit: A medium sized spiny tree.

Root : Tap, branched.

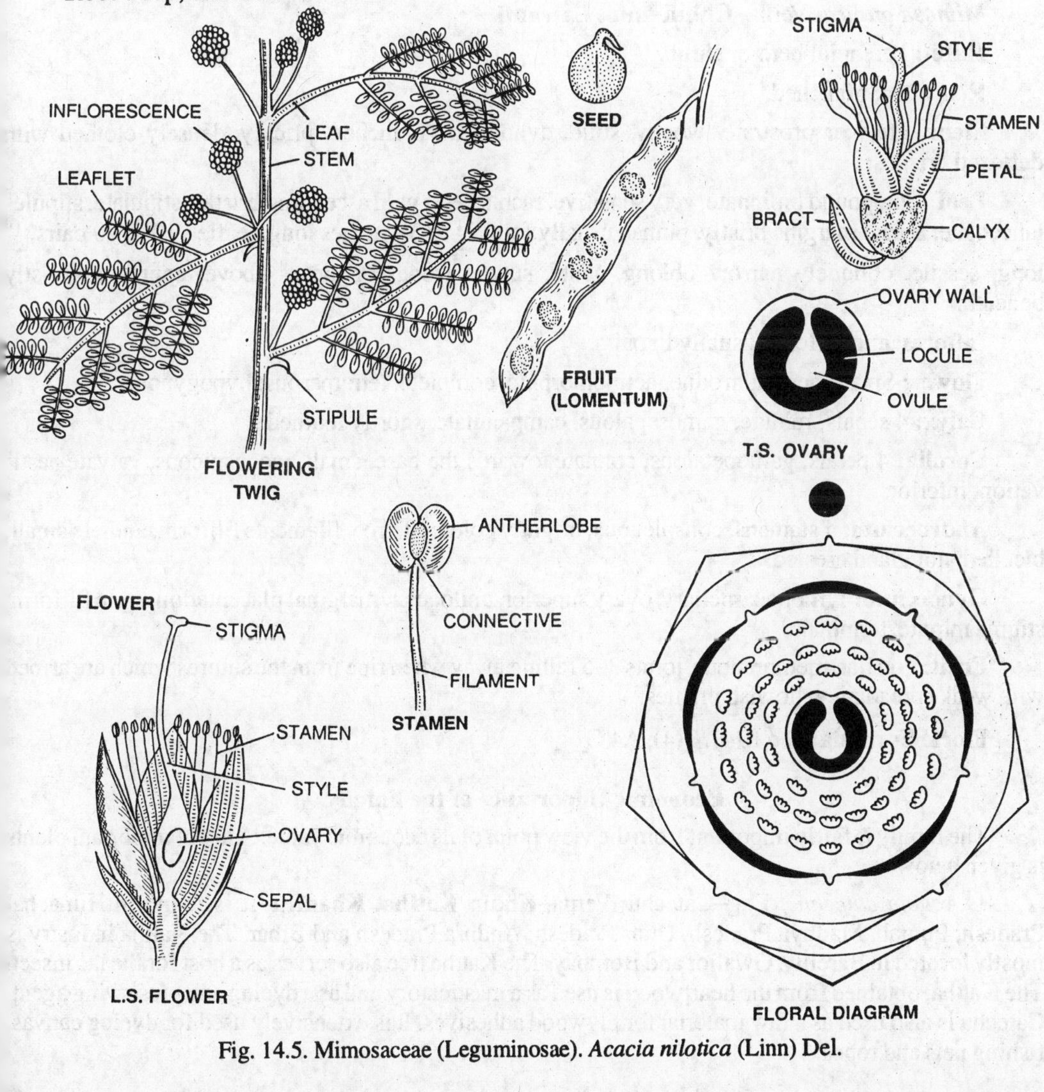

Fig. 14.5. Mimosaceae (Leguminosae). *Acacia nilotica* (Linn) Del.

Stem : Erect, woody, solid, cylindrical, branched.

Leaf : Compound, bippinnate, stipulate, stipules modified into spines, petiolate, rachis possesses glands, leaflet oval, entire, obtuse, unicostate reticulate venation.

Inflorescence: The flowers are arranged in compound cymose heads.

Flower: Sessile, bracteate, actinomorphic, hermaphrodite, hypogynous, complete, fragrant small yellow.

Calyx : Sepals 4 or 5, minute, gamosepalous, campanulate, valvate aestivation, inferior.

Corolla : Petals 4 or 5, gamopetalous, valvate aestivation, inferior.

Androecium : Stamens indefinite, polyandrous, long filaments, minute anthers, introrse, yellow.

Gynoecium: Monocarpellary, ovary superior, unilocular, marginal placentation, style long, stigma minute.

Fruit : A lomentaceous pod.

Floral Formula : ⊕ ⚥ K(4 – 5), C(4 – 5), A∝ $G\underline{1}$.

Mimosa pudica, Verna–**Chhui–mui, Lajvanti.**

Habit: Perennial herb or shrub.

Root: Tap, branched.

Stem: Erect or prostrate, woody, solid, cylindrical, branched, prickly, densely clothed with deflexed bristles.

Leaf: Compound bipinnate, very sensitive, rachis beset with ascending bristles stipulate, stipules lanceolate, striate margins bristly, pinnae usually four, $2\frac{1}{2}$ to $3\frac{1}{3}$ inches long, leaflet, 12 to 20 pairs, $\frac{1}{2}$" long, sessile, obliquely narrow–oblong, acute, subcoriaceous, glabrous above, appressed–bristly beneath.

Inflorescence: Heads usually in pairs.

Flower: Small, hermaphrodite, actinomorphic, complete, tetramerous, hypogynous.

Calyx: 4 sepals, minute, gamosepalous, campanulate, shortly toothed.

Corolla: 4 petals, gamopetalous, connate towards the base, small, inconspicous, valvate aestivation, inferior.

Androecium: 4 stamens, conspicuous, brightly coloured, free, filaments filiform, anthers small, bicelled, not gland crested.

Gynoecium: 1, monocarpellary, ovary superior, unilocular, marginal placentation, style filiform, stigma minute, terminal.

Fruit: Pod flat, membranous, joints 3–5 falling away when ripe from the sutures which are armed with weak spreading yellowish bristles.

Floral Formula: ⊕ ⚥ K(4), C(4), A4, $G\underline{1}$.

Economic Importance of the Family

The family is fairly important from the view point of its economic value. A list of important plants is given below:

1. *Acacia catechu ;* Eng–Catechu; Verna–**Khair, Kattha, Khadira**–It is found in Himachal Pradesh, Punjab, Madhya Pradesh, Uttar Pradesh, Andhra Pradesh and Bihar. The Kattha industry is mostly located in Bareilly, Gwalior and Bombay. The Kattha tree also serves as a host for the lac insect. The Kattha, obtained from the heartwood is used as a masticatory and as a dyeing and preserving agent. Catechu is also used as a raw material for plywood adhesives. It is extensively used for dyeing canvas, fishing nets and ropes.

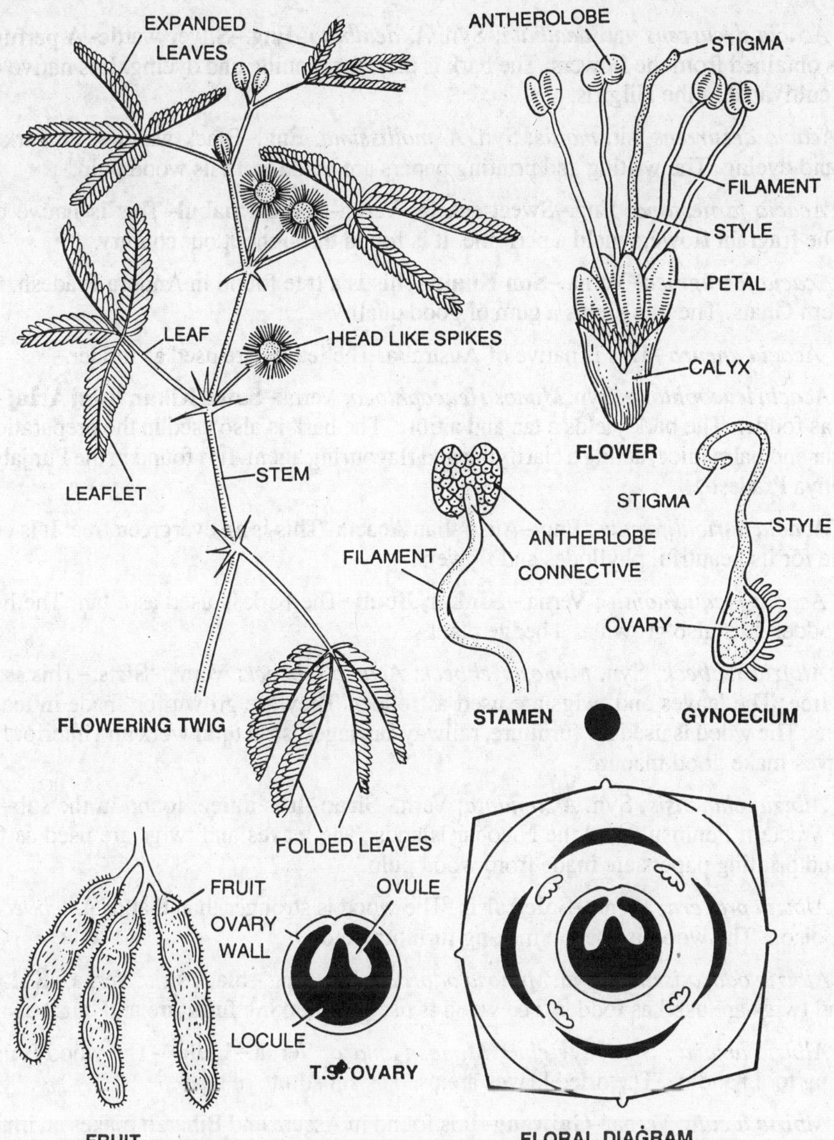

Fig. 14.6. Mimosaceae (Leguminosae). *Mimosa pudica* Linn.

2. *Acacia concinna ;* Eng–Soap pod; Verna–**Shikakai, Banritha**–The fruits (pods) are used as a detergent for washing hairs, silk and woollen fabrics. It is found along the West Coast, Andhra Pradesh, Assam and Bihar.

3. *Acacia nilotica;* Syn. *A. arabica;* Eng.–Acacia; Verna–**Babul, Kikar**–This is a tree. The pods (fruits) are used as fodder. The bark is used as a tan. The gum obtained from the bark is of good quality, which is used in confectionery. Baskets are made from young, green twigs. The twigs are used as tooth greenbrushes(Datun). The bark is employed in tanning.

4. *Acacia senegel*; Syn. *A. verek;* Verna.–**Kumta**–This is native of North Africa but now found in the Punjab and Rajasthan. The stem bark, yields a gum which is used in the textile, mucilage, polish, paste and confectionery industries. This is also used as a glaze in painting.

5. *Acacia pycnantha;* Eng.–Goldern wattle–This is native of Australia, but now cultivated in the Nilgiris. The extract of bark is used as **Kattha.**

6. *Acacia decurrens* var. *dealbata*; Syn. *A. dealbata*; Eng.–Silver wattle–A perfume of good quality is obtained from the flowers. The bark is used for tanning and dyeing. It is native of Australia but now cultivated in the Nilgiris.

7. *Acacia decurrens* var. *mollis*; Syn. *A. mollissima.* Eng.–Black wattle–The bark is used for tanning and dyeing. The writing and printing papers are made from its wood pulp.

8. *Acacia farnesiana;* Eng.–Sweet acacia; Verna–**Gandh babul**–This is native of the West Indies. The fragrant flowers yield a perfume. It is found throughout our country.

9. *Acacia ferruginea;* Verna.–**Son Khair**–This is a tree found in Andhra Pradesh, Gujarat and the Western Ghats. The tree yields a gum of good quality.

10. *Acacia aneura*–This is native of Australia. The leaves are used as fodder.

11. *Acacia leucophloea;* Syn. *Mimosa leucophloea;* Verna–**Safed Kikar, Ronj Arinj**–The leaves are used as fodder. The bark yields a tan and a fibre. The bark is also used in the preparation of spirits from sugar and palm juice, acting a clarifying and flavouring agent. It is found in the Punjab, Rajasthan and Madhya Pradesh.

12. *Acacia auriculiformis* ; Eng.–Australian Acacia–This is an evergreen tree. It is cultivated as an avenue for its beautiful phyllodes and flowers.

13. *Acacia jacquemontii* ; Verna.–**Kinkar, Bouli**–The bark is used as a tan. The branches are used as fodder. It is also grown as a hedge plant.

14. *Albizia lebbeck*; Syn. *Mimosa lebbeck*; *Acacia lebbeck*; Verna.–**Siris.**–This is a common roadside tree. The leaves and twigs are used as fodder. They are grown for shade in tea and coffee plantations. The wood is used for furniture, railway carriages, structural work and interior fittings. The fallen leaves make good manure.

15. *Albizia chinensis*; Syn. *A stipulata*; Vern.–**Siran**–It is a tree, found in the sub–Himalayan tract, the Western Peninsula and the Nicobar islands. The leaves and twigs are used as fodder. The writing and printing papers are made from wood pulp.

16. *Albizia procera;* Verna.–**Safed siris**–The wood is stronger than that of *A. lebbeck.* The bark yields a poison. The wood is used in making furniture, etc.

17. *Albizia odoratissima;* Syn. *Mimosa odoratissima*; Eng.–black siris; Verna.–**Kala siris.**–The leaves and twigs are used as fodder. The wood is used for making furniture and burnt as fuel.

18. *Albizia amara ;* Syn. *A. wightii; Mimosa amara;* Verna.–**Lallei.**–The wood is used for fuel and making tool handles. The dried leaves are used as substitute of soap.

19. *Albizia lucida*; Verna.–**Galwang**–It is found in Assam and Bihar. It makes an important host of lac insect.

20. *Mimosa pudica*; Eng.–Sensitive plant; Verna.–**Lajwanti, Chhui–mui.**–A spreading under-shrub, grown as an ornamental.

21. *Mimosa rubicaulis*; Verna.–**Shiah–Kanta, Aila**–This is a shrub, grown as a hedge plant.

22. *Neptunia oleracea*; Syn. *N.prostrata*; Verna.–**Lajalu**–A common aquatic herb. It is used as a vegetable.

23. *Prosopis juliflora;* Syn. *Mimosa juliflora;* Verna. –**Kabuli kikar**–This is a tree. It is native of Mexico and Central America. It is grown as a hedge plant. It is also being planted to check the extension of the desert. The fruits are used as fodder. This is also planted as windbreak.

24. *Prosopis spicigera*; Verna–**Chhonkar Jand**–This is grown as a hedge plant.

25. *Entada phaseoloides*; Syn. *E. scandens*; *Lens phaseoloides*; Verna.–**Barabi, Chian, Gilla**–This is a large climbing shrub. The seeds are edible. They are also used as a substitute of soap.

26. *Xylia xylocarpa*; Syn. *Xylia dolabriformis* ; Verna–**Jambu**–This is a tree found in Madhya Pradesh, Andhra Pradesh and orissia. The wood is used for construction work, railway–sleepers, ship–building and bridges.

27. *Pithecolobium dulce*; Syn. *Mimosa dulcis; Inga dulcis*; Eng.–Manila tamarind; Verna.–**Jangal jalebi**–This is a tree cultivated throughout India for its edible fruits. They are also grown as hedge plants.

Family–Papilionaceae

The family Papilionaceae is the largest of the three families of the order Leguminales. This includes about 375 genera.

Distribution: The members of the family are being distributed in the temperate regions of both northern and southern hemispheres. In our country the Papilionaceae is represented by many important genera commonly found in hills and plains. Usually the herbs belonging to this family appear and flower in winter season in the Indian plains.

Habit: Generally the plants are herbs or shrubs (often climbing), very rarely trees. Sometimes they are hydrophytes and xerophytes. Most of the members are cultivated in the fields which yield pulses.

Root: The root is tap, branched, bearing nodules containing nitrogen fixing bacteria (*Rhizobia),* and therefore, the plants are being used for green manuring.

Stem: The stem is erect or climbing (it climbs by means of tendrils), branched, angular or cylindrical herbaceous or woody.

Leaves: The leaves are alternate, opposite or whorled, usually compound (digitate or pinnate), rarely even pinnate, sometimes simple, stipulate, stipules occur at the base of the petiole (foliaceous and large), sometimes secondary stipules arise at the base of individual leaflets.

Inflorescence: The inflorescence is racemose, the raceme, the spike, and the contracted raceme or head.

Flower: The flowers are usually pedicellate, zygomorphic, irregular, hermaphrodite (bisexual), complete, perigynous and papilionaceous.

Calyx: It consists of sepals, gamosepalous (united sepals), equal or unequal, below the disc united in a tubular calyx, 5 toothed or five lobed or bi–labiate (the two upper and three lower may unite). Aestivation ascending imbricate.

Corolla: The corolla consists of 5 petals (unequal), the uppermost and the largest petal is known as **standard** or **vexillum;** the two free lateral petals are known as **wings** or **alae**; the anterior pair of united petals is termed **keel** or **carnia**, this encloses stamens and pistil. They are of various colours. Venation is conspicuous. Aestivation descending imbricate.

Androecium: The stamens are usually 10, inserted on a disc below the calyx, they may be in two bundles (diadelphous) of 9 + 1 or 5 + 5, or in one bundle (monadelphous), rarely free. Anthers bi–celled, dorsifixed, dehiscence by longitudinal slits.

Gynoecium: Carpel one, free, ovary superior, stalked or sessile, unilocular, marginal placentation, style, bent and short, hairy, stigma simple.

Fruit: legume or pod, splitting along both dorsal and ventral sutures. In some species it is indehiscent lomentum.

Seeds: Many, exalbuminous, usually flattened.

Pollination: Entomophilous(insect pollinated)

Floral formula: ·|· ⚥ K(5), C 1+2+ (2), A(9)+1, $\underline{G1}$.

Bentham and Hooker(1862)	**Engler and Prantl (1931)**	**Hutchinson(1959)**
Phanerogams	Phanerogams	Angiospermae
Dicotyledones	Dicotyledones	Dicotyledones
Polypetalae	Archichlamydeae	Lignosae
Calyciflorae	Rosales	Leguminales
Leguminosae	Papilionaceae	
Papilionatae		

Some important plants have been described here in semi–technical language.

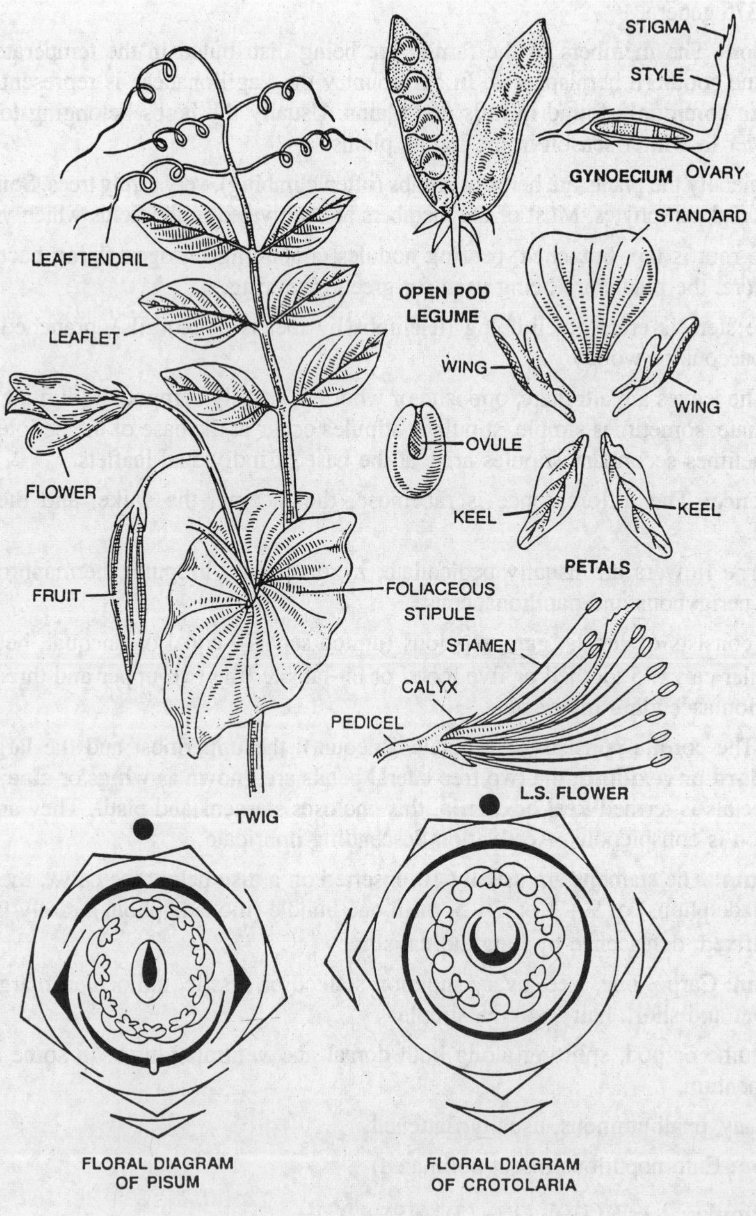

Fig. 14.7. Papilionaceae(Leguminosae)–*Pisum sativum* Linn.

Pisum sativum; Verna.–**Matar.**

Habit : Annual herb; cultivated.

Root : Tap, branched, with nodules containing nitrogen fixing bacteria (*Rhizobium radicicola).*

Stem : Herbaceous, weak, climbing with the help of leaf tendrils, cylindrical, branched, smooth, glaucous.

Leaves: Alternate, compound, imparipinnate, stipulate (stipules large foliaceous, ovate, semi-cordate, irregularly toothed at the base), leaflets 4 or 6, the common rachis ends in a branched tendril; the leaflets entire, smooth, net veined, oval to oblong, mucronate tips, green and glaucous, the terminal leaflet is always a tendril.

Inflorescence: Racemose, flowers arranged in axillary racemes or solitary.

Flower: Pedicellate, zygomorphic, irregular, hermaphrodite, papilionaceous, white or pink, complete, hypogynous to perigynous, bracteate or ebracteate.

Calyx: 5 sepals, gamosepalous, campanulate calyx tube, teeth long or the upper short; sepaloid, ascending imbricate aestivation.

Corolla: 5 petals, 1 standard, 2 wings, 2 keels united, keels shorter than wings and enclose the pistil and stamens; corolla papilionaceous, white or pink in colour; descending imbricate (vexillary) aestivation; inferior.

Androecium: 10 stamens in two bundles (diadelphous) of 9+1, nine stamens unite at the base and form a tube around ovary tenth is posterior and free; anthers bi–lobed, basifixed, introrse, dehiscence by longitudinal splitting.

Gynoecium: Carpel one (monocarpellary); ovary superior, unilocular; marginal placentation; ovules many; style bent and long, stigma simple; terminal and hairy; ovary also hairy.

Fruit: A legume (pod), broad.

Seeds: Rounded, uniform, white.

Generic Characters: Leaflets are even in number; terminal leaflet is represented by a tendril; stamens diadelphous (9 + 1); pods normal, flattened, not partitioned ; style dialated from the base towards the apex, hairy on one side; calyx lobes leafy.

Floral Formula: ·|· ⚥ K(5), C 1+2+(2), A(9)+1, $\underline{G}$1

Arachis hypogea; Verna–**Moongfali.**

Habit: Low annual herb; usually cultivated for its benefical fruits, used in several ways.

Root: Tap and branched; lateral branches are fewer near the surface of earth; root hairs usually absent; bacterial nodules present.

Stem: The central axis of the stem is erect. On the basis of the branching the plants can be subdivided into two groups(*i*) prostrate type, (*ii*) bush type. In the former, the laterals are prostrate from the very beginning, but in the latter the laterals in the beginning have a tendency towards erectness, but in later stages they also become prostrate as the former. The stem is cylindrical, hairy, short internodes, becomes somewhat angular and hollow as it becomes older.

Leaves: Pinnately compound, stipulate, usually the stipules are one inch long, linear and fused half of their length to the petiole; petiolate petiole long, slender, with a distinct swelling (pulvinus.); leaflets four, arranged in two opposite and equal pairs. 1 to 2½ inches long, entire, obovate to obtuse, netveined, pubescent.

Inflorescence: Flowers arranged either solitary or in groups of three in leaf axils. Mostly the flowers develop towards the base of the plant, and sometimes buried in the soil.

Flower: Pedicellate (small pedicels), pedicel usually hairy, zygomorphic, irregular, hypogynous, bisexual, complete, papilionaceous.

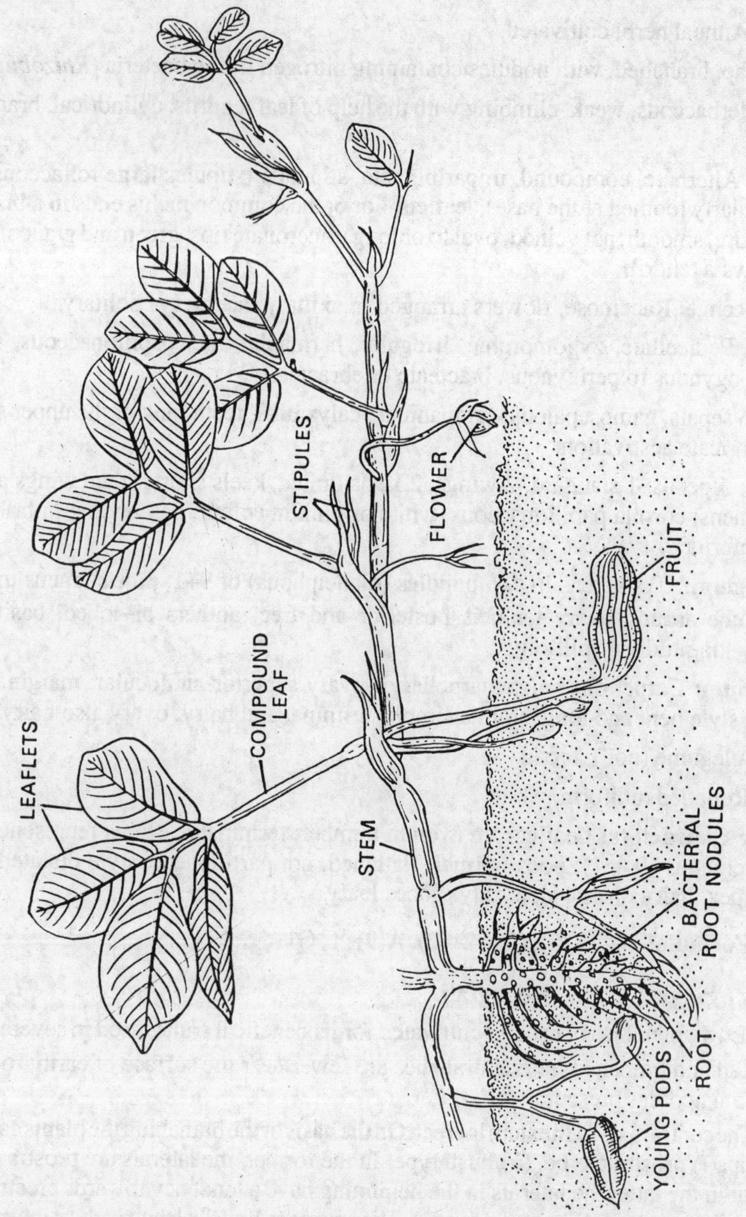

Fig. 14.8. Papilionaceae–*Arachis hypogea* (Verna–**Moongfali**)–A, a complete plant; B, floral diagram.

Calyx: 5 petals, gamosepalous, the sepals unite to form a calyx tube approximately one inch long; the free lobes are short and irregular.

Corollà : 5 petals, free, one standard, two wing and two keels united, united keels form a beak like structure: the corolla is inserted on the rim of the calyx tube, yellow, papilionaceous.

Androecium: 10 stamens, monadelphous, filaments fused about two–thirds of their length. Sometimes only nine stamens. Out of ten, two stamens are sterile and their filaments are fused with the filaments of the neighbouring stamens. In the remaining eight, fertile stamens four have bi–celled, elongated anthers and dehisce by longitudinal slits, while the remaining four have rounded and elongated anthers.

Gynoecium: Carpel one (monocarpellary), ovary sessile, unilocular, marginal placentation, 1 to 5 ovules, the ovary wall is somewhat constricted between each ovule; style curved and hairy, stigma simple and terminal.

Fruit: Oblong, large, indehiscent, ridged (due to these ridges the surface of pod becomes somewhat reticulate). It contains 1 to 3 seeds, constrictions present between the seeds.

Seeds: Elongated cylindrical or elongated and ovoid, short beaked, rounded or diagonally flattened. The embryo is covered by a thin papery, reddish seed coat.

Development of the fruit: After the fertilization the ovary elongates and turns down towards the earth. As soon as the gynophore penetrates the ground upto 2 to 5 cm., the ovary begins to develop and swell. The position of the ovary becomes horizontal within the soil, and gradually it develops into a pod.

Generic characters: Leaflets four, stamens monadelphous; pods short, woody, constricted between seeds, ripening and developing underground.

Floral Formula: ·|· ⚥ K(5), C 1+2+(2), A 10, $\underline{G1}$.

Phaseolus sp.

Habit: An annual herb; cultivated for their edible seeds and pods. It is cultivated in rainy season. Mainly grown in the tropical regions of the world.

Root: Tap and branched. The roots possess tuburcles containing nitrogen fixing bacteria (*Rhizobium radicicola*) and therefore, the plants are used for green manuring.

Stem: Erect, one to three feet high, densely pubescent somewhat angular, herbaceous, green branched.

Leaves: Alternate, petiolate, stipulate, stipules 1/4 to 1/2 inch long attached near the base, compound (trifoliate), leaflets dark green, two to four inches long, entire, acuminate, pubescent on both the sides, net veined.

Inflorescence: Racemose, flowers are being arranged in clustered axillary racemes.

Flower: Short peduncled, zygomorphic, irregular, hermaphrodite, perigynous, complete, bracteate(the bracts usually small and deciduous), bracteolate (bracteoles usually conspicuous and persistent); small and yellow.

Calyx : 5 sepals, gamosepalous, campanulate; the lowest lobe usually longest, two uppermost lobes are somewhat united.

Corolla: 5 petals, it is much exerted, the keel is being spirally coiled with terminal horn–like appendage; descending imbricate aestivation.

Androecium: 9 + 1 stamens, diadelphous, anthers uniform.

Gynoecium: One carpel (monocarpellary), ovary superior, unilocular, sessile many ovuled, marginal placentation; style hairy, filiform, twisted round with the keel; stigma oblique and terminal.

Fruit: Legume or pod; linear rarely oblong, green, very often separate between the seeds, usually 1½ to 2 inches long.

Seeds: Exalbuminous, with thick cotyledons, green (*Phaseolus aureus*–**Moong**) or black (*P. mungo*–**Urd**).

Generic Characters : Leaflets three in number; stamens diadelphous; style hairy; inflorescence racemose in clusters;calyx lobes equal to the tube; keel of corolla coiled.

Floral Formula: ·|· ⚥ K(5), C 1+2+(2), A 9+1, $\underline{G1}$.

Cajanus cajan **;** Verna.–**Arhar.**

Habit: It is an erect branched shrub 4 to 10 feet high; cultivated for edible seeds, in tropical and sub–tropical regions.

Root: Tap, branched with bacterial nodules.

Stem: Erect, much–branched, solid, oody, green pubescent.

Leaves: Alternate, petiolate, compound, trifoliate (unipinnate, unijugate), gland dotted below; leaflets are lanceolate to narrow elliptical, two to four inches long, acuminate, softly pubescent on both surfaces, net veined, stipulate, the stipules are minute and caducous.

Inflorescence: Racemose, corymbose raceme or a terminal panicle.

Flower: Pedicellate, zygomorphic, irregular, hermaphrodite, complete, perigynous, yellow or orange in colour but brownish at the back.

Calyx: Two upper lobes united, calyx tube campanulate, toothed or lobed, teeth short.

Corolla: 5 petals, polypetalous; the standard is large and broad, auricled; the keel is obtuse and incurved; descending imbricate aestivation; yellow or pink in colour.

Androecium: Stamens 9+1, diadelphous; anthers uniform, bicelled.

Gynoecium: Carpel one (monocarpellary), ovary superior, pubescent, unilocular, few ovuled, sessile, placentation marginal; style long; thickened, much incurved, beardless; capitate, terminal stigma.

Fruit: Pod or legume (lomentaceous pod), many, compressed, with diagonal depressions, two or three inches long by half an inch broad, pubescent and beaked.

Seeds: Three to five in a single pod; small sized, orbicular and compressed one edge flattened; brown coloured with a white small hilum.

Generic Characters: Leaves gland dotted beneath, leaflets lanceolate oblong, leaflets two inches or more; pods with depressed lines between the seeds; ovules 3 to 5; seeds not strophiolate.

Floral Formula: ·|· ⚥ K(5), C1 + 2+ (2), A9 + 1, $\underline{\text{G1}}$

Crotalaria sp.

Habit: The plants are either herbs or shrubs; cultivated for fibre (*Crotalaria juncea*–Indian sunn hemp–Verna–**San**) and seeds; some are ornamentals.

Root: Tap and branched; nodulated, nodules contain nitrogen fixing bacteria.

Stem: Erect, branched, cylindrical, pubescent, green, sometimes grooved.

Leaves: Alternate, petiolate, simple or compound, if compound it is trifoliate rarely 5 to 7 foliate; digitate (usually all leaflets born at the apex of the petiole); stipulate (minute stipules); leaflets are linear or oblong, entire, shining silky hairs on both the sides; net veined.

Inflorescence: Flowers arranged in solitary or racemose inflorescences; the flowers are being arranged in either terminal, lateral or leaf opposed racemes.

Flower: Pedicellate (short peduncle), zygomorphic, irregular, hermaphrodite, complete, perigynous, bracteate, usually large and showy.

Calyx: 5 sepals, gamosepalous, sub–bilabiate, the upper lip bifid, the lower cordate trifid; the teeth being linear or lanceolate.

Corolla: 5 petals, polypetalous; standard usually broad and shortly clawed; the wings obovate oblong; the keel is broad and nearly equal to wing in size, petals of keel are united, incurved and beaked; papilionaceous; descending imbricate aestivation; usually yellow coloured, sometimes pink and purplish; big and showy.

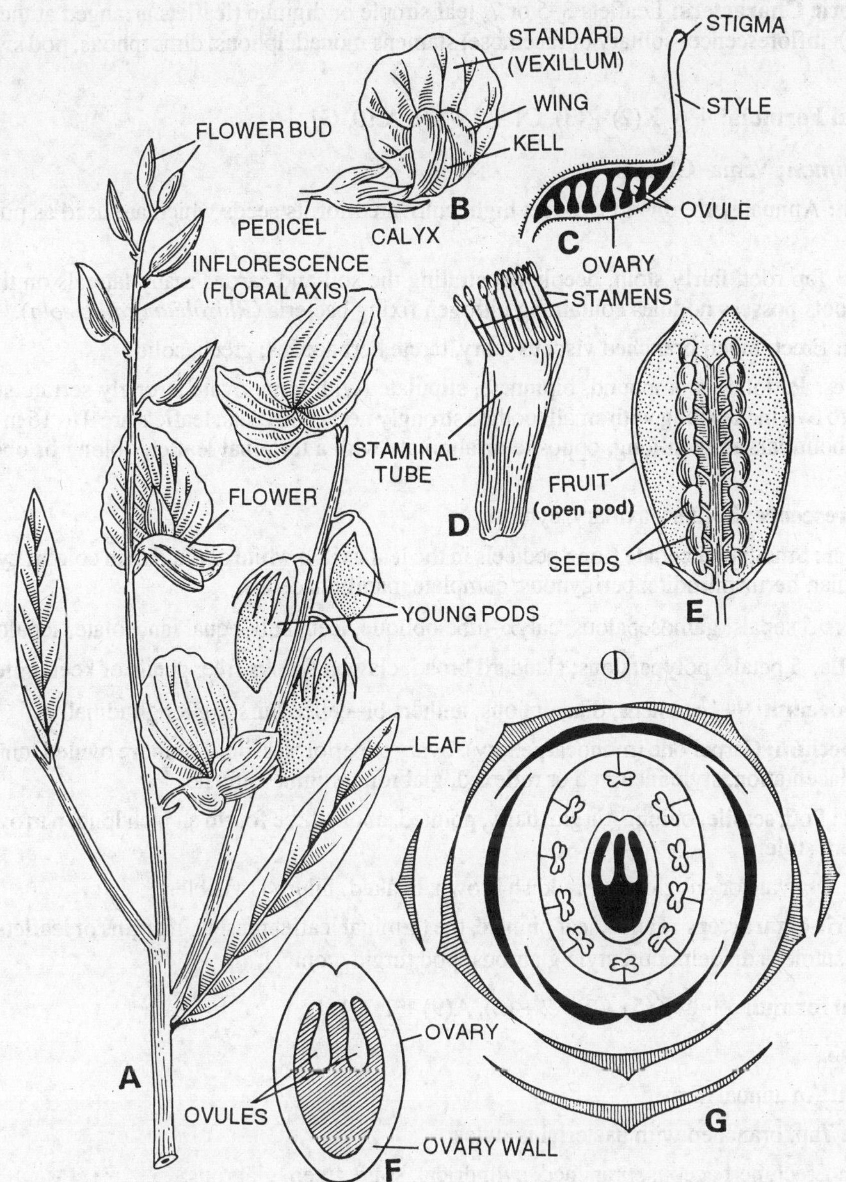

Fig. 14.9. Papilionaceae–*Crotalaria juncea.* A, flowering twig with young pods; B, flower; C, L.S. gynoecium; D, androecium; E, dehisced fruit; F, T.S. of ovary; G, floral diagram.

Androecium: 10 stamens, monadelphous, connate, the anthers are 2–celled, dimorphous, dehiscence by longitudinal slits.

Gynoecium: One carpel (monocarpellary), ovary superior, unilocular, many ovuled sessile; style long, strongly incurved or reflexed, somewhat bearded (hairy); the stigma is small and oblique; placentation marginal.

Fruit: Legume or pod, globose or oblong sessile, turgid or inflated, with many loose seeds.

Seeds: Exalbuminous, small, usually flat, different coloured.

Generic Characters: Leaflets 3, 5 or 7; leaf simple or digitate (leaflets arranged at the apex of the petiole); inflorescence–solitary or racemose; stamens monadelphous; dimorphous; pod swollen or inflated.

Floral Formula: ·|· ⚥ K(2) + (3), C1 + 2 + (2), A 10, $\underline{G1}$.

***Cicer arietinum*; Verna–Chana.**

Habit: Annual herb, one to two feet high, cultivated for its seeds which are used as pulses and forage.

Root: Tap root, fairly stout, deeply penetrating the soil and carries many laterals on the upper part. The roots possess nodules containing nitrogen fixing bacteria (*Rhizobium radicicola*).

Stem: Erect, much branched viscous hairy, terete herbaceous, green, solid.

Leaves: Petiolate; compound, bipinnate, stipulate (with large ovate strongly serrate stipules); rachis one to two inches long with small toothed strongly nerved leaflets; leaflets are 9 to 16 in number and each about half an inch long, opposite or alternate with a terminal leaflet, oblong or obovate in outline.

Inflorescence: Solitary axillary, cymose.

Flower: Small; pedicellate (long pedicels in the leaf axils); white or pinkish in colour; zygomorphic, irregular, hermaphrodite, perigynous, complete, papilionaceous.

Calyx : 5 sepals, gamosepalous, calyx–tube oblique, teeth sub–equal lanceolate, sepaloid.

Corolla: 5 petals, polypetalous; standard broad; clawed; wings free; petals of keel united.

Androecium: 9+1 stamens, diadelphous; anthers bi–celled bursting longitudinally.

Gynoecium: Carpel one (monocarpellary), ovary superior, sessile, 2 or more ovuled, unilocular, marginal placentation, style incurved or reflexed, glabrous, stigma capitate.

Fruit : Pod, sessile, oblong, turgid, hairy, pointed, about three fourth an inch long, narrowed into the persistent style.

Seed: Obovate or subglobose, reddish brown, beaked; hilum is visible.

Generic Characters : Herbs, leaf pinnate, the terminal leaflet present, margins of leaflets serrate all–round; stamens diadelphous, style glabrous; pod turgid, conical.

Floral formula: ·|· ⚥ K(5), C1 + 2 + (2), A(9) + 1, $\underline{G1}$.

Melilotus sp.

Habit: An annual herb.

Root: Tap, branched with bacterial nodules.

Stem: Erect, herbaceous, branched, cylindrical, solid green, glabrous.

Leaf: Compound trifoliate, alternate, cauline, petiolate, stipulate (stipules linear), obovate, acuminate, margins serrate, mucronate apex, unicostate reticulate venation.

Inflorescence: Racemose, typical raceme.

Flower: Pedicellate, bracteate, minute, hermaphrodite, zygomorphic (irregular), complete, yellow, hypo or perigynous.

Calyx: Five sepals, gamosepalous, nearly equal lobes, lanceolate, valvate aestivation, odd sepal posterior.

Corolla: Five petals, polypetalous, yellow, one standard, two wings, two innermost used forming keel, aestivation descending imbricate.

Androecium: Ten stamens, diadelphous, filaments of 9 stamens fused forming one bundle, tenth stamen free (9+1), anthers dorsifed.

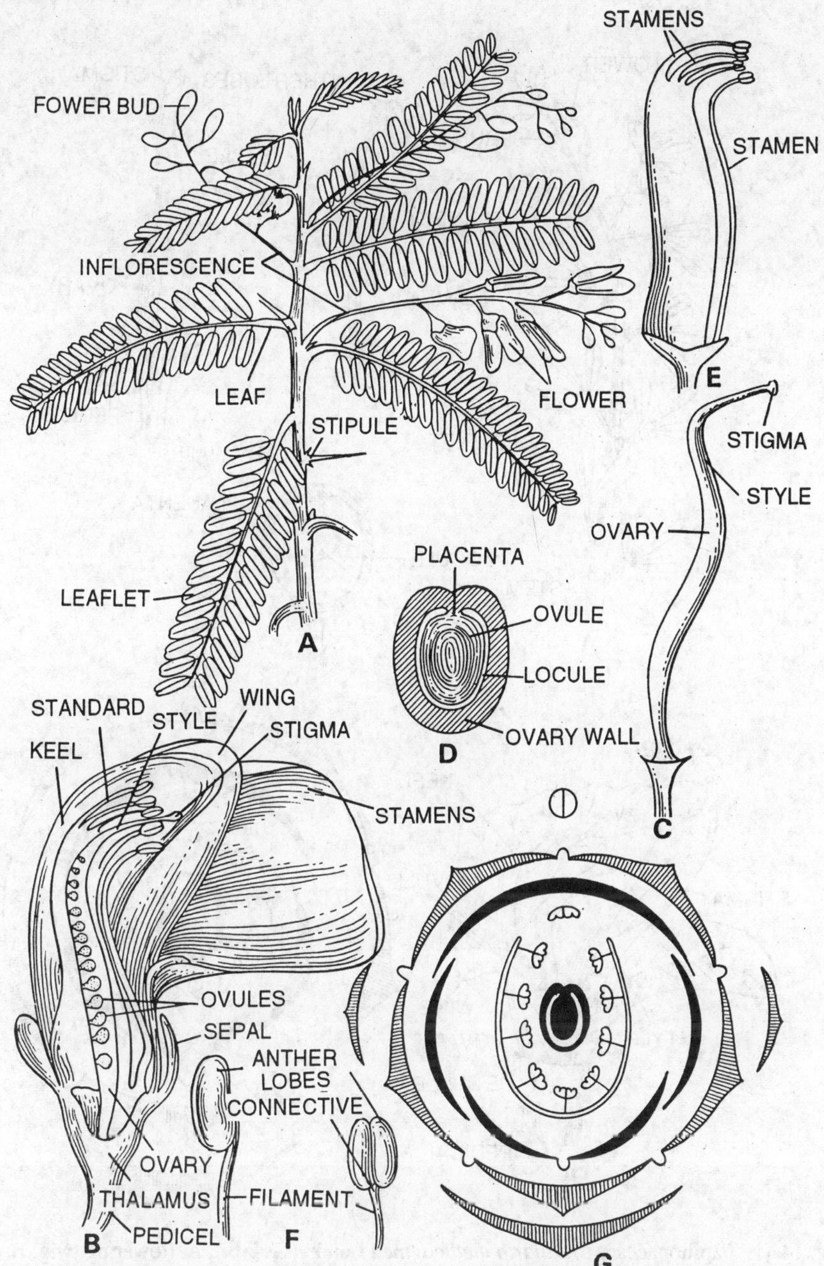

Fig. 14.10. Papilionaceae. *Sesbania sesban* (L.) Merr. **(Jaynti)**. A, flowering twig; B, L.S. of flower; C, gynoecium; D, T.S. of ovary; E, androecium; F, stamens; G, floral diagram.

Gynoecium: One carpel (monocarpellary), ovary superior, unilocular, marginal, placentation.
Fruit: A pod.

Floral Formula: ·|· ⚥ K(5), C1 + 2 + (2), A(9) + 1, $\underline{G1}$.

Sesbania sesban (L) Merr.; Verna **Jayanti, Jait. Stem**–herbaceous and woody, aerial, erect, cylindrical, branched, solid, smooth, green. **Leaf**–cauline, ramal, alternate, stipulate, leaf base pulvinus, compound, unipinnate; pinna oblong, entire, mucronate, unicostate reticulate. **Inflorescence**–racemose, axillary raceme. **Flower**–bracteolate, pedicellate, complete, zygomorphic, hermaphrodite,

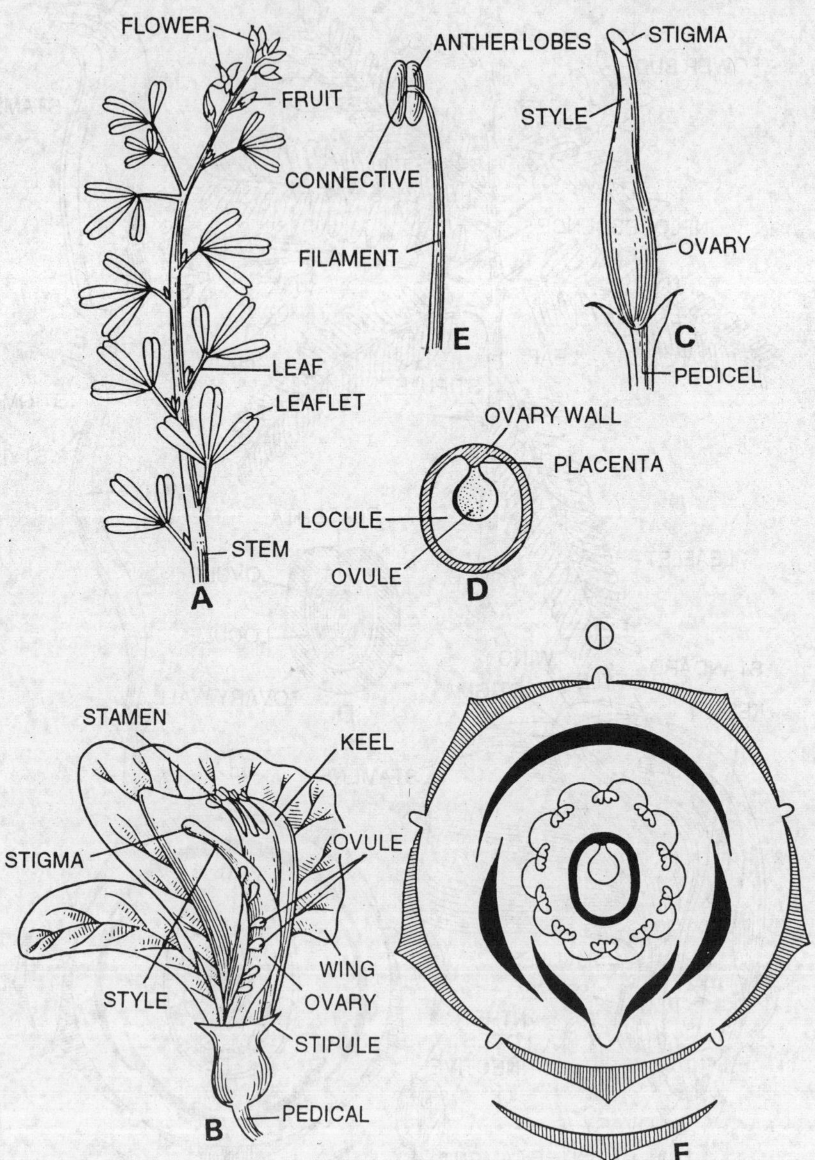

Fig. 14.11. Papilionaceae. *Crotalaria medicaginea* Lamk. (**Gulabi**). A, flowering twig; B, L.S, of flower; C, gynoecium; D. T.S. of ovary; E, stamen; F, floral diagram.

pentamerous, perigynous, cyclic, **Calyx**–5, gamosepalous, persistent, green, valvate. **Corolla**–5, polypetalous, vexillary, papilionaceous, yellow. **Androecium**–(9)+1, diadelphous, dithecous, dorsifixed, introrse. **Gynoecium**–monocarpellary, half–inferior, unilocular, marginal, style long, stigma capitate. **Fruit**–legume pod. **Floral formula**–Br Brl ·|· ⚥ K5, C5, A(9) + 1, G1.

Crotalaria medicaginea Lamk; Verna **Gulabi. Stem**–herbaceous, aerial erect, branched, cylindrical, solid, pubescent, green. **Leaf**–cauline, ramal, alternate, stipulate, palmately compound trifoliate, petiolate, leaf base pulvinus, leaflet obovate, entire, hairy. **Inflorescence**–racemose, raceme. **Flower**–pedicellate, complete, zygomorphic, hermaphrodite, pentamerous, perigynous, cyclic. **Calyx**–5, gamosepalous, green, valvate. **Corolla**–5, polypetalous, vexillary, papilionaceous, yellow.

Androecium–(10), monadelphous, dithecous, dorsifixed, introrse. **Gynoecium**–1, monocarpellary, half inferior, unilocular, many ovules, marginal, style long, stigma capitate. **Fruit**–legume, pod, **Floral formula–** Br Brl ·|· ⚥ $K_{(5)}$, C_5, $A_{(10)}$, G1–

Economic Importance of the Family

This family is of great economic value. A list of few very important plants is given below :

1. *Phaseolus mungo;* Eng.–black gram; Verna.–**Urd** this is cultivated as a pulse crop in Uttar Pradesh, Madhya Pradesh, Punjab and Bengal. The seeds are rich in proteins and used as **'dal'.**

2. *Phaseolus aureus;* Syn. *P. mungo; P. radiatus;* Eng.–green gram; Verna.–**Mung.** This is grown as pulse crop in Uttar Pradesh, Madhya Pradesh, Bihar, Rajasthan and Bengal. The seeds are eaten as pulse. They are rich in proteins.

3. *Phaseolus vulgaris*; Eng.–Kidney bean; Verna.–**Vilayati sem**–They are cultivated throughout India. The tender pods are used as vegetable.

4. *Phaseolus aconitifolius;* Syn. *P. trilobus; Dolichos dissectus;* Eng. Moth bean; Verna.–**Moth**– This is a herb. It is native of India. It is grown as a pulse crop in Andhra Pradesh, Uttar Pradesh, Bihar, Maharashtra the Punjab and Bengal. The seeds are rich in proteins and eaten as **'dal'.**

5. *Phaseolus lunatus;* Eng.–Lima bean; Verna.–**Lobia**–This is native of tropical America but now grown in Northern India. The young pods are used as vegetable and the seeds are eaten as pulse.

6. *Phaseolus calcaratus;* Eng.–Rice bean. This is introduced from Philippines but now cultivated in Tamil Nadu, Assam, Bengal and the Punjab. The seeds are used as a pulse.

7. *Phaseolus trilobus;* Eng.–Wild gram; Verna.–**Mugani**–It is used as a fodder.

8. *Cajanus cajan;* Syn. *Cajanus indicus*; Eng.–Pigeon pea; Verna.–**Arhar.**–This is a shrub, grown chiefly in Uttar Pradesh, Bihar, Madhya Pradesh, Andhra pradesh, Maharashtra and Karnataka. The immature and ripe seeds are used as human food. The leaves and twigs are used as cattle fodder. The enzyme urease is obtained from it which is used for estimation of urea in blood, urine, etc. The baskets are made from thin straight branches. The plants also serve as a host for the lac insects.

9. *Glycine max* (Linn.) Merr.; Eng. Soybean; Verna. **Bhat, Ramkurthi**–Soybean makes a good food. The bean has a high oil content, and the kernel is rich in proteins. The oil can be used for cooking. Oil–cake is used as cattle feed. Soybean flour or cooked bean is vary good for diabetic patients.

10. *Cicer arietinum;* Eng.–Gram; Verna–**Chana**–This is a herb. It is native of South Europe, but now commonly grown in Uttar Pradesh, the Punjab, Rajasthan, Bihar and Madhya Pradesh. The seeds are edible. The seeds are eaten raw as well as roasted. The vegetative parts of the plant are cooked as vegetable and used as fodder.

11. *Pisum sativum* var. *arvense;* Syn. *P. arvense;* Eng. Garden pea; Verna–**Mattar**–It is native of South Europe but now cultivated in our country chiefly in the Punjab, Uttar Pradesh, Delhi and Himachal Pradesh. The seeds and buds are eaten as vegetable. They are quite rich in proteins.

12. *Lens culinaris*; Syn. *L. esculenta; Ervum lens; Eng.* Lentil; Verna–**Masur**–This is native of South West Asia but now cultivated in our country chiefly in Andhra Pradesh, Uttar Pradesh, Rajasthan and Bengal for its edible seeds. The seeds are consumed as pulse. The leaves and stems are used as fodder.

13. *Dolichos lablab;* Syn. *Lablab niger*; Eng. Lablab; Verna–**Sem**–This is a twining herb. It is cultivated throughout our country. The green pods and seeds are consumed as vegetable.

14. *Dolichos biflorus;* Eng. Horse gram; Verna–**Kulthi**–This is a trailing herb. It is cultivated throughout our country as fodder crop.

15. *Canavalia gladiata;* Syn. *Dolichos gladiatus;* Eng.–Jack bean; Verna–**Bara sem**–It is cultivated throughout our country. The pods are used as vegetable. The plant is also used as green manure.

16. *Lupinus albus;* Eng–white lupine; Verna–**Turmas**–This is a herb, grown for fodder and green manue.

17. *Cyamopsis tetragonoloba*; Eng.–Cluster bean; Verna–**Gwar**–This is a herb or small shrub. It is cultivated throughout our country. The pods are eaten as vegetable and used as fodder. The seeds are the source of a gum which is used in food, paper and textile industries.

18. *Melilotus alba;* Eng–White sweet clover. This is a herb. It is native of Eurasia, but now grown in Northern India. It is used as fodder.

19. *Melilotus indica*; Syn. *M. parviflora*; Eng.–Sweet clover; Verna–**Banmethi, Senji**–This is native of Europe and Asia but now grown in the Punjab and Uttar Pradesh as a winter fodder crop.

20. *Medicago sativa;* Eng.–**Alfalfa;** Verna–**Vilayati Gawuth**–This herb is used as fodder.

21. *Medicago denticulata;* Eng.–Toothed, bur clover; Verna–**Maina**–This is a common herb and used as fodder.

22. *Trifolium alexandrium;* Eng.–Egyptian clover : Verna–**Berseem**–This is a herb grown as green manure and fodder crop.

23. *Trifolium resupinatum;* Eng.–Persian clover; Verna–**Shaftal**–It is used as green fodder.

24. *Trifolium pratense;* Eng–Red clover; Verna–**Trepatra.** This is used as fodder. A yellow dye is also extracted from the flowers.

25. *Trifolium repens;* Eng.–White clover; Verna–**Shaftal** It is used as fodder. This is grown in the temperate Himalayas and the Nilgiris.

26. *Trigonella foenum–graecum;* Verna–**Methi**–This is a herb. It is native of South Europe, but now grown chiefly in Northern India. The leaves are used as vegetable and fodder. The seeds are used as spice and condiment. The seeds are also used medicinally as carmintive and tonic.

27. *Vicia faba*; *Faba vulgaris*; Eng.–Broad bean, Verna–**Bakla**–This is a herb commonly grown in North Western India. It is used as fodder. The seeds are eaten as vegetable.

28. *Vigna sinensis;* Eng–Cowpea. Verna–**Lobia, Chowli**–It is cultivated throughout our country. The young pods and seeds are used as a vegetable and pulse.

29. *Abrus precatorius;* Verna–**Geumchi, Rati**–This is a climbing shrub. The seeds are used as weights by goldsmiths. They are also used as beads for necklaces.

30. *Butea monosperma;* Syn. *B. frondosa;* Eng.–Flame of the forest; Verna–**Dhak, Palas, Tesu**-This is a small tree. This serves as a host for the lac insect. The flowers yield a yellow dye. The leaves are used as substitute for plates chiefly on festive occasions. The ripe fruits are edible

31. *Indigofera tinctoria*; Eng.–Indigo; Verna–**Nil**–This is a common small shrub, grown as green manure crop. The plant is of great medicinal value.

32. *Arachis hypogea;* Eng.–Groundnut; Verna–**Mungphali.**This is a native of Brazil but now widely grown in South India, Maharashtra and Uttar Pradesh. The nuts (seeds) are used for the preparation of peanut butter. They are also eaten after roasting. The filtered refined oil is used for cooking and in making margarine. The oil cake is used as fodder. The protein obtained from groundnuts is used in the manufacture of ardil, a synthetic fibre.

33. *Aeschynomene indica;* Verna–**Phulan**–The soft wood is used for making sola hats. It is found in Kashmir, Assam, Bengal and South India.

34. *Aeschynomene aspera*; Eng. Sola pith; Verna–**Sola**–This is a tall herb, found in wet places in Bengal and South India. The soft wood is used for making sun–hats, toys, models, artificial flowers, etc. It is used in the manufacture of swimming jackets and life–belts. The leaves are edible.

35. *Alhagi pseudalhagi*; Syn. *A. maurorum A. camelorum* ; Eng.–Camel's thorn; Verna–**Jawasa**–This is a spiny shrub of the deserts. It is used as camel fodder. The twigs are used in making screens.

36. *Dalbergia sissoo;* Verna–**Shisham**–it is found in the Punjab, Uttar Pradesh, Bengal and Assam. The wood is used for making furniture and carving. The pulp is used for making writing and printing papers.

37. *Dalbergia latifolia*; Verna–**Kala Shisham**–This is a tree found in Bengal, Bihar, Madhya Pradesh and Western Peninsula. The wood is used for cabinet work and furniture.

38. *Dalbergia melanoxylon;* Eng.–African black wood. It is cultivated in Tamil Nadu, Maharashtra and Bengal. The wood is used for musical instruments, walking sticks and fancy articles.

39. *Desmodium triflorum;* Verna–**Kubaliya**–It is grown as a soil, binder and for fodder.

40. *Crotalaria juncea;* Eng.–Sunn hemp; Verna–**San**–This is a tall shrub, cultivated throughout our country. The stem fibre is used for making ropes, cordage, mats and paper. The flowers and fruits are eaten as vegetable. The green stems and leaves make good manure.

41. *Crotalaria mucronata;* Syn. *C. striata;* Verna–**San**–This is a herb used as a cover crop and green manure.

42. *Crotalaria retusa*; Verna–**Ghungbunian**–This is cultivated all over India. A fibre is obtained from its stem bark, which is used for making canvas and cordage.

43. *Crotalaria burhia;* Verna–**Khip, Sis, Booi**–This is a xerophytic under shrub, found in the Punjab, Uttar Pradesh, Rajasthan and Gujarat. The stem fibre is used for making ropes and cordage. The branches and leaves are used as fodder.

44. *Lathyrus odoratus*; Eng.–Sweet pea; Verna–**Phul mattar**–It is a native of Italy, cultivated as an ornamental.

45. *Lathyrus aphaca;* Verna.–**Jangli matar**–This is a climbing herb. It is grown in the Punjab, Uttar Pradesh, Bengal and Madhya Pradesh. It is used as a fodder.

46. *Lathyrus sativus*; Verna.–**Khesari**–It is cultivated in the Punjab, Bihar, Baroda, Assam and Orissa. It is used as a fodder. The seeds are eaten as vegetable. The ripe seeds are source of adulteration in pulses.

47. *Erythrina indica* var. *parcellii;* Eng.–Indian coral tree. This is a small tree, grown for its beautiful flowers.

48. *Erythrina blake.* This is a small tree, grown for its beautiful scarlet flowers.

49. *Erythrina crista–galli*; Eng.–Cockspur coral tree. It is grown as an ornamental.

50. *Erythrina variegata* var. *orientalis;* Eng.–Coral tree; Verna–**Pangra, Nasut**–This is a tree found in the Sundarbans, the Andaman and Nicobar Islands, Assam, Tamil Nadu and Orissa. The tender pods and seeds are eaten as vegetable. The bark fibre is used for making cordage. The flowers yield a red dye. It is cultivated as a hedge plant and also employed to support the black pepper vine.

51. *Erythrina suberosa;* Verna.–**Panga**–This is a small tree found in the Punjab and Rajasthan. A fibre is obtained from its bark which is used for cordage.

52. *Clitoria ternatea;* Eng.–Butterfly pea. A twiner with beautiful blue flowers; grown in the gardens.

53. *Sesbania bispinosa;* Syn. *S. aculeata;* Eng–Dhencha; Verna.–**Jayanti**–This is a tree. Its stem fibre is used for making ropes.

54. *Sesbania cannabina*; Syn. *S. aculeata* var. *cannabina;* Verna.–**Dhaincha**–This is a tall herb, grown for its fibre. The fibre is used in making fishing–nets. The plant is also used as a green manure crop.

55. *Sesbania grandiflora*; Syn. *Robinia grandiflora*; *Agati grandiflora;* Verna.–**Basna**–This is a tree. It is native of tropical Asia but now grown in Assam, Bengal, Andhra Pradesh, Baroda and Tamil Nadu. The flowers, fruits and leaves are eaten as vegetable. The plant is also used as a support for the betel vine.

56. *Sesbania sesban*; Syn. *S. aegyptiaca;* Verna–**Jayanti**–This is a large shrub, cultivated throughout our country. The stem fibre is used for making ropes and cordage. The leaves are used as fodder.

57. *Pongamia pinnata*; Syn. *P. glabra*; Eng.–Pongam; Verna.–**Karanja**–This is a tree, commonly found near banks of streams in both the Peninsulas, West Bengal and Travancore. An oil is extracted from the seeds which is used for soap making and as an illuminant. It is also used in the treatment of skin diseases and rheumatism. The leaves are used as manure. The fruits are edible.

58. *Alysicarpus longifolius*; Verna–**Jangli gailia**–This is a herb, used as fodder.

59. *Astragalus heratensis.* This is a large shrub. The stem is the source of gum Katira gabina, which is used for glazing.

60. *Astragalus prolixus.*This is an undershrub, found in the Punjab. The stem is the source of gum tragacanth, which is used in cosmetics, calico–printing and confectionery.

61. *Astragalus strobiliferus*; Verna.–**Garmezu**–The gum obtained from the stem is used in confectionery. It is found in Kashmir.

62. *Calopogonium mucunoides.* This is a herb, introduced in the Western Ghats. It is used as fodder and also prevents soil erosion.

63. *Castanospermum australe*; Eng.–Chestnut. This is an introduced tree from Australia, the seeds are edible.

Systematic Relationships : According to Bentham and Hooker and Engler the family Leguminosae consists of three sub–families, *i.e.,* Mimosoideae, Caesalpinioideae and Papilionatae (Lotoideae). Many botanists have treated these sub–families as three distinct families, *i.e.,* the Mimosaceae, the Caesalpiniaceae and the Papilionaceae. Authors who accept them as separate families treat the trio as a single order. Hutchinson has placed these three families within the order Leguminales. The characters that distinguish one sub–family from the other are more apparent than the characters which bond them together. Many botanists bond them together only because of conservatism and tradition. Sen (1943) has pointed out, that as regards their wood anatomy there are no clear cut lines which separate the three sub–families. He also advocated that there was no cytological data in the support or denial of the elevation of each sub–family to the rank of family. The Mimosoideae of the three sub–families are accepted to be most primitive and the Papilionatae to be most evolved.

15

ORDER–CASUARINALES

According to Hutchinson this is the eighteenth order of the phylum Angiospermae, subphylum Dicotyledones and division Lignosae. This order contains a single family the Casuarinaceae. This family has been described here in detail. Bentham and Hooker placed the family Casuarinceae in their seventh series, the Unisexuales along with eight other families, of class Dicotyledones and division Monochlamydeae or Incompletae. Engler and Prantl have placed the single family Casuarinaceae in their first order Verticellatae of class Dicotyledoneae and sub-class Archichlamydeae. They have considered the family to be the primitive most among the Dicotyledones.

Male flowers spicate, female capitate; no perianth; male with only one stamen; seed solitary, no endosperm; leaves completely reduced.

Family–Casuarinaceae

This is a monogeneric family. The single genus *Casuarina* contains about 40 species.

Distribution: The species of *Casuarina* are chiefly found in Australia. The genus is also found in tropical Asia, the Mascarene Islands and the Islands of the Pacific Ocean. *Casuarina equisetifolia* is quite common in the coastal regions of both the Peninsulas of our country. *Casuarina suberosa* is found in the Nilgiris.

Habit: The habit of the plant resembles with that of *Ephedra* (a gymnosperm) in having the jointed or switch-like branches which bear alternating whorls of scaly leaves. These scale-like leaves unite at the base forming a sheath.

The plants are evergreen profusely branched shrubs or trees. They possess long, slender, cylindrical and grooved branches.

Leaves: The number of leaves in a whorl varies from species to species. It ranges from 4 to 12. The internodes are furrowed.

Anatomically the structure is quite peculiar. The ridges contain palisade tissue; the sunken stomata are found to be situtated in the furrows. The transverse section of a young branch shows two rings of vascular bundles arranged alternately. The stem possesses peculiar xerophytic features. The wood contains vessels, tracheids, wood parenchyma (xylem parenchyma) and fibres.

Flowers: The flowers are unisexual and the perianth is absent.

Male Flowers: The male flowers arise at the ends of the branches and are found to be arranged in erect catkin-like spikes. The flowers are arranged in whorls. Each whorl of flowers develops in the axil of a bract. This bract, and the other bracts situated at a node form a sheath which protects the young flowers. A pair of bracteoles and two small perianth leaves are also found. There is a single central stamen, showing the tendency to split. The filament is short and bears a two-celled anther which

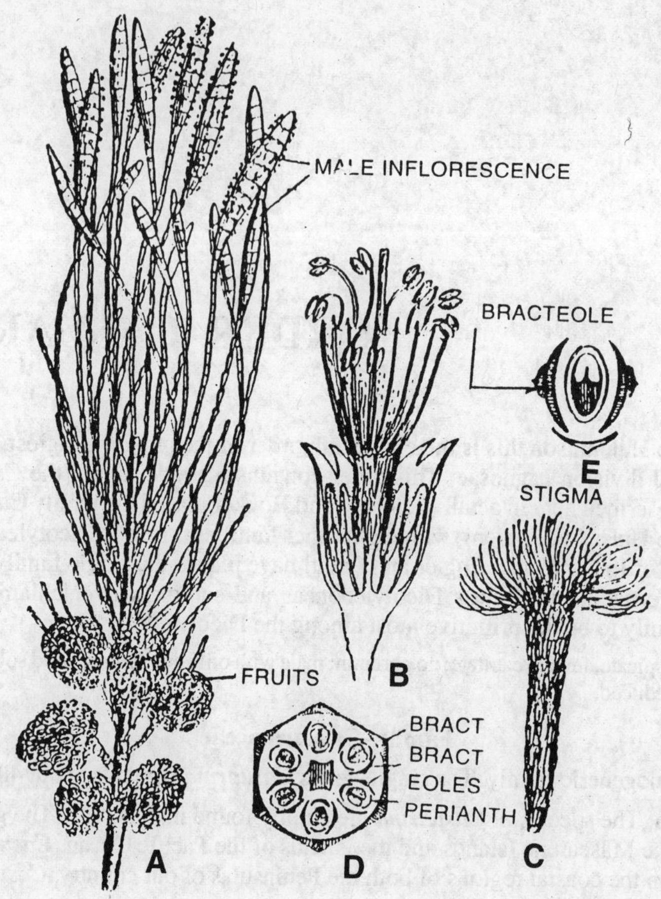

Fig. 15.1. Casuarinaceae–*Casuarina equisetifolia.* A, a twig with male inflorescences and fruits; B, male inflorescence showing two whorls; C, female inflorescence with long stigmas; D, floral diagram of a whorl of male flowers; E, floral diagram of female flower.

dehisces longitudinally. The filament elongates lengthwise before dehiscence and the perianth which first covers the anther is pushed off. The open anther is then pushed above the sheath.

Female Flowers: The female flowers are borne in dense heads at the end of short lateral branches. Each flower is found single in the axil of a bract and two bracteoles. The flower is without perianth and consists of two carpels which are syncarpous. In the beginning the ovary is bilocular, but later on by the suppression of the posterior chamber it becomes unilocular. The posteior locule becomes empty and anterior one bears two ovules arranged on a parietal placenta. There are two long stigmas.

Fruit: The fruit is one-seeded winged nut protected by the hardened bracteoles. The bracteoles become woody and make a five valved capsule which envelops the compressed nut-like fruit having a membranous wing. Several fruits form a cone like structure.

Seeds: The seeds are non-endospermic. They are winged and being enclosed within woody bracteoles. Each seed bears a straight embyro with an upwardly directed radicle and two large flat cotyledons.

Pollination: It takes place through the agency of insects.

Floral Formulae: ♂ P0, A1

♀ P0, G(2).

Bentham & Hooker (1862)	**Engler & Prantl (1931)**	**Hutchinson (1959)**
Phanerogams	Phanerogams	Angiospermae
Dicotyledones	Dicotyledones	Dicotyledones
Monochlamydeae or	Archichlamydeae	Lignosae
Incompletae	Verticellatae	Casuarinales
Unisexuales	Casuarinaceae	Casuarinaceae
Casuarineae		

Economic Importance of the Family

The family is of little economic value.

1. *Casuarina equisetifolia;* Syn. *C. muricata;* Eng.–Sheoak; Verna–**Jangali jhau–A** tall evergreen tree planted as an avenue and also used for fuel in the coastal regions of both the Peninsulas. The bark is used as a tan. The wood is very hard and is known as 'iron wood'. The bark of the tree possesses medicinal properties and used to cure diarrhoea and dysentery.

2. *Casuarina suberosa*; Eng.–Swamp Oak. It is found in the Nilgiris. The bark is used for tanning.

Systematic Relationships: Engler and Prantl have put the family Casuarinaceae in the order Verticellatae. Engler has treated the family to be most primitive among Dicotyledons. Rendle and Wettstein retained this family within the Amentiferae and thought it to be most primitive angiosperm derived from the Ephedraceae. Bessey, Hallier, Hutchinson and Tippo have recognised the family to be an advanced one. Hutchinson (1926) placed the family at the top of the Amentiferae and derived the group from the Hamamelidales. Tippo (1938) considered the family to be advanced because of certain anatomical evidences.

16

ORDER–URTICALES

According to Hutchinson this is the nineteenth order of his phylum Angiospermae, subphylum Dicotyledones and division Lignosae. The order comprises of six families, *i.e.*, Ulmaceae, Cannabiaceae, Moraceae, Urticaceae, Barbeyaceae and Eucmmiaceae, of which Moraceae and Urticaceae have been described in detail in the present text. Bentham and Hooker have placed the Urticaceae in their seventh series the unisexuales of class Dicotyledones and division Monochlamydeae, or Incompletae, together with eight other families. However, they have not mentioned the Moraceae in their system. Engler and Prantl have included the Ulmaceae, Moraceae and Urticaceae in 12th order the Urticales of class Dicotyledoneae and subclass Archichlamydeae.

Ovary usually 1-locular; ovule 1, rarely 2, erect or pendulous; endosperm present; leaves stipulate, mostly alternate.

Family—Moraceae

There are about 71 genera and 1,000 species in this family.

Distribution: The members of this family are widely distributed in the tropical and sub-tropical regions of the world. *Ficus* is the largest genus of the family containing about 800 species, which vary in their habit too much. In our country the family is represented by several important members, such as, *Ficus religiosa* (Verna–**Pipal**), *F. bengalensis* (Verna–**Bargad**), *F. rumphii* (Verna–**Pakar**), *F. glomerata* (Verna–**Gular**), *F. elastica*(Verna–**Bor**), *Artocarpus heterophyllus* syn. *A. integrifolia* (Verna–**Kathal**), *Morus alba* (Verna–**Tut**) and many others.

Habit: They are mostly trees or shrubs. The herbs are very rare, *e.g., Dorstenia.* Many species of *Ficus* are epiphytic and sometimes they form such a tight network of roots round the stem of the host plant that ultimately they straggle it. The family is distinguished by the presence of latex, which is found in long sacs, especially in the secondary cortex or the phloem. Certain species of *Ficus* are big trees, *e.g., Ficus religiosa, F. bengalensis* ,etc. *Ficus scandens* (Verna.–**Chhanchri**) is a creeping shrub with roots given out from the stem and branches. *Ficus hispida* (Verna.–**Kathumbar**)is a large shrub or small tree. Many species of *Ficus* possess prop roots arising from their stems and branches. These roots reach down in the soil and give support to the tree. The species of *Artocarpus* and *Morus* are also trees.

Leaves: The leaves are simple, alternate, rarely opposite, entire, serrate or lobed margins and stipulate. The stipules vary in their form and position. In the species of *Morus*, the stipules are lateral and small whereas in *Ficus carica,* (Verna–**Anjir**) the stipules are intrapetiolar. In *Ficus elastica,* the pair of intrapetiolar stipules may well be seen to unite and forming a cap round the younger leaves. As the leaves become old and are unfolded, the stipules fall off and a circular scar remains there.

Inflorescence: The inflorescence is cymose or racemose but frequently modified. In the species of *Ficus,* the cymes coalesce and form a fleshy hollow axis bearing the flowers on the interior surface. Here the receptacle forms a hollow cavity, with an apical opening possessing scales and the inflorescence is known as **hypanthodium.** In each hypanthodium the female flowers are found to be developed at the base of the cavity and the male flowers near its apical region. In *Morus* the inflorescence is **catkin** where the male and female flowers develop on different shoots.

Flowers: The flowers are unisexual, regular (actinomorphic), incomplete, small and hypogynous.

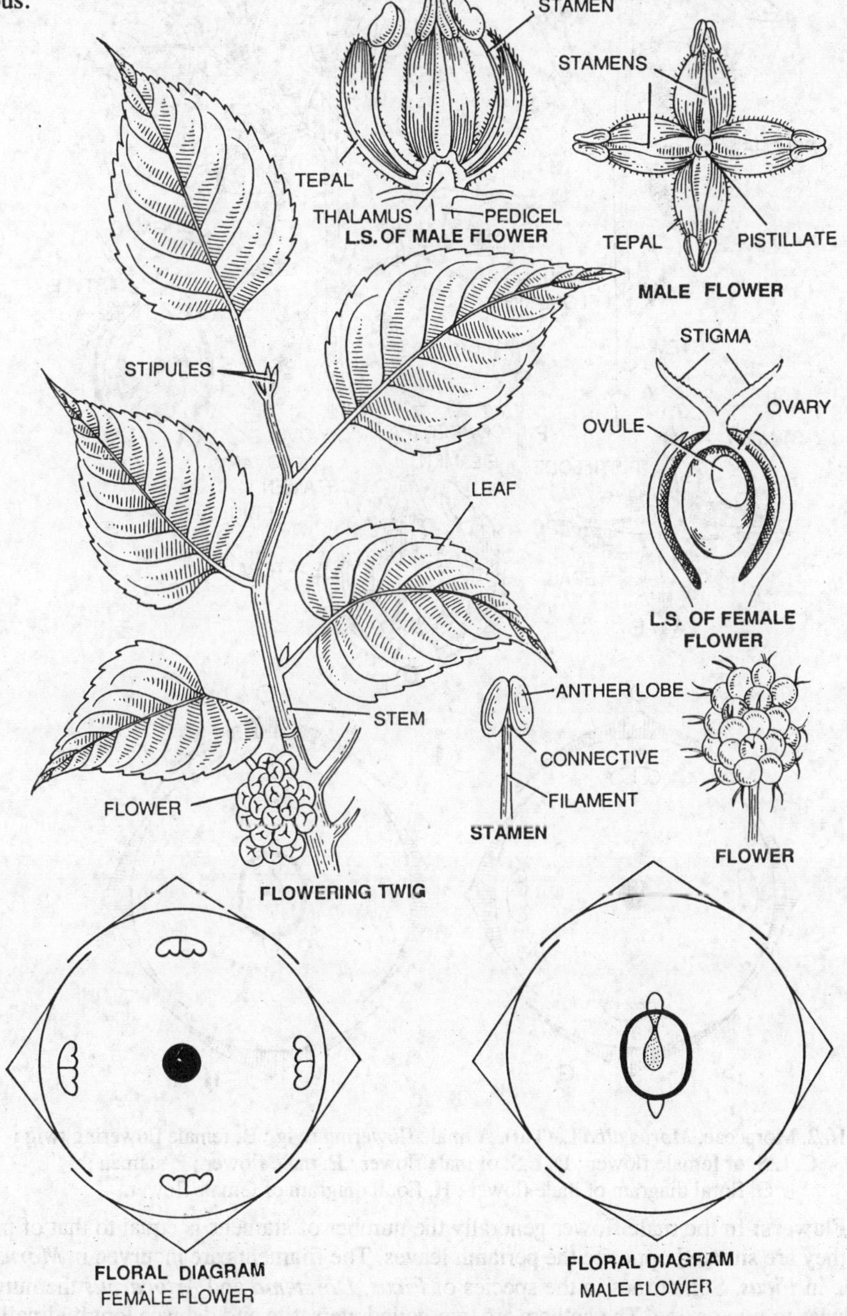

Fig. 16.1. Moraceae. *Morus australis* Poir. (syn. *M. indica*).

Perianth: The perianth consists of four perianth leaves which are occasionally united. They are arranged in two whorls each consisting of two tepals. In rare cases the number of tepals may be reduced to two or increased to five or six tepals. The perianth is persistent.

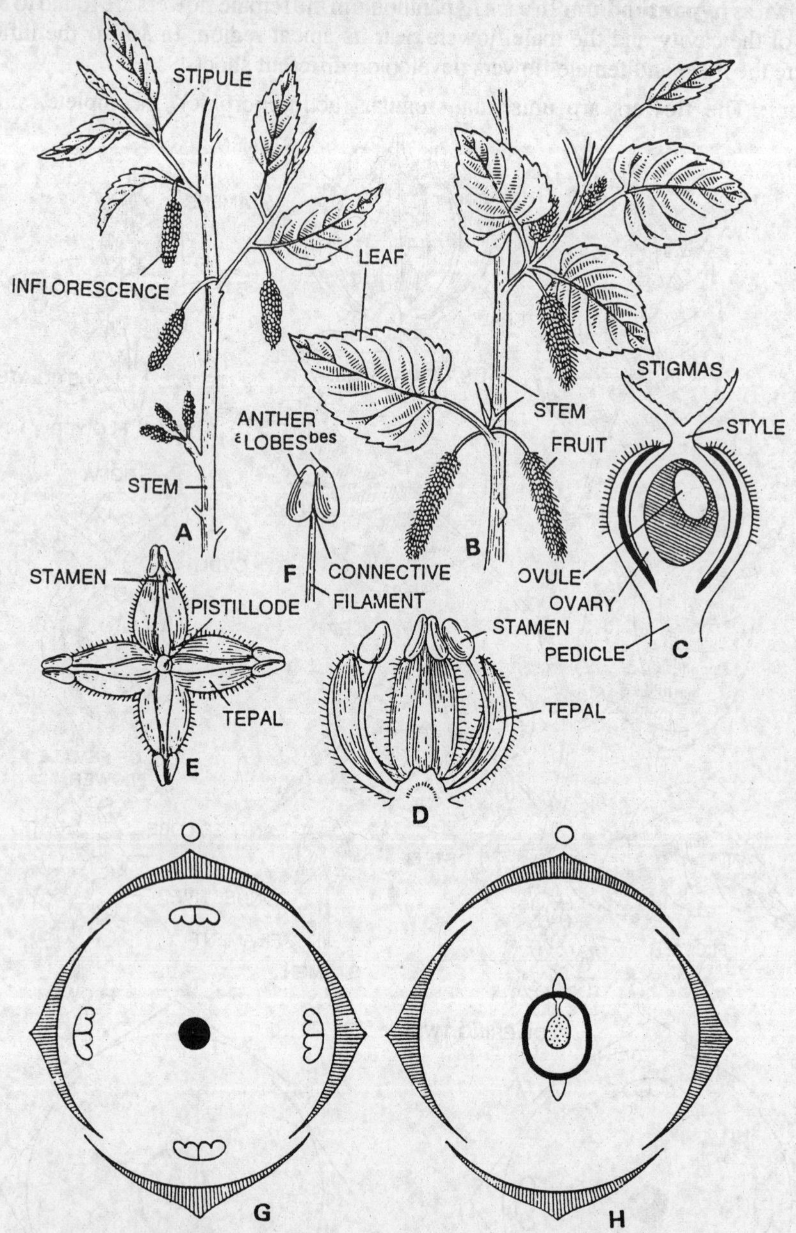

Fig. 16.2. Moraceae. *Morus alba* L. (Tut). A, male flowering twig ; B, female flowering twig ; C, L.S. of female flower ; D, L.S. of male flower ; E, male flower ; F, stamen ; G, floral diagram of male flower ; H, floral diagram of female flower.

Male Flowers: In the male flower generally the number of stamens is equal to that of perianth leaves, and they are situated opposite the perianth leaves. The filaments are incurved in *Morus* sp. or straight, *e.g.*, in *Ficus*. Sometimes in the species of *Ficus, Dorstenia* and *Artocarpus* the number of stamens reduces to two or one. The anthers are two-celled, versatile and dehisce longitudinally.

Female Flowers: In the female flower there are two medianly placed carpels, the hinder one of which shows various degrees of abortion. In many cases it is represented only by a style similar to that of the anterior carpel. In *Artocarpus* the hinder carpel is sometimes developed, while in *Morus* it is represented by an empty ovary chamber. Sometimes the second style is either represented by a small protuberance or altogether absent, and there is no remnant or indication of the hinder carpel, *e.g., Chlorophora.* The ovary is unilocular with a pendulous more or less curved ovule.

Fruit: The fruit is generally a drupe, the perianth becomes fleshy and surrounds it. It is sometimes an achene. The single fruits are frequently aggregated together by the union of neighbouring flowers to form a pseudocarp. In *Morus*, the individual fruits may be seen with their enclosing perianth, but in *Artocarpus,* the fruits, perianth leaves and axis become fleshy and so united that one solid mass is resulted. The fruit is called **sorosis** in the case of *Morus* and *Artocarpus* while **syconus** in the case of *Ficus.*

Seeds: Albuminous or exalbuminous with a bent or curved embryo.

Pollination: Usually takes place through the agency of inisects (entomophily). In many *Ficus* species the pollination is effected by a gall-wasp which lays its eggs in the ovaries of the female flowers and when it insists the flower it transfers the pollen.

Floral Formulae

Male flower : ⊕ ♂ P2 + 2, A4 or 2–1

Female flower : ⊕ ♀ P2 + 2, G$(\underline{2})$ or (2)

Bentham & Hooker(1862)	**Engler & Prantl(1931)**	**Hutchinson(1959)**
Phanerogams	Phanerogams	Angiospermae
Dicotyledones	Dicotyledones	Dicotyledones
Monochlamydeae or	Archichlamydeae	Lignosae
Incompletae	Urticales	Urticales
Unisexuales	Moraceae	Moraceae

Economic Importance of the Family
(*Moraceae*)

The family is fairly important from the point of view of its economic value. A list of certain important plants is given below:

1. *Artocarpus heterophyllus*; Syn. *A. integra; A. integrifolia*; Eng.–Jack tree or jack fruit; Verna.–**Kathal**–This is a tree, cultivated for its fruits in Uttar Pradesh, Bengal, Bihar, Andhra Pradesh and Orissa. The fruit is eaten as vegetable and made into pickle. The bark is used for tanning. The yellow dye is obtained from the heart-wood. The leaves are used for skin diseases. The ripe fruit is laxative.

2. *Artocarpus lakoocha*; Eng.–Monkey jack; Verna.–**Berhal**–The ripe, yellow coloured fruits are edible. It is a tree, cultivated in Uttar Pradesh, Khasia hills and Western Ghats.

3. *Artocarpus communis*; Syn. *A. altilis; A. incisa*; Eng.–Bread fruit.–This is a small tree grown in Maharashtra and South India. The fruits are eaten as vegetable.

4. *Artocarpus chaplasha*; Verna.–**Chaplash**–The strong wood of this tree is used for making furniture, tea boxes and packing cases. The tree is found in Assam, Bengal and the Andaman Islands.

5. *Ficus benghalensis*; Eng.–Banyan; Verna.–**Bargad**–This is a large shade giving tree. This is supposed to be the sacred tree of Hindus. The wood is quite durable under the water. The leaves are used as fodder for cattle and elephants. A fibre is obtained from the bark which is used for coarse cordage. The milky juice is applied in rheumatic pains externally. The bark is used for the treatment of dysentery and diarrhoea. The trees are grown as hedge plants by the sides of the roads. This is found throughout India.

6. *Ficus altissima*; Verna.–**Gadgubar**–This is a tree, commonly found in Assam, the Khasia hills and the Andaman Islands, The coarse cordage is made from its aerial roots. The inner bark is used for making coarse paper.

7. *Ficus religiosa*; Verna.–**Pipal**–The Hindus regard the tree very sacred and grow it near the temples. The bark is astringent. The seeds are cooling and the fruits are laxative. The leaves and young shoots are given as purgative. The bark is used for tanning. The tree is grown as an avenue.

8. *Ficus carica*; Eng.–Fig., Verna.–**Anjir**–This is a small tree, cultivated in the Punjab, Uttar Pradesh, Rajasthan, Andhra Pradesh and Maharashtra for its edible frutis. The fresh fruits are rich in vitamins. The dried fruits are used as laxative. The latex of fig is used in several ways. It also possesses medicinal properties.

9. *Ficus elastica*; Eng.–India rubber; Verna.–**bor**–This is a big tree, native of India. It is found in the Assam and Khasia hills. The rubber is prepared from the latex of the bark of the tree. The tree is also grown as a shade tree and an ornamental.

10. *Ficus glomerata*; Verna.–**Gular**–This is a tree producing edible frutis. The leaves used as fodder and the bark is used for tanning. The unripe fruits are used as vegetable. The wood of the tree is supposed to be sacred and burnt in hawans. The root is used as a remedy for dysentery. The milky juice is used for the treatment of piles and diarrhoea. Also serves as a host for the lac insect.

11. *Ficus cunia*; Verna.–**Khewnau**–This is a tree. The bark yields a fibre which is used for making cordage. The trees are found in the Punjab, Bihar, Khasia hills and Manipur.

12. *Ficus hispida*; Verna.–**Kathumbar**–This is a large shrub or a small tree. The fruits are edible. They are found throughout our country.

13. *Ficus lacor*; Syn. *F. infectoria*; Verna–**Plaksha**–This is tree. Commonly found in Northen India, Madhya Pradesh and Western Peninsula. Usually the tree is grown as an avenue. This also serves as a host plant for lac insect.

14. *Ficus rumphii*; Verna.–**Pakar**–This is a tree. The leaves and branches are used as fodder for cattle and elephants. The bark is used as an antidote for snake bite. The fruits are edible.

15. *Ficus palmata*; Syn. *F. virgata* ; Verna.–**Anjiri**–This is a tree. The fruits are edible. The young shoots are used as vegetable. The latex may be used in the preparation of curd.

16. *Broussonetia papyrifera*; Eng–Paper–Mulberry–This is a small tree found in Northern India. A strong lustrous fibre is obtained from the bark. This fibre is used for making writing paper, umbrellas and paper lanterns. The pulp is used for making newsprint paper.

17. *Morus alba*;,Eng.–White mulberry ; Verna.–**Tut**–This is a tree, cultivated in the Punjab, Uttar Pradesh, Kashmir and the North Western Himalayas. The fruits (catkins) are edible. They are very sweet and full of vitamins. The hockey–sticks are made from its wood.

18. *Morus serrata*; Verna.–**Kimu**–The fruits are edible. The leaves are used as fodder. The trees are found in the Western Himalayas.

19. *Morus austrails*; Syn. *Morus acidosa*; *Morus indica*; Eng.–Common mulberry; Verna–**Tut**–The fruits are edible. The leaves are the source of food of silk worms. The trees are commonly found in the sub–Himalayan tracts.

Family—Urticaceae

There are about 41 genera and 500 species in this family (Rendle).

Distribution: Usually the members of the family are common in tropical regions but several species, *e.g., Urtica dioica*, are quite common in temperate climates. About one third of the total number of species are found in America. *Parietaria debilis* is widely distributed and commonly found in Siberia to New Zealand and from North America to Argentina. *Urtica hyperborea* is found at 15,000 feet elevation in the Himalayas. The species of *Parietaria* are chiefly found in temperate regions and rarely in tropics.

Habit: Generally the plants are herbs and under–shrubs. The shrubs and trees are rare. Climbing habit is also found. The milky latex, which is commonly found in Moraceae is absent from this family. Stinging hairs often present on the leaves and stems of various plants, *e.g., Urtica, Laportea* etc.

From the viewpoint of the anatomy, the cystoliths (calcium carbonate crystals) are found in enlarged epidermal cells which vary in their size and shape from species to species. The bast fibres are quite long, *e.g.,* in *Boehmeria nivea* (Ramie) an individual fibre reaches upto nine inches in length.

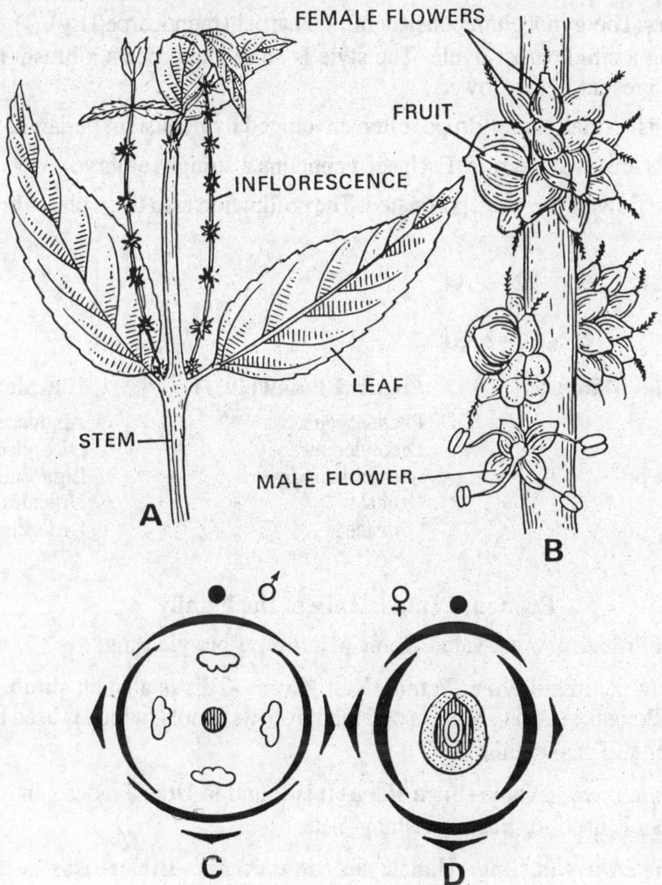

Fig. 16.3. Urticaceae. *Boehmeria* sp, A, a twig with inflorescence; B, close–up view of inflorescence showing male and female flowers; C and D, floral diagrams of male and female flowers of *Urtica* sp.

Leaves: The leaves are simple, alternate or opposite, stipulate except in *Parietaria* where stipules are absent. Sometimes the leaves may be alternate on the stem and opposite in the inflorescences on same plant. Stinging hairs are very often present on the leaves.

Inflorescence: The inflorescence is of cymose type. The flowers are sessile and so arranged that the inflorescence appear either to be condensed heads (pseudoheads) or catkin as in *Boehmeria*. In certain cases the inflorescence reduces to a single flower.

Flowers: Usually the flowers are unisexual, and monoecious or dioecious. Rarely the flowers are hermaphrodite, *e.g.,* in *Parietaria*. The flowers are incomplete, when unisexual and complete when bisexual. They are actionomorphic (regular) and hypogynous. In male flowers the rudimentary pistil is present whereas in female flowers the staminodes may be present. These facts suggest that the unisexual condition has been derived from the bisexual condition.

Perianth: Usually it consists of four perianth leaves arranged in two whorls of two each. The perianth leaves are green, free or more or less united especially in the female flowers.

Male Flowers: Usually the androecium consists of four stamens, in rare cases they are 3–5 in number. The stamens are situated opposite the perianth leaves. In bud condition, the filaments are bent inwards but on maturity they spread backwards and outwards. The anthers are two celled and release the pollen in a tiny cloud when mature. The rudiment pistil is often present in the flower.

Female Flowers: The gynoecium consists of one carpel (monocarpellary). The ovary is unilocular, superior having a single erect ovule. The style is simple and bears a brush–like stigma. The staminodes are often present in the flower.

Fruit: The fruit is a small nut or drupe, often enveloped by persistent perianth.

Seeds: The seeds are endospermic. Each seed contains a straight embryo.

Pollination: The flowers are wind pollinated. The pollination also takes place through the agency of insects.

Floral Formulae: ⊕ ♂ P2+2, A4

⊕ ♀ P2+2, $\underline{G1}$.

Bentham& Hooker (1862)	**Engler & Prantl (1931)**	**Hutchinson (1959)**
Phanerogams	Phanerogams	Angiospermae
Dicotyledones	Dicotyledoneae	Dicotyledones
Monochlamydeae or	Archichlamydeae	Lignosae
Incompletae	Urticales	Urticales
Unisexuales	Urticaceae	Urticaceae
Urticaceae		

Economic Importance of the Family

The family is of little economic value. Some plants are fibre yielding.

1. *Boehmeria nivea*; Eng.–Ramie; Verna.–**Kan Khura**–This is a small shrub. It is commonly grown in Assam and Bengal. A strong fibre is obtained from its shoots, which is used in making sacks, cordage, thread, paper and gas–mantles.

2. *Boehmeria macropylla*; Verna–**Bara siaru**–It is found in Uttar Pradesh and Bihar. The fibre obtained from the stem is used for making fishing–nets.

3. *Girardinia heterophylla*; Eng.–Himalayan nettle; Verna.–**Bichu**–This is a perennial herb commonly found in Kashmir, Assam and the Khasia hills. The bark fibre is used for making rough cloth, ropes and cordage.

4. *Pilea muscosa*, grown as an ornamental in the gardens for its beautiful foliage.

5. *Urtica dioica*, Verna.–**Bichhu ghas**–very commonly found in the hills of Himalayas. The stem yields a storng fibre which is used for many purposes. The plant possesses some medicinal properties and the decoction of it is given in jaundice.

6. *Maoutia puya*. Verna–**Pua, Poi**–This is a shrub found in Uttar Pradesh, Bihar and Assam. The stem fibre is used in making fishing nets and twines.

Systematic Relationships: Engler and Prantl and Hutchinson put the families Moraceae and Urticaceae in the Order Urticales. Within the order, the family Urticaceae has been considered to be most advanced.

17

ORDER–THYMELAEALES

According to Hutchinson this is the twenty first order of the phylum, Angiospermae, subphylum, Dicotyledones and division Lignosae. This order comprises of six families, *i.e.*, Gonystylaceae, Aquilariaceae, Geissolomataceae, Penaeaceae, Thymelaeaceae and Nyctaginaceae. The Nyctaginaceae is the highest evolved family within this order. This family has been described here in detail. Bentham and Hooker have placed the family Nyctagineae along with other six families in their first series Curvembryae of class Dicotyledones and division Monochlamydeae or Incompletae. Engler and Prantl put the family Nyctaginaceae along with eight other families in the seventeenth order–the Centrospermae of class Dicotyledoneae and sub-class Archichlamydeae.

Mostly apetalous and Monocarpellary, ovules few to solitary, flowers often capitate; endosperm present or absent; calyx imbricate or valvate; stipules absent or minute and glandular.

Family–Nyctaginaceae

There are about 30 genera and 300 species in this family.

Distribution: The members of this family are found in the tropical and sub-tropical region on northern and southern hemispheres. The plants are especially found in America. In our country the family is represented by several genera, such as *Bougainvillaea, Mirabilis, boerhaavia,* etc.

Habit: The plants may be herbs, shrubs or trees. The species of *Boerhaavia* are herbs. *Mirabilis jalapa* is a well-known beautiful shrub. The species of *Bougainvillaea* are climbers, *Bougainvillaea spectabilis,* a native of Brazil is a beautiful climber.

Needle-like raphides of calcium oxalate are found in the various organs of the plant. Anatomically, the structure of the stem is anomalous. The cambium develops outside the original ring of bundles from which now closed bundles and conjunctive tissue are produced. The phloem in the new bundles is less developed.

Leaves: Usually the leaves are opposite, entire, simple and exstipulate. The opposite leaves of a pair are usually very unequal in size.

Inflorescence: The inflorescence is of cymose type. In *Mirabilis,* the dichasial cyme has a tendency to become monochasial cyme in the higher branches, in this case, each flower is surrounded by an involucre of flve sepaloid bracts. In *Bougainvillaea,* there are three large, coloured, petaloid bracts which surround a group of three flowers. Each flower becomes adnate at the base of each petaloid bract. In *Boerhaavia* and certain other plants, the involucral bracts are reduced to teeth or scale-like structures. In *Mirabilis,* the flowers are arranged in cymes of three, of which only the middle flower develops. There remains an involucre of five parts at its base, which really belongs to the cyme, but looks exactly like the calyx of the flower.

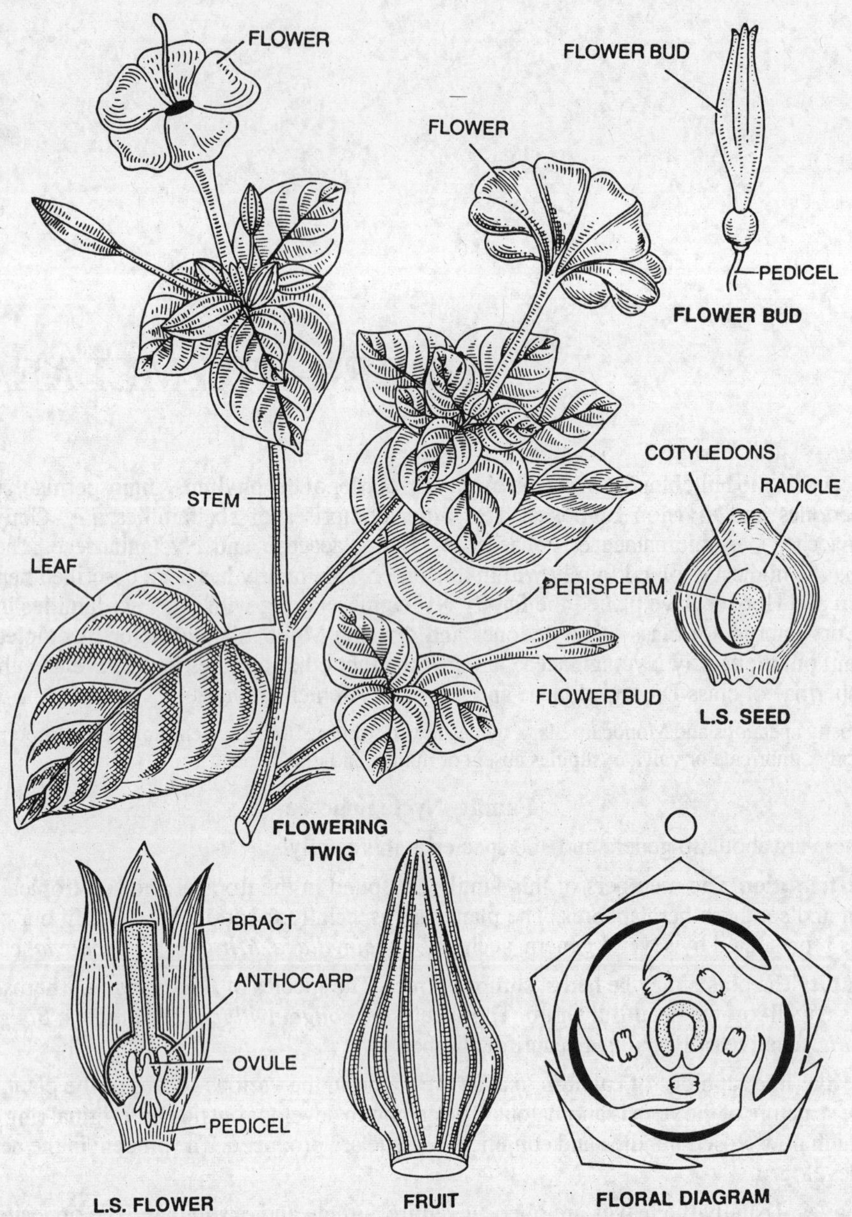

Fig. 17.1. Nyctaginaceae–*Mirabilis jalapa* Linn.

Flowers: The flowers are bisexual (hermaphrodite) but sometimes as in *Pisonia*, by suppression of stamens or pistil they become unisexual. They are regular (actinomorphic) and hypogynous.

Perianth: Usually it consists of five petaloid perianth leaves. It is tubular and remains persistent in the fruit. The persistent portion of the perianth which envelops the fruits becomes leathery, hardened or fleshy. The mucilage or glandular hairs develop on this portion and serve as a means of dissemination, *e.g.*, in *Pisonia.*

Androecium: Usually it consists of five stamens alternating with the perianth leaves. The number of stamens is very variable. It ranges from 1 to 30. The filaments are often unequal in their length.

Gynoecium: It consists of a single carpel. The ovary is free, superior and unilocular. It bears a long simple style and a single basal erect ovule. The ovule may be anatropous or campylotropous.

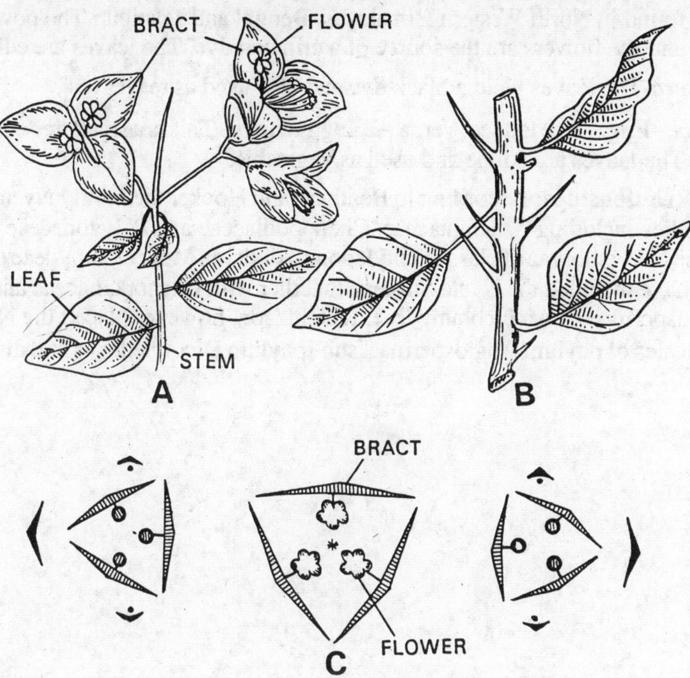

Fig. 17.2. Nyctaginaceae–*Bougainvillaea;* A, a flowering shoot; B, a vegetative shoot; C, floral diagram of inflorescence.

Fruit: The fruit is an achene which remains enclosed within the persistent perianth.

Seeds : The seeds are endospermic. It contains a large, straight, curved or folded embryo. The cotyledons are unequal in size.

Pollination: It takes place through the visiting insects. The petaloid bracts attract the insects and the visiting insects effect the cross pollination.

Floral formula: Br ⊕ ⚥ P5, A 5, <u>G1</u>

Bantham&Hooker (1962)	**Engler&Prantl (1931)**	**Hutchinson (1959)**
Phanerogams	Phanerogams	Angiospermae
Dicotyledones	Dicotyledones	Dicotyledones
Monochlamydeae or	Archichlamydeae	Lignosae
Incompletae	Centrospermae	Thymelaeales
Curvembryae	Nyctaginaceae	Nyctaginaceae
Nyctagineae		

Economic Importance of the Family

From the point of view of its economic importance, the family is of little value. Usually the plants are grown as ornamentals in the gardens.

1. *Bougainvillaea spectabilis*; Eng.–Glory of the garden; Verna–**Baganvilas–**A woody, scandent or straggling shrub of South America; very commonly grown in gardens as ornamental.

2. *Bougainvillaea glabra*–This is a native of Brazil grown in the gardens as ornamental.

3. *Bougainvillaea buttiana*–This is a climbing, spiny shrub, grown in the gardens as ornamental.

4. *Boerhaavia diffusa*-Verna.-**Santh**-the root of the plant possesses medicinal properties. It is used in asthama, jaundice, anaemia, etc. This is also an antidote to snake bite.

5. *Mirabilis jalapa*-Eng.-Four o-clock plant, Verna.–**Gulabbas**–This is ornamental herb or small shrub, commonly found in North Western Himalayas, Bengal and Manipur. The powdered seeds are used in cosmetics and the flowers are the source of a crimson dye. The leaves are edible.

6. *Neea theifera*-The leaves yield a black dye and also used as tea leaves.

7. *Pisonia alba*–Eng.–Tree lettuce, Verna.–**Bhaggachura**–This is a small tree cultivated in South India and Bengal. The leaves are edible and used as vegetable.

Systematic Relationships: According to Bentham and Hooker the family Nyctagineae together with six other families including Amarantaceae, Chenopodiaceae and Polygonaceae has been placed in the first series the Curvembryae of class Dicotyledones, division Monochlamydeae or Incompleteae. Engler and Prantl have included the Nyctaginaceae together with Chenopodiaceae and Amarantaceae in the other Centrospermae of Archichlamydeae. Hutchinson, however placed the Nyctaginaceae in the order Thymelaeales of phylum Angiospermae, sub–phylum Dicotyledones and division Lignosae.

18

ORDER–CAPPARIDALES

According to Hutchinson this is the twenty-fourth order of his phylum Angiospermae, sub phylum dicotyledones and division Lignosae. The order comprises of three families, *i.e.,* Capparidaceae, Moringaceae and Tovariaceae. The Capparidaceae is the primitive most family within the order. This family has been described in this text in detail. Bentham and Hooker have placed the family Capparidaceae along with other eight families in their order the Parietales of class Dicotyledones, division Polypetalae and series Thalamiflorae. Engler and Prantl put the family Capparidaceae along with other five families in their nineteenth order the Rhoeadales of class Dicotyledoneae, and sub-class Archichlamydeae.

Hypogynous or subperigynous; bisexual; petals present; syncarpous with parietal placentation; carpels 2; stamens numerous to few, mostly equal; endosperm absent; embryo variously folded; flowers actinomorphic to zygomorphic; leaves rarely stipulate.

Family—Capparidaceae

There are about 45 genera and 700 species in this family.

Distribution: The members of this family are mainly distributed in the warmer (tropical) parts of the world. The plants occur mostly in dry regions. Several shrubby species of *Capparis* occur in the Mediterranean region. Several genera, such as *Capparis, Gynandropsis, Cleome, Crataeva,* etc., are found in our country.

Habit : They may be herbs, shrubs or trees. *Gynandropsis pentaphylla* is an annual rainy season herb, *Capparis decidua (C. aphylla)* is a xerophytic shrub, *Crataeva nurvala(C. religiosa)* is a tree. Sometimes the plants are climbing, *e.g.,* species of *Maerua.*

Root : Tap and branched.

Stem: Erect, branched, woody or herbaceous, solid, cylindrical, etc.

Leaves : The leaves may be simple or palmately compound. The stipules may or may not be present. in *Capparis spinosa* the stipules are modified into spines. In certain species, *e.g., Capparis decidua* (*i.e., C. aphylla*) the leaves are altogether absent.

Inflorescence : Usually the inflorescence is of racemose type. The flowers of *Cleome, Gynandropsis,* etc., are arranged in racemes; in *Maerua* and *Capparis decidua,* they are arranged in corymbs; in *Capparis sepiaria,* the flowers are found to be arranged in umbels; in *Crataeva nurvala* they are found to be arranged in corymbose clusters. Rarely the flowers are solitary, *e.g., Niebuhria.*

Flowers : The flowers are pedicellate and bracteate. The bracteoles are absent. From the point of view of the structure of the flower, the family occupies an intermediate position between Papaveraceae and Cruciferae. The flowers are hermaphrodite, actinomorphic (sometimes zygomorphic, *e.g., Capparis decidua*), complete, regular or irregular and hypogynous.

Calyx : It consists of four sepals, arranged in two whorls. Each whorl consists of two sepals similar to that of Cruciferae. The sepals are free (polysepalous). The sepals of *Capparis decidua* are unequal in size. In this case the posterior sepals form a hood-like structure. The aestivation is either valvate or imbricate.

Corolla: It consists of 4 petals, polypetalous. The petals are arranged diagonally similar to that of Cruciferae. Usually regular but sometimes irregular, *e.g.*, in *Capparis decidua.* The aestivation may be imbricate, *e.g.*, in *Cleome* or valvate, *e.g.*, in *Crataeva.*

Androecium: The number of stamens vary much in different species. In different species of *Cleome* the number of stamens ranges from four to indefinite. The most reduced type of flower is *Cleome tetrandra* where there are only four stamens in *Cleome spinosa*, there are six stamens; in *C. viscosa,* the number of stamens varies from 12 to 24 and in *C. chelidonii,* the number of stamens is idenfinite. In *Gynandropsis,* the number of stamens is six; in *Capparis, Maerua* and *Crataeva,* the stamens are indefinite. This way, the number of stamens ranges from 4 to indefinite in the different species of the family. The stamens are never tetradynamous in Capparidaceae.

The characteristic of this family is the development of an internode between petals and stamens, *i.e.,* androphore or between stamens and pistil, *i.e.,* gynophore. The androphore and gynophore are known together as gynandrophore. In *Gynandropsis pentaphylla*, both the gynophore and androphore are well developed. In *Capparis, Crataeva* only gynophore is present.

The filaments are filiform, the anthers are basifixed and dithecous.

Gynoecium: It is situated on a gynophore. It consists of two carpels (bicarpellary) syncarpous. Rarely the number of carpels is more than two. The ovary is superior and unilocular. Sometimes it becomes multilocular by the ingrowth of parietal placentas. The placentation is parietal. The ovules are many and campylotropus. The style is either very short or absent.

Fruit: In *Cleome* it is siliqua with replum, in *Capparis* it is berry, in *Rogdsia*, it is drupe and in *Emblingia* it is a nut. In majority of cases the fruit is siliqua.

Seeds: The seeds are non-endospermic. They are reniform. Each seed contains a large variously folded embryo.

Pollination: It takes place by means of insects (*i.e.*, entomophilous). Dichogamy and protandry are found. The nectar is also secreted which attracts insects.

Floral Formula: ⊕ ⚥ K 2+2, C 4, A 4– ∝ , G (2).

Bentham& Hooker (1862)	**Engler & Planil (1931)**	**Hutchinson (1959)**
Phanerogams	Phanerogams	Angiospermae
Dicotyledones	Dicotyledones	Dicotyledones
Polypetalae	Archichlamydeae	Lignosae
Thalamiflorae	Rhoeadales	Capparidales
Parietales	Capparidaceae	Capparidaceae
Capparidaceae.		

Description of some important plants in semi-technical language.

1. *Capparis spinosa* Linn.; Verna., **Kabra**.

Habit : Erect or climbing shrubs, sometimes small trees.

Root : Tap and branced.

Stem: Erect, branched, solid, rough, woody, spiny, usually armed with a pair of stipular thorns at the base of petioles.

Leaves: Simple, rarely altogether absent, alternate, stipulate, stipules spiniscent, acute apex, reticulate venation.

Inflorescence: Cymose or racemose. In *C. aphylla* manyflowered corymbs, in *C. sepiaria* many flowered umbels or umbellate corymbs, and in *C. horrida* solitary axillary cyme.

Flower : Pedicellate, hermaphrodite, actinomorphic or zygomorphic, ebracteate, tetramerous, hypogynous and complete.

Calyx : Four sepals, arranged in two series, polysepalous, imbricate aestivation, inferior. In *C. aphylla*, the sepals unequal, the inner being saccate.

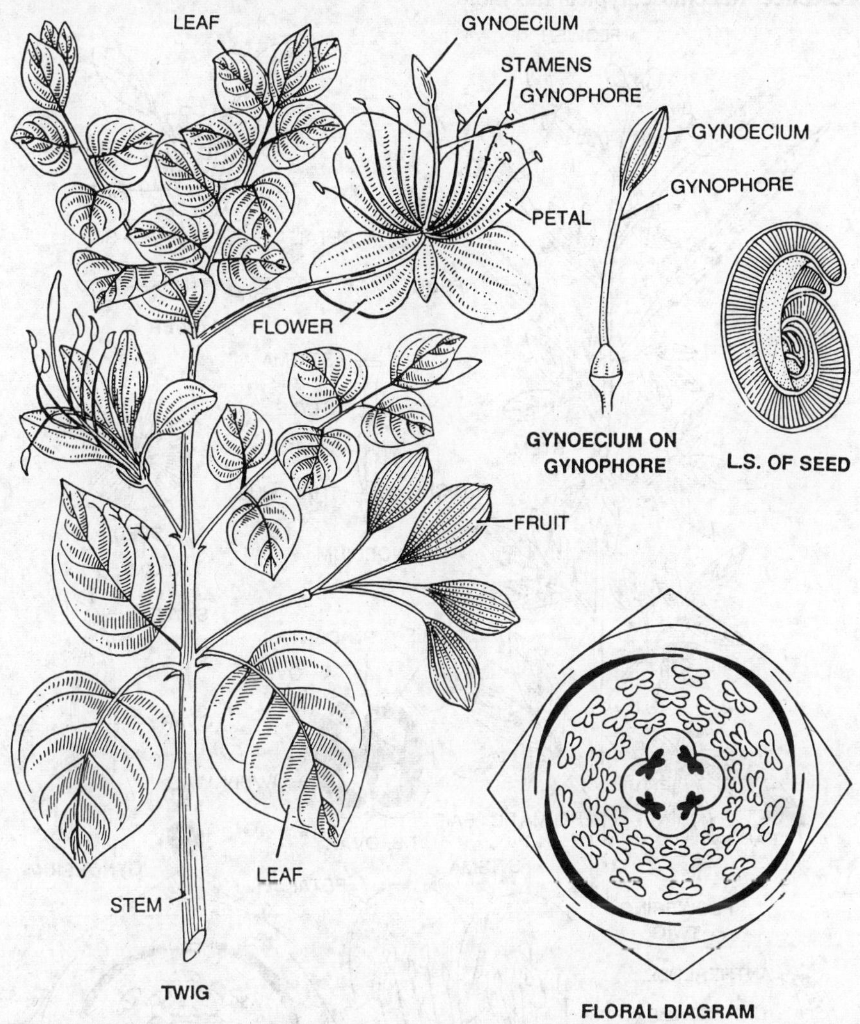

Fig. 18.1. Capparidaceae–*Capparis spinosa* Linn.

Corolla: Four petals, polypetalous, sessile.

Androecium : Many, polyandrous, inserted on the disc at the base of the long gynophore, anthers dorsifixed, introrse, inferior.

Gynoecium : Two carpels (bicarpellary), syncarpous, ovary superior, unilocular, parietal placentation, many ovules on each placenta, ovary being situated on a long stalk (gynophore), stigma sessile.

Fruit : A berry.

Floral Formula: ·|· or ⊕ ⚥ K 2 + 2, C 4, A ∞, $G\underline{(2-6)}$.

2. ***Cleome viscosa.***

Habit: An annual rainy season herb, weed, 1-3 feet high.

Root: Tap, branched.

Stem: Erect, glandular pubsecent, branched, solid.

Leaves: 3-5 foliate, leaflets sessile, ovate or obovate.

Inflorescence: Racemose, typical raceme.

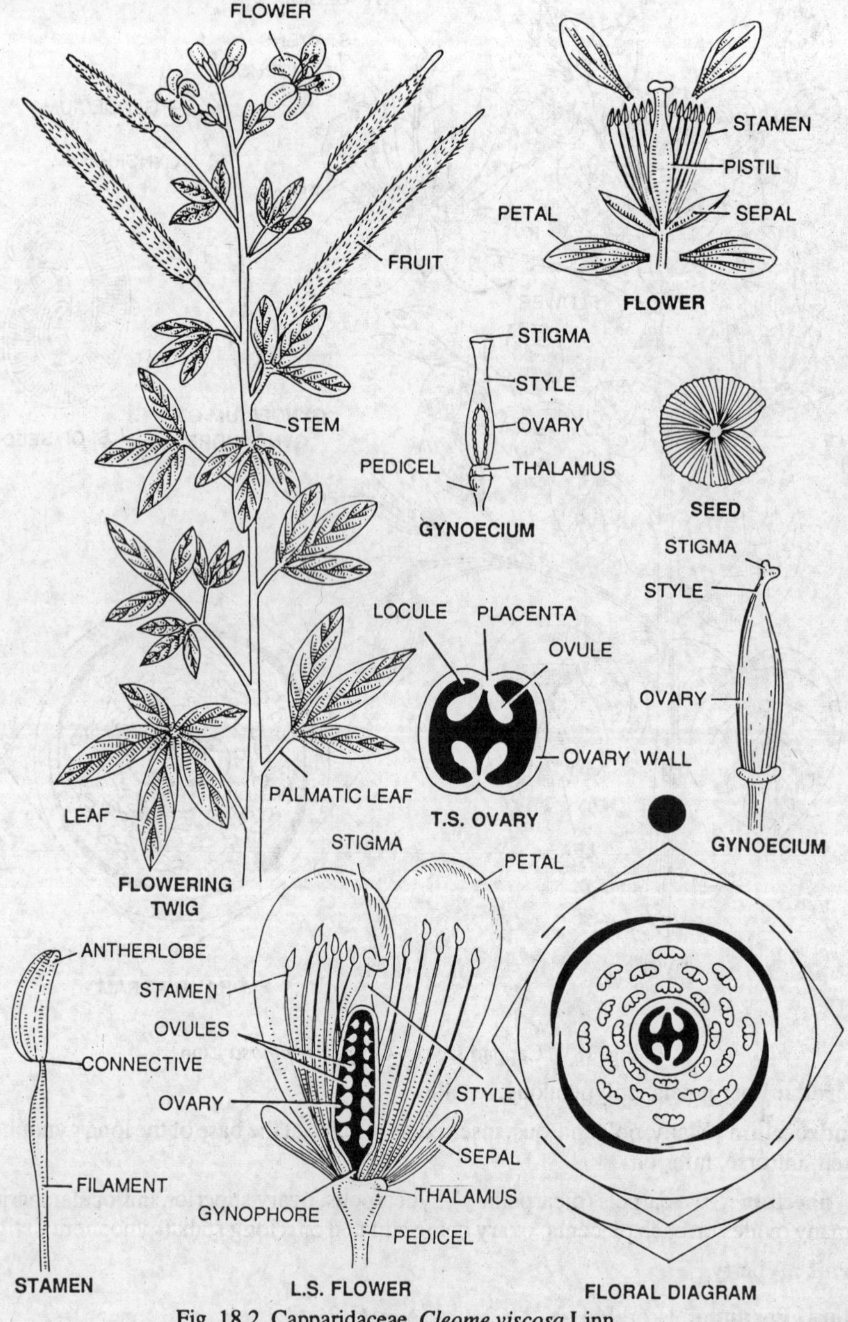

Fig. 18.2. Capparidaceae. *Cleome viscosa* Linn.

Flower : Pedicellate (long pedicels), ebracteate, hermaphrodite, yellow, actinomorphic, complete.

Calyx : Four sepals, arranged in two whorls, polysepalous, ovate, acute, inferior valvate.

Corolla: Four petals, polypetalous, two proximate and two spreading, obtuse, clawed, yellow, inferior, valvate aestivation.

Androecium: Many stamens (12-24), polyandrous, filaments long, inferior, anthers introrse, basifixed.

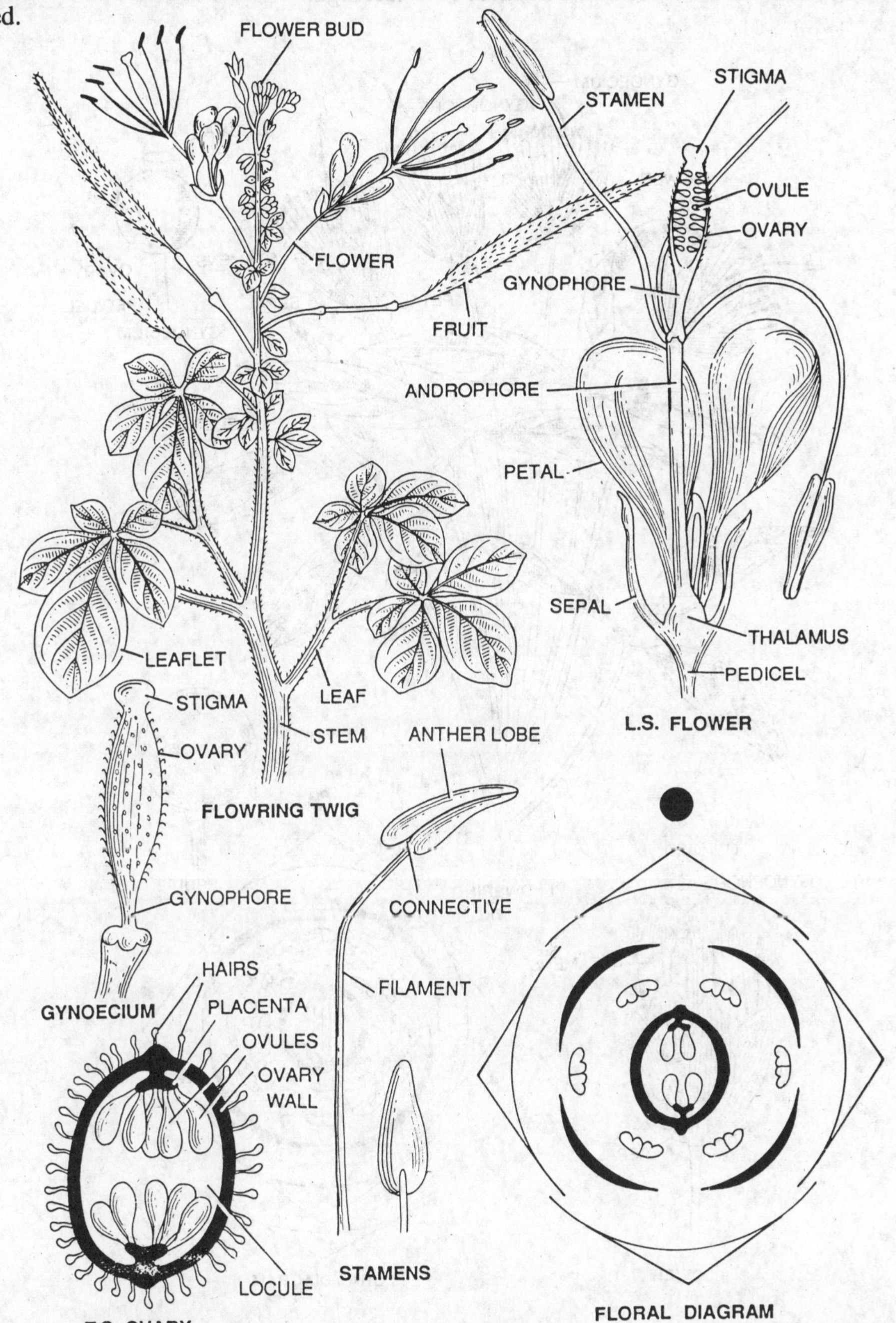

Fig. 18.3. Capparidaceae. *Cleome gynandra* L. Syn. *Gynandropsis pentaphylla* (Verna.–**Hurhur**)

Gynoecium: Two carpels (bicarpellary), syncarpous, ovary superior sessile or sub-sessile, unilocular, placentation parietal, many ovules on two parietal placentas, style short, stigma round.

Fruit : Oblong or linear capsule.

Floral Formula: ⊕ ⚥ K 2 + 2, C4, A ∝, G(2).

3. ***Gynandropsis pentaphylla***; Verna. **Hulhul.**

Habit: An annual rainy season wild herb, 1 – 3 feet high.

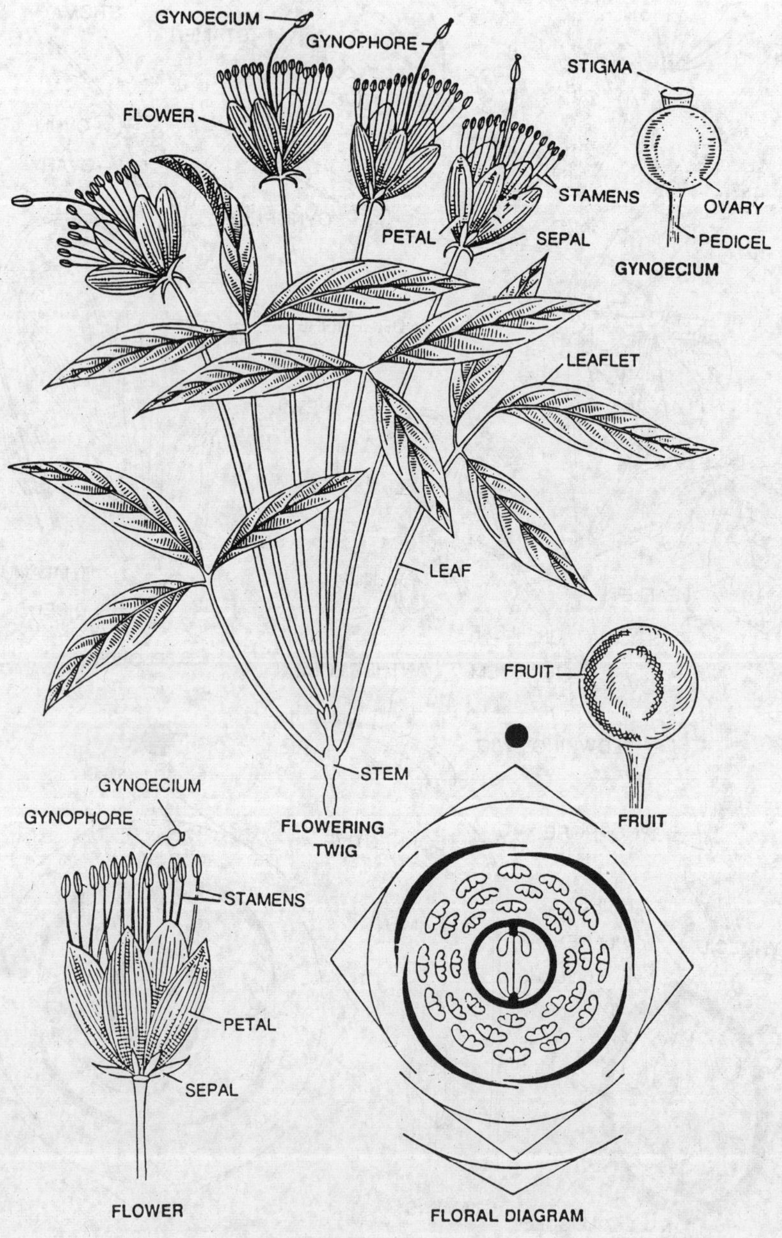

Fig. 18.4. Capparidaceae–*Crataeva religiosa* Linn.

Root: Tap, branched.

Stem: Erect, cylindrical, branched, glandular pubescent, solid.

Leaf: Palmately compound, alternate, petiolate, pentafoliate, leaflets sessile, unequal obovate, entire or finely serrate margin, leaflet's venation unicostate reticulate.

Inflorescence: Racemose, terminal corymboracemes.

Flower: Pedicellate, hermaphrodite, actinomorphic, complete bracteate (bracts foliaceous and trifoliate), hypogynous, white or purple, distinct gynandrophore present.

Calyx: Four sepals, polysepalous, arranged in two whorls of two each, imbricate aestivation.

Corolla: Four petals, polypetalous, long clawed, white, valvate aestivation.

Androecium: Six stamens, polyandrous, androphore present, anthers purple, introrse, filaments adnate below to the slender gynophore.

Gynoecium: Two carpels (bicarpellary), syncarpous, superior ovary, stalked or being situated on an elongated gynophore, unilocular, parietal placentation, ovules many, stigma capitate and sessile.

Fruit : Stalked elongate capsule.

Floral formula: Br ⊕ ⚥ K 2 + 2, C4, A 6, G(2).

4. ***Crataeva religiosa.***

Habit: A small spreading unarmed tree.

Stem : Erect, woody, branched, cylindrical, solid.

Leaves : Compound, alternate, deciduous, appearing with the flowers, petiolate (long petioles), 3-foliate, leaflets 3 - 6" long, ovate or ovate lanceolate, acuminate, entire, glabrous, pale beneath.

Inflorescence : Racemose, flowers arranged in corymbose clusters at the ends of the new growths.

Flower Large, yellow or purplish, hermaphrodite, actinomorphic (regular), pedicellate, hypogynous, complete.

Calyx: Four sepals, polysepalous, inferior, inserted on the edge of a large lobed disk, valvate or imbricate.

Corolla: Four petals, polypetalous, long clawed, open in bud, claw half as long as the limb, ovate or oblong, obtuse or acute.

Androecium: Stamens many, polyandrous, adnate to the base of the gynophore, stamens longer than the petals, filaments long, anthers bicelled, introrse.

Gynoecium: Two carpels (bicarpellary) ovary superior, on a slender stalk, unilocular, parietal placentation, ovules, many on two parietal placentas.

Fruit: Berry, 1-2" diameter situated on a thick woody gynophore, pericarp hard.

Floral formula : ⊕ ⚥ K, 4, C 4, A ∞ G (2).

Economic Importance of the Family

The family is of little economic importance. Some of the plants of economic value are given here.

1. *Capparis decidua;* Syn. *C. aphylla*; Verna.—**Karil**–This is a shrub without leaves. It is found in the Western Peninsula, Punjab, Rajasthan and Uttar Pradesh. The flowers, buds and the fruits are used as vegetable. The pickle is prepared of green berries. The bark is used in cough, asthama and inflamation. The root is useful in fever and rheumatism. The wood is hard and heavy and is used for making tool-handles.

2. *Capparis grandis*; Verna.—**Puchaonda**–A small tree. It is found in Rajasthan and Western Ghats. It yields an oil which is used as an illuminant.

3. *Capparis sepiaria*; Verna.–**Kanthari**–It is branched, leafy, climbing shrub with recurved spines. The plants are planted as hedge plants. It is also used medicinally in skin diseases.

4. *Capparis spinosa*; Verna.–**Kabra**–The flower buds are used as condiment. The root bark is used medicinally in the treatment of rheumatic pains.

5. *Capparis zeylanica* ; Syn. *C. horrida*; Verna.–**Gitoran**–A thorny shrub used as hedge plant. The fruits are edible and used as vegetable.

6. *Cleome spinosa*; A strongly scented herb, usually grown as an ornamental.

7. *Cleome viscosa*; Verna.– **Hurhur**-A common rainy reason herb with sticky shoot and yellow flowers. The seeds are used in curries. The seeds and leaves are also used medicinally. The seeds are stimulant and carminative.

8. *Cleome icosandra*; Verna–**Hurhuria**– The seeds are carminative. The leaves possess medicinal properties. They are used for curing ulcers and wounds.

9. *Crataeva nurvala*; Syn. *C. religiosa*; Verna. –**Barana**–This is a tree. The bark and leaves possess medicinal properties. The pulp of this plant is used as an adhesive and mixed with cement to strengthen it. The combs are prepared from its wood. The plants are commonly planted near Muslim tombs.

10. *Gynandropsis pentaphylla*; Verna. –**Hulhul**–The plants are annual rainy season white flowered herbs. The leaves, roots and stems possess several medicinal properties.

11. *Maerua* sp. The plants are of ornamental value. They are climbing shrubs. The roots used as tonic and stimulant.

Systematic Relationships : According to Norris (1941) the Resedaceae and the Capparidaceae are supposed to be most primitive families of the order Rhoeadales and this was also noted that the Cruciferae and Papaveraceae of the same order were derived by parallel evolution from a common ancestral group resembling the present Resedaceae and Capparidaceae. Dr. V. Puri also holds the similar views. He has shown that the parietal placentation in the families Moringaceae, Curciferae and Capparidaceae has been derived from the axile type. Hutchinson has separated the Capparidaceae as a distinct order Capparidales with herbaceous or woody plants. On the basis of the scanty development of endosperm, he has separated the family from rest of the Rhoeadales. Bentham and Hooker have placed the Capparideae in their second order the Parietales of Dicotyledones.

19

ORDER–VIOLALES

According to Hutchinson this is the twentysixth order of the phylum Angiospermae, sub-phylum Dicotyledones and division Lignosae. The order consists of a single family, *i.e.*, Violaceae. This family has been described in this text in detail. Bentham and Hooker have included the family Violaceae together with other eight families including Papaveraceae, Cruciferae, Capparidaceae. etc., in the second order–the Parietales of class Dicotyledones, division Polypetalae and series Thalamiflorae. Engler and Prantl have put the family Violaceae along with twenty eight other families in their twenty seventh order–the parietales of class Dicotyledoneae and subclass Archichlamydeae.

Hypogynous to perigynous; bisexual; rarely unisexual; petals present, syncarpous with parietal placentation; stamens several to few; endosperm present or absent; embryo straight or curved; flowers mostly zygomorphic; leaves with stipules.

Family–Violaceae

There are about 16 genera and 850 species in this family.

Distribution: The genera are widely distributed in temperate and tropical regions, but they are mainly found in the tropical regions of the world. The genus *Viola* is the largest of the family and contains about half of the species which are mainly found in the temperate zones and hilly tracts of the northern hemisphere. A few species of *Viola* are cultivated in our country for their ornamental value.

Habit: The plants are annual or perennial herbs. The under shrubs or shrubs are also found, and occasionally they may attain the tree-like habit. *Viola tricolor,* also known as garden Pansy is an ornamental annual herb. *Viola patrinii* possess sweetscented flowers and grow in Himalayas as perennial herbs. The species of *Ionidium* are either herbs or undershrubs. *Alsodeia roxburghii* and *A. lanceolata* are undershrubs. *Alsodeia bengalensis* and *A. racemosa* are large shrubs and occasionally may attain the size of a small tree. *Alsodeia maingayi* is a small tree. The species of *Anchietea* are climbers.

Leaves : The leaves are simple and stipulate. They are scattered and rarely opposite as in the species of *Hybanthus.* In the species of *Viola* the stipules become large and leaf-like. Mostly the leaves are entire and undivided but sometimes they are pinnately divided. The leaves of the species of *Alsodeia* are distichous.

Inflorescence: In majority of cases the inflorescence is of cymose type, *i.e.,* solitary axillary. In some cases it is racemose type. Sometimes two flowers arise in each axil. In *Viola tricolor* in addition to the flower a vegetative shoot also comes out from the same axil.

Flower: The flowers are bracteate and bracteolate. The bracts and bracteoles are well developed.

They are hermaphrodite and usually zygomorphic, but sometimes actinomorphic, *e.g., Alsodeia,* complete and hypogynous. In *Viola* the flowers are medianly zygomorphic, because of the presence of the large anterior petal which is spurred.

Pentamery prevails below the gynoecium. There are five sepals, five petals and five stamens arranged in alternating whorls. The parts of these whorls are usually free, but sometimes connate at the base.

Calyx: It consists of five sepals. The sepals are free, green in colour and generally uniform in size. The sepals are generally persistent and the aestivation is imbricate in the bud. In *Viola,* the sepals are appendiculate, *i.e.,* they are produced below the point of insertion.

Corolla : It consists of five petals. The petals are variously coloured. Usually they are large and either equal or unequal in size. They are polypetalous. The aestivation is either imbricate or contorted in the bud. In *Viola,* the anterior petal is large in size and spurred.

Androecium: It consists of five stamens, which are hypogynous. They are alternating with the petals. The anthers are erect and form a ring-like structure around the style. The anthers are introrse, twocelled and dehisce longitudinally. The anthers possess apical appendages. The stamens possess short filaments. In *Viola* and other zygomorphic flowers the connective of the anterior pair of anthers is spurred.

Gynoecium: It consists of three to five carpels. The ovary is unilocular and superior. The placentation is parietal. The ovules are many and anatropous. The style is simple, short and terminal. The stigma is very often hooded.

Fruit : The fruit is a capsule usually splitting into three boatshaped valves. Rarely the fruit may be a berry.

Seed : The seed is small, hard and shiny. They are endospermic and possess straight embryo. In climbing genera such as *Agatea* and *Anchieta*, the seeds are winged.

Pollination: It is usually of entomophilous type. In *Viola*, flowers are specially adapted for insect pollination. They are generally pollinated by bees, butterflies and moths. In *Viola biflora*, the flowers are pollinated by *Diptera.* The lower petal is spurred and forms the landing place for the insects. The anthers of the lower stamens possess nectar secreting processes which store the nectar. The connectives of all the five stamens form a membranous appendage around the style. This is a conical chamber which receives the pollen grains when the anthers dehisce. The stigma comes out of this cone and closes the opening to the flower. An insect first approaches the stigma and slightly raises it so that an opening is formed in the anther cone from which the pollen falls on the proboscis of the insect. This insect visits the other flowers and the pollens are transferred on its stigma, thus effecting the cross pollination.

Floral Formula: Br ·|· ⚥ K 5, C 5, A 5, G $\underline{(3)}$.

Bentham & Hooker (1862)	**Engler & Prantl (1931)**	**Hutchinson (1959)**
Phanerogams	Phanerogams	Angiospermae
Dicotyledones	Dicotyledones	Dicotyledones
Polypetalae	Archichlamydeae	Lignosae
Thalamiflorae	Parietales	Violales
Parietales	Violaceae	Violaceae
Violaceae		

Description of ***Viola tricolor*** in semi-technical language:–

Habit : Annual herb, ornamental, cultivated.

Root : Tap, branched.

Stem: Erect, branched, glabrous, angular, green, herbaceous, fistular.

Leaf: Simple, cauline and ramal, alternate, stipulate, (foliaceous stipules), petiolate, oblong, margins with rounded teeth, unicostate reticulate venation.

Inflorescence: Cymose, solitary axillary.

Flower : Large, showy (violet), pedicellate, bracteate, bracteolate (two bracteoles), hermaphrodite, zygomorphic, hypogynous, complete.

Calyx : Five sepals, polysepalous, green, lanceolate, appendiculate, quincuncial aestivation, inferior.

Corolla: Fives petals, polypetalous, petals unequal, violet, dark streaks present, the anterior or lower petal is produced backward into a spur and collects honey, imbricate aestivation.

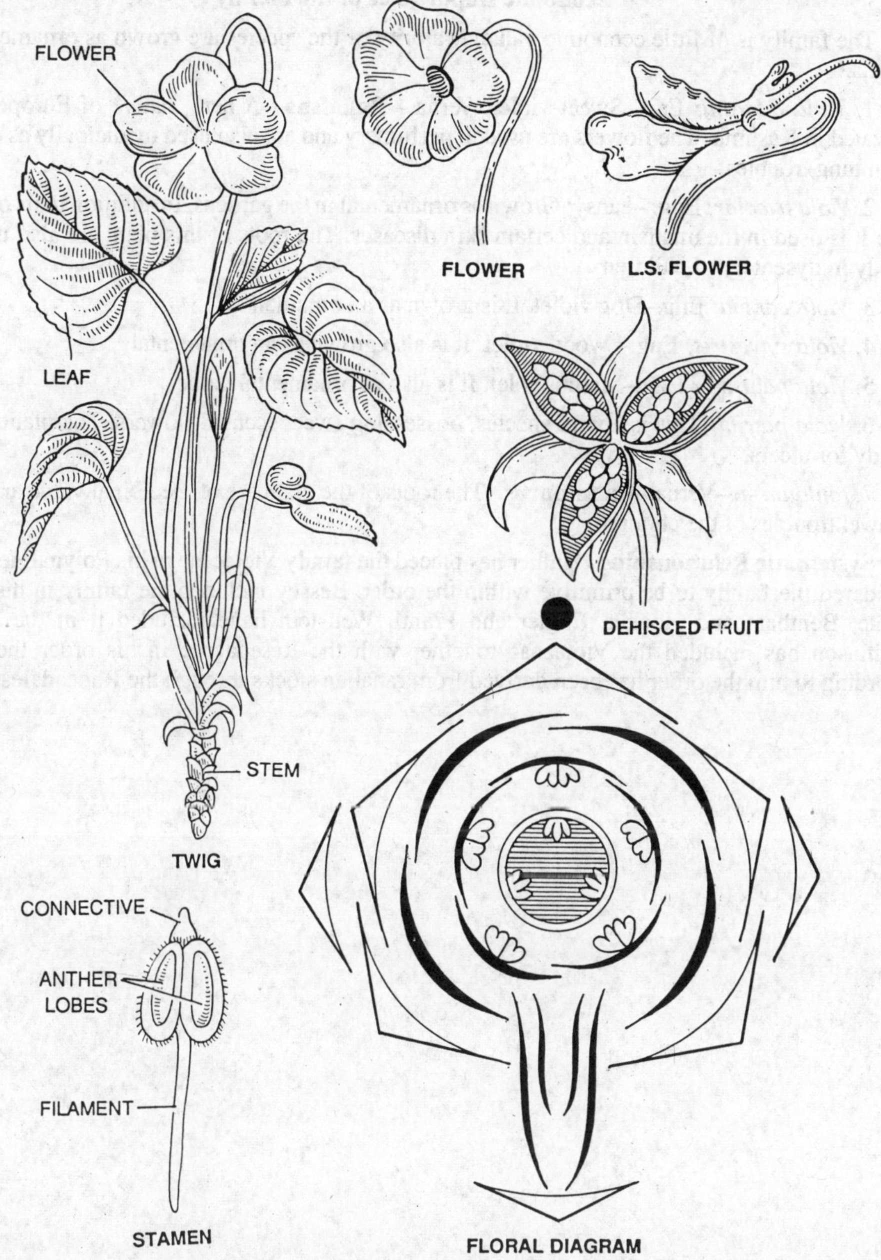

Fig. 19.1: Violaceae–*Viola tricolor* Linn. Eng., Pansy.

Androecium: Five stamens, alternating with the petals, the stamens form a ring like structure round the ovary and style, filaments short, anthers introrse, connate, connectives of the two lower often spurred at the base, which project into the spur of the anterior petal.

Gynoecium: Three carpels, syncarpous, ovary superior, unilocular, parietal placentation, many ovules attached to parietal placentas, style short, tip straight or oblique, stigma lobed.

Fruit: 3-valved loculicidal capsule.

Floral Formula: Br Brl ⫮ ⚥ K 5, C 5, A 5, G (3).

Economic Importance of the Family

The family is of little economic value. Majority of the species are grown as ornamental in the gardens.

1. *Viola odorata;* Eng.–Sweet violet; Verna.– **Banafsha**– A herb, native of Europe, but now cultivated in Kashmir. The flowers are used in perfumery and are also used medicinally as demulcent and in lung troubles.

2. *Viola tricolor;* Eng. –Pansy; Grown as ornamental in the gardens. In addition to its ornamental value it is used in rheumatism and certain skin diseases. The roots of this plant are also used as the remedy in dysentery of children.

3. *Viola canina;* Eng.–Dog violet. It is grown as an ornamental.

4. *Viola sylvestris;* Eng.– Wood violet. It is also grown as an ornamental.

5. *Viola palustris;* Eng.–March violet. It is also of ornamental value.

6. *Viola patrini;* This is a wild species, possessing sweet scented flowers. The plant is used as remedy for ulcers.

7. *Ionidum* sp.–Verna.–**Ratanpurus** –The roots of the plant are of medicinal value and are used in bowel troubles of the children.

Systematic Relationships : Hallier has placed the family Violaceae in his Polygalineae. He has considered the family to be primitive within the order. Bessey included the family in his Guttales. Rendle, Bentham and Hooker, Engler and Prantl, Wettstein have included it in the Parietales. Hutchinson has included the Violaceae together with the Resedaceae in his order the Violales. According to him the order has been derived from ranalian stocks through the Rhoeadales.

20

ORDER–CUCURBITALES

According to Hutchinson this is the thirteenth order of the phylum Angiospermae, subphylum Dicotyledones and division Lignosae. The order comprises of four families, *i.e.,* Cucurbitaceae, Begoniaceae, Datiscaceae and Caricaceae. The Cucurbitaceae is the primitive most and the Caricaceae is the most evolved within the order. These two families have been described in this text in detail. Bentham and Hooker have placed the families Cucurbitaceae and Caricaceae (under the name Passifloreae) along with other five families in the thirteenth order–Passiflorales of class Dicotyledones, division Polypetalae and series Thalamiflorae. Engler and Prantl have included the family Caricaceae along with other twenty eight families including Violaceae in their twenty ninth order–the Parietales of class Dicotyledonae and subclass Archichlamydeae. They have placed the single family Cucurbitaceae in their ninth order–Cucurbitales of class Dicotyledoneae and subclass Sympetalae.

Constantly unisexual; ovary inferior with parietal placentation; endosperm scanty or often present; corolla sometimes sympetalous; stipules present or absent.

Family—Cucurbitaceae

The family Cucurbitaceae includes about 100 genera and 850 species.

Distribution: The members belonging to this family are mainly tropical and sub-tropical in distribution. Only few genera are found in temperate regions. In our country the family is represented by a large number of cultivated plants especially for their edible fruits, *e.g., Citrullus vulgaris* (Water Melon–**Tarbooz**), *Cucumis melo* (Melon–**Kharbooza), *Cucumis sativus*** (Cucumber– **Khira**), *Luffa aegyptiaca* (**Ghia tori), *Lagenaria vulgaris*** (**Lauki**), *Cucurbita maxima* (Pumpkin–**Kaddu**), etc.

Habit: The majority are annual or perennial, trailing or decumbent vines, usually climbing by means of tendrils. Often some plants reach a great size. *Acanthosicyos* is a thorny leafless shrub, the two species of *Dendrosicyos* from the Socotra island of the Arabian coast are small trees but the stems are soft and not woody.

Root : Tap and branched.

Stem: Herbaceous, climbing, hollow, often 5-angled and either glabrous, hairy or prickly (bi-collateral vascular bundles present).

Leaves: The leaves are simple and in most cases are palmately lobed or divided, usually exstipulate, large and long petioled. The arrangement on the stem is alternate. The petioles are often hollow. The tendrils may be simple or branched and may twist to the right and to the left at different points along their axes. There is much controversy about the morphological nature of the tendrils.

Morphology of the Tendrils: According to some workers these tendrils are believed to be leaves, or that their branches are leaves and the main portion stem. But some botanists consider them as

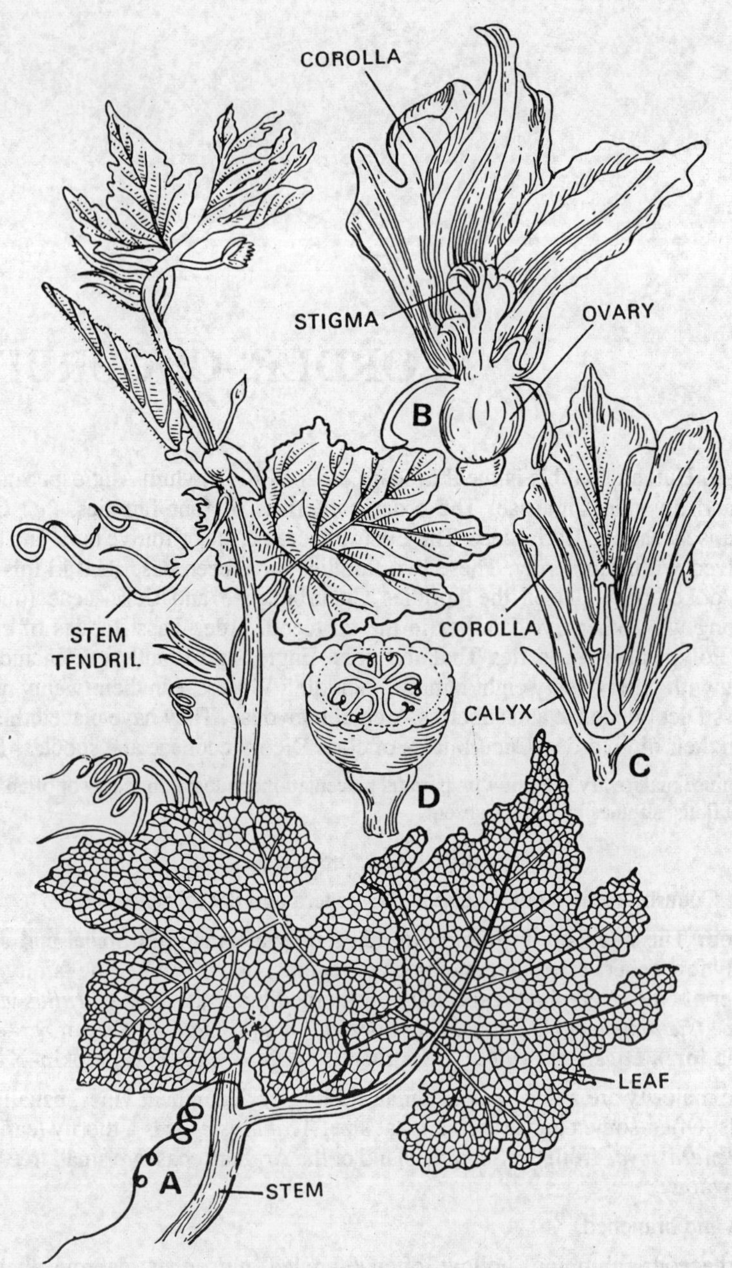

Fig. 20.1. Cucurbitaceae–*Cucurbita moschata*–A, a twig with leaves and tendrils; B, female flower cut lengthwise; C, male flower cut lengthwise; D, fruit cut transversely showing parietal placentation.

bracteoles and according to some they are stipules. We can conclude that in different genera, they are of different origin. According to Braun, Eichler and others, they are modified bracteoles of the flowers, which have been pushed out of position.

Inflorescence: Usually cymose solitary axillary, sometimes panicle, more rarely racemose or sub-umbellate.

Flowers : They are typically regular and unisexual (rarely bisexual) and either monoecious or dioecious. Actinomorphic, epigynous, pentamerous but with the tricarpellary pistil by reduction and the stamens united into two pairs. Calyx and corolla inserted on a variously shaped often elongated hypanthium which is constricted above the ovary. Male and female flowers are described here separately.

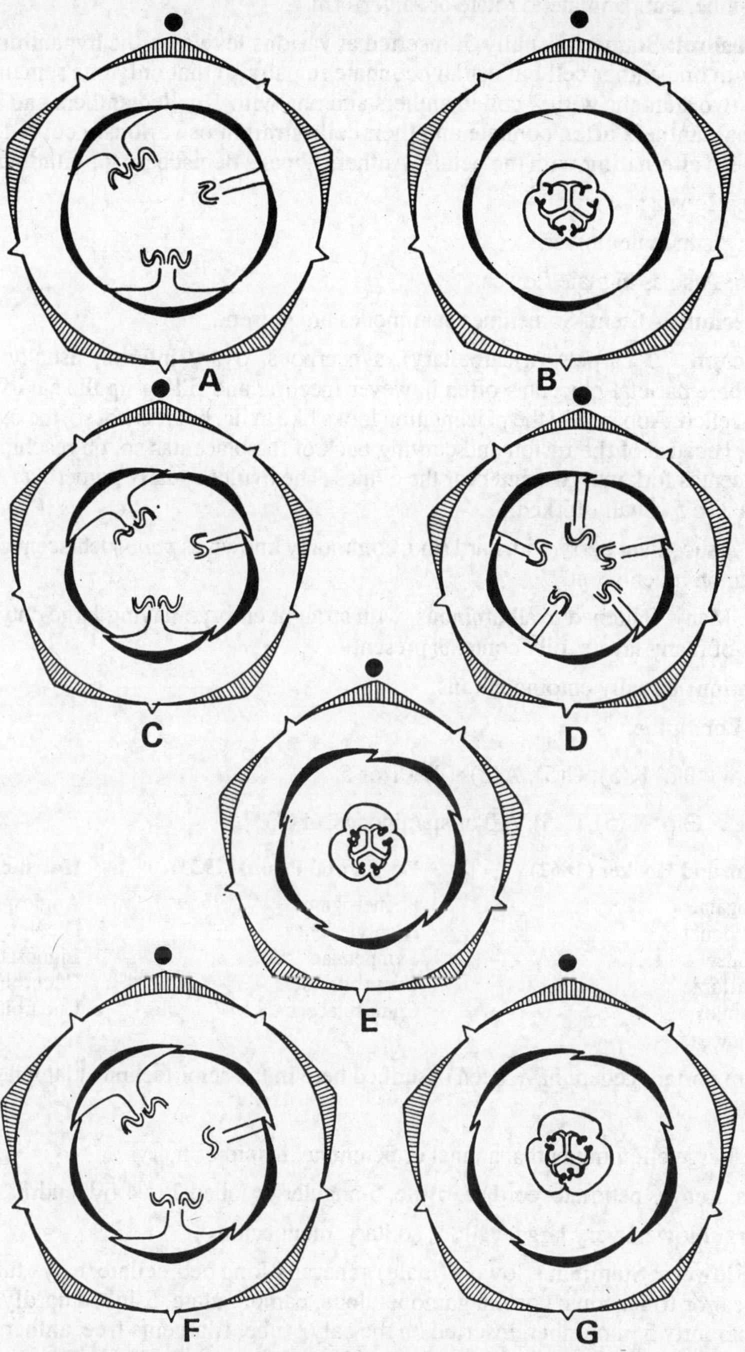

Fig. 20.2. Floral diagrams. A and B, male and female–Cucurbitaceae (typical); C, D and E *Luffa*, male early, late and female; F and G *Lagenaria*, male and female.

Male Flower :–

Calyx: 5 lobed, segments generally narrow and pointed petaloid, campanulate, gamosepalous. Aestivation is usually imbricate quincuncial or valvate.

Corolla: 5 petals mostly sympetalous or free often deeply 5-lobed, valvate or imbricate, inserted on the calyx-tube, campanulate to rotate or salverform.

Androecium: Stamens usually 5, inserted at various levels on the hypanthium, extrorse anthers each with one anther cell but usually connate in pairs so that only one remains free (so there appear to be two stamens with 2-celled anthers and one with 1-celled anthers and are described as three stamens), anthers often connate and their cells straight or variously curved or twisted (sigmoid). Stamens alternating with the petals. Anthers always dehisce by longitudinal slits.

Female Flower :

Calyx : As in male flower.

Corolla: Also as in male flower.

Androecium: Absent. Sometimes staminodes are present.

Gynoecium : 3 carpels (tricarpellary), syncarpous, ovary inferior, usually unilocular (one celled), the three parietal placentas often however meeting and filling up the cavity of the ovary or ultimately 3-celled. Apparently the placentation looks like axile, but it is not so, the ovules do not born in the centre, because of the fusion and curving back of the placentas so, on reaching the carpellary. walls the placentas fork into two and bear the ovules. The ovules usually numerous, anatropous, style short, stigma 3 to 5 usually forked.

Fruit : A succulent berry with hard rind, commonly known as *pepo*, dehiscent often explosively or more common indehiscent.

Seeds : Many, flattened exalbuminous, with straight embryo having large cotyledons, the seed coat consists of many layers, oily contents present.

Pollination: Usually entomophilous.

Floral Formulae:

Male : ⊕ ♂ K(5), C(5), A(2)+(2)+ 1 or 5

Female : ⊕ ♀ K (5), C(5), A 0 or staminodes, $G (\overline{3})$

Bentham and Hooker (1862)	**Engler and Prantl (1931)**	**Hutchinson(1959)**
Phanerogams	Phanerogams	Angiospermae
Dicotyledones	Dicotyledones	Dicotyledones
Polypetalae	Sympetalae	Lignosae
Thalamiflorae	Cucurbitales	Cucurbitales
Passiflorales	Cucurbitaceae	Cucurbitaceae
Cucurbitaceae		

A few important genera have been described here in the semi-technical language :

***Cucurbita* sp.**

Habit: Large climbing herbs, annual or perennial, hispid or hairy.

Leaves: Leaves petiolate, cordate, ovate, 5-angular or lobed 2 to 4 fid tendrils.

Flowers: Flowers very large, yellow, solitary, monoecious.

Male Flowers: Staminate flowers (male) generally long pedicellate; calyx tube campanulate, 5-lobed, linear or foliaceous; corolla gamopetalous, campanulate, 5-lobed hardly half way down; stamens apparently 3 in number, inserted on the calyx tube, filaments free, anthers connate, linear and united, one 1-celled, two 2-celled cells conduplicate, ovary rudimentary.

Female Flowers: Pistillate flowers (female) are short peduncled; calyx and corolla as in male;

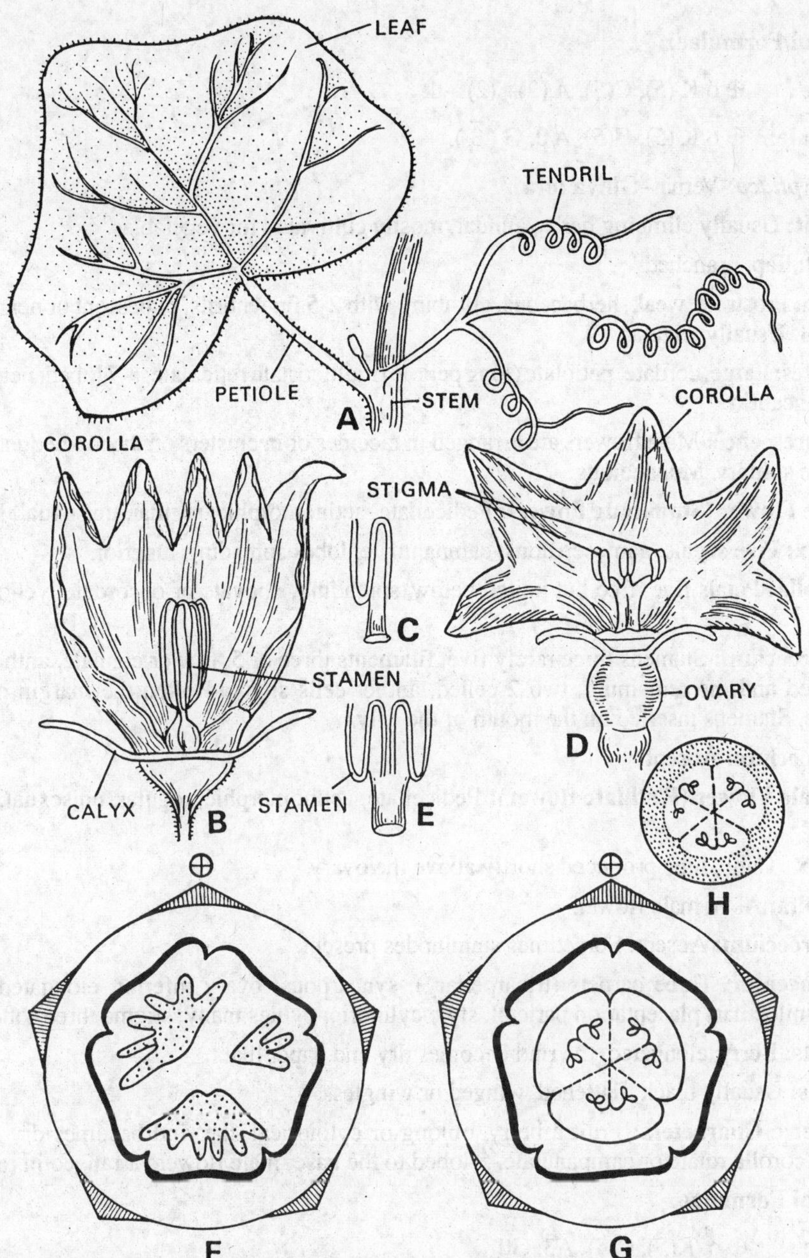

Fig. 20.3 : Cucurbitaceae—*Cucurbita*. sp. A, a twig ; B, L.S. male flower ; C, single lobed stamen; D, female flower ; E, double lobed stamen ; F, floral diagram of male flower ; G, floral diagram of female flower ; H, T.S. ovary.

ovary oblong; 1-celled with 3 to 5 or many ovuled placentae; style short and thick with 3 to 5 stigmas; large epigynous cushion shaped 5-lobed disc.

Fruit: A large pepo.

Seeds: Ovoid or oblong, flattened smooth, margin present or absent.

Generic Characters: Fruit a pepo, with hard firm rind, indehiscent; corolla campanulate (bell shaped), 5-lobed to, or just below the middle; anthers connate.

Floral Formulae:

Male : ⊕ ♂ K (5), C(5), A (2)+ (2) + 1.

Female: ⊕ ♀ K (5), C(5), A 0, G ($\overline{3}$).

Luffa aegyptiaca; Verna.–**Ghiya torai.**

Habit: Usually climbing herbs, annual, mostly cultivated for vegetables.

Root: Tap, branched.

Stem: Prostrate, weak, herbaceous, climbing with 2-5 fid tendrils, pubescent or nearly glabrous, sap present. Usually angular.

Leaves: Large, cordate, petiolate (long petiole) multicostate reticulate, 5-7 lobed, nearly glabrous to slightly scabrid.

Inflorescence: Male flowers are arranged in racemes or in clusters on the top peduncles. Female flowers are solitary. Monoecious.

Male Flower (Staminate flower): Pedicellate, actinomorphic, regular, unisexual, incomplete.

Calyx: Five sepals, gamosepalous, campanulate, lobes spreading superior.

Corolla: Petals five, free but united below, spreading, obovate or obcordate, yellow or white, superior.

Androecium: Stamens three rarely five, filaments three or 5, free or connate, anthers extrorse, one 1-celled and the remaining, two 2-celled, anther cells sigmoid often the margin of the broad connective. Stamens inserted on the mouth of the calyx.

Gynoecium: Absent.

Female Flower (Pistillate flower): Pedicellate, actinomorphic, regular, unisexual, epigynous, incomplete.

Calyx : Calyx tube produced shortly above the ovary.

Corolla: As in male flower.

Androecium: Absent. Sometimes staminodes present.

Gynoecium: Three carpels (tricarpellary), syncarpous; ovary inferior, elongated, angled or grooved, unilocular; placentation parietal, style cylindric; ovules many; stigma three-lobed.

Fruits: Berry, elongated, the rind becomes dry and paper like.

Seeds: Usually black, flattened, winged or wingless.

Generic Characters: Fruit a berry, oblong or cylindrical, the rind becomes dry and papery, dehiscent; corolla rotate or campanulate, 5-lobed to the base; male flowers arranged in racemes.

Floral Formulae:

Male : ⊕ ♂ K(5), C(5), A3, G0.

Female : ⊕ ♀ K(5), C(5), A0, G($\overline{3}$).

Lagenaria vulgaris; Verna. **Ghiya, lauki.**

Habit: Large, long and running annual vine. Usually cultivated for its fruits, which are eaten as vegetable.

Root: Tap, branched.

Leaves: Large, 6 to 12 inches across, ovate or orbicular to reniform, cordate, dentate, lobes absent or 3 to 7 lobed, petiolate (long petiole), two glands at the apex of the petiole. The tendrils are usually bifid.

Inflorescence: Cymose, solitary, the flowers are monoecious rarely dioecious.

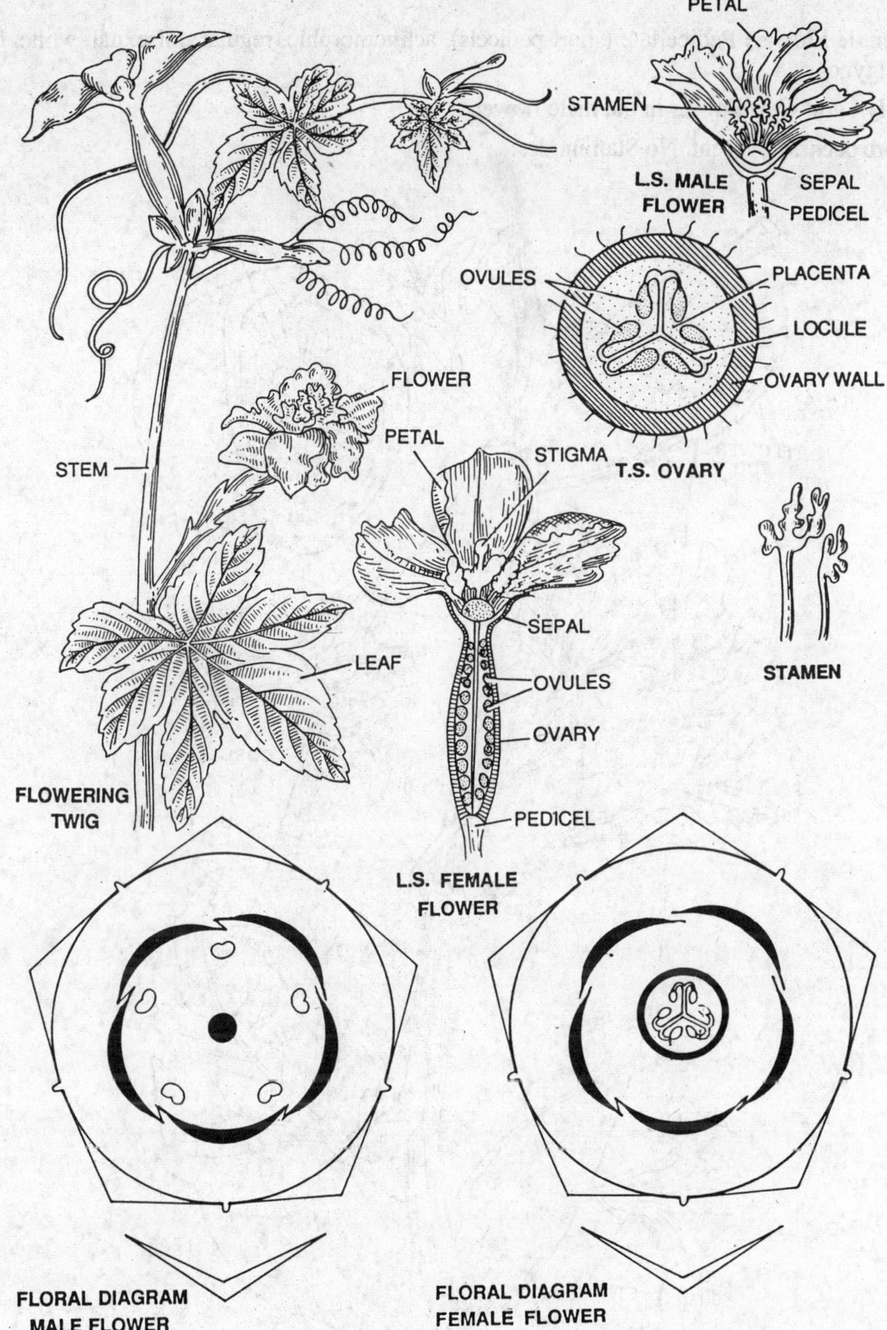

Fig. 20.4. Cucurbitaceae–*Luffa cylindrica* (Linn.) Roem., syn. *L. aegyptiaca* Mill.

Male flower: Pedicellate (long pedicels), actinomorphic, regular, unisexual, incomplete, white, showy and short-lived.

Calyx: 5 narrow sepals, calyx tube funnel shaped, subcampanulate, superior.

Corolla: petals, free, obovate or obcordate, white, superior.

Androecium: Stamens 3 (apparently), 3 filaments, free; stamens inserted on the calyx tube, anthers included and connate, one 1-celled, two 2-celled, conduplicate cells.

Female Flower: Pedicellate (short pedicels), actinomorphic, regular unisexual, white, incomplete, epigynous.

Calyx and Corolla: As in the male flower.

Androecium: Absent. No Staminodes.

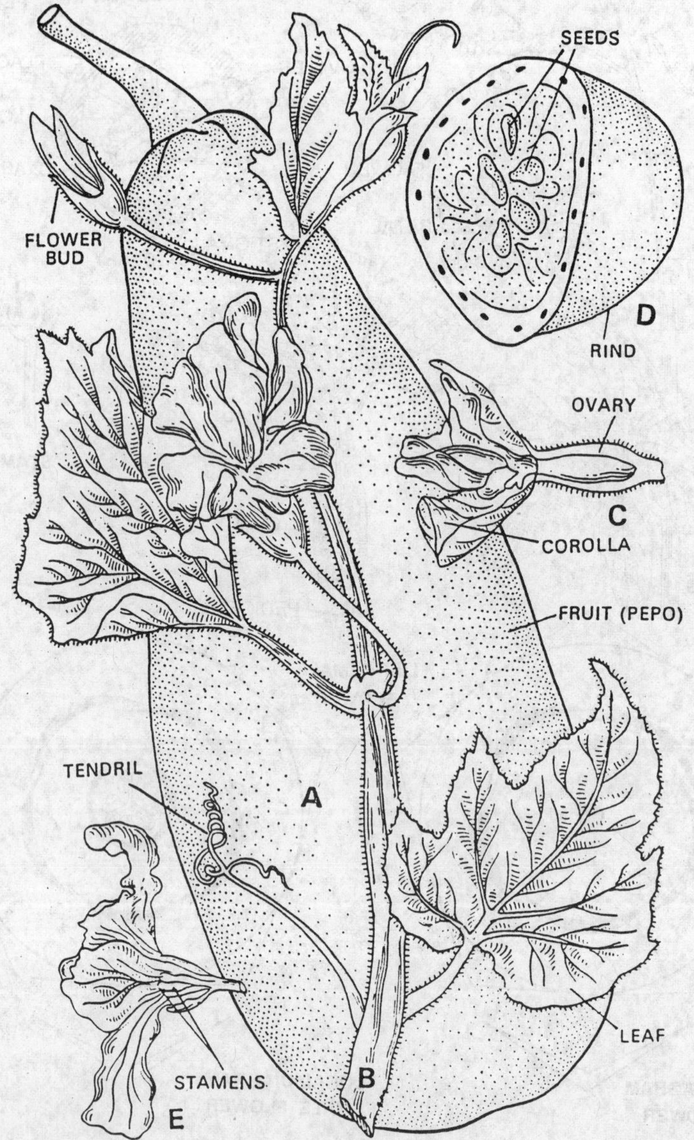

Fig. 20.5. Cucurbitaceae–*Lagenaria vulgaris;* A, fruit; B, a twig; C, L.S. female flower; D, fruit ; E. L.S. male flower.

Gynoecium: 3 carpels, syncarpous; inferior ovary, ovate to cylindrical; style short and stout; 3 bifid stigmatic lobes; ovules many, horizontal; placentation parietal; ovary unilocular.

Fruit: A pepo, varies in length from 3 inches to 3 feet. The rind becomes hard and woody, they are oblong or bottle shaped. Indehiscent.

Seeds: Many, white, oblong, flattened distinctly margined, furrows and ridges (longitudinal) also present on the seeds.

Generic Characters : Fruit a pepo, with hard firm rind, indehiscent; corolla campanulate, 5-lobed just beyond the middle, male flowers solitary; sepals small, entire, erect or spreading; tendrils branched, usually 2-fid; leaves not lobed.

Floral Formulae:

Male : ⊕ ♂ K5, C5, A3, G0.

Female : ⊕ ♀ K5, C5, A0, $G(\overline{3})$.

Economic Importance of the Family
(*Cucurbitaceae*)

Economically the family is fairly important. They possess edible fruits. The flowers buds and fruits are also eaten as vegetables. Some plants are of medicinal value. A list of important plants is given here :

1. *Cucurbita maxima;* Verna.—**Sitaphal**—This is grown for its large edible fruits which are used as vegetable.

2. *Cucurbita moschata;* Eng.—Squash ; Verna.—**Mitha Kaddu**—The fruits are edible. They are eaten as vegetable.

3. *Cucurbita pepo* ; Eng.—Pumpkin; Verna—**Safed Kaddu, Kumra**— The Plant is commonly cultivated in Northern India. The fruits are edible.

4. *Lagenaria siceraria;* Syn. *L. vulgaris* ; Eng.—Bottle gourd; Verna—**Lauki**—This is a climbing or trailing herb. It is cultivated throughout our country. The fruits are eaten as vegetable.

5. *Luffa cylindrica*; Syn. *Luffa aegyptiaca; Momordica cylindrica;* Eng.—Vegetable sponge; Verna—**Ghia torai**—It is cultivated throughout the country for its fruits which are used as vegetable. The dried fruits yield a spongy substance which is used as a bath sponge.

6. *Luffa acutangula;* Syn. *Cucumis acutangulus*; Eng.—Vegetable sponge; Verna.—**Kali torai**– This is a trailing herb cultivated throughout the country. The fruits are eaten as vegetable. The fibrous material obtained from the dried fruits, is used as a substitute for bath sponges.

7. *Luffa echinata;* Verna.—**Chagarabela**—It is found in Bihar, Bengal, Gujarat and Dehradun. The fruits possess medicinal properties and are used to cure dropsy. They are also given as purgative.

8. *Momordica charantia;* Eng.—Bitter gourd; Verna—**Karela**—This is climbing or trailing herb. This is cultivated throughout the country for its fruits which are eaten as vegetable.

9. *Momordica tuberosa;* Syn. *Luffa tuberosa*, Verna.—**Kadavanchi**—This is a trailing or climbing herb, found mainly in Maharashtra and Tamil Nadu. The fruits are eaten as vegetable.

10. *Momordica cochinchinensis;* Verna—**Bhat Karela**—This is cultivated throughout India for the fruits which are used as vegetable.

11. *Momordica dioica;* Eng.—small bitter gourd; Verna—**Jangli Karela**—This is a climbing shrub. The fruits are used as vegetable.

12. *Momordica balsamina;* Eng.—Balsam apple; Verna—**Mokha**—This is climbing or spreading herb. The fruits are edible.

13. *Momordica subangulata*—This is a climbing shrub, grown as an ornamental.

14. *Trichosanthes anguina;* Eng.—Snake gourd; Verna—**Chichenda**-This is a climbing or trailing herb, cultivated throughout the country. The fruits are eaten as vegetable.

15. *Trichosanthes dioica;* Verna-**Parwal**-It is cultivated commonly in Northern India, Assam and Bengal. The fruits are eaten as vegetable. The leaves are also consumed as vegetable.

16. *Trichosanthes cucumerina;* Verna—**Jangli chachinda, Rambel**—The fruits are edible and eaten as vegetable.

17. *Cucumis melo;* Eng.—Musk melon; Verna—**Kharbooza**—It is cultivated in Northern India. The fruits are edible.

18. *Cucumis melo;* var. *agrestis;* Eng.—small gourd; Verna—**Meki, Takmak**—The fruits are edible.

19. *Cucumis melo* var. *momordica*; Verna—**Phunt, Kachri**—The fruits are edible.

20. *Cucumis melo* var. *utilissimus;* Verna.—**Kakri**—The fruits are edible.

21. *Cucumis sativus;* Eng.-Cucumber; Verna.—**Khira**—The fruits are edible. They are used in salad.

22. *Cucumis anguria;* Eng.—Gherkin—This is a creeping herb, introduced from Canada. The fruits are edible.

23. *Benincasa hispida;* Syn. *B. cerifera;* Eng. —Ash gourd; Verna.—**Petha**—This is a trailing or climbing herb. The fruits are edible. Sweets are prepared from the pulp of fruits.

24. *Citrullus vulgaris;* Eng.—Water melon; Verna—**Turbooz**—This is native of tropical Africa but now cultivated throughout our country for its large edible fruits.

25. *Citrullus vulgaris* var. *fistulosus;* Eng.—Round gourd; Verna—**Tinda**—The fruits are used as vegetable. It is cultivated chiefly in Northern India.

26. *Citrullus colocynthis*; Eng.—Colycynth; Bitter apple; Verna—**Indrayan**—This is a trailing herb. The fruits are used as purgative. This is native of Africa but now found throughout our country.

27. *Coccinia cordifolia*; Syn. *C. indica;* Verna.—**Kanduri**—This is a twining or spreading shrub. The fruits are used as vegetable. It is cultivated in Assam, Bihar, Bengal, Orissa, Maharashtra and Tamil Nadu.

Family—Caricaceae

There are only 3 genera and about 270 species in this family.

Distribution: The plants of this family are found in the tropics of the world. This family is important only on account of *Carica papaya,* which is widely cultivated in the tropical regions for its fruit. The genus *Carica* comprises of 40 species.

Habit: The plant is a small, soft wooded tree. The stem is stout and generally unbranched, a feature uncommon in Dicotyledons but common in Monocotyledons, *e.g.,* palms. The tissues are permeated with a laticiferous system which contains the powerful proteolytic enzyme, *papain.*

Leaves: The leaves are large and palmately lobed. The petioles are long and fistular. They are exstipulate and arranged alternately or spirally usually forming a crown at the terminal end of the stem.

Flowers: The flowers are usually unisexual and rarely bisexual. Usually the flowers are regular and pentamerous. From the point of view of their morphology the flowers are grouped into four types. They are as follows:

Male Flowers: These flowers develop on male plants. The flowers are sessile and unisexual. Sometimes the terminal flowers of the inflorescence are either female or bisexual. The number of stamens is 10. They are arranged in two whorls of 5 each. The rudimentary pistils may be present.

The flowers are arranged in clusters on pendant racemes or richly branched panicles.

Female Flowers: These flowers develop on female plants. They are subsessile. The ovary is large and globose. Style short with fine stigmas. The fruit is globose or ovoid.

The flowers are arranged in few flowered corymbs in leaf axils.

Long-fruited Flowers: The flowers are like the female flowers. The corolla is united and 10 stamens present at the base of petals. The ovary is elongated and the fruit is cylindrical in shape.

The flowers are arranged in 5-6 flowered corymbs in the axils of the leaf.

Polygamous Flowers: Such flowers are of two types. One type bears 10 sessile stamens found

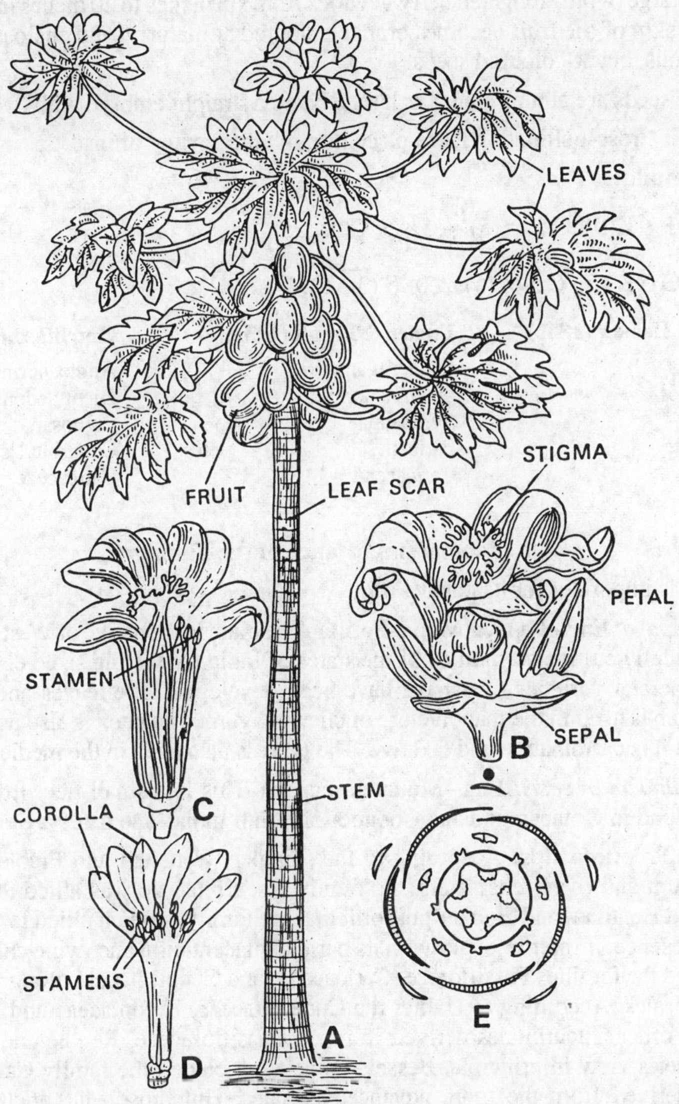

Fig. 20.6. Caricaceae-*Carica papaya* (Verna—**Papita**): A, a female tree bearing fruit; B, female flower ; C, male flower; D, male flower cut open; E, floral diagram (after Rendle).

at the throat of the corolla. The flowers of other type bear 5 stamens with long filaments, found at the base of the ovary. The corolla tube is very minute.

Calyx: 5 sepals, gamosepalous. It is very minute in all the flowers of above given types.

Corolla: It consists of 5 petals. The petals may be united or free. Aestivation twisted in bud. The male flowers bear long corolla tube and the female ones the short tube.

Androecium: In male flowers it consists of 10 stamens arranged in two whorls of 5 each. sometimes the stamens of inner whorls are suppressed. They are epipetalous. The filaments may be short or long. The anthers are dithecous. They dehisce by longitudinal slits.

Gynoecium: In female flowers usually it consists of 5 carpels, syncarpous. The ovary is unilocular. The placentation is parietal. The ovules are many.

Fruit: It is large ovoid elongated berry. It varies from six inches to 20 inches in length and weighs upto 10 kg. The skin of the fruit becomes orange coloured at maturity and encloses orange coloured flesh and numerous, black-coloured seeds.

Seeds: The seeds are albuminous. Each seed bears a straight embryo with two large cotyledons.

Pollination: Cross-pollination takes place through the agency of insects.

Floral Formulae:

$\oplus$ ♂ K(5), C 5 or (5), A 5 + 5, G 0

$\oplus$ ♀ K (5), C 5 or (5), A 0, G $\overline{(3-5)}$.

Bentham & Hooker (1862)	**Engler & Prantl (1931)**	**Hutchinson (1959)**
Phanerogams	Phanerogams	Angiospermae
Dicotyledones	Dicotyledones	Dicotyledones
Polypetalae	Sympetalae	Lignosae
Thalamiflorae	Parietales	Cucurbitales
Passiflorales	Caricaceae	Caricaceae
Passifloreae		

Economic Importance of the Family

The family is important for its papaya fruit.

1. *Carica papaya;* Eng.–Papaya; Verna–**Papita**–The plant is the native of West Indies and Central America. Now chiefly cultivated in most of the states of India for the big sized edible fruits. A large number of commercial varieties of papaya have been developed. The leaves and the unripe fruits contain latex which is used in the manufacture of chewing gum. The latex is also used for tenderizing meat. The ripe fruit is carminative and laxative. The latex is also used in the medicine.

2. *Carica candamarcensis:* Eng.–Mountain papaya–This is a small tree. It is native of South America. Now found in Ootacamund and Coonoor of South India. The fruits are edible.

Systematic Relationships: According to Pax, Engler, Wettstein and Eichler, on the basis of corolla, androecium and ovary characters, the family Cucurbitaceae was allied to Campanulaceae. Robert Brown and Bentham and Hooker put forth that the family is much allied to the Passifloraceae because of the presence of inferior ovary with its parietal placentation and ovule characters. Bentham and Hooker placed the families Passifloreae (Caricaceae) and Cucurbitaceae along with five others in the order Passiflorales. According to Hallier the Cucurbitaceae, Bigoniaceae and Datiscaceae comprised a separate order Cucurbitales derived from the Passifloraceae. Rendle and Hutchinson later adopted the Hallier's view in principle. Bessey, however, included the family Cucurbitaceae in the order Loasales derived from the more advanced Rosales, Hutchinson has included the families Cucurbitaceae and Caricaceae along with other two families in the order Cucurbitales.

21

ORDER–CACTALES

According to Hutchinson this is the thirty first order of the phylum Angiospermae, subphylum Dicotyledones and division Lignosae. The order contains the single family–the Cactaceae which has been described in this text in detail. Bentham and Hooker have included the families Cactaceae and Ficoideae in their fourteenth order–the Ficoidales of class Dicotyledones, division polypetalae and series Calyciflorae. Engler and Prantl have put the single family Cactaceae in their twenty eighth order–the Opuntiales of class Dicotyledones and subclass Archichlamydeae.

Succulent or woody; sepals, petals, and stamens mostly numerous and in several series on a tubular axis; ovary inferior, 1-locular with parietal placentas.

Family—Cactaceae

There are about 130 genera and over 1200 species in this family (Lawrence).

Distribution: The plants show remarkable adaptation to desert conditions, and therefore they are chiefly distributed in the dry and desert regions. The members of this family are chiefly distributed in the dry districts of the warmer parts of America, especially of Mexico, including the temperate and colder part. Many species are found in the interior of Brazil. In North America *Opuntia* is found as far north as 59° latitude and South America the species are found as far south of Patagonia. The plants are found on the mountains from 3000 to 4000 metres elevation. The species of *Opuntia* and *Echinocactus* remain exposed to the frost throughout the year on the high mountains of Colorado. Several genera are found in the West Indies. The epiphytic genera such as *Rhipsalis, Phyllocactus* and *Epiphyllum* are found in the tropical forests of America. The species of *Opuntia* are very commonly found in Southern Europe and warmer parts of the old World. The species of *Rhipsalis,* are found in tropical Africa and Madagascar. The species of *Opuntia* are commonly found round the Mediterranean. Some cacti are found as far as the north of Columbia. The species of *Opuntia* and *Pereskia* are found in our country.

Habit: The cacti show varied and remarkable adaptations to a xerophytic mode of life. They are succulent plants. They range in size from small plants, *Mammilaria,* to large, columnar trees such as *Carnegiea gigantea* which reach a height of 70 feet by 2 feet girth. The genera, *Rhipsalis, Phyllocactus* and *Epiphyllum* live as epiphytes on forest trees.

Stem: The stems are fleshy and of various shapes. They are often greatly enlarged and cylindrical, flattened or fluted. The stems are frequently constricted and jointed. They are columnar, conical or subglobose angular, very often ribbed and warty. They are simple or branched. Sometimes they are rope like and flat like leaves. The stem of *Pereskia* possesses normal dicotyledonous growth, which is much-branched and woody. The leaves of *Pereskia* are broad and fleshy. In *Opuntia,* sometimes the narrow fleshy leaves are found but generally they fall early. In other genera the leaves are either very small or undeveloped. In all the genera but the species of *Pereskia* the assimilation takes place by the simple or branched, flattened or columnar stem; the tubercles are found to be developed upon the stem. The spines are developed upon the stem which are borne upon the tubercle. The spines are generally found at the top of the tubercle. The spines are very much varied in size and form. The spine bearing area at the tip of the tubercle or along the ridges or edges of the columnar or flattened stem is known as the **areole.** The tubercles may unite in vertical rows to form ribs. The spines correspond to leaves. The transitional structures between spines and leaves which are found in *Opuntia* and *Echinopsis* indicate the leaf character. Darbishire, while working on *Mamillaria elongata* considered the tubercle as the leaf base and the spines as modified portions of the leaf blade.

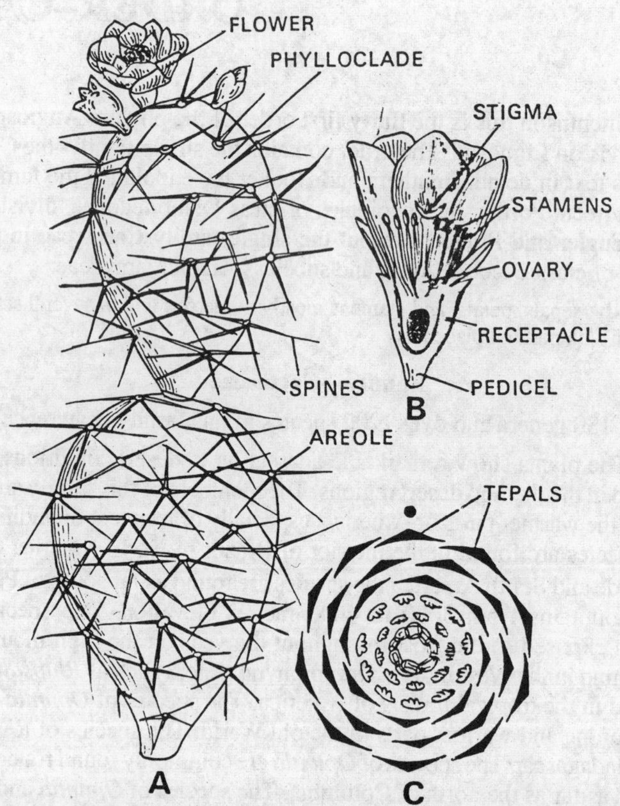

Fig. 21.1. Cactaceae—A, Phylloclade of *Opuntia tuna* showing spines, flower, bud and young fruit; B, flower of same cut lengthwise; C, floral diagram of *Opuntia* sp.

Schumann, however, could not compare the spines with leaves and regarded the spines simply the emergences. In *Opuntia*, apart from the spines, certain other slender, barbed bristles are also present on the inner side of the apex of the axillary shoot. These bristles are known as **glochidia** which are supposed to be homologous with the spines.

The spines are of various forms. The spines of *Pilocereus* are weak and hair like, whereas the spines of *Mamillaria,* are featherlike. The spines of *Leuchtenbergia* are flat and papery. The spines of the species of *Pereskia* or hooks which help in climbing. Ganong found the nectar secreting glands to be present among the spines of the species of *Opuntia, Mamillaria, Cereus* and others, which are supposed to be metamorphosed spines. The spines perform various functions. They protect the plants from stray animals, and reduce transpiration. In *Mamillaria elongata* they act as a protective screen which protects the underlying tissues from the strong sunlight.

Anatomical Characters: The epidermis possesses thick cuticle and sunken stomata. A collenchymatous hypodermis is found beneath the epidermis, and next to it the chlorophyllous cells are present. The pith and cortex are much developed. Usually they consist of large round pitted water storing cells. Very often these cells contain mucilage. The xylem is made of ringed and spiral tracheids, however, there are no annual rings. All the tissues retain the capacity of growth for a long time, and therefore, the plants may be reproduced vegetatively and the shoots may be grafted on stems of other species.

Inflorescence: Usually the flowers are solitary and develop in the axils or at the ends of the tubercles. In *Pereskia*, the inflorescence is branched producing a terminal panicle. In the species of *Pereskia* the lateral panicles are also produced, which arise from the axils of the upper leaves. In *Rhipsalis,* several flowers develop from an areole successively.

Flower : Usually the flowers are large and showy. They are bisexual, actinomorphic or sometimes becoming zygomorphic by curving of the perianth, stamens and pistil (*e.g.*, in *Ephiphyllum*) and epigynous.

Perianth: Usually the perianth is not clearly distinguished into calyx and corolla. The number of the floral leaves is generally very large. The floral leaves are united below developing into a long tube. In *Rhipsalis* and *Opuntia* the tube is quite short and the floral leaves are almost free. The perianth leaves are spirally arranged. The lower floral leaves often resemble sepals being green and thick while the upper ones resemble the petals which are delicate and white or coloured. In the species of *Echinopsis* and *Echinocactus,* the overlapping scales bearing tufts of hairs in their axils are found upon the perianth tubs. These scales become larger above and pass into the petaloid leaves. The larger floral leaves do not bear hairs or spines in their axils.

Androecium: The androecium consists of indefinite number of stamens arising from the perianth tube. They are either arranged spirally or in groups. Very often a ring of short stamens surrounds the throat of the tube. The stamens are epiphyllous. The anthers are basi or dorsifixed and two celled, dehiscing by means of longitudinal slits. The pollens are very small, yellow, spherical, smooth and with three small germ pores.

Gynoecium: Usually consists of 4 – ∝ carpels which are syncarpous. The ovary is inferior, unilocular, more or less sunken in the floral axis, parietal placentation. The parietal plancentas being arranged on the interior of the ovary wall. The number of placentas correspond in number with the stigmas. The style is simple, the ovules are many each with two integuments.

Fruit: Generally the fruit is fleshy, one celled berry with numerous seeds. In *Echinocactus* the fruit is dry. The fruits are often sweet and tasty and therefore they are undoubtedly disseminated by animals.

Seeds: The seeds are endospermic or non-endospermic having straight or curved embryo.

Floral Formula: ⊕ or ·|· ⚥ P ∝, A ∝, G $(\overline{4 - \propto})$.

The Cactaceae has been sub-divided into three sub-families. They are– 1. Pereskioideae, 2. Opuntioideae and 3. Cereoideae.

1. Pereskioideae: They possess flat and somewhat fleshy leaves. The spines are present. The stalked flowers are found to be arranged in panicles. The only genus–*Pereskia.*

2. Opuntioideae: The plants are succulent. They possess flattened jointed stems. The leaves are small cylindrical to subulate and caducous. The areoles are found with glochidia. The flowers are rotate. The best known genera are–*Opuntia, Pereskiopsis* and *Nopalea.*

3. Cereoideae: The plants are succulent. The leaves are reduced to minute scales. The areoles are without glochidia. The flowers are funnel or salver shaped except in *Rhipsalis.* The important genera are–*Cereus, Echinocactus, Phyllocactus* ,etc.

Bentham & Hooker (1862)	**Engler & Prantl (1931)**	**Hutchinson (1959)**
Phanerogams	Phanerogams	Angiospermae
Dicotyledones	Dicotyledones	Dicotyledones
Polypetalae	Archichlamydeae	Lignosae
Calyciflorae	Opuntiales	Cactales
Ficoidales	Cactaceae	Cactaceae
Cactaceae		

Economic Importance of the Family

The family is of little economic value. Most of them are grown as ornamentals. They are commonly known as cacti. Some important plants are given below:

1. *Opuntia dillenii*; Eng.–Prickly pear; Verna–**Nagfani**–This is a thorny bush. It is native of South America. It is generally grown as a hedge plant. The fruits and stem are edible. The fruits are useful in gonorrhoea. The stem is given in whooping cough. The latex is used as purgative. The plant is antidote in snake-bite.

2. *Cereus grandiflorus*; Eng.–Night blooming Cereus–This is a tall cactus, grown as an ornamental. It possesses large, attractive, fragrant flowers which open at night and wither before morning. The stem is also used medicinally to cure dropsy and cardiac affections.

3. *Pereskia aculeata*; Eng.–West Indian gooseberry, Lemon vine–This is a woody vine with edible fruits.

4. *Nopalea.cochinellifera*; Eng–Cochinal cactus–A scarlet dye known as cochineal is prepared from the cochineal insects which feed upon this plant. These insects also feed upon *Opuntia tuna.*

Systematic Relationships: The order Cactales is also known as Ficoidales (Bentham and Hooker) and Opunitiales (Engler and Prantl). The systematic relationship of the group is doubtful. According to Bentham and Hooker the order is much related to the Passiflorales. Engler places the order near to the Myrtiflorae. Rendle has followed Engler. Hutchinson is of opinion that the view of Bentham and Hooker is more correct and relates the order Cactales to the Passifloraceae through the Cucurbitales.

Engler puts the family Cactaceae in his order Opuntiales. According to him this has been derived from the Parietales, Bessey and Hutchinson placed the Cactaceae near the Cucurbitaceae. Wettstein, Hallier, Buxbaum(1948), Maheshwari (1945) and Martin (1946), hold the view, on the basis of the embryology, floral morphology and anatomy that the family must be transferred to a position within or near the Centrospermae.

22

ORDER–TILIALES

According to Hutchinson this is the thirty second order of the phylum Angiospermae, sub-phylum Dicotyledones and division Lignosae. This order comprises of six families, *i.e.,* Dirachmaceae, Scytopetalaceae, Tiliaceae, Sterculiaceae, Peridiscaceae and Bombacaceae, of which Tiliaceae and Sterculiaceae have been described in the Present text in detail. Bentham and Hooker have placed the families Sterculiaceae and Tiliaceae along with one more family, *i.e.,* Malvaceae in the sixth order–the Malvales of class Dicotyledones division Polypetalae and series Thalamiflorae. Engler and Prantl have included these families along with other six families including Malvaceae in their twenty sixth order–the Malvales of class Dicotyledoneae and sub-class Archichlamydeae.

Hypogynous, actinomorphic; bisexual or unisexual; syncarpous with axile placentation; stamens rarely few and then opposite the petals, free to partially united; 2-1 locular; endosperm copious, rarely absent; embryo fairly large, straight or curved; calyx mostly valvate; leaves alternate, stipulate; indumentum often stellate; trees, shrubs or shrublets.

Family—Tiliaceae

There are about 41 genera and 400 species in this family (Rendle).

Distribution: The majority of plants of this family are found in the tropical regions. Some plants are found in temperate regions. The most important genus which is found away from the tropics is *Tilia* with about ten species. This is found in the northern regions of both hemispheres. In our country the family is represented by *Corchorus, Tilia,* etc.

Habit: The plants are usually trees or shrubs.

Leaves: Generally the leaves are alternate, entire, dentate or lobed stipulate. The stipules serve as organs for protection of leaves (*e.g.,* in *Tilia*), and as the leaves become unfold they fall off.

Mucilage cells are found both in the pith and cortical regions of the plant.

Inflorescence: Cymose and often very complex (*e.g.,* in *Tilia* and *Triumfetta*). In *Tilia,* there is a leaf bud in between the main stem and the inflorescence and the foliage leaf apparently subtends both a floral axis and a vegetative bud.

Flowers: The flowers are regular, actinomorphic, hermaphrodite, complete, penta or tetramerous and hypogynous. Rarely the unisexual flowers develop on monoecious plants.

Calyx: Usually it consists of five sepals, rarely three or four sepals, polysepalous, sometimes gamosepalous (connate at the base). Valvate aestivation.

Corolla: Usually it consists of five petals, polypetalous. The glands are found at the base of petals. The aestivation is imbricate in the bud. They are usually coloured but sometimes green.

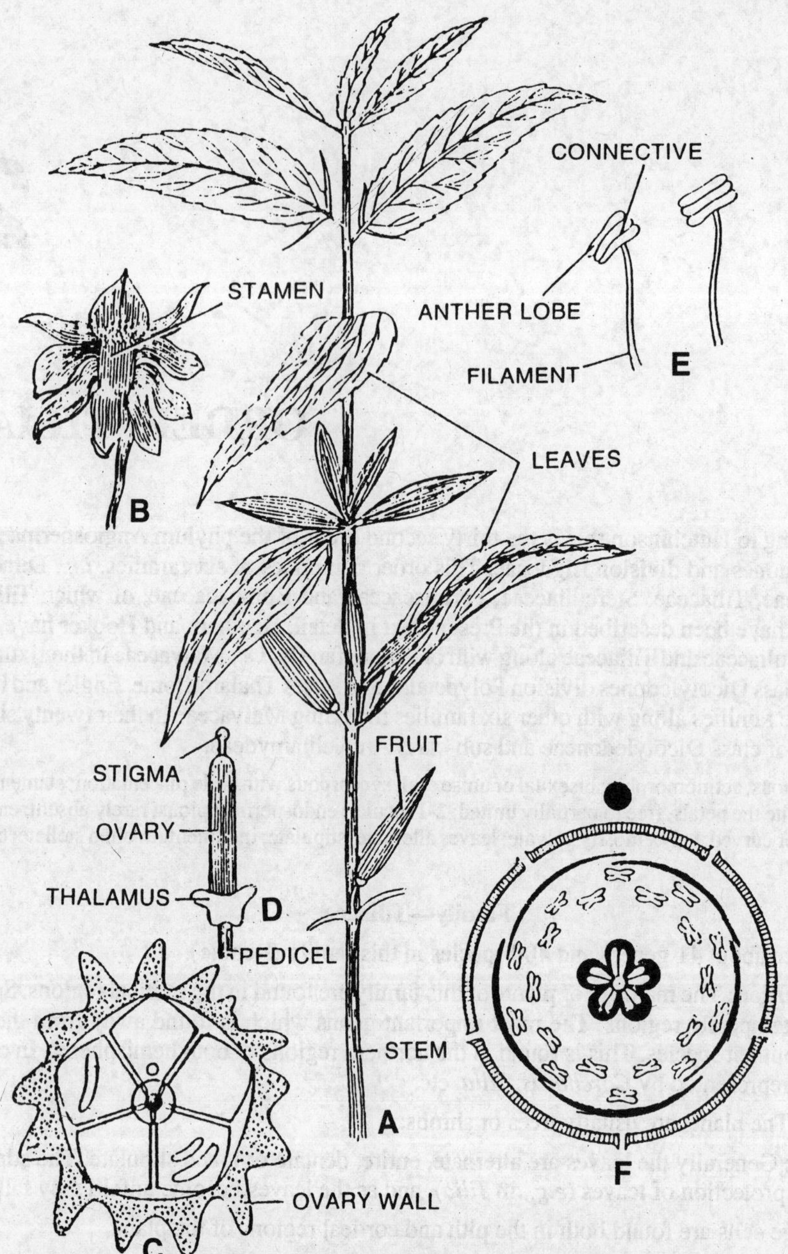

Fig. 22.1. Tiliaceae. *Corchorus* sp.–A, twig ; B, flower ; C, T.S. of ovary ; D, gynoecium; E, stamens; F, floral diagram

Androecium: The number of stamens varies from ten to indefinite. The stamens are quite free or connate at the base only. The stamens remain inserted at the base of the petals. In *Grewia,* they are raised above the corolla and are situated on an **androgynophore.** The anthers are two celled and basifixed. They dehisce by means of apical pores or longitudinal slits.

Gynoecium: The number of carpels ranges from two to many (2 -∝). The ovary is superior and bears two to many locules. One or more ovules are found in the inner angle. Placentation is axile. A simple style bears a lobed or capitate stigma.

Fruit: The fruit may be dehiscent or indehiscent. In *Tilia,* the fruit is a globose indehiscent nutlet having one and rarely two seeds. In the other genera the fruit is loculicidal capsule (*e.g.,* in *Corchorus*).

In *Sparmannia* winged cocci are found. In *Grewia,* the fruit is indehiscent and drupaceous. The spines are developed on the capsule and cocci of *Triumfetta, Sparmannia* and others which help in the dispersal of fruits by animals.

Seeds: The seeds are endospermic. The embryo is straight.

Pollination: The pollination is effected by the agency of insects. The bees and flies are attracted by the nectar and strong scent of the flowers.

Floral Formula: ⊕ ⚥ K 5 or (5), C 5, A ∝ , G ($\underline{2 - \propto}$).

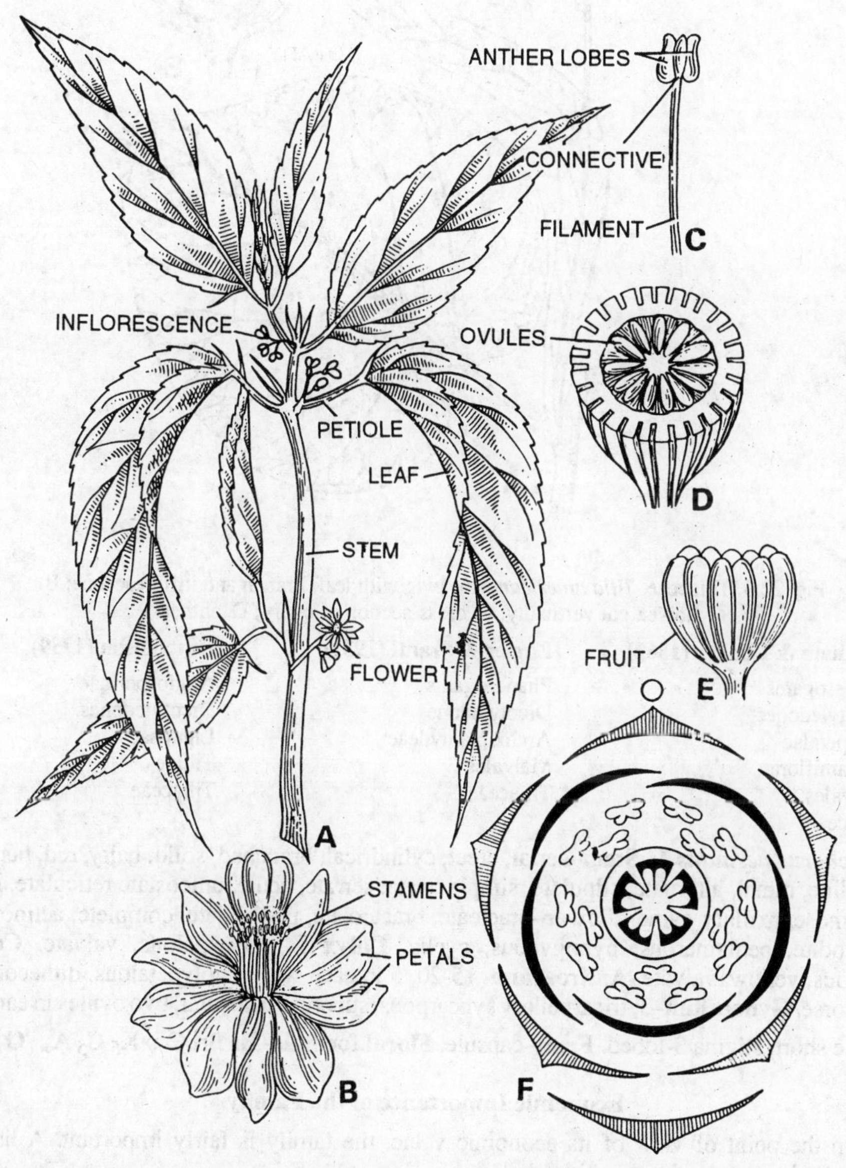

Fig. 22.2. Tiliaceae. *Corchorus capsularis*. A, flowering twig ; B, flower top view ; C, stamen ; D, T, S. of ovary ; E, fruit-capsule ; F, floral diagram.

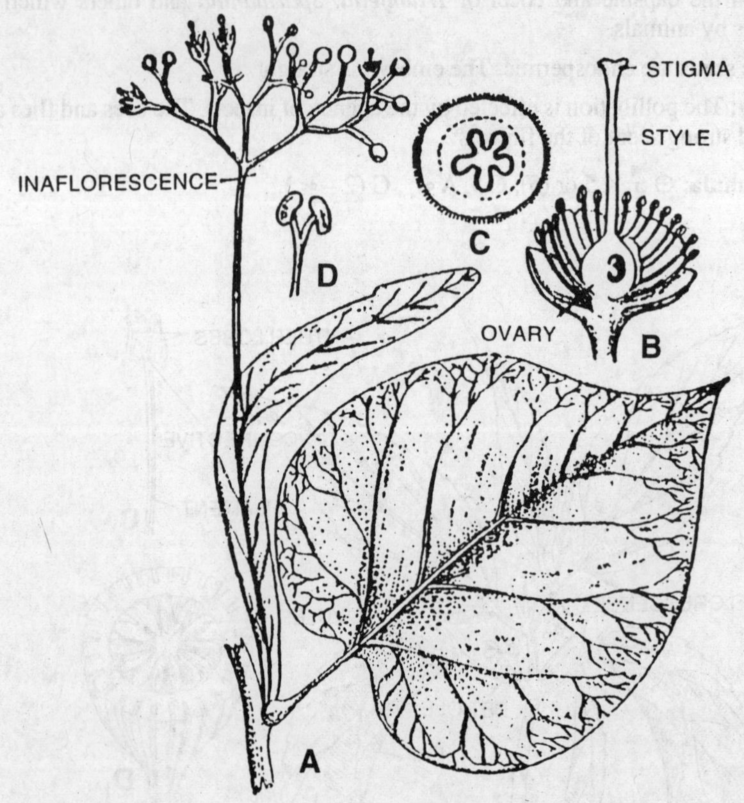

Fig. 22.3. Tiliaceae–*Tilia americana.* A, twig with leaf, branch and inflorescence; B, flower. cut vertically; C, cross-section of ovary; D, anther.

Bentham & Hooker (1862)	**Engler & Prantl (1931)**	**Hutchinson (1959)**
Phanerogams	Phanerogams	Angiospermae
Dicotyledones	Dicotyledones	Dicotyledones
Polypetalae	Archichlamydeae	Lignosae
Thalamiflorae	Malvales	Tiliales
Malvales	Tiliaceae	Tiliaceae
Tiliaceae		

***Corchorus aestuans* L. Stem**–aerial, erect, cylindrical, branched, solid, hairy, red, herbaceous. **Leaf**–cauline, ramal, alternate, stipulate, simple ovate, serrate, acute, unicostate reticulate. **Inflorescence**–cymose, axillary cyme. **Flower**–bracteate, bracteolate, pedicellate, complete, actinomorphic, hermaphrodite, pentamerous, hypogynous, cyclic. **Calyx**–5, polysepalous, valvate. **Corolla**–5, polypetalous, yellow, valvate. **Androecium**–15-20, 5 in each whorl, polypetalous, dithecous, dorsifixed, introrse. **Gynoecium**–3, tricarpellary syncarpous, superior, trilocular, two ovules in each locule, axile, style short, stigma 3-lobed. **Fruit**–capsule. **Floral formula**–Br Brl $\oplus$ ⚥ K_5 C_5 A_∞ $G_{\underline{(3)}}$

Economic Importance of the Family

From the point of view of its economic value, the family is fairly important. A list of few economically important plants is given below:

1. *Corchorus capsularis*; English–Jute, White jute; Verna.–**Pat, titapat, Narecha**–This is native of India. This is a shrub, cultivated in West Bengal, Assam, Bihar, Uttar Pradesh and Orissa. The fibre,

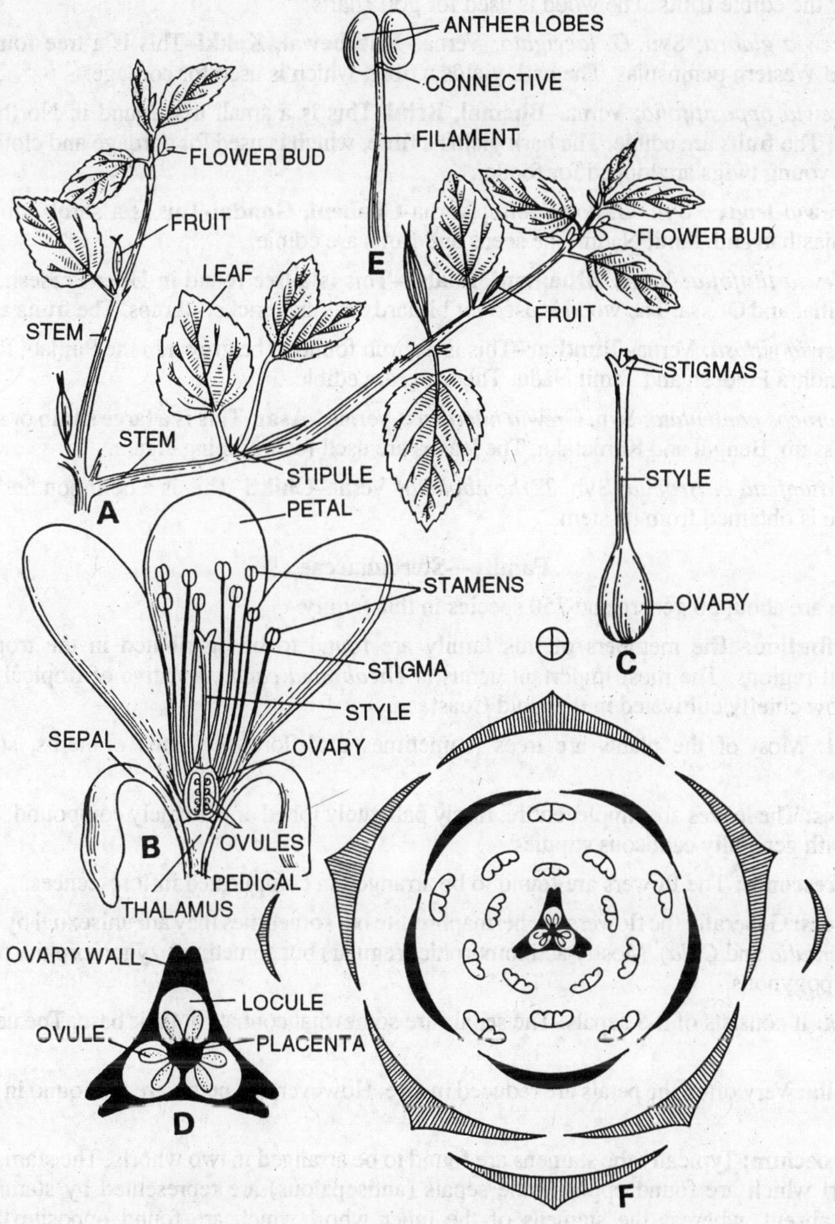

Fig. 22.4 : Tiliaceae. *Corchorus aestuans* L.A, flowering twig, B, L.S. of flower ; C, gynoecium ; D, T.S. ovary ; E, stamen ; F, floral diagram.

known as jute, obtained from the stem is used for rough weaving, jute bags, twine, covers for cotton bales, carpets, coarse cloth and in paper making. Jute sticks are used as a suitable raw material for production of active carbon which is used as a medicinal charcoal. Jute stick charcoal has high adsorptive power for gases and liquids.

2. *Corchorus olitorius*; Eng.–Tossa jute; Verna.–**Koshta, Mithapat**–This is also native of India, grown in West Bengal, Assam, Bihar, Orissa and Uttar Pradesh. The fibre obtained from the stem, is used for rough weaving and coarse cloth.

3. *Grewia asiatica;* Syn. *G. elastica* ; Hindi–**Phalsa**–A small tree, cultivated throughout the country for the edible fruits. The wood is used for golf shafts.

4. *Grewia glabra;* Syn. *G. laevigata;* Verna.–**Kathbewal, Kakki**–This is a tree found in the Eastern and Western peninsulas. The bark yields a fibre, which is used for cordage.

5. *Grewia oppositifolia*; Verna.–**Bhemal, Briul**–This is a small tree found in North Western Himalayas. The fruits are edible. The bark yields a fibre, which is used for cordage and clothing. The leaves and young twigs are lopped for fooder.

6. *Grewia tenax ;* Syn. G. *populifolia*; Verna-**Chabeni, Gondni**-This is a shrub found in the Punjab, Rajasthan and Tamil Nadu. The seeds and fruits are edible.

7. *Grewia tiliifolia;* Verna.–**Dhamani, Phalsa**–This is a tree found in Utter Pradesh, Madhya Pradesh, Bihar and Orissa. The wood is used for billiard cues and cricket stumps. The fruits are edible.

8. *Grewia villosa;* Verna–**Tamthar**–This is a shrub found to be grown in the Punjab, Rajasthan, Gujarat, Andhra Pradesh and Tamil Nadu. The fruits are edible.

9. *Microcos paniculata;* Syn. *Grewia microcos;* Verna. –**Asar**. This is a large shrub or small tree found in Assam, Bengal and Karnataka. The leaves are used for wrapping cigars.

10. *Triumfetta bartramia*; Syn. *T. rhomboidea;* Verna–**Chikti**–This is a common herb. A soft, glossy fibre is obtained from its stem.

Family—Sterculiaceae

There are about 50 genera and 750 species in this family.

Distribution: The members of this family are found to be distributed in the tropical and sub-tropical regions. The most important genus is *Theobroma cacao,* a native of tropical America which is now chiefly cultivated in the Gold Coast.

Habit: Most of the plants are trees (sometimes cauliflorous) shrubs or herbs, sometimes climbers.

Leaves: The leaves are simple, entire, rarely palmately lobed or palmately compound, alternate, stipulate with generally caducous stipules.

Inflorescence: The flowers are found to be arranged in complicated inflorescences.

Flowers: Generally the flowers are hermaphrodite but sometimes they are unisexual by abortion, (*e.g.,* in *Sterculia* and *Cola*). Mostly actinomorphic (regular) but sometimes zygomorphic, pentamerous and hypogynous.

Calyx: It consists of 3-5 sepals. The sepals are somewhat connate at their base. The aestivation is valvate.

Corolla: Very often the petals are reduced in size. However, the petals are not found in *Sterculia* and *Cola.*

Androecium: Typically the stamens are found to be arranged in two whorls. The stamens of the outer whorl which are found opposite the sepals (antisepalous) are represented by staminodes or altogether absent, whereas the stamens of the inner whorl which are found opposite the petals (antipetalous) are fertile. The filaments are more or less united to form a tube (monadelphous).

Gynoecium: It consists of 4-5 carpels, syncarpous; the superior ovary is 4-5 loculed; the carpels are generally antipetalous and contain two to numerous anatropous ovules; the placentation is axile; sometimes the ovary is found to be carried up along with the stamens above the petals by the development of **androgynophore,** the styles are as many as the number of carpels which are either free or united together.

Fruits: Generally the fruits are dry, which very often separate into cocci.

Seeds: They are endospermic. Each seed contains an embryo with two folded, leaf-like cotyledons.

Pollination: The pollination usually takes place through the agency of insects.

Floral Formula: ⊕ ⚥ K 3 — 5, C, A 5, G, $\underline{(4-5)}$.

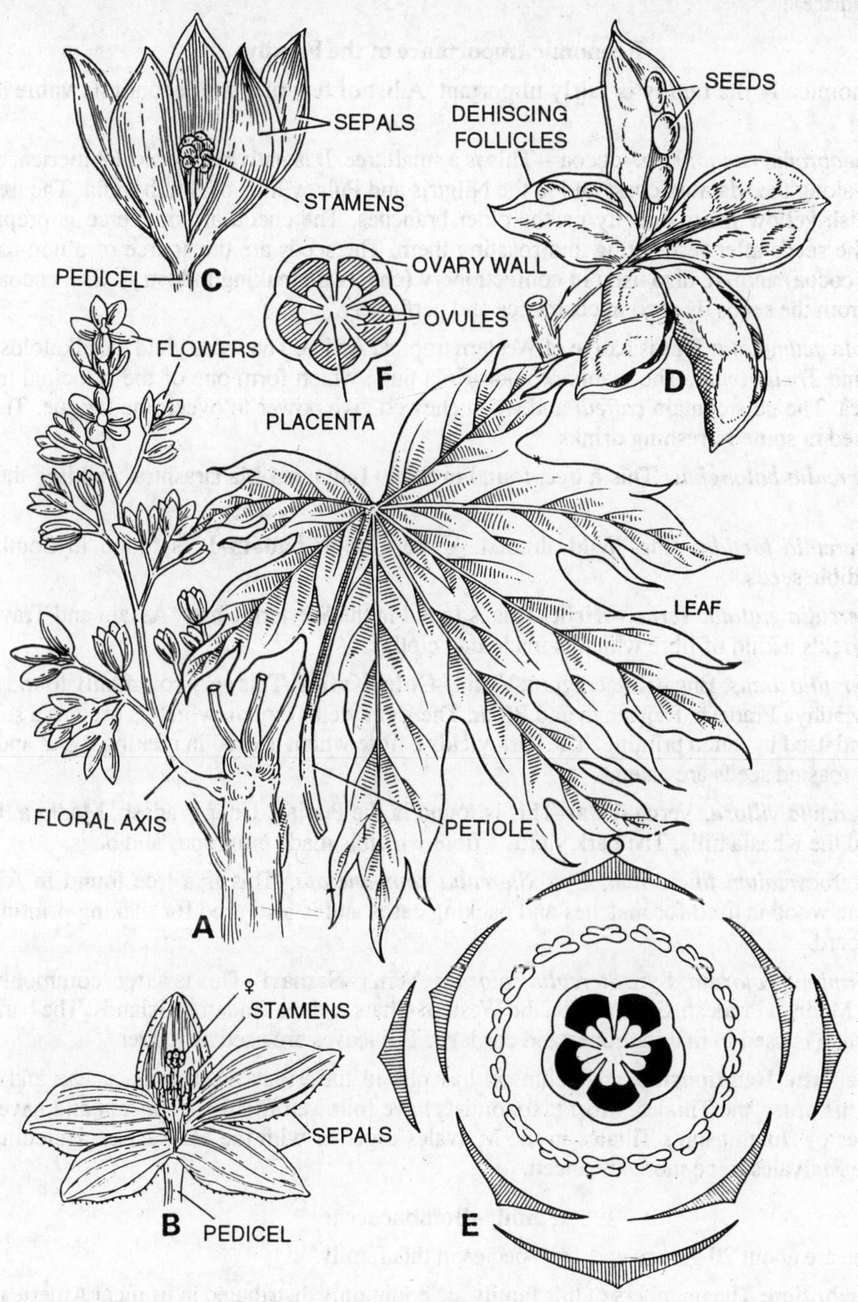

Fig. 22.5. Sterculiaceae—*Sterculia villosa*. A, flowering twig and leaf; B, flower top view ; C, flower lateral view ; D, fruit—dehiscing follicles; E, floral diagram ; F, T.S. of ovary.

Bentham & Hooker (1862)	**Engler & Prantl(1931)**	**Hutchinson(1959)**
Phanerogams	Phanerogams	Angiospermae
Dicotyledones	Dicotyledones	Dicotyledones
Polypetalae	Archichlamydeae	Lignosae
Thalamiflorae	Malvales	Tiliales
Malvales	Sterculiaceae	Sterculiaceae
Sterculiaceae		

Economic Importance of the Family

Economically the family is fairly important. A list of few plants of economic value is given here.

1. *Theobroma cacao;* Eng.–cocoa—This is a small tree. It is native of tropical America, but now cultivated along the Malabar coast and in the Nilgiris and Pulney hills of South India. The tree bears large reddish-yellow fruits directly on the older branches. The cocoa of commerce is prepared by grinding the seeds after fermenting and roasting them. The seeds are the source of a non-alcoholic beverage (cocoa) and are also used in confectionery (chocolate making industry). The cocoa butter, obtained from the seeds, is used in cosmetics and perfumery.

2. *Cola acuminata*; This is native of Western tropical Africa. The seeds yield the alkaloids known as *Thein* and *Theobromin*. The plants provide Kola nuts, which form one of the principal trades in West Africa. The nuts contain *caffein* and when chewed give power to overcome fatigue. The seeds are also used in some refreshing drinks.

3. *Sterculia balanghas;* This a tree, found in South India and Maharashtra, yielding the edible seeds.

4. *Sterculia foetida;* Eng. Wild almond; Verna.–**Jangli badam**–It is found in South India, yielding edible seeds.

5. *Sterculia guttata: Verna.*–**Hirik**–This is found in the Western Ghats, Assam and Travancore. The bark yields a kind of fibre which is made into cloth.

6. *Sterculia urens;* Eng.–Kateeragum; Verna.–**Gulu, Katira**–This very commonly found in Uttar Pradesh, Madhya Pradesh, Rajasthan and Bihar. The stem yields a gum, which is used as a substitute for gum and used in calico printing. The bark yields a fibre which is used in making ropes and coarse cloth. The roasted seeds are eaten.

7. *Sterculia villosa;* Verna–**Udal**–This is found in the Punjab, Uttar Pradesh, Madhya Pradesh, Assam and the Khasia hills. The bark yields a fibre which is made into ropes and bags.

8. *Pterocymbium tinctorium;* Syn. *Sterculia campanulata;* This is a tree found in Andaman islands. The wood is used for matches and packing cases and is also used for making printing paper and cardboard.

9. *Firmiana colorata;* Syn. *Sterculia colorata;* Verna.–**Samari** –This is a tree, commonly found in Assam, Madhya Pradesh, South India, the Western Ghats and the Andaman islands. The bark yields a fibre, which is used in making ropes and cordage. The leaves are used as fodder.

Systematic Relationships: Hutchinson has placed the Tiliaceae, Bombacaceae and Sterculiaceae in his order, the Tiliales. Most taxonomists have followed Eichler's concept and have placed the families of Hutchinson's Tiliales in the Malvales together with the Malvaceae. Hutchinson has treated the Malvales to be more advanced.

Family–Bombacaceae

There are about 20 genera and 140 species in this family.

Distribution: The members of this family are commonly distributed in tropical America, Africa and Asia. The common plants in India are represented by *Bombax ceiba*(silk cotton tree), *Adansonia digitata* (Baobab tree) and *Eriodendron pentandrum.*

Habit: The plants are tall trees with thick trunks and spreading branches.

Leaves: The leaves are alternate, simple or palmately compound, deciduous, often with slime or mucilage containing cells and the vesture of stellate hairs or peltate scales; stipules caducous.

Inflorescence: Cymose, solitary or fascicled in leaf axils or situated opposite a leaf.

Flowers: Generally the flowers are actinomorphic (rarely slightly zygomorphic), bisexual, large and showy, commonly bracteate, often appearing before the leaves; the perianth sometimes subtended by an involucre.

Calyx: Sepals 5, distinct or basally connate, valvate.

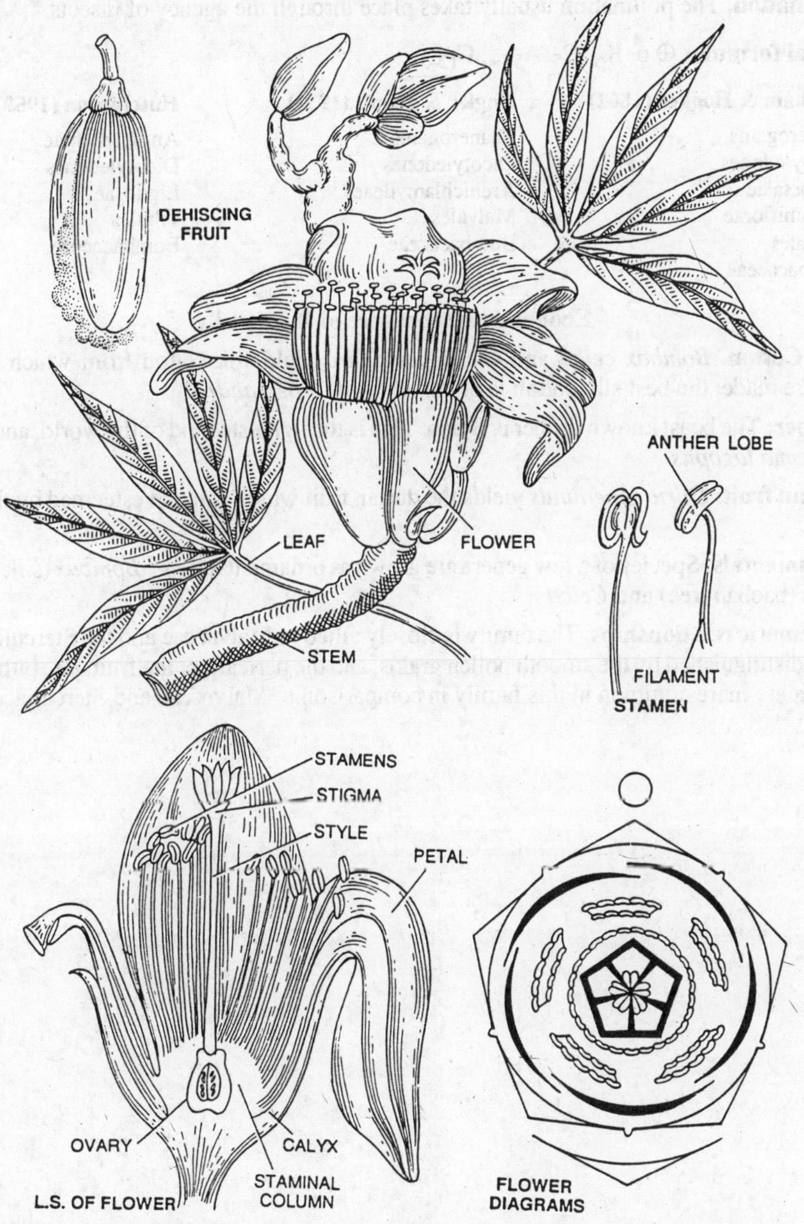

Fig. 22.6 : Bombacaceae. *Bombax ceiba;* Eng., silk cotton tree; Verna. **Semal.**

Corolla: Petals 5, or occasionally absent, contorted in bud, inbricate and free.

Androecium: Stamens 5-many, distinct or monadelphous, the anthers often one celled (monothecous), dehiscing longitudinally, the pollen grains smooth; staminodes often present.

Gynoecium: Carpels are 2-5, connate in a superior 2-5 loculed ovary, with 2-many anatropous ovules in each locule; placentation axile, style 1, capitate or lobed, the stigmas 1-5.

Fruit: A loculicidal capsule, sometimes an indehiscent pod or berry like.

Seed: Smooth, sometimes embedded in a pith like tissue or in a woolly proliferation of the pericarp, occasionally arillate, the endosperm scant to absent.

Pollination. The pollination usually takes place through the agency of insects.

Floral formula: ⊕ ⚥ K_5, C_5, $A_{5-\infty}$, $\underline{G_{(2-5)}}$.

Bentham & Hooker (1862)	Engler & Prantl (1931)	Hutchinson (1959)
Phanerogams	Phanerogams	Angiospermae
Dicotyledones	Dicotyledones	Dicotyledones
Polypetalae	Archichlamydeae	Lignosae
Thalamiflorae	Malvales	Tiliales
Malvales	Bombacaceae	Bombacaceae.
Bombacaceae		

Economic Importance of the Family

Silk Cotton. *Bombax ceiba* and *Cieba pentandra* yield silk cotton from which pillows and cushions are made; the best silk cotton is obtained from *C. pentandra.*

Timber: The beast known timber is **Balsa.** This is the lightest wood of the world, and is obtained from *Ochroma lagopus.*

Durian fruit: *Durio zebethinus* yields the durian fruit which is much esteemed by the people of Myammar.

Ornamentals: Species of a few genera are grown as ornamentals, *e.g., Bombax* (silk cotton tree), *Adansonia* (baobab tree) and *Ceiba.*

Systematic relationships. The family is closely allied to Malvaceae and the Sterculiaceae, from which it is distinguished by the smooth pollen grains, and the pericarp of the fruit, *i.e.*, pithy to woolly. Staminodes are more common in this family in comparison to Malvaceae and Sterculiaceae.

23

ORDER–MALVALES

According to Hutchinson this is the thirty third order of the phylum Angiospermae, subphylum Dicotyledones and division Lignosae. This order consists of a single family–the Malvaceae, which has been described in this text in detail. Bentham and Hooker have included the family Malvaceae along with other two families, *i.e.,* Sterculiaceae and Tiliaceae in the sixth order–the Malvales of class Dicotyledones, division Polypetalae and series Thalamiflorae. Engler and Prantl have placed the family Malvaceae along with seven families including Tiliaceae and Sterculiaceae in their twenty sixth order–the Malvales of class Dicotyledoneae and subclass Archichlamydeae.

Mostly as in Tiliales, but mainly herbaceous or softly woody and fibrous, calyx always valvate, anthers split in two and therefore 1-locular.

Family—Malvaceae

There are about 82 genera and 1,500 species in this family.

Distribution: The members of this family are cosmopolitan in their distribution. They are specially found in tropical regions of the world. Several species of this family have been reported from our country, *e.g., Hibiscus, Sida, Malvastrum, Malva, Abutilon, Gossypium,* etc. Several of them are wild and some are commonly cultivated in gardens, *e.g., Althaea rosea.* However, the most important member of the family is the genus *Gossypium,* from which various cottons are obtained.

Habit: The plants may be herbs, shrubs or trees. A few plants are annual or sometimes perennial herbs, *e.g., Abutilon, Malva, Malvastrum, Hibiscus* sp., etc., they may be under shrubs, *e.g., Sida cordinifolia, S. carpinifolia,* etc. The species of *Hibiscus,* may be herbs, shrubs or even small trees. *H. rosa sinensis* is a large ornamental shrub with beautiful red flowers. *II. mutabilis* is a small ornamental tree.

Root: Tap and branched.

Stem: The stem is erect, herbaceous or woody, branched, cylindrical, solid usually with stellate hairs. Mucilage sacs occur abundantly in the tissues.

Leaves: The leaves are simple, alternate, round or palmately divided, stipulate, usually palmately veined. The leaves bear stellate hairs. Mucilage sacs are also found.

Inflorescence: The inflorescence is of cymose type, very rarely racemose. This may be solitary axillary terminal, or compound cyme.

Flower: The flowers are usually pedicellate, bracteolate (the bracteoles are arranged in a whorl known as epicalyx), actinomorphic regular, hermaphrodite, complete, hypogynous and pentamerous.

Calyx: It consists of five sepals, gamosepalous, *i.e.,* the sepals are united to each other, sepaloid, inferior, and the aestivation is valvate.

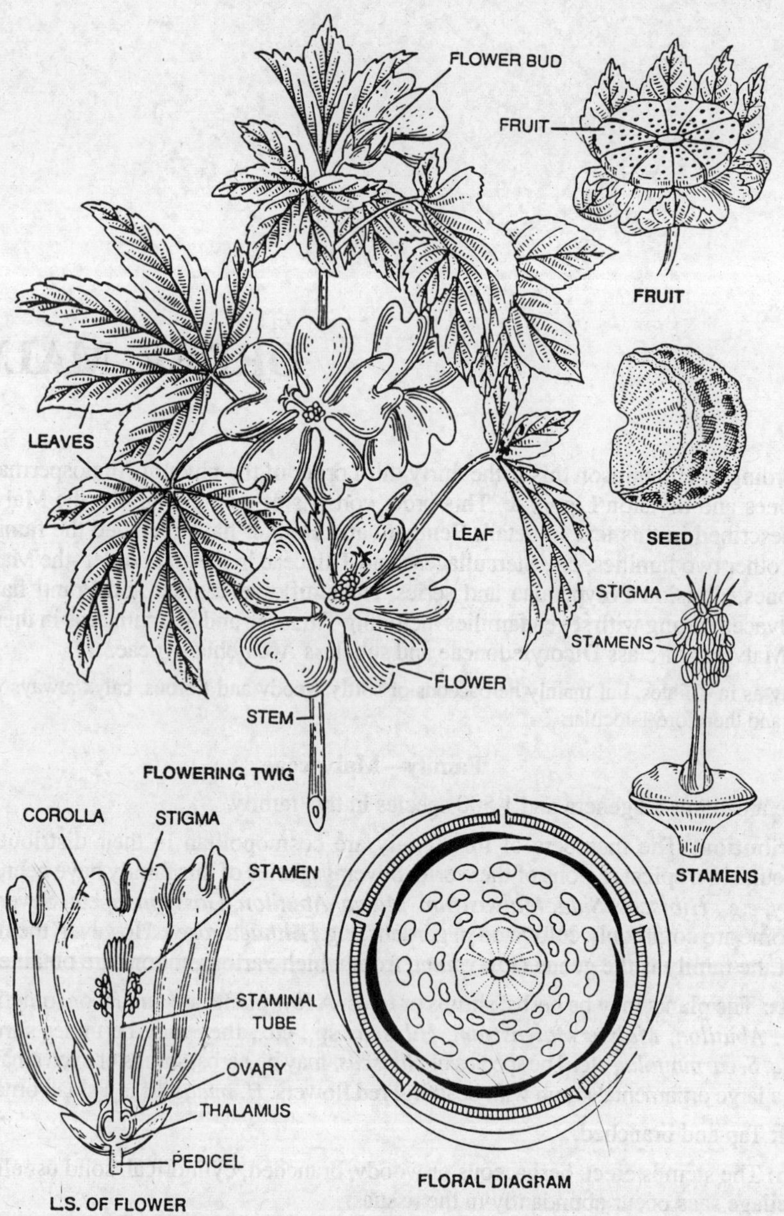

Fig. 23.1: Malvaceae. *Malva sylvestris* Linn., Eng., Common mallow; Verna. **Vilayati Kangi.**

Corolla: It consists of five petals, polypetalous but the petals are inferior and the aestivation is twisted. The petals are variously coloured.

Androecium: It consists of indefinite stamens, the stamens are found arranged on a staminal tube. The stamens are monadelphous. The staminal tube remains united at the base to the petals, and therefore, the condition is known to be epipetalous. The filaments are short and the anthers are monothecous, *i.e.,* onecelled and dorsifixed. The anthers dehisce transversely.

The stamens are derived from profuse branching of five antipetalous stamens. The outer whorl has been lost, but however, represented in *Hibiscus* by five staminodes.

Gynoecium: It consists of 5 to indefinite carpels (polycarpellary), syncarpous. The ovary is superior and multilocular. Each loculus bears one to many ovules. The placentation is axile. The styles are united but the stigmas are free. The style passes through the staminal tube. The number of stigmas is as much as the number of carpels. In *Hibiscus* the carpels are five, in *Althaea rosea* the number is ten and in *Abutilon* the number of carpels is indefinite.

Fruit: The fruit is loculicidal capsule or schizocarp. It splits into a large number of mericarps. The number of mericarps is equal to the number of carpels. In *Hibiscus* and *Gossypium* the fruit is loculicidal capsule.

Seed: The seeds are usually buried in a hairy covering formed from the testa. The reniform or ovoid in shape. The embryo possesses large much folded cotyledons. The endosperm is scanty.

Pollination: Entomophilous.

Floral Formula : $\oplus$ ⚥ K (5), $\widehat{C5\,A}$ (∞), G $(\underline{5-\infty})$

Bentham & Hooker (1862)	Engler& Prantl (1931)	Hutchinson (1959)
Phanerogams	Phanerogams	Angiospermae
Dicotyledones	Dicotyledones	Dicotyledones
Polypetalae	Archichlamydeae	Lignosae
Thalamiflorae	Malvales	Malvales
Malvales	Malvaceae	Malvaceae.
Malvaceae		

Description of some important plants in semi-techinical language.

1. ***Hibiscus rosa-sinensis.***

Habit: Ornamental perennial shrub.

Root: Tap, branched.

Stem : Erect, woody, branched, cylindrical, glabrous and solid.

Leaf: Simple, alternate, petiolate, stipulate, ovate, serrate, glabrous, apex acuminate, multicostate reticulate venation.

Inflorescence: Cymose, solitary axillary.

Flower: Pedicellate, hermaphrodite, large, showy, pentamerous, actinomorphic, complete, hypogynous.

Epicalyx: 5-7 bracteoles constitute epicalyx below calyx.

Calyx: Five sepals, gamosepalous, green, inferior, valvate aestivation:

Corolla: Five petals, polypetalous, slightly united below, adnate to the staminal tube, inferior twisted aestivation.

Androecium: Stamens indefinite, monadelphous, stamens form a staminal tube round the style, epipetalous, anthers monotheous, reniform, basifixed.

Gynoecium: Five carpels (pentacarpellary), syncarpous ovary superior, pentalocular, axile placentation, style long and passes through the staminal tube ending in five distinct rounded stigmas.

Fruit: In other species of *Hibiscus, loculicidal* capsule.

Floral Formula: $\oplus$ ⚥ K(5), $\widehat{C5, A}$(∞), G($\underline{5}$)

2. ***Abutilon indicum;*** Verna–**Kangi.**

Habit: Annual or perennial herb or undershrub.

Root: Tap, branched.

Stem: Erect, branched, herbaceous, solid, hairy, green.

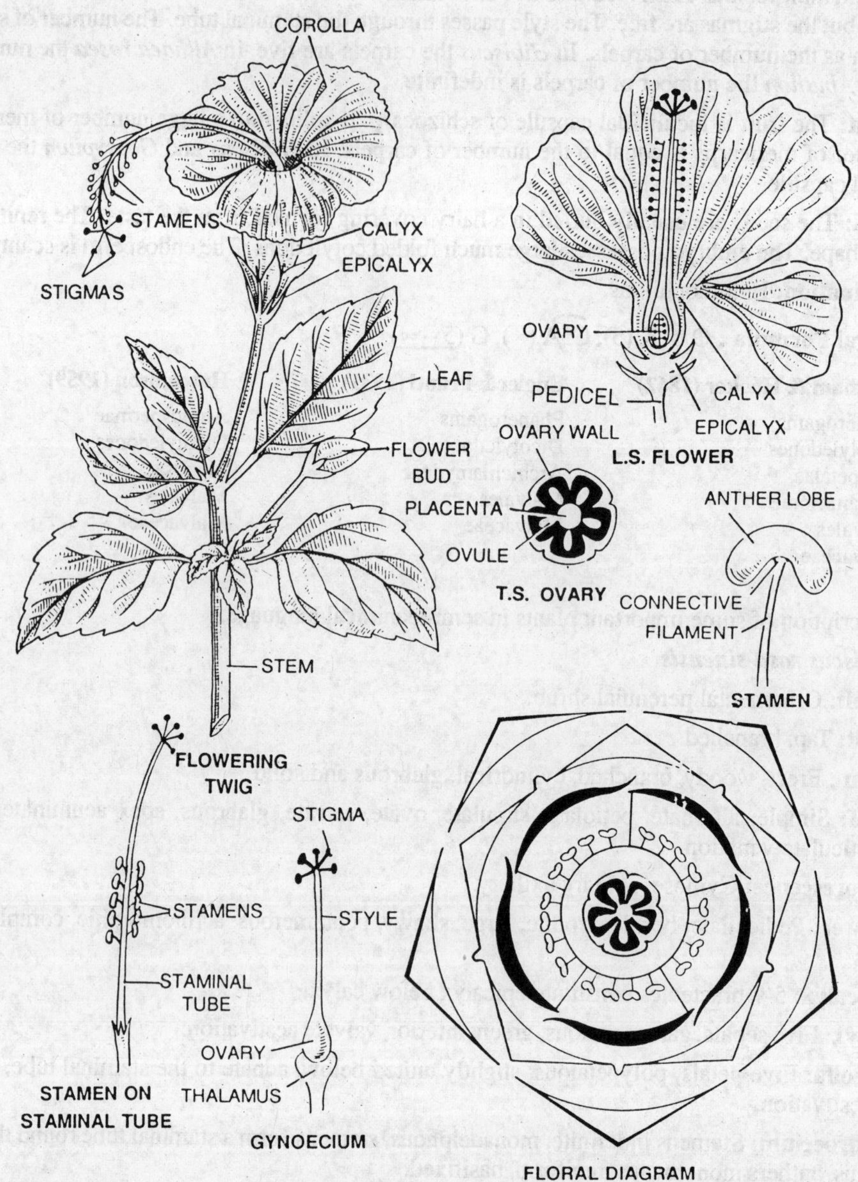

Fig. 23.2. Malvaceae. *Hibiscus rosa sinensis* Linn. Eng., China shoeflower; Verna. **Gurhal.**

Leaves: Simple, alternate, petiolate (petiole usually longer than blade), broadly cordate, acute or acuminate apex, irregularly toothed, pale, minutely pubescent on both surfaces, entire, multicostate reticulate venation.

Inflorescence: Cymose, solitary axillary.

Flower: Pedicellate (pedicels longer than the petioles and jointed near the top), one inch diameter, hermaphrodite, actinomorphic, complete, hypogynous.

Calyx: Five sepals, gamosepalous, shallow, ovate, acute, persistent, valvate aestivation.

Corolla: Five petals, polypetalous, free above, connate below and adnate to the staminal tube, orange, yellow, petals spreading imbricate aestivation.

Androecium: Indefinite stamens, staminal tube dividing at the top into numerous anther bearing filaments, monadelphous, anthers one-celled, basifixed, dehisce longitudinally.

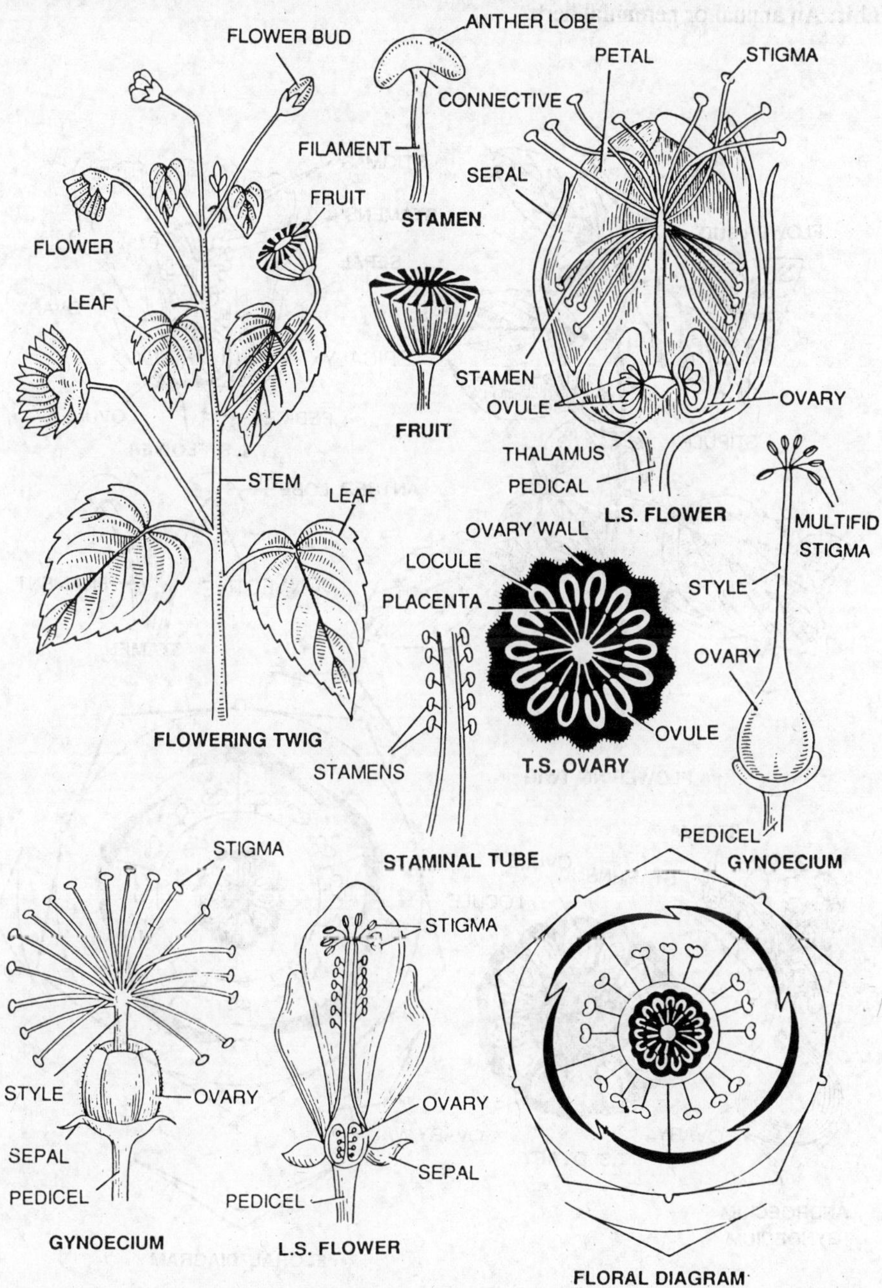

Fig. 23.3 : Malvaceae—*Abutilon indicum* (Linn) Sweet. Eng. Country mallow, Verna. **Kangi**

Gynoecium: 15-20 carpels (polycarpellary), carpels longer than the calyx, with short spreading awns, hairy, syncarpous, ovary superior, multilocular, axile placentation, many ovules in each locule.

Floral Formula: ⊕ ⚥ K(5), C 5, A (∝) G($\underline{\infty}$).

Note. The genus *Sida* is closely allied to *Abutilon.*

Malvastrum tricuspidatum A. Gray.

Habit: An annual or perennial herb.

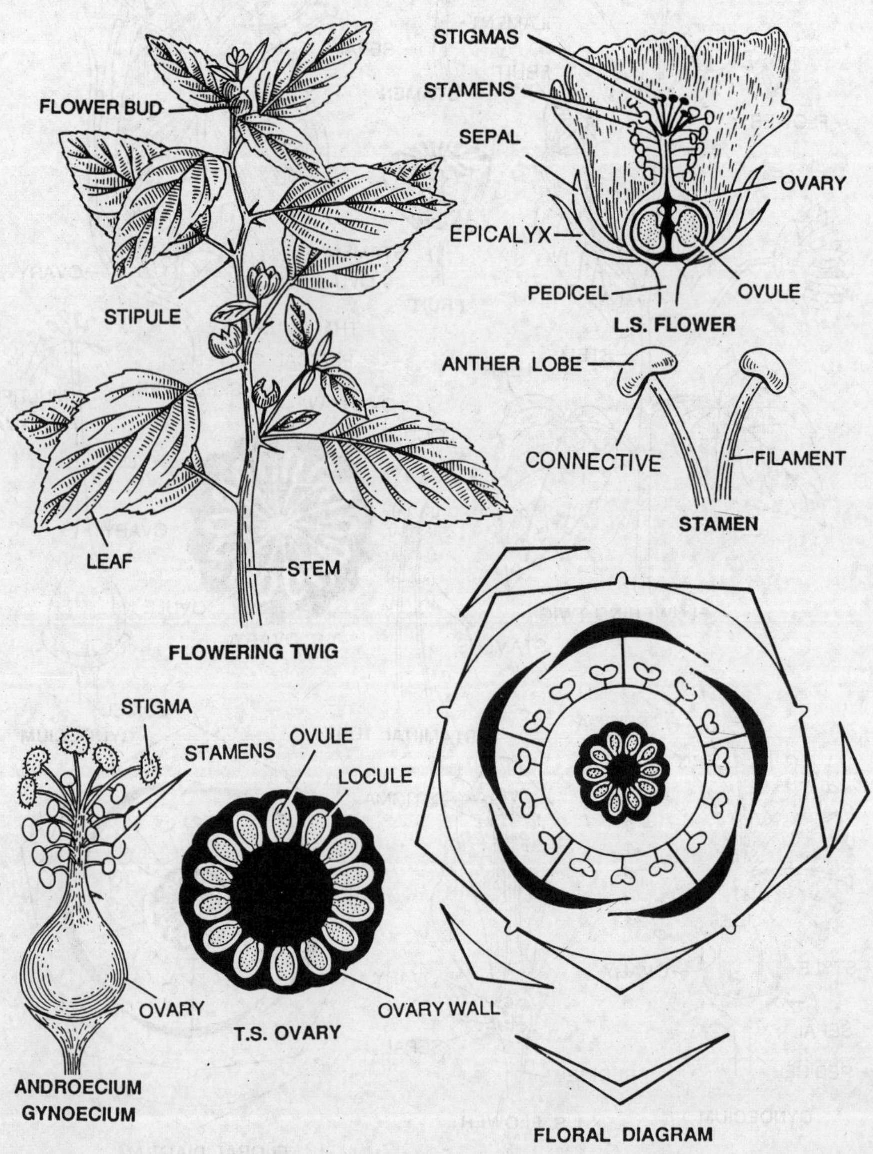

Fig. 23.4 : Malvaceae. *Malvastrum tricuspidatum* A. Gray.

Root: Tap, branched.

Stem: Erect, aerial, herbaceous, cylinderical, branched, solid, hairy, green.

Leaf: Cauline and ramal, petiolate, stipulate, stipules small and free-lateral, alternate, simple, ovate-lanceolate, serrate, acute, glabrous, unicostae reticulate.

Inflorescence: Cymose, solitary axillary.

Flower: Bracteate, bracteolate, 3 small bracteoles, pedicellate, actinomorphic, hermaphrodite, hypogynous, pentamerous, cyclic.

Calyx: 5 sepals, gamosepalous, valvate aestivation, persistent.

Corolla: 5 petals, polypetalous, twisted, yellow.

Androecium: Indefinite, monadelphous, a staminal tube is formed around the style, epipetalous, monothecous, reniform, basifixed, extrorse.

Gynoecium: Many carpels, syncarpous, ovary superior, multilocular, axile placentation, stigmas as many as carpels, style within staminal tube.

Fruit : A capsule.

Floral formula: Br. Brl. $\oplus$ ⚥ K(5), C 5, A, $\propto$ G ($\underline{\propto}$).

Economic Importance of the Family

The family is of great economic value as it contains valuable fibre yielding plants. some plants are of medicinal value and some are highly ornamental. The important plants are as follows:

1. *Abutilon asiaticum;* Eng. Don (Verna–**Kanghi**)– A shrub, cultivated for its stem fibre, which is used for ropes and cordage.

2. *Abutilon indicum;* (Eng.-Indian Abutilon; Verna-**Kanghi**) The stems, on retting, yield a fibre, which is used for making ropes.

3. *Abutilon polyandrum;* The stem fibre is used for manufacturing ropes and cordage. Found in North Western India, the Western Peninsula and the tropical Himalayas.

4. *Abutilon theophrasti ;* Syn. *A. avicennae* Gaertn. (Eng.–China jute; Verna–**Nahani khapat**)- A herb. Its stem fibre called China jute; is used in rug making. Found in North-West India, Kashmir and Bengal.

5. *Althaea rosea;* (Eng.–Hollyhock; Verna–**Gulkhera**). An ornamental herb. Its flowers yield a red dye.

6. *Abelmoschus esculentus;* Syn. *Hibiscus esculentus;* (Eng.–Lady's finger; Verna–**Bhindi**)–The fruits are used as vegetable and stalks are sometimes used for making fibres. Native of Tropical Africa, now cultivated throughout India.

7. *Abelmoschus manihot;* (Syn. *Hibiscus manihot*: Verna-**Ran bhindi**)– The fibre, obtained from the stem, is used for making wrapping cloth. Found in Bengal, the Western Ghats, South Kanara and Travancore.

8. *Abelmoschus moschatus;* Syn. *Hibiscus abelmoschus;* (Verna–**Mushkdana**)–Source of a fibre, which is used for cloth. The seed-oil is used in perfumery. Found in the hilly regions of Deccan and Karnatak.

9. *Gossypium arboreum;* Syn. *G. indicum, G. neglectum;* (Eng.–Tree cotton; Verna–**Kapas, Rui**)-A shrub cultivated all over India as a fibre plant. The fibre obtained from the surface of the seeds, is utilized in the manufacture of textiles of all types, rubber-tyre fabrics and other woven products. The stalks are used for making paper. The cotton seed and the seed cake are used as fodder and the oil from the seed is edible.

10. *Gossypium arboreum* var. *nadam;* Syn. *G. barbadense;* (Eng.–Egyptian cotton; Verna–**Kapas, Rui**).-It is native of tropical America, grown mainly in Malabar and South Kanara. The fibre obtained from the surface of the seeds, is utilized in the manufacture of textiles and the seed oil is edible.

11. *Gossypium herbaceum;* (Eng–Asiatic cotton; Verna–**Kapas**)–Cultivated mainly in Ma-

harashtra, Tamil Nadu, Andhra Pradesh and Karnataka. The fibre obtained from the seeds is used in making cloth and other woven products. The oil–cake is used as fertilizer and fodder.

12. *Gossypium hirsutum;* Syn. *G. mexicanum* (Eng.–Upland cotton; Verna–**Kapas**)– Grown chiefly in the Punjab, Uttar Pradesh, Rajasthan, Madhya Pradesh, Tamil Nadu and Maharashtra. The fibre obtained from the seeds, is utilized in the manufacture of textiles, rubber–tyre fabrics and other woven products. The cotton seed and cake are used as fodder and the seed oil is edible.

13. *Hibiscus cannabinus;* (Eng.–Kenaf, Verna–**Patsan**)–A herb grown in West Bengalm, Andhra Pradesh, Assam, Bihar and the Punjab. The fibre obtained from the stem, is used for making ropes cordage and paper pulp and the seed oil is edible.

14. *Hibiscus collinus;* A small tree grown as an ornamental.

15. *Hibiscus ficulneus;* Syn. *Abelmoschus ficulneus;* Verna–**Ran Bhindi**–A herb, found in the Punjab, Bengal and South India. The mucilaginous extract of the green stem is used for clarifying sugar juice and the seeds are used for flavouring purposes.

16. *Hibiscus urcatus* Roxb.–Found in Assam, Madhya Pradesh, Orissa, Andhra Pradesh, Bengal and Karnataka. The leaves are edible and the stem fibre is used for flavouring purposes.

17. *Hibiscus macrophyllus;* Roxb. (Verna–**Baiza**)–A small shrub, found in Bengal and the Eastern Peninsula. The stem fibre is used for ropes and cordage.

18. *Hibiscus mutabilis*; (Eng.–cotton rose; Verna–**Guliajaib**)–A bushy shrub; native of China. The bark yields a fibre, which is used for ropes and cordage.

19. *Hibiscus rosa–sinensis;* (Eng.–Rose of China; Verna–**Gurhal**)–A common ornamental shrub. the flowers yield a dark purplish dye, which is used for making shoe–polishes.

20. *Hibiscus sabdariffa;* (Verna–**Patwa**)– A shrub native of the West Indies, now grown in Andhra Pradesh, Bihar, the Punjab, Assam and Tamil Nadu. The stem fibre is utilized for making ropes and cordage and the red fleshy epicalyx is edible.

21. *Hibiscus schizopetalous;* (Eng–coral Hibiscus)–A shrub, native of East Africa, grown as an ornamental for its orange–red flowers.

22. *Hibiscus surattensis;* (Verna–**Ranbhindi**)–A herb cultivated mainly in Bengal. The leaves are edible and the stem fibre is used for rough cloth and cordage.

23. *Hibiscus syriacus* ; (Verna–**Swet jaha**)–A shrub native of China and Japan, now cultivated all over India for its edible flowers.

24. *Hibiscus tiliaceus* ; Syn. *Paritium tiliaceum;* (Eng.–Cuban bast; Verna–**Chelwa**)–Commonly found along the coasts of both the Peninsulas and in the Sunderbans. The leaves are edible and the bark fibre is used for ropes and mats.

25. *Hibiscus vitifolius;* (Verna–**Ban kapas**)–Found in Northern India and Madhya Pradesh. The bark is the source of strong fibre.

26. *Malachra capitata*; (Eng.–cork wood: Verna–**ranbhindi**)–A herb, found throughout the hotter parts of the country. The stem fibre is used for cloth, twines and ropes.

27. *Urena lobata;* Syn. *U. sinuata;* (Verna–**Vilayati san**)–A small shrub, commonly found in Bengal. The stem fibre is a substitute for jute, and is used for cordage ropes and coarse fabrics. The mucilaginous seeds are used for making soap.

Systematic Relationships: According to different taxonomists the arrangement of the families within the order Malvales varies. Engler, Wettstein and Rendle include the common families, Tiliaceae, Bombacaceae, Sterculiaceae and Malvaceae in this order. Hutchinson, however, separated the Malvales to include only the Malvaceae and included the Tiliaceae, Bombacaceae, and Sterculiaceae in his new order the Tiliales. According to Rendle the Malvales show the affinities with the Guttiferae in having hypogynous, regular and pentamerous flowers. He also thinks that on the other hand the Malvales are related to Euphorbiales in some respects. Bessey expanded Eichler's concept of Malvales and included the Balanopsidaceae, Ulmaceae, Moraceae and Urticaceae within this order.

24

ORDER–MALPIGHIALES

According to Hutchinson this is the thirty fourth order of the phylum Angiospermae, subphylum Dicotyledones and division Lignosae. The order comprises of twelve families. They are–Ixonanthaceae, Malpighiaceae, Humiriaceae. Linaceae, Irvingiaceae, Huaceae, Ledocarpaceae, Erythroxylaceae, Ctenolophonaceae, Lepidobotryaceae, Balanitaceae and Zygophyllaceae. Of which only Linaceae has been described in the present text in detail. Bentham and Hooker have placed the family Linaceae along with other ten families including Rutaceae and Meliaceae in their seventh order the Geraniales of class Dicotyledones. division Polypetalae and series Disciflorae. Engler and Prantl have included the family Linaceae along with other nineteen families including Rutaceae, Meliaceae and Euphorbiaceae in the twenty third order–the Geraniales of class Dicotyledoneae and subclass Archichlamydeae.

Hypogynous; actinomorphic to subzygomorphic; bisexual; syncarpous with subapical placentation; ovules few; stamens usually definite in number; endosperm mostly absent. Often climbers with opposite leaves.

Family—Linaceae

There are about 9 genera and 200 species in this family.

Distribution: The members of this family are found throughout the world, primarily in temperate regions of both northern and southern hemispheres. In our country the family is chiefly represented by the genus *Linum usitatissimum* (Linseed or Flax). This is also known as flax family. In India linseed is extensively grown especially in the Benares division, in Bundelkhand and in the Sub Himalayan tract.

Habit: Mostly annual herbs, sometimes shrubs, trees also occur.

Root: Usually tap and branched.

Leaves: Alternate or opposite, rarely whorled, simple, entire, stipulate or exstipulate (stipules sometimes intrapetiolar).

Inflorescence; Cymose, dichasial cyme or cincinnus (sometimes appearing racemose).

Flower: Hermaphrodite (bisexual), actinomorphic, complete, hypogynous, mostly pentamerous.

Calyx: Usually 5 sepals, rarely four, the sepals may be free or connate at the base (rarely 4-parted with 3-fid lobes), imbricate aestivation.

Corolla: Usually 5 petals; rarely four, the petals contorted in bud, free, often clawed and when so the claw naked or crested; imbricate or contorted aestivation; arranged, alternately to sepals, usually coloured blue, white or red, early deciduous.

Androecium: Stamens 5, arranged alternately with the petals, the filaments are connate at the base and form a ring below the ovary, outside the ring nectar secreting glands are present, sometimes the five stamens are alternated with 5 or 10 tooth like staminodes, the anthers dithecous (2-celled), introrse, always dehisce by longitudinal slits.

Gynoecium: Usually it consists of 5 carpels (rerely 3-4), 5 locules or falsely 10-loculed by the intrusion of carpel midribs, each locule contains 2 ovules; the ovules are pendulous and anatropous; placentation axile, the ovary superior, pistil one; there are as many styles as many ovary locules, styles are free, filiform and at the terminal end of each style there is a capitate stigma.

Fruit: It may be a drupe or septicidal capsule, if drupe it is surrounded by persistent calyx; the fruit splits septicidally into several valves corresponding to the number of the carpels present.

Seeds: Albuminous or exalbuminous; embryo nearly as long as the seed, straight, rarely incurved, cotyledons usually broad and superior radicle.

Pollination: Usually the flowers are cross pollinated through the agency of insects, sometimes self pollination also occurs.

Floral Formula : ⊕ ⚥ K 5, C 5, A 5, G(2 - 5)

Bentham and Hooker (1862)	**Engler and Prantl (1931)**	**Hutchinson (1959)**
Phanerogams	Phanerogams	Angiospermae
Dicotyledones	Dicotyledones	Dicotyledones
Polypetalae	Archichlamydeae	Lignosae
Disciflorae	Geraniales	Malpighiales
Geraniales	Linaceae	Linaceae

Linum usitatissimum; Verna–**Alsi**

Habit: Annual herb 2-4 feet high, cultivated for the seeds and fibres.

Root: Tap and branched.

Stem: Erect, usually corymbosely branched towards the upper side.

Leaves: Simple, alternate, exstipulate, linear or lanceolate, sub 3-nerved, entire.

Inflorescence: Cymose, flower arranged in broad cymes.

Flower: Pedicellate, hermaphrodite, complete, actinomorphic, hypogynous, blue or sometimes white, one inch across.

Calyx: 5 sepals, ovate acuminate, 3-nerved, glandular, margins white, ciliate or not, aestivation quincuncial.

Corolla: 5 petals, polypetalous, free, contorted aestivation arranged alternately to sepals, claw-shaped, blue or white.

Androecium: 5 stamens alternating with 5 minute staminodes; filaments connate at the base, nectar secreting glands are present on the ring below the ovary, the anthers dithecous, introrse and versatile.

Gynoecium: 5 carpels, ovary pentalocular, two ovules in each locule; styles usually free; stigmas linear, clavate.

Fruit: Septicidal capsules, capsules hardly exceeding the sepals, edges of valves ciliate.

Floral formula: ⊕ ⚥ K 5, C 5, A 5, G (5).

Generic characters: Herbs, perfect, stamens 5, styles 5, capsule 5-celled.

Economic Importance of the Family

1. *Linum usitatissimum* (Eng.–Linseed or Flax; Verna–**Alsi, tisi, Bijri**)–It is grown for its seeds only, the flax is not manufactured in India. It has got some medicinal properties for the colic of ponies. In other countries the cloth **linen** is prepared out of its bast fibres. There are many species of *Linum*

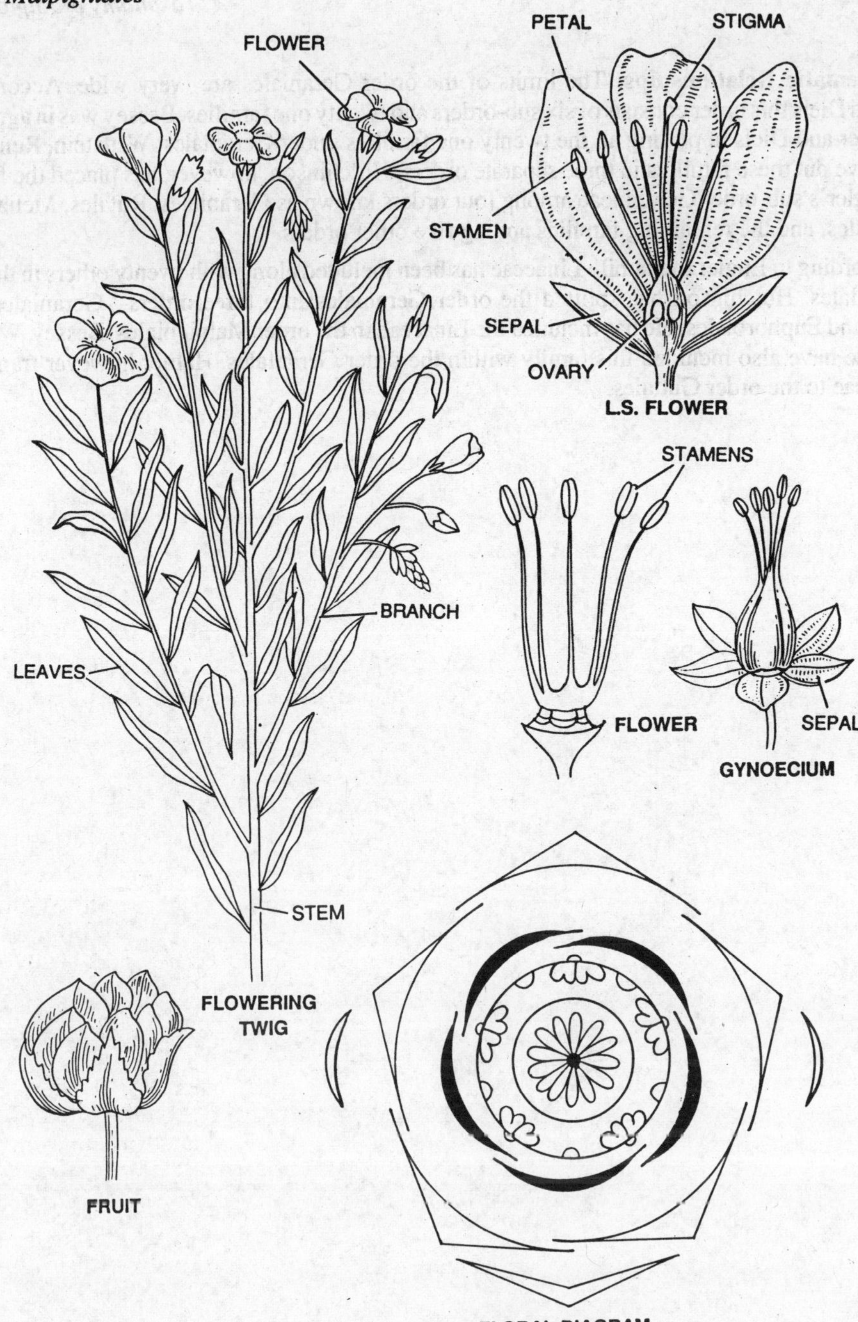

Fig. 24.1. Linaceae- *Linum usitatissimum* Linn. A, twig; B, L.S. of flower; C, flower without calyx and corlla; D, gynoecium; E, L.S. of seed; F, fruit; G, floral diagram.

possessing beautiful flowers of different colours, and they are grown for the same purpose in the gardens. It is chiefly cultivated in Madhya Pradesh, Uttar Pradesh, Maharashtra, Bihar and Rajasthan.

2. *Linum grandiflorum* var. *rubrum;* Eng.–Flowering flax-This is cultivated as an ornamental for its beautiful flowers.

3. *Reinwardtia indica* (Verna–**Basant**)–This possesses some medicinal qualities for cattle diseases.

Systematic Relationships: The limits of the order Geraniales are very wide. According to Engler and Diels this order consists of six sub-orders and twenty one families. Bessey was in agreement with Engler and Diels in putting all the twenty one families under Geraniales. Wettstein, Rendle and Hallier have put these families in three separate orders. Hutchinson, however, has placed the families of the Engler's sub-order Geranineae among four orders known as Geraniales, Rutales, Meliales and Malpighiales, and the remaining families among five other orders.

According to Engler the family Linaceae has been included along with twenty others in the order the Geraniales. Hutchinson has splitted the order Geraniales into three orders—Geraniales, Malpighiales and Euphorbiales. He has included the Linaceae in the order Malpighiales. Bessey, Wettstein and Rendle have also included this family within the order Geraniales. Hallier however transferred the Linaceae to the order Guttales.

25

ORDER–EUPHORBIALES

According to Hutchinson this is the thirty fifth order of the phylum Angiospermae, sub phylum Dicotyledones and division Lignosae. The order consists of a single family, *i.e.*, the Euphorbiaceae which has been described in the present text in detail. Bentham and Hooker have included the Euphorbiaceae along with other eight families including Urticaceae and Casuarinaceae in their seventh series—the Unisexuales of class Dicotyledones, and division Monochlamydeae or Incompletae. Engler and Prantl have included the family Euphorbiaceae along with other nineteen families including Linaceae, Rutaceae and Meliaceae in their twenty third order—the Geraniales of class Dicotyledoneae and sub-class Archichlamydeae.

Hypogynous; actinomorphic; unisexual; petals usually absent, if present sometimes quite sympetalous(*Jatropha*); syncarpous with 1-2 ovules pendulous from the inner angle; stamens various; seeds often with a conspicuous caruncle; endosperms mostly copious; stipules mostly present.

Family—Euphorbiaceae

There are about 283 genera and 7,300 species in this family.

Distribution: The plants of this family are found throughout the world. However, they are not found in arctic regions. In our country the family is represented by several genera such as, *Euphorbia, Ricinus, Phyllanthus, Croton, Pedilanthus,* etc. In the desert regions of Africa and elsewhere the family is represented by cactus-like plants of different species of *Euphorbia.* Heath-like *Euphorbias* are quite common in Australia. In Britain only two genera, *i.e., Euphorbia* and *Mercurialis* are found, which are represented by sixteen and two species respectively.

Habit: The plants exhibit great variation in their habit. The plants may be herbs, shrubs or trees. *Euphorbia hirta, E. thymifolia, E. helioscopica, E. peplus; E. heterophylla, E. cristata, E. elegans; Phyllanthus niruri, Croton* sp., *Acalypha indica,* etc. are annual or prennial herbs. *Euphorbia pulcherrima, E. splendens,* are beautiful shrubs. *Pedilanthus* sp., and *Jatropha* sp., are shrubby plants. *Euphorbia royleana, E. tirucalli* are cactus like shrubs. *Ricinus communis* (**Arand**) is a tall annual and becomes small tree-like in habit. The tree habit of the family is represented by *Phyllanthus emblica* (**Amla**), *Bischofia javanica, Putranjiva roxburghii,* etc. *Havea brasiliensist* (rubber tree) is a large tree 60 to 100 feet in height and 8-12 feet in girth. Species of the genus *Tragia* are tropical climbers. Majority of the members of the family possess large laticiferous vessels which contain latex.

Root: Tap and branched. The roots of *Manihot utilissima* and *M.palmata* are tuberous and rich in starch.

Stem: Herbaceous or woody, erect, very rarely climbing as in a tropical genus *Tragia.* The species of *Xylophylla* possess flat phylloclades. The stem is branched. It may be cylindrical, angular or flat. Usually solid but sometimes hollow as in *Ricinus communis.* Many stems possess spines. In many

Euphorbia sp., the stems become fleshy, green and cactus like in appearance.

Leaves: The form and position of leaves are variable. The arrangement is usually alternate but sometimes they are opposite, *e.g., Euphorbia hirta*. In *Pedilanthus* the leaves are arranged alternately in the lower region of the plant whereas opposite in the floral region. Usually the leaves are simple but in some they are deeply incised, *e.g., Ricinus, Manihot,* etc. In many *Euphorbias* the leaves are scaly and caducous. In many cases the leaves are reduced to spines. In few cases the leaves are replaced by cladodes. Usually the leaves are stipulate. In *Jatropha* sp., the stipules become branched and hair-like. In many *Euphorbia* sp., they are represented by glands or spines.

Inflorescence: The inflorescence varies greatly. It may be racemose or cymose or sometimes complex. In *Euphorbia,* the inflorescence is peculiar but very characteristic and known as **cyathium.** This is the modification of a cyme. In cyathium inflorescence a large number of male flowers each

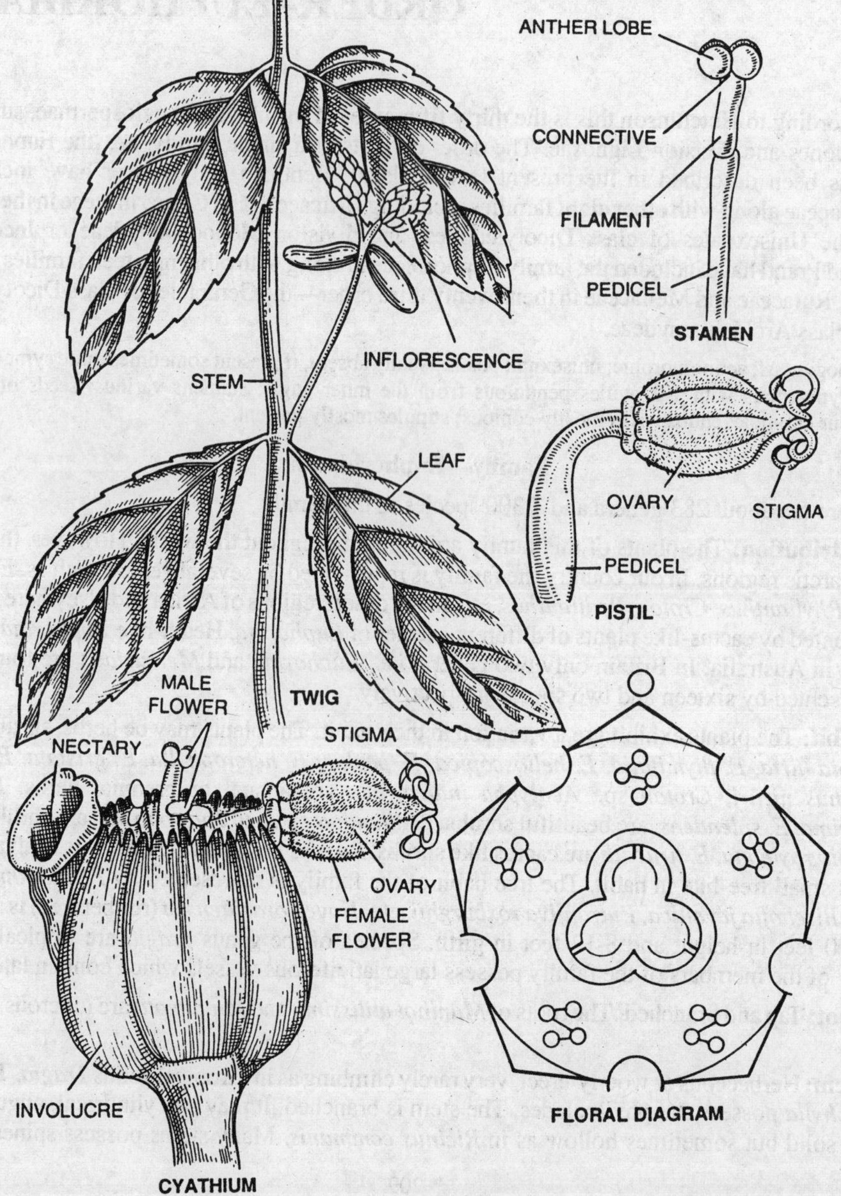

Fig. 25.1 : Euphorbiaceae. *Euphorbia* spp., plant with latex and cyathium inflorescence.

represented by a stalked stamen are found arranged around a central stalked female flower. The female flower consists of gynoecium only. The complete inflorescence looks like a single flower. The bracts are being arranged like a perianth. The bracts are so united that they form a cup-like structure. In *Acalypha* the inflorescence is catkin type. In *Croton* and *Ricinus* the flowers are arranged in terminal racemes. In *Jatropha* the inflorescence is of cymose type and the flowers are arranged in terminal cymes. In *Manihot* the flowers are being arranged in racemes.

Flowers: The flowers are always unisexual. They are much reduced and may be monoecious or dioecious. In *Euphorbia* sp., each male flower is represented by a single stalked stamen. The flowers are incomplete, regular, actinomorphic and hypogynous.

Perianth: Occasionally, both calyx and corolla are present, *e.g., Croton.* In majority of cases either calyx or corolla or both are absent. In *Ricinus communis* the calyx is present and the corolla absent. In *Euphorbia hirta* both the whorls of calyx and corolla are absent. In *Jatropha* sp., both calyx and corolla are present. In *Acalypha indica* the perianth is represented by four minute sepaloid petals. In *Phyllanthus* only sepaloid perianth is present. In *Euphorbia* the perianth is absent or represented by tiny scaly structures.

The perianth consists of 4 to 5 petals. The calyx and corolla consists of 4 or 5 sepals or petals. The aestivation is valvate or imbricate.

Androecium: The number of stamens varies from one to many. Usually as many stamens are present as many perianth leaves. In *Euphorbia* a single stalked stamen represents a single male flower. In *Ricinus* sp., usually five stamens are present, each stamen is profusely branched. In *Jatropha* they are aranged in two whorls each of five stamens. In many the stamens are indefinite, *e.g., Croton.* The filaments may be free or united. The anthers are dithecous. They dehisce either by apical pores or by transverse or longitudinal slits.

Gynoecium: Three carpels (tricarpellary), syncarpous; the ovary is trilocular, superior. Each locule contains one or two pendulous, anatropous ovules. The placenation is axile.

Fruit: The fruits are schizocarpic. The fruits break violently and dehisce into one seeded cocci. Such type of fruit is termed **regma** which is characteristic of *Ricinus* sp. The sp., of *Trewia* and *Bridelia* bear drupe fruit. *Phyllanthus emblica* also bears drupe.

Seeds: The seed is endospermic. In *Ricinus* caruncle develops from the micropyle. The cotyledons either lie flat or are folded within the endosperm.

Pollination: Usually entomophilous, *i.e.,* through the agency of insects. Only cross-pollination takes place. In many species the leaves and bracts become coloured and showy to attract the insects. Sometimes anemophily is also found.

Floral Formulae: The floral formulae of different genera are as follows:

Euphorbia — ♂ K0, C0, A1

♀ K0, C0, G(3).

Ricinus— ⊕ ♂ K5, C0, A5 (branched)

⊕ ♀ K3, Co, G(3).

Croton— ⊕ ♂ K5, C5, A ∝

⊕ ♀ K5, C5, G(3).

Bentham & Hooker (1862)	**Engler & Prantl (1931)**	**Hutchinson (1959)**
Phanerogams	Phanerogams	Angiospermae
Dicotyledones	Dicotyledones	Dicotyledones
Monochlamydeae	Archichlamydeae	Lignosae
Incompletae or	Geraniales	Euphorbiales
Unisexuales	Euphorbiaceae	Euphorbiaceae
Euphorbiaceae		

A few important plants have been described here in semi-technical language:

1. ***Ricinus communis;*** Eng.–Castor oil plant; Verna.–**Arand.**

   Habit: Tall annual or perennial shrub or small tree.

   Root: Tap and branched.

   Stem: Erect, cylindrical, branched, solid, glabrous.

   Leaf: Alternate, simple, petiolate, broad, palmately lobed, lobes usually 7-9, serrate, multicostate reticulate venation.

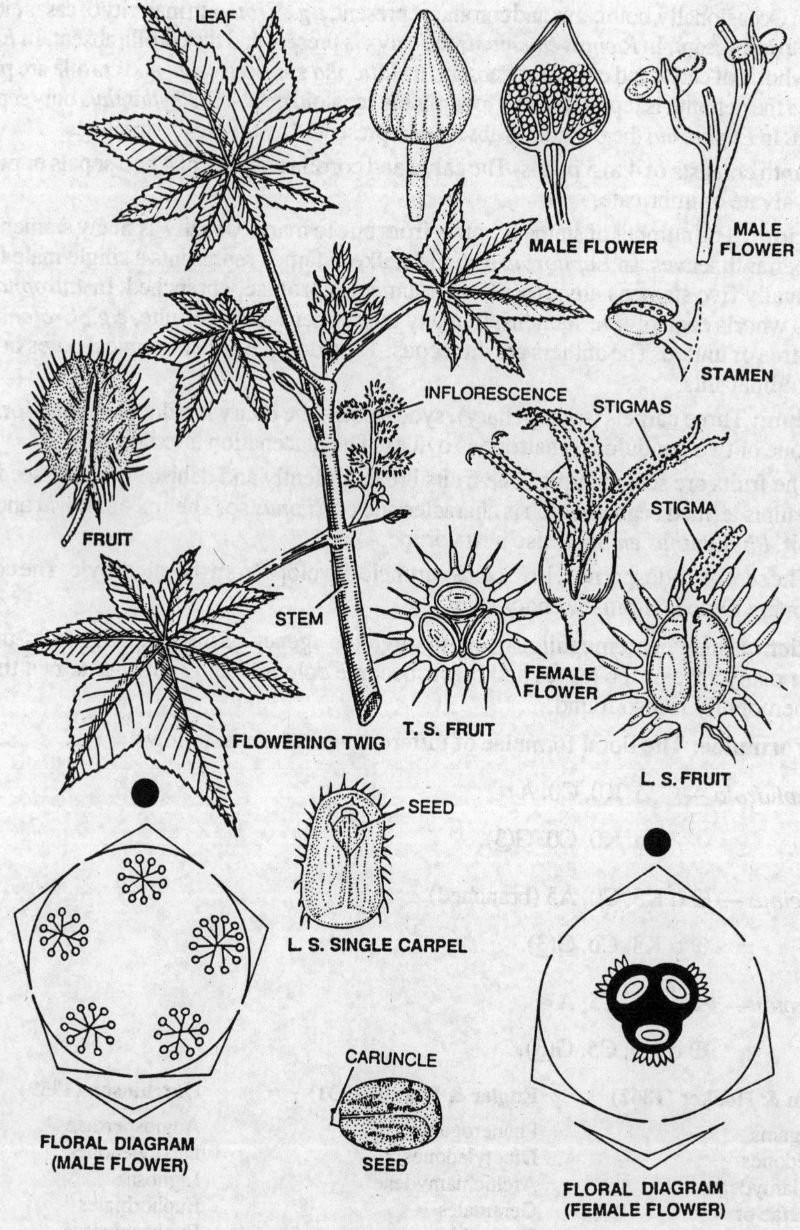

Fig. 25.2. Euphorbiaceae—*Ricinus communis* Linn., Eng., Castor oil plant; Verna. **Arand**

Inflorescence: Terminal panicled cyme.

Flower: Pedicellate, ebracteate, unisexual, monoecious, male flowers towards lower portion, female flowers towards apical portion of inflorescence, actinomorphic, incomplete, hypogynous.

Male Flowers:

Calyx: 5 or 4, membranous, polysepalous, green, inferior, valvate aestivation.

Corolla: Absent.

Androecium: 5 stamens, each stamen profusely branched, anther cells borne on ultimate branchlets, anthers basifixed.

Female Flowers:

Calyx: 3 to few or more, membranous, caducous, spathaceous, valvate aestivation.

Corolla: Absent.

Gynoecium: 3 carpels (tricarpellary), syncarpous, ovary superior, trilocular, covered with spiny outgrowths, axile placentation, single large ovlue in each locule, three styles, three bifid stigmas.

Fruit: A schizocarpic capsule or regma, splitting into three cocci, spiny outgrowths on fruit.

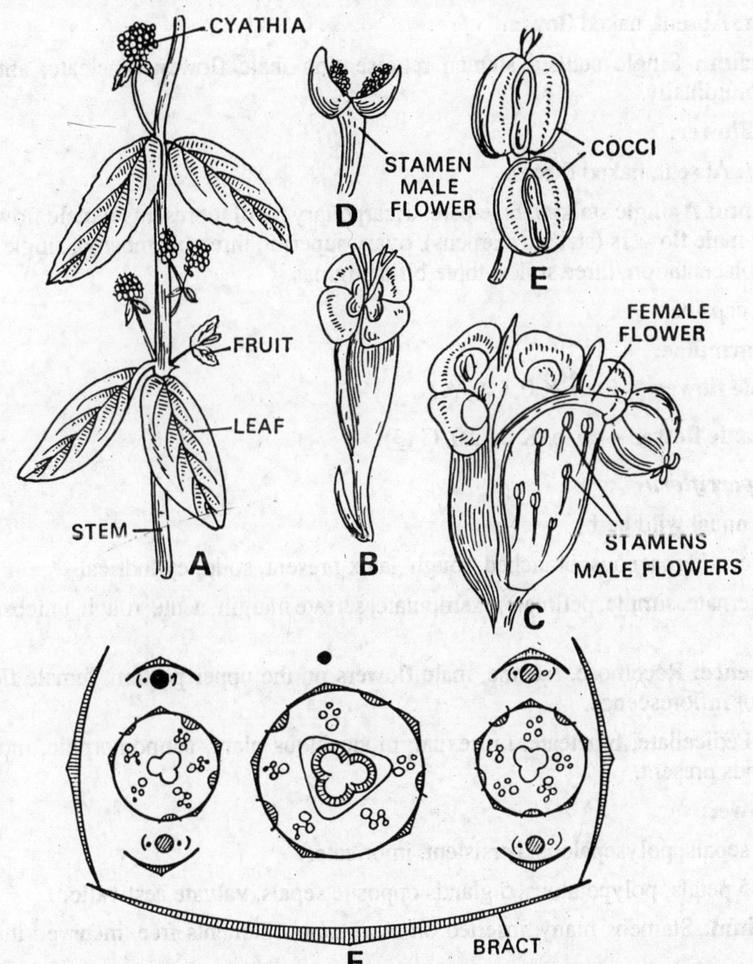

Fig. 25.3. Euphorbiaceae. *Euphorbia* sp. A, a flowering twig; B, cyathium; C, cyathium cut open; D, stamen; E, fruit-capsule showing cocci; F, floral diagram showing three cyathia.

Floral Formula :

Male : ⊕ ♂ K 5, C 0, A 5 (branched)

Female: ⊕ ♀ K 3, C 0, G (3).

2. ***Euphorbia hirta.***

Habit: Annual wild herb.

Stem: Erect, usually unbranched (sometimes branched), herbaceous, cylindrical, solid, covered with yellow crisped hairs.

Leaf: Simple, opposite, superposed, subsessile (short petiole), acute, oblong, lanceolate, serrulate, stipulate, stipules caducous, unicostate reticulate venation.

Inflorescence: Large number of cyathia densely crowded and arranged in peduncled axillary cymes.

Flowers: Pedicellate, unisexual, monoecious, enclosed within minute involucre of bracts forming cupular structure.

Male Flower:

Perianth: Absent, naked flower.

Androecium: Single stalked stamen representing male flower, bracteate, anthers 2-celled dehiscing longitudinally.

Female Flower:

Perianth: Absent, naked flower.

Gynoecium: A single stalked, bracteate, tricarpellary pistil represents female flower, it remains surrounded by male flowers (stalked stamens), ovary superior, three-chambered, single ovule in each loculus, axile placentation, three styles, three bifid stigmas.

Fruit: A capsule.

Floral Formulae:

Male flower— ⊕ ♂ K 0, C 0, A I

Female flower— ⊕ ♀ K 0, C 0, G (3).

3. ***Croton sparciflorus***

Habit: Annual wild herb.

Stem: Erect, herbaceous, branched, rough, latex present, solid, cylindrical.

Leaf: Alternate, simple, petiolate, exstilpulate, serrate margin, acute, rough, unicostate reticulate venation.

Inflorescence: Racemose, raceme, male flowers on the upper portion, female flowers on the lower portion of inflorescence.

Flower: Pedicellate, bracteate, unisexual, monoecious plant, actinomorphic, incomplete, hypogynous, glands present.

Male Flower:

Calyx: 5 sepals, polysepalous, persistent, imbricate.

Corolla: 5 petals, polypetalous, 5 glands opposite sepals, valvate aestivation.

Androecium: Stamens many, inserted on receptacle, filaments free, incurved in bud, anthers adnate.

Female Flower:

Calyx : As in male flowers.

Corolla: Absent.

Gynoecium: 3 carpels (tricarpellary), syncarpous, 3 styles ending in 3 bifid stigmas, ovary superior, three chambered, single ovule in each loculus, axile placentation.

Fruit : Indehiscent capsule,

Floral Formulae:

Male flower- ⊕ ♂ K 5, C 5, A ∝

Female flower : ⊕ ♀ K 5, C 0, G (3).

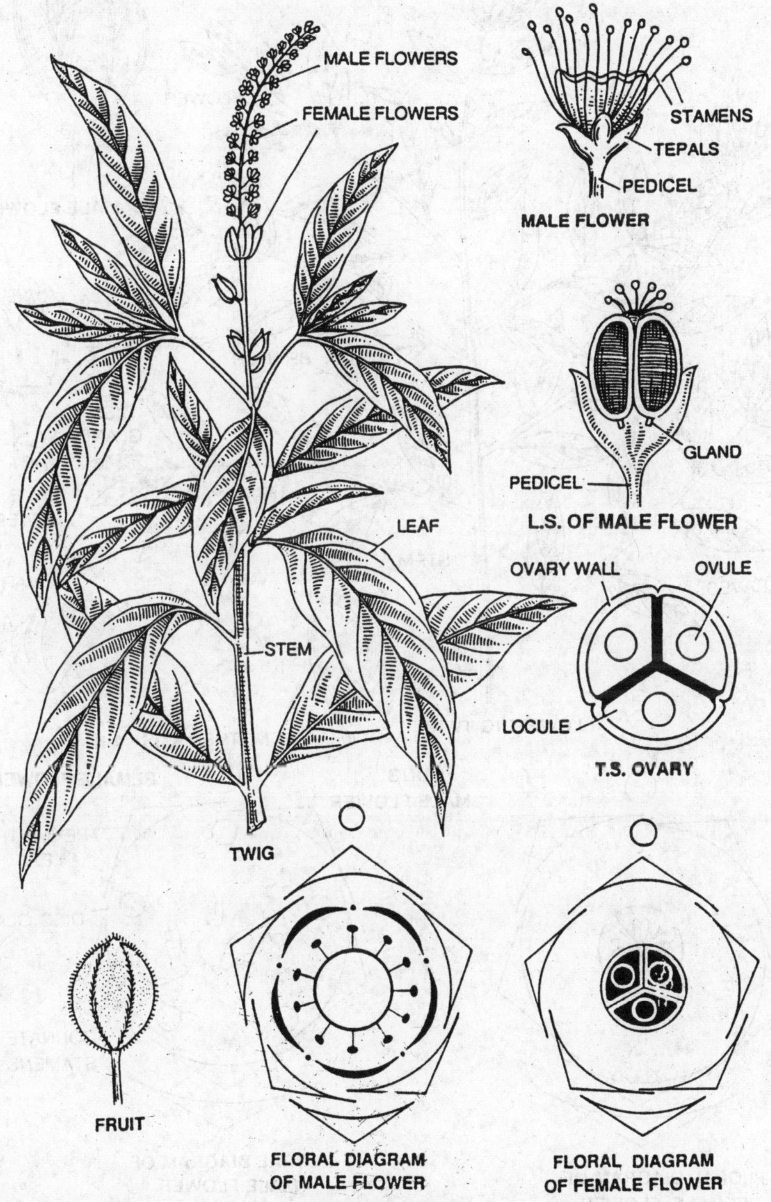

Fig. 25.4. Euphorbiaceae. *Croton sparciflorus.*

4. ***Phyllanthus reticulata;*** Verna. **Panjoli.**

Habit: An annual herb.

Stem: Erect, branched, herbaceous, green, solid, cylindrical $\frac{1}{2}$ to 2 feet in height.

Leaf: Simple, alternate, subsessile, elliptic-oblong, rounded obtuse, glaucous beneath, stipulate, distichous often resembling pinnate compound leaves, $\frac{1}{4}$ – $\frac{1}{2}$ inch long.

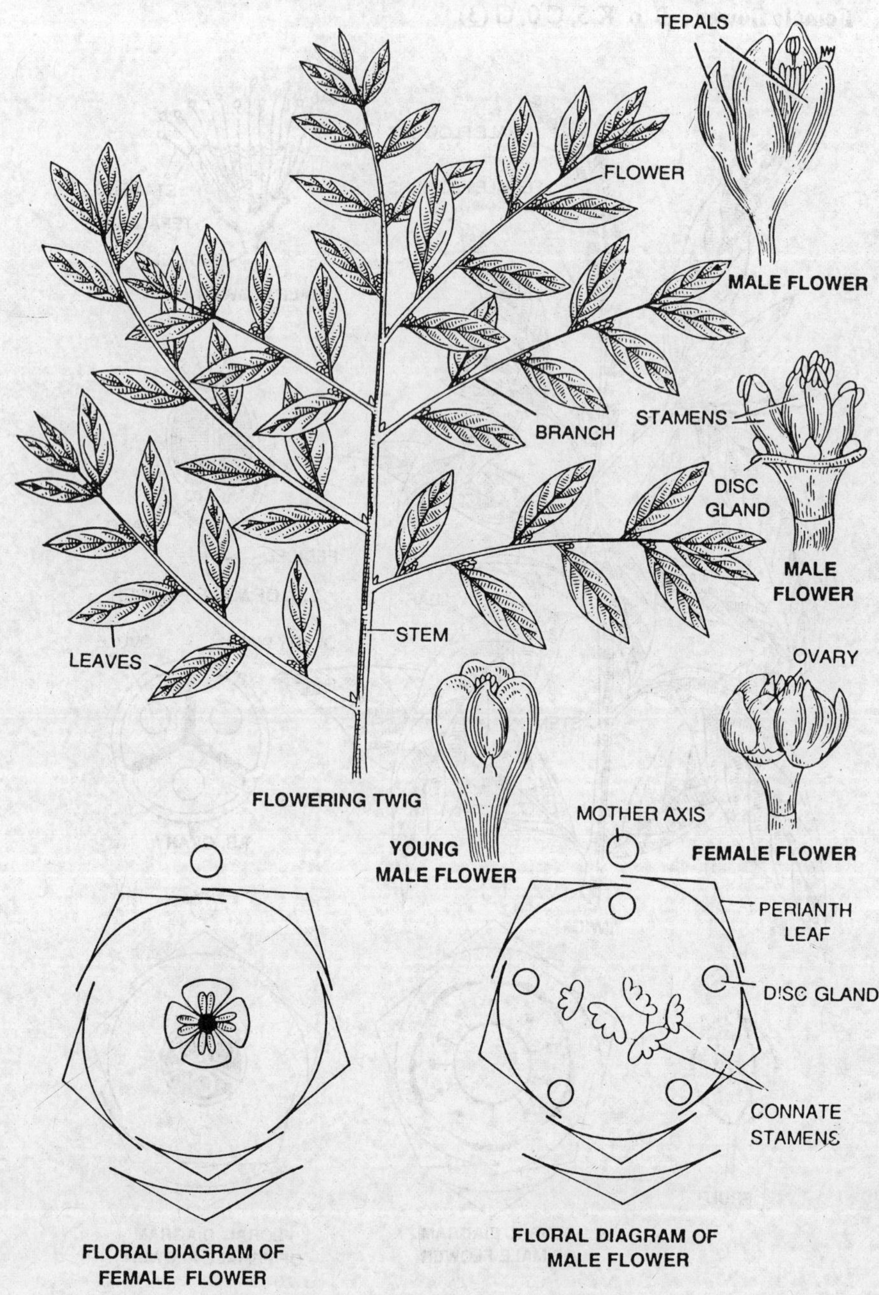

Fig. 25.5. Euphorbiaceae. *Phyllanthus reticulata* Poir.

Inflorescence: Axillary flowers.

Flower: Minute, unisexual, trimerous, actinomorphic, bracteate, shortly pedicellate.

Male flower:

Perianth: 6 petals (perianth leaves), arranged in two whorls of three each, polyphyllous, sepaloid, imbricate aestivation, disc of minute glands inside the perianth.

Androecium: 3 stamens, polyandrous, filaments short, borne on a short column.

Female Flower:

Perianth: As in male flower.

Gynoecium: 3 carpels (tricarpellary), syncarpous, ovary superior, three chambered, each loculus having two ovules, axile placentation, stigma trifid.

Fruit: Schizocarpic regma, opening into three cocci.

Floral Formula: **Male :** ⊕ ♂ P3 + 3, A 3

Female :– ⊕ ♀ P3 + 3, G($\underline{3}$).

5. ***Acalypha indica***

Habit: An annual herb.

Stem: Erect, branched, herbaceous, solid cylindrical, hairy.

Leaf: Alternate, long petiolate, stipulate, rhomboid ovate, serrate obtuse, multicostate reticulate venation.

Inflorescence: Racemose, axillary spikes, male flowers towards the apex of spike, female flowers towards lower portion of spike in clusters of 3-5 subtended by a leafy bract.

Flower: Sessile, unisexual, monoecious plant, male flower ebracteate, female flower bracteate, minute incomplete, hypogynous.

Male Flower:

Perianth: 4, minute sepaloid perianth leaves, polyphyllous, valvate aestivation.

Androecium: 8 stamens, polyandrous, filaments short.

Female Flower:

Perianth: 3 - 4 sepaloid perianth leaves, polyphyllous, imbricate aestivation.

Gynoecium: 3 carpels (tricarpellary), syncarpous, ovary superior, 3-chambered, axile placentation.

Fruit: Capsule, often one seeded.

Floral Formula: **Male –** ⊕ ♂ P 4, A8.

Female– ⊕ ♀ P3-4, G($\underline{3}$).

6. ***Jatropha gossypifolia*** L. Verna. **Bherenda.**

Habit: A cultivated shrub.

Stem: Erect, branched, solid, cylindrical, woody, hairy, gland dotted.

Leaf: Simple, alternate, cauline, petiolate, petiole spiny, stipulate, stipules spiny, entire, acuminate, bright green on dorsal side, greenish violet on ventral side, unicostate reticulate venation.

Inflorescence: Cymose, polychasial cyme.

Flowers: Monoecious, actinomorphic, unisexual, incomplete, bracteate.

Calyx: 5 sepals, gamosepalous, petaloid, reddish green, valvate.

Corolla: 5 petals, polypetalous, twisted aestivation.

Androecium: 10 stamens, arranged in two whorls of five each, monadelphous, anthers dorsifixed.

Female Flowers: Pedicellate, bracteate, unisexual, incomplete, hypogynous.

Calyx: As in male flower.

Corolla: As in male flower.

Gynoecium: 3 carpels (tricarpellary), syncarpous, ovary superior, three-chambered, axile placentation, three styles, three bifid stigmas.

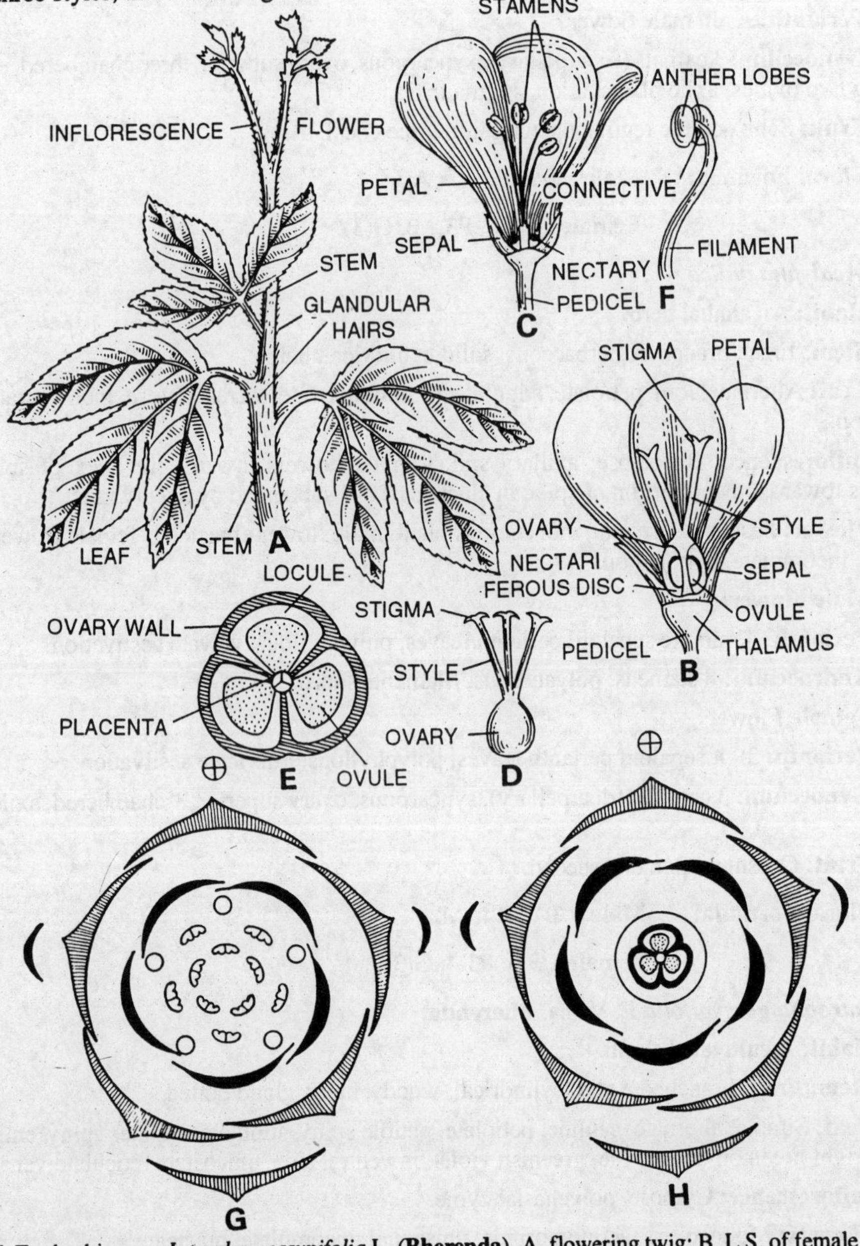

Fig. 25.6. Euphorbiaceae. *Jatropha gossypifolia* L. (**Bherenda**). A, flowering twig; B, L.S. of female flower; C, L.S. of male flower; D, gynoecium; E, T.S. of ovary; F, stamen; G, floral diagram of male flower; H, floral diagram of female flower.

Fruit: Three valved capsule.

Floral Formulae: **Male** : $\oplus$ ♂ K (5), C5, A(5 + 5)

Female : $\oplus$ ♀ K (5), C5, G(3).

7. ***Jatropha gossypifolia*** L. Eng. Bellyache bush; Verna. **Bherenda. Stem**–woody, aerial, erect, cylindrical, branched, solid, hairy (glandular hairs). **Leaf**–cauline and ramal, exstipulate, simple, lobed, petiolate, hairy (glandular hairs), leaf lobe elliptic-ovate, serrulate, acute, glaucous, multicostate reticulate, coriaceous. **Inflorescence**–cymose, panicled cyme. **Male flower**–bracteate, bracteolate,

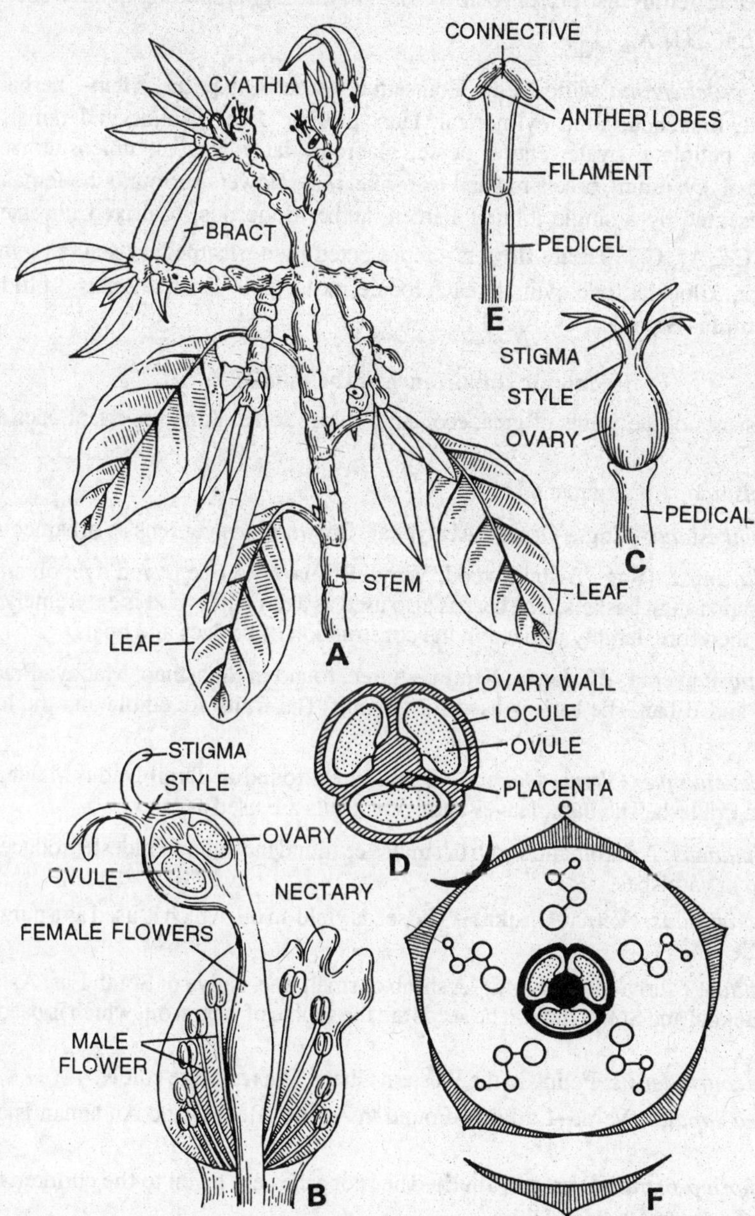

Fig. 25.7. Euphorbiaceae. *Euphorbia pulcherrima* Willd. (**Lal patta**). A, flowering twig; B, L.S. of cyathium; C, female flower; D, T.S. of ovary; E, male flower represented by a stamen; F, floral diagram of cyathium (inflorescence).

pedicellate, unisexual, incomplete, actinomorphic, staminate, pentamerous, cylic. **Calyx**–5, polysepalous, quincuncial, persistent. **Corolla**–5, polypetalous, twisted, red, 5 nectariferous discs present alternating petals. **Androecium**–5 + 5, filaments short, dithecous, dorsifixed, introrse. **Gynoecium**–absent, **Floral formula**–Br Brl ⊕ ♂ K_5, C_5, A_{5+5}, G_0. **Female flower**–Bracteate, bracteolate, pedicellate, unisexual, incomplete, actinomorphic, pistillate, pentamerous, hyogynous. **Calyx**–5, polysepalous, quincuncial, persistent. **Corolla**–5, polypetalous, twisted. **Androecium**–absent. **Gynoecium**–3, tricarpellary, syncarpous, superior, trilocular, single ovule in each locule, style 3, stigmas 3 bifid, nectariferous disc present below ovary. **Fruit**–regma splitting in three cocci. **Floral formula**–Br Brl ⊕ ♀ K_5, C_5, A_0, $G_{(\underline{3})}$.

8. ***Euphorbia pulcherrima*** willd., Eng. Poinsettia; Verna. **Lal patta. Stem**– herbaceous or woody, aerial, erect, branched, solid cylindrical, latex present. **Leaf**–cauline and ramal, simple, aternate, extipulate, petiolate, ovate, entire, acute, glabrous, latex present, unicostate reticulate. **Inflorescence**–special, cyathium, bracts red and leaf–like, male flowers surround the female flower. **Male flower**–represented by a single jointed stamen, anther dithecous, basifixed introrse. **Floral formula**–Br ♂ K_0, Co, A_1, G_0, **Female flower** – represented by a tricarpellary stalked gynoecium, syncarpous, superior, trilocular, one ovule in each locule, axile, style short, stigmas 3 bifid. **Fruit**–capsule. **Floral formula**–Br, ♀ K_0, C_0, A_0, $G_{(\underline{3})}$.

Economic Importance of the Family

The family consists of the plants of great economic value. Some of the important ones are given below:

1. *Acalypha hispida;* An ornamental herb.

2. *Acalypha wilkesiana;* (Eng. –Garden Acalypha). Grown in the gardens as ornamental plant.

3. *Bischofia javanica;* (Eng.–Bishop wood; Verna–**Bhillar**)– A tree. A red dye, obtained from the bark is used to stain rattan baskets. The bark is also used as a tan. The wood is exteremely resistant to water effect and therefore, largely utilized in the construction of bridges and boats.

4. *Bridelia retusa*; (Verna.–**Ekdania, Khaja**)–A tree, found in Rajasthan, Madhya Pradesh, the Western Peninsula and Bihar. The bark is used for tanning. The fruits are edible and the leaves are used as fodder.

5. *Cleistanthus collinus;* (Verna–**Garari**)–A small tree, found in Tamil Nadu, Malabar, Bihar, Orissa and Madhya Pradesh. The bark, leaves and green fruits are used as tan.

6. *Croton aromaticus;* An aromatic shrub or small tree found in Andhra Pradesh produces a gum–resin, which is used in varnishes.

7. *Croton oblongifolius;* (Verna–**Chuka**) – The seeds yield an oil, which is used as a purgative and also as an insecticide.

8. *Croton tigllium;* (Verna–**Jamalgota**)–A shrub or small tree, native of South East Asia but cultivated in Assam, Bengal and South India. The seeds are the source of croton oil, which is used as a purgative.

9. *Baccaurea courtallensis:* Found in the Western Ghats. The fruits are edible.

10. *Baccaurea sapida;* (Verna–**Lathua**)–Found in Assam, Bengal and Andaman Islands. The fruits are edible.

11. *Chrozophora prostrata* (Verna–**Subali**)–The root ashes are given to the children for cough treatment. The seeds are used as a purgative.

12. *Baliospermum montanum;* (Verna–**Danti**)–The seeds are used as a strong purgative. The decoction of leaves is used in asthma.

13. *Euphorbia milii;* Syn. *E. splendens;* A small climbing shrub; native of Madagascar. Grown in gardens as ornamental.

14. *Euphorbia pulcherrima;* Syn. *Poinsettia pulcherrima;* Eng–Poinsettia; (Verna.–**Lal Patta**)–A shrub, native of Central America, Grown as an ornamental.

15. *Euphorbia tirucalli;* (Verna.–**Tohar**)–Succulent spineless small tree, native of Africa. The roots are used for poisoning fish and birds.

16. *Euphorbia hirta;* (Verna.–**Dudhi**)–The plant is of medicinal value and used in many diseases of children and adults.

17. *Euphorbia thymifolia;* (Verna.–**Chhoti dudhi**)–The juice of leaves and seeds is used as a strong purgative. The juice is also used for remedy of ringworms and other skin diseases. It is an antidote for snake bite.

18. *Euphorbia royleana;* (Verna.–**Thar**)–The plants are used in hedges. The latex is used medicinally in several ways.

19. *Euphorbia antiquorum;* (Verna.–**Tridhara, sehund**)–The decoction of stem is used as a remedy of gout. The juice of the plant is also used as a strong purgative. The root bark is also used as a purgative.

20. *Glochidion zeylanicum;* (Verna.–**Kumbalm**)–The leaves are used in itches.

21. *Securinega leucopyrus;* Syn. *Phyllanthus leucopyrus; Flueggea leucopyrus* (Verna.–**Hartho**)–A large shrub or small tree, found in Uttar Pradesh, the Punjab, Maharashtra and Tamil Nadu. The fruits are edible.

22. *Jatropha curcas;* (Verna.–**Safed arand**)– A shrub or small tree, native of tropical America, now cultivated in Travancore. The seed oil is used for manufacturing candles, soaps and as a lubricant and for illumination. The seed oil is also used as a purgative. The tender shoots are edible.

23. *Jatropha gossypifolia;* (Verna.–**Bherenda**)–A shrub, native of Brazil. Cultivated as an ornamental.

24. *Jatropha hastata;* Grown as an ornamental for bright crimson flowers.

25. *Jatropha padagrica;* (Eng.–Gouty stemmed Jatropha)–Native of Panama, grown for its bright scarlet flowers.

26. *Manihot esculenta;* Syn. *M. utilissima;* (Eng.–Tapioca; Verna.–**Sakarkand**)– A small shrub; native of Brazil, now grown in Kerala, Tamil Nadu and Karnataka. The tapioca tubers are exploited commercially to obtain starch, sago, semolina and flour.

27. *Hevea brasiliensis;* (Eng.-Para rubber; Verna.-**Rabar**) – A tall tree; native of Brazil;, now grown in Kerala, Tamil Nadu, Karnataka and North–Eastern Assam. The latex, obtained from the bark of the tree, is used for preparing rubber, which is used for tyres and inner tubes, waterproof clothing and various electrical goods.

28. *Mallotus philippinensis;* (Verna.–**Kamala, Rauni**)–A small tree found commonly in Bengal, Madhya Pradesh, Maharashtra and Orissa. The red dye, obtained from the surface of the fruits, is used for dyeing silk. Also used medicinally to remove thread worms and Ascaris.

29. *Cicca acida;* Syn. *Phyllanthus acidus; Averrhoa acida;* (Eng.–Stargoose-berry; Verna.-**Hariphul**)- A small tree cultivated in Bengal and South India for the edible fruits. The leaves are edible. The bark is used as a tan.

30. *Emblica officinalis*; Syn. *Phyllanthus emblica*; (Verna.-**Amla**)- A common tree with edible fruits. The fruits are also used in diarrhoea and dysentery. The bark, leaves and fruits are used in dyeing and tanning. The wood yields excellent charcoal. The pickle or jam is prepared from the fruits. The fruit is very rich in vitamin C.

31. *Emblica fischeri*; A small tree, found in South India. The fruits are edible.

32. *Kiganelia reticulata;* Syn. *Phyllanthus reticulatus;* (Verna.-**Panjoli**)- A climbing shrub, commonly found in Northern India. The roots are the source of a red dye.

33. *Putranjiva roxburgii*; (Eng.-Child-life tree; Verna.-**Putranjiva**)-A tall tree, grown as a hedge plant. The nuts are made in rosaries. The Hindus believe that if the hard stones of the fruits are made into rosaries and placed around the neck of the children, they keep them in good health. The seeds also yield an oil which is used for burning purposes. The leaves are used as fodder. The leaves, fruits and stone of fruits are used medicinally in colds and fevers.

34. *Ricinus communis*; (Eng.-castor-oil plant; Verna–**Arand**)–A small tree, cultivated chiefly in Andhra Pradesh, Maharashtra, Karnataka and Orissa. The seeds are the source of castor-oil, which is mainly used as a lubricant and as a purgative. It is also used for transparent soap, textile soap, typewriter-inks, perfume, aromatics, varnishes and paints. The seed cake is used as a fertilizer. The writing and printing-papers are made of the wood-pulp. Castor stems are used for strawboards and cheap wrappings.

35. *Tragia involucrata*; (Verna–**Barhaita**)–The roots and fruits are used medicinally.

Systematic Relationships: Bentham and Hooker have included nine families in the series Unisexuales. They are Euphorbiaceae, Balanopseae, Urticaceae, Plantanaceae, Leitnerieae, Juglandeae, Myricaceae, Casuarineae and Cupuliferae.

Hallier included the family Euphorbiaceae in his Passionales, Rendle and Hutchinson placed the family in the order Euphorbiales.

26

ORDER–OCHNALES

According to Hutchinson this is the thirtyseventh order of the phylum Angiospermae, sub-phylum Dicotyledones and division Lignosae. The order consists of six families, *i.e.,* Strasburgeriaceae, Ochnaceae, Rhodolaenaceae, Sphaerosepalaceae, Dipterocarpaceae and Ancistrocladaceae. The family Dipterocarpaceae has been described in this text in detail. Bentham and Hooker have included the family Dipterocarpeae together with other five families, *i.e.,* Elatineae, Hypericineae, Guttiferae, Ternstroemiaceae and Chlaenaceae in the fifth order–the Guttiferales of class Dicotyledones, division Polypetalae and series Thalamiflorae. Engler and Prantl have put the family Dipterocarpaceae along with twentyeight other families in their twentyseventh order–the Parietales of class Dicotyledoneae and sub-class Archichlamydeae.

Hypogynous; bisexual; petals present; syncarpous with axile placentation, stamens 5 to many in one to several whorls; carpels usually, syncarpous, superior ovary; fruit mostly surrounded by persistent winged calyx segments; exendospermic, resin ducts or oil cells present in wood; leaves with stipules.

Family—Dipterocarpaceae

There are about 25 genera and 50 species in this family.

Distribution: The members of this family are mainly confined to the tropics of the Old World. Most of the species are found in Malaysia and Philippines. In India *Shorea robusta* (**sal**) is ditributed in Assam, Terai, Madhya Pradesh, Bihar (Chota-Nagpur) ascending upto 500 m forming the extensive sal forest. *Monoporandra* is found only in Shri Lanka. Several species of *Dipterocarpus* occur in Bangladesh and Tripura (India)

Habit: The plants are trees (*e.g. Shorea robusta* **Sal**) and rarely shrubs. They are usually large trees with high erect stems.

Branched system of resin ducts usually lined with epithelium are found in almost all members of the family. In many members resin, balsam, camphor, etc., are found in ducts.

Leave : The leaves are simple, entire and coriaceous; stipules small and sometimes surround the internode. The petiole is locally thickened.

Inflorescence : The flowers are found to be arranged mostly in paniculate terminal or axillary spikes or racemes.

Flower : The floral axis may be broad, saucer-shaped or concave; bisexual, hypogynous pentamerous, complete.

Calyx : 5 sepals, free, unequal, imbricate or valvate; calyx segments are persistent winglike and enclose the fruit.

Corolla. 5 petals, imbricate, polypetalous.

Androecium: Stamens 5, 10, 15 or indefinite. Only in *Monoporandra* stamens are 5. In *Hopea pierrei* Roxb., stamens are 15 in three whorls, of which those of the outer and inner whorls are antisepalous, the middle five are antipetalous. In *Hopea recopei* Roxb., stamens are monadelphous. Filaments short usually, rarely long; anther lobes unequal, connective produced at the apex very often(*e.g.* in *Shorea, Stemenoporus,* etc.)

Gynoecium. Carpels 3 usually, syncarpous, 3-loculed superior ovary, rarely ovary unilocular; two or more ovules in each locule; axile placentation. In *Marquesia* the ovary is one-celled with parietal placentation. The stigma almost always exceeds the stamens.

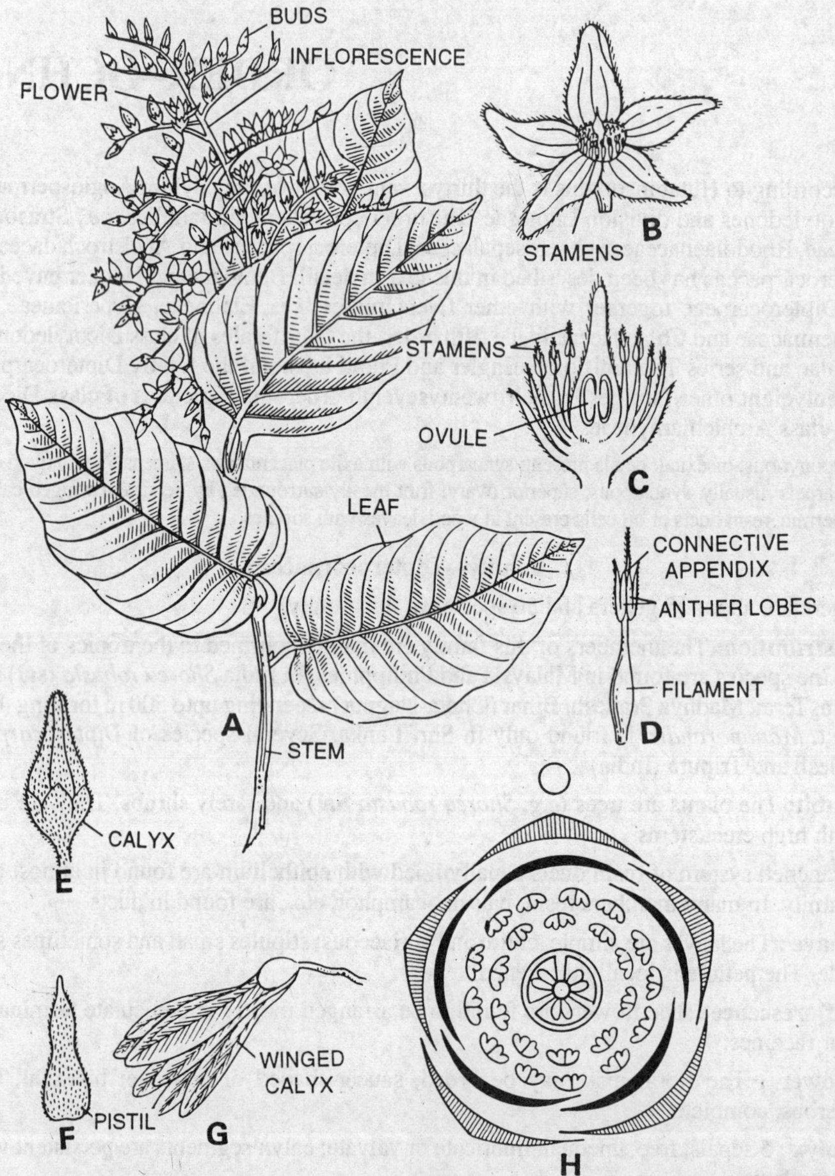

Fig. 26.1. Dipterocarpaceae–*Shorea robusta* Gaertn. **(sal)**. A, flowering twig with panicle inflorescence; B, flower; C, L.S. of flower; D, stamen with connective appendix; E, flower bud; F, gynoecium; G, persistent winged calyx; H, floral diagram.

Fruit. It may be leathery, woody indehiscent nut-like, mostly surrounded by persistent winged calyx segments.

Seed. The seed is non-endospermic.

Pollination. Entomophily. In many species the flowers are pendulous so that cross pollination takes place by means of insects. Collection of flowers in inflorescence, sweet scent etc., also attract the insects.

Floral formula. $\oplus$ ⚥ $K_5, C_5, A_{5-\infty}, G_{(3)}$

Bentham & Hooker (1862)	Engler & Prantl (1931)	Hutchinson (1959)
Phanerogams	Phanerogams	Angiospermae
Dicotyledoneae	Dicotyledoneae	Dicotyledones
Polypetalae	Archichlamydeae	Lignosae
Thalamiflorae	Parietales	Ochnales
Guttiferales	Dipterocarpaceae	Dipterocarpaceae
Dipterocarpeae		

Economic Importance of the Family

The useful products obtained from the plants of this family are as follows:

1. In the species of *Dipterocarpus* resins, balsam, oils, camphor-like substances, turpentine, etc., are produced; yellow camphor is obtained from *Dryobalanops aromatica* Gaertn., *Shorea koordarsii* yields turpentine and resin.

2. **Fats and Oils.** The oily seeds of *Shorea aptera* yield fats and oils.

3. **Food.** The starchy seeds of *Vateria acuminata, Doona* and *Vatica* are used as a source of starch.

4. **Timber.** *Shorea robusta* (**sal**) is a source of good timber.

5. **Copal.** *Vateria indica* L., is a source of gum-resin known as 'Copal'.

Systematic Relationships. The genera *Ancistrocladus* and *Lophira* formerly placed by Bentham and Hooker in the family Dipterocarpeae have been shifted to the families Ancistrocladaceae and Ochnaceae respectively by E. Gilg which shows that the family Dipterocarpaceae has affinity with these two families. The family Guttiferae stands close to the family Dipterocarpaceae where it differs in opposite leaves and exstipulate condition. Hutchinson (1969) postulates dilleniaceous ancestry of this family because of many primitive characters.

27

ORDER–MYRTALES

According to Hutchinson this is 40th order of the phylum Angiospermae, subphylum Dicotyledoneae and division Lignosae. He included seven families in this order. They are Myrtaceae, Lecythidaceae, Rhizophoraceae, Sonneratiaceae, Punicaceae, Combretaceae and Melastomataceae, of which the Myrtaceae and the Combretaceae have been described in this text in detail. Bentham, and Hooker have also included these two families along with other four families in their 12th order–the Myrtales of class Dicotyledones, division polypetalae and series Disciflorae. Engler and Prantl have placed these families along with other seventeen families in the 29th order–the Myrtiflorae of class Dicotyledoneae and subclass Archichlamydeae.

Mostly epigynous with opposite often gland dotted leaves; stamens becoming specialized and dissimilar; some in maritime habitats; porous anthers; calyx becoming valvate.

Family—Myrtaceae

There are about 80 genera and 3,000 species in this family.

Distribution: The members of this family are mostly found in tropical regions. Chiefly they are found in Australia and tropical America. In our country the family is represented by several important genera, such as, *Syzygium, Eucalyptus, Psidium, Callistemon,* etc. The main centre of the distribution of this family is Australia.

Habit: Generally the plants are moderate-sized or small trees, or shrubs; the herbs are rare. The *Eucalyptus* species are giant trees. Species of *Eucalyptus* reach 300 to 400 feet in height with a girth at the base of 90 feet. *Syzygium cumini*(**Jumun**) is a tree where as *Psidium guajava* (**Amrood**) is a smallsized tree. *Syzygium aromaticum*(**Laung**) is a shrub which attains a height of about 12 feet. *Careya herbaceae* is an undershrub.

Root: Tap and branched.

Stem: Erect, branched, woody, solid.

Leaves: Mostly the leaves are opposite and sometimes alternate, *e.g., Eucalyptus.* The petioles are short. They are simple, entire, evergreen and exstipulate. Many Australian xerophytic species possess acicular (needle-like) leaves. The leaves contain numerous oil glands which secrete ethereal oils.

Inflorescence: The inflorescence is of cymose type. Sometimes the flowers are found solitary in leaf axils, *e.g., Myrtus communis.* Rarely the inflorescence is of racemose type. In *Syzygium fruticosum* the flowers are found to be arranged in panicled cymes. In *Syzygium mappaceum* the inflorescence is of corymbose cyme type. In *Syzygium aqueum* the flowers are either solitary or they are found to be arranged in cymes. In *Psidum guajava* 1-3 flowers are found in each axil. In *Callistemon,* the flowers are found to be arranged in spikes. The inflorescence looks like a bottle-brush,

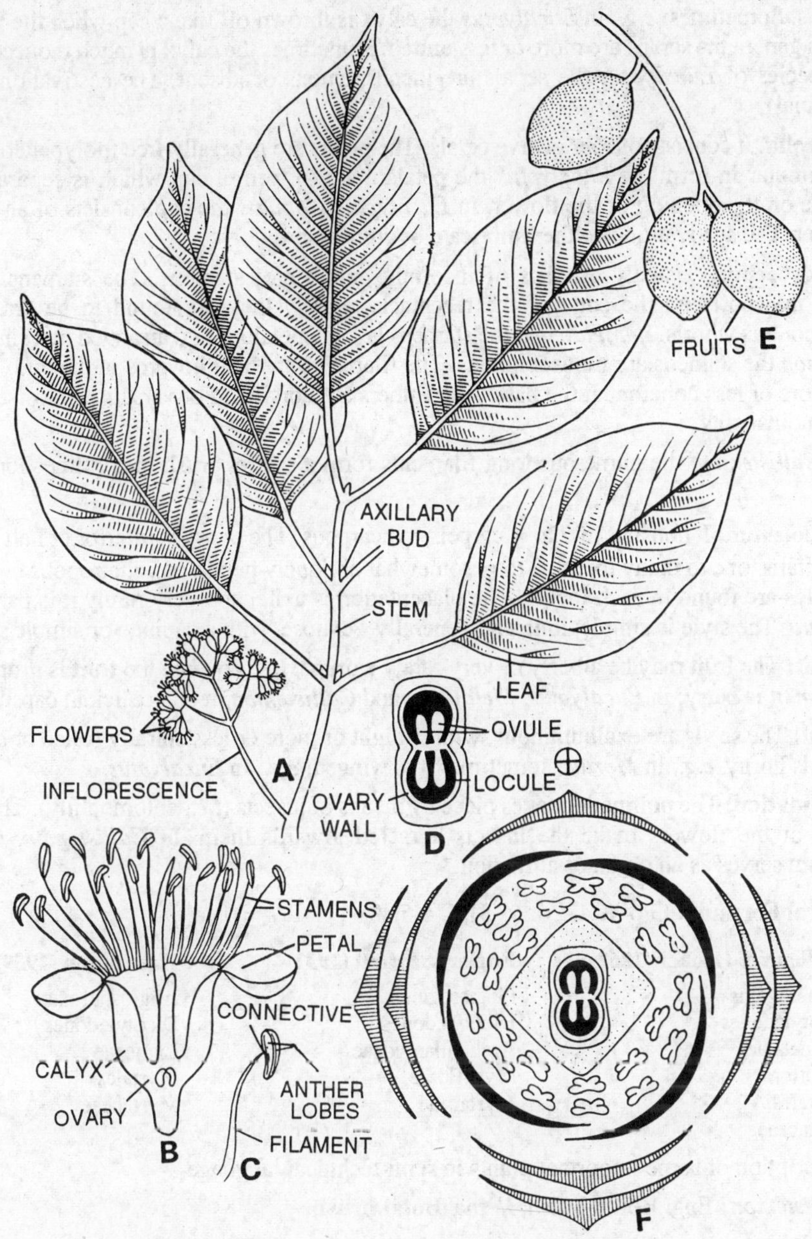

Fig. 27.1. Myrtaceae—*Syzygium cumini.* A, flowering twig ; B, L.S.of flower; C, stamen; D.T.S. of ovary; E, fruits; F, floral diagram.

and therefore the plant is named as a bottle brush tree. In *Eucalyptus teretocornis,* the flowers are arranged in axillary umbels which may be simple or panicled.

Flowers: The flowers are bracteate, bracteolate (usually two bracteoles), sometimes ebracteolate; actinomorphic, hermaphrodite, regular, complete, cyclic, epi or perigynous. The receptacle is either united with the ovary giving rise to complete epigynous condition, *e.g., Myrtus communis* and *Psidium guajava,* or the union may not be completed, which results in perigynous condition.

Calyx: The calyx normally consists of either four or five sepals, alternating with four or five free petals. Sometimes, *e.g.*, in *Eucalyptus* the calyx is thrown off like a cap when the flower opens. In certain genera the sepals are more or less united. Sometimes the calyx is much reduced and in most of the species of *Eucalyptus* the sepals are inconspicuous or absent. The aestivation is imbricate (quincuncial).

Corolla: It consists of four or five petals. The petals are generally free (polypetalous) and more or less circular in form. In *Eucalyptus* the petals unite to form a cap which is separated from the receptacle on the opening of the flower. In *Eucalyptus globulus* the cap consists of an inner and an outer layer. The aestivation is either imbricate or quincuncial.

Androecium: Usually consists of indefinite number of stamens. The stamens are free and arranged in whorls on the edge of the receptacle. Rarely they are found to be arranged in ob-diplostemonous whorls. In *Melaleuca* the stamens are antepetalous and arranged in a single whorl. In *Callistemon* the stamens are monadelphous. The filaments are bent inwards in the bud. They may be free or more or less coherent at the base. The anthers are dorsifixed or versatile, introrse and dehisce by longitudinal slits.

In *Callistemon,* the numerous long filaments form a scarlet brush many times longer than the corolla.

Gynoecium: It consists of 2 to ∝ carpels, syncarpous. The ovary is inferior or half inferior. The ovary contains one to many loculi, 2 to ∝ somewhat obliquely pendulous anatropous or campylotropous ovules are found in each locule. The placentation is axile, and very rarely it is parietal, *e.g., in Rhodamnia.* The style is simple, long and generally flexuose, with a capitate or simple stigma.

Fruit: The fruit may be a berry or very rarely a drupe. In *Syzygium* the fruit is drupe-like berry. In *Psidium,* it is berry; in *Eucalyptus, Melaleuca* and *Callistemon,* it is loculicidal capsule.

Seed: The seeds are exalbuminous, with straight or more or less spirally rolled or bent embryo. The testa is horny, *e.g.*, in *Myrtus,* sometimes it is winged, *e.g.*, in *Eucalyptus.*

Pollination: The pollination takes place by means of insects (*i.e.*, entomophily). The colour and the scent of the flowers make the insects attracted towards them. In *Callistemon,* the coloured androecium serves as an organ of attraction.

Floral Formula: ⊕ ⚥ K 4-5 or (4-5), C 4-5, A ∝, $G\ \overline{(2-\infty)}$

Bentham & Hooker (1862)	Engler & Prantl (1931)	Hutchinson (1959)
Phanerogams	Phanerogams	Angiospermae
Dicotyledones	Dicotyledones	Dicotyledones
Polypetalae	Archichlamydeae	Lignosae
Disciflorae	Myrtiflorae	Myrtales
Myrtales	Myrtaceae	Myrtaceae
Myrtaceae		

Description of some important plants in semi-technical language—

1. *Callistemon;* Eng. Bottle-brush; Verna. **Botal brush.**

Habit: A small tree.

Root: Tap, branched.

Stem: Erect, woody, cylindrical, branched, hairy, solid.

Leaf: Simple, cauline, petiolate, petiole very short, lanceolate, acute apex, unicostate reticulate venation, inframarginal veins.

Inflorescence: Spike.

Flower: Sessile, bracteate, bracts leafy, green, bracteolate, two bracteoles, hermaphrodite, actinomorphic, complete, epigynous.

Calyx: 5, gamosepalous, green, superior, quincuncial aestivation.

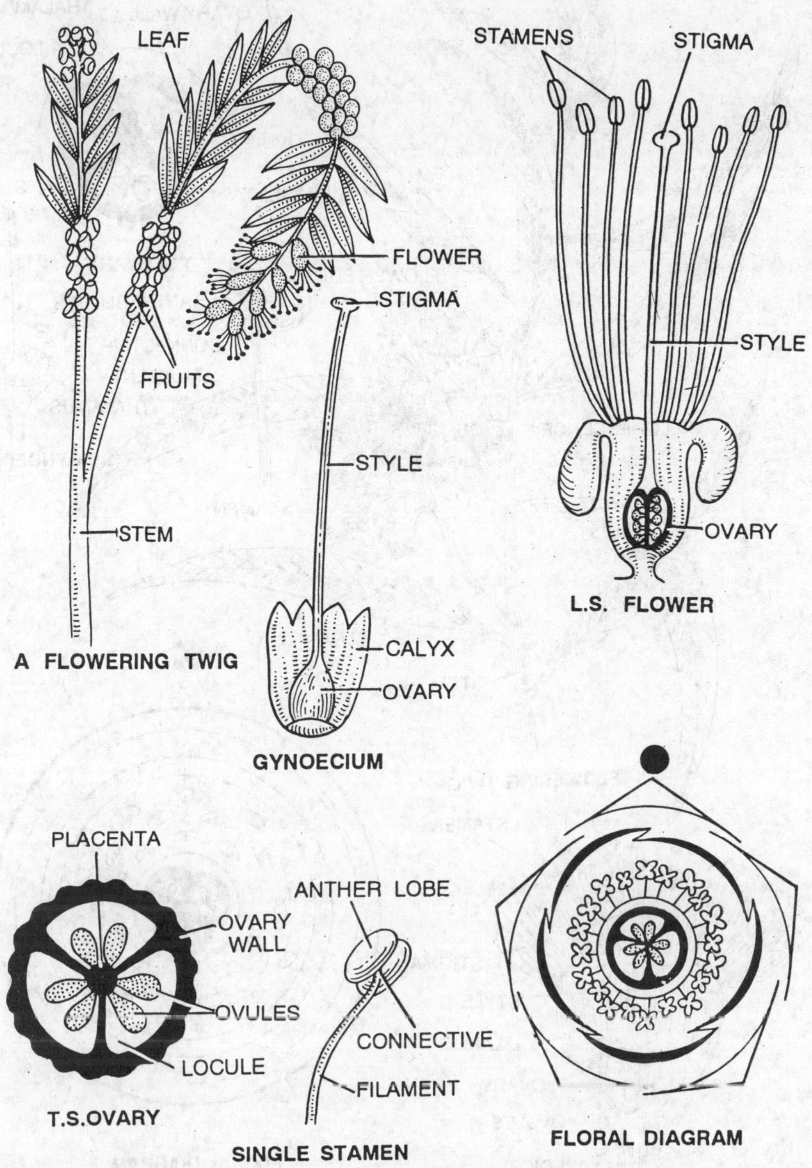

Fig. 27.2. Myrtaceae. *Callistemon citrinus* (Curtis) Skeels.

Corolla: 5, polypetalous, superior, whitish green, quincuncial aestivation.

Androecium: Indefinite stamens, monadelphous, scarlet red, filaments very long, slightly connate at the base, anthers dorsifixed, bicelled, dark brown, dehisce longitudinally.

Gynoecium: The number of carpels varies from 2-4 even in the different flowers of the same spike, syncarpous, ovary inferior, 3 or 4 chambered according to the number of carpels, axile placentation, one or two ovules in each locule.

Floral Formula: ·|· ⚥ K (5), C 5 A_{∞}, $G_{\overline{(2-4)}}$.

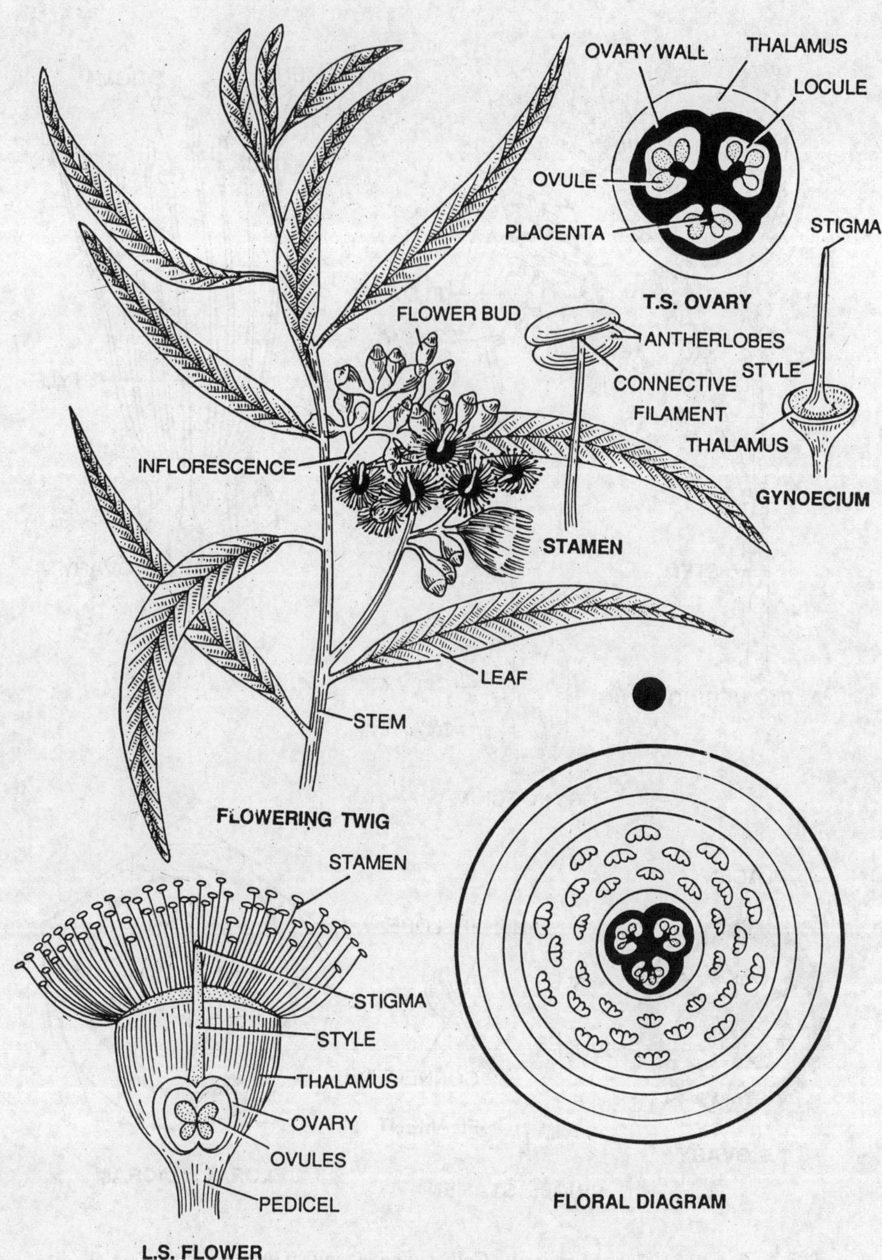

Fig. 27.3. Myrtaceae. *Eucalyptus citriodora* Hook.

2. ***Eucalyptus citriodora*** **Hook**; Eng. Lemon-scented gum; Verna.–**Safeda**.

Habit: A large tree with whitish bark.

Root: Tap, branched.

Stem: Erect, woody, branched, cylindrical, solid, glabrous, white.

Leaves: Simple, cauline, alternate, petiolate, exstipulate, gland dotted, aromatic, entire, glaucous, apex acute, unicostate reticulate venation.

Inflorescence: Cymose umbellate clustered cyme.

Flower: Pedicellate, ebracteate, complete, hermaphrodite, actionomorphic, epigynous.

Calyx: 5 sepals, gamosepalous, reduced, fused, forming a cap-like structure on the corolla, superior.

Corolla: 5 spetals, gamopetalous, forming a corolla cap covered by calyx cap, the so formed calyptra falls off when flower opens, superior.

Androecium: Stamens indefinite, polyandrous, attached on the rim of calyx cap, folded inwards in bud condition, anthers dorsifixed, versatile, bicelled, small, dehisce longitudinally.

Gynoecium: 3, tricarpellary, syncarpous, ovary inferior, trilocular, style single, stigma minute, terminal, axile placentation.

Fruit: A capsule.

Floral Formula: ⊕ ⚥ K (5), C 5, A_∞, G $(\overline{3})$.

2. *Psidium guajava* Linn.; Verna. **Amrood;** Eng. Guava.

Habit: Medium sized tree with whitish bark.

Root: Tap, branched.

Stem: Erect, aerial, woody, branched, cylindrical, solid, glabrous, white or brown.

Leaves: Simple, cauline, alternate, petiolate, exstipulate, gland dotted, aromatic, entire, glaucous, apex ovate, unicostate reticulate venation.

Inflorescence: Cymose, solitary axillary.

Flower: Pedicellte, bracteate, complete, hermaphrodite, actinomorphic, epigynous, pentamerous, cyclic, white.

Calyx: 5 sepals, gamosepalous, reduced, fused.

Corolla: 5 petals, gamopetalous, forming a corolla cap covered by calyx cap, the so formed calyptra falls off when flower opens, superior.

Androecium: Stamens indefinite, polyandrous, attached on the rim of calyx cap, folded inwards in bud condition, anthers dorsifixed, versatile, bicelled, small, dehisce longitudinally.

Gynoecium: Penta-carpellary, syncarpous, ovary inferior, pentalocular, style, stigma minute, terminal, axile placentation.

Fruit: Berry.

Floral formula: Br ⊕ ⚥ K (5), C 5, A ∞, G $(\overline{5})$.

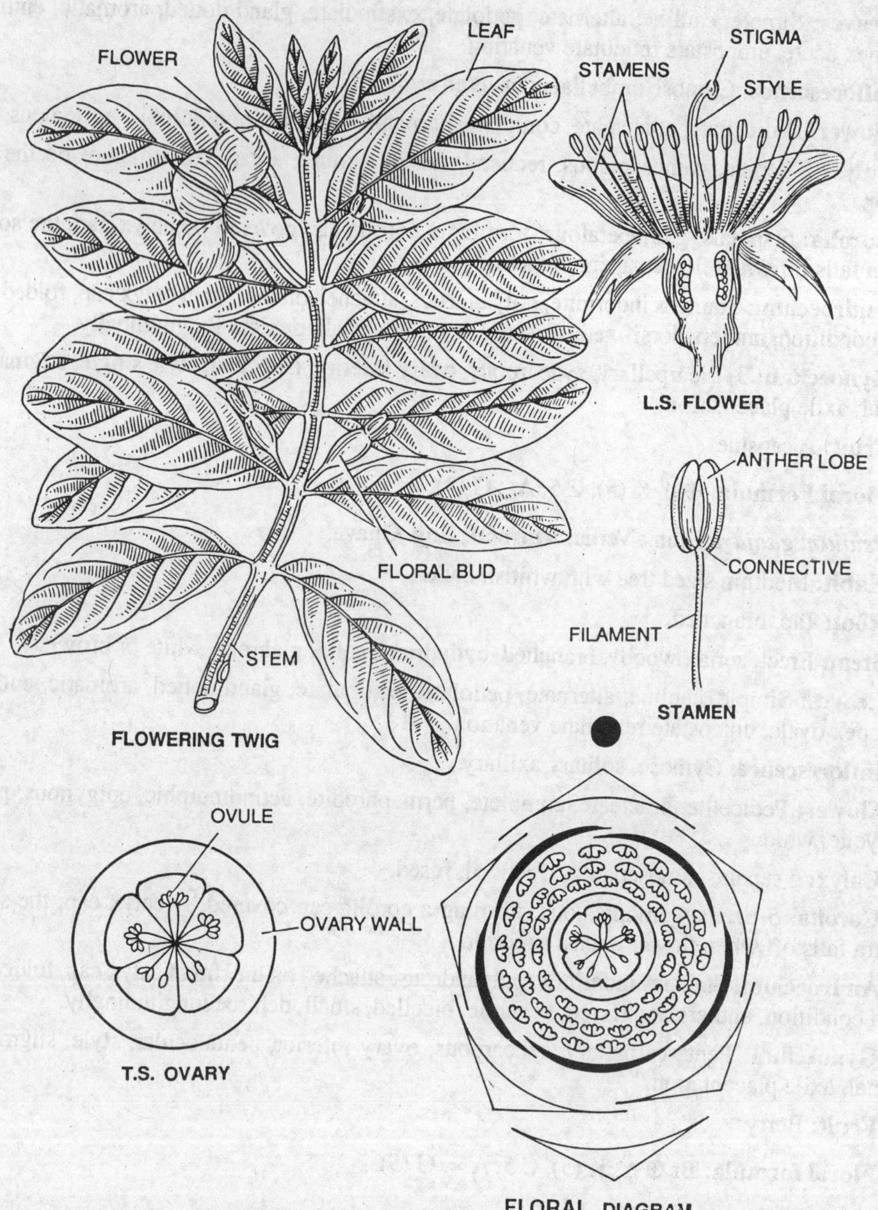

Fig. 27.4. Myrtaceae. *Psidium guajava* Linn.; Eng., guava; Verna, **amrood.**

Economic value. The fruits are edible. It is aromatic, sweet, juicy and highly flavoured. It contains acid, sugar and pectin. It is one of the richest sources of vitamins A, B, C and of ascorbic acid. It is commonly used for making jams, jellies and pastes. Bark of root is astringent and used in diarrhoea of children; fruit is laxative. Leaves are astringent and used for bowels and for wounds and ulcers.

Economic Importance of the Family

Economically, the family is fairly important. Some plants are of medicinal value while some produce edible fruits rich in vitamins. Certain plants are grown as ornamentals in the gardens. A list of few important plants is given below:

1. *Psidium cattleyanum;* Syn. *P. littorale;* Eng.–Strawberry guava –This is a small tree grown for its edible fruits.

2. *Psidium guajava*; Eng. Guava; Verna. **Amrood**–This is a small tree, chiefly cultivated in Uttar Pradesh, the Punjab, Bihar, Maharashtra and Andhra Pradesh. The fruits are rich in vitamins. The fruit is a laxative. The bark of the root is given in the diarrhoea of children. The bark and leaves are used for tanning.

3. *Syzygium aqueum;* Syn; *Eugenia aquea; Eugenia javanica;* Eng.–Watery rose apple; Verna.-**Lal jumrool**–This is a small tree, grown for its edible fruits. It is found in Khasia hills and Sikkim.

4. *Syzygium aromaticum;* Syn. *Eugenia aromatica; E. caryophyllata;* Eng.–Clove; Verna.-**Laung**–The dried flower buds are used as spice. They are also used medicinally as carminative, stimulant and in flatulence. An oil, known as 'clove oil' is extracted from the dried buds, which is given in many diseases. The cloves are eaten with betel.

5. *Syzygium cumini;* Syn. *Myrtus cumini; Eugenia jambolana;* Eng.-Javaplum; Verna.–**Jamun**–A large tree cultivated throughout India for the edible fruits. The vinegar is prepared from the juice of its fruits, which is very useful for stomach troubles. The seeds are used as fodder. The bark is used in tanning.

6. *Syzygium fruticosum;* Syn. *Eugenia fruticosa* ; Eng.–Wild jamun; Verna–**Jangli Jamun**–it is grown as a road–side tree. The fruits are edible.

7. *Syzygium jambos;* Syn. *Eugenia jambos;* Eng.–Rose apple; Verna.–**Gulabjaman**–the plants are grown for edible fruits. They are commonly cultivated in Asam, Bihar, Andhra Pradesh. Tamil Nadu, Maharashtra and Bengal.

8. *Syzygium malaccense;* Syn. *Eugenia malaccensis;* Eng.–Mountain apple; Verna–**Malaya jam**–The fruits are edible. It is introduced from Malacca.

9. *Syzygium samarangense;* Syn. *Eugenia javanica* ; Eng. Wax jambu; Verna.–**Jamul**–The fruits are edible. The plants are commonly found in the Andaman and Nicobar islands.

10. *Eugenia cuspidata;* Verna–**Jamoah**–This is grown as an avenue tree.

11. *Eugenia uniflora;* Eng.–Surinamcherry–The plants are grown in the hills in South India for their edible fruits.

12. *Eucalyptus citriodora;* Eng.–Lemon–scented gum.–An essential oil is extracted from its leaves which is used in perfumery. They are trees and grown in Northern India, Kerala and Nilgiris.

13. *Eucalyptus globulus;* Eng.–Tasmanian blue gum–This is a large tree. The eucalyptus oil is extracted from its dried leaves. This oil is used to cure asthma and bronchitis, and is also used in perfumery. The wood is used for fuel. The plants are grown for afforestation in the marshy places as an anti–malarial measure. The braille paper is manufactured from its pulp. The tree is native of Australia, but now grown chiefly in the Nilgiris, the Anamalis, the Pulney hills, Shimla hills and at Shillong.

The *Eucalyptus* tree is the principle source of timber in Australia. The timber is used for bridge, railway sleepers, building materials, plywood industries, etc.

14. *Eucalyptus rostrata*; Eng.–Murray red gum. The tree is used for afforestation in marshy areas.

15. *Callistemon citrinus;* Syn. *Callistemon lanceolalus*; Eng.–Bottle–brush; Verna–**Lal botal brush**–This is grown in the gardens as an ornamental. This a large shrub or small tree.

16. *Callistemon rigidus;* Eng.–Bottle–brush.–This is also grown as an ornamental, and introduced in India from East Africa.

17. *Melaleuca leucadendron;* Eng.–Cajeput tree; Verna–**Kayaputi.**–This is a tree. Usually grown as a hedge plant. An essential oil is extracted from its leaves and twigs. The oil is known as cajeput oil, which is used as an insecticide, especiaily for expelling mosquitoes.

Family—Combretaceae

There are about 17 genera and 500 species in this family (Rendle).

Distribution: The members of this family are commonly found in tropical and sub–tropical regions of the world. In our country the family is represented by few important genera, such as *Terminalia, Combretum, Anogeissus* ,etc.

Habit: The family consists of trees (*e.g., Terminalia chebula, T. bellerica,* etc.), shrubs and climbers (*e.g. Quisqualis indica*).

Leaves: The leaves are simple, alternate, sub–opposite or opposite, sometimes ternate (in threes), petiolate, entire, exstipulate.

Inflorescence: The flowers are usually arranged in spikes or racemes, the racemes are often panicled.

Flower: The flowers are bracteolate, hermaphrodite or unisexual by reduction, actinomorphic with a tendency to zygomorphy, pentamerous and sometimes tetramerous, epigynous.

Calyx: The calyx–tube consists of 4–5 lobes, the tube remains adnate to the ovary and is produced above it, generally valvate aestivation.

Corolla: It consists of 4–5 petals alternating with the sepals or sometimes altogether absent.

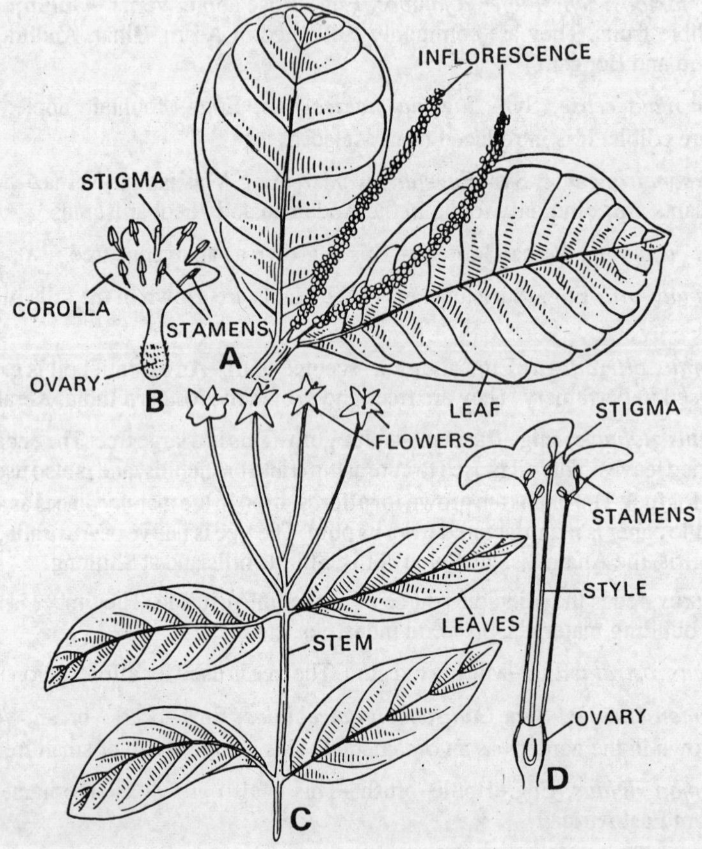

Fig. 27.5. Combretaceae-A, a twig of *Terminalia catappa* with leaves and inflorescences; B, flower of same; C, a flowering twig of *Quisqualis indica*; D, flower in vertical section of same.

Androecium: The stamens are 4–5 or 8–9, situated on the calyx; generally the number of stamens is twice the number of sepals; they are found in two series the lower opposite the sepals and the upper alternating the sepals; sometimes the upper series is found to be very much reduced or absent; stamens are bent inwards in the bud condition.

Gynoecium: Monocarpellary, ovary one chambered, inferior, angled; ovules (anatropous) two to five rarely more; style simple, one, long, filiform bearing a simple pointed and rarely capitate stigma.

Fruit: Two to five angled or winged, coriaceous (leathery) or drupaceous, one seeded, indehiscent.

Seeds: The seeds are non–endospermic (exalbuminous), the cotyledons are spirally folded.

Pollination: Generally the pollination takes place through the agency of insects.

Floral Formula: $\oplus$ ⚥ K (4–5), C(4–5) or 0, A 4–9, $G_{\bar{1}}$.

Bentham & Hooker (1862)	Engler & Prantl (1931)	Hutchinson (1959)
Phanerogams	Phanerogams	Angiospermae
Dicotyledones	Dicotyledones	Dicotyledones
Polypetalae	Archichlamydeae	Lignosae
Disciflorae	Myrtiflorae	Myrtales
Myrtales	Combretaceae	Combretaceae
Combretaceae		

Economic Importance of the Family

Economically the family is fairly important. It consists of several timber yielding and medicinal plants. Some are ornamentals.

1. *Terminalia arjuna* ; Syn. *Terminalia glabra*; *Pentaptera arjuna*; Verna–**Arjun**–This is a tree commonly found in Madhya Pradesh, Bihar and the Western Peninsula. The bark and ashes are used in dyeing and tanning. The gum, bark, leaves, and fruits are medicinal. The heartwood is brown with dark coloured streaks and very hard, but not very durable. Also grown as an avenue.

2. *Terminalia bellirica;* Syn. *Myrobalanus bellirica*; Verna–**Bahera**–This is a tree grown as avenue. The fruit constitutes one of the **myrobalans** (an adhesive) of commerce. It is used in dyeing and tanning, and largely exported. Native ink is prepared from it, and it is also used medicinally, as also the oil extracted from the kernels. The fruits are used to increase the potency of spirits. The kernels are largely eaten by natives. The wood is used for various purposes. In Northern India it is usually steeped for sometime in water in order to make it more durable when used for house building.

3. *Terminalia bialata*; Eng.–White chuglam.–This is found in the Andaman islands. The wood is used for cabinet work, furniture and interior fittings.

4. *Terminalia catappa*; Eng. Tropical almond; Verna–**Janglibadam**.–This is cultivated in gardens in the hotter parts of the country. It is a large, very handsome tree with buttressed trunk, and the large leaves turn crimson red before falling. The kernels (seeds) are eaten and the bark and fruits are used for tanning. The fruits also yield a black dye.

5. *Terminalia chebula*; Verna–**Harara**–Commonly found in Northern India. In Northern India, the tree does not attain to any great size, but large trees upto 100 feet in height, are often met with in south of Nerbudda. The most important product of this tree is the dried fruit, which constitute the black **myrobalan** of commerce, one of the most valuable of Indian tanning materials. The fruit is also much valued for medicinal purposes. The kernel is eaten, and the leaves are used as good fodder for cattle.

6. *Terminalia citrina*; Verna–**Harira**–This is found in Assam and the Khasia hills. The fruits are used for tanning.

7. *Terminalia manii;* Eng.–Black chuglam–This is found in the Andaman and Nicobar islands. The wood is used for making fishing rods.

8. *Terminalia myriocarpa*; Eng.–Hollock–This is commonly found in Assam. The wood is used for making tea–boxes and plywood.

9. *Terminalia procera;* Verna.–**Badam**–This is found in the Andaman islands. The wood is used for making furniture and interior fittings.

10. *Terminalia tomentosa;* Verna–**Asan**–The bark and fruits are used for tanning and dyeing. The leaves afford good fodder for cattle, and in Madhya Pradesh they are much used for feeding the tassar silkworm. The bark is also used medicinally. The dark brown and streaked heart wood is largely used for building purposes and it makes excellent fuel and charcoal.

11. *Anogeissus acuminata*; Eng.–Yon; Verna–**Dhaura**–This is a source of timber, which can be used with advantage where strength and shock resisting ability are required. The handles of tools are made from its wood.

12. *Anogeissus latifolia*; Eng.–Gum ghatti; Verna–**Bakia**, **Dhaura**–This is a deciduous tree found in the sub–Himalayan tract, Madhya Pradesh, the Western Peninsula and the Nilgiris. The gum obtained from the stem is known as 'ghati–gum', is used in calicoprinting and the leaves are used for tanning. The wood is used for making wheels.

13. *Anogeissus pendula*; Verna–**Dhao**–This is a tree grown as an avenue on road sides.

14. *Quisqualis indica;* Eng.–Rangoon creeper, Verna–**Rangoon-ki-bel**–This is a beautiful climbing shrub. This is native of Java and Malaya. This is grown as an ornamental in the gardens. The tender shoots are edible.

Systematic Relationships: Eichler, Engler and Rendle have treated the order Myrtales as Myrtiflorae. Engler and Diels have included about twenty–three families in this order, Bentham and Hooker, Bessey and Hutchinson designated the order, the Myrtales Hutchinson has splitted the Engler's Myrtiflorae into three orders, *i.e.*, Thymelaeales, Lythrales and Myrtales.

28

ORDER-SANTALALES

According to Hutchinson this is the forty-third order of the phylum Angiospermae, sub-phylum Dicotyledones and division Lignosae. This order includes five families, *i.e.*, Loranthaceae, Grubbiaccae, Santalaceae, Myzodendraceae and Balanophoraceae. The Loranthaceae is the primitive most within the order and has been described in the present text in detail. Bentham and Hooker have included the family Loranthaceae along with other two families, *i.e.*, Santalaceae and Balanophoraceae in their sixth series – Achlamydosporeae of class Dicotyledones and division Monochlamydeae or Incompletae. Engler and Prantl put the family Loranthaceae along with other seven families in the fourteenth order the Santalales of class Dicotyledoneae and sub-class Archichlamydeae.

Flowers mostly epigynous; calyx valvate or open or absent; stamens opposite to the corolla-lobes when latter present; endosperm present; embryo straight.

Family—Loranthaceae

Distribution: The members of this family are being distributed in temperate and tropical regions of the world. The only genus found in Britain is *Viscum. Viscum album* is popularly known as **Mitsletoe.** *Loranthus* is the largest genus of the family which comprises of about 500 species. The second largest genus is *Phoradendron* which comprises of about 135 species. In our country the family is represented by *Viscum* sp. and *Dendrophthoe falcata* **(Banda).**

Habit: The plants are either herbs or shrubs. They live on tree branches in semiparasitic state. They remain attached to their hosts by suckers which have been variously interpreted. But according to most of botanists they have been thought to be adventitious roots greatly modified as organs of absorption. A few species are found to be grown on the ground, but majority of them live upon the trees. Though the semiparasitic plants bear green leaves, yet they obtain a considerable amount of their nourishment from the tissues of the host tree, on which they live.

Root: The roots are represented by suckers or haustoria which derive their nourishment from the host tree. Such roots branch within the host tissue as found in *Viscum album*.

Stem: The stem develops sympodially. In Mistletoe (*Viscum album*), the stem forms a dichasium and here each successive axis bears a pair of persistent leathery green leaves. The nodes of the stems are usually swollen; dichotomously branched, woody, glabrous, solid, cylindrical and green in colour.

Leaves: The leaves are simple, entire, persistent, coriaceous (leathery), thick, opposite or whorled rarely alternate, exstipulate, glabrous. Sometimes the leaves are very much reduced and represented by scales.

Inflorescence: There are many variations in the inflorescence. The flowers may be solitary or may be arranged in dichasia. Each flower remains situated in the axil of a bract. The flowers may also

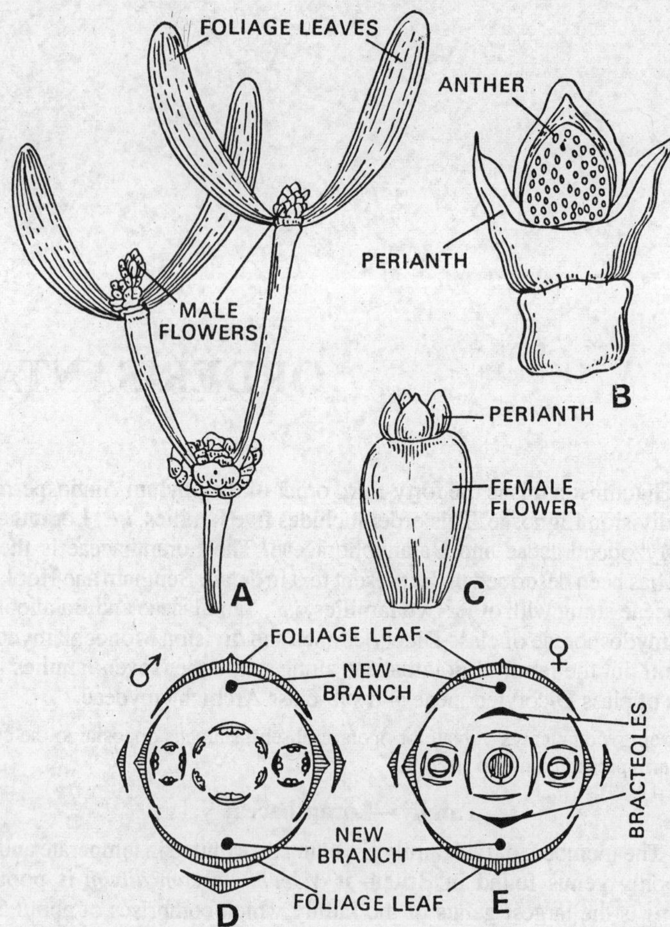

Fig. 28.1. Loranthaceae. *Viscum album*. A, a male shoot; B, male flower cut lengthwise; C, female flower; D, floral diagram of male flower bearing branchlet; E, floral diagram of female flower bearing branchlet.

found to be arranged in a raceme, spike or a panicle. The spike of *Dendrophthoe* and *Phoradendron* consists of two or more nodes with decussating bracts. In *Viscum*, the male and female flowers are found to be arranged in dichasial cyme which ends in the leafy shoot. When the flowers are pedicellate, the bract adnates to the pedicel.

Flower: The flowers are actinomorphic with a tendency to zygomorphy. They are either hermaphrodite or unisexual and dioecious. They are small and green but the flowers of Loranthoideae are large and brightly coloured. They are complete or incomplete and epigynous. They are dimerous or trimerous in their plan.

Perianth: There is a cupular receptacle and from its edge a perianth of generally two similar 2 - 3 merous whoris arises. The perianth leaves may be free or united. The perianth leaves of subfamily Viscoideae are small and sepal like (green), whereas of subfamily Loranthoideae, they are large and brilliantly coloured (petaloid). In Loranthoideae, a slightly toothed or irregular rim known as **calyculus** is present below the perianth and sometimes this may be treated as calyx and the perianth proper as corolla.

Androecium: The stamens are equal in number to the segments of the perianth and situated opposite the perianth leaves. They are more or less united with the perianth, segments however, in

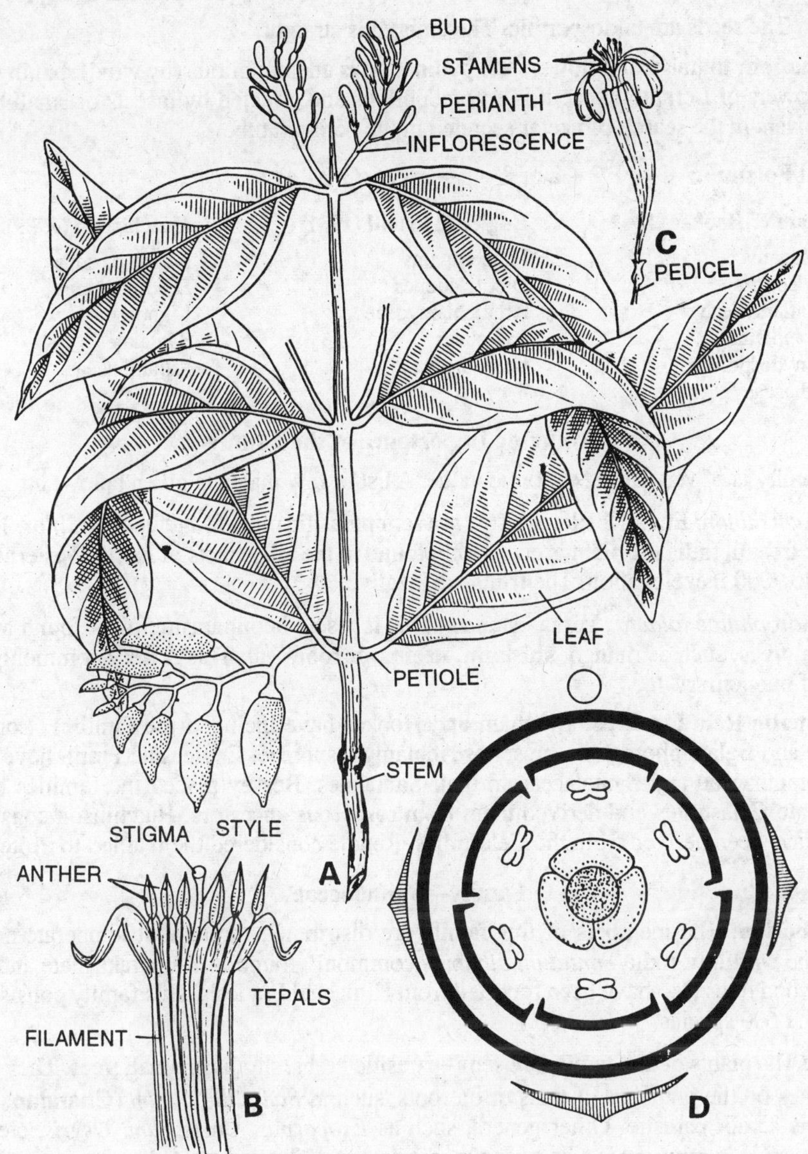

Fig. 28.2. Loranthaceae. *Dendrophthoe falcata* A, flowering twig; B, perianth with stamens, style and stigma; C, tubular flower; D, floral diagram.

Viscum, this union is complete. Here the petals and stamens arise as one structure from the floral axis on the front of which a large number of pollen sacs are produced.

Gynoecium: It consists of 3-4 syncarpous carpels. The ovary is inferior unilocular with a large central placenta having many ovules. The ovules are not differentiated from the placenta. The placentation is basal. In *Viscum*, a group of the sporogenous cells is produced in the placenta, and the ovary-cavity is hardly being distinguished from it. Generally one such sporogenous cell develops into an embryo sac.

Fruit: The fruit is a pseudocarp, the ovary unites with the receptacular cup developing a berry-like or more rarely a drupaceous fruit. The fruit may be compared to that of apple.

Seeds: The seeds are endospermic. The embryo is straight.

Pollination: In unisexual flowers the pollination is anemophilous (by wind), but in the brightly coloured flowers of Loranthoideae it is entomophilous and effected by insects or small birds which visit the flowers in the search of nectar secreted in the perianth tube.

Floral Formula: $\oplus$ ⚥ P 2 + 2 or 2 + 2 + 2, A 6, G$(\overline{3-4})$.

Bentham & Hooker (1862)	Engler & Prantl (1931)	Hutchinson (1959)
Phanerogams	Phanerogams	Angiospermae
Dicotyledones	Dicotyledones	Dicotyledones
Monochlamydeae or Incompletae	Archichlamydeae	Lignosae
Achlamydosporeae	Santalales	Santalales
Loranthaceae	Loranthaceae	Loranthaceae

Economic Importance of the Family

The family is of very little economic value. A list of few plants is given below:

1. *Viscum album;* Eng.–Mistletoe–This is a semiparasitic plant found upon the branches of trees, such as Oak tree. In India the plant is commonly found in the hilly tracts of North Western Himalayas from 4000 to 9000 feet elevation. The fruits are laxative.

2. *Dendrophthoe falacta;* Verna.–**Banda**–This is also a hemiparasitic plant found to be grown upon several trees, such as mango, **shisham, neem, amrood,** etc. This is very commonly found on the plants of our country.

Systematic Relationships: Bentham and Hooker have included the families, Loranthaceae, Santalaceae and Balanophoreae in the series Achlamydosporeae. Engler and Prantl have placed the families Loranthaceae and Santalaceae in their Santalales. Bessey placed the families of Engler's Santalales into Celastrales and derived them from rosaceous ancestors. Hutchinson considered that Santalales have been derived from the Celastrales. Rendle considered them allied to Proteales.

Family—Santalaceae

Distribution: The members of this family are distributed throughout temperate and tropical regions of the world. In India *Santalum album* is commonly found in Karnataka state and Mysoore. The sandalwood trees also have been reported from Himachal Pradesh. This family consists of about 26 genera and 600 species.

Habit: The plants of this family are semi–parasitic herbs, shrubs or small trees. They are mostly semi–parasites on the branches of trees or on roots, such as *Santalum album* (**Chandan**, the sandal wood tree) is a root parasite. Other genera, such as *Exocarpus, Choretrum, Osyris*, etc., are root parasites. *Thesium humifusum* is a herbaceous root parasite. They are root parasites in seedling stage. The seedlings get themselves attached to the host by suckers.

Haustoria: As the members of this family are semi–parasites, they remain attached to the hosts by means of haustoria. The haustoria penetrate the roots of the host (another angiosperm) as far as xylem, as a result the vessel of the host and that of the parasite become organically jointed, the inter–communicating vascular elements are known as **'phloeotracheids'**. The haustoria may be simple or compound. The compound types occur in the species attacking hosts having hard roots which are difficult to penetrate. In both types of haustoria glandular activity is marked on the surface of the haustoria, the glands secrete some substances that help the haustoria for penetration. The structure of a mature haustorium is like an inverted flask. The vessels of the haustorium become opposed to those of the host and finally organic connection between the vessels of the host and parasite is established. The haustoria undergo secondary thickening also.

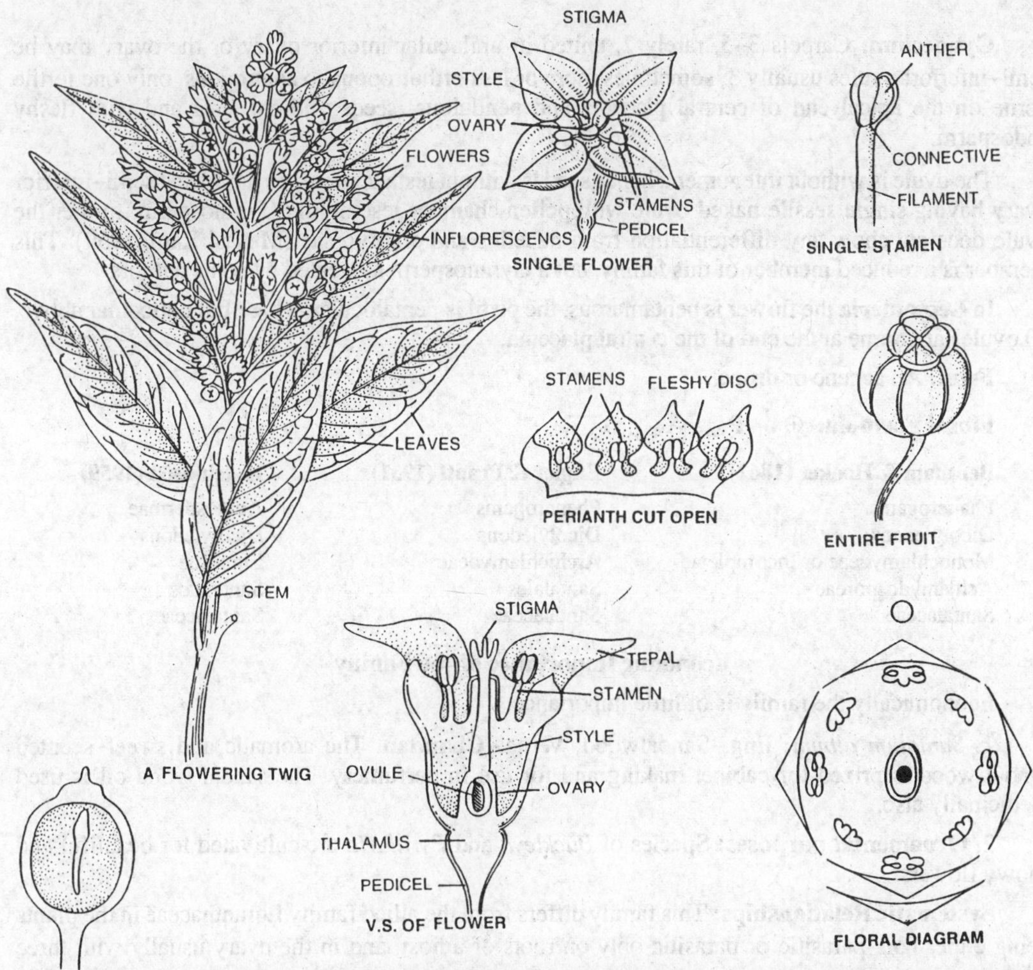

Fig. 28.3. Santalaceae. *Santalum album.*

Stem: Erect, branched, solid, cylindrical, glabrous green when young, brown or grey when old.

Leaves: The leaves are simple, entire, leathery, opposite, sometimes alternate and exstipulate. Net veined with acute apex and petiolate.

Inflorescence: There are many variations in the inflorescence. The flowers may be solitary or arranged in dichasia. Usually a raceme, spike, head.

Flower: The flowers are small, actinomorphic, haplochlamydous, bisexual or unisexual by abortion of one of the essential members. All transitions from bisexual to unisexual staminate or pistillate flowers are found occurring in this family; dioecious plants with unisexual flowers are represented by *Osyris pyrularia;* pentamerous or trimerous. In *Leptomeria* the flower is pentamerous. Nectar secreting disc is often present in many genera, such as in *Leptomeria* five lobed epigynous disc is present, epigynous or hypogynous.

Perianth: Tepals are usually 4–5, sometimes 3–6, united at the base and valvate in aestivation, either sepaloid or petaloid; *Comandra umbellata* has pentamerous perianth. The perianth of *Osyris* and others are trimerous.

Androecium: Stamens are as many as perianth members (tepals) and opposite to them and are epiphyllous, *i.e.*, adnate to the perianth leaves; filaments are short, anthers dithecous, dehiscing vertically; at the time of the bursting of anther cells the anther cells fuse, the pollens are discharged through vertical pores.

Gynoecium: Carpels 3–5, rarely 2, united in unilocular inferior ovary or the ovary may be semi–inferior; ovules usually 3, sometimes more or less orthotropous to anatropous, only one fertile borne on the apical end of central placenta and pendulous; seeds without testa and with fleshy endosperm.

The ovule is without integument, hence seed is without testa. *Exocarpus* has naked semi–inferior ovary having single sessile naked ovule with pollen chamber resembling Gymnosperm, further the ovule does not show any differentiation from nucellus and embryo sac (Manasi Ram, 1959). This member is a reduced member of this family, not a Gymnosperm as treated by some authors.

In **Leptomeria** the flower is pentamerous, the pistil is pentalocular at base but unilocular above, 50 ovules are borne at the end of the central placenta.

Fruit: An achene or drupe.

Floral Formula: $\oplus$ ⚥ $P_{4-5}, A_{4-5}, G_{\overline{(3-5)}}$.

Bentham & Hooker (1862)	**Engler & Prautl (1931)**	**Hutchinson (1959)**
Phanerogams	Phanerogams	Angiospermae
Dicotyledons	Dicotyledons	Dicotyledons
Monochlamydeae or Incompletae	Archichlamydeae	Lignosae
Achlamydosporeae	Santalales	Santalales
Santalaceae	Santalaceae	Santalaceae

Economic Importance of the Family

Economically the family is of little importance.

1. *Santalum album;* Eng.–Sandalwood; Verna.–**Chandan.** The aromatic and sweet–scented sandal wood is prized for cabinet making and for use in perfumery. The sandal wood oil is used medicinally also.

2. **Ornamental purposes:** Species of *Buckleya* and *Pyrularia* are cultivated for beautiful and showy flowers.

Systematic Relationships: This family differs from the allied family Loranthaceae in the plants being either non–parasitic or parasitic only on roots of a host, and in the ovary usually with three ovules.

The family Santalaceae is most closely allied to family Loranthaceae (these families were united by Baillon). Because of the 4–5 merous flower and less complicated floral and fruit morphology, this family is supposed to be less advanced. The Santalaceae were divided into 3 tribes by Engler and Diels on the basis of ovary position and the character of the disc.

29

ORDER–EBENALES

According to Hutchinson this is the forty-sixth order of the phylum Angiospermae, sub-phylum Dicotyledones and division Lignosae, the order comprises of three families, of which Sapotaceae and Ebenaceae have been described in the present text in detail. Bentham and Hooker have also placed the family Sapotaceae along with two others, *i.e.*, Ebenaceae and Styraceae in their sixth order–the Ebenales of class Dicotyledones, division Gamopetalae and series.Inferae. Engler and Prantl have put the family Sapotaceae along with three others, *i.e.*, Ebenaceae, Symplocaceae and Styraceae in their fourth order– the Ebenales, of class Dicotyledoneae and sub class Sympetalae.

Sympetalous; corolla-lobes imbricate; stamens epipetalous or rarely hypogynous, 1-4 times as many as the corolla-lobes; petaloid staminodes often present; anthers opening lengthwise; ovary superior; ovules 1-2 in each loculus, axile; seeds often with shining hard testa and conspicuous hilum. Trees or shrubs, leaves alternate, rarely with stipules.

Family—Sapotaceae

There are about 40 genera and 600 species in this family (Rendle).

Distribution: The members of the family are widely distributed in the tropical regions of the world. In our country the family is represented by some important genera, *e.g., Mimusops, Achras, Palaquium, Sideroxylum* ,etc.

Habit: The plants are mostly trees and rarely shrubs. *Achras sapota* is a tall tree, reaching 75 feet in height.

The miky latex is found in the pith, cortical region and in the leaves. The leaves, flowers and fruits are usually covered with unicellular two-armed hairs.

Leaves: The leaves are simple, alternate, rarely sub-opposite, petiolate, coriaceous, entire, exstipulate or very caducous.

Inflorescence: The flowers are solitary or in cymose inflorescences. In certain cases the inflorescence is found to be situated above the scars of fallen leaves or on old stems.

Flowers: Generally hermaphrodite, actionomorphic, bracteolate, the bracteoles when found are small and caducous, complete and hypogynous.

Calyx: The number of sepals shows considerable variation, *e.g.*, 2 + 2, 3 + 3, 4 + 4 or 5. The sepals are somewhat connate at the base (gamosepalous). The aestivation is imbricate.

Corolla: Generally the number of petals corresponds to the number of sepals, the petals alternate the sepals. Mostly the petals are found to be arranged in a single whorl, but sometimes they are arranged in two whorls. The petals are basally united. In *Mimusops*, the number of petals apparently increase be-

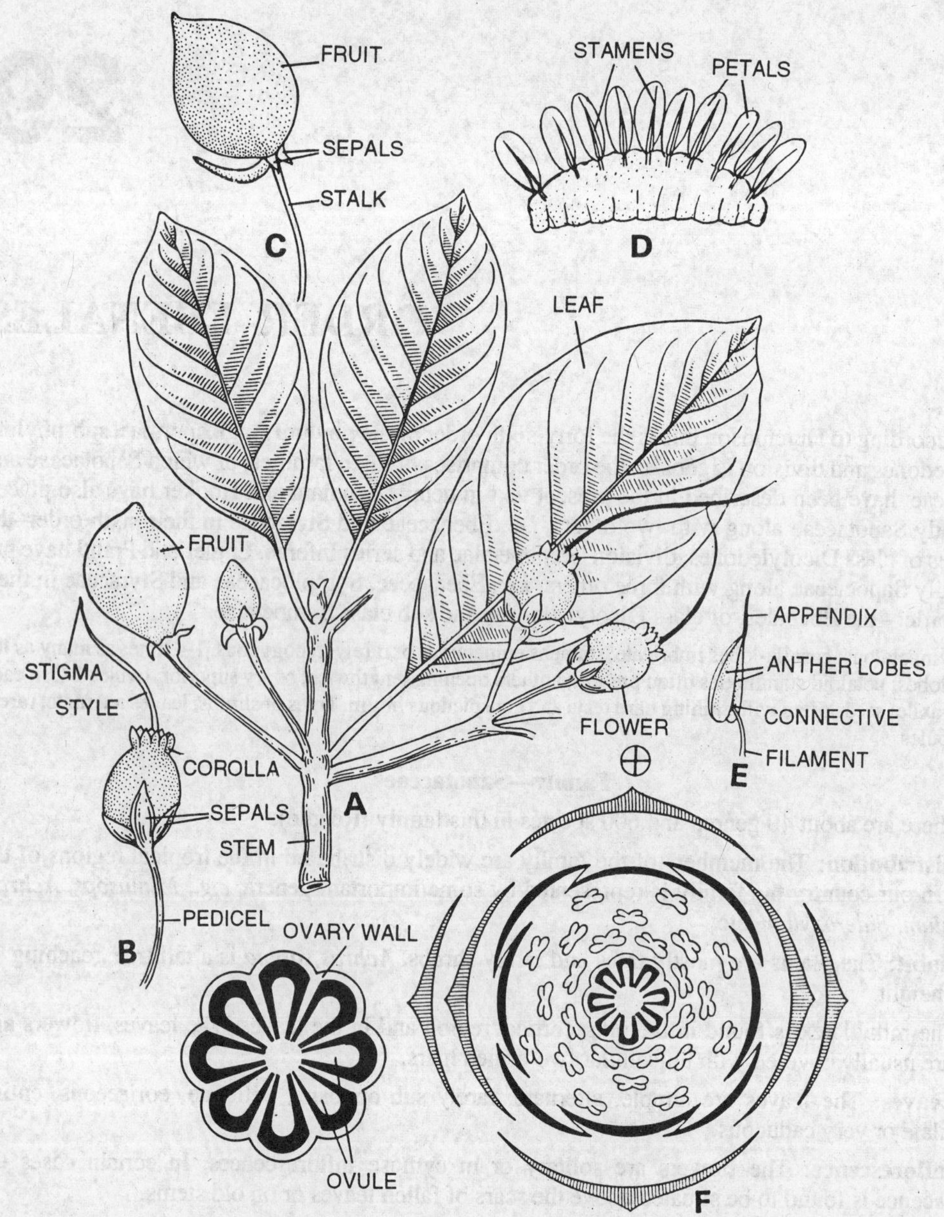

Fig. 29.1. Sapotaceae–*Madhuca indica* (**Mahua**). A, flowering twig; B, flower; C, fruit; D, stamens; E, single stamen; F, floral diagram; G. T. S. of ovary.

cause of the development of dorsal outgrowth from the base of each petal. These outgrowths resemble the petals or be much cut. In other genera each segment is found with a pair of lateral outgrowths.

Androecium: The stamens are epipetalous, *i.e.,* situated on the corolla tube. Generally they are found to be arranged in two or three whorls, each of 4-5 stamens. Generally the inner whorl consists of fertile stamens and the stamens of remaining whorls are being reduced to staminodes or sometimes altogether suppressed. The staminodes (when present) remain alternate to stamens, filaments usually short; anthers oblong-lanceolate, connective often produced, anthers 2-celled, mostly extrorse, dehiscing longitudinally.

Gynoecium: The gynoecium consists of as many carpels or double as many as the number of stamens in a whorl; the ovary is superior, tetra, penta or multichambered; the placentation is axile; one anatropous ovule in each loculus; style; simple stigma inconspicuous.

Fruit: The fruit is berry; the outer layer is brown and sometimes sclerenchymatous. The berries are sticky and this nature helps in their distribution by birds.

Seeds: The seeds are endospermic or non-endospermic when the cotyledons are flat, when non-endospermic the cotyledons are large and fleshy, the testa is hard and shiny.

Pollination: Usually entomophilous.

Floral Formula: ⊕ ⚥ K 4-12, C 4-12, A 4-5, or ∝, G $\underline{(4-5)}$.

Bentham & Hooker (1862)	Engler & Prantl (1931)	Hutchinson (1959)
Phanerogams	Phanerogams	Angiospermae
Dicotyledones	Dicotyledoneae	Dicotyledones
Gamopetalae	Sympetalae	Lignosae
Inferae	Ebenales	Ebenales
Ebenales	Sapotaceae	Sapotaceae
Sapotaceae		

Economic Importance of the Family

From the viewpoint of its economic value the family is fairly important. It contains some important fruit trees. Some are grown for their foliage and fragrant flowers. A list of few important plants is given below:

1. *Mimusops hexandra;* Syn. *Manilkara hexandra;* Verna.–**Khirni**– This is a tree cultivated in Delhi, Madhya Pradesh, Gujarat, Maharashtra and Andhra Pradesh. The wood is hard, tough and very durable, and is much used for oil presses, house building and in turnery. The bark is used for medicinal purposes. The fruits are edible, and an oil is extracted from the seeds.

2. *Mimusops elengi* ; Verna–**Maulsari**–This is a tree. Generally grown in Northern India, Western Peninsula and South India. The wood is used in the same way as that of *M.hexandra.* The small fragrant flowers are made into garlands, and also yield **Maulsari attar.** The ripe fruits are edible. An oil is extracted from the seeds. The bark is much employed in native medicine.

3. *Mimusops littoralis.* The bark yields a dye.

4. *Manilkara kauki*; Syn. *Mimusops kauki;* Verna–**Khirni**–It is a tree, grown in Northern India, Andhra Pradesh and Malabar. The fruits are edible.

5. *Achras zapota;* Syn. *Achras sapota; Sapota achras; Manilkara zapota;* Eng.–Sapodilla; Verna–**chiku**–This is a tree. It is native of tropical America but commonly cultivated in Bengal, Tamil Nadu and Maharashtra for the edible fruits. The milky juice is known as chicle and used in manufacture of chewing gum. The tender leafy tips are eaten.

6. *Palaquium ellipticum;* Syn. *Bassia elliptica; Dichopsis elliptica;* Eng.–India gutta parcha tree; Verna–**Halganne.**–This is a tall tree found in the Western Ghats, North Kanara and Tamil Nadu. An oil is extracted from the seeds, which is used in the manufacture of soap and as an illuminant.

7. *Diploknema butyracea;* Syn. *Bassia butyracea; Madhuca butyracea;* Eng.–Bassia fat. Indian butter tree; Verna–**Phalwara.** –This is a large tree, found in Uttar Pradesh and Bihar. The flowers are edible and the seed oil is used for cooking. The bark is used for dyeing.

8. *Madhuca indica;* Syn. *Madhuca latifolia*; *Bassia latifolia;* Verna.–**Mahua.** – This is a large tree, generally grown in Eastern Uttar Pradesh and Bihar. The flowers are used as vegetable and also are the source of alcohol. An oil is extracted from the seeds, which is used for cooking and soap–making. The oil cake is used as a manure.

9. *Madhuca longifolia;* Syn. *Bassia longifolia;* Eng.–Monira fat; Verna.–**Mohua**–This is a tree, cultivated in Uttar Pradesh, Bihar, Andhra Pradesh, Karnataka, Bengal and Maharashtra. The flowers are used in the manufacture of alcohol. An oil is extracted from the seeds, which is used for cooking purpose, the bark is used as a tan.

10. *Madhuca malabarica;* Syn. *Bassia malabarica*; This is a tree found in Western Peninsula. The flowers are edible and the seedoil is used for soap–making and cooking.

Systematic Relationships: According to Engler and Diels the order Ebenales consists of two sub orders and seven families. Hallier has divided this order into two orders known as the Sapotales and the Santalales. Hutchinson has also put forth two orders–the Ebenales and the Styracales but on wholly different morphological grounds. According to Copeland the Ebenales have been derived from the family Theaceae. According to Wettstein the order Ebenales has affinities with the orders Malvales, Geraniales and Rutales.

Family—Ebenaceae

Distribution: The members of this family are commonly distributed in India, Malay–Peninsula and Africa. *Maba* and *Diospyros* are Indo–Malayan genera, *Royena* and *Euclea* are African, and *Tetraclis* is endemic in Madagaskar. This family consists of 5 genera and about 325 species.

Habit: Usually trees or shrubs with very hard wood in stems. The heartwood is often black, red or green; the plants lack milky sap; secretary cells with tannin are found in the tissues of stem.

Leaves: The leaves are alternate, rarely opposite or whorled, simple, entire, coriaceous, exstipulate. Secretory cells containing tannin–like substances occur in the tissues of leaves.

Inflorescence: Cymose or solitary axillary.

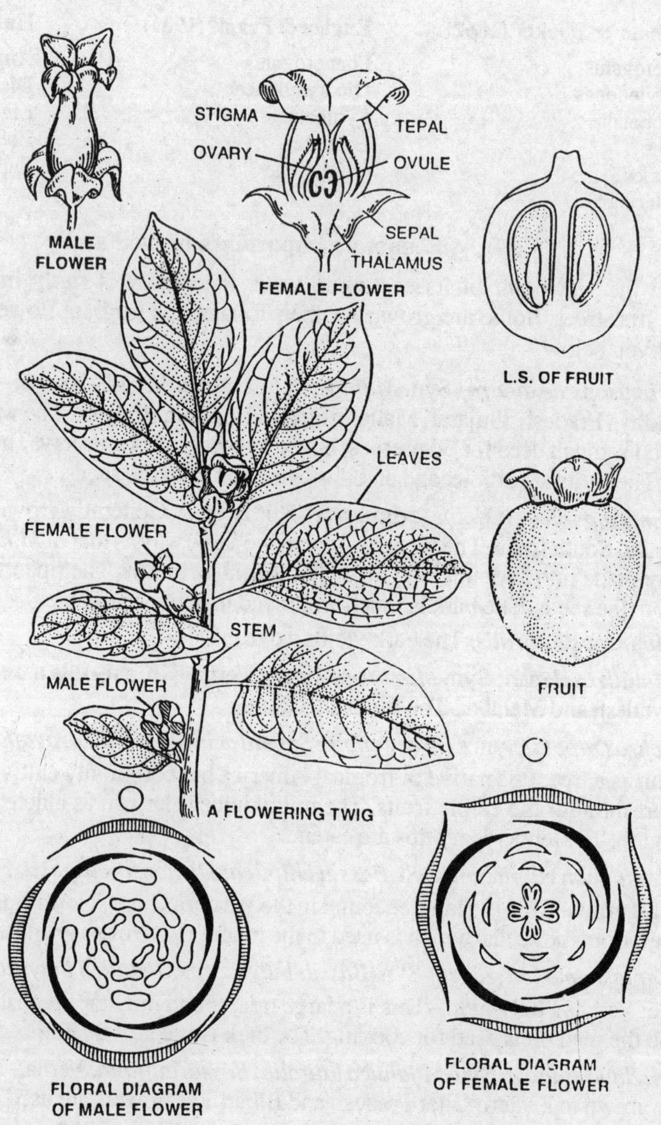

Fig. 29.2. *Diospyros kaki.*

Flowers: Flowers are unisexual, usually in dioecious plants, rarely flowers are bisexual and polygamous, actionomorphic, 3 – 7 merous, hypogynous; staminate flowers mostly more abundant than pistillate.

Calyx: Sepales 3 – 7, usually 4 – 5, united in persistent lobed calyx (gamosepalous).

Corolla: Petals 3 – 7, as many as sepals, gamopetalous, united in an urceolate, tubular, campanulate or semi–funnel–shaped corolla, lobes imbricate or contorted in bud.

Androecium: Stameus epipetalous or hypogynous, of same number as corolla lobes, *i.e.*, 3–7 or 2–3 times as many, distinct or united in pairs, the anthers dithecous (2–celled), introrse, dehiscing longitudinally; staminodes usually present in pistillate (female) flower, similarly pistillode (rudimentary pistil) may be present in a staminate (male) flower. The frequent occurrence of numerous stamens is the result of branching of original staminal primordia. In some species of *Diospyros* the dehiscence of anthers is by apical pores.

Gynoecium: Carpels 2–16, usually 2 – 8, syncarpous, superior ovary, with locules as many as carpels; false septum frequently develops in the ovary; ovules 2–1, integumens 2, ovules pendulous and anatropous; the styles and stigmas 2 – 8, styles distinct or basally connate, placentation axile. Usually the number of carpels corresponds to that of the sepals and the carpels are opposite to the sepals, but the ovarian chambers become double the number of carpels actually present due to the development of false septa in the ovary, such as in *Diospyros* carpels are 4 – 5, the ovary is 8 – 10 loculed.

Fruit : A bery with one or many seeds.

Seed: Seed with straight embryo and very hard, copious endosperms.

Floral Formula: $\oplus$ ♂ or ⚥ $K_{(4-5)}, C_{(4-5)}, A_{4-5}, G_{\underline{(2-16)}}$

Secretory cells containing tannin–like substance occur in the tissues of leaves and stems in almost all genera. Various kinds, of hairs are present on the leaves.

Differences from Sapotaceae: The Ebenaceae is distinguished from the Sapotaceae by the absence of milky sap, the unisexual flowers, the multilocular superior ovary, and by the ovules typically in pairs and always with two integuments.

Economic Importance of the Family

The family is of much economic importance.

1. **Timber.** All species of *Diospyros* yield useful timbers. The best known species is *D.ebenum* (Eng. ebony; Verna. **Abnus**) that yields true ebony. True ebony is the black heartwood of *D. ebenum* obtained by stripping off the peripheral light coloured wood of felled trees. *D. tomentosa* and *D. melanoxylon* are also ebony yielding plants. The heartwood of most species is, however, only streaked or mottled with black, and of these variegated ebonies the best known are *D. marmorate* and *D. quaesita.*

The heartwoods of ebony trees are extremely hard and tough. They are fancy and ornamental woods and have long been esteemed for carving, turning, musical instruments, umbrella handles, walking sticks, ornamental cabinets, furniture making, etc.

D. virginianum is widely used for weaving shuttles and is the standard wood for the heads of golf clubs.

2. **Edible fruits:** *Diospyros kaki* (Eng. Japanese Persimmon; Verna. **Halwa tendu**). A large tree; the fruits are orange colored berries 3" in diameter, and are eaten fresh or dried.

3. **Bidi industry:** The leaves of *Diospyros melanoxylon* (**tendu**) are used for wrapping **bidis.** Trees are commonly found in Madhya Pradesh forests.

Systematic Relationships. Ebenaceae is very much allied to Sapotaceae. Engler and Diels treated the order as composed of two sub–orders: 1. Sapotineae, and 2. Diospyrineae, and seven families of which two belong to Sapotineae, they are Sapotaceae and Hoplestigmataceae; and five belong to Diospyrineae, they are Ebenaceae, Diclidantheraceae, Symplocaceae, Styracaceae and Lissocarpaceae. According to Hallier this order is not a natural taxon. He has accepted each of the suborders as orders, the first as the Sapotales and the second as Santalales to which he added also the Loranthaceae, Olacaceae and Alangiaceae. He traced the origin of both the orders Sapotales and Santalales from the family Linaceae. Bessey put the order Ebenales, Ericales and Primulales very close to each other having relationship between them. According to him the Geraniales gave rise to both Ebenales and Ericales on parallel line. Hutchinson (1959) has placed the order Ebenales near Rhamnales from which he traced the origin. Again Hutchinson (1969) has placed the Ebenales next to Myrsinales. Rendle divided the order Ebenales into two suborders, *i.e.*, Sapotineae (single family Sapotaceae) and Diospyrineae (three families–Ebenaceae, Styracaceae and Symplocaceae). The division of Ebenales into two suborders suggests two divergent lines of evolution. However, Copeland (1938) supported the origin of Ebenales from Theales.

30

ORDER–RUTALES

According to Hutchinson this is the forty seventh order of the phylum Angiospermae, subphylum Dicotyledones and division Lignosae. This order comprises of four families, *i.e.*, Rutaceae, Simaroubaceae, Burseraceae and Averrhoaceae, of which the Rutaceae has been described in the present text in detail. Bentham and Hooker placed the family Rutaceae along with other ten families including Lineae and Meliaceae in their seventh order–the Geraniales of class Dicotyledones, division Polypetalae and series Disciflorae. Engler and Prantl have also placed the Rutaceae along with nineteen other families including Linaceae and Meliaceae in the twenty third order–the Geraniales of class Dicotyledoneae and Subclass Archichlamydeae.

Hypogynous to slightly perigynous; bisexual or unisexual; apocarpous to syncarpous with axile, basal or apical placentation; ovules mostly few; stamens definite in number, free; petals contorted to valvate; disc mostly conspicuous; leaves often gland-dotted; endosperm present.

Family—Rutaceae

There are about 140 genera and 1300 species in this family.

Distribution: They are widely distributed in warm temperate regions. They are also found in tropical regions. In our country the family is represented by several important genera such as, *Citrus, Aegle, Feronia, Murraya, Ruta,* etc.

Habit: Usually the plants are trees or shrubs. Herbs are rarely found. Some of the shrubs are climbing and xerophytic. *Aegle marmelos* (**Bel**) and *Feronia elephantum* (**Kaith**) are large sized trees. *Murraya exotica* is an ornamental shrub with scented white flowers. Sometimes the shrub becomes tree-like in habit. Species of *Citrus* are small trees. They are grown in the gardens for their fruits rich in vitamins. *Ruta graveolens* is a strong-smelling herb which is a common garden shrub. The plant of *Paraminya grifithii* is a woody climber. This possesses strong curved spines.

It is believed that *Citrus aurantium* (**Narangi**) probably originated in Northern India and Assam. It has been cultivated since 1500 B.C.

Root: Tap and branched.

Stem: Usually erect, sometimes climbing, branched, woody or rarely herbaceous, cylindrical, solid, green when young, grey when becomes old.

Leaves: The leaves may be either simple or compound. They are usually alternate and rarely opposite. They are exstipulate. The leaves are dotted with glands. These glands contain volatile oils, which give typical smell to the leaves. In most of the *Citrus* species the leaves possess winged petioles. According to many authors the leaves with winged petioles are regarded as unifoliate palmate type of compound leaves. In early stages actually the leaf possesses three leaflets but on its maturity two of them fall off. In certain Australian genera such as *Citrus, Feronia,* etc., the leaves are reduced to spines. Such plants are extreme xerophytes.

Inflorescence: The inflorescence consists of several variations in different species. Generally it is cymose, but sometimes it is racemose. In some cases the flowers are solitary axillary. In *Murraya exotica* the flowers are arranged in either axillary or terminal corymbs. In *Citrus* sp., the flowers are either solitary axillary or arranged in racemes.

Flower: The flowers are pedicellate, complete, hermaphrodite, regular or occasionally slightly zygomorphic, white or yellow, hypogynous. A disc is present beneath the ovary. Usually pentamerous and rarely tetramerous, or trimerous. Sometimes the flowers are unisexual, *e.g., Evodia, Zanthoxylum,* etc.

Calyx: There are usually five sepals. Sometimes only four sepals are found. Usually polysepalous but in the genera with zygomorphic flowers the calyx becomes either tubular or united. The aestivation is either imbricate or quincuncial.

Corolla: There are usually five petals. Sometimes only four petals are present. The corolla is polypetalous. However, in an Australian species *Correa speciosa,* the petals become united attaining campanulate shape. The petals are coloured–white, yellow or red. The petals are inferior. The aestivation is imbricate.

Androecium: Usually ten or eight stamens are present in obdiplostemonous condition. Sometimes the number of stamens decreases or increases to indefinite. In *Citrus* sp., and *Aegle marmelos* the stamens are indefinite. In such cases they are arranged in irregular bundles, *i.e.,* polyadelphous condition. In *Murraya exotica* there are ten stamens arranged in two whorls of five each.

One whorl of stamens is alternating and the other antipetalous. The filaments are free. The anthers are bilobed, basifixed and introrse.

Gynoecium: The gynoecium consists of 3, 4, or five carpels. Rarely it consists of a fewer number of carpels or indefinite carpels. The carpels are fused together, *i.e.,* syncarpous. In most of cases the placentaion is axile. The ovary is tri to multilocular. Each locule contains one or more anatropous ovules. In *Feronia,* however, the placentation is parietal. In this a large number of ovules are arranged in several rows on the parietal placentas.

Fruit: The fruits are of various kinds in different genera. They may be berry, drupe or capsule. In *Murraya exotica* the fruit is berry. The *Citrus* fruits are known as hespiridium (a berry). In *Acronychia,* the fruit may be capsule or drupe.

Seeds: The seeds are endospermic or non-endospermic. Each seed contains a large straight very often curved embryo.

Pollination : Usually entomophily, *i.e.,* by means of the agency of insects. The nectar secreting disc found beneath the ovary is main source of attraction for the insects. The coloured and scented flowers also attract the insects.

Floral Formula: ⊕ ⚥ K 5, or (5), C 5, A 10, G $\underline{(3-5)}$.

Bentham & Hooker (1862)	**Engler & Prantl (1931)**	**Hutchinson (1959)**
Phanerogams	Phanerogams	Angiospermae
Dicotyledones	Dicotyledones	Dicotyledones
Polypetalae	Archichlamydeae	Lignosae
Disciflorae	Geraniales	Rutales
Geraniales	Rutaceae	Rutaceae
Rutaceae		

Description of some important plants in semi-technical language:

1. ***Citrus aurantium.***

Habit: A small tree.

Root: Tap and branched.

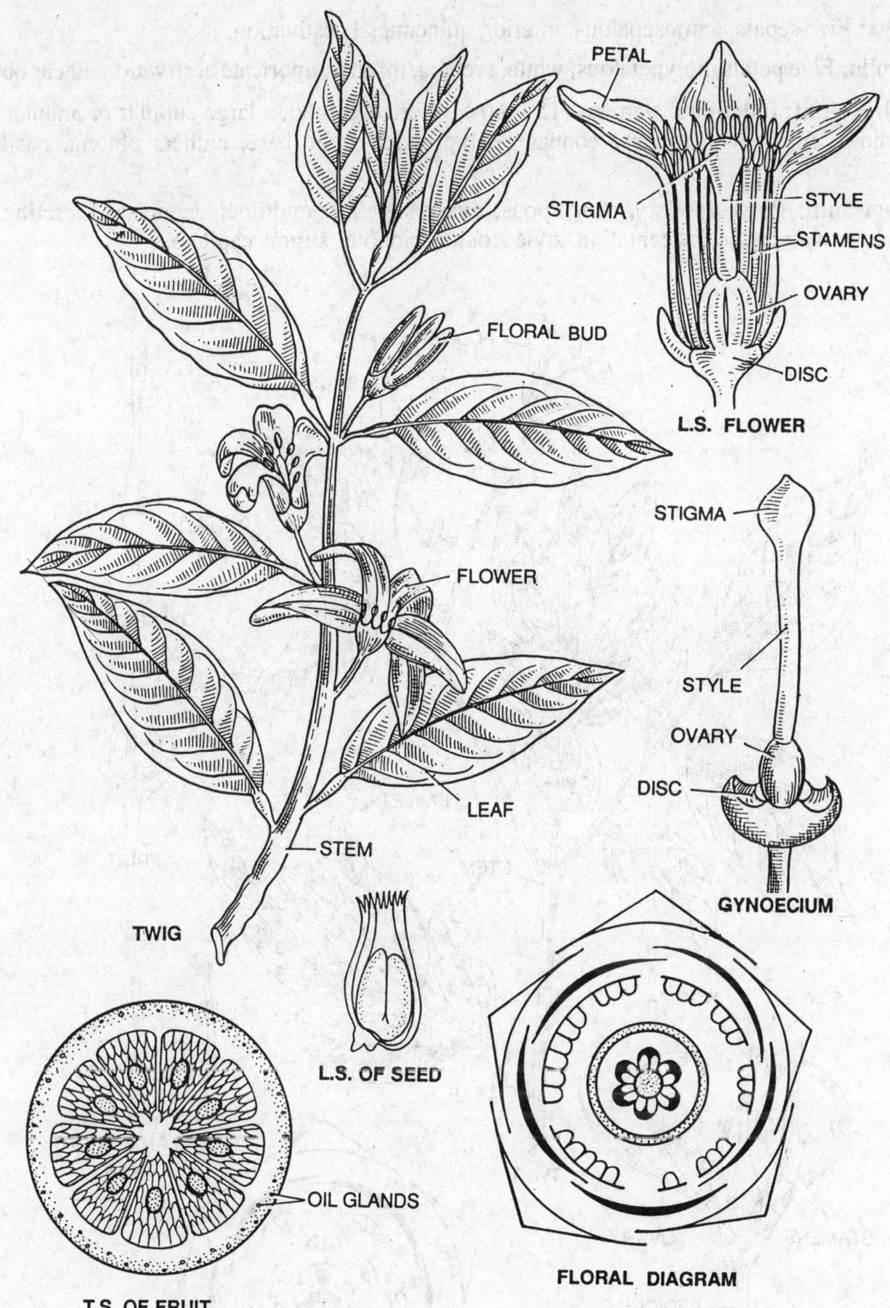

Fig. 30.1: Rutaceae. *Citrus aurantium* Linn., Eng. Orange; Verna. **Narangi.**

Stem: Erect, woody, branched, cylindrical, glabrous, solid.

Leaf: Cauline, alternate, petiolate (petiole winged and flattened), stipulate, coriacious, persistent, more or less serrate, acute or obtuse, glabrous, unicostate reticulate venation.

Inflorescence: Cymose, solitary axillary or in small cymes.

Flower: Pedicellate, bracteate, hermaphrodite, white or pinkish, scented, actinomorphic, complete, hypogynous.

Calyx: Five sepals, gamosepalous, inferior, quincuncial aestivation.

Corolla: Five petals, polypetalous, white, scented, inferior, imbricate aestivation, linear oblong.

Androecium: Indefinite, stamens (15 – 60), inserted around, a large cupular or annular disc, polyadelphous, filaments variously connate, compressed at the base, anthers oblong, basifixed, introrse.

Gynoecium: Polycarpellary, syncarpous, ovary superior, multilocular, nectar secreting disc present below ovary, axile, placentation, style stout, deciduous, stigma capitate.

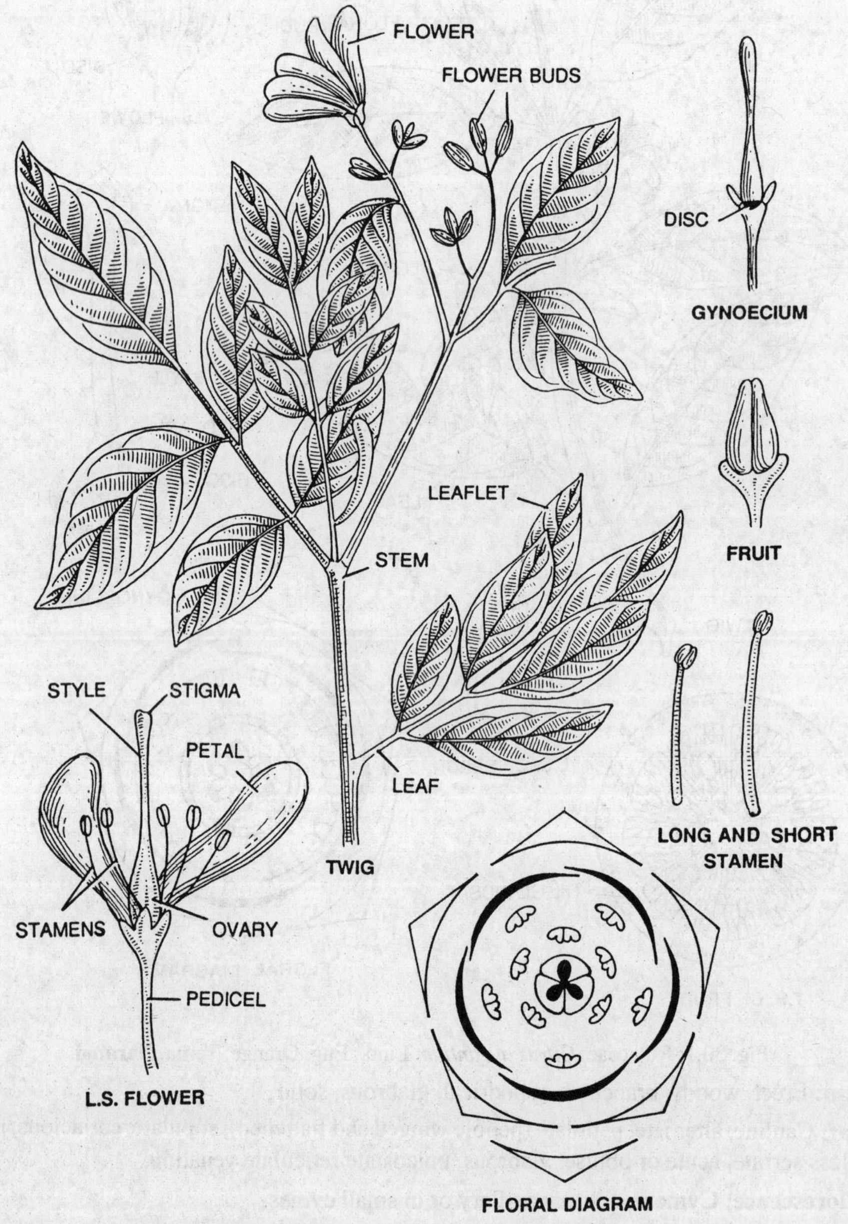

Fig. 30.2. Rutaceae. *Murraya exotica* Spreng.

Fruit: Berry (hesperidium).

Floral Formula: ⊕ ⚥ K (5), C 5, A ∞ , G ($\underline{\infty}$)

2. ***Murraya exotica*** ; Verna–**Kamini**

Habit: Evergreen large shrub.

Root: Tap, branched.

Stem: Erect, branched, cylindrical, woody, green, solid.

Leaf: Compound, pinnate, exstipulate, alternate, leaflets 3 – 7, oval, entire, smooth, coriaceous, alternate, petioled, unicostate reticulate venation, notched at the tip, dark green, shining above, oblique at the base.

Inflorescence: Cymose, short terminal and axillary corymbocymes.

Flower: Pedicellate; ebracteate, pentamerous, hermaphrodite, complete, hypogynous, white, very fragrant, ½" diameter.

Calyx: Five sepals, five-fid or five partite, gamosepalous, minute, acute, inferior, quincuncial aestivation.

Corolla: Five petals, polypetalous, white, fragrant, oblong lanceolate, spreading above, inferior, imbricate aestivation.

Androecium: Ten stamens, inserted round an elongated disc, alternate filaments shorter, obdiplostemonous, anthers bi-celled, introrse, polyandrous.

Gynoecium: Two or three carpels, syncarpous, ovary, superior, two or three locular, axile placentation, two ovules in each chamber, style filiform, disk present below ovary.

Fruit: Ovoid berry, orange red when ripe.

Floral Formula: ⊕ ⚥ K (5), C 5, A 5 + 5, G ($\underline{2-3}$).

3. ***Aegle marmelos*** Corr. Verna. **Bel**; Eng. Bael, Bengal quince.

Habit: Large spinous tree.

Root: Tap and branched.

Stem: Erect, aerial, branched woody, cylinderical, solid, strong thorns present.

Leaf: Palmately compound, trifoliate, exstipulate, petiolate, alternate, gland dotted, each leaflet ovate-lanceolate, acute, entire, glabrous, net-veined, unicostate reticulate, lateral sessile, terminal stalked, gland dotted.

Inflorescence: Cymose, axillary panicles.

Flower: Pedicellate, bracteate, hermaphrodite, actinomorphic, complete, hypogynous, yellowish white, disc present beneath ovary, pentamerous, sweet scented, cyclic.

Calyx: 5 sepals, polysepalous, valvate or imbricate, pubescent, deciduous.

Corolla: 5 petals, polypetalous, imbricate, yellowish white.

Androecium: Indefinite stamens, polyadelphous, polyandrous, filaments short, fascicled, basi-fixed, anther dithecous, introrse, elongate.

Gynoecium: 10, indefinite carpels, syncarpous, axile placentation, ovary superior, multilocular, each locule contains one or more ovules, style short and stout stigma capitate, sticky.

Fruit: Berry, large, globose oblong or pyriform, rind woody, grey or yellow, pulp orange-coloured, sweet.

Floral formula: Br. ⊕ ⚥ K 5, C 5, A ∞, G ($\underline{\infty}$)

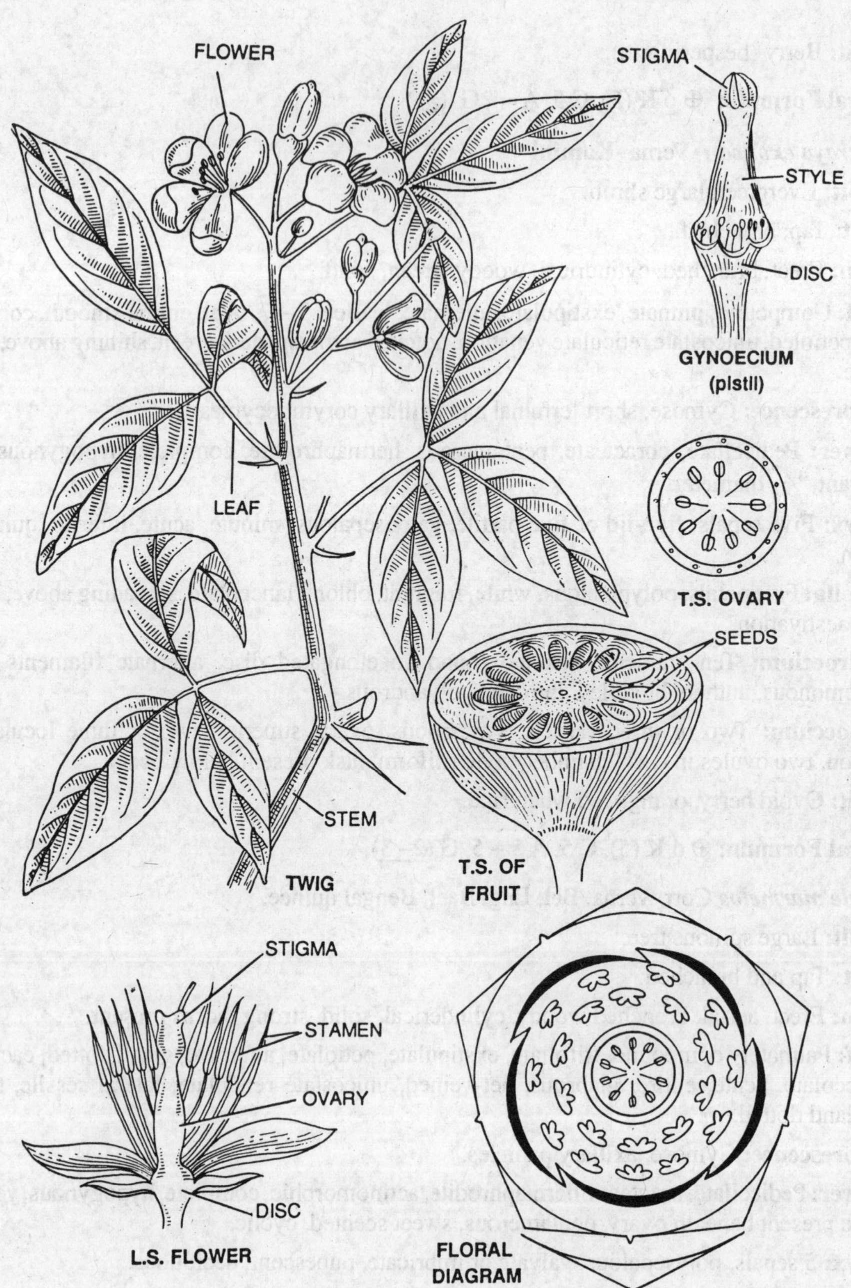

Fig. 30.3. Rutaceae. *Aegle marmelos* Corr.; Eng. Bengal quince; Verna., **bell.**

Economic value. The ripe fruits are edible. The unripe fruits are used as an astringent, stomachic, and also in the treatment of diarrhoea. The mucilaginous substance, secreted round the seeds, is used as a cement and is also employed as a varnish.

Economic Importance of the Family

The family is fairly important from economic viewpoint as it consists of many fruit yielding and medicinal plants. Some important plants are enlisted below:

1. *Citrus aurantifolia* ; Syn. *Limonia aurantifolia; Limonia acidisima; Citrus lima; C. medica var. acida;* (Eng.–Lime; Verna.–**Kaghzi-nimbu**). A shrub or small tree, cultivated nearly all over the country. The oil distilled from the peel is mainly used in confectionery, pharmaceuticals and toilet preparations. The fruits are used for making jam, jellies, pickles and alcoholic drinks. It is a good source of vitamin C.

2. *Citrus aurantium;* (Eng.–Sour orange; Verna.–**Khatta**)–Not grown commercially anywhere in India but found scattered along with other *Citrus*. The fruits are edible. The leaves are the source of an essential oil, which is used in the preparation of cosmetics, perfumes and confectionery. Rich in vitamin C.

3. *Citrus limettioides; C. medica* var. *limetta;* (Eng.–sweet lime; Verna.–**Mitha-nimbu**)–They are mainly cultivated in Northern India and Madhya Pradesh for their edible frutis. They are also used as root stocks for sweet oranges. An essential oil is extracted from the leaves by means of distillation. This oil is known as 'Petitgrain oil' which is used in perfumery, confectionery and cosmetics for flavouring purposes. Rich in vitamin C.

4. *Citrus limon;* Syn. *C. medica* Var. *limonum, C. limonum.* (Eng.–Lemon, Verna–**Baranimbu**)–They are grown commonly in the Punjab and Uttar Pradesh by their popular name 'galgal'. In Maharashtra and Deccan the fruits are known as 'Pat nimbu'. The fruits are employed in the preparation of pickles and lemon squash. Good source of vitamin C.

5. *Citrus maxima ;* (Eng.–Shadlock, Verna.–**Chakotra**)–Grown in Tamil Nadu, Karnataka, the Punjab and Uttar Pradesh. The fruits are edible. This is a good source of vitamin C.

6. *Citrus medica*; (Eng.–Citron, Verna.–**Bara nimbu**)–They are grown on the Khasia hills of Assam and sub-mountainous Himalayan ranges in Garhwal, Kumaon in the Uttar Pradesh. They are also found in the Punjab and South India. The fruits are edible and rich in vitamin C.

7. *Citrus paradisi;* Syn. *C. racemosa*; (Eng.–Grape fruit)–They are commonly grown in the Punjab, Uttar Pradesh, Madhya Pradesh and Bengal. The fruits are edible and rich in Vitamin C.

8. *Citrus reticulata;* Syn. *C. nobilis* (Eng.–Mandarin ; Verna–**Santara**)–They are grown in Khasia and Darjeeling hills. They are also cultivated in the Punjab, Uttar Pradesh, Madhya Pradesh, Maharashtra and South India. The fruits are edible and rich in vitamin C. The essential oil is extracted by means of distillation which is used in the manufacture of confectionery, perfumery and cosmetics, toilet products and pharmaceutical preparations.

9. *Citrus sinensis*; (Eng.–Sweet orange; Verna– **Musambi**)–They are grown for their edible fruits in South India, Maharashtra, the Punjab, Uttar Pradesh, Rajasthan and Tamil Nadu. Orange oil is extracted from the fruit peel.

10. *Clausena dentata*; A small tree found in the Western Ghats and Assam. The fruits are edible.

11. *Aegle marmelos*; (Verna–**Bel**) – A tall tree. The fruits are edible when ripe. The unripe fruits are used as an astringent, stomachic and in the treatment of diarrhoea. The medicines of stomach troubles are prepared from the pulp of ripe fruits. The mucilaginous substance secreted around the seeds is used as a cement and is used as a varnish. The pulp of the fruit is a good laxative. The pulp is often used to strengthen mortar. The seeds yield a dye. The leaves are used as fodder. The leaves are employed in the worship of Lord Shiva by Hindu ladies.

12. *Feronia limonia;* Syn. *F. elephantum;* (Eng.–Elephant apple; Verna–**Kaith**) – A tree found all over India. The fruits are with hard rind, large, globose and edible. The pulp of the fruits is eaten in many ways. It is also used in the treatment of diarrhoea and dysentery. The leaves are used as fodder. A gum is obtained from the trunk and branches of the tree.

13. *Fortunella japonica*; Syn. *Citrus japonica*; (Eng.–Kumquat)–This is a shrub or a small tree. It is native of East Asia, but now grown mainly in Northern India. The fruits are edible. The shrub is ornamental.

14. *Fortunella margarita;* Syn., *Citrus margarita*–This is an ornamental shrub possessing edible fruits.

15. *Acronychia pedunculata;* A small tree. The leaves yield an essential oil. The tender leave are used as condiment. It is found in Dehraduan, West Bengal and Khasia hills of Assam.

16. *Glycosmis pentaphylla;* (Verna-**Ban-nimbu**)–The ripe fruits are edible. The branches are used as tooth-brushes.

17. *Luvunga scandens;* (Verna–**Long–phul**)–A large, thorny climbing ornamental shrub, with beautiful scented white flowers. They are found in Bengal and Assam. The fruits yield an essential oi which is used in perfumery and pharmaceuticals.

18. *Murraya paniculata;* Syn. *Murraya exotica;* (Verna–**Kamini**)–This is an ornamental shrub possessing white scented flowers. The fruits are edible.

19. *Murraya koenigii*; (Verna–**Mitha neem**)– A large shrub or small tree. Cultivated in both Northern and South India. The leaves are used for flavouring curries.

20. *Ruta graveolens;* (Verna.–**Sadab**)–A herb, cultivated throughout country. The leaves are used as condiment and for flavouring curries. It also possesses medicinal properties.

21. *Toddalia asiatica;* (Verna.–**Kanj**)–A shrub. It is found in sub-tropical Himalayas. The roo yields a yellow dye. The leaves and fruits are edible. The pickle is prepared from the fruits.

22. *Zanthoxylum alatum;* (Verna.–**Tejbal**)– A small tree found in sub-tropical Himalayas. The seeds are used as tonic. The walking sticks are made from its wood. The branches are used as tooth brushes.

23. *Fagara budrunga*; Syn. *Zanthoxylum rhesta;* (Verna–**Badrang**)–A large tree. It is found in Western Ghats and West Bengal. An aromatic oil is extracted from its seeds. The fruits are used as spice. The wood is used in making shuttles.

24. *Fagara oxyphylla;* (Verna–**Timar**)–A shrub found in the hills of Kumaon and Assam. The tender shoots are used as vegetable.

Systematic Relationships: Hutchinson has placed the family Rutaceae along with three other families in the order–Rutales. Earlier this view was put forth by Rendle and accepted by Wettstein in principle.

31

ORDER–MELIALES

According to Hutchinson this is the 48th order of the phylum Angiospermae; Subphylum Dicotyledones and division Lignosae. The order consists of a single family–the Meliaceae. The family has been described here in detail. Bentham and Hooker included the family along with ten other families including Lineae and Rutaceae in their seventh order–the Geraniales of class Dicotyledones, division Polypetalae and series Disciflorae. Engler and Prantl have included the Meliaceae along with nineteen other families including Linaceae and Rutaceae in the twenty third order–the Geraniales of class Dicotyledoneae and subclass Archichlamydeae.

They are mainly tropical; trees, shrubs or climbers, rarely herbs; leaves simple or compound ,rarely stipulate; flowers hypogynous to slightly perigynous, mostly hermaphrodite; sepals generally imbricate; petals contorted to valvate, free or connate near the base; disc present; stamens often completely connate into a tube with the anthers inside; ovary superior, syncarpous; styles free or connate; seeds endospermic or non-endospermic; embryo straight or curved.

Family—Meliaceae

This is a small family consisting of about 50 genera and 800 species. The members of this family are found to be distributed in tropical regions. In our country the family is represented by *Azadirachta indica* and some other plants.

Habit: The plants are mostly trees or shrubs. The wood of this family emits a characteristic smell.

Root: Tap and branched.

Ste : Erect, branched, solid, woody with characteristic smell.

Leaves: Usually the leaves are alternate, petiolate, pinnately compound or decompound, rarely simplc and cxstipulatc.

Inflorescence: It is of cymose type. In many cases the axillary cymose panicles are present, *e.g.*, *Azadirachta indica.*

Flowers : The flowers are actinomorphic, hermaphrodite, rarely unisexual, regular, complete, rarely incomplete and hypogynous.

Calyx: It consists of 4 or 5 sepals, poly or gamosepalous. When gamosepalous, the sepals are connate at the base. The sepals are small in size. Mostly the aestivation is imbricate and rarely valvate.

Corolla: It consists of 4 or 5 petals, rarely 3 – 8 petals are also found, polypetalous, rarely they are connate or adnate to the staminal tube. The aestivation is contorted, valvate or imbricate.

Androecium: Usually it consists of 8 – 10 stamens. In rare cases the number of stamens is 5 or indefinite. Generally the stamens are united and form a long or short staminal tube, *i.e.,* monadelphous condition. In *Cedrela*, the stamens are distinct. Sometimes the filaments become united, in such cases the anthers are sessile on the staminal tube. Generally a disc is found in between ovary and stamens. The anthers are bilobed. The anthers dehisce by longitudinal slits.

Gynoecium: It consists of 2 to 5 carpels, syncarpous. The ovary is superior; it is 2 – 5 loculed one, two or more ovules in each loculus; the placentation is axile. The ovules are pendulous and anatropous. The style is simple with capitate or lobed stigma.

Fruit: Berry, capsule or rarely drupe.

Seeds: The seeds are albuminous or exalbuminous and winged.

Pollination: Anemophily and entomophily.

Floral Formula: ⊕ ⚥ K 4 – 5 or (4 – 5), C 4 – 5, A (8 – 10), G $(\underline{2-5})$.

Bentham & Hooker (1862)	**Engler & Prantl (1931)**	**Hutchinson (1959)**
Phanerogams	Phanerogams	Angiospermae
Dicotyledones	Dicotyledones	Dicotyledones
Polypetalae	Archichlamydeae	Lignosae
Disciflorae	Geraniales	Meliales
Geraniales	Meliaceae	Meliaceae
Meliaceae		

*1. **Azadirachta indica*** Juss.; Syn. *Melia azadirachta;* Eng. Margosa tree; Verna. **Neem.** The Neem tree is found to be distributed in tropical regions. Common shade giving tree of N. India with antiseptic medicinal properties.

Habit: A large tree with blackish bark.

Root: Tap and branched.

Stem: Erect, aerial, branched, solid, woody with characteristic smell.

Leaves: Cauline and ramal, alternate, petiolate, compound, bipinnate and imparipinnate, unicostate reticulate, leaflet coarsely serrate.

Inflorescence: Axillary cymose panicle.

Flower: Bracteate, bracteolate, pedicellate, complete, actinomorphic, hermaphrodite, pentamerous, hypogynous, cyclic.

Calyx: Sepals 5, polysepalous, sepals connate at the base, small, aestivation valvate or imbricate.

Corolla: Petals 5, polypetalous, aestivation imbricate.

Androecium: Stamens 10, monadelphous, forming a cylindrical staminal tube, anthers dithecous, introrse, basifixed.

Gynoecium: Carpels 5–7, syncarpous, ovary superior, 5 – 7, loculed one or two ovules in each locule, placentation axile, ovule pendulous and anatropous, style long and slender, capitate stigma.

Fruit: Drupe.

Floral formula : Br Brl. ⊕ ⚥ K 5, C 5, A (10), G $(\underline{5-7})$.

2. ***Melia azedarach*** Linn.; Eng. Pride of India; Verna. **Bakain.**

Habit : A tree

Root : Tap, branched.

Stem : Erect, aerial, woody, solid, cylinderical, branched, smooth and green.

Leaf : Cauline and ramal, exstipulate, leaf base pulvinus, alternate, compound, bipinnate imparipinnate, pinnae ovate-lanceolate, serrate, acute, oblique, glabrous, pinna unicostate reticulate.

Inflorescenece : Cymose, axillary panicle cyme.

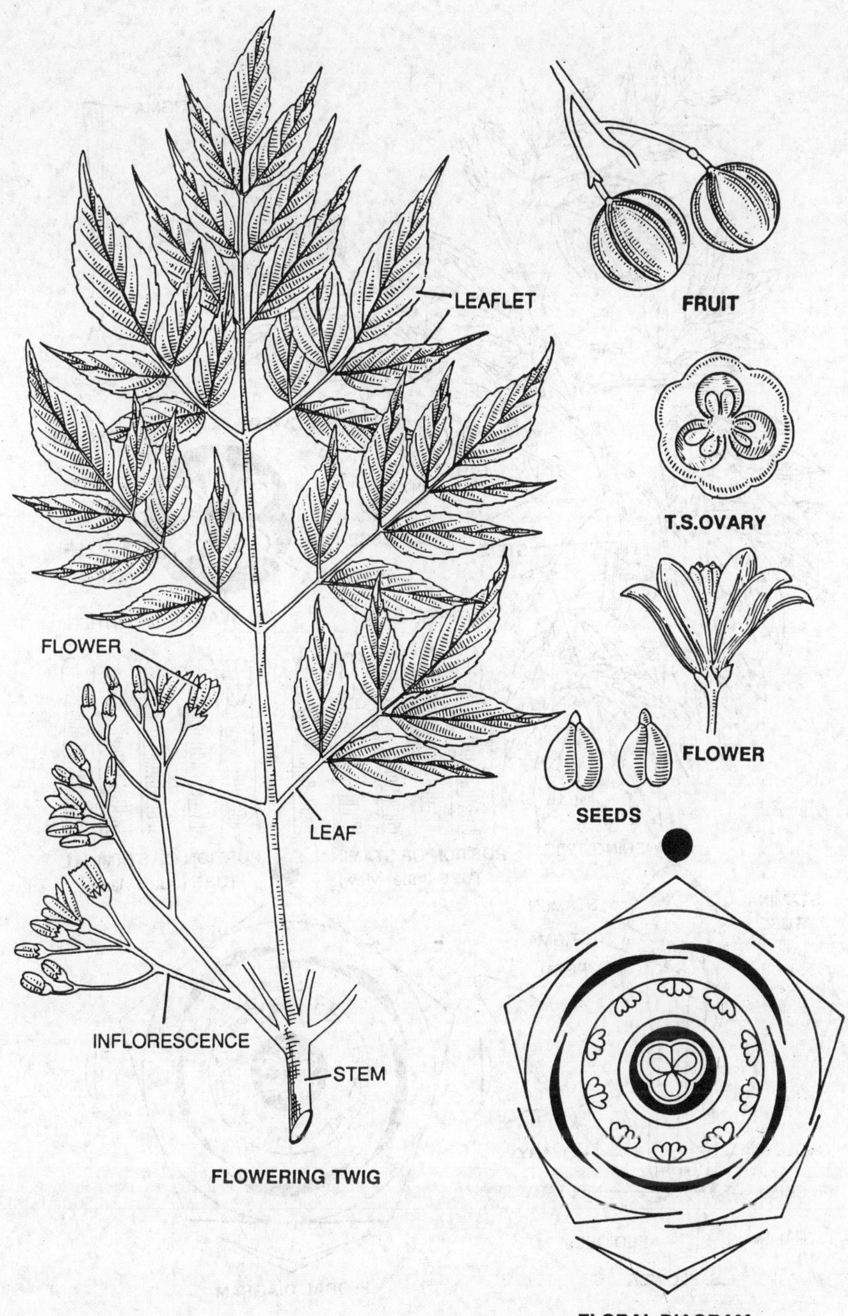

Fig. 31.1 : Meliaceae. *Azadirachta indica* Juss; Eng. Margosa tree; Verna, **neem.**

Economic Value. All parts of the plant used medicinally. The leaves are placed in suit–cases to repel insects and to preserve woollens. An extract of the leaves is used in tooth-pastes and soaps. The seed-oil is used as an antiseptic and for burning purposes. The oilcake, obtained from the seeds, is used as a fertilizer and manure.

Flower : Pedicellate, bracteate, bracteolate, actinomorphic, hermaphrodite, complete, pentamerous, hypogynous, cyclic.

Calyx : 5 sepals, polysepalous, connate at base, valvate, acute, hairy, green.

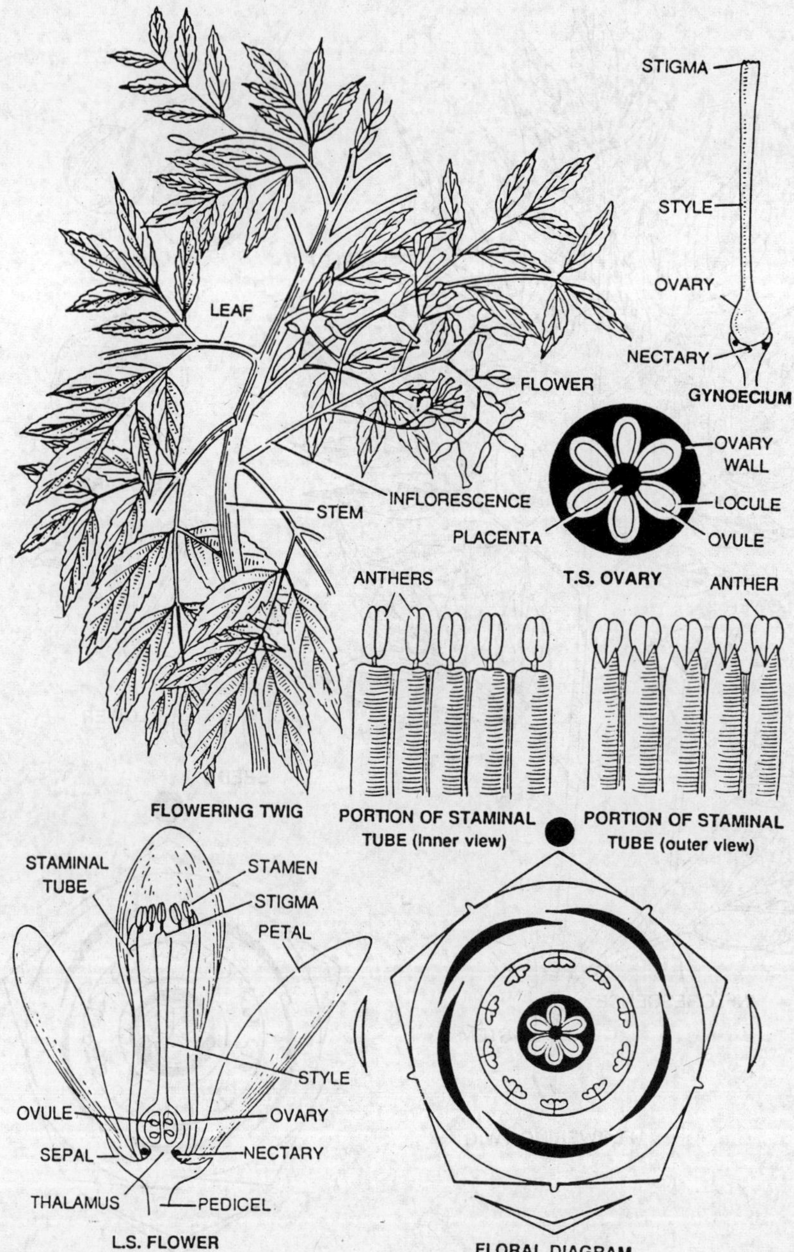

Fig. 31.2. Meliaceae. *Melia azedarach* Linn.; Eng. Pride of India; Verna. **Bakain.**

Economic Value. Root bark, fruit flowers and leaves are used medicinally. Flowers and leaves are applied as poultice to relieve nervous headaches. Juice of leaves is used internally as anthelmintic, diuretic and emmenagogue. Seeds are prescribed in rheumatism. Oil is used as antiseptic. Fig. 31.2)

Corolla : 5 petals, polypetalus, imbricate or quincuncial aestivation, purple white.

Androecium : 10, monadelphous, a staminal tube encloses the ovary, tube cylindric dialated at apex and base, ten-toothed at apex, anthers inserted near apex, anothers inserted near apex, dithecous (bicelled), basifixed, introse.

Gynoecium : 5-8 capels, syncarpus, ovary superior, 5-8 loculed, one to two ovules in each locule, axile placentation, style slender and long, stigma capitate, 5-lobed, nectariferous disc present beneath the ovary.

Fruit : A drupe.

Floral formula : Br. Brl. ⊕ ⚥ K 5, C 5, A (10), G ($\underline{2}$ – $\underline{8}$)

Economic Importance of the Family

The family is of little economic value. Some plants yield timber while some are of medicinal value. A list of some important plants is given below:

1. *Azadirachta indica;* Syn. *Melia azadirachta;* Eng.–Margosa tree; Verna–**Neem**–This is a large tree, usually found in the upper Gangetic plains of our country. The nim oil is extracted from the pulp of the fruits, which is used in the manufacture of margosa soap and several skin ointments. The oilcake, obtained from the seeds, is used as a fertilizer and manure. Almost all the parts of the tree are of medicinal value. The leaves are placed in the suit cases to repel insects and to preserve woollens. Decoction of leaves is antiseptic and used to wash ulcers and wounds. An extract of the leaves is used in the manufacture of tooth pastes and soaps. The seed oil is used as an antiseptic. This oil is also burnt in the earthen lamps. Dry flowers are used as a tonic. The bark is used as antiseptic. The young branches are used as *dataun.* The gum, bark, leaves and seeds are used in snake-bite. The wood makes a good timber. The branches are burnt as fuel. In summers people take shelter under the shade of neem tree.

2. *Toona ciliata;* Syn. *Cedrela toona;* Eng.–Red cedar; Verna–**Toon**–This is a tree usually grown in the Punjab, Assam, Bihar, the western Ghats, Maharashtra, Madhya Pradesh and the Nilgiris. This is a road side tree. The flowers yield a yellowish red dye, which is used for dyeing cotton. The bark is used medicinally as astringent, tonic, etc. The wood is used for furniture, tea chests, shuttles and cigar boxes. The bark is also used as a tan. This is a good timber tree and sometimes known as 'Indian Mahogany'.

3. *Melia azedarach;* Eng.–Pride of India; Verna. **Bakain–**A small tree, grown as roadside or hedge plant. It is used medicinally; the juice of the leaves is used as an anthelmintic and the seeds in rheumatism.

4. *Swietenia mahagoni;* Eng.–West Indian Mahogany; Verna. – **Mahogani**–This is a tree valued for its timber. It is cultivated as a timber crop in Tamil Nadu; this is also grown as a hedge plant.

5. *Soymida febrifuge;* Eng.–Indian red-wood; Verna. **Rohan**–This is a tree, found in the Western Peninsula, Rajasthan and Bihar. A red fibre is obtained from the bark, which is used for making ropes. The bark is used as a tan. The bark is also used in medicines for the treatment of diarrhoea and dysentery. The reddish-brown wood is strong and durable and used as timber.

Systematic Relationships: Hallier included the family Meliaceae in his Terebinthales and he thought it to have been derived from Rutaceae. On the basis of the characters, that the leaves were not gland dotted and the stamens were connate into a tube, he segregated the single family Meliaceae into the order Meliales.

32

ORDER–SAPINDALES

According to Hutchinson this is the forty-ninth order of the phylum Angiospermae, subphylum Dicotyledones and division Lignosae. The order consists of eleven families. They are : 1. Melianthaceae, 2. Sapindaceae, 3. Podoaceae, 4. Sabiaceae, 5. Anacardiaceae, 6. Aceraceae, 7. Hippocastanaceae, 8. Staphyleaceae, 9. Akaniaceae, 10. Julianiaceae and 11. Didiereaceae of which the Anacardiaceae has been described in the present text in detail. Bentham and Hooker have placed the family Anacardiaceae along with other four families in their tenth order, the Sapindales of class Dicotyledones, division Polypetalae and series Disciflorae. Engler and Prantl have included the family Anacardiaceae along with twenty other families in their twenty fourth order, the Sapindales of class Dicotyledoneae and subclass Archichlamydeae.

Flowers often unisexual; petals mostly present, free; stamens more or less perigynous; ovules 1– 2 in each ovary loculus; endosperm mostly absent; embryo often curved or crumpled; trees or shrubs.

Family—Sapindaceae

There are about 158 genera and 2230 species in this family.

Distribution: The members of this family are mainly distributed in the tropical and subtropical parts of the world. In India, the family is represented by 24 genera and 72 species. The plants are commonly found in the North Eastern and North Western Himalayas. The common examples of this family are–*Aesculus indica* (horse chest nut), *Litchi chinensis* (litchi), *Sapindus mukorossi* (soap nut), *Acer* (sugar maple), *Nephelium longama, Cardiospermum helicacabum,* etc. *Aphania danura* is a small tree found in Sunderbans.

Habit. The plants are trees (*e.g., Litchi, Aesculus*), shrubs or climbers by tendrils (*e.g., Cardiospermum*).

Anomalous Secondary Growth in Stem : Anomalous secondary growth takes place in many climbing genera (*e.g., Serjania, Paullinia.* etc.). This type of anomalous thickening takes place as a result of unusual position of primary vascular bundles.

Leaves : Alternate rarely opposite (e.g., in *Acer*),usually compound, pinnate, sometimes simple (*e.g.,* in *Cardiospermum*), exstipulate, or stipules sometimes present in climbing species (*e.g.,* in *Melianthus*)

Inflorescence : Cymose or racemose. The small flowers are found to be arranged in racemose inflorescence or in paniculate cymes.

Flower : The flowers are abliquely zygomorphic or actinomorphic, hermaphrodite or unisexual, hypogynous, polygamous or polygamodioecious, pentamerous, rarely tetramerous. Unilateral, annular or glandular extrastaminal disc present. The flowers may be bracteate and bracteolate.

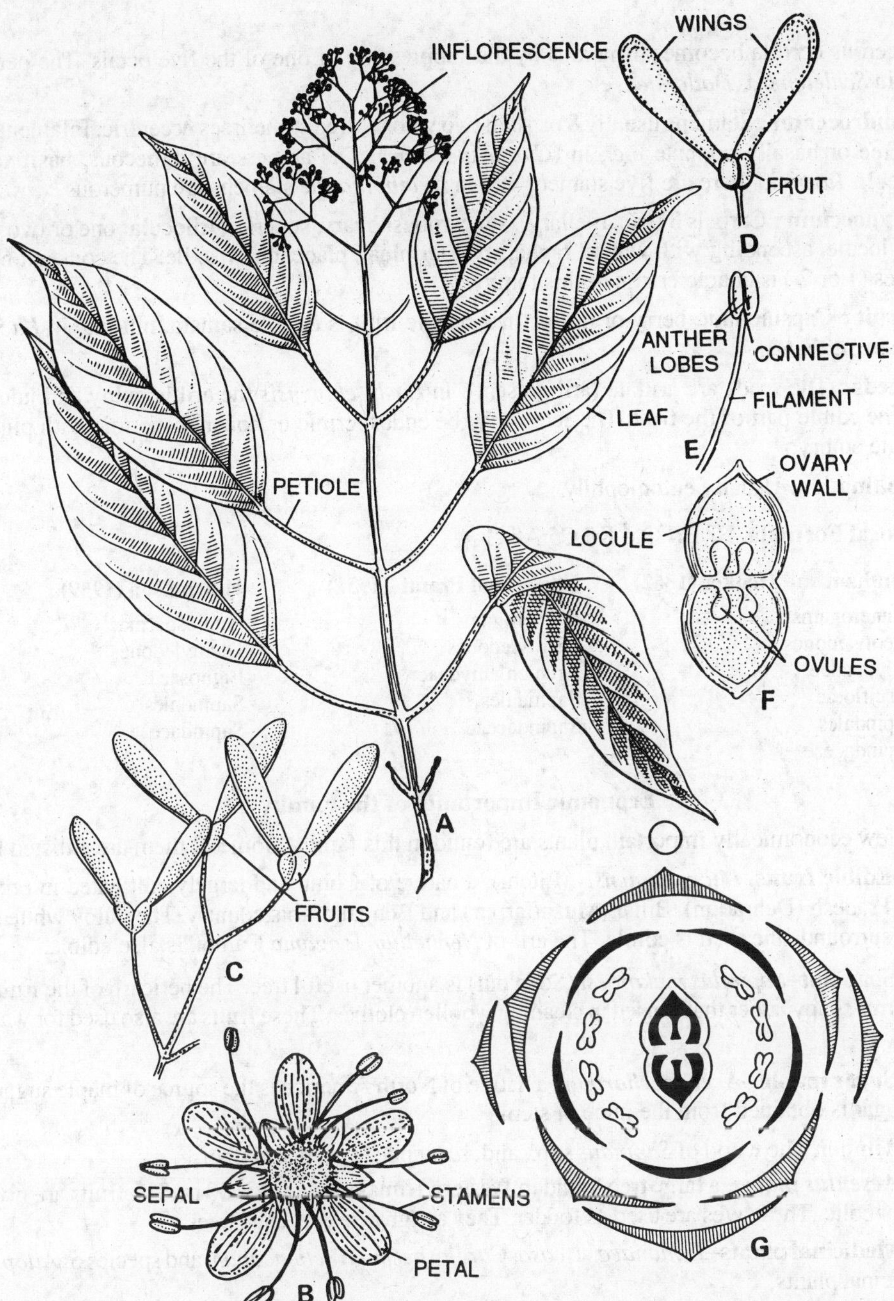

Fig. 32.1. Sapindaceae. *Acer oblongum*. A, flowering twig; B, flower top view; C, fruits; D, fruit with wings; E, stamen; F, T.S. of ovary; G, floral diagram.

The actionomorphic flowers are found in *Sapindus, Litchi* ,etc., whereas zygomorphic flowers with eccentric androecium are found in *Cardiospermum, Erioglossum* etc. Both unisexual and bisexual flowers are found in the same individual (polygamous) or bisexual and male or bisexual and female flowers (polygamo-dioecious) may occur in separate individuals.

Calyx : Sepals usually 5, free and valvate or imbricate, in actinomorphic flowers sepals become 4 by the fusion of 3rd and 5th sepals.

Corolla : Petals 4 or 5, free, provided with scaly or hairy appendages. In actinomorphic flowers the 5-merous corolla becomes 4-merous by the suppression of one of the five petals. The petals are absent in *Schleichera, Dodonaea*, etc.

Androecium : Stamens usually 8 or 10 in two whorls, free sometimes eccentric. Filaments often hairy, free or basally connate *e.g.*, in (*Cardiospermum*). The anthers are dithecous, basifixed and introrse. In *Turpinia* there are five stamens and in *Deinbollia* the stamens are numerous.

Gynoecium : Carpels 3 (tricarpellary), syncarpous, ovary, superior, trilocular one or two ovules in each locule, ascending with ventral raphe, style terminal, placentation axile. The constant number of ovules (1 or 2) is characteristic of this family.

Fruit : Capsule, nut, berry or drupe. In *Acer* the fruit is double samara, nut-like in *Litchi* and capsule in *Sapindus.*

Seeds : The seeds are arillate (aril present). In *Litchi chinensis* the aril is pulpy and juicy, and makes the edible part of the fruit. The seeds may be endospermic or nonendospermic with plicate or convolute embryo.

Pollination. Usually entomophily.

Floral Formula. Br Brl ⊕ ⚥ K5 C5 A8 $G_{(\underline{3})}$

Bentham and Hooker (1862)	Engler and Prantl (1931)	Hutchinson (1959)
Phanerogams	Phanerogams	Angiospermae
Dicotyledones	Dicotyledones	Dicotyledones
Polypetalae	Archichlamydeae	Lignosae
Disciflorae	Sapindales	Sapindales
Sapindales	Sapindaceae	Sapindaceae
Sapindaceae		

Economic Importance of the Family

A few economically important plants are found in this family. Some of them are enlisted here.

1. **Edible fruits.** *Litchi chinensis* (Litchi), a native of China and largely cultivated in orchards of Uttar Pradesh (Dehradun), Bihar (Muzaffarpur) and Bengal of this country. The pulpy white sweet aril that surrounds the seed is edible. The aril of *Nephelium longama* Camb., is also edible.

2. **Soap nut**–*Sapindus mukorossi* (Soap nut) is another useful tree. The pericarp of the fruit with water forms soapy lather that is used in cleaning woollen clothes. These fruits are also used for washing hair.

3. **Sugar maple** –*Acer saccharinum* a native of North America is the source of maple sugar. The maple sugar is obtained from the juice of stem.

4. **Timber.** The wood of *Sapindus* spp., and *Acer* spp., is used as timber.

5. ***Aesculus indica,*** a large tree found in Western Himalayas upto 3000 m. The fruits are given to horses for colic. The leaves are used as fodder. They are grown as avenue trees.

6. **Medicinal plants**–*Dodonaea viscosa, Cardiospermum helicacabum* and species of *Allophylus* are medicinal plants.

The family Sapindaceae because of the characteristic of unisexuality in flowers, presence of definite number of ovules with ventral raphe, tricarpellary ovary and arillate seeds has close affinity with family Euphorbiaceae. This family is related to Anacardiaceae in habit and floral characters, and therefore, they are grouped together in the order Sapindales (Hutchinson, 1959).

Family—Anacardiaceae

There are about 73 genera and 600 species in this family.

Distribution: The plants of this family are mainly found in tropical regions. Although a few members are found in Southern Europe. In our country the family is represented by very important genera such as *Mangifera indica* (mango–**Aam**), *Anacardium occidentale* (cashew nut–**Kaju**),

Anacardium occidentale though a native of Brazil but now cultivated in South India. *Pistacia vera* (Verna–**Pista**) is native of West Asia also cultivated in North India for its edible fruits.

Habit: They are mainly trees or shrubs. The resin canals are present in the wood.

Roots: Tap and branched.

Stem : The stem is erect, branched, woody, cylindrical and solid.

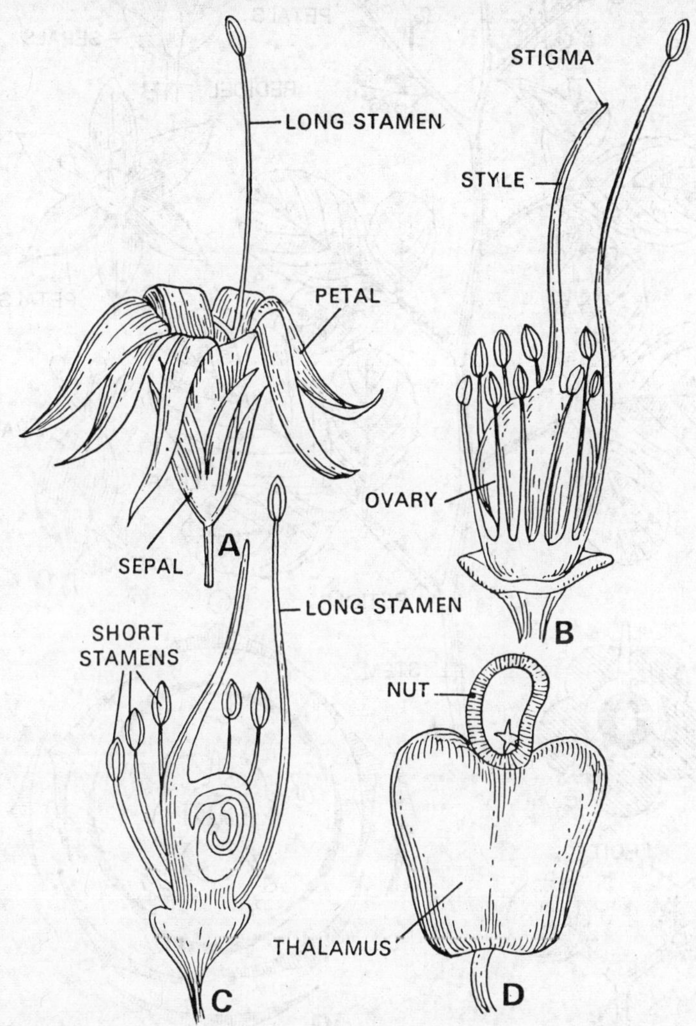

Fig. 32.2. Anacardiaceae, *Anacardium occidentale* ; A, flower; B, flower with petals removed; C, flower in vertical section; D, fruit in vertical section.

Leaves: The leaves may be simple, *e.g., Mangifera* sp., or compound (*e.g., Rhus, Spondias*, etc.). They are usually alternate.

Rarely the leaves are opposite (*e.g., Dobinea,* sp.). They are usually petiolate and exstipulate.

Inflorescence: The flowers are arranged in terminal or axillary panicles.

Flower: The flowers are hermaphrodite, but sometimes become unisexual by reduction of any of the sexes. They are usually regular (actinomorphic) and pentamerous. They may be hypo, peri or epigynous.

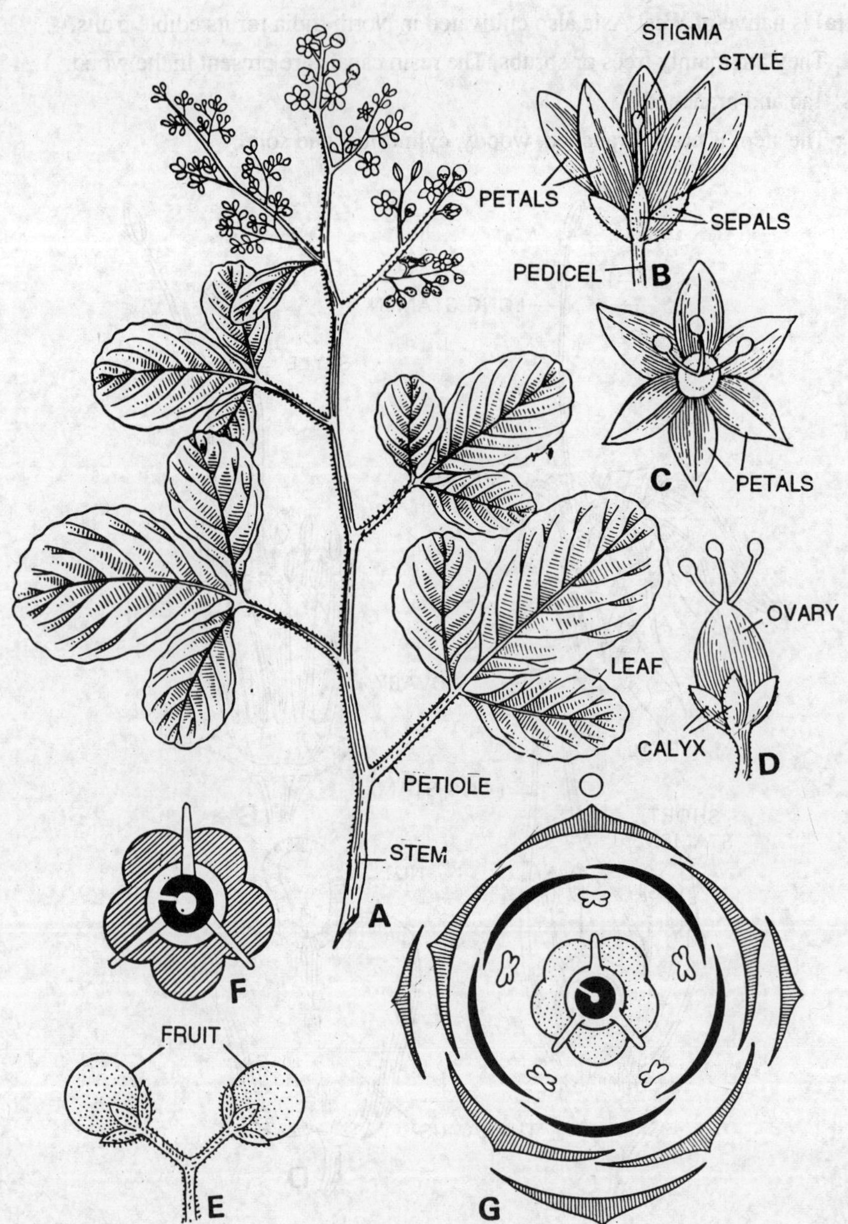

Fig. 32.3. Anacardiaceae. *Rhus parviflora*. A, flowering twig; B, flower lateral view; C, flower top view; D, gynoecium; E, fruits; F, T. S. of ovary; G, floral diagram.

Calyx: It consists of 3 to 5 sepals, gamosepalous. Sometimes the sepals are adnate to the ovary wall.

Corolla: It consists of 3 to 7 petals. Sometimes the corolla is altogether absent, polypetalous but sometimes united at the base.

Androecium: It consists of 10 stamens arranged in two whorls of five each. Very often the stamens are double the number of the petals. In *Buchanania* sp., 10 stamens are present in two whorls of five each. In *Anacardium* sp., however, one stamen is fertile and the remaining 6 to 9 stamens are represented by staminodes. In *Mangifera* sp., there are five stamens but sometimes only one of them remains functional. They are antisepalous in position. The stamens may be free or united at the base. They arise from the edge of an intrastaminal disc. The anthers are dithecous. They split by longitudinal slits.

Gynoecium: Usually it consists of three carpels (tricarpellary), syncarpous . Out of three carpels one remains functional. The ovary is superior (sometimes inferior) and contains a single, solitary, pendulous ovule. Almost in all cases only one ovule matures and becomes seed. In *Mangifera, Anacardium* sp., only one carpel is present. The placentation is basically axile. Each locule contains a single ovule. The style is short, one rarely more in number. The stigmas are as many as the number of carpels.

Fruit: The fruit is usually a drupe.

Seeds: The seed is non-endospermic or possesses very little endosperm. The embryo bears fleshy cotyledons.

Pollination: Entomophilous, *i.e.*, through the agency of insects.

Floral formula : ⊕ ⚥ K3—5, C3—7, A5+5, G($\underline{3}$)

Bentham & Hooker (1862)	**Engler & Prantl (1931)**	**Hutchinson (1959)**
Phanerogams	Phanerogams	Angiospermae
Dicotyledones	Dicotyledoneae	Dicotyledones
Polypetalae	Archichlamydeae	Lignosae
Disciflorae	Sapindales	Sapindales
Sapindales	Anacardiaceae	Anacardiaceae
Anacardiaceae		

Mangifera indica Linn. Verna. **Aam;** Eng. Mango tree.

Mangifera indica Linn. (Vern. **Aam;** Mango tree). A large evergreen tree with widely spreading branches and dark coloured bark.

Distribution : Tropical Himalayas from Garhwal to the Bhutan and Khasia hills, up to 3, 000 feet, Bihar and on the hills of Central and South India cultivated throughout the greater part of India, and in most tropical countries. Flowers during March and April, and the fruit ripens during the early part of the rainy season.

Habit: Medium or large tree; resin canals present in wood.

Root: Tap and branched.

Stem: Erect, aerial, branched, woody, cylindrical and solid.

Leaves: Simple, alternate, petiolate, exstipulate, acute, entire, unicostate reticulate, glabrous, thick or coriaceous, young leaves soft and purple.

Inflorescence: Flowers arranged in panciles.

Flower: Pedicellate, bracteate, hermaphrodite, actinomorphic, regular, complete, pentamerous, hypogynous, cyclic, yellowish white.

Calyx : 5 sepals, gamosepalous. sepals adnate to the ovary wall.

Corolla : 5 petals, polypetalous, quincuntial or imbricate aestivation.

Androecium: 5 stamens, one functional or perfect stamen, antisepalous, stamens arise from the edge of an intrastaminal disc; anthers dithecous, introrse, dehiscence by longitudinal slits, basifixed.

Gynoecium: 3 carpels (tricarpellary), syncarpous, only one carpel functional; ovary superior, placentation basically axile; each locule contains a single pendulous ovule; style short, stigma lobed.

Fruit: Drupe.

Seed: Exalbuminous, embryo bear fleshy cotyledons.

Floral formula : Br. Brl. ⊕ ⚥ K 5, C 5, A 1, G ($\underline{1}$).

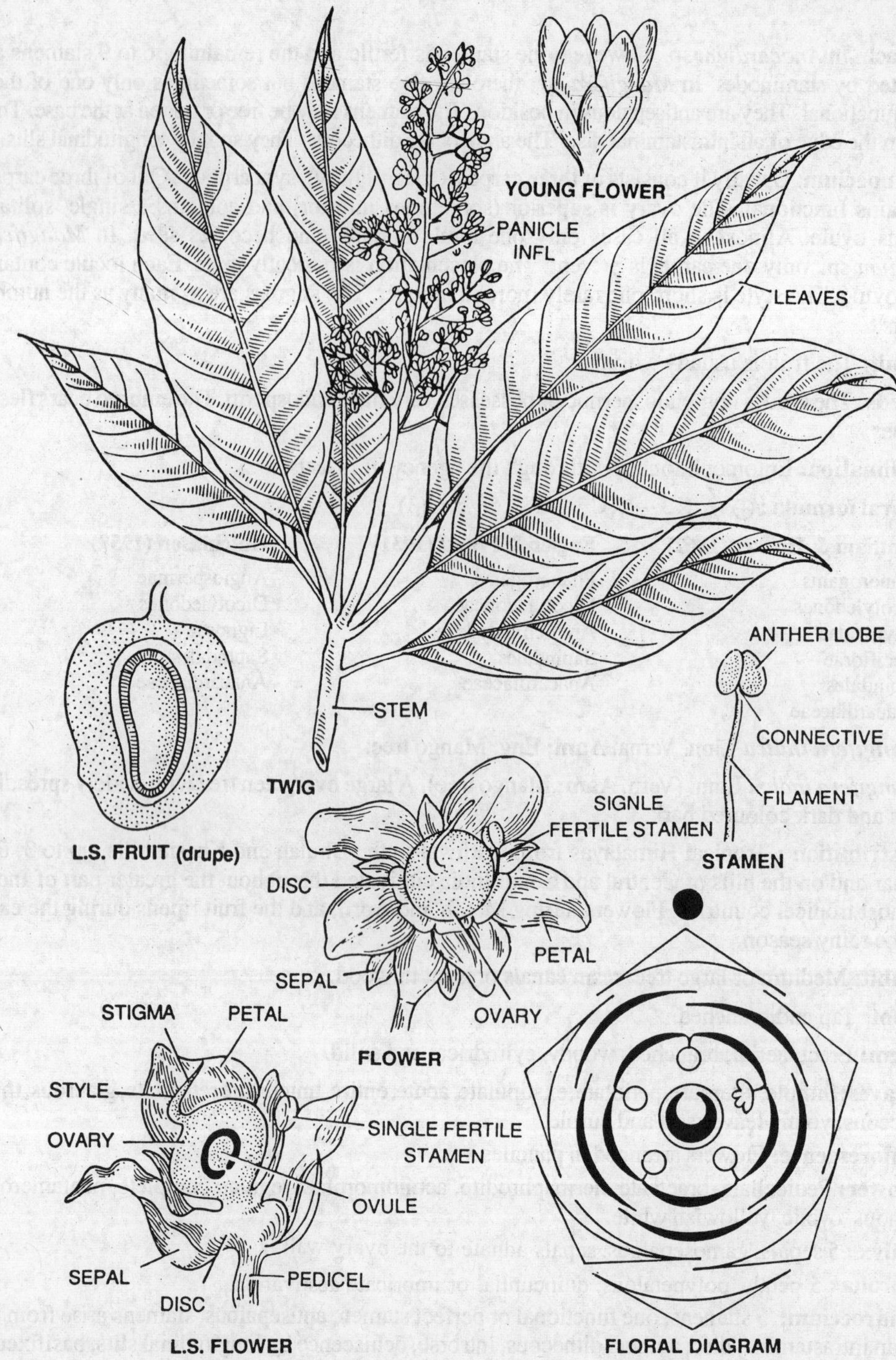

Fig. 32.4. Anacardiaceae. *Mangifera indica* Linn.; Eng. mango tree; Verna., **aam.**

Econimic value. The mango fruit is one of the most highly prized dessert fruits of the tropics. It has a rich, luscious, aromatic flavour and a delicious taste in which sweetness and acidity are delightfully blended. Ripe fruits constitute a rich source of vitamin A. Young and unripe fruits are usually acidic and used in pickles, chutney, *amchur* and culinary preparations.

Economic Importance of the Family

The family is fairly important from the economic viewpoint as it consists of several plants yielding valuable fruits. The economic value of some of very important plants is given here:

1. *Anacardium occidentale;* Eng.–Cashew nut; Verna.–**Kaju**–A small tree native of Brazil but now cultivated in Kerala, Malabar, Karnataka, Tamil Nadu and Andhra Pradesh. The thalamus known as the cashew apple and the true fruit known as the cashew nut are edible. The oil obtained from the shells is used in the manufacture of varnishes, typewriter rolls, industrial flooring tiles, gums, inks, oil, cloth, paints, water proof paper, acid and alkali resistant cements, cements, laminating resins and rubber-compounding resins. The oil is used for treating cracks on the soles of feet and also used for producing synthetic cation-exchange resin for use in the treatment of water required for boilers, air conditioning units, ice plants, textile industries, etc.

2. *Buchanania angustifolia* – Eng. Buchanan's mango, Verna–**Piyala**–A tree of Madhya Pradesh and Western Peninsula. The fruits are edible.

3. *Buchanania lancifolia*–Found along the Eastern coast of India and the Andaman Islands. The fruits are edible.

4. *Buchanania langan* ; Syn. *B. latifolia*; Eng. Cuddapah almond; Verna–**Chironji**–The bark is used for tanning. The fruits as well as seeds are edible and a gum, obtained from the stem, is employed for printing cloth and dyeing. Except the arid regions of N.W. India, it is commonly found throughout the country. The gum is used medicinally for diarrhoea.

5. *Mangifera indica*; Eng.–mango; Verna–**Aam**– A tree, grown extensively in U.P., the Punjab, Maharashtra, Andhra Pradesh, West Bengal and Tamil Nadu. The fruits are edible. The best varieties of mango are obtained by grafting.

6. *Pistasia khinjak* ; Syn. *P. integerrima;*Eng. – China turpentine tree: Verna–**Kakra**–A tree found in the N.W. Himalayas and the Punjab. The leaves are used as fodder and the galls and bark are used for tanning.

7. *Pistacia vera*; Eng. Pistachia; Verna.–**Pista**–Native of West Asia. The plants are cultivated in Northern India for its edible seeds.

8. *Lannea coromandelica*; Syn. *L. grandis;* Eng.*Odina wodier*; Verna–**Jhingan**–A tree commonly found throughout India. The young leaves are eaten with rice. The wood pulp is used for manufacturing paper and board. The gum, obtained by tapping the stem bark, is used in confectionery and as a mucilage for making ink. The boiled leaves are applied for local swellings and pains on body. The decoction of bark is used for toothache.

9. *Rhus chinensis;* Syn. *R. semialata*; Verna.–**Tatri**–A tree found in the Khasia and Naga hills. The fruits are edible.

10. *Rhus mysurensis;* Verna, – **Dasni**–Found in Rajasthan, the Punjab and Andhra Pradesh. The fruits are edible and the stem bark is used as a tan.

11. *Rhus punjabensis*; Verna–**Tatri**–Found in the North Western Himalayas from Kashmir to Kumaon. The fruits are edible and the leaves are used for tanning.

12. *Rhus succedanea;* Eng. Japanese wax tree; Verna–**Kakrasingi**–Found in the temperate Himalayas, Kashmir, Sikkim and the Khasia hills. A liquid obtained from the mesocarp of the fruit, is used in varnishes , ointments and furniture-polish. The galls, formed on the leaves, are used for tanning. The seeds yield a wax, which is used for candles.

13. *Rhus vernicifera*; Eng–Lacquer tree–Native of China, now found in Kumaon and Garhwal. A liquid, obtained from the bark, is used for varnishes. Also produces an oil which is used for candles.

14. *Semecarpus anacardium*; Syn. *Anacardium orientale;* Verna–**Bhilawa**–A tree, found in the Punjab, Assam, the Khasia hills, Madhya Pradesh and the Western Peninsula. A black resin, obtained

from the pericarp of the fruit, is used as marking ink to write on cloth, the seed oil is used for protection against white ants and for floor dressing. The nuts are used for tanning.

15. *Spondias cytherea*; Syn. *S. dulcis*, Eng.–Golden apple; Verna–**Amra**–A tree; the fruits are edible.

16. *Spondias mombin*; Syn. *S. lutea*; Eng. – Hogplum. Grown in S. India; the fruits are edible.

17. *Spondias pinnata*; Syn *S. mangifera*; *Mangifera pinnata;* Eng.–*Hogplum; Verna*–**Ambara, Amara** – cultivated in the Punjab, Maharashtra, Bengal and Assam for the edible fruits. The bark is used as a tan.

Systematic Relationships: Most modern taxonomists divide the order Sapindales into two orders, *i.e.,* the Sapindales and the Celastrales.

All the taxonomists include the family Anacardiaceae in the order Sapindales. According to Hallier the Amentiferous families, such as Juglandaceae, Fagaceae, Urticaceae, etc., have been derived from Anacardiaceae, and the Anacardiaceae itself has been derived from Rutaceae. Hutchinson has treated the Anacardiaceae as one of the more advanced members of the Sapindales.

33

ORDER–LOGANIALES

According to Hutchinson this is the 50th order of the phylum Angiospermae, subphylum Dicotyledones and division Lignosae. This order comprises of seven families, *i.e.* Potaliaceae, Loganiaceae, Buddleiaceae, Antoniaceae, Spigeliaceae, Strychnaceae and Oleaceae, of which Oleaceae has been described here in detail. Bentham and Hooker have placed the family Oleaceae in Gamopetalae, series Bicarpellatae and Order Gentianales along with other five families, *i.e.*, Salavadoraceae, Apocynaceae, Asclepiadaceae, Loganiaceae and Gentianaceae. Engler and Prantl have placed the Oleaceae in the order–Contortae of subclass Sympetalae, along with other four families, *i.e.*, Loganiaceae, Gentianaceae, Apocynaceae and Asclepiadaceae.

Plants with commonly opposite simple or pinnately compound exstipulate leaves; flowers bisexual, rarely unisexual by abortion, hypogynous, actinomorphic; corolla 5, gamopetalous, imbricate, contorted, convolute, rarely valvate; stamens adnate to or near corolla base, 2 - 5; carpels 2, ovary 2 - 4 loculed; herbs, shrubs and trees, climbing or erect.

Family—Oleaceae

Distribution: This family consists of 32 genera and about 500 species, distributed in temperate and paleotropical regions of Asia and East Indies. The important genera of the family are *Olea, Jasminum, Fraxinus, Forsythia, Menodora, Syringa, Ligustrum, Osmanthus, Chionanthus, Forestiera,* etc.

Habit: Trees or shrubs, sometimes lianes, scandent or twining. In olives (*Oleaeuropaea*) and ash (*Fraxinus*). They are trees, while the genus *Jasminum* is a twining or scandent shrub.

Leaves: The leaves are opposite, simple or compound, exstipulate. In *Jasminum*, the leaf axil subtends several buds arranged one above the other, these buds are accessory buds. In *Syringa*, the scale of the winter bud represents undeveloped leaf and is considered as leaf sheath phyllode (Arber, 1920). The peltate hairs occur on the leaf epidermis; besides there are extra floral nectaries in the leaves of many genera.

Inflorescence: Terminal or axillary cymes, panicles or sometime racemes. In *Jasminum*, the inflorescence is typical cymose, often dichasium; in *Syringa.* the inflorescence is compound raceme.

Flower: Flowers are bisexual, hypogynous, rerely unisexual; unisexual flowers occur in *Fraxinus* and in some species of *Olea*; actinomorphic, with small pedicel.

Calyx: Sepals 4, valvate, rarely absent; in some species of *Fraxinus* both sepals and petals are absent; in some species of *jasminum* sometimes more than 5 sepals are present.

Corolla: Petals 4 or 5, connate in salver-shaped or other forms of corolla or rarely free, imbricate sometimes valvate, rarely absent; in *Fraxinus excelsior* corolla is absent; in some species of *Jasminum* sometimes more than 5 petals are present.

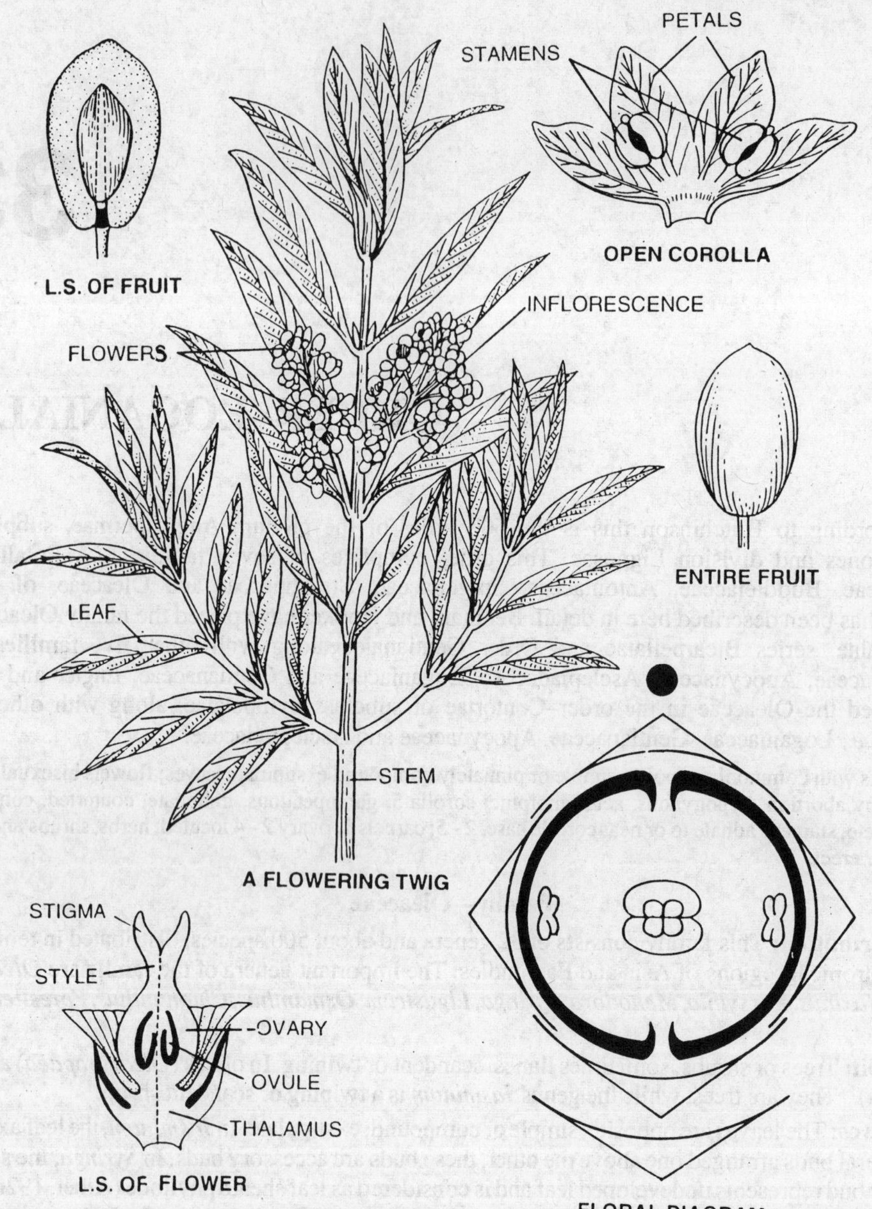

Fig. 33.1. Oleaceae. *Olea europaea*

Androecium: Stamens usually 2, sometimes 4, anther dithecous; 4 stamens are in *Hesperelaea* and *Tessarandra;* distinct, often apiculate by extension of the connective, dehiscing longitudinally.

Gynoecium: Carpels 2, pistil 1, ovary superior, two locules, placentation axile, two ovules in each locule; ovules anatropous, ascending or pendulous; style 1 or absent, simple; stigmas 1–2, bilobed.

Fruit: A berry (*e.g.*, in *Ligustrum*), drupe, (*e.g.*, in *Olea, Jasminum,* loculicidal capsule (*e.g.*, in *Syringa, Forsythia*), a circumscissile capsule (*e.g.*, in *Menodora*), or a samara (as in *Fraxinus*)

Seed: Seeds are with or without endosperm; endosperm when present, it is fleshy; straight embryo.

Floral Formula: $\oplus$ ⚥ K_4, C_{4-5}, A_2, $G_{\underline{(2)}}$.

Bentham & Hooker (1862)	**Engler & Perantl (1931)**	**Hutchinson (1959)**
Phanerogams	Phanerogams	Angiospermae
Dicotyledones	Dicotyledones	Dicotyledones
Gamopetalae	Sympetalae	Lignosae
Bicarpellatae	Contortae	Loganiales
Gentianales	Oleaceae	Oleaceae
Oleaceae		

Economic Importance of the Family

The family is of considerable economic importance.

1. **Oil and food.** *Olea europaea* (Eng. Olive; Verna—Zaitun). The olive is a source of food; the oil expressed from the fruit pulp is of high value. The oil is golden yellow, clear and odourless; it is used chiefly as salad or cooking oil; the inferior grades of oil with greenish tinge are used for soap making and as lubricants; the oil cake is used as cattle feed.

2. **Timber trees.** *Fraxinus excelsior* (ashwood tree). Valuable timber, the ashwood is obtained from this tree; The wood is used for cabinet work, and for high quality furniture.

3. **Ornamentals.** Species of *Jasminum, Syringa, Ligusturm, Chionanthus, osmanthus,* etc, are grown as ornamentals. In most of the plants, the flowers are fragrant.

Systematic Relationships : The family is supposed to be related to Loganiaceae, and has been put under order Loganiales. Hutchinson established relationship of Oleaceae with Rubiaceae and Melastomaceae.

34

ORDER–APOCYNALES

According to Hutchinson this is the fifty first order of the phylum Angiospermae, sub-phylum Dicotyledones and division Lignosae. This order comprises of only four families, *i.e.,* Plocospermaceae, Apocynaceae, Periplocaceae and Asclepiadaceae. The Apocynaceae and the Asclepiadaceae have been described in the present text in detail. Bentham and Hooker have placed the Apocynaceae and the Asclepiadaceae along with other four families in the seventh order– the Gentianales of class Dicotyledones, division Gamopetalae and series Bicarpellatae. Engler and Prantl have put these families along with three others in the fifth order– the Contortae of class Dicotyledones and sub-class Sympetalae.

Corolla sympetalous; actinomorphic; stamens epipetalous, alternate with the corolla lobes; pollen becoming waxy and collecting into masses as in orchids; ovary superior, carpels-2, free or becoming free in fruit; styles united.; ovules mostly numerous; a parietal or on the septa; seeds usually with endosperm and straight embryo. Leaves opposite, simple; no stipules.

Family—Apocynaceae

There are about 300 genera and 1300 species in this family.

Distribution: The members of this family are found throughout the world, but they are more commonly met with in the tropical regions.

Habit: There is a great variation in the habit of the plants of this family. They may be herbs, erect or twining shrubs or trees. *Vinca rosea* (Verna–**Sadabahar**) is a perennial herb; *Vallaris solanacea* (Verna–**Ramsar**) is a large twining shrub; *Nerium indicum* (*N. odorum*) is a large shrub with beautiful red or white flowers; *Thevetia peruviana* or *Thevetia nerifolia* (Verna–**Pili kaner**) is a large shrub or a small tree; *Plumeria acutifolia* is a small sized tree and *Alstonia scholaris* is a medium sized tree. In some genera, the stem becomes tuber like, *e.g., Adenium.* The species of *Landolphia* and *Clitandra* are climbing shrubs.

The latex is present in most of the genera.

Root: Tap and branched.

Stem: usually erect, branched, solid, glabrous rarely tuberlike and thick.

Leaves: The leaves are simple, petiolate, usually opposite decussate. In rare cases the leaves are alternate or even whorled (*e.g., in Nerium odorum, Alstonia,* etc.) Usually the leaves are exstipulate and very rarely they may be stipulate.

Inflorescence: Usually the inflorescence is of cymose type. It is very rarely solitary as in *Vinca.* In *Carissa,* the flowers are found to be arranged in corymbose cymes. In *Plumeria*, the flowers are arranged in terminal cymes. In *Alstonia,* the flowers are found to be arranged in umbellate branched panicled cymes. In *Rauvolfia,* the flowers are arranged in umbellate or corymbose cymes.

Flowers: The flowers are pedicellate, bracteate, bracteolate, hermaphrodite, actinomorphic, regular, sometimes slightly, zygomorphic, complete, hypogynous and pentamerous. In rare cases the flowers are tetramerous with reduction to two in the pistil.

Calyx: Usually it consists of five sepals, gamosepalous. The calyx is generally divided almost to the base. The aestivation is quincuncial.

Corolla: Usually the corolla consists of five petals, gamopetalous. It is generally salver or funnel shaped. The corolla tube usually possesses hairy appendge or scales. The aestivation is contorted.

Androecium: It consists of five stamens alternating with the petals. The stamens are situated on the tube or the throat of the corolla (*i.e.,* epipetalous). The filaments are short; anthers introrse, polyandrous or connate and often adhere to the stigma. The antherlobes are sometimes empty at their base and prolonged into spines.

Gynoecium: It consists of two carpels. The carpels may be free (apocarpous) or connate (syncarpous); superior, sometimes partly inferior as in *Plumeria*. The style is simple and the stigma is thick and often bilobed. Rarely the number of carpels exceeds, *i.e.,* 3 to 5. Usually a nectar secreting disc is situated beneath the gynoecium. In syncarpous gynoecium, the ovary may be unilocular with parietal plcentation or it may be bilocular with axile placentation. In the case of separate ovaries the placentation is marginal. The ovary is superior or half–inferior. Numerous ovules are found to be situated on parietal placentas or in two chambered ovaries on marginal walls.

Fruit: In the case of free ovaries, the fruit is a pair of follicles. Sometimes the fruits of separate ovaries are fleshy and indehiscent, or may be one seeded, *e.g., Cameraria.* In the case of syncarpous ovary, usually the fruit is indehiscent, fleshy and berry-like, *e.g.,* in *Landolphia.* In *Cerbera,* it may be a drupe. This fruit is coconut like and distributed by means of water currents. In certain genera, possessing syncarpous ovaries a two-valved capsule is found, *e.g.,* in *Aspidosperma* and *Allamanda.*

Seed: In dry fruits the seeds are generally winged, *e.g.,* in *Plumeria.*Sometimes the seed bears a tuft of hairs at the base, *e.g., in Kickxia,* and sometimes at both ends, *e.g.,* In *Strophanthus.* The embryo is straight, with or without endosperm.

Pollination: Generally it takes place through the agency of insects.

Floral Formula : $\oplus$ ⚥ K (5), $\overparen{C(5)\ A(5)}$, G$\underline{(2)}$

Bentham & Hooker (1862)	**Engler & Prantl (1931)**	**Hutchinson (1959)**
Phanerogams	Phanerogams	Angiospermae
Dicotyledones	Dicotyledones	Dicotyledones
Gamopetalae	Sympetalae	Lignosae
Bicarpellatae	Contortae	Apocynales
Gentianales	Apocynaceae	Apocynaceae
Apocynaceae		

Description of important plants in semi-technical language:

1. ***Vinca rosea*; Verna.–Sadabahar.**

Habit: A perennial herb.

Stem: Erect, cylindrical, branched, solid, reddish green, glabrous.

Leaf: Cauline, simple, opposite, decussate, petiolate, exstipulate, obovate, entire, glabrous, mucronate apex, unicostate reticulate venation.

Inflorescence: Cymose, flowers arranged in axillary pairs.

Flower: Pedicellate, bracteate, hermaphrodite, actinomorphic, complete, pink, hypogynous.

Calyx: 5, polysepalous, glandular, green, inferior, quincuncial aestivation.

Corolla: 5, gamopetalous forming corolla tube, throat of corolla tube hairy forming a corona, contorted aestivation.

Androecium: 5, free, epipetalous, alternate to petals, almost sessile, anthers dorsifixed, connivent round the stigma, yellowish.

Gynoecium: 2 carpels (bicarpellary), syncarpous, carpels united above in the region of style and stigma, ovaries free (apocarpous below), with single style and stigma, ovaries superior, nectar secreting

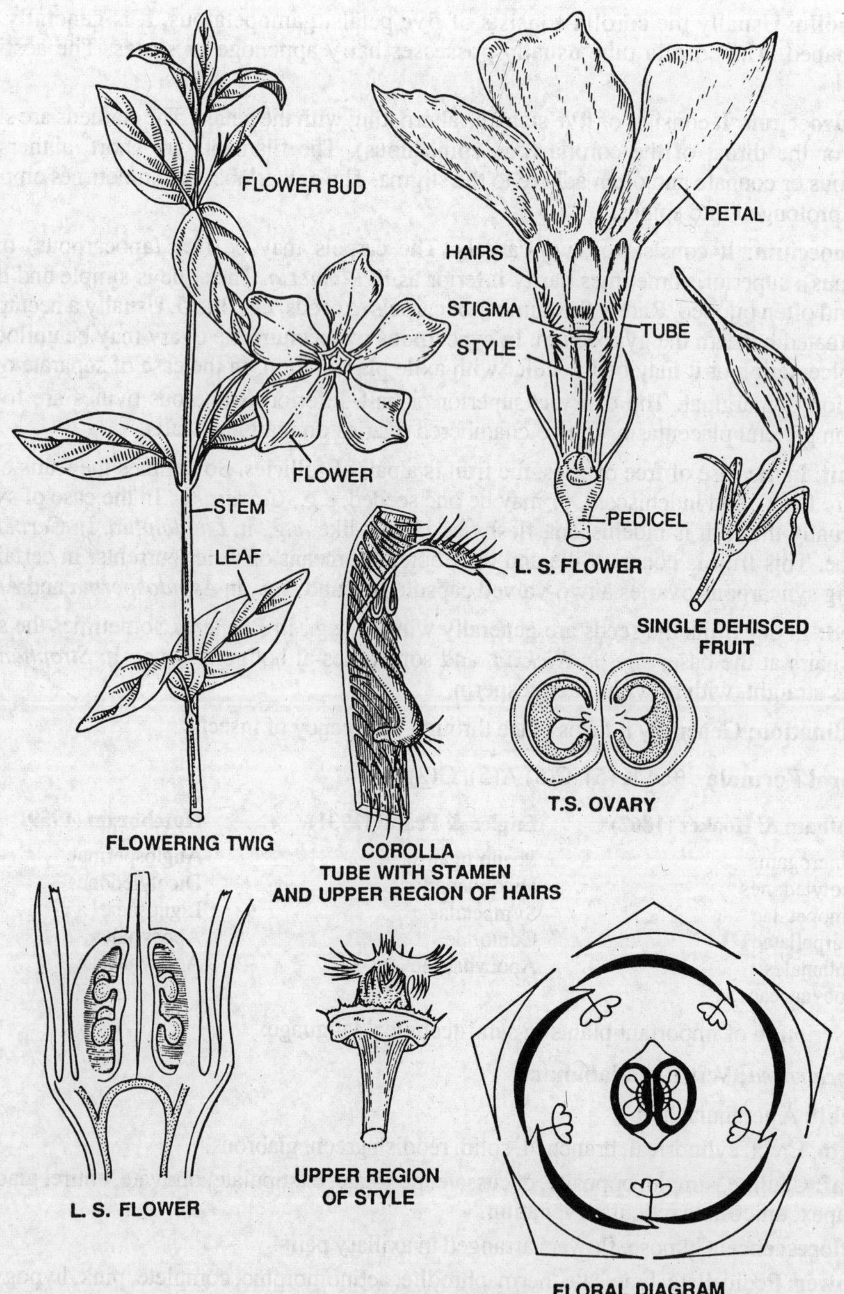

Fig. 34.1. Apocynaceae. *Lochnera rosea;* Syn. *Vinca rosea* Linn.

disc present beneath ovaries, unilocular, marginal placentation, glands present alternating with the carpels, style filiform, stigma thickened, dumb-bell shaped.

Fruit: A pair of elongated follicles.

Floral Formula: ⊕ ⚥ K 5, $\widehat{C\ 5, A}$ 5, G (2).

2. ***Plumeria acutifolia;*** Verna.–**Goburchampa.**

Habit: A large shrub or small tree.

Stem: Erect, solid, branched, woody, latex present, glabrous, green, cylindrical.

Leaf: Simple, whorled, petiolate, margin smooth, apex acute, unicostate reticulate venation, large.

Inflorescence: Cymose.

Flowers: Pedicellate, bracteate, hermaphrodite, actinomorphic, complete, hypogynous.

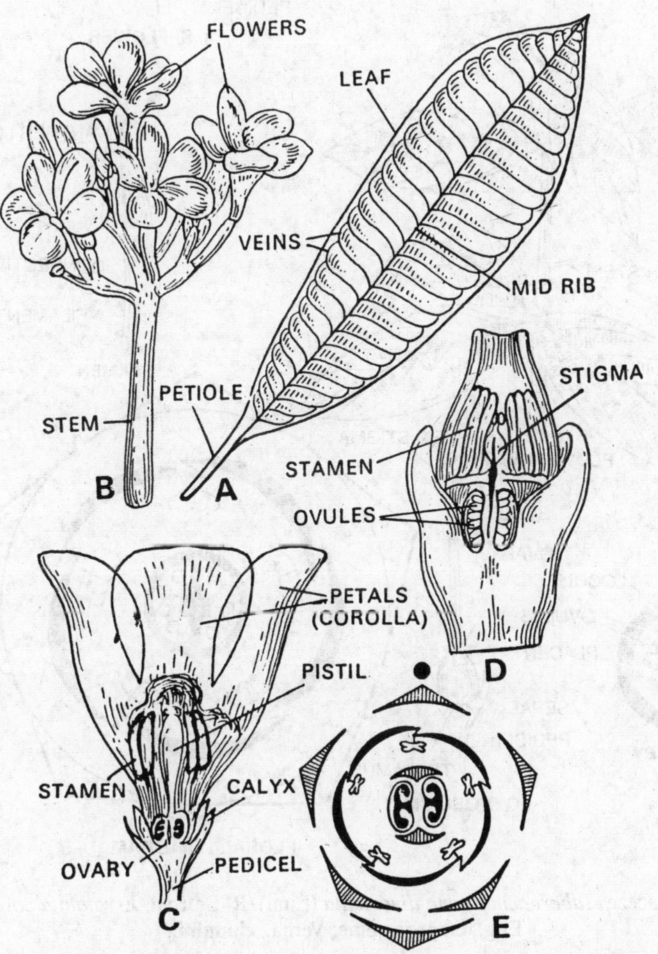

Fig. 34.2. Apocynaceae. *Plumeria* sp. A, a leaf; B, inflorescence; C, flower cut lengthwise; D, lower portion of flower, in longitudinal section, E, floral diagram.

Calyx: 5, polysepalous, imbricate aestivation.

Corolla : 5, gamopetalous forming a corolla tube, twisted aestivation.

Androecium: 5, free, epipetalous, included in the corolla tube, introrse, two-celled, basifixed.

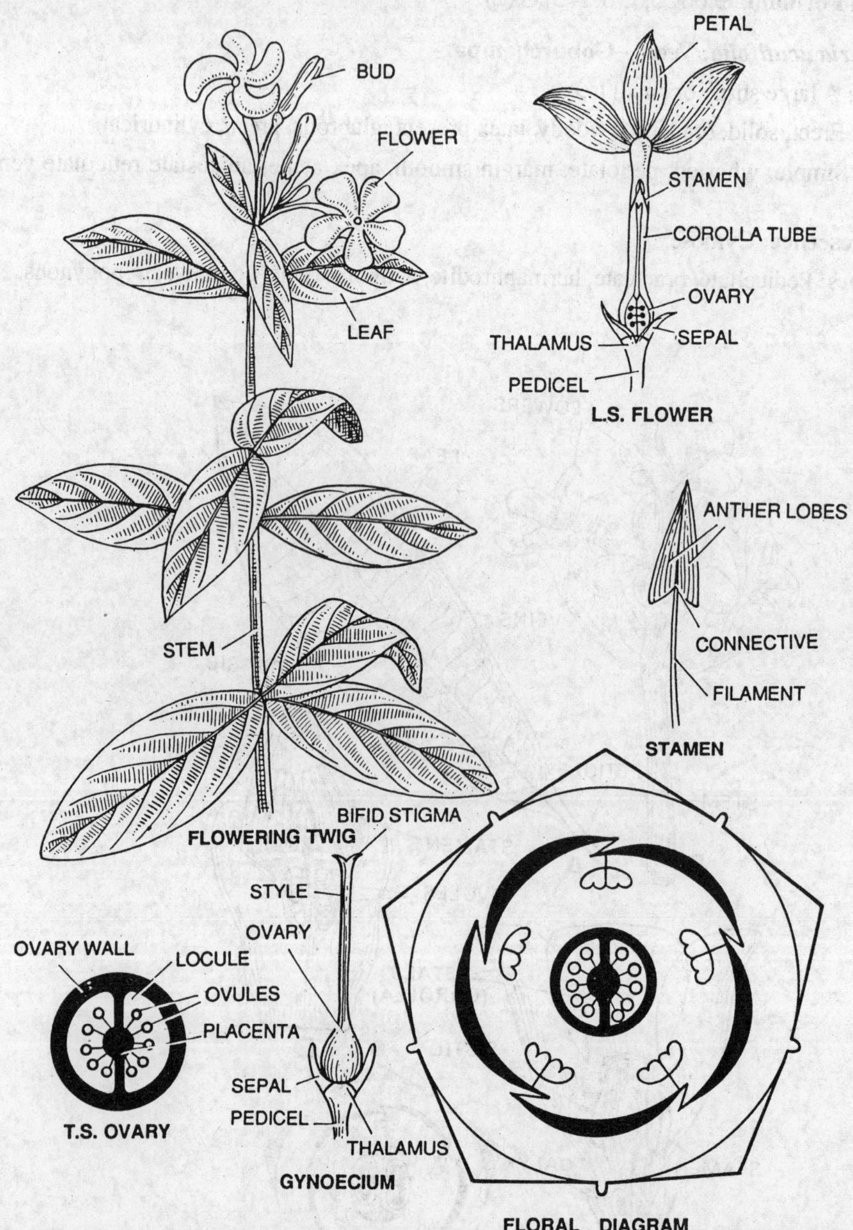

Fig. 34.3. Apocynaceae. *Tabernaemontana divaricata* (Linn). R. Br. Syn. *Ervatamia coronaria* Stapf.; Eng., crape jasmine; Verna., **chandni.**

Economic value. Grown as an ornamental. The red pulp around seed is used as a dye. Wood is refrigerant. Milky juice is used for diseases of eye. Root is acrid, bitter, used as local anodyne and chewed for relief of toothache.(Fig. 34.3)

Gynoecium: 2(bicarpellary), syncarpous, ovary superior or partly inferior, bilocular, axile placentation, style one, stigma simple.

Floral Formula: $\oplus$ ⚥ K 5, $\widehat{C(5), A\ 5}$, G $(\underline{2})$.

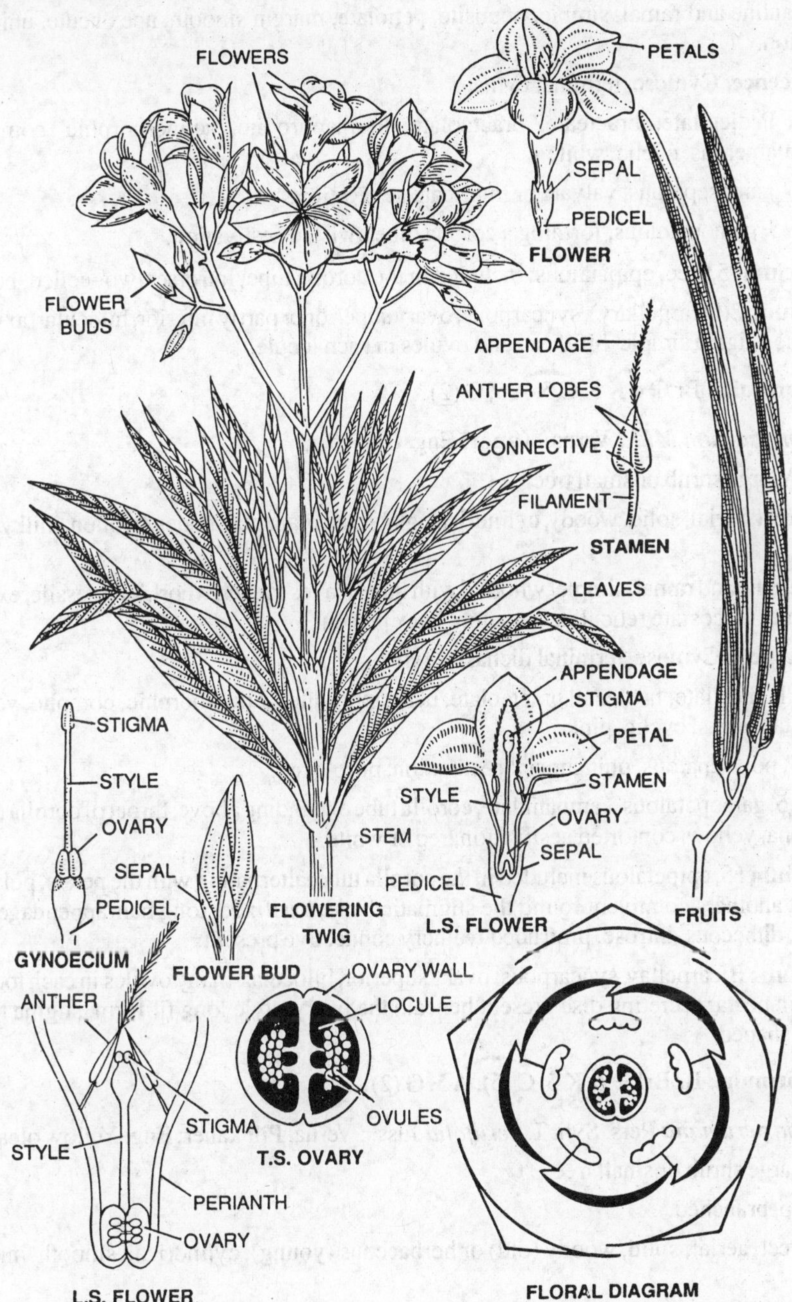

Fig. 34.4. Apocynaceae. *Nerium indicum* Mill.; Eng., oleander; Verna. **kaner.**

Economic value. A beautiful shrub; grown as a hedge plant. Plant is poisonous. Root is powerful resolvent and attenuant; used externally. Oil prepared from root bark is used in skin diseases and leprosy. (34.4)

1. ***Tabernaemontana divaricata*** (Linn); R. Br.; Verna. **Chandni;** Eng. Crape jasmine.

Habit: A large shrub.

Stem: Erect, solid, branched, woody, latex present, glabrous, green, cylindrical.

Leaf: Cauline and ramal, simple, opposite, petiolate, margin smooth, apex acute, unicostate reticulate venation.

Inflorescence: Cymose, terminal cyme.

Flower: Pedicellate, bracteate, bracteolate, hermaphrodite, actinomorphic, complete, hypogynous, pentamerous, cyclic, white.

Calyx: 5, gamosepalous, valvate or imbricate aestivation.

Corolla: 5, gamopetalous, forming a corolla tube, twisted aestivation.

Androecium: 5, free, epipetalous, included in the corolla tube, introrse, two–celled, basifixed.

Gynoecium: 2(bicarpellary), syncarpous, ovary superior or partly inferior, bilocular, axile placentation, style one, stigma simple, bifid, several ovules in each locule.

Floral formula : Br ⊕ ⚥ K 5, $\overline{C\ 5,\ A}$ 5, G $(\underline{2})$.

2. ***Nerium indicum Mill.;*** Verna. **Kaner;** Eng. Oleander.

Habit : A large shrub or small tree.

Stem : Erect, aerial, solid, woody, cylindrical, glabrous, branched, nodes swollen, milky latex present.

Leaf : Cauline and ramal, simple, whorled with three leaves in each whorl, sub-sessile, exstipulate, linear-lanceolate; unicostate reticulate venation, latex present.

Inflorescence : Cymose, terminal dichasial cyme.

Flower : Pedicellate, braceate, bracteolate, hermaphrodite, actinomorphic, complte, yellow, hypogynous, pentamerous, cyclic, pink.

Calyx : 5, polysepalous, quincuncial aestivation, purple red.

Corolla : 5, gamopetalous, campanulate, corolla tube expanding above, throat of corolla tube hairy forming a corona, yellow, contorted aestivation, red or white.

Androecium : 5, epipetalous included in the corolla tube, alternating with the petals, polyardrous, filaments short anothers connivent round the stigmatic head, basifixed, long hair appendages present on their apices, dithecous, introse, protruded feathery connective present.

Gynoecium : Bicarpellay, syncarpous, ovary superior, bilocular, many ovules in each loculus, axile placetation, a nectar secreting disc present beneath the ovary, style long filiform, stigma thickened and dumb-bell shaped.

Floral Formula : Br Brl ⊕ ⚥ K 5, $\overline{C\ (5),\ A}$ 5, G $(\underline{2})$.

3. ***Thevetia peruviana*** Pers. Syn. T. *neriifolia* Juss. : Verna. Pili kaner; Eng. Yellow oleander.

Habit : Large shrub or small tree.

Root : Tap, branched.

Stem : Erect, aerial, solid, woody (old) or herbaceous (young), cylindrical, smooth, milky latex present.

Leaf : Cauline and ramal, exstipulate, sub-sessile, spiral, linear-lanceolate, entire, acute, glabrous, unicostate reticulate, latex present.

Inflorescence : Cymose, axillary dichasial cyme.

Flower : Bracteate, bracteolate, pedicellate, complete, actinomorphic, regular, hermaphrodite, pentamerous, hypogynous, cyclic.

Calyx : 5 sepals, polysepalous, quincuncial aestivation, persistent, green.

Corolla : 5 petals gamopetalous, twisted aestivation, bell-shaped (infundibuliform), yellow fringed coronary outgrowths present.

Androecium : 5 stamens, polyandrous, epipetalous, stamens inserted at the throat of corolla, filament short anther sagittate, dithecous, basifixed and introrse.

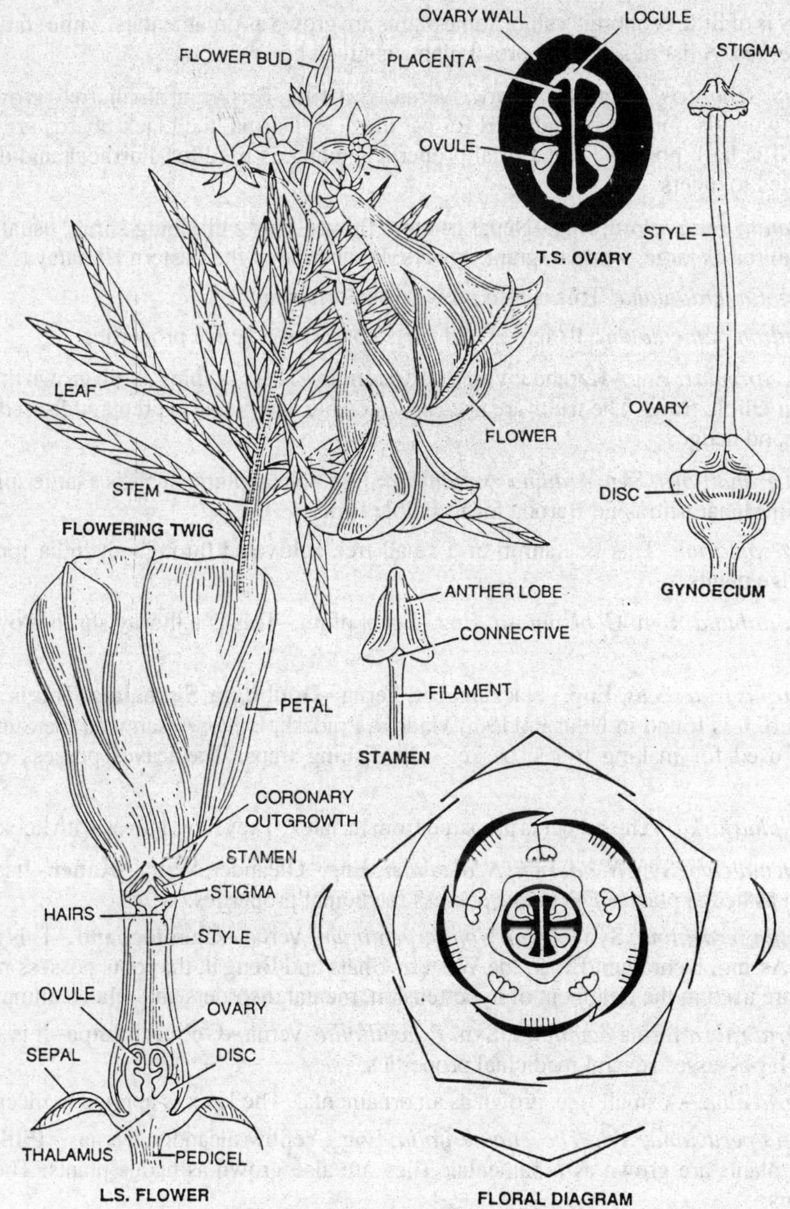

Fig. 34.5. Apocynaceae. *Thevetia peruviana* Pers. , Syn. *T. neriifolia.* Juss.; Eng., yellow oleander; Verna., **pili kaner.**

Economic value. The bark is bitter, cathartic, febrifuge and useful in different kinds of intermittent fevers. Milky juicy and seeds are poisonous. (Fig. 34.5)

Gynoecium : 2 capels (bicarpellary), syncarpous, ovary superior, bilocular, two ovules in each locule, axile placentation, style long, stigma umbrella-shaped, 5-lobed nectar secreting disc present.

Fruit : Drupe.

Floral Formual : Br Brl ⊕ ⚥ K 5, C (5), A 5, G (2).

Economic Importance of the Family

The family is of little economic value. Some plants are grown as ornamentals, while some possess medicinal properties. A list of some important plants is given here:

1. *Alstonia scholaris;* Eng.–Dita bark; Verna.–**Satwin.** This is a small tree grown as an ornamental. Its wood is quite light and used for carvings. In Burma, the black boards are prepared from its wood. The bark possesses medicinal properties, which is used for diarrhoea and dysentery. Its latex is applied to ulcers.

2. *Beaumontia grandiflora*; Eng.–Nepal trumpet flower–It is a climbing shrub, usually grown as an ornamental for its large, white fragrant flowers. It is native of the Eastern Himalayas.

3. *Beaumontia jerdoniana;* This is also grown as an ornamental.

4. *Anodendron paniculatum;* Its leaves and roots possess medicinal properties.

5. *Carissa carandas;* Eng.–Karanda; Verna.–**Karaunda**–This is a spiny shrub grown throughout India for its sour edible fruits. The fruits are used as vegetable and pickle is prepared from them. The plant makes a good hedge.

6. *Carissa grandiflora;* Syn. *Arduina grandiflora;* Eng.–Natal plum.–This is a large spiny shrub usually grown in Maharashtra and Baroda for its edible fruits.

7. *Carissa spinarum.* This is a shrub or a small tree cultivated throughout India for fragrant flowers and hedge plants.

8. *Carrisa arduina;* Syn. *C. bispinosa*; Eng.–Natal plum.–This is a thorny shrub grown for its edible fruits.

9. *Ichnocarpus frutescens;* Eng. –blackcreeper; Verna.–**Dudhilata, Siamalata.** This is a twining ornamental shrub. It is found in Uttar Pradesh, Madhya Pradesh, Bihar, Assam and the Sundarbans. The stems are used for making ropes, baskets and fishing traps. The leaves possess medicinal properties.

10. *Landolphia kirkii.*–The rubber is prepared from its latex. They have leaves with hook tendrils.

11. *Nerium indicum*; Syn. *N. odorum; N. oleander;* Eng.–Oleander; Verna.–**Kaner.**–It is a shrub. They are grown as hedge plants. The plants possess medicinal properties.

12. *Rauvolfia serpentina;* Syn. *Ophioxylon serpentinum;* Verna.–**Chhotachand.**–This is a small shrub found in Assam, Dehradun, Bihar, the Western Ghats and Bengal, the roots possess medicinal properties and are used in the treatment of hypertension, mental disorders and related ailments.

13. *Plumeria rubra* forma *acutifolia*; Syn. *P. acutifolia;* Verna.–**Goburchampa**–It is grown as an ornamental. It possesses several medicinal properties.

14. *Plumeria alba.*–A small tree, grown as an ornamental. The latex is applied to ulcers.

15. *Thevetia peruviana*; Syn. *Thevetia nerifolia*; Eng.–Yellow oleander; Verna. – **Pilikaner.**–It is a shrub. The plants are grown as ornamental. They are also grown as hedge plants. The latex is highly poisonous.

16. *Wrightia tinctoria ;* Syn. *W. rothii; Nerium tinctorum;* Verna.–**Dudhi.**–A tree, found in Rajasthan, Madhya Pradesh and Tamil Nadu. A blue dye is obtained from its flowers and fruits. The fruits are edible. The bark and seeds possess medicinal properties.

17. *Wrightia tomentosa;* Syn. *W. mollissima; Nerium tomentosum;* Verna.–**Dharauli**–The seeds

and roots yield a yellow dye. The leaves and fruits are edible. Its soft wood is used for carvings. The bark and roots are used as an antidote for snake bite.

18. *Aganosma dichotoma;* Syn. *Echites dichotoma;* Verna.–**Malati.**–This is a climbing shrub, grown as an ornamental in the gardens.

19. *Allamanda cathartica;* Eng.–Allamanda–This is a beautiful climbing shrub, grown as an ornamental in the gardens. It is native of Central America and Brazil.

Family–Asclepiadaceae

According to Rendle there are about 280 genera and over 1800 species in this family.

Distribution: The members of this family are mainly found in the tropical regions of the world. They occur specially in the drier parts of Central and South America. In our country the family is represented by several genera, such as, *Asclepias, Calotropis, Ceropegia,* etc. *Asclepias curassavica* is commonly found in sub-Himalayan tracts; *Calotropis gigantea* is commonly found in Northern India; *Calotropis procera* is quite common in the Punjab, Uttar Pradesh and Andhra Pradesh. The species of *Ceropegia* are commonly found in South India.

Habit: The plants of this family are mostly erect herbs or woody climbers but some are succulent. The perennial herbs are quite common in the dry regions of South and Central Africa. Some plants perennate by means of tuber-like rhizomes. Large shrubs and trees are rarely found. *Hoya* is a plant of succlent habit, it possesses fleshy stem and leaves. *Stapelia* is a native of South Africa. It is cactus like in habit possessing thick and fleshy stems; the leaves are reduced to thorns or scales. *Dischidia* is an epiphytic plant; it climbs upon its host by means of adventitious roots, the leaves are succulent and covered with a coating of wax. *Cryptostegia grandiflora* is a climbing shrub. The species of *Hemidesmus, Gymnema, Pergularia,* etc., are also climbing shrubs. *Leptadenia spartium* is a shrub. *Calotropis procera,* (Verna.–**Ak**) is a xerophytic shrub and *Calotropis gigantea* (Verna.–**Safed Ak**) is a large shrub. *Asclepias curassavica* is a perennial herb. The species of *Ceropegia, Pentatropis* etc., possess underground tuber like parts. *Tylophora indica* is a herb. Most of the plants possess xerophytic characters and certain are extreme xerophytes.

Anatomically the plants of this family resemble with that of Apocynaceae. They also possess bicollateral bundles and the latex. According to Busich (1913), the endophytic mycorhiza is frequently found in the roots of succulent plants of this family.

The species of *Pentatropis* possess tuberous roots; whereas *Ceropegia tuberosa* possesses tubers, by means of which, they perennate.

Leaves: The leaves are simple, sub-sessile and exstipulate. In many cases the leaves are fleshy and covered with a coating of wax, *e.g., Calotropis.* In *Stapelia,* the leaves are very much reduced and represented by spines and scales.

Inflorescence: The inflorescence is of two types, *i.e.,* (1) cymose and (2) racemose. Generally the cymose type of inflorescence is found. In such cases the axis ends in a terminal flower, while branching occurs in the axils of the two branches forming a dichasial cyme, but one of the two branches of the dichasium grows faster than the other and ultimately one of the branches is completely suppressed and the inflorescence becomes monochasial cyme. In certain cases the inflorescence is of racemose type. This is either a raceme or a umbel. Sometimes the flowers are found to be arranged in umbelliform cymes. When the inflorescence is axillary only one flower is found at each node; the other leaf either remains completely sterile or possesses a vegetative shoot. In *Calotropis, Asclepias* the flowers are being arranged in umbellate cymes. In *Hemidesmus,* the flowers are found to be arranged in axillary cymes.

Flowers: The flowers are pedicellate, bracteate, hermaphrodite, actinomorphic, rarely zygomorphic, *e.g.,* in *Ceropegia,* complete. The general plan of the flower is pentamerous with three regularly alternating pentamerous whorls of calyx, corolla and androecium, however, the number of carpels is reduced to two in the gynoecium. Usually the flowers are small in size, but the flowers of *Ceropegia, Stapelia* and *Stephanotis* are quite large in size.

Calyx: It consists of five sepals, which are either free (polysepalous) or somewhat connate at the base; with the odd sepal posterior. The aestivation is quincuncial or rarely valvate.

Corolla: It consists of five united petals (*i.e.*, it is gamopetalous). The petals are spreading (*i.e.*, rotate), but in *Stephanotis* the corolla tube is long, forming a salver-shaped corolla. In *Ceropegia* the corolla is pitcher-like in appearance (zygomorphic). The aestivation is contorted and rarely valvate. Sometimes the petaloid appendages arise either from the corolla or from the back of the stamens. Very often these hairs or appendages are found inside or at the mouth of the corolla forming the **corolline corona.**

Androecium: It consists of five stamens. The anthers are united laterally, giving rise to a five sided blunt cone, which is usually attached on the inside to the stigma head, except in the sub-family Periplocoideae. With the result of this union of anthers and pistils, the **gynostegium** is formed. The pollen grains of each half anther associated in tetrads are found in a sac like structure known as **pollinium.** The anthers are almost quite sessile. In the typical members of the family the **translator** consists of two parts–(1) **corpusculum** and (2) a **pair of arms.** The corpusculum is attached to the margin of stigma head between the anthers and the pair of arms is attached to the pollinia of the adjacent anther halves.

In the members of the sub family Periplocoideae the translator consists of a spoon or funnel-shaped structure in between each anther, terminating below in a sticky disc. The granular pollen grains from one half of each of the two adjacent anthers, on maturity, pour into the concave receptacle of the translator.

Certain appendages or horn-like projections are given out from the backs of the anthers, are known as **cuculi,** which secrete and store honey. These appendages form a corona known as **staminal corona.**

Gynoecium: It consists of two carpels. The ovaries remain free, but styles unite to form a common swollen stigma-head. This stigma-head may be flat or somewhat conical and sometimes even beaked. The receptive surfaces of the stigma are situated on the edge or on the underside of it. The ovary is superior. The placentation is marginal with many ovules.

Fruit: Generally it is a pair of follicles. Sometimes there is only one follicle, because of the suppression of the other.

Seed: Usually the seeds are flat, ovate-oblong and are crowned by a fruit of hairs. These hairs facilitate the dispersal of the seeds by wind. The embryo is large.

Floral Formula: ⊕ ⚥ K5, $\widehat{C(5), A5}$, G($\underline{2}$).

There are two sub-families, which may be distinguished as follows:

1. Periplocoideae–The pollen grains are granulur and arranged in tetrads. The translators are spoon or funnel shaped.

2. Cynanchoideae–Pollen grains are in twos, found in pollinia. The translators are differentiated into a corpusculum and a pair of arms.

Pollination: The pollination takes place strictly by means of the agency of insects. As an insect visits the flower, the sticky basal disc of translator turns outwards and adheres to the head of the insect visitor. The whole translator with its pollinia or pollen contents is carried over by the insect from one flower to another flower where the pollen grains are deposited on the stigmatic surfaces on the underside of the stigma-head. In this sub family Cynanchoideae, where the pollinia are found the insect visiting the flower in the search of nectar pushes its leg in the slit left between the adjacent anthers, and when it pulls out the leg, it comes in contact with the notched base of the corpusculum, and this way the complete translator with two pollinia is dragged out. At first the pollinia are quite apart from each other, but as the hygroscopic arms of the translator are dried up they become close to each other. If this insect visits another flower, the pollinia are readily pushed into one of the slits, and will slip into the stigmatic chamber, thus effecting cross–pollination. Chiefly the pollination is by two–winged flies.

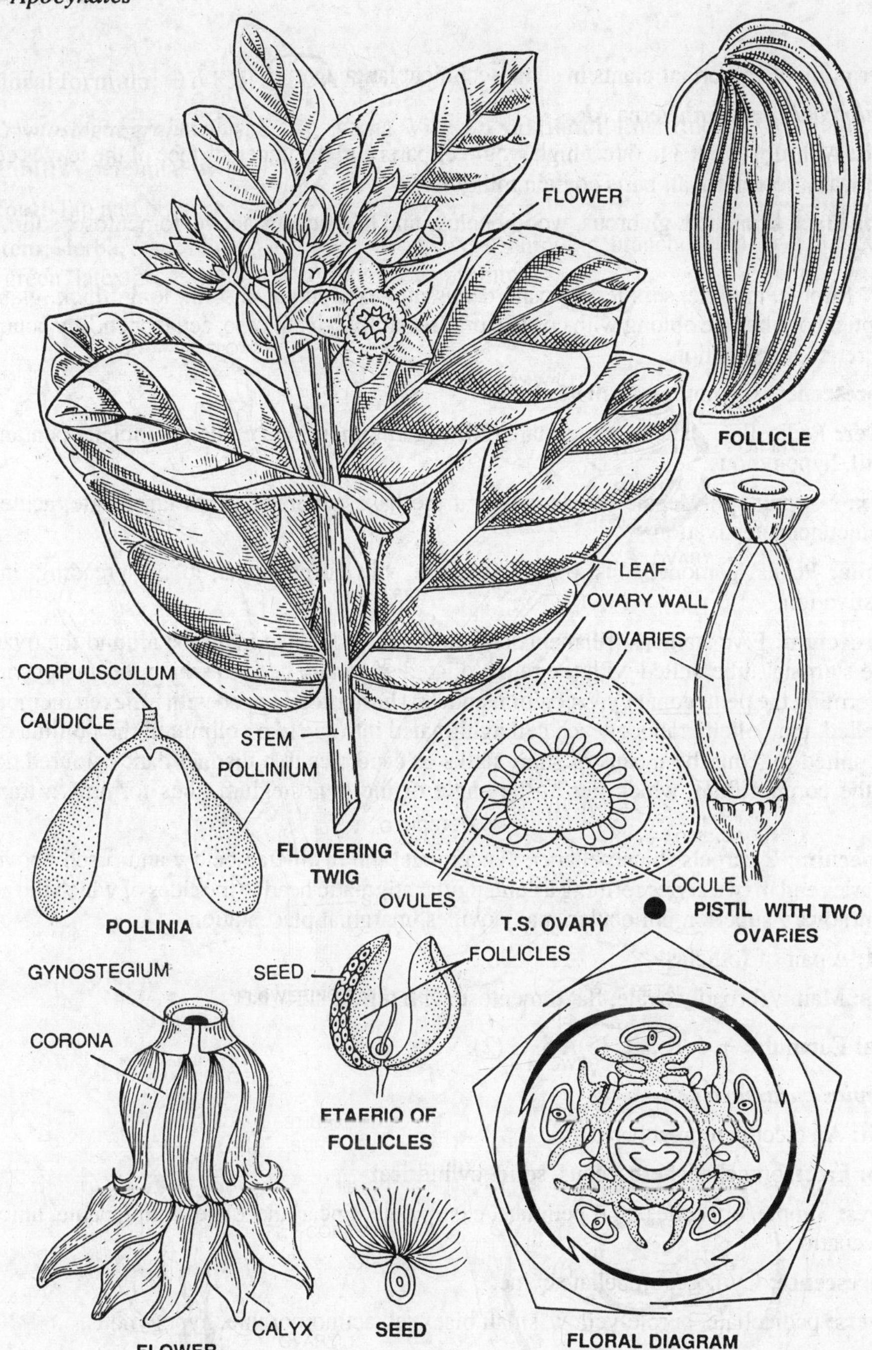

Fig. 34.6. Asclepiadaceae.*Calotropis procera* **(Ak.)**

Bentham & Hooker (1862)	**Engler & Prantl (1931)**	**Hutchinson (1959)**
Phanerogams	Phanerogams	Angiospermae
Dicotyledones	Dicotyledoneae	Dicotyledones
Gamopetalae	Sympetalae	Lignosae
Bicarpellatae	Contortae	Apocynales
Gentianales	Asclepiadaceae	Asclepiadaceae
Asclepiadaceae		

Description of important plants in semi–technical language:

1. ***Calotropis procera;*** Verna.**Ak.**

Habit: A shrub, about 3 to 6 feet high, younger parts and the under surface of the leaves covered with white waxy secretion, all parts contain milky latex.

Stem: Erect, branched, glabrous, woody below and herbaceous above, tomentose, solid, cylindrical.

Leaf: Simple, cauline, sessile, opposite decussate, exstipulate, 2-5 in., long, thick, glaucous–green, elliptical or obovate oblong with cordate or often amplexicaul base, acute or shortly acuminate, unicostate reticulate venation.

Inflorescence: Cymose, umbellate cyme.

Flower: Pedicellate, bracteate, hermaphrodite, actinomorphic, regular, complete, pentamerous except pistil, hypogynous.

Calyx: 5 sepals, polysepalous or connate at the base, greenish, lobes lanceolate, acute apex, interior quincuncial aestivation.

Corolla: Petals, gamopetalous, pink or whitish with purple spots, lobes spreading, inferior, twisted aestivation.

Androecium: Five stamens, filaments connate in a fleshy staminal tube around the ovary, the apex of the staminal tube united with the much-dilated stigmatic head to which the anthers are also coherent, forming the pentagonal gynostegium; anthers short, broad tipped with inflexed membranous flaps, bi-celled, the pollen grains of each cell aglutinated into sac like pollinium; the pollinia of each anther are united together by means of short stalks or caudicles to a distinct dark coloured dot-like structure, the corpusculum, which lies at the angle of the gynostegium, thus forming a translator apparatus.

Gynoecium: 2 carpels (bicarpellary), syncarpous; the pistil free below and fused above; two distinct ovaries end in two styles forming a pentangular stigmatic head to the sides of which the anthers are coherent; ovary superior, unilocular, many ovules, marginal placentation.

Fruit: A pair of follicles.

Seeds: Mainly, broadly ovate, flat tomentose with tuft of silky hairs.

Floral Formula: $\oplus$ ⚥ K 5, $\overbrace{C\ (5) A}$ 5, G $(\underline{2})$.

2. ***Asclepias curassavica.***

Habit: An erect perennial herb.

Stem: Erect, branched, herbaceous, solid, cylindrical.

Leaves: Simple, opposite, short petiolate, exstipulate, linearlanceolate, entire, acute, unicostate reticulate venation.

Inflorescence: Cymose, umbellate cyme.

Flowers: pedicellate, purple-yellow, small bisexual, actinomorphic, hypogynous.

Calyx: 5 sepals, polysepalous, imbricate aestivation.

Corolla: 5 petals, gamopetalous, twisted aestivation.

Androecium: Similar to that of *Calotropis procera.* The filaments attached to the base of the corolla are united to form a column around the pistil, the anthers being pressed close to the styles; each stamen bears on the side away from the pistil an appendage which enfolds the anther like a hood; these appendages collectively form a corona.

Gynoecium: 2 carpels (bicarpellary), syncarpous ovary superior.

Fruits: Follicle.

Floral formula: ⊕ ⚥ K 5, $\overset{\frown}{C(5), A\,5}$, G $(\underline{2})$.

3. *Cryptostegia grandiflora* R. Br.; Verna. **Vilayati vakhandi**; Eng. Rubber vine.

Habit: A perennial twiner.

Root: Tap and branched.

Stem: Herbaceous, lower part woody, aerial, weak, climbing, branched, twiner, solid, cylindrical, rough, green, latex present, nodes and internodes prominent.

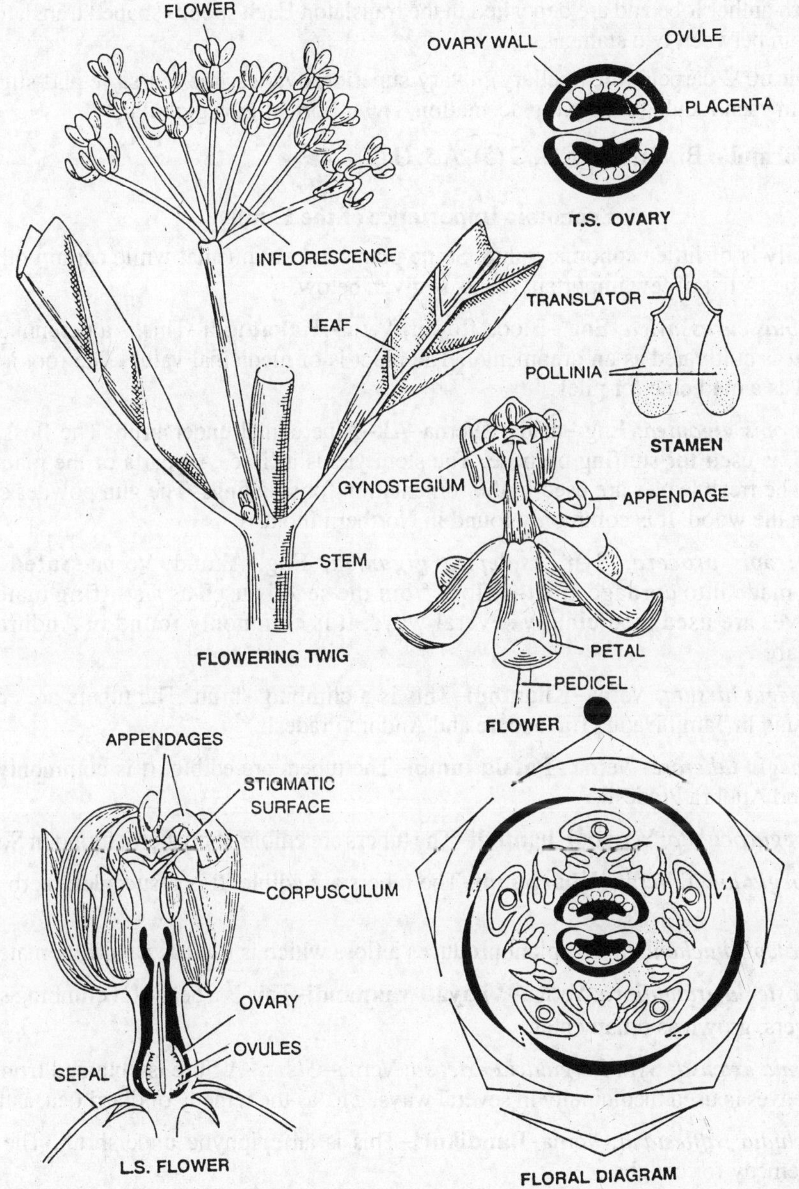

Fig. 34.7. Asclepiadaceae. *Asclepias curassavica* Linn. Eng. Blood flower ; Verna. **Kukatundi.**

Leaf: Ramal and cauline, opposite decussate, exstipulate, simple, petiolate elliptic ovate, entir acute, coriaceous, glaucous, unicostate reticulate.

Inflorescence: Cymose dichasial cyme.

Flower: bracteate, bracteolate, pedicellate, actinomorphic, regular, hermaphrodite, pentamerou hypogynous, cyclic.

Calyx: 5 sepals, polysepalous, quincuncial aestivation, green, margins membranous.

Corolla: 5 petals, gamopetalous, twisted, coronary outgrowths present, violet.

Androecium: 5 stamens, polyandrous, epipetalous, eithecous, introrse; pollen grains are bein shed from each anther lobe and are deposited in the translator. Each spoon–shaped translator is foun to be situated in between two stamens.

Gynoecium: 2 carpels (bicarpellary), ovary superior, ovaries free but style and stigma fusec many ovules in each locule, marginal placentation, style short, stigma globular.

Floral formula: Br, Brl ⊕ ⚥ K 5, $\widehat{C\,(5),\,A}$ 5, G $\underline{(2)}$.

Economic Importance of the Family

The family is of little economic value. Some plants are ornamental while certain others are o medicinal value. A list of few important plants is given below:

1. *Asclepias curassavica*; Eng.–Blood flower; Verna–**Kakatundi**–This is a perennial herb or small shrub. It is cultivated as an ornamental. The plant is of medicinal value. The root is used as purgative and as a medicine for piles.

2. *Calotropis gigantea;* Eng.–madar; Verna–**Ak**–A perennial undershrub. The floss, obtaine from the seed, is used for stuffing purpose. The stem yields a fibre. All parts of the plant are use medicinally. The fresh leaves are used in the fomentation for swellings. The gun powder charcoal i prepared from the wood. It is commonly found in Northern India.

3. *Calotropis procera;* Syn. *Asclepias gigantea;* Eng. Akund; Verna–**Safed Ak.**–Th stem fibre is made into cordage and the floss from the seed is used as a stuffing material. Th stem and leaves are used medicinally several ways. It is commonly found in Andhra Prades and the Punjab.

4. *Ceropegia hirsuta;* Verna–**Khantodi**–This is a climbing shrub. The tubers are edible. It i commonly found in Tamil Nadu, Travancore and Andhra Pradesh.

5. *Ceropegia tuberosa*: Verna.–**Patala tumbi**–The tubers are edible. It is commonly found i Tamil Nadu and Andhra Pradesh.

6. *Ceropegia oculata*; Verna.–**Khantodi**–The tubers are edible. It is a herb, found in South India

7. *Ceropegia lawii*; Verna–**Kharpundi**–The tubers are edible. It is a succulent herb. Found i South India.

8. *Cryptolepis buchanani*; The plant produces a floss which is used as a stuffing material.

9. *Cryptostegia grandiflora*: Verna.–**Vilayati-vaknandi**–This is a beautiful climbing shrub wit red–rose flowers, grown as ornamental.

10. *Daemia extensa*; Syn. *Pergularia extensa*; Verna–**Utsan**–A fibre is obtained from its stem The juice of leaves is used medicinally in several ways, *e.g.,* as the remedy of diarrhoea, asthama, etc

11. *Dischidia rafflesiana*; Verna–**Bandikuri**–This is an epiphytic undershrub. The roots ar chewed as a remedy for coughs.

12. *Gymnema sylvestre*; Verna.–**Gurmar**–This is a large climbing shrub. The flowers are foun to be arranged in umbellate cymes. The roots are used as antidote for snake-bite. The leaves are use as remedy for diabetes.

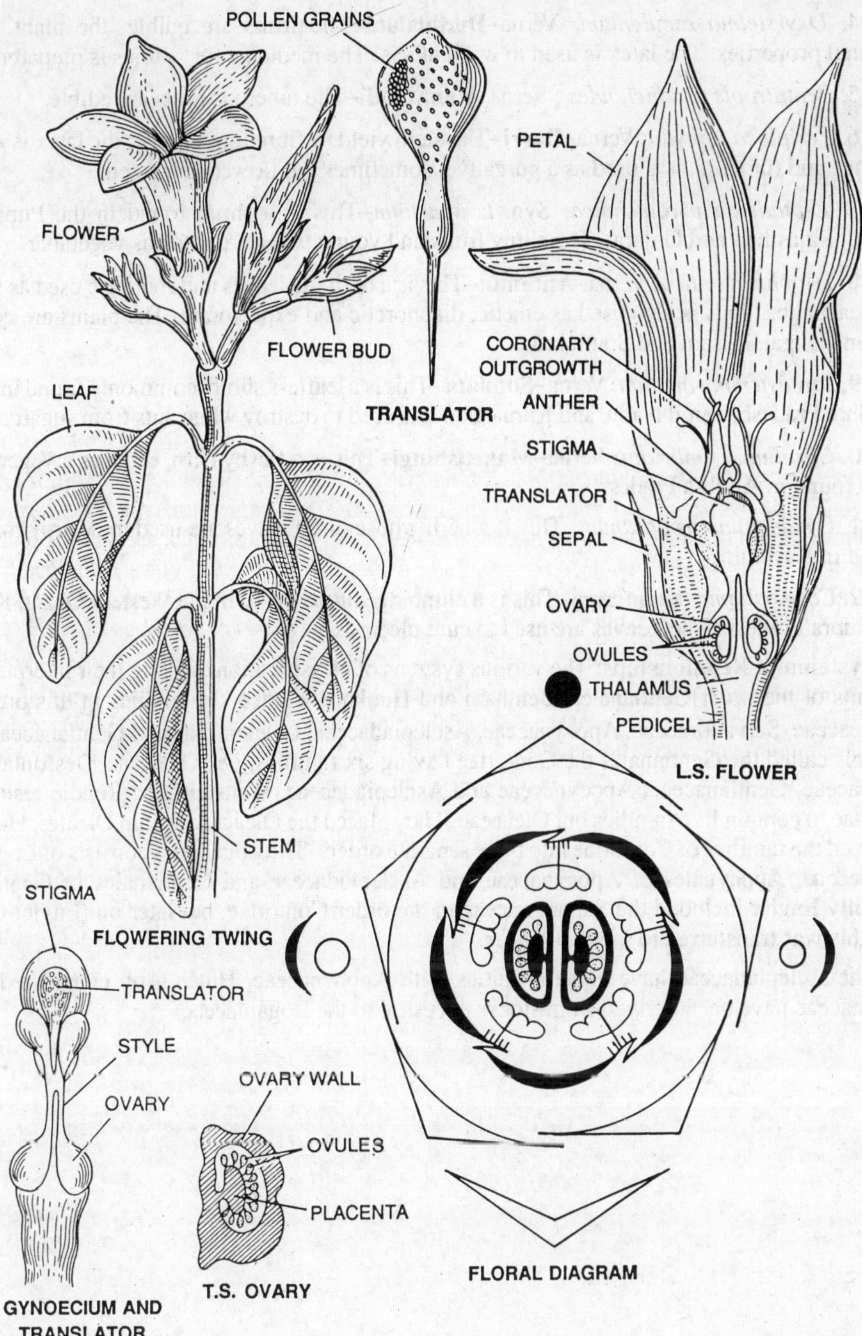

Fig. 34.8. Asclepiadaceae. *Cryptostegia grandiflora* R. Br., rubber vine; Verna., **vilayati vakhandi.**

Economic value. Grown as an ornamental for its beautiful violet-coloured flowers. The leaves are toxic.

13. *Hemidesmus indicus*; Eng. –Indian sarsaparilla; Verna.–**Magrabu**–This is a twining shrub. The roots are used as a tonic, which is useful in loss of appetite, fever, skin diseases, as blood purifier, rheumatism and snake-bite. The tonic (sarsaparilla) is diuretic diaphoretic and demulcent.

14. *Oxyrstelma esculentum;* Verna–**Dudhialata**–The fruits are edible; the plant possesse medicinal properties. The latex is used to wash ulcers. The medicine for scabies is prepared from it.

15. *Pentatropis cynanchoides* ; Verna–**Kanathodi**–The tuberous roots are edible.

16. *Periploca aphylla*; Verna–**Barri**–The stem yields a fibre for cordage; the latex is applied t swellings and tumours, it is used as a purgative; sometimes the flowers are eaten.

17. *Leptadenia pyrotechnica;* Syn. *L. spartium*–This is a shrub found in the Punjab, Utta Pradesh, Rajasthan and Gujarat. The slimy fruits and young twigs are eaten as vegetable.

18. *Tylophora indica*; Verna–**Antamul**–This is a herb. Its leaves and roots are used as substitut for ipecacuanha. This is also used as emetic, diaphoretic and expectorant. The plants are commonl found in Bengal, Assam and South India.

19. *Sarcostemma acidum;* Verna–**Somlata**–This is a leafless shrub, commonly found in dry area of Andhra Pradesh, Tamil Nadu, and Karnataka. It is used to destroy white ants from sugarcane fields

20. *Caralluma fimbriata;* Verna–**Makedshingi**–This is a fleshy herb, eaten as a vegetable. It i mainly found in Andhra Pradesh.

21. *Cynanchum arnottianum;* This is a herb whose dried leaves are used to destroy maggots. I is found in Kashmir.

22. *Cosmostigma racemosum*–This is a climbing shrub, found in the Western Ghats, Karnatak and Andhra Pradesh. The Leaves are used to cure ulcerous sores.

Systematic Relationships: The various systems of classification differ in their interpretation o the limits of the order Gentianales. Bentham and Hooker included six families in this order. The are–Oleaceae, Salvadoraceae, Apocynaceae, Asclepiadaceae, Loganiaceae and Gentianaceae. Engle and Diels called the Gentianales the Contortae having six families, *i.e.,* Oleaceae, Desfontaineaceae Loganiaceae, Gentianaceae, Apocynaceae and Asclepiadaceae. Wettstein and Rendle restricted th Contortae to contain five families but Oleaceae. They placed the Oleaceae within Oleales. Hutchinso has placed the families of Contortae into three separate orders. The Loganiales consists of Loganiacea and Oleceae; Apocynales of Apocynaceae and Asclepiadaceae and Gentianales of Gentianaceae Originally Engler included the Salvadoraceae in the order Contortae, but later on (Engler and Gilg 1924) this was transferred to the Sapindales.

The Aclepiadaceae have close affinities with Apocynaceae, Hutchinson considered that th Apocynaceae have been derived from stocks ancestral to the Loganiaceae.

35

ORDER–RUBIALES

According to Hutchinson this is the fifty second order of the phylum Angiospermae, subphylum Dicotyledones and division Lignosae. This order consists of only two families, *i.e.,* Dialypetalanthaceae and Rubiaceae. The Rubiaceae has been described in the present text in detail. Bentham and Hooker placed the family Rubiaceae along with one more, *i.e.,* Caprifoliaceae, in the first order, the Rubiales of class Dicotyledones, division Gamopetalae and series Inferae. Engler and Prantl have included the Rubiaceae along with four other families in the 8th order– the Rubiales of class Dicotyledones and sub-class Sympetalae.

Corolla sympetalous, more or less zygomorphic; stamens 5, rarely more, epipetalous, alternate with the corolla-lobes; anthers free, opening lengthwise; ovary inferior; axile placentation, rarely parietal; style 1; ovules numerous to solitary; seeds with endosperm. Leaves opposite, simple, entire, stipules inter-or intrapetiolar.

Family—Rubiaceae

There are about 400 genera and 5,300 species in this family.

Distribution: The family is cosmopolitan in distribution but the great majority of the species are found in tropical regions. The species of *Galium, Oldenlandia,* etc., are frequently found in the temperate regions. Certain species of *Galium* are found even in the Arctic zone or high elevations on tropical mountain ranges. The species of *Nertera* are found to be spread along the Andes. The genus *Comprosma* is commonly found in the south temperate and Antarctic regions, *Guettarda speciosa* is quite common round the Indian ocean from Africa to Malacca. The family is represented by several genera in our country, *e.g., Cinchona, Coffea, Adina, Hamelia, Ixora, Galium, Gardenia, Mussaenda, Rubia, Morinda* and many others.

Habit: The plants show great variations in habit. Majority of plants are either trees or shrubs. The herbaceous plants are also common and mainly they are confined to certain tribes. The members of the tribe **Galieae** are exclusively herbaceous. The tendency towards the formation of herbaceous forms is widespread, as certain herbaceous genera, such as, *Bouvardia* sp., is found in the typically woody tribe **Cinchoneae.** The climbing habit is also found in the family. Some climbers are herbaceous, whereas certain others are shrubby twiners, *e.g., Manettia. Uncaria* is a typical example of hook climbers. However, *Galium aparine* possesses stiff recurving hairs, which help in climbing on the supporting vegetation. The epiphytes also occur, *e.g., Myrmecodia. Myrmecodia* possesses a tuber-like stem developed from the swollen hypocotyl. This tuberlike stem contains a number of communicating galleries which harbour the ants. Such type of plants are known as **myrmecophilous** but the exact relation of ants to plant is not clear.

In the species of *Nauclea, Duroia,* etc., the ants live just beneath the inflorescence in the swollen hollow internodes. It is thought that the ants protect the flowers from other harmful insects. In the species of *Remijia* and *Duroia,* the ants live in the flask-shaped swellings situated on the leaf base Some important trees of the family are, *Anthocephalus cadamba* (Verna–**Kadam**), *Adina cordifolia* (Verna–**Haldu**), the species of *Cinchona, Mitragyna parviflora,* etc., *Coffea arabica, Hamelia patens* many species of *Ixora, Gardenia, Mussaenda,* etc., are medium sized or large shrubs. Certain species of *Galium, Rubia, Oldenlandia,* etc, are beautiful herbs.

Leaves: They are simple, stipulate, opposite decussate, entire or rarely toothed.

The stipules may be inter or intrapetiolar. In the former case, the stipules are situated at the side of leaf-base and thus lie in between the leaves; whereas in the latter case the stipules are situated in front of the leaf base in between the petiole and the stem. Sometimes the stipules unite to each other and then to the petiole, forming a sheath like structure around the stem. In certain species of the genus *Gardenia,* the stipules form a conical cap like structure which protects the apex of the stem, and is thrown off as the bud develops. Sometimes the stipules split into bristle-like structures, and at the top of each bristle a resin secreting gland develops, *e.g.,* in *Pentas.* In the case of the stipules of *Coffea, Gardenia,* etc., such glands are situated at the base of the stipules. In the case of *Galium* and allied genera, the leaf like (foliaceous) stipules are found. In such cases the stem apparently bears the whorls of leaves at the nodes, and the leaves may only be distinguished from the stipules by the presence of axillary buds.

Inflorescence: Usually the inflorescence is of cymose type, *i.e.,* a dichasial or a panicled cyme. Sometimes the small-flowered dichasia are aggregated into dense globose heads. In these heads the inferior ovaries of the flowers unite together, *e.g.,* in *Sarcocephalus, Morinda,* etc. In *Gardenia, Randia,* and some other plants the inflorescence consists of solitary terminal flowers. In *Coffea arabica* the flowers are found to be arranged in axillary cymes, whereas in *Ixora subsessilis,* they are arranged in subsessile corymbose cymes. The flowers of *Oldenlandia* are being arranged in panicled cymes. In *Anthocephalus indicus* and *Adina cordifolia* the flowers are found to be arranged in globose heads. In the species of Australian *Pamax,* the two flowers of a pair are so united, that the two possess a common calyx.

Flowers: The flowers are hermaphrodite, usually actinomorphic and either tetramerous or pentamerous, with five or four stamens and two carpels. They are complete and epigynous. Sometimes the flowers become unisexual by means of reduction, *e.g., Coprosma.* The flowers of *Henriquezia,* are however zygomorphic.

Calyx: The calyx consists of five or four sepals, gamosepalous. In the case of *Mussaenda, Warscewiezia,* etc., one or rarely more of the sepals become leaf like and white, yellow or variously coloured. The development of such leaf-like sepal increases the attractiveness of the inflorescence and attraction for the insects. In certain cases the sepals increase in size only after fertilization *e.g., Nematostylis, Alberta* ,etc. The aestivation is valvate.

Corolla: The corolla consists of five or four petals, gamopetalous. The corolla is tubular, campanulate or rotate. In *Henriquezia,* the corolla is more or less zygomor-

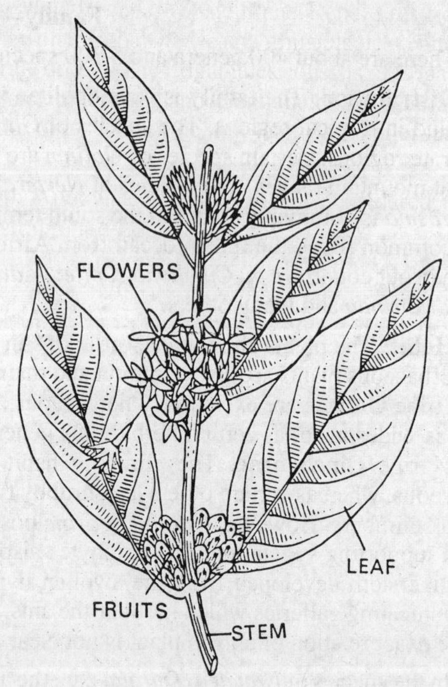

Fig. 35.1 Rubiaceae–a twig of *Coffea arabica* with leaves, flowers and fruits.

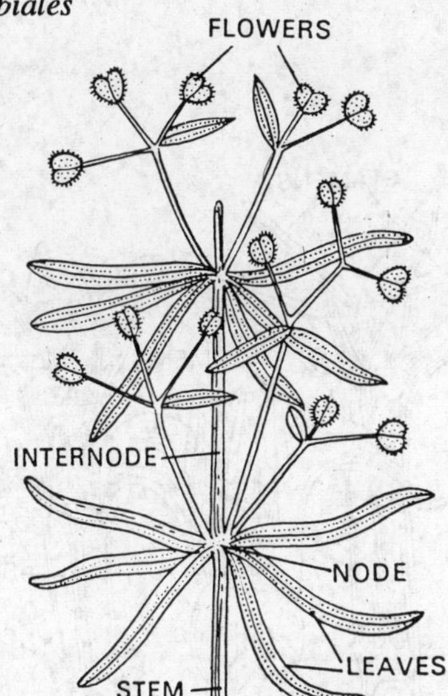

Fig. 35.2. Rubiaceae—A flowering twig of *Galium* Sp.

phic and bilabiate. The aestivation is either contorted or valvate.

Androecium: It consists of five or four stamens. The stamens are epipetalous, *i.e.*, they are inserted at or near the throat of the corolla tube. The stamens alternate with the petals. The anthers are bicelled, introrse and dehisce by longitudinal slits and sometimes by apical pores. In irregular corollas, the stamens are of unequal length and bent. In *Oldenlandia,* hereroandry occurs (*i.e.,* the stamens are of different length).

Gynoecium: Usually the gynoecium consists of two carpels, syncarpous, Generally the ovary is inferior, but in the Australian genus *Synaptantha,* it is half inferior. However, *Gaertnera* and *Pogamea,* the ovary is superior. Generally the ovary is bilocular but in *Gardenia,* it is unilocular, with parietal placentation. In general, the placentation is axile. One to many ovules are found in each locule of the ovary. The style is simple with globose stigma.

Fruit: The fruit is a drupe, capsule or berry. They may be dry or fleshy. It is a fleshy berry *e.g.,* in the case of *Randia, Mussaenda,* etc., it is a drupe, *e.g.,* in *Coffea* and others; it is a capsule, *e.g.,* in *Silvianthus, Ophiorrhiza,* etc. The fruits of the species of *Galium* are sticky, because of the presence of small stiff recurving hairs.

Seeds: Generally the seeds are small and more or less winged, endospermic; the flat cotyledons of embryo lie face to face.

Pollination: The pollination is usually effected by means of insects (*i.e.,* entomophilly). The nectar, the flag-like sepals attract the insects.

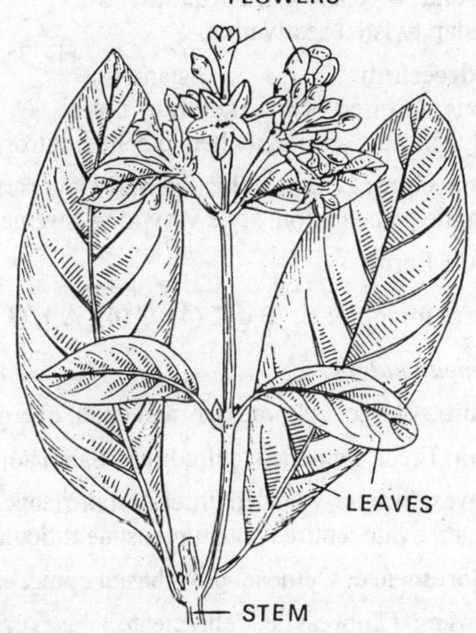

Fig. 35.3. Rubiaceae—A flowering twig of *Cinchona* sp.

Floral formula: $\oplus$ ⚥ K (4–5), $\overset{\frown}{\text{C (4–5), A5}}$, G$(\overline{2})$.

Bentham &Hooker (1862)	Engler & Prantl (1931)	Hutchinson (1959)
Phanerogms	Phanerogams	Angiospermae
Dicotyledones	Dicotyledones	Dicotyledones
Gamopetalae	Sympetalae	Lignosae
Inferae	Rubiales	Rubiales
Rubiales	Rubiaceae	Rubiaceae
Rubiaceae		

Description of important plants in semi-technical language.

1. Ixora coccinea Linn; Verna. **Rangan, Rookmini;** Eng. Jungle flame *Ixora.*

Habit: A large cultivated ornamental shrub.

Stem: Erect, branched, solid, cylindrical, green, herbaeous or somewhat woody.

Leaves: Cauline and ramal, simple, opposite, sessile, stipulate, ovate, smooth entire, acute, unicostate reticulate venation.

Inflorescence: Cymose, corymbose cyme.

Flower: Bisexual, actinomorphic, pedicellate, bracteate, bracteolate, complete, epigynous, tetramerous.

Calyx: 4 sepals, gamosepalous, superior, valvate aestivation.

Corolla: 4 petals, gamopetalous, corolla tubular, twisted aestivation.

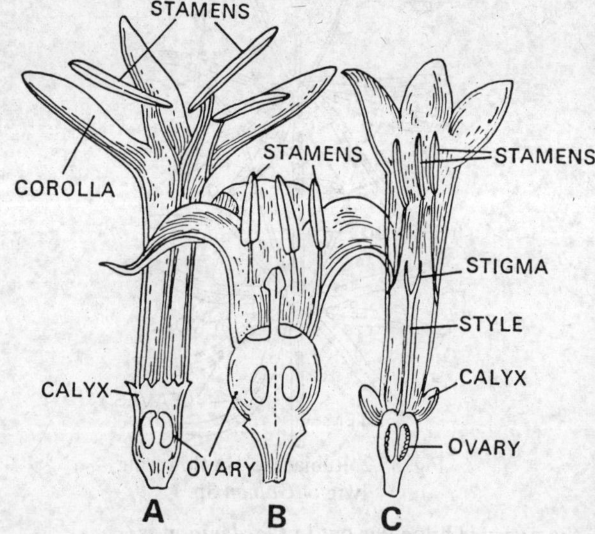

Fig. 35.4 : Rubiaceae—flowers in vertical sections : A, *Coffea arabica;* B, *Rubia tinctorum* ; C, *Cinchona* sp.

Androecium: 4 stamens, epipetalous, inserted at the mouth of corolla, polyandrous, sessile, anthers bicelled, introrse, dorsifixed.

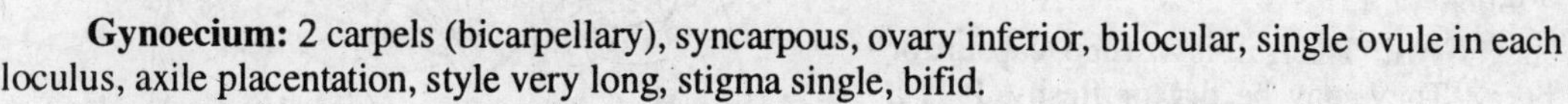

Gynoecium: 2 carpels (bicarpellary), syncarpous, ovary inferior, bilocular, single ovule in each loculus, axile placentation, style very long, stigma single, bifid.

Fruit: Berry.

Floral formula: Br ⊕ ⚥ K (4), $\overparen{C (4), A 4}$, G $(\overline{2})$.

2. ***Hamelia patens.***

Habit: A large, cultivated, ornamental, evergreen shrub.

Stem: Erect, branched, cylindrical, hairy, solid, woody.

Leaves: Simple, whorled (three or more leaves at each node), cauline, petiolate, stipulate, stipules interpetiolar, ovate, entire acute, unicostate reticulate venation.

Inflorescence: Cymose, polychasial cyme, each branch helicoid.

Flowers: Subsessile, ebracteate, bisexual, actinomorphic, complete, orange coloured, epigynous.

Calyx: 5 sepals, gamosepalous, petaloid, reddish, adnate to the ovary wall, superior, valvate aestivation.

Corolla:5 petals, gamopetalous, tubular, orange coloured, superior, twisted aestivation.

Androecium: 5 stamens, polyandrous, alternate to the petals, filaments short, epipetalous, anthers basifixed, long, bicelled, introrse.

Gynoecium: 5 carpels (pentacarpellary), syncarpous, ovary inferior, pentalocular, several ovules in each loculus, axile placentation.

Fruit: Berry.

Floral Formula: ⊕ ⚥ K(5), $\overparen{C(5) A}$ (5), G$(\overline{5})$.

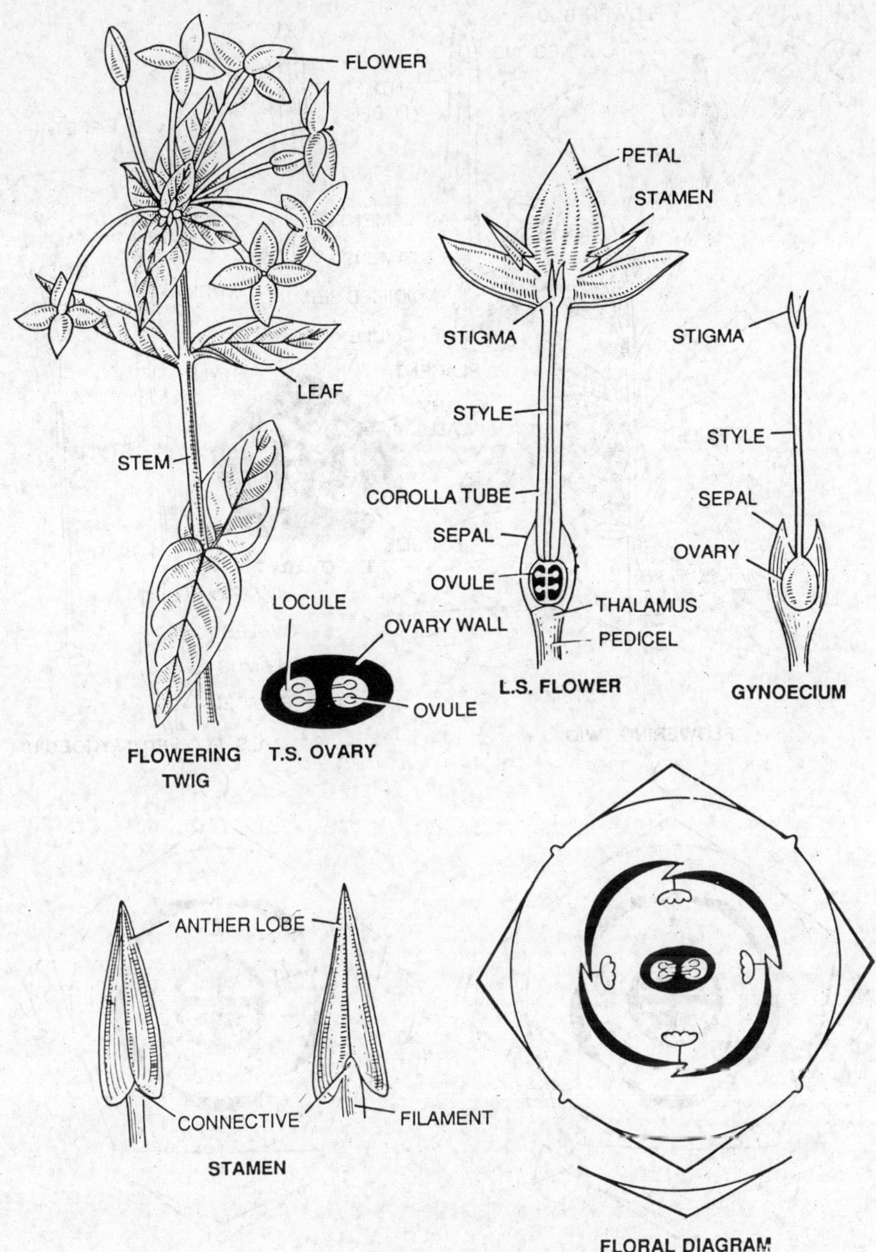

Fig. 35.5. Rubiaceae. *Ixora coccinea* Linn.; Eng. Jungle flame Ixora; Verna. **Rangan, rookmini.**

Economic value. Grown as a hedge plant.

2. ***Oldenlandia corymbosa*** L., Eng. Daman-paper. **Root**–simple, tap. **Stem**–herbaceous, aerial, prostrate, branched, solid, angular, green. **Leaf**–cauline and ramal, opposite decussate, stipulate, interpetiolar stipules, simple, sessile, lanceolate, entire, acute, pubescent, unicostate reticulate. **Inflorescence**–cymose, axillary dichasial cyme. **Flower**–Bracteate, bracteolate, pedicellate, actinomor-

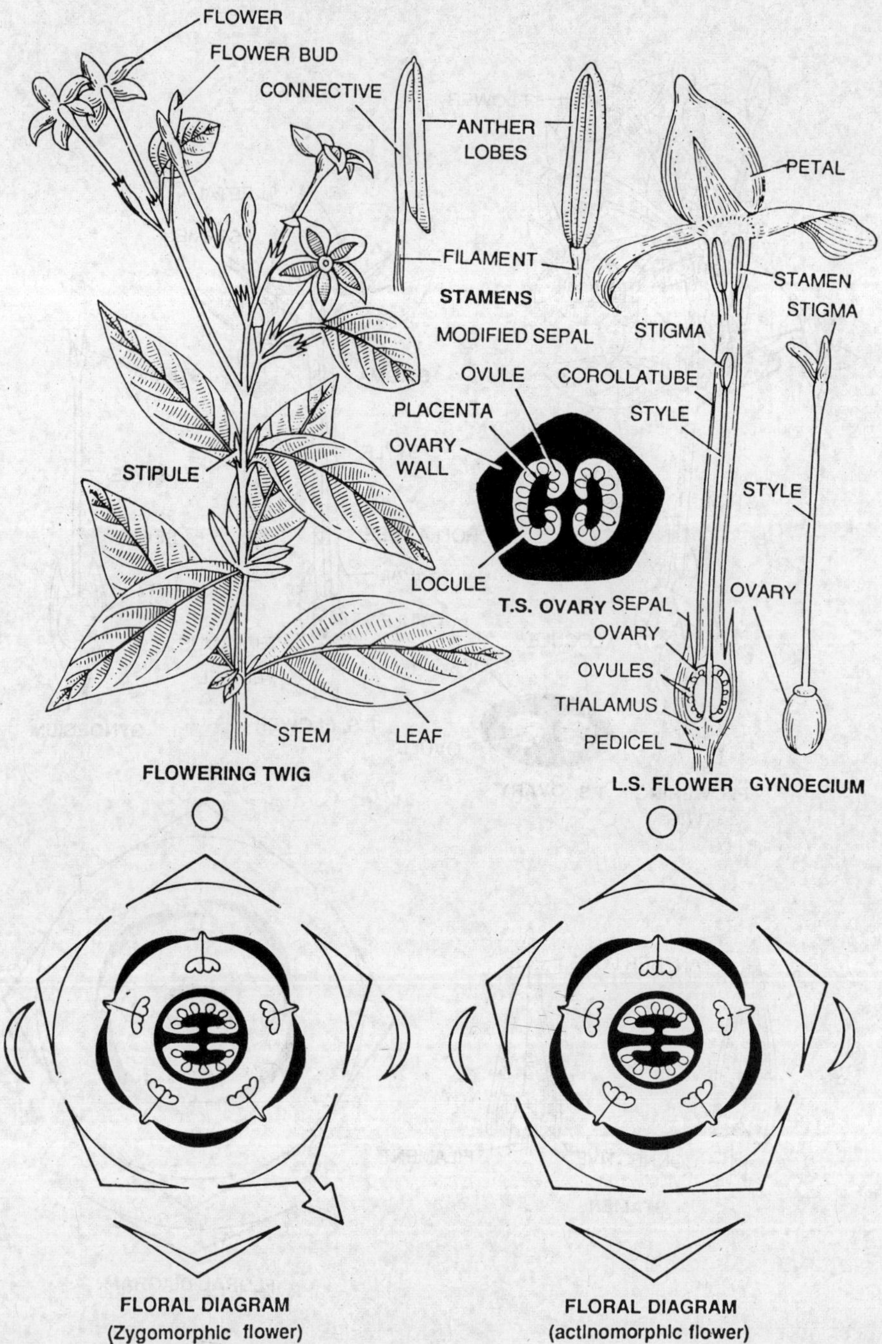

Fig. 35.6. Rubiaceae. *Mussaenda luteola* Delile; Eng., advertising flag plant.

Economic value. The plant is grown in gardens as an ornamental.

phic, hermaphrodite, complete, tetramerous, epigynous, cyclic. **Calyx**–4, gamosepalous, valvate, persistent. **Corolla**–4, gamopetalous, valvate, viloet. **Androecium**–4, polyandrous, epipetalous, filaments short, dithecous, basifixed introrse. **Gynoecium**–2, bicarpellary, syncarpous, inferior, bilocular, many ovules in each locule, axile, style short, stigma simple and globular. **Fruit**–berry.

Floral formula–Br Brl $\oplus$ ⚥ $K_{(4)}$, $\overset{\frown}{C_{(4)}, A_4}$, $\overline{G}_{(2)}$.

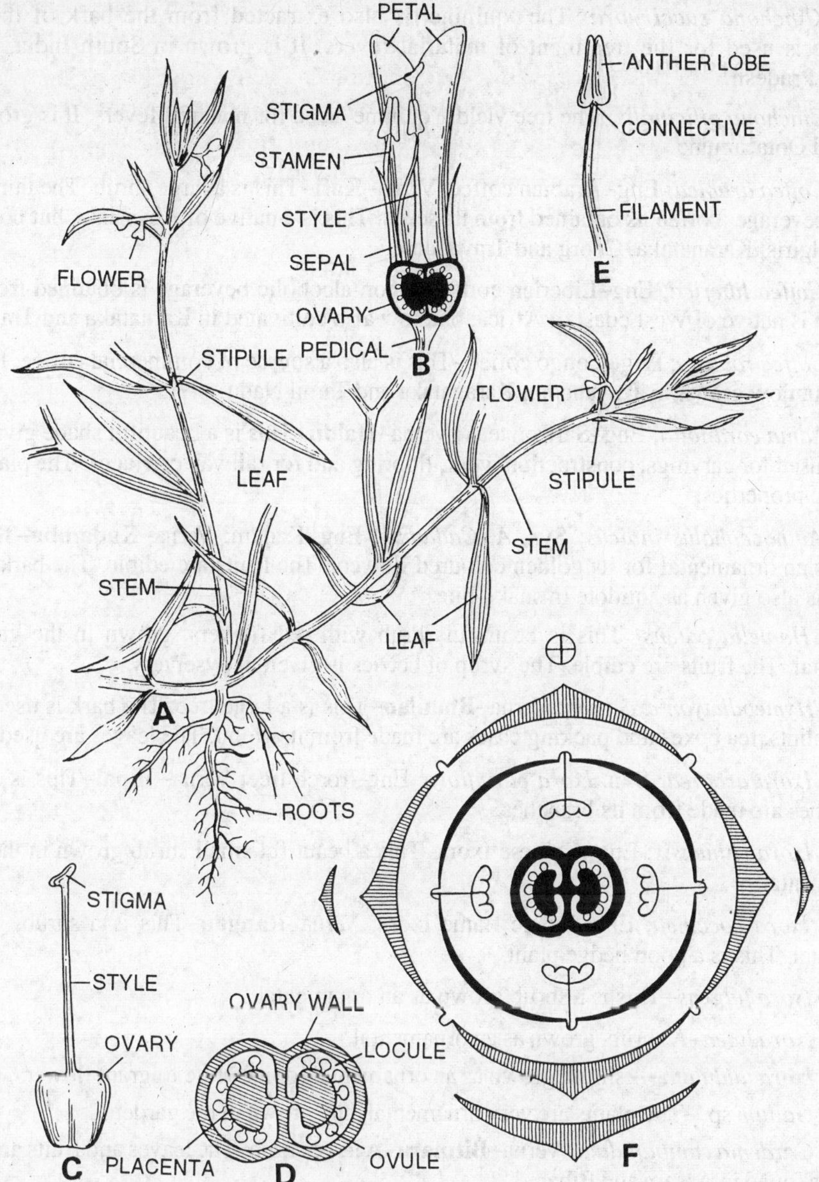

Fig. 35.7. Rubiaceae—*Oldenlandia corymbosa* L.A. flowering plant ; B, L.S. of flower ; C, gynoecium ; D, T.S. of ovary ; E, stamen ; F, floral diagram.

Economic Importance of the Family

From the point of view of its economic importance, the family is of great value. A list of few important plants is given below:

1. *Cinchona calisaya;* Eng.–Quinine; Verna.-**Cinchona**–This is a tree. The famous medicine of malaria known as quinine is obtained from the bark of this tree. The tree is native of South America, but found in the Nilgiris and Sikkim.

2. *Cinchona ledgeriana;* The medicine known as 'quinine' which is used for malarial fevers is also extracted from the bark of the tree. This is grown in Bengal, the Khasia hills and South India.

3. *Cinchona succirubra;* The 'quinine' is also extracted from the bark of this tree. The medicine is used for the treatment of malarial fevers. It is grown in South India, Sikkim and Madhya Pradesh.

4. *Cinchona officinalis;* The tree yields 'quinine' used for malarial fevers. It is grown in South India and Ootacamund.

5. *Coffea arabica;* Eng–Arabian coffee; Verna–**Kafi**–This is a large shrub. The famous non–alcoholic beverage 'coffee' is obtained from its seeds. This is a native of Abyssinia, but now cultivated in the Nilgiris, Karanataka, Coorg and Travancore.

6. *Coffea liberica*; Eng–Liberian coffee–A non-alcoholic beverage is obtained from its seeds. The bush is native of West coast of Africa, but now also cultivated in Karnataka and Travancore.

7. *Coffea robusta*: Eng–Congo coffee–This is also a source of commercial coffee. It is native of Congo but now grown in Travancore, Karanataka and Tamil Nadu.

8. *Adina cordifolia*; Eng–Saffron teak; Verna–**Haldu**–This is a beautiful shade giving tree. The wood is used for carvings, construction work, flooring and for railway carriages. The plant possesses antiseptic properties.

9. *Anthocephalus indicus*; Syn. *A. Cadamba*; Eng–Kadam; Verna–**Kadamba**–This is a tree grown as an ornamental for its golden coloured flowers. The fruits are edible. The bark is used as a tonic. It is also given as antidote in snake-bite.

10. *Hamelia patens;* This is beautiful shrub with red flowers, grown in the gardens as an ornamental. The fruits are edible. The syrup of berries is given in dysentery.

11. *Hymeodictyon excelsum*; Verna–**Bhaulan**–This is a large tree. The bark is used in tanning. Match splints, tea boxes and packing cases are made from its wood. The leaves are used as fodder.

12. *Ixora arborea;* Syn. *Ixora parviflora*; Eng–Torch tree; Verna–**Jilpai**–This is a small tree. The torches are made from its branches.

13. *Ixora chinensis*; Eng–Chinese Ixora–This a beautiful small shrub grown in the gardens as an ornamental.

14. *Ixora coccinea*; Eng–Jungle flame Ixora; Verna–**Rangan**–This is a shrub, grown as an ornamental. This is a good hedge plant.

15. *Ixora fulgens*–This is a shrub grown as an ornamental.

16. *Ixora lutea*–A shrub, grown as an ornamental.

17. *Ixora nudulata*–A shrub, grown as an ornamental for its white fragrant flowers.

18. *Galium* sp. –The plants are very ornamental, and grown in the gardens.

19. *Gardenia campanulata;* Verna–**Bitmara**–A large shrub. The leaves and fruits are edible. The plants are found in Assam and Bihar.

20. *Gardenia gummifera;* Eng–Gummy cape jasmine; Verna–**Dikamli**–The plant yields a gum-resin. It is used as carminative and stimulant, and also given in dyspepsia. The fruits are edible.

21. *Gardenia jasminoides*; Verna–**Gandharaj**–This is an ornamental shrub grown in the gardens. A yellow dye is obtained from the pulp of its fruits. An essential oil is extracted from its flowers, which is used in making of perfumes.

22. *Gardenia latifolia ;* Verna–**Papra**–This is a small tree, usually grown as hedge plant.

23. *Gardenia resinifera;* Verra–**Dikamli**–This is a large ornamental shrub with fragrant flowers.

24. *Mussaenda glabrata;* Verna–**Bedina**–This is a small tree, grown as an ornamental in the gardens. The plant is used for asthma, fevers and dropsy.

25. *Mussaenda luteola*; This is a shrub of ornamental value.

26. *Mitragyna parviflora*; Verna–**Kadam**–This is a tree found throughout our country. A fibre is obtained from the bark which is used for cordage. The furniture, agriculture implements and carvings are prepared from its wood. The leaves are used as a fodder.

27. *Morinda angustifolia* ; Verna–**Banhardi**–This is a large climbing shrub. The roots yield a yellow dye. The plants are found in the Khasia hills, Assam and Sikkim.

28. *Oldenlandia umbellata*; Eng–Indian madder; Verna–**Saya**–This is a herb, commonly found in Bengal, Orissa and South India. A red dye is obtained from the bark of its root.

29. *Oldenlandia corymbosa*; Verna–**Daman paper**–The decoction is used as a remedy for dysentery and cholera.

30. *Randia dumetorum;* Syn. *Gardenia dumetorum*; Verna–**Mainphal**–This is a large shrub or a small tree. The fruits are edible. The pulp of the fruits is used to cure dysentery. The plants are commonly found in Tamil Nadu, Maharashtra, Gujarat and Uttar Pradesh.

31. *Randia uliginosa*; Syn. *Gardenia uliginosa*; Verna– **Pindalu**–The fruits are edible and used for dyeing purposes. The leaves are used as fodder. The unripe fruits are used as a remedy for dysentery and diarrhoea. The plants are commonly found throughout the country except in the Northern India.

32. *Rubia cordifolia*; Verna–**Manjit**–This is a herb. A dye is extracted from its roots.

33. *Rubia khasiana*; A red dye is obtained from its roots.

34. *Rubia sikkimensis*–A red dye is obtained from its roots. It is used for dyeing both cotton and hair.

35. *Rubia tinctorum*; Verna–**Bacho**–The roots of this plant also yield a red dye which is used for dyeing cloth. The plant is also used as fodder plant.

Systematic Rlationships: Bentham and Hooker included two families, *i.e.*, the Caprifoliaceae and the Rubiaceae within the order Rubiales, Engler and Diels included five families in the order. They are–Rubiaceae, Adoxaceae, Dipsacaceae, Caprifoliaceae and Valerianaceae. Hutchinson (1926) however, has placed the Rubiaceae and the Caprifoliaceae of Engler and Diels to his Asterales, but later on (1948) he modified his own view and placed these three families in a new order, the Valerianales. According to him the Rubiales and the Valerianales are not at all related phylogenetically. According to him the Rubiales are derived from Loganiaceae and the Valerianales from Saxifragaceae.

36

ORDER–BIGNONIALES

According to Hutchinson this is the fifty third order of the phylum Angiospermae, subphylum Dicotyledones and division Lignosae. This order comprises of four families, *i.e.,* Coleacaceae, Bignoniaceae, Pedaliaceae and Martyniaceae of which the Pedaliaceae has been described in the present text in detail. Bentham and Hooker have placed the family Pedalineae along with seven others including Scrophulariaceae, Acanthaceae, etc., in the ninth order–the Personales of class Dicotyledones, division Gamopetalae and series Bicarpellatae. Engler and Prantl have put the family Pedaliaceae along with nineteen other families including Convolvulaceae, Verbenaceae, Labiatae, Solanaceae, Scrophulariaceae, Acanthaceae, etc., in the sixth order–the Tubiflorae to class Dicotyledones and sub-class Sympetalae.

Corolla sympetalous, more or less zygomorphic, stamens, 5, 4, or 2, epipetalous, alternate with the corolla lobes, fifth reduced to a staminode; ovary superior, 2-or 1-locular; ovules numerous; seeds without endosperm, often winged; leaves opposite, mostly compound; stipules very rare..

Family—Bignoniaceae

There are about 120 genera and 650 species in this family.

Distribution : The family has its main distribution in South America; some genera are tropical African, e.g., *Kigelia, Crescentia* etc. In India the family is represented by about 15 genera and 40 species which occur mostly in Western and Southern India and a few in Himalayas. Species of *Millingtonia* and *Oroxylum* occur frequently in India; many American and African genera have been introduced in India and they are now naturalised *e.g., Tabebuia, Kigelia, Pyrostegia, Spathodea,* etc. *Tecoma stans* (yellow bells), *Kigelia pinnata* (sausage tree), *Millingtonia hortensis*(Indian cork tree) are some well known Indian species.

Habit: Most plants of this family are represented by trees and shrubs. Most of the shrubs are climbers by tendrils or twiners, rarely root climbers. *Millingtonia hortensis* L., *Kigelia pinnata DC., Pajanelia rheedei* DC., are examples of very large trees. The examples of tendril climbers are represented by species of *Pyrostegia, Bignonia,* etc., twiners by *Tecomaria, Pandorea,* etc, and that of root climbers by *Tecoma radicans.* The species of *Campsis* (*e.g., Campsis radicans*) climb with the help of aerial roots.

Anomalous Secondary Thickening in Stems. In twining plants anomalous secondary thickening in the stems is of universal occurrence. Here the cambium inspite of its normal position adds secondary phloem and secondary xylem in very much varying quantities that is in some segments much greater amount of xylem than of pholem and in others more phloem than xylem–as a result a characteristic structure with ridged and furrowed xylem cylinder is formed.

Leaves : The leaves are opposite, compound pinnate and exstipulate. The leaves are sometimes rather frequently 2-3 times pinnate *e.g., Oroxylum indicum, Millingtonia hortensis,* etc.

Tendrils : The tendrils are the metamorphosed terminal leaflets of a compound pinnate leaf, hence the tendrils are typical leaf tendrils. In *Bignonia* the tendril is grapple-like with three claws which become thick after grasping.

Inflorescence: The flowers are arranged in racemose or cymose inflorescence, either dichasium or helicoid cyme or panicle of dichasia or clustered.

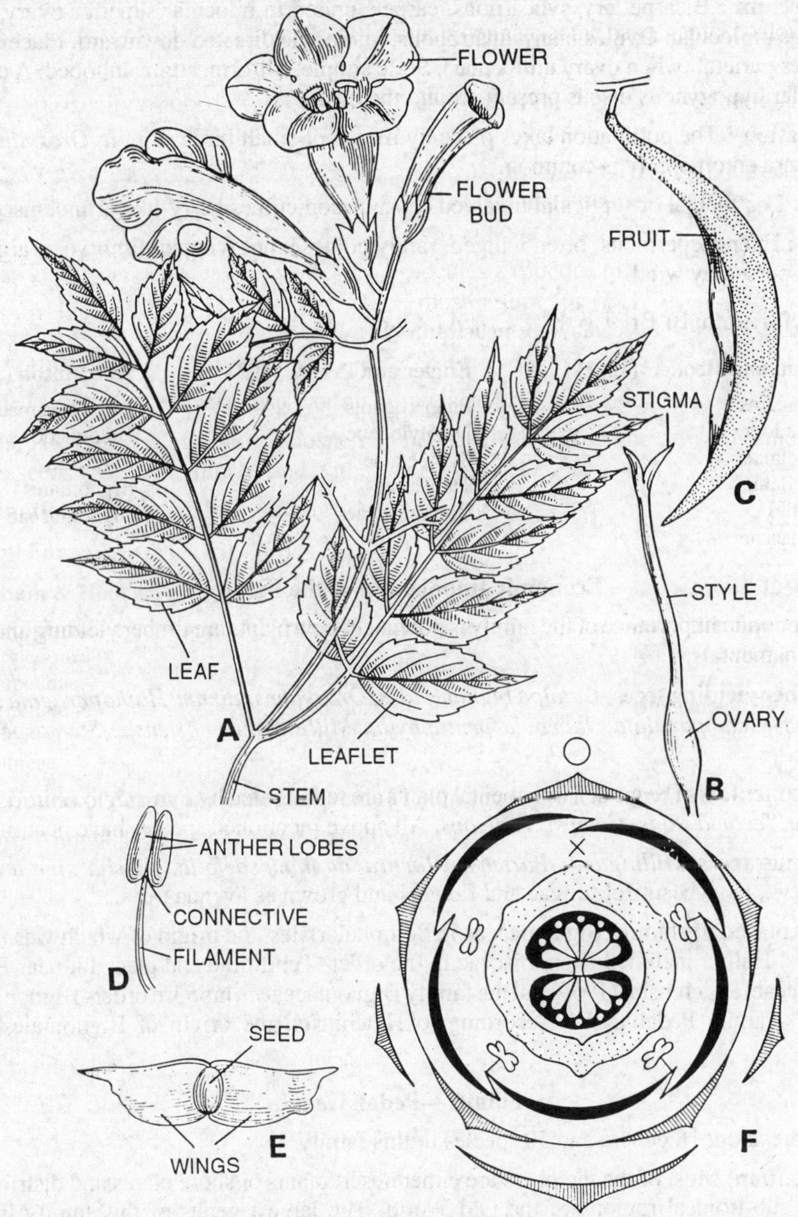

Fig. 36.1. Bignoniaceae. *Campsis radicans*. A, flowering twig; B, gynoecium; C, fruit; D, stamen; E, winged seed; F, floral diagram.

Flowers : The flowers are showy, zygomorphic, hermaphrodite and hypogynous.

Calyx : Sepals 5, united in 5 lobed or toothed, sometimes bilabiate irregular calyx, lobes imbricate or rarely valvate in bud *e.g., Pyrostegia*. Spathaceous calyx is present in *Dolichandrone*.

Corolla : Petals 5, united in funnel-shaped or bell-shaped (campanulate) or sometimes bilabiate corolla; aestivation of petals is descending imbricate.

Androecium: Stamens 4, didynamous, the posterior fifth stamen often reduced to staminode or absent, inserted on the basal part of corolla tube; anthers dithecous, anther lobes are placed little above the other; dehiscence of anther lobes is longitudinal.

Gynoecium : Bicarpellary, syncarpous, carpels united in bilocular superior ovary, sometimes ovary may be unilocular. Ovules many, anatropous, micropyle directed downward. Placentation axile or sometimes parietal (when ovary unilocular). Style simple, filiform, stigma bilobed. A conspicuous fleshy annular hypogynous disc is present around the ovary.

Pollination : The pollination takes place by means of small birds (*e.g.*, in *Oroxylum indicum*). In other genera entomophily is common.

Fruit : Loculicidal or septicidal bivalved capsule, sometimes berry-like or indehiscent.

Seeds : Exendospermous, often winged, rarely comose, not winged. Embryo straight. Winged seeds are dispersed by wind.

Floral formula. Br Brl ·|· ⚥ $K_{(5)}$, $\overparen{C_{(5)}, A_4}$, $G_{(2)}$.

Bentham and Hooker (1862)	Engler and Prantl (1931)	Hutchinson (1959)
Phanerogams	Phanerogams	Angiospermae
Dicotyledones	Dicotyledones	Dicotyledones
Gamopetalae	Sympetalae	Lignosae
Bicarpellatae	Tubiflorae	Bignoniales
Personales	Bignoniaceae	Bignoniaceae
Bignoniaccae		

Economic Importance of the Family

The economic importance of the family is not much. Few plants are timber yielding and cultivated as garden ornamentals.

1. **Timber yielding trees**–*Catalpa bignonioides, Oroxylum indicum, Haplophragma adenophyllum, Spathodea campanulata, Tabebuia pentaphylla, Millingtonia hortensis, Stereospermum personatum,* etc.

2. **Ornamentals**–The garden ornamental plants are represented by *Pyrostegia venusa, Spathodea campanulata, Tecoma stans, Campsis radicans.* All above mentioned species have beautiful flowers.

3. **Avenue trees**–*Millingtonia hortensis, Jacaranda mimosaefolia, Kigelia pinnata, Tabebuia spectabilis,* etc., have beautiful foligae and flowers and grown as avenue trees.

Bessey placed the family Bignoniaceae in Scrophulariales, the origin of which was traced from Polemoniales. Hallier included Bignoniaceae in the order–Tubiflorae and placed it near Pedaliaceae and Acanthaceae. Hutchinson (1959) put the family Bignoniaceae in his 53rd order–Bignoniales which also includes family Pedaliaceae. According to Hutchinson the origin of Bignoniales was from Loganiales.

Family—Pedaliaceae

There are about 16 genera and 50 species in this family.

Distribution: Most of the members are either desert plants or shore plants and distributed in the tropical and sub-tropical regions of the Old World. The largest genus of this family is *Sesamum* containing about 16 species of Africa and India. in our country the family is represented by *Sesamum indicum* (**Til**).

Habit: Mostly annual or perennial herbs, rarely shrubs.

Root: Usually tap and branched.

Leaves: Opposite or the uppermost sometimes alternate, simple, entire or lobed, exstipulate, mucilaginous glandular hairs present.

Inflorescence: Cymose, solitary axillary or simple axillary dichasial cyme (usually 3-flowered cymes). Rarely the flowers are fascicled or in racemes.

Flower: Pedicellate, hermaphrodite (bisexual), zygomorphic, irregular, complete, hypogynous (epigynous in *Trapella*). The bracts may or may not be present.

Calyx: Consists of 5 sepals (sometimes 4), sepals connate at the base, gamosepalous, aestivation imbricate.

Corolla: Usually 5-lobed and somewhat bilabiate corolla forms a broad tube, gamopetalous, aestivation imbricate.

Androecium: Consists of 4 stamens, didynamous (only 2 in *Trapella*), sometimes the fifth posterior stamen is represented by a small staminode, epipetalous stamens, the anthers are dithecous and dehisce by longitudinal slit, introrse. Sometimes the number of stamens is reduced to two only.

Gynoecium: Usually 2 carpels, syncarpous, locules 2 (but due to the formation of false septum, the number of locules is doubled, *i.e.,* 4), ovary mostly superior (inferior in *Trapella,* an aquatic) placentation axile, the ovules one to many on each placenta, ovules anatropous. Style usually one, slender filiform. The stigma 2-lobed.

Fruit: Either a loculicidal capsule or nut, often spiny, winged, hooked or thorny.

Seeds: Albuminous with thin endosperm and a small straight embryo. Seeds smooth. In Indian species the seeds are exalbuminous and without wings.

Pollination: Entomophilous (through the agency of insects).

Floral Formula : ·|· ⚥ K (5), $\overbrace{C(5)\ A}$ 2+2, G($\underline{2}$).

Bentham & Hooker (1862)	**Engler & Prantl (1931)**	**Hutchinson (1959)**
Phanerogams	Phanerogams	Angiospermae
Dicotyledones	Dicotyledones	Dicotyledones
Gamopetalae	Sympetalae	Lignosae
Bicarpellate	Tubiflorae	Bignoniales
Personales	Pedaliaceae	Pedaliaceae
Pedalineae		

Description of ***Sesamum indicum***; Verna–**Til,** in semitechnical language.

Habit: Annual herb 1 to 2 feet high, cultivated.

Root: Tap and branched.

Stem: Erect, pubescent, profusely branched from the base, angular.

Leaves: Oblong or ovate, 3 to 5 inches, lower leaves opposite, lobed, with serrated margins, upper leaves alternate, narrow, oblong, the upper surface glabrous, the lower surface pubescent.

Inflorescence: Cymose, solitary axillary.

Flower: Pedicellate (small pedicels), hermaphrodite, zygomorphic, hypogynous, compele, pinkish in colour, ebracteate.

Calyx: 5 sepals (small in size about ¼ inch long), gamosepalous, sepals lanceolate, acute, hairy, aestivation imbricate or valvate.

Corolla; 5 petals, gamopetalous, 2 lipped, pinkish in colour with yellow markings, each petal about 1½ inches long, corolla 5-lobed, the upper lip consists of two equal sized lobes while the lower lip includes three petals out of which the central one is largest, corolla tubular-ventricose with subgibbous throat which is hairy outside.

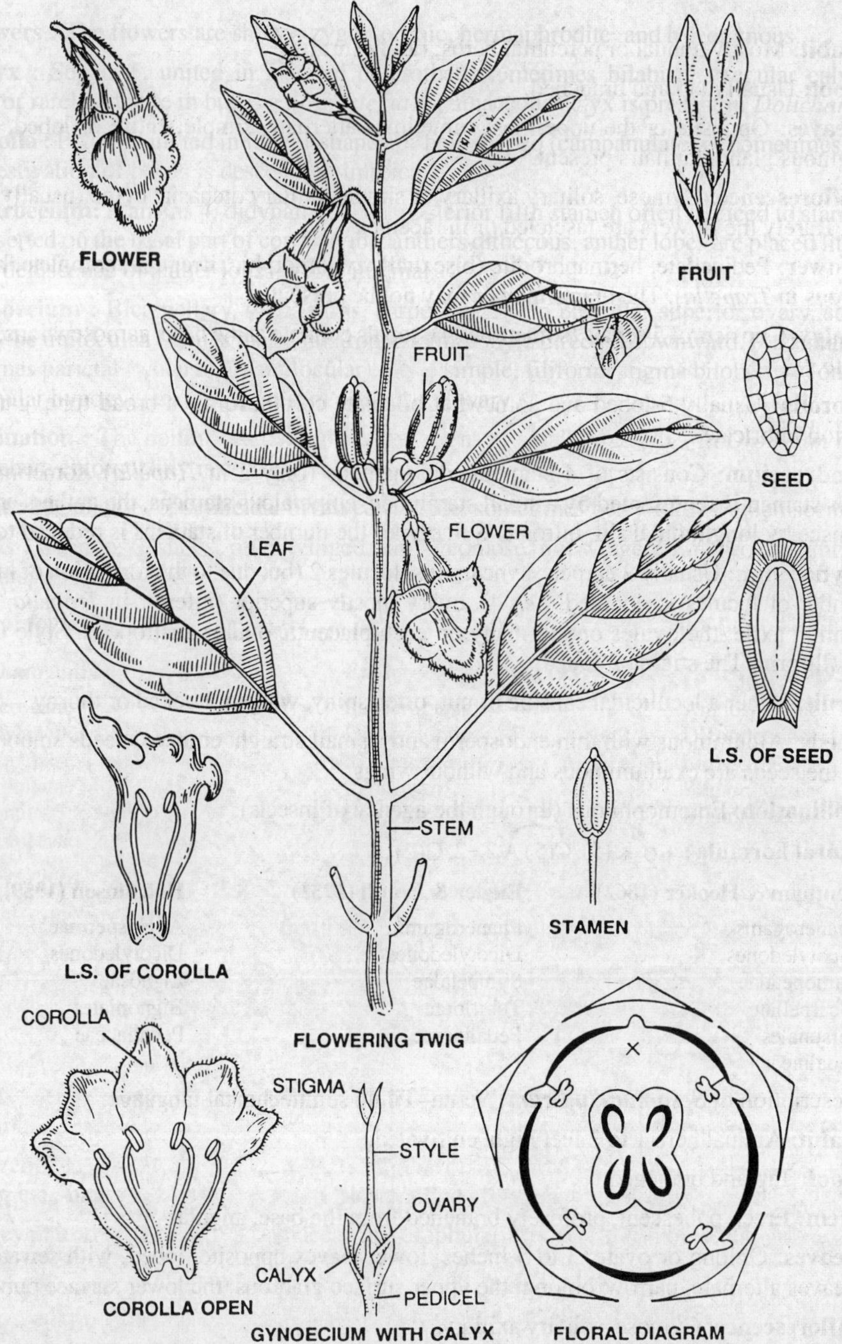

Fig. 36.2. Pedaliaceae. *Sesamum indicum* Linn; Eng., sesame; Verna.**til.**

Androecium: 4 stamens, didynamous, epipetalous. Instead of fifth stamen usually a staminode is present, anthers 2-celled (dithecous) sagittate, introrse and dehisce by longitudinal slit.

Gynoecium: 2-carpels (bicarpellary), syncarpous, ovary superior, bilocular (four locules formed due to false septation), placentation axile, style filiform, stigmas 2 and lobed.

Fruit: Loculicidal capsule, one inch long, erect, hispid, beak short, valves separating half way down.

Seeds: Small and many, smooth, albuminous with straight embryo, black and white.

Floral Formula: ·|· ⚥ K (5) $\widehat{C(5)\ A}$ 2 + 2 G ($\underline{2}$).

Generic characters: Stems and leaves pubescent and with sticky glandular hairs. Inflorescence solitary axillary. Corolla bilabiate, upper lip consists of two equal sized lobes, the lower lip includes three petals. Corolla ventricose with hairy subgibbous throat, two carpels but due to false septation four chambers are formed. Fruit loculicidal capsule. Seed, small black or white.

Economic Importance of the Family

1. *Sesamum indicum* (Verna–**til, tili, tila**); In India it is extensively cultivated during the rainy season, but chiefly as a mixed crop. In many parts of India it is grown as pure crop. There are two distinct varieties, black seeded and white seeded; the former yields the best oil, which is largely used in India for multipurpose uses. The oil is burnt in lamps, used for the preparation of scented oils, employed in anointing the body and in the manufacture of soap and perfumes. The seeds of white seeded variety are often eaten, chiefly in the form of sweet meats.

2. *Pedalium* sp. (Verna–**Bara gokhru**); It is used in the preparation of many Ayurvedic Medicines.

Systematic Relationships: Bentham and Hooker have included the family Pedaliaceae along with seven others in the order–Personales. Engler and Prantl included the families of Bentham and Hooker's Personales within their order–the Tubiflorae. Hutchinson, however, puts the family Pedaliaceae in the order–Bignoniales which is not related to his Personales.

37

ORDER–VERBENALES

According to Hutchinson this is the fifty-fourth order of the phylum Angiospermae, subphylum Dicotyledones and division Lignosae. The order comprises of five families, *i.e.* Ehretiaceae, Verbenaceae, Stilbeaceae, Chloranthaceae and Phrymaceae. The Verbenaceae has been described here in detail. Bentham and Hooker have placed the Verbenaceae along with three other families including Labiatae in the tenth order–the Lamiales of class Dicotyledones, division Gamopetalae and series Bicarpellatae. Engler and Prantl put the family Verbenaceae along with nineteen other families including Convolvulaceae, Labiatae, Solanaceae, Scrophulariaceae, Pedaliaceae, Acanthaceae, etc., in the sixth order–the Tubiflorae of class Dicotyledones and sub-class Sympetalae.

Corolla sympetalous, actionomorphic to zygomorphic; stamens 5, 4, or 2, epipetalous; ovary superior, entire bicarpellate; ovules solitary or paired; style terminal on the ovary. Leaves alternate, opposite or verticillate; no stipules.

Family—Verbenaceae

There are about 80 genera and 800 species in this family.

Distribution: The family is almost entirely tropical or subtropical in distribution. *Verbena* has 80 species which are mainly distributed in the tropical and extratropical regions of America. Only *Verbena officinalis* is found in Britain. In our country the family is represented by several genera, *e.g.*, *Verbena, Lantana, Duranta, Clerodendron, Tectona,* etc.

Habit: The plants may be herbs, shrubs or trees. The species of *Verbena, Lippia*, etc., are annual herbs. The species of *Lantana, Clerodendron, Duranta*, etc., are shrubs. *Tectona grandis* is a large tree reaching the height of 80-120 feet. Some plants of the family are climbers. The species of *Lantana, Clerodendron, Vitex*, etc., are good examples of climbers. *Avicennia* is a mangrove shrub found in tropical shores of both hemispheres.

Stem: The stems or branches are usually quadrangular and possess spines.

Leaves: Usually the leaves are opposite or whorled, only a few species possess alternate leaves. Generally they are simple and very rarely compound, *e.g.*, in *Vitex* they are 3-7 foliate. The leaves may be entire or divided and exstipulate.

Inflorescence: The inflorescence shows great variation. This may be either racemose or cymose. The flowers may be arranged in racemes, spikes or head or they may be axillary. The well developed coloured bracts are often present. Dichasial cymes are usually present which very often pass into monochasial cymes, *e.g.*, in *Vitex* and *Clerodendron.* The axillary cymes are aggregated in panicles and showy inflorescence.

Flower: The flowers are usually pentamerous, rarely showing reduction in the calyx, *e.g.*,

Physopsis, which possesses a regular tetramerous flower. In species of *Vitex* by means of reduction or abortion a two-lipped calyx is formed. Generally the flowers are hermaphrodite, rarely unisexual by means of abortion or reduction; zygomorphic, sometimes actinomorphic complex, hypogynous.

Calyx: Five sepals, gamosepalous, valvate aestivation. In the fruiting stage the calyx becomes much altered and enlarged. Sometimes it becomes bladder like and encloses the fruit and sometimes it becomes brightly coloured. Rarely the number of sepals ranges from 4 to 8.

Corolla: Usually it consists of five petals. It is tubular often curved with a spreading limb, rarely campanulate, lobes equal, unequal or more or less two-lipped, aestivation imbricate.

Androecium: It consists of four fertile stamens which are didynamous, sometimes the number is two, the stamens remain inserted on the corolla tube and alternate with its lobes, the fifth posterior is usually represented by a staminode or altogether absent. A pentamerous androecium (five stamens) is very rare as found in *Tectona* and *Geunsia* where the flower remains pentamerous throughout, sometimes there are only two fertile stamens and the remaining three are staminodes, *e.g.*, in *Ozera.* The filaments are free, anthers dorsifixed, two-celled, dehiscence longitudinal and introrse.

Gynoecium: Usually it consists of two carpels. Rarely there are four or five carpels as found in *Duranta* and *Geunsia* respectively. The posterior carpel becomes aborted in *Lippia, Lantana* and allied genera. A false septum develops from the middle of each carpel inwards dividing the two carpels into four one ovuled chambers. In the young ovary the placentation is parietal, but later on it becomes axile.

Fruit: Usually the fruit is drupe. The exocarp is fleshy or dry and the endocarp is hard. There are usually division forming two or four one seeded stones. In *Verbena* the fruit consists of four nutlets.

Seeds: The seeds are exalbuminous, ovate and sessile. The seed coat is thin and membranous. The embryo is straight and the radicle small directed towards the micropyle.

Pollination: The pollination takes place through the agency of insects.

Floral formula: $\cdot|\cdot$ ⚥ K (5), $\widehat{C(5), A}$ 2 + 2 or 5, G $(\underline{2})$

Bentham & Hooker (1862)	Engler & Prantl (1931)	Hutchinson (1959)
Phanerogams	Phanerogams	Angiospermae
Dicotyledones	Dicotyledonae	Dicotyledones
Gamopetalae	Sympetalae	Lignosae
Bicarpellatae	Tubiflorae	Verbenales
Lamiales	Verbenaceae	Verbenaceae
Verbenaceae		

1. ***Duranta repens*** Linn.; Eng. Pigeon berry.

Habit: Shrub, largely grown as hedge plants.

Root: Tap and branched.

Stem: Erect, herbaceous to woody, branched, solid, cylindrical, glabrous, spiny.

Leaves: Cauline and ramal, opposite decussate, exstipulate, sub-sessile, simple, leaf base swollen, dentate, acute, ovate, unicostate reticulate, glabrous.

Inflorescence: Panicled terminal and axillary racemes.

Flower: Bracteate, bracteolate, sub-sessile, hermaphrodite, zygomorphic, complete, pentamerous, hypogynous, cyclic.

Calyx: 5 sepals, gamosepalus, valvate.

Corolla: 5 petals , gamopetalous, imbricate, tubular, violet coloured.

Androecium: 5 stamens, free, epipetalous, didynamous; anthers dithecous, basifixed, introrse, dehiscing longitudinally.

Gynoecium: 4 carpels, syncarpous, ovary superior, tetralocular, axile placentation, style simple, stigma lobed.

Fruit : Drupe.

Floral formula : Br, Brl ·|· ⚥ K (5), $\widehat{C(5), A4}$, G(4).

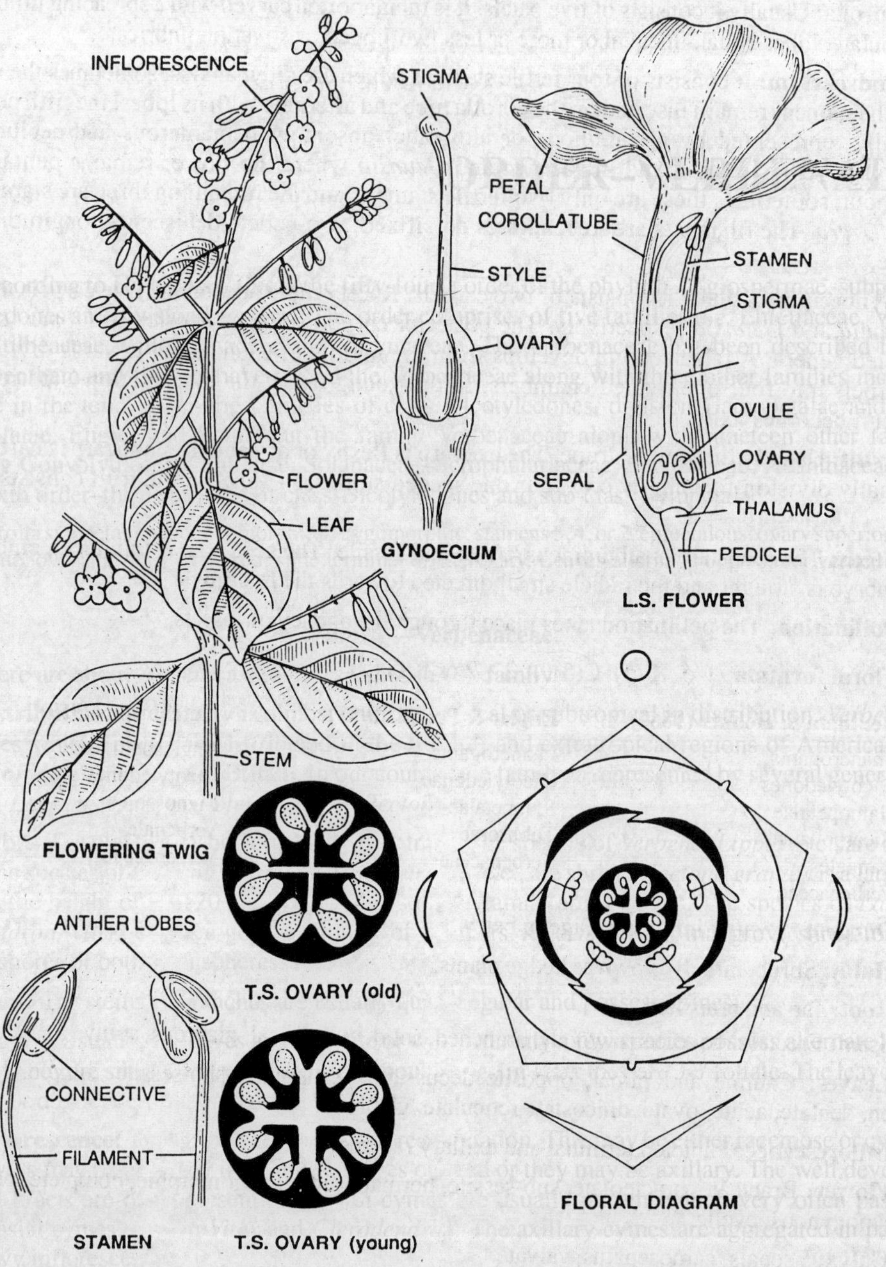

Fig. 37.1 : Verbenaceae. *Duranta repens* Linn; Eng., pigeon berry.

Economic value: Grown as a hedge plant. Juice of fruits to kill mosquito larvae. The juice can be used as a larvicide in ponds and swamps.

2. ***Lantana indica*** Roxb.; Verna. **Kuri.**

Habit : A perennial shurb.

Root : Tap and branched.

Stem : Erect, aerial, lower part woody, terminal branches herbaceous, branched, guardrangular, solid, hairy (stiff hairs), green.

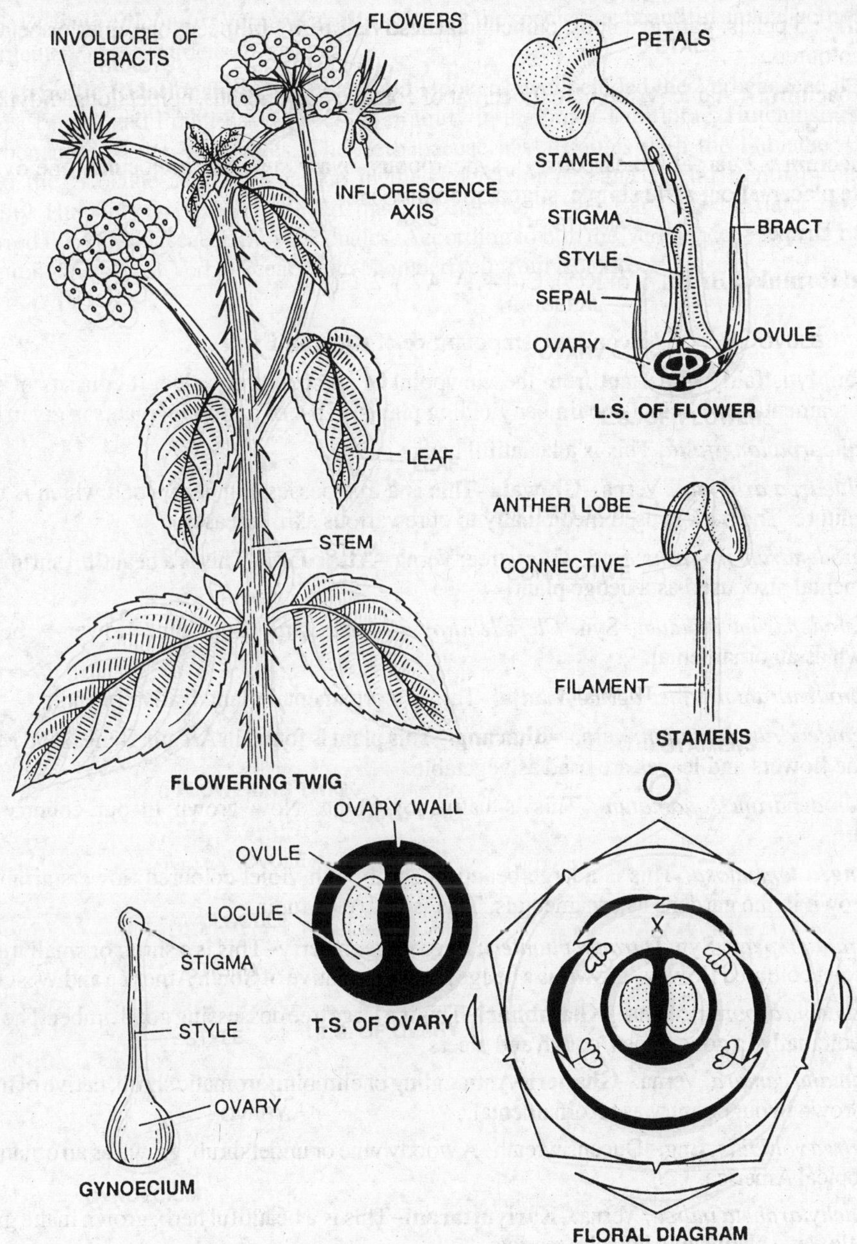

Fig. 37.2. Verbenacceae. *Lantana indica* Roxb; Verna. **Kuri.**

Economic value: The leaves are used as a cure for snake-bite.

Leaf : Cauline and ramal, exstipulate, opposite, decussate, simple, ovate, crenate, acute, rough surface, coriaceous, unicostate reticulate.

Inflorescence : Umbellate cyme.

Flower : Bracteate, bracteolate, sessile, zygomorphic, irregular, hermaphrodite, pentamerous, hypogynous, complte, cyclic.

Calyx : 5 sepals, gamosepalus, valvate, green.

Corolla : 5 petals, gamopetalous, quincuncial aestivation, 4/1 bilipped, one anterior petal large, variously coloured.

Androecium : 4 stamens, found in the throat of corolla, polyandrous, epipetalous, didynamous, dithecous, introrse.

Gynoecium : 2 carpels (bicarpellary), syncarpoous, ovary superior, biulocular, one ovule per locule, axile placentation, style simple, stigma globular.

Fruit : Drupe.

Floral formula : Br Brl ·|· ⚥ K(5), C(4 + 1), A2 + 2, G($\underline{2}$)

Economic Importance of the Family

The family is fairly important from the viewpoint of its economic value. It consists of several important ornamental, medicinal and timber yielding plants. A list of important plants is given below·

1. *Callicarpa longifolia*; This is a beautiful hedge plant.

2. *Callicarpa arborea ;* Verna.–**Ghiwala**–This is a tree possessing hard wood which is used in making furniture. The bark is used medicinally to cure various skin diseases.

3. *Clerodendrum fragrans;* Eng.–Glory tree; Verna.–**Attardana**–This is a beautiful shrub grown as an ornamental also, used as a hedge plant.

4. *Clerodendrum indicum*; Syn. *Clerodendron siphonanthus;* Eng.–tube flower–A beautiful shrub, grown as an ornamental.

5. *Clerodendrum inerme;* Verna.–**Lanjai**–This is a ornamental plant. Native of India.

6. *Clerodendrum serratum;* Verna.–**Bharangi**–This plant is found in Assam, Bengal and Madhya Pradesh. The flowers and leaves are used as vegetable.

7. *Clerodendrum squamatum;* This is native of China. Now grown in our country as an ornamental.

8. *Congea tomentosa;* This is a large beautiful shrub with violet coloured flowers arranged in racemes, grown in the gardens as ornamentals. This is native of India.

9. *Duranta repens;* Syn. *Duranta plumieri*; Eng.–Pigeon berry–This is a shrub or small tree with beautiful violet coloured flowers, grown as a hedge plant. It is native of South America and West Indies.

10. *Gamelna arborea;* Verna.–**Khambhari**–This is a large tree possessing good timber. The leaves are used medicinally in gonorrhoea, cough and ulcers.

11. *Lantana camara*; Verna.–**Ghaneri**–A straggling or climbing aromatic shrub, native of tropical America. Grown in our country as an ornamental.

12. *Petrea volubilis*; Eng.–Queens wreath–A woody vine or undershrub, grown as an ornamental, native of tropical America.

13. *Stachytarpheta indica;* Verna.–**Kariyuttarani**–This is a beautiful herb, grown in the gardens for its blue flowers. Native of tropical America.

14. *Tectona grandis*; Eng.–Teak; Verna. –**Sagwan**–A large tree found in the Western Ghats, Tamil Nadu, Madhya Pradesh, Orissa, Karnataka and Bihar. The wood is used for making furniture, cabinet work and construction work. The bark yields a yellow dye which is used for staining baskets.

15. *Vitex altissima;* Eng.–Tall chaste tree–A tree found in South West India. The wood is used for building purposes and cabinet work.

16. *Vitex negundo;* Eng.–Chinese chaste tree; Verna.–**Nirgandi, Samaalu**–A shrub or small tree, grown for reclamation of forest land. The branches are used for manufacturing baskets.

17. *Vitex peduncularis*; This is a tree. A drug is obtained by the infusion of leaves which is used to cure black-water fever.

18. *Verbena officinalis*; Verna. **Karaita**–Most of the species are beautiful annual herbs and grown as ornamentals in the gardens.

Systematic Relationships: Bentham and Hooker have included the Verbenaceae in the order-Lamiales. Engler and Prantl have placed this family in the order-Tubiflorae. Hutchinson has placed the Verbenaceae in the Verbenales. The Verbenaceae has affinities with the Labiatae. Bessey has included the Labiatae and Verbenaceae in the order-Lamiales and so did Bentham and Hooker. Originally Hutchinson (1926) included the Verbenaceae in the Lamiales but later on (1948) he transferred the Verbenaceae in the Verbenales. According to him the Verbenaceae and the Labiatae are quite unrelated and the Verbenaceae have been derived from Rubiaceae.

38

ORDER–RANALES

According to Hutchinson this is the fifty-fifth order of the phylum Angiospermae, subphylum Dicotyledones and division Herbaceae. This order comprises of seven families, *i.e.,* Paeoniaceae, Helleboraceae, Ranunculaceae, Nymphaeaceae, Podophyllaceae, Ceratophyllaceae, Cabombaceae, of which Ranunculaceae and Nymphaeaceae have been described in the present text in detail. Bentham and Hooker have considered the order–Ranales as the primitive most. According to them this is the first order of class Dicotyledones, division Polypetalaae and series Thalamiflorae. They have placed eight families in this order a including Ranunculaceae, Magnoliaceae, Annonaceae, Nymphaeaceae. The Ranunculaceae has been supposed to be most primitive within the order. Engler and Prantl have placed the family Ranunculaceae along with other seventeen families including Nymphaeaceae, Magnoliaceae, Annonaceae, etc., in their eighteenth order–the Ranales of class Dicotyledones and subclass Archichlamydeae.

Hypogynous to rarely perigynous; bisexual; hemicyclic to rarely cyclic; apocarpous; petals usually present; stamens free, often numerous seeds rich in endosperm; embryo minute. Herbs to softly woody; some aquatics.

Family—Ranunculaceae

There are about 35 genera and 1500 species in this family.

Distribution: They are chiefly found in the temperate and other colder parts of the world. In our country, the plants are chiefly found in the hilly regions. The species of *Clematis* are found in the Western Himalayas up to 5,000 feet. *Thalictrum javanicum* is found to be distributed from Shimla to Sikkim hills and also in Khasia hills of Assam. *Thalictrum foliolosum* grows in the Himalayas upto 9000 feet. The species of *Ranunculus* are found in the Indian plains. In Britain the family is represented by several genera, such as, *Ranunculus, Clematis, Anemone, Thalictrum, Paeonia, Aquilegia, Delphinium, Aconitum,* etc.

Habit: Generally the plants are annual or perennial herbs. They perennate either by means of rhizomes or by tuberous roots, *e.g., Ranunculus, Aconitum, Paeonia,* etc. In many plants the primary root perishes and is replaced by adventitious roots. These roots become tuberous and contain reserve food material in them, *e.g., Ranunculus, Aconitum, Paeonia,* etc. A few species are annuals. Undershrubs, and shrubs are also found in this family. Usually the species of *Clematis* are climbing undershrubs. The species of *Naravelia* are also climbing shrubs.

Leaves: The leaves are alternate, exstipulate and more or less divided. In some cases the leaves may be opposite or even radical. In *Clematis,* the leaves are opposite and compound. They possess twining petioles. In *Clematis aphylla* the whole leaf becomes modified into a tendril. In the case of *Naravelia,* the leaves are compound and trifoliate, here the terminal leaflet becomes modified into a tendril. In *Anemone* the leaves are radical. The leaf base usually broadens into a sheath which sometimes elongates into a pair of lateral stipules, *e.g.*, in *Thalictrum.* The leaves of *Thalictrum foliolosum* are pinnately decompound with auricled sheaths. The leaf lamina is sometimes entire. It is narrow in some species of *Ranunculus,* however, in *Ranunculus ficaria* and *Caltha* sp., it is cordate. Heterophylly is found in the aquatic species of *Ranunculus.* In such cases the submerged leaves are very much dissected whereas the floating ones are simply lobed. In *Delphinium* the leaves are palmately lobed and much dissected. The leaves of *Clematis* are climbing.

Inflorescence: In most cases the inflorescence is of cymose type, *e.g., Ranunculus.* However, in some species of *Ranunculus* the floweres are found to be arranged in panicles. The flowers are often solitary and terminal, *e.g., Anemone* sp. In *Anemone vitifolia* the flowers are arranged in decompound cymes. In *Aconitum* and *Delphinium,* the inflorescence is of racemose type. In *Clematis cadmia* the inflorescence is solitary axillary. In the species of *Nigella,* in the beginning a terminal flower is formed but later on the raceme develops.

Flower: The flowers are ebracteate rarely bracteate, hermphrodite and mostly actinomorphic (regular), rarely they are zygomorphic (irregular) such as *Delphinium* and *Aconitum,* hypogynous and sometimes a disc is found below the gynoecium, *e.g., Paeonia.* The flowers are usually developed on an elongated receptable. The floral parts being arranged wholly or partly spirally.

Perianth: In most of the flowers of this family, the perianth is not distinguished into calyx and corolla. It is simple, petaloid and variously coloured. However, in *Ranunculus* there are two distinct whorls of calyx and corolla, each consisting of five sepals and five petals respectively. The perianth is often associated with nectar secreting structures of various forms. In *Ranunculus,* the petals possess pocket like nectaries at their bases. The nectaries are supposed to be the modification of the petals. In certain genera the nectaries develop in the association of sepals, stamens or even carpels. The number of perianth leaves varies from 4 to 20.

Androecium: Usually the number of stamens is indefinite. The stamens are free (polyandrous). They are hypogynous and arranged spirally. Mostly the anthers are extrorse basifixed and dithecous. They dehisce laterally.

Gynoecium: It consists of indefinite carpels (polycarpellary), the carpels are free, *i.e.*, apocarpous. In *Delphinium* the number of carpels is reduced to one. In *Aconitum* sp., there are three to five carpels. In *Nigella,* there are 5 to 8 carpels, which are more or less united (*i.e.,* syncarpous). In each ovary the number of ovules ranges from one to many. The ovules are anatropous. The placentation is either marginal (*e.g., Delphinium*), or basal (*e.g., Ranunculus*).

Fruit: The fruit is usually either an etaerio of achenes, *e.g., Ranunculus* or an etaerio of follicles, *e.g., Aconitum.* In *Actaea,* it is a berry. In *Nigella* it becomes capsule, because of the fusion of carpels. In some of the cases the achenes possess long persistent feathery styles which help in the distribution of the fruits by means of wind, *e.g., Clematis* and *Anemone.* The seeds of *Helleborus* are disseminated by ants. The ants are attracted by the oil-containing swellings on the raphe. The fruits of certain species of *Ranunculus* possess hooks and are distributed by animals and birds.

Seeds: The seeds are endospermic. Each seed contains only endosperm with a very small embryo.

Pollination: The pollination takes place either by means of insects (entomophily) or by wind (anemophily). In *Ranunculus* and others the flowers are generally protandrous. In *Thalictrum* and *Helleborus* the flowers are protogynous, whereas, in *Anemone* and *Trollius* the flowers are homogamous. The flowers of some species of *Thalictrum* are wind pollinated but most of the other genera are insect pollinated. The flowers of *Delphinium* and *Aconitum* are pollinated by long tongued bees.

Floral Formulae:

Ranunculus: ⊕ ⚥ K 5, C 5, A $\propto$, G $\underline{\propto}$

Delphinium: ·|· ⚥ K 5, C (4), A $\propto$, G $\underline{1}$

Bentham & Hooker (1862)	**Engler & Prantl (1931)**	**Hutchinson (1959)**
Phanerogams	Phanerogams	Angiospermae
Dicotyledones	Dicotyledones	Dicotyledones
Polypetalae	Archichlamydeae	Herbaceae
Thalamiflorae	Ranales	Ranales
Ranales	Ranunculaceae	Ranunculaceae
Ranunculaceae		

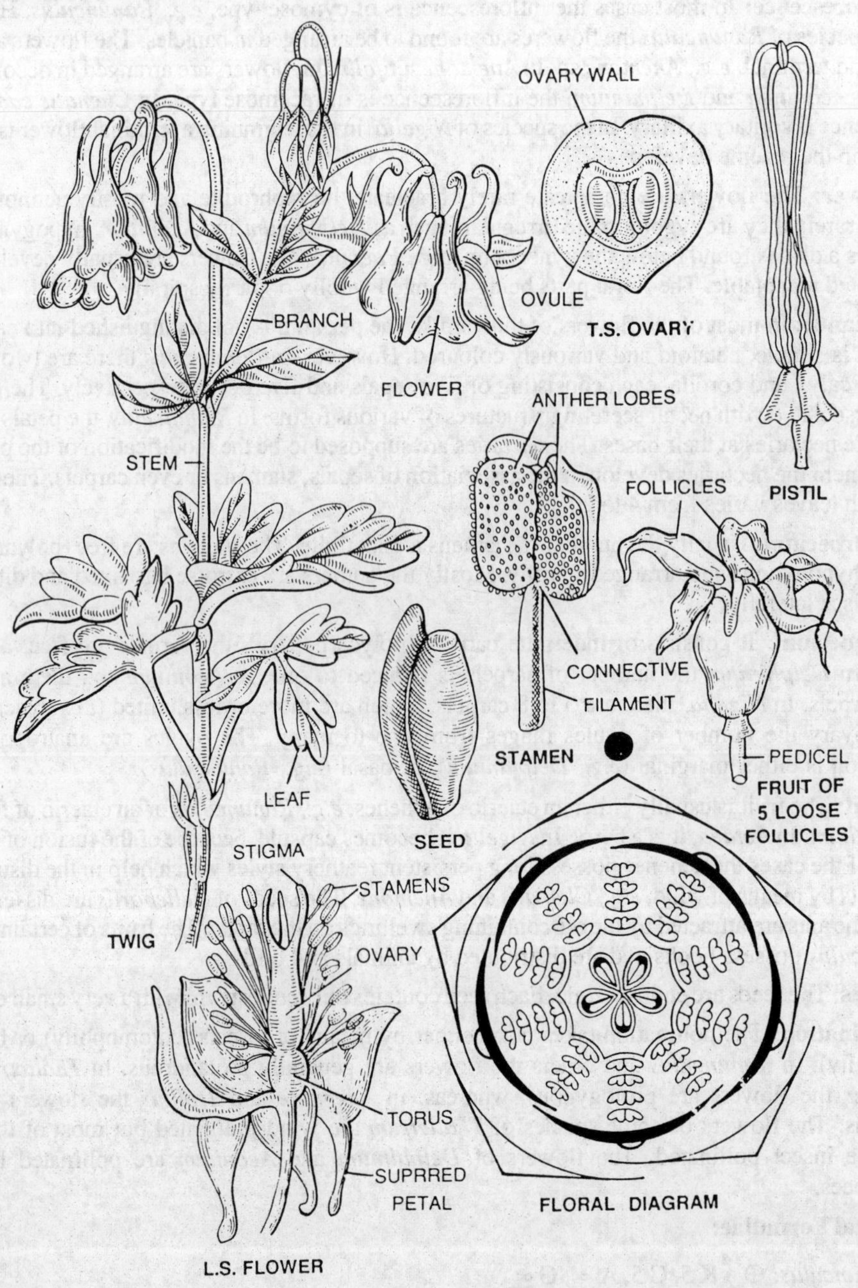

Fig. 38.1. Ranunculaceae. *Aquilegia vulgaris* (Linn.) Thunb.

Economic value. Grown as on ornamental. It is poisonous.

Description of some important plants in semi-technical language:

1. ***Aquilegia vulgaris*** Linn.

Habit: Annual or perennial herbs.

Root : Tap and branched.

Stem: Erect, herbaceous, aerial green, branched, cylinderical glabrous.

Leaf: Cauline and ramal, simple, petiolate, much dissected, entire, lobes with acute apices, stipulate, membranous stipules fused with leaf base, multicostate reticulate venation.

Inflorescence: Cymose.

Flower: Pedicellate, bracteate, hermaphrodite, complete, zygomorphic, hypogynous, nectar secreted in spurs of petals, pentamerous.

Calyx: 5 petals, polysepalous, petaloid, imbricate aestivation.

Corolla: 5 petals, polypetalous imbricate aestivation, nectar secreted in spurs of petals.

Androecium: Indefinite, polyandrous, filament long, anthers dithecous, extrorse, basifixed.

Gynoecium: 5 carpels, apocarpous, carpels are connate in a five-chambered ovary, ovary superior.

Fruit: Follicle, made up of 5 loose follicles.

Floral formula: Br. Brl. ·|· ⚥ K 5, C 5, A ∞ G $\underline{5}$.

2. ***Clematis paniculatea*** Thunb.

Habit : A climber, annual or perennial.

Root : Tap and branched.

Stem : Herbaceous, aerial weak, climbing with the help of tendrilar petioles, branched, solid, smooth and green.

Leaf : Cauline and ramal, simple, petiolate, petiole tendrilar, exstipulate, opposite decussate, ovate, acute, entire, glabrous, unicostate reticulate, coriaceous.

Inflorescence : Cymose, dichasial cyme.

Flower : Pedicellate, bracteate, bracteolate, complete, actinomorphic, hermaphrodite, tetramerous, hypogynous, hemicyclic and scented.

Calyx : 4 sepals, polysepalous, valvate, petaloid, white.

Corolla : Absent.

Androecium : Indefinite, polyandrous, filamens long, monothecous, extrorse, basifixed.

Gynoecium : 4-6 carpels, apocarpous, ovary superior, unilocular, basal placentation, style short, stigma simple, hairy outgrowths arise from ovary.

Fruit : Achene.

Floral formula : Br. Brl ⊕ ⚥ K 4, C 0, A ∞, G $\underline{4\text{-}6}$.

3. ***Ranunculus scleratus***: Verna–**Shim**.

Habit: Annual herb.

Root: Tap, branched.

Stem: Erect, reduced in early stages, herbaceous, green, glabrous, solid, branched, aerial with nodes and internodes.

Leaf: Simple, petiolate, radical with long petioles, tripartite or trilobed, each lobe further divided with obovate, cuneate segments, membranous stipules fused with the leaf base, multicostate reticulate venation.

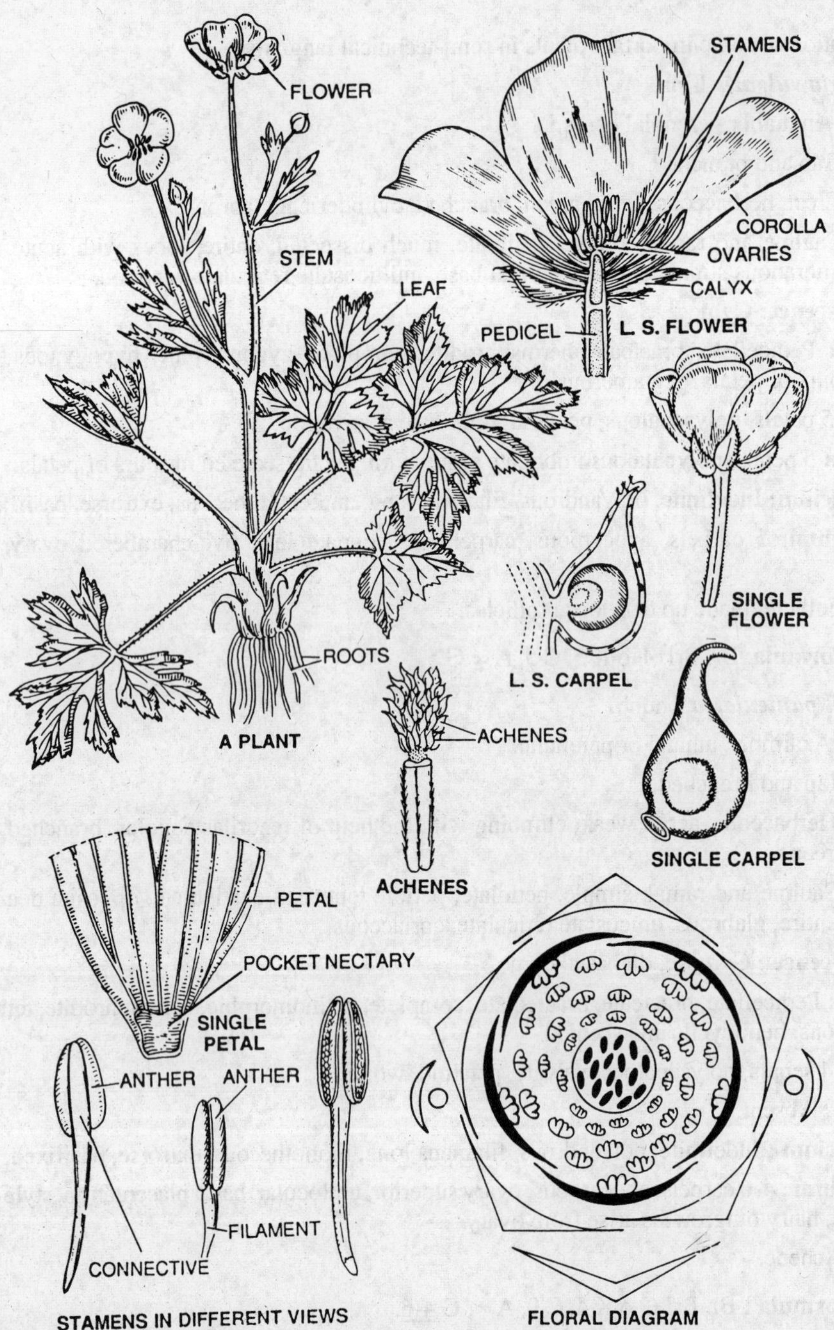

Fig. 38.2. Ranunculaceae–*Ranunculus sceleratus* Linn.

Inflorescence: Cymose.

Flower: Pedicellate, bracteate, hermaphrodite, complete, actinomorphic, yellow and hypogynous.

Calyx: Five, polysepalous, petaloid, quincuncial, *i.e.*, imbricate in bud condition.

Corolla: Five, polypetalous, yellow, oblong, pocket-shaped nectary present at the base of each petal, imbricate.

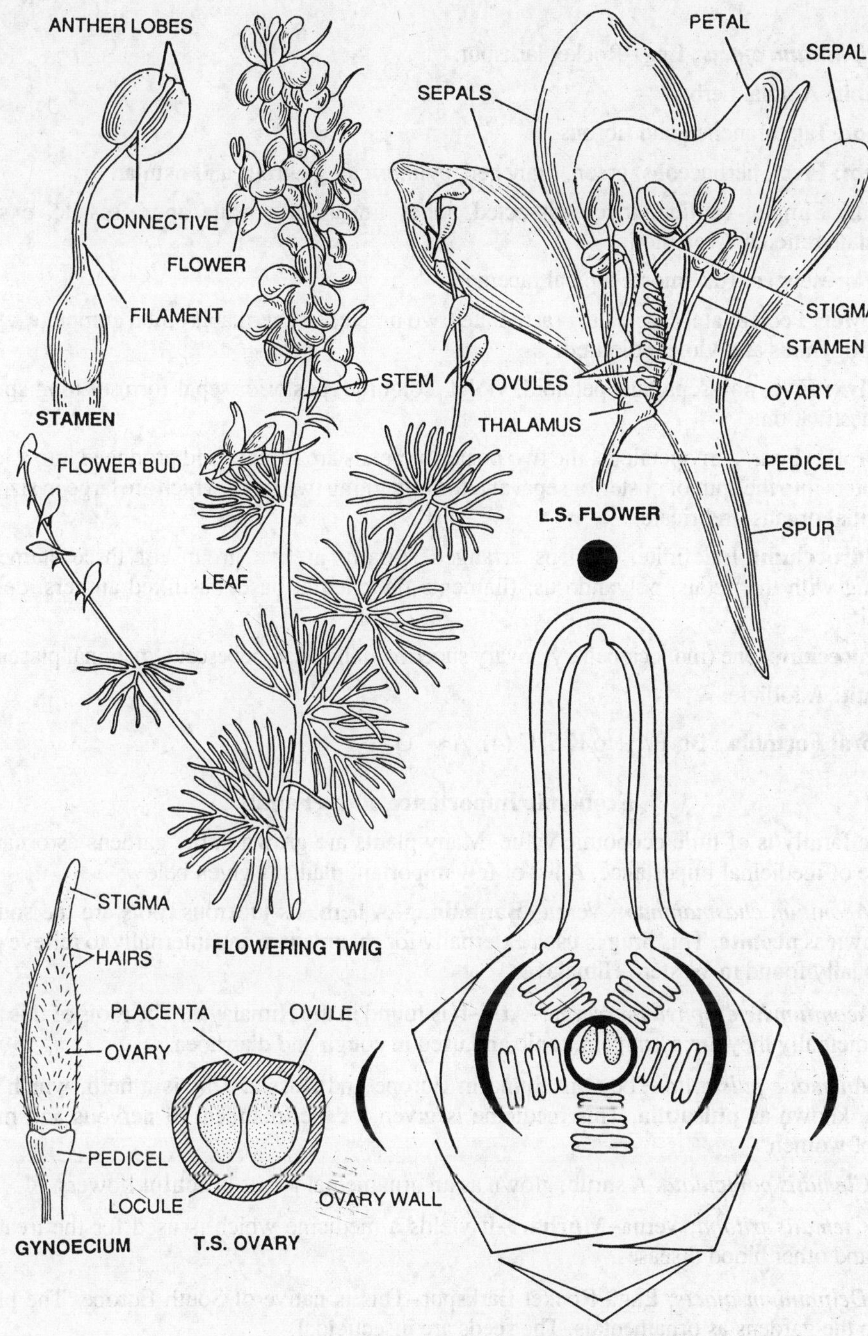

Fig. 38.4. Ranunculaceae. *Delphinium ajacis* Linn.

Androecium: Indefinite, the stamens arranged spirally, polyandrous, filaments long, anthers extrorse, elongated, yellow and basifixed.

Gynoecium: Indefinite (polycarpellary), apocarpous, ovary superior, unilocular, basal placentation.

Fruit: The fruit is an etaerio of achenes

Floral Formula : ⊕ ⚥ K 5, C 5, A ∞, G $\underline{\infty}$

4. ***Delphinium ajacis*;** Eng.–Rocket larkspur.

Habit: Annual herb.

Root: Tap, branched and fibrous.

Stem: Erect, herbaceous, green, branched, cylindrical, glabrous and fistular.

Leaf: Simple, cauline, much dissected, lobes linear with acute apex, sessile, exstipulate, multicostate reticulate venation.

Inflorescence: Racemose, typical raceme.

Flower: Pedicellate, bracteate, bracteolate (two bracteoles, hermaphrodite, complete, zygomorphic, hypogynous and violet coloured.

Calyx: Five, polysepalous, petaloid, violet coloured, posterior sepal forms a long spur, quincuncial aestivation.

Corolla: Four, gamopetalous, the two posterior petals are smaller and produced into a long spur which enters into the spur of posterior sepal and the remaining two lobes which are large in size enclose the essential organs, imbricate.

Androecium: Indefinite, stamens arranged spirally in five groups of three stamens each alternating with the petals, polyandrous, filaments flattened at base, basifixed anthers, dehisce by lateral slits.

Gynoecium: One (monocarpellary), ovary superior, unilocular, pubescent, marginal placentation.

Fruit: A follicle.

Floral Formula : Br, Brl ·|· ⚥ K 5, C (4), A∞ , G$\underline{1}$.

Economic Importance of the Family

The family is of little economic value. Many plants are grown in the gardens as ornamentals. Some are of medicinal importance. A list of few important plants is given below:

1. *Aconitum chasmanthum*; Verna.–**Banbalnag**–A herb. Its tuberous roots are the source of a drug known as **aconite.** This drug is used externally for rheumatism and internally to relieve pain and fever; usually found in Western Himalayas.

2. *Aconitum heterophyllum*; Verna.–**Atis**–It is found in the Himalayas. The roots of this plant are used medicinally they are astringent, tonic and used in cough and diarrhoea.

3. *Anemone pulsatilla*–They are found in Europe and Russia. This is a herb, which yields a medicine known as **pulsatilla.** This medicine is given for the treatment of nervous and menstrual trouble of women.

4. *Clematis paniculata*; A shrub, grown as an ornamental for its beautiful flowers.

5. *Clematis triloba*; Verna–**Murhari**–It yields a medicine which is used for the treatment of leprosy and other blood diseases.

6. *Delphinium ajacis*; Eng.–Rocket Larkspur–This is native of South Europe. The plants are grown in the gardens as ornamentals. The seeds are insecticidal.

7. *Delphinium brunonianum*; Eng.–Musk larkspur; Verna.–**Laskar**. The juice of leaves is used as an insecticide. This is found in the Western Himalayas.

8. *Delphinium caeruleum*; Verna.–**Dhakangu**–The root is used as an insecticide to kill maggots in the wounds of goats.

9. *Delphinium denundatum;* Verna–**Nirbisi**–The roots are used as a tonic.

10. *Nigella damascena*; Eng.–Love-in-a mist.– A herb, grown as an ornamental in the gardens. It is native of South Europe.

11. *Nigella sativa;* Eng.–Black cumin; Verna.–**Kala jira, Kalonji**–The seeds are used as condiment and spice. The seeds mixed with sesamum oil are used externally in eruption of skin and for scorpion sting. The seeds are stimulant, carminative and diuretic. It is a herb. The plants are cultivated in the Punjab, Assam, Bihar and Bengal.

12. *Ranunculus aquatilis*; Verna.–**Jal dhania**–It is used medicinally in the treatment of rheumatic pain and asthma.

13. *Ranunculus ficaria;* Eng.–Pilewort–The plant is native of Europe. It makes a good medicine of piles. The patent medicine of piles, known as 'pilewort ointment' is prepared from this plant.

14. *Thalictrum foliolosum;* Verna.–**Mamira**–The roots are used as a tonic. The plants are found in the Himalayas.

15. *Thalictrum javanicum*; This plant is the alternate host of brown rust of wheat, caused by *Puccinia recondita.*

Family—Nymphaeaceae

There are about 8 genera and one hundred species in this family.

Distribution: The plants of the family are widely distributed in tropical and north temperate regions of the world. The plants are aquatic in habit. They are either water or marsh-plants with generally sub-merged and floating leaves, and large, beautiful solitary flowers. In our country, the family is represented by few important genera, such as *Nymphaea, Victoria, Euryale, Nelumbium*(*Nelumbo*), etc.

Habit: As already stated, the plants are aquatic herbs.

Stem: The stem is a rhizome which is short, thick and erect. In *Victoria* sp., this survives for few years, whereas in *Euryale* it lives only for one year. In *Nuphar* and *Nymphaea*, the stem is branched rhizome, which creeps in mud and may live for a long time.

Leaves: Usually the leaves are petiolate, simple, floating and large in size. They are peltate or ovate in shape. Heterophylly prevails in the species of *Cabomba,* where the floating leaves are ovate or peltate, and the submerged leaves are finely dissected. In majority of cases the leaves possess shining and smooth surface, but the lower surface of the leaves of *Victoria* and *Euryale* is prickly. The leaves of *Victoria regia* are usually very large in size. Sometimes they are six-feet in diameter and look like big saucers. The leaves of *Cabomba* are sessile, otherwise they possess long petioles.

The plants possess the following chief anatomical features:

1. The vascular bundles possess much reduced woody elements.

2. The vascular bundles have no cambium and remain scattered in the parenchyamatous tissue and thus they resemble to that of monocotyledonous plants.

3. Laticiferous vessels are present.

4. Stellate hairs are very commonly found in the angles of the walls of large, intercellular spaces.

Inflorescence: The inflorescence is of cymose type. The flowers are solitary and being found on long pedicels.

Flower: The flower is large, showy, variously coloured, pedicellate, hermaphrodite, actinomorphic, complete, hypogynous to epigynous, acyclic, hemicyclic or cyclic.

Perianth: It consists of six to indefinite free perianth leaves and usually differentiated into a calyx and corolla.

Calyx: Usually it consists of 3-5 sepals or indefinite sepals. The sepals are green in colour, but in *Nuphar,* they become petaloid and larger than the petals.

Corolla: It consists of 3 to indefinite free petals (polypetalous). Usually the petals are large, coloured and showy. In *Nuphar* they are very small and scale-like.

Androecium: It consists of 3 to indefinite number of free stamens. When they are few in number

(3-6) their arrangement is cyclic, but when they are indefinite, the arrangement becomes acyclic. The anthers are bicelled and introrse. Very often the filaments project out beyond the anther sacs as sterile appendages. The anthers dehisce by longitudinal slits.

Gynoecium: It consists of 3 to indefinite number of carpels. They may be apocarpous or syncarpous. The ovary may be superior or inferior. The ovary is unilocular having many ovules being found on parietal placenta (parietal placentation).

Fruit: The fruits are variable. They may be follicles, berries or etaerio of indehiscent nutlets.

Seed : The seeds are endospermic. They are often arillate. The embryo is well developed with fleshy cotyledons.

According to Lawrence, the family has been divided into three sub-families. These sub-families are:

1. Cabomboideae.
2. Nelumbonoideae.
3. Nymphaeoideae.

Sub-family 1. **Cabomboideae**–There are two genera, known as *Cabomba* and *Brasenia* in this sub-family. Sepals-3; petals 3; stamens–3–6; carpels 2–3 apocarpous. The fruit is usually 3 seeded pod. The seed is endospermic and also possesses abundant perisperm.

Floral Formulae:

$$\oplus \text{⚥} K3, C3, A3\text{-}6, G\underline{2\text{–}3}.$$

OR

$$\oplus \text{⚥} P3+3, A\,3\text{-}6, G\underline{2\text{–}3}.$$

Sub-family 2. **Nelumbonoideae**–There is single genus *Nelumbo* or *Nelumbium* in this sub-family. The leaves are large, peltate and long petiolate. Sometimes the long petiolate leaves reach high above the surface of water. Perianth consists of indefinite, perianth leaves, which may be white or pink. The stamens are indefinite which are arranged spirally below the obconical receptacle. Several carpels are found on the upper flat surface of this receptacle. Each carpel remains embedded in a round pit. Each carpel possesses a pendulous ovule on the side towards the axis and matures in an indehiscent nut with a very hard pericarp, completely filled up by the embryo.

Floral Formula: $\oplus$ ⚥ $P \infty, A \infty, G \infty$.

Sub-family 3. **Nymphaeoideae**–There are five genera in this sub-family. The carpels are united into a single many chambered ovary which bears a flat radiating stigma above it. The flowers may be hypogynous, perigynous.

Nymphaea–This genus contains about 32 species. The most important Indian species is *N.lotus* (Verna.–**Chhota Kamal**). The species are large aquatic herbs. Root stock is creeping. In *N. lotus* the root stock is short, erect roundish and tuberous. Leaves and flowers are floating. Flowers are expanded large, solitary and situated on long radical scapes. Sepals 4, and adnate to the base of the disk. Petals are found in many series, successively transformed into stamens. All stamens are adnate to the disk Filaments are petaloid. Anthers are small, linear and introrse. Ovaries many in one series, sunk in the fleshy disk and with it forming a many-celled ovary crowned by connate radiating stigmas. Ovules many. Fruit–a spongy berry, which ripe under water. Seeds–small and buried under pulp.

Floral Formula : $\oplus$ ⚥ $K\,4, X \infty, A \infty, G\,(\underline{5 - \infty})$.

The genera, *Euryale* and *Victoria* possess completely epigynous flowers and the prickles are copiously developed on the surface of the organs.

The genera, *Euryale* and *Victoria* possess completely epigynous flowers and the prickles are copiously developed on the surface of the organs.

Pollination: Usually takes place through the agency of insects. Both self and cross pollination are found.

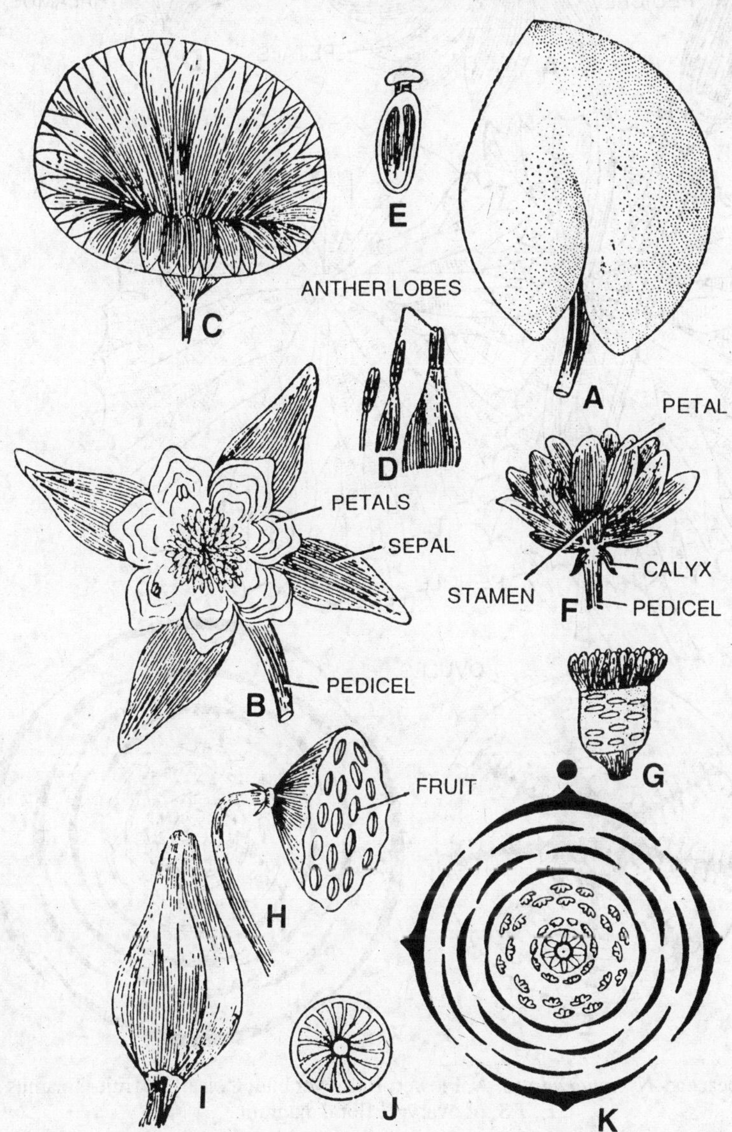

Fig. 38.4. Nymphaeaceae: A, leaf of *Nymphaea* sp. ; B, flower of *Nymphaea;* C, leaf of *Nelumbo;* D, *Nymphaea*–stamens of various sizes from the innermost to the outermost whorl; E, N*elumbo*-single carpel separated with ovary cut longitudinally; F, *Nelumbo*–flower in vertical section; G, *Nymphaea*–ovary with scars of stamens and petals and bearing the stylar processes; H, Nelumbo-receptacle in ripe fruit; I, flower bud of *Nymphaea;* J, *Nymphaea*–T.S. of ovary; K; floral diagram of *Nelumbium* sp.

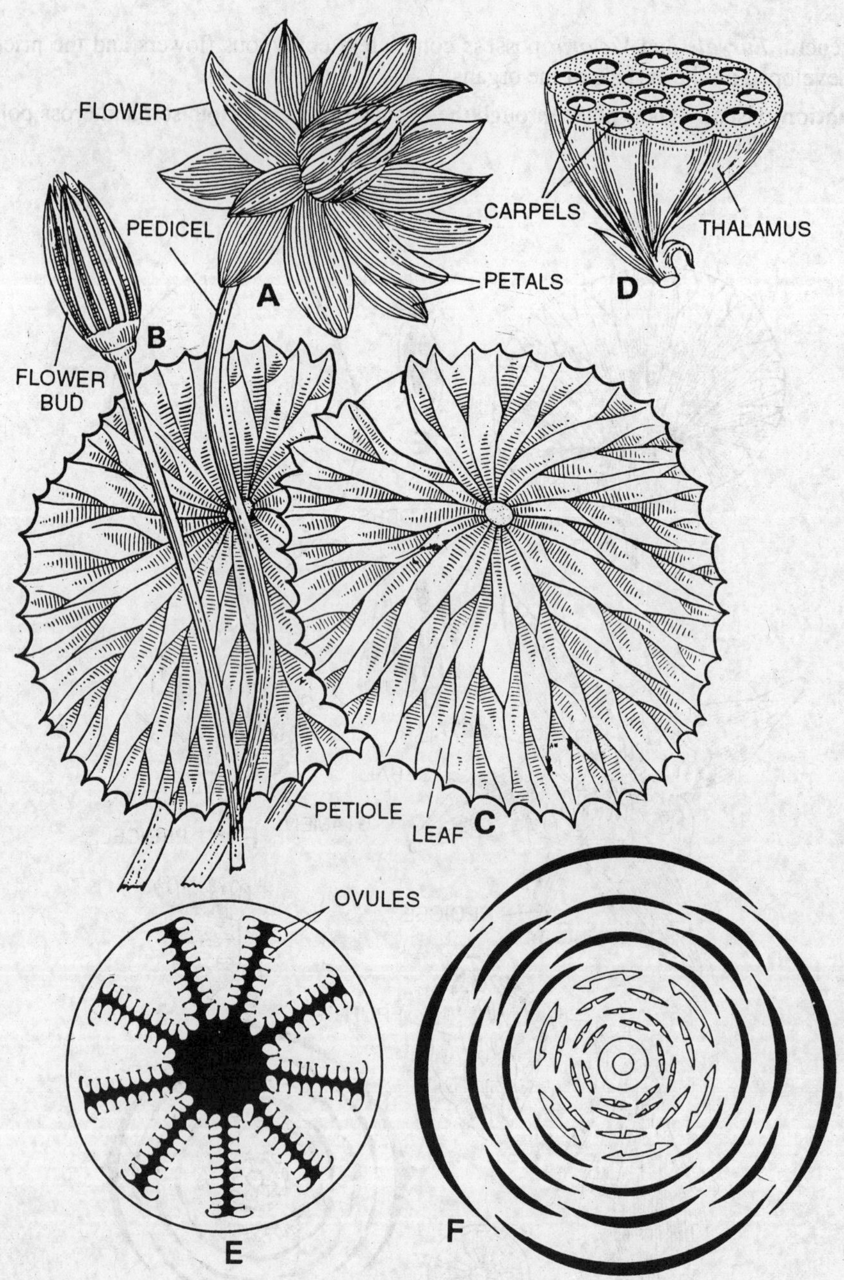

Fig. 38.5. Nymphaeaceae–*Nymphaea lotus*. A, Flower; B, flower bud; C, leaf; D. fruit-thalamus with carpels; E, T.S. of ovary; F, floral diagram.

Bentham and Hooker (1862)	**Engler and Prantl (1931)**	**Hutchinson (1959)**
Phanerogams	Phanerogams	Angiospermae
Dicotyledones	Dicotyledones	Dicotyledones
Polypetalae	Archichlamydeae	Herbaceae
Thalamiflorae	Ranales	Ranales
Ranales	Nymphaeaceae	Nymphaeaceae
Nymphaeaceae		

Economic Importance of the Family

The family is of some economic value. Mostly ornamentals. Some plants are medicinal and some are edible. A list of few important plants is given below:

1. *Euryale ferox*; Eng–Gorgan nut; Verna.–**Makhana**–This is a prickly aquatic herb, found commonly in Kashmir, Bihar, Assam, Manipur, Tripura, Bengal and Uttar Pradesh. The seeds are roasted and eaten.

2. *Nelumbo nucifera;* Syn. *Nelumbium speciosum*; Eng.–East Indian lotus; Verna.–**kamal**–A common aquatic herb whose rhizomes, seeds and young leaves are eaten as vegetable. The mature leaves are used as substitute for feast plates. This is highly ornamental aquatic plant.

3. *Nymphaea alba;* Eng.–European white water lily; Verna.–**Pandharen-kamal**–An aquatic herb. This is native of Europe and North Africa but also found in Kashmir. The root-stocks are edible.

4. *Nymphaea esculenta*; Eng.–Indian white waterlily: Verna.–**Chhota sundi.**This is a floating aquatic herb whose tubers and seeds are edible.

5. *Nymphaea nouchali;* Syn. *N.pubescens;* Eng.–White lotus; Verna.–**Kamal kakri, Neelophal.** This is a floating, aquatic, ornamental herb. This is native of Central Africa. The rhizomes and flowers are edible.

6. *Nymphaea stellata*; Eng.–Blue lotus of India; Verna.–**Nilkamal.** This is native of South East Asia. The rhizomes and seeds are edible. Also grown as an ornamental.

Systematic Relationships: Hutchinson has divided the order Ranales into two–1. The Ranales include herbaceous type and 2. The Magnoliales include the arborescent forms (trees, etc.). He has treated the Magnoliales to be more primitive. The systematic position of the order Ranales is a subject of controversy. According to most of the taxonomists they are supposed to be primitive. According to Bentham and Hooker they are most primitive and reduced series derived from more elaborate ancestral types. This is also believed that most of the living angiosperms have been evolved from this series. They are supposed to be most primitive because of the presence of free carpels and the indefinite number of floral parts being inserted spirally on the receptacle.

39

ORDER–PIPERALES

According to Hutchinson this is the 58th order of the phylum Angiospermae, subphylum Dicotyledones and division Herbaceae. This order comprises of only three families, *i.e.*, Piperaceae, Saururaceae and Chloranthaceae, of which Piperaceae has been described here in detail. Bentham and Hooker have placed the family Piperaceae along with three other families (*i.e.*, Chloranthaceae, Myristiceae and Monimiaceae) in the fourth series–the Micrembryae of class Dicotyledones and division Monochlamydeae or Incompletae. Engler and Prantl have placed the Piperaceae along with other families among the second order–the Piperales of class Dicotyledones and sub-class Archichlamydeae.

Usually no perianth; ovary mostly superior; placentas parietal to subaxile; endosperm copious; embryo minute.

Family—Piperaceae

There are about 10 to 12 genera and more than 1300 species in this family.

Distribution: They are widely distributed in the tropical regions of both hemispheres. Only a few species are extra-tropical in distribution. About 700 species have been included in the genus *Piper.* The plant of chief economic importance is *Piper nigrum* (Verna.–**Kali Mirch**), a native of Indo-Malaya where it is chiefly cultivated. *Piper longum* and *Piper betle* are chiefly cultivated in our country.

Habit: They are generally shrubs or herbs, very rarely trees. The species of *Piper* which are over 100 in number are generally shrubs, often climbing, rarely trees or herbs. The species of *Peperomia*, which are over 600 in number are annual or perennial herbs, often fleshy.

A large number of geophilous species which are hardly 2-5 cm. in height and bearing a short thickened rhizome are found in Mexico, Central America, in the mountains of South America upto the height of 13,000 feet a.s.1.

Stem: The stem is generally succulent and herbaceous. They are very often provided with jointed or swollen nodes. The stem (*e.g.*, in *Peperomia*) possesses distinct and scattered vascular bundles like that of monocot stems. However the bundles are open. The resin and oil secreting sacs are generally found both in the epidermis and ground tissue.

Leaves: The leaves are alternate, rarely opposite or whorled, petiolate, entire, stipulate or exstipulate, when the stipules are present. They are adnate to the petiole. They are palmately or penninerved and often fleshy. The leaves are of pungent taste. In the fleshy leaves of *Peperomia* there is a well developed aqueous tissue beneath the upper epidermis. Resin or oil containing sacs are generally found to be scattered both in the epidermis and ground tissue.

Inflorescence: The flowers are generally arranged in catkinlike terminal or axillary spikes. Sometimes (*e.g.*, in *Piper*) the spikes are found to be situated opposite the leaves.

Flower: The flowers are naked which are often more or less sunk in the fleshy axis of the inflorescence (spike) on which they are closely arranged. The flowers are very minute bracteate, usually hermaphrodite or in some species unisexual.

Perianth: It is altogether absent. The flowers are naked.

Androecium: The number of stamens varies from one to ten. Most of the flowers are of trimerous type having two whorls, each of three stamens (*e.g.,* in *Piper amalago*). The trimerous flowers are considered the primitive ones from which the other types of flowers have been derived. In certain other species of *Piper* the two posterior stamens of the inner whorls are absent. Sometimes the whole inner whorl is found to be absent. In *Piper nigrum* there are only two stamens, the posterior stamen

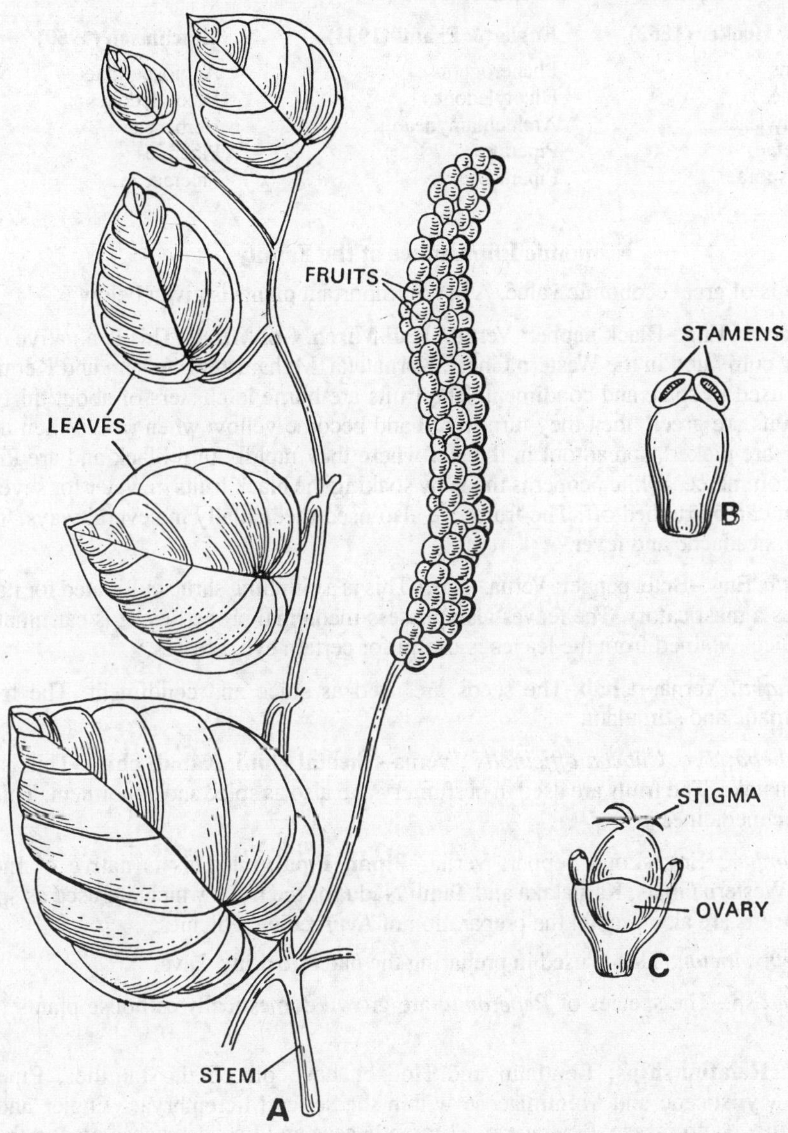

Fig. 39.1. Piperaceae—*Piper nigrum* (Verna. **Kali Mirch**) : A, a twig with leaves and fruits; B, male flower; C female flower.

of the inner whorls has also been aborted. This arrangement of stamens is constant feature in *Peperomia*, and this has been considered to be a constantly reduced form. The filaments are usually distinct, the anthers are two-celled and dehisce by longitudinal slits.

Gynoecium: The number of carpels varies from 2-5. The number of carpels in primitive forms is probably three. The ovary is superior, one chambered with a solitary basal orthrotropous ovule. The style may be absent or one. In *Piper,* the stigmas are as many in number as that of carpels. In *Peperomia* the stigma is simple often brushlike and lateral.

Fruit: The fruit is a small drupe but described very erroneously as a berry.

Seeds: The seeds are small, endospermic with a minute embryo.

Floral Formula: ⚥ P 0, A 1-10, G $(\underline{2\text{-}5})$.

Bentham & Hooker (1862)	Engler & Prantl (1931)	Hutchinson (1959)
Phanerogams	Phanerogams	Angiospermae
Dicotyledones	Dicotyledones	Dicotyledones
Monochlamydeae or Incompletae	Archichlamydeae	Herbaceae
Achlamydosporae	Piperales	Piperales
Piperaceae	Piperaceae	Piperaceae

Economic Importance of the Family

The family is of great economic value. A list of important plants is given below:

1. *Piper nigrum*; Eng.–Black pepper; Verna.–**Kali Mirch, Gol Mirch**–This is a native of Indo–Malaya, but now cultivated in the Western Ghats, Karnataka, Maharashtra, Assam and Kerala for its fruits which are used as spice and condiment. The fruits are borne in clusters of about thirty. In the first stage the fruits are green, then they turn to red and become yellow when ripe. When the fruits become red they are picked, spread out in the sun where they rapidly turn black and are known as black pepper of commerce. White pepper is made by soaking the black fruits in water for several days till the outer coat can be rubbed off. The fruits are also used medicinally in several ways. They are useful in cholera, headache and fever.

2. *Piper betle*; Eng.–Betel pepper; Verna.–**Pan**–This is a climbing shrub cultivated for its leaves, which are used as a masticatory. The leaves also possess medicinal properties. It is carminative and stimulant. The juice obtained from the leaves is useful for certain eye diseases.

3. *Piper chaba*; Verna.–**Chab**–The seeds are used as spice and condiment. The fruits are carminative, aromatic and stimulant.

4. *Piper cubeba*; Syn. *Cubeba officinalis* ; Verna–**Sheetal chini, Kabab chini**–This is chiefly cultivated in Karnataka. The fruits are used in perfumery and also as spice and condiment. This is also used in Ayurvedic medicines.

5. *Piper longum*; Eng.–Long pepper; Verna.–**Pipar, Pipaluml**–This is native of India and cultivated in the Western Ghats, Karnataka and Tamil Nadu for the fruits which are used as spice and condiment. The fruits are also used in the preparation of Ayurvedic medicines.

6. *Piper methysiticum*: This is used in preparing the narcotic drink, kava.

7. *Peperomia* sp.–The species of *Peperomia* are grown domestically as house plants for their beautiful foliage.

Systematic Relationships: Bentham and Hooker have placed the families, Piperaceae, Chloranthaceae, Myristiceae and Monimiaceae within the series Micrembryae. Engler and Prantl included the families Saururaceae, Piperaceae, Chloranthaceae and Lacistemaceae among the Piperales. It has been thought by many workers such as, Hallier, Bessey and Hutchinson, that the Piperaceae, which of undeterminate origin, is not one of the most primitive, dicotyledons but probably is an independent and terminal offshoot of ranalian ancestors.

40

ORDER–RHOEADALES

According to Hutchinson this the fifty-ninth order of the phylum Angiospermae, subphylum Dicotyledonés and division Herbaceae. The order consists of only two families, *i.e.,* Papaveraceae and Fumariaceae. Both the families have been described in the present text in detail. The other authors (Bentham and Hooker, Engler and Prantl) have treated these two families as a single–the Papaveraceae. Bentham and Hooker have placed the Papaveraceae along with other eight families including Cruciferae, Capparideae, Violaceae, etc., among the second order–the Parietales of class Dicotyledones, division Polypetalae and series Thalamiflorae. Engler and Prantl have placed the family Papaveraceae along with other five families including Capparidaceae, Cruciferae, etc., in the nineteenth order–Rhoeadales of class Dicotyledones and Subclass Archichlamydeae.

Petals present, becoming zygomorphic; stamens numerous to few; syncarpous, with parietal placentatión; endosperm copious; embryo minute.

Family—Papaveraceae

There are about 27 genera and 250 species in this family.

Distribution: The members of the family are commonly found in the sub-tropical and temperate regions of the Northern hemisphere. Some genera are found in America, some in Asia whereas some are found in South Africa. The plants of the family are very uncommon in the Southern hemisphere.

Habit: Commonly the plants are annual herbs: *e.g., Papaver.* Some are undershrubs. *e.g., Argemone*. Shrubs are very rare *e.g. Dendromecon* and trees are extremely rare *e.g., Bocconia.*

The plants usually possess a milky or coloured latex. In *Argemone* the latex is yellow coloured whereas in *Hunnemannia,* it is watery.

Root: Tap and branched.

Stem: Usually erect (*e.g., Argemone*) herbaceous, cylindrical, branched, solid, green.

Leaves: Simple, alternate, entire or more or less lobed or cut margins (*e.g., Argemone*); sometimes dissected.

Inflorescence: Cymose or Racemose. In the species of *Macbaya* the flowers are arranged in panicled racemes.

Flower: Large, showy, attractive, bisexual, actinomorphic, complete, hypogynous.

Perianth: Biseriate or triseriate.

Calyx: 2 or 3 deciduous sepals; usually polysepalous; aestivation twisted or imbricate.

Corolla: The number of petals is usually double the number of sepals. Usually the petals are arranged in two whorls. They are variously coloured. The petals very quickly fall off. The aestivation is imbricate.

Androecium: Usually indefinite stamens arranged in several alternating whorls. Polyandrous, anthers bilobed, dehisce longitudinally, extrorse.

Gynoecium; 2 to indefinite carpels; syncarpous; the ovary is superior and unilocular; the placentation is parietal. The ovules are anatropous. In *Papaver*, the placentation becomes superficial.

Fruit: The fruit is a capsule or nut. The capsule opens by valves or pores.

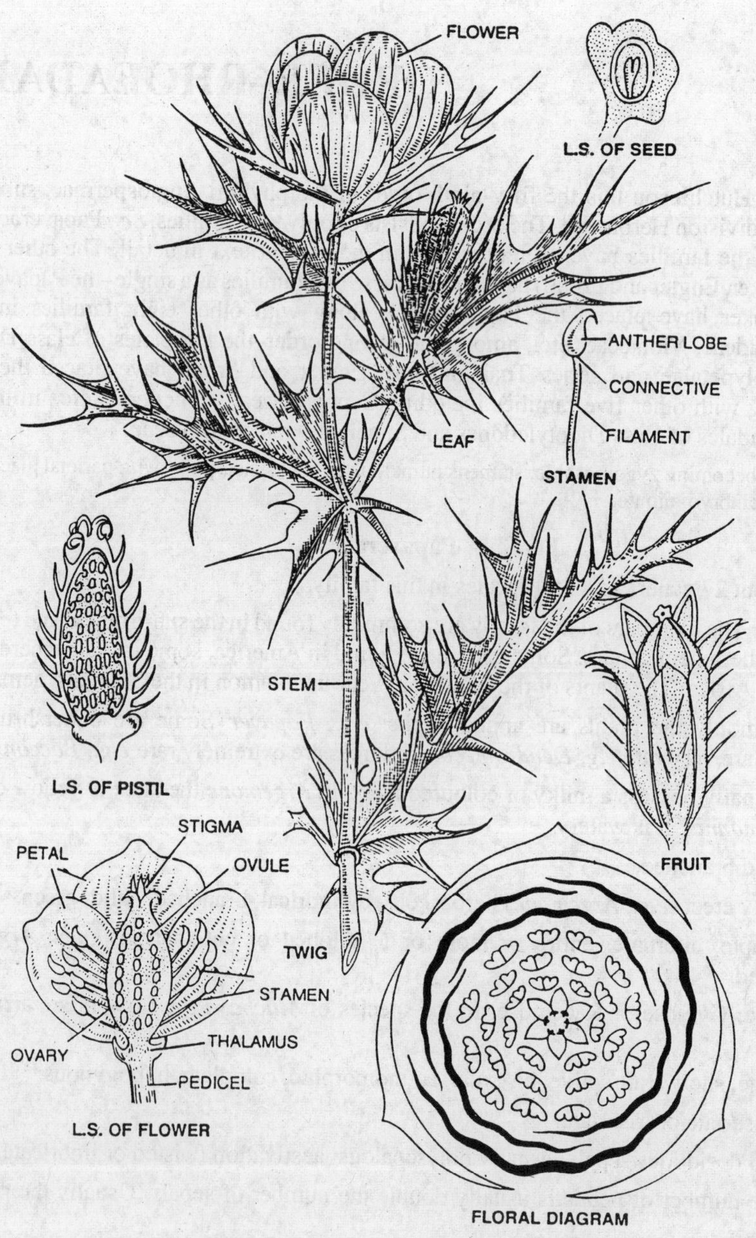

Fig. 40.1. Papaveraceae. *Argemone mexicana* Linn.

Seeds : The seeds are small and endospermic with a crested smooth raphe.

Pollination: The pollination takes place through the agency of insects.

Floral Formula: ⊕ ⚥ K 2, C 2 + 2, A ∝ , G ($\underline{2 - \infty}$)

Bentham & Hooker (1862)	**Engler & Prantl (1931)**	**Hutchinson (1959)**
Phanerogams	Phanerogams	Angiospermae
Dicotyledones	Dicotyledones	Dicotyledones
Polypetalae	Archichlamydeae	Herbaceae
Thalamiflorae	Rhoeadales	Rhoeadales
Parietales	Papaveraceae	Papaveraceae
Papaveraceae		

Description of important plants in semi-technical language.

1. ***Argemone mexicana***; Verna.–**Shialkanta.**

Habit: An annual prickly herb with yellowish latex.

Root: Tap, branched.

Stem: Erect, woody below, cylindrical solid, branched spiny, greenish, yellow latex.

Leaf: Simple, alternate, cauline, sessile, exstipulate prickly deeply cut, with spiny teeth, unicostate reticulate venation.

Inflorescence: Cymose, solitary axillary.

Flowers: Pedicellate, ebracteate, large, yellow, hermaphrodite, actinomorphic, complete, hypogynous.

Calyx: Three, polysepalous, green, caducous, horned at the top, twisted aestivation.

Corolla: Six, polypetalous, petals arranged in two whorls, of three each (petals of inner whorl slightly smaller), yellow, inferior, imbricate aestivation.

Androecium: Indefinite, polyandrous, stamens arranged in several alternating whorls, filaments long, anthers extrorse, bicelled basifixed.

Gynoecium: Four to six carpels (tetracarpellary to hexacarpellary), syncarpous, ovary superior, unilocular, parietal placentation, several ovules on each placenta, style absent, stigmas as many as the number of carpels.

Fruit :A spiny capsule.

Floral Formula: ⊕ ⚥ K 3, C3 + 3, A ∝, G ($\underline{4 - 6}$).

2. ***Papaver somniferum*** Linn.; Verna.–**Afim, Post;** Eng. Opium poppy.

Habit : An annual herb with latex.

Root: Tap, branched.

Stem: Erect, herbaceous, green, sometimes branched, cylindrical, hairy and fistular, latex.

Leaves: Simple, alternate, sessile, sheathing base, exstipulate, lobed, unicostate reticulate venation.

Inflorescence: Cymose, solitary or terminal.

Flower: Pedicellate (long pedicels), ebracteate, actinomorphic, hermaphrodite, complete, hypogynous, red or white large, showy.

Calyx: Two sepals, polysepalous, caducous, hairy, inferior, twisted aestivation.

Corolla: Four petals, polypetalous, arranged in two whorls of two each, inferior, often crumpled or rolled, in the bud aestivation, imbricate.

Androecium: Indefinite stamens arranged in several whorls, polyandrous anthers extrorse, basifixed, bi-celled, dehisce by longitudinal slits.

Gynoecium: Polycarpellary, syncarpous, ovary superior, unilocular, parietal placentation, many ovules on each placenta, stigma persistent, discoid with stigmatic rays.

Fruit: A capsule dehiscing by minute pores.

Floral Formula: ⊕ ⚥ K 2, C 4, A ∝ G, ($\underline{\infty}$).

3. *Eschscholtzia californica* Camb. mb.

Habit: An annual herb.

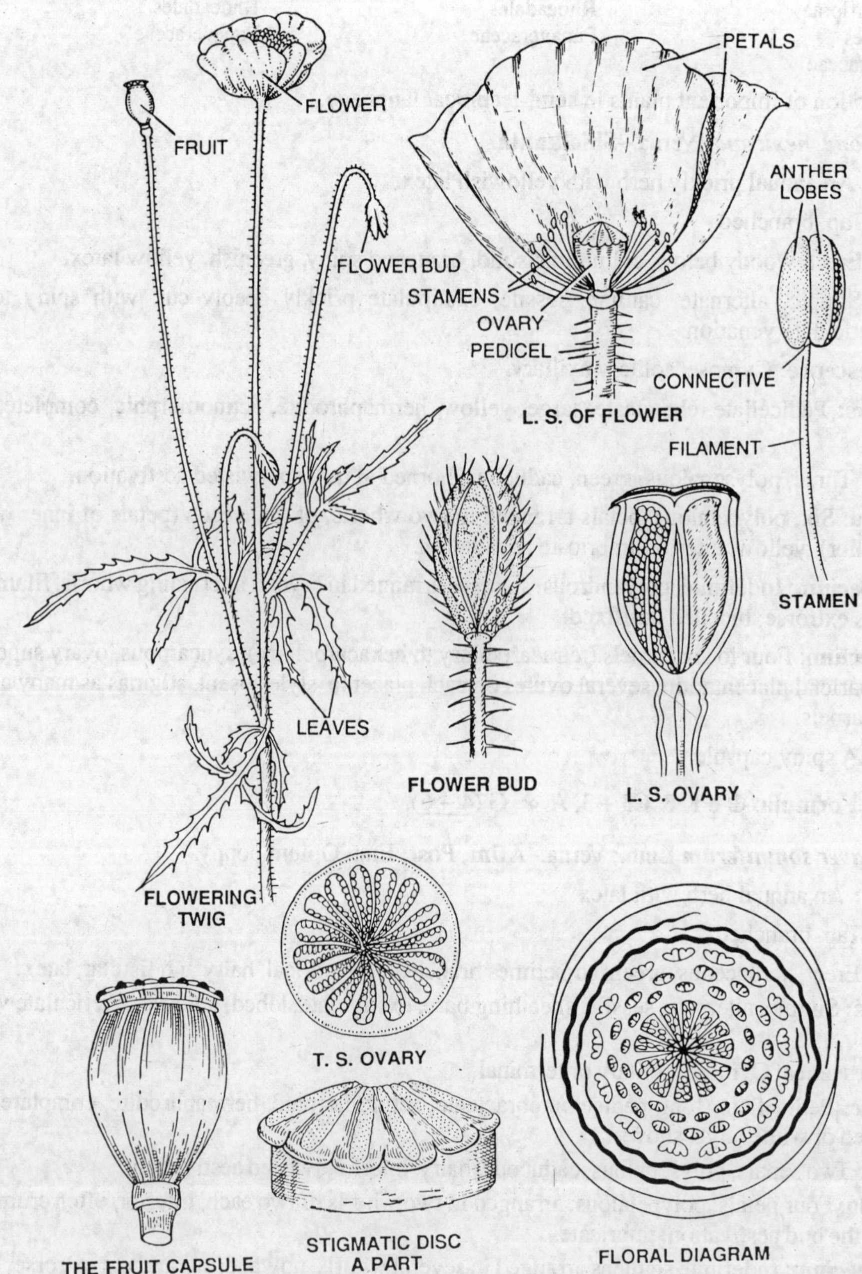

Fig. 40.2. *Papaver somniferum* Linn.

Root: Tap, branched.

Stem: Erect, herbaceous, aerial, branched, solid, green, watery, juice present.

Leaf: Cauline and ramal, simple, much dissected, alternate, sessile, exstipulate, sheathing at base, acute, glabrous, unicostate reticulate.

Inflorescence: Cymose, solitary axillary.

Flower: Pedicellate, ebracteate, complete, actinomorphic; hermaphrodite, hypogynous, yellow, dimerous, cyclic.

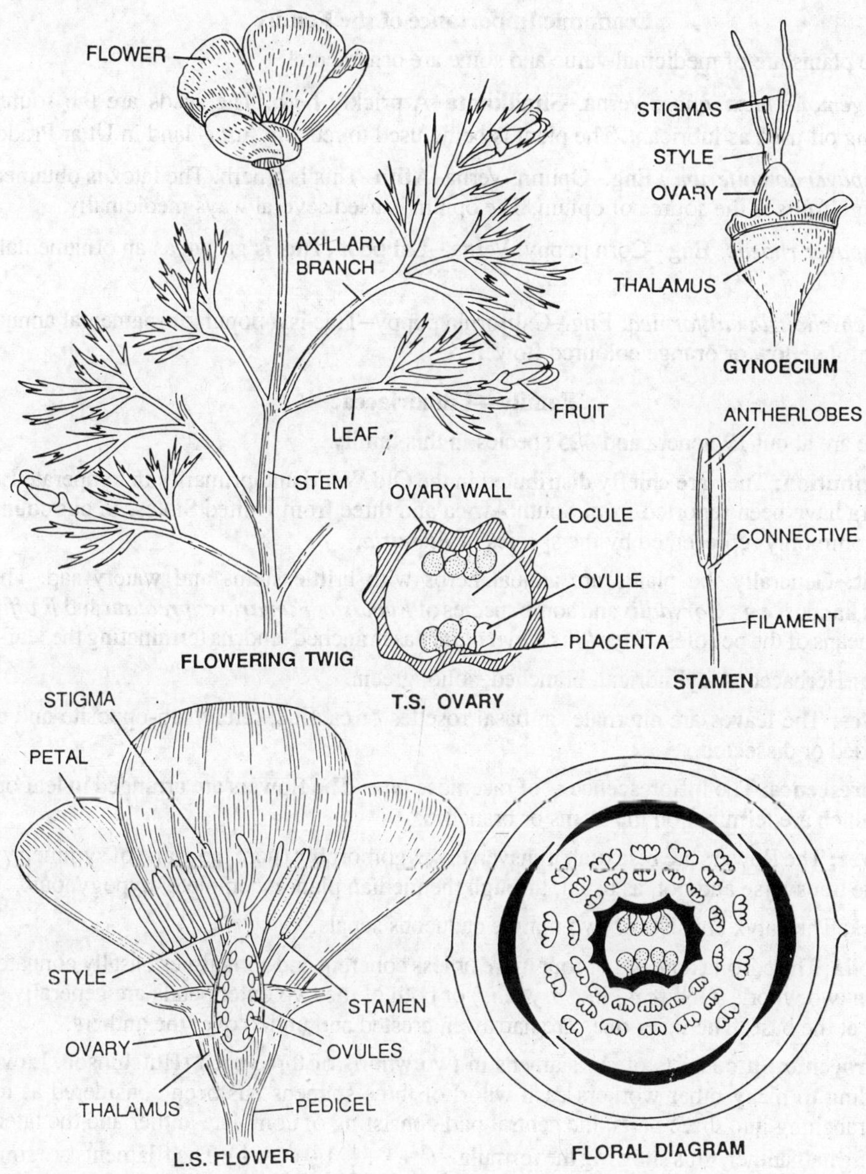

Fig. 40.3. Papaveraceae. *Eschscholtzia californica* Chamb.

Economic value. A popular annual ornamental herb with beautiful yellow or orange coloured flowers.

Calyx: 2 sepals, arise from a cup-like structure of floral axis, fused together (gamosepalous), caducous.

Corolla: 4 petals (2 + 2), polypetalous, imbricate, yellow.

Androecium: Indefinite, polyandrous, filaments short, anthers dithecous, basifixed, extrorse.

Gynoecium: 2 carpels, syncarpous, ovary half superior–half inferior, unilocular, parietal placentation, style short, stigma 2, each bifid and linear, one is shorter than the other.

Fruit: Capsule.

Floral formula: ⊕ ⚥ K 2, C 2 + 2, A∝, G (2).

Economic Importance of the Family

Some plants are of medicinal value and some are ornamental:

1. *Argemone mexicana* ; Verna.–**Shailkanta**–A prickly herb. The seeds are the source of a semi–drying oil used as lubricant. The plant is being used to reclaim 'usar' land in Uttar Pradesh.

2. *Papaver somniferum ;* Eng.–Opium; Verna–**Afim**–This is a herb. The latex is obtained from the immature fruits is the source of opium. The opium is used several ways medicinally.

3. *Papaver rhoeas;* Eng.–Corn poppy; Verna.–**Lal post**–This is grown as an ornamental in the gardens.

4. *Eschscholtzia californica*; Eng.–California poppy–This is a popular ornamental annual herb with beautiful yellow or orange coloured flowers.

Family—Fumariaceae

There are about 19 genera and 425 species in this family.

Distribution: They are chiefly distributed in the Old World and primarily in temperate Eurasia. Four genera have been reported from South Africa and three from United States. In our country the family is commonly represented by the species of *Fumaria.*

Habit: Generally the plants are annual herbs with brittle stems and watery sap. They are sometimes lianous, *e.g., Corydalis* and some species of *Fumaria. Fumaria capreolata* and *F. officinalis* climb by means of the petioles. *Corydalis claviculata* has branched tendrils terminating the leaf-stalks.

Stem: Herbaceous, cylindrical, branched, solid, green.

Leaves: The leaves are alternate, in basal rosettes or cauline, rarely sub-opposite and usually much divided or dissected.

Inflorescence: The inflorescence is of racemose type. The flowers are arranged in leaf opposed racemes which are terminal on the stems or branches.

Flower: The flowers are bisexual, transversely zygomorphic (here the plane of symmetry passes through the transverse and not, as usual, through the median plane), complete, hypogynous.

Calyx: The calyx consists of two minute caducous sepals.

Corolla: The corolla consists of four more or less coherent and sometimes basally connate petals arranged in two whorls. One (*e.g.,* in *Corydalis*) or both of the two outer petals are generally saccate or spurred at the base. The inner ones are narrower, crested and united over the anthers.

Androecium: It consists of six stamens in two whorls of three each (Hutchinson, Lawrence), but according to many other workers each whorl of three stamens has been considered as a single stamens branching into three parts, the central part consisting of complete anther and the lateral part each bearing half anther, thus showing the formula – $(\frac{1}{2}+1+\frac{1}{2})(\frac{1}{2}+1+\frac{1}{2})$. The filaments are somewhat winged and coherent to connate for much of their length. Usually one or two nectar secreting glands found to be present at the base of the androecium.

Gynoecium: The gynoecium consists of two carpels (syncarpous); the ovary is superior, unilocular; the placentation is parietal, the ovules are two-many; the style is one and slender; the stigma is one, sometimes bi-lobed.

Fruit: The fruit is a capsule or a nut. In *Fumaria,* the fruit is an indehiscent one–seeded nut.

Seeds. The seeds are endospermic with a minute embryo.

Pollination: The pollination takes place through the agency of insects.

Floral Formula: ·|· ⚥ K 2, C 2+2, A (3)+(3) or $(\frac{1}{2}+1+\frac{1}{2})(\frac{1}{2}+1+\frac{1}{2})$, G $\underline{(2)}$.

Bentham & Hooker (1862)	Engler & Prantl (1931)	Hutchinson (1959)
Phanerogams	Phanerogams	Angiospermae
Dicotyledones	Dicotyledones	Dicotyledones
Polypetalae	Archichlamydeae	Herbaceae
Thalamiflorae	Rhoeadales	Rhoeadales
Parietales	Papaveraceae	Fumariaceae
Papaveraceae	Fumarioideae	

Discription of *Fumaria parviflora* and *Corydalis cava* in semi-technical language : 1. ***Fumaria parviflora.***

Habit: An annual herb, wild, weed.

Root: Tap, branched.

Seed: Herbaceous, branched, glabrous, solid, scandent, watery juice, pale green.

Leaves: Simple, alternate, much dissected, pale green.

Inflorescence: Racemose, raceme.

Flower: Small, bracteate, pedicellate, zygomorphic, hermaphrodite, hypogynous, complete, purplish.

Calyx : 4 sepals, minute, caducous.

Corolla : Four petals, polypetalous, two petals of outer whorl dissimilar, one of them flat or concave and other spurred petal contains nectary.

Androecium: According to Lawrence and Hutchinson six stamens, united into two bundles one on each side of the pistil. According to other authors two stamens, each stamen three-partite, middle segment of complete anther, and two lateral segments bear half anther each. The stamen towards the spurred petal produces a nectariferous spur which runs into the spur of the petal.

Gynoecium: Two carpels (bicarpellary), syncarpous, ovary superior, unilocular, two ovules on each placenta, parietal placentation, style slender, stigma lobed.

Fruit: A capsule or a single-seeded nut.

Floral Formula: Br ·|· ⚥ K 2, C 2 + 2, A $(\frac{1}{2}+1+\frac{1}{2})(\frac{1}{2}+1+\frac{1}{2})$, G $\underline{(2)}$.

2. *Corydalis cava* (Linn.) Schweigg.; Verna. **Bhutkesi.**

Habit: Annual herbs.

Root: Tap, branched.

Stem: Herbaceous, aerial, branched, glabrous, solid, cylinderical, scandent, watery juice present, green.

Leaf: Alternate, petiolate, simple, much dissected, each lobe acute at apex, entire, glabrous, exstipulate, leaf base sheathing.

Inflorescence: Racemose, raceme.

Flower: Bracteate, bracteolate, pedicellate, hermaphrodite, zygomorphic, hypogynous, complete, dimerous.

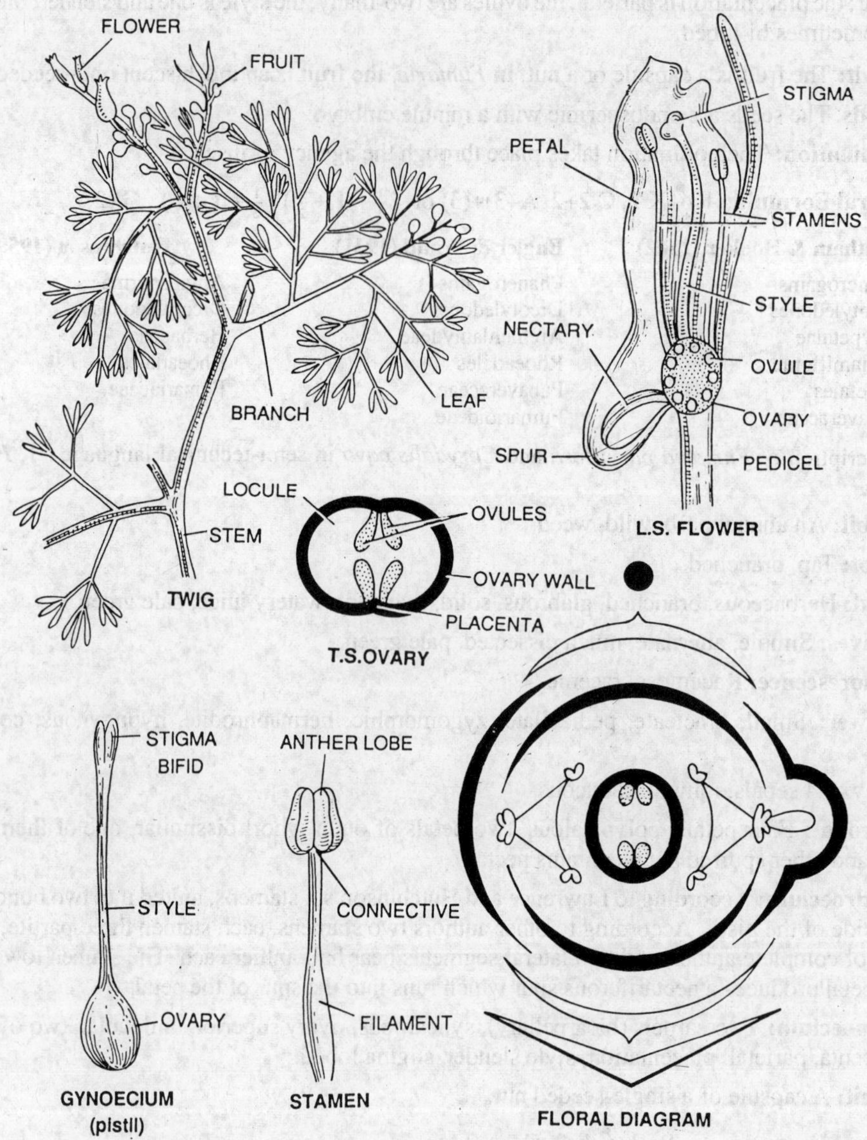

Fig. 40.4. Fumariaceae (Papaveraceae).*Fumaria parviflora* Lamk. ,Verna.–**Pit papra**.

Calyx: 4 sepals, minute, caducous, placed anterio-posteriorly.

Corolla: 4 petals, polypetalous, two petals of outer whorl dissimilar, one flat or concave other saccate or spurred, inner hooded and envelop the stamens like a cap, the spurred petal contains nectar, petals of inner whorl are smaller and placed anterio-posteriorly.

Androecium: 6 stamens, arranged in two series, outer two with dithecous anthers, inner four with monothecous anthers; each stamen branches off into three, two of the lateral branches bear monothecous anthers ; stamens are coherent in two bundles; one or two nectar glands at the base of androecium, or stamens in two groups and each group consists of $(\frac{1}{2} + 1 + \frac{1}{2})$ stamens, basifixed, extrorse.

Gynoecium: 2, bicarpellary, syncarpous, ovary superior, unilocular, parietal placentation, two ovules on each placenta, style slender, stigma lobed.

Fruit: Capsule or nut.

Floral formula : Br. Brl ·|· ⚥ K2, C 4, A 6, G $(\underline{2})$ OR Br. Brl ·|· ⚥ K 2, C 2+2, A $(\frac{1}{2}+1+\frac{1}{2})$ $(1/2+1+\frac{1}{2})$, G $(\underline{2})$.

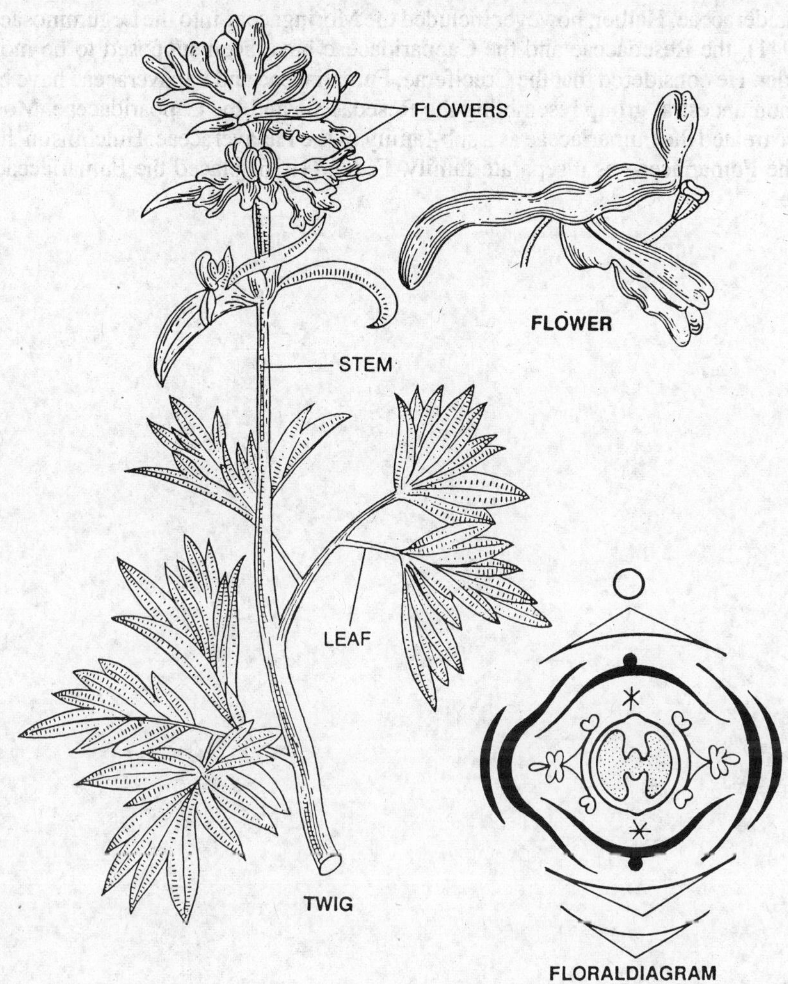

Fig. 40.5. Fumariaceae (Papaveraceae). *Corydalis cava* (Linn.) Schweigg : Verna. **bhutkesi.**

Economic value. A medicinal plant. The root contains a principle, *Corydalia.* The roots are supposed to be tonic, diuretic, and alternative, and are prescribed in syphilitic and cutaneous affections.

Economic Importance of the Family

The family is of little economic value. Some important plants are given below:

1. *Dicentra officinalis*; Eng.–Bleeding heart. This is a highly ornamental garden plant.
2. *Fumaria parviflora* ; Verna–**Pit papra**–This is a common winter weed. It is used medicinally.

3. *Fumaria officinalis* –This plant possesses medicinal properties. It is diuretic and laxative.

4. *Fumaria indica* – The plant is diuretic

5. *Corydalis govaniana*; Verna–**Bhut kesi**–This is a plant of medicinal importance. It is used as diuretic and tonic.

Systematic Relationships: According to Engler and Diels the order Rhoeadales comprises of seven families, *i.e.* Papaveraceae, Capparidaceae, Cruciferae, Tovariaceae, Resedaceae, Moringaceae and Bretschneideraceae. Hallier, however included the Moringaceae into the Leguminosae. According to Norris (1941), the Resedaceae and the Capparidaceae have been supposed to be most primitive within the order. He considered that the Cruciferae, Fumariaceae and Papaveraceae have been derived from a common ancestral group resembling the Resedaceae and the Capparidaceae. Most European botanists have treated the Fumariaceae as a sub-family of the Papaveraceae. Hutchinson, however, has recognized the Fumariaceae as a separate family. Fedde (1936) placed the Fumariaceae within the Papaveraceae.

41

ORDER–BRASSICALES (CRUCIALES)

According to Hutchinson this is the sixtieth order of the phylum Angiospermae, sub-phylum Dicotyledones, and division Herbaceae. This order consists of a single family the Cruciferae. This family has been described here in detail. Bentham and Hooker have placed the family Cruciferae along with eight other families including Papaveraceae, Capparidaceae, Violaceae, etc., among the second order–the Parietales of class Dicotyledones, division Polypetalae and series Thalamiflorae. Engler and Prantl have placed the Cruciferae along with other five families including Papaveraceae, Capparidaceae, etc., in the nineteenth order–the Rhoeadales of class Dicotyledones and sub-class Archichlamydeae.

Stamens reduced to 6, 4 long and 2 short, petals 4; ovary usually divided by a false septum; no endosperm.

Family—Cruciferae (Brassicaceae)

There are about 350 genera and 2500 species in the family.

Distribution: The members of this family are cosmopolitan in their distribution but majority of them are common in north temperate regions. Some species are sub-arctic. A large number of important vegetable crops as well as many garden and wild flowers are included in this family. The family is represented by several important species in our country, such as *Brassica campestris, Brassica nigra, Raphanus sativus, Iberis amara, Senebiera didyma* and several others. Usually the plants grow in winter season in our country.

Habit: Mostly the plants are annual, bienninal or perennial herbs. Sometimes they are small shrubs. Majority of the species, of *Brassica* are annual, biennial or sometimes perennial herbs. The species of *Raphanus* may be annual or biennial herbs. However, majority of the plants are annuals of a life cycle of only a few weeks. Usually the biennials, *e.g.*, Turnip (**Shalgam**) develop swollen tap roots which store enough of nourishment. *Farsetia* is a small branched undershrub.

Root: Usually tap and branched. The tap root modifications–fusiform root, *e.g., Raphanus sativus,* napiform root, *e.g., Brassica rapa,* are also found. Such roots become swollen and store nourishment in them.

Stem: Usually tap stem is herbaceous. However, in *B. oleracea* var. *caulorapa* (knol khol, **Ganth-Gobhi**). The stem becomes corm like and very much thickened, which is eaten as vegetable. In *Raphanus sativus* (Radish, **Muli**), the stem becomes very much condensed.

Leaves: The leaves are simple, alternate, exstipulate and possess simple or branched hairs. They are usually radial or cauline, sub-sessile or sessile, lyrate. When radical they are found in rosettes, *e.g., Raphanus sativus.*

Sometimes the bulbils develop in the axils of the upper leaves which act as vegetative structures for propagation.

Inflorescence: The inflorescence is generally of racemose type and very often may be a raceme, a corymb or corymboraceme. Bracts and bracteoles are absent. They may be present only in rare cases.

Flowers: Mostly the flowers are ebracteate, pedicellate, regular (actinomorphic), hermaphrodite, complete, cruciform and hypogynous.

Calyx: The calyx is polysepalous and consists of four sepals. The sepals are arranged in two whorls of two each. The lateral inner sepals are sometimes pouched at the base, which serve as nectar containers. The aestivation is imbricate.

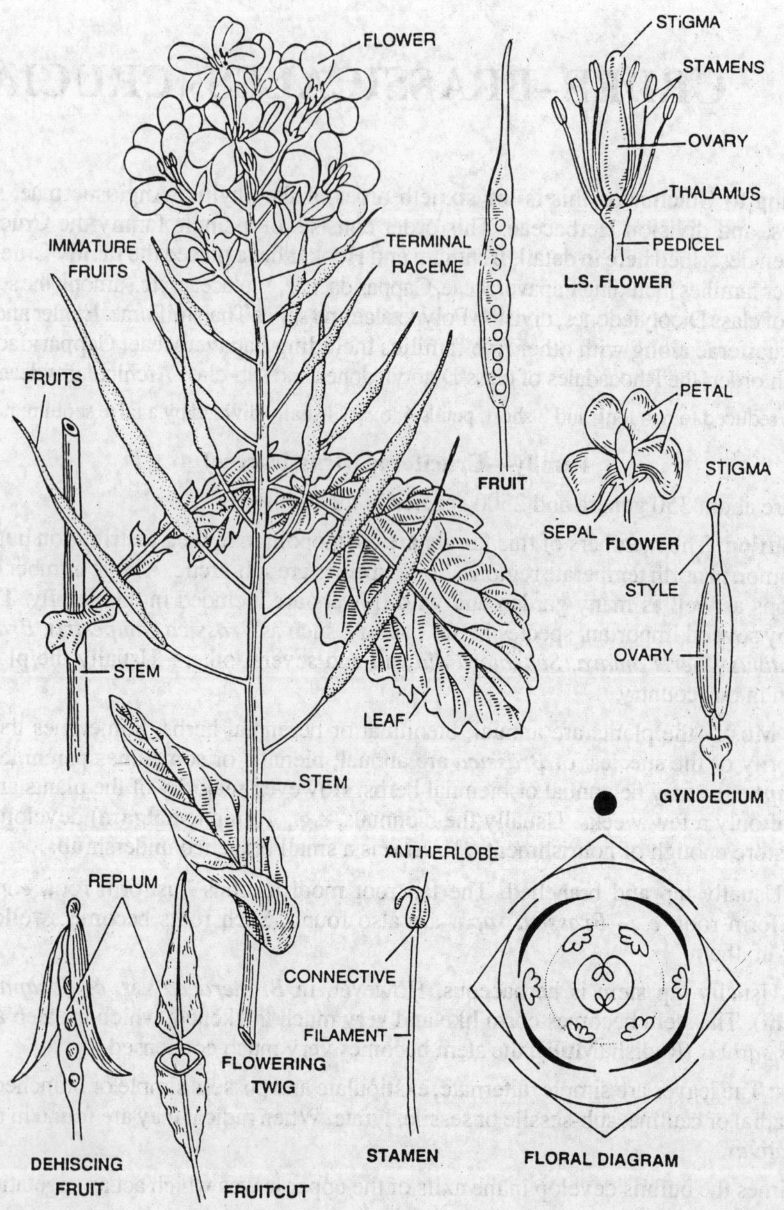

Fig. 41.1. Cruciferae (Brassicaceae). *Brassica campestris* Linn.; Eng. Yellow mustard; Verna. **Sarson.**

Corolla: The corolla is polypetalous and cruciform. It consists of four petals. The petals are found in one whorl. They are alternate to the sepals. Each petal is usually differentiated into a narrow claw and a broad expanded limb. In *Iberis amara* (candytuft) the flowers become zygomorphic by the enlargement of the two outer petals. In *Senebiera* sp., the petals are represented by four minute lobes.

Androecium: Usually the androecium consists of six stamens arranged in two whorls. The outer two stamens are short whereas the inner four are long. This condition of the stamens is known as **tetradynamous** condition and considered to be characteristic of the family. Nectaries are developed as small green glands at the base of the two short stamens. In *Senebiera,* there are only two stamens, the anthers are bilobed; basifixed and introrse.

Gynoecium: It consists of two carpels (bicarpellary), syncarpous. The ovary is superior, unilocular but becomes bilocular because of the development of the **false septum** or **replum** from the ingrowths of the parietal placentas. The placentation is parietal. Many anatropous or campylotropous ovules develop from the parietal placentas. The style is short with two lobed stigmas.

Fruit: The fruit is either siliqua or silicula. Sometimes the fruit is lomentum (lomentaceous siliqua) as in *Raphanus sativus.*

Pollination: The pollination is usually entomophilous, *i.e.,* through the agency of insects. The honey secreting nectaries found at the base of short stamens and anterior sepals are the main centres of the attraction of the insects. The flowers arranged in corymbs also attract the insects and cross pollination is effected. In such cases several flowers are pollinated by one insect at one time.

Floral Formula: ⊕ ⚥ K 2 + 2, C 4, A 2 + 4, G $\underline{(2)}$.

Bentham & Hooker (1862)	Engler & Prantl (1931)	Hutchinson (1959)
Phanerogams	Phanerogams	Angiospermae
Dicotyledones	Dicotyledoneae	Dicotyledones
Polypetalae	Archichlamydeae	Herbaceae
Thalamiflorae	Rhoeadales	Cruciales (Brassicales)
Parietales	Cruciferae	Cruciferae or
Cruciferae		Brassicaceae

Description of some important plants in semi-technical language.

1. ***Brassica campestris*** **Linn.**; Verna. **Sarson;** Eng. **Yellow mustard.**

Brassica campestris Linn., very largely grown during the cold season, especially in Uttar Pradesh. The oil (**Karwa tel**) is yielded from its seeds.

Habit: Annual herb, cultivated.

Root: Tap, branched.

Stem: Erect, herbaceous, cylindrical, slightly hairy, branched, smooth and green.

Leaf: Cauline and ramal, simple, alternate, sessile, glabrous, upper leaves entire, lower leaves lyrate, exstipulate, unicostate reticulate venation.

Inflorescence: Racemose, typical raceme.

Flower: Pedicellate, ebracteate, hermaphrodite, actinomorphic, yellow, hypogynous, complete, tetramerous, cyclic.

Calyx: Four sepals, polysepalous, arranged in two whorls of two each, the sepals of inner whorl are longer, imbricate aestivation.

Corolla: Four petals, polypetalous, each petal consists of a limb and claw, cruciform, alternating the sepals, imbricate or valvate.

Androecium: Six stamens, polyandrous arranged in two whorls, the two stamens of outer whorl are smaller than the four stamens of inner whorl (*i.e.,* tetradynamous condition), anthers bicelled, basifixed, introrse.

Gynoecium: The carpels (bicarpellary), syncarpous, ovary superior, unilocular but becomes bilocular because of the development of false septum (replum)m parietal placentation, style short and stigma bilobed.

Fruit: Siliqua.

Floral Formula: ⊕ ⚥ K 2 + 2, C 4, A 2 + 4, G (2).

Economic value. The tender leaves, and shoots are used as vegetable. Ground mustard seeds are used as a condiment. The seeds yield a fatty oil, which is generally used for cooking purposes.

2. ***Senebiera didyma*** Pers.

Senebiera didyma Pers. (Lesser Wart-cress). An established weed of cultivation during the cold season. Distribution–plains of N.India.

Habit: Annual herb, wild, weed.

Root: Tap, branched.

Stem: Prostrate, aerial, weak, trailing, cylindrical, branched, branches arranged in rosettes, green, glabrous, herbaceous, solid.

Leaf: Radical, cauline and ramal, compound, exstipulate, unipinnate and imparipinnate, pinnae with cut margins, petiolate, petiole base hairy, alternate, leaf surface glabrous, reticulate venation.

Inflorescence: Racemose, leaf opposed typical raceme.

Flower: Minute, greenish, pedicellate, ebracteate, hermaphrodite, actinomorphic, hypogynous, complete, tetramerous, cyclic.

Calyx: Four sepals, polysepalous, arranged in two whorls of two each, sepals spreading, short, larger than petals, inferior, valvate aestivation.

Corolla: Very much reduced (rudimentary), represented by four minute whitish scaly structures alternating the sepals, valvate.

Androecium: Two stamens, polyandrous, median, anterior-posterior, anthers basifixed, filaments long tapering at apex and broad at base, dithecous, introrse.

Gynoecium: Two carpels (bicarpellary), syncarpous, ovary superior, unilocular but becomes bilocular due to formation of replum, parietal placentation, one ovule on each locule, style reduced, stigma discoid.

Fruit: Compressed silicula.

Floral formula: ⊕ ⚥ K 2 + 2, C 4, A 2 , G (2).

3. ***Iberis amara*** Linn.; Eng. Candytuft.

Habit: Annual ornamental herb, cultivated.

Stem: Herbaceous, erect, aerial, branched, green, solid, rough surface.

Leaf: Cauline and ramal, alternate, sometimes opposite, sessile, exstipulate, simple, margin somewhat dissected, acute, glabrous, unicostate reticulate.

Inflorescence: Racemose, corymb.

Flower: Ebracteate, pedicellate, complete, irregular (zygomorphic), hermaphrodite, tetramerous, hypogynous, white, cyclic.

Calyx: 4, polysepalous, in two whorls of 2 each, imbricate, petaloid, boat shaped.

Corolla: 4, polypetalous, valvate, 2 anterior petals large, 2 posterior petals small, each petal consists of a claw and limb, cruciform.

Androecium: 6, free stamens (polyandrous), tetradynamous –2 outer lateral short, remaining 4 anterio-posterior long, dithecous, dorsifixed, introrse.

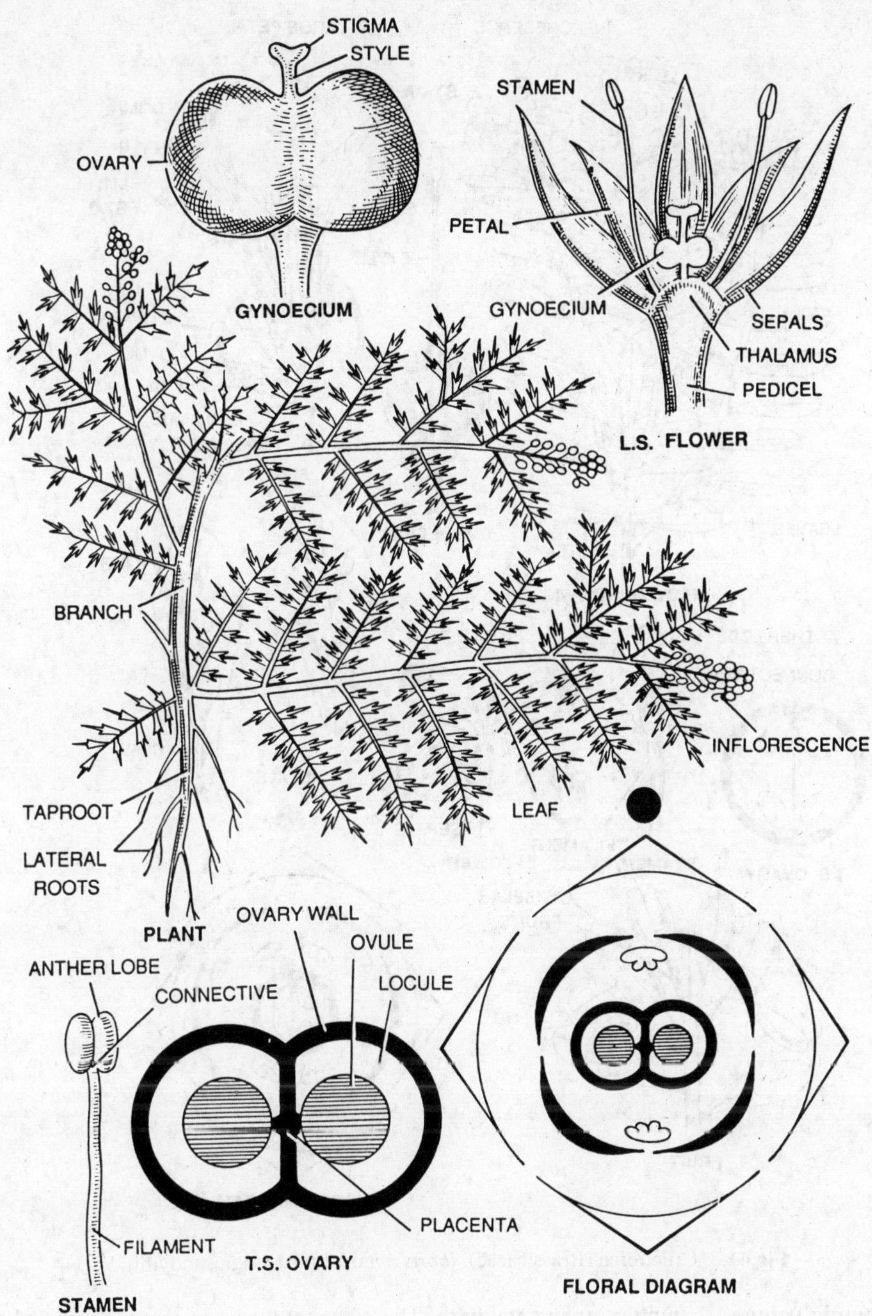

Fig.41.2: Cruciferae (Brassicaceae). *Senebiera didyma* Pers.; Eng. lesser wart cress.

Gynoecium: Bicarpellary, syncarpous, ovary superior, unilocular when young, at maturity becomes bilocular because of the development of false septum, parietal placentation, style long, stigma globular.

Fruit: Silicula.

Floral Formula: ⊕ ⚥ K 2+2, C 4, A 2+4, G($\underline{2}$).

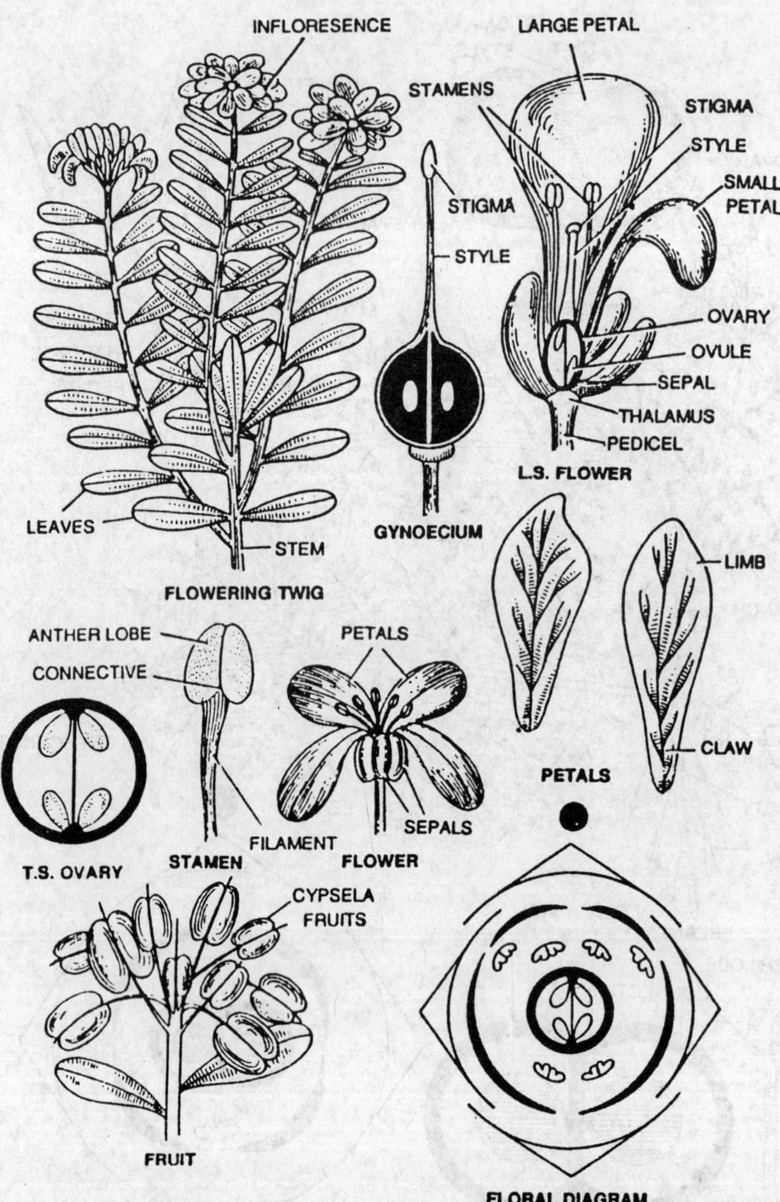

Fig.41.3. Cruciferae (Brassicacae). *Iberis amara* Linn.; Eng., candytuft.

Economic value. A common ornamental herb. The plants are used in rheumatism and gouts. Seeds are used in asthma and bronchitis.

Economic Importance of the Family

The family is fairly important from the economic point of view. Most of the plants contain sulphur compounds. The seeds of many plants yield vegetable oil of multipurpose use.

1. *Brassica campestris* var. *dichotoma* (Verna.–**Kali Sarson**) cultivated as an oil yielding crop, mostly in the Punjab.

2. *Brassica campestris* var. *sarson;* (Verna.–**Sarson**)–An oil-seed crop grown mainly in Uttar Pradesh, the Punjab, Bihar and Assam. The oil is used for cooking and burning purposes, and the oil cake as a cattle feed. The tender leaves and shoots are used as vegetable.

3. *Brassica campestris* var. *toria* (Verna.–**Toria**)–An oil-yielding crop grown in Uttar Pradesh, West Bengal and the Punjab. The oil is edible and the oil-cake is used as cattle and manure.

4. *Brassica hirta*; Syn. *B. alba* (Eng.–White Mustard; Verna.–**Safed Rai**)–The young leaves and tender shoots are used as vegetable. The seeds yield fatty oil.

5. *Brassica juncea* (Verna.–**Rai**)–The seed oil is used for cooking purposes. Cultivated in the Punjab, West Bengal and Uttar Pradesh.

6. *Brassica juncea* var. *cuneifolia*; Syn. *B. rugosa* var. *cuneifolia* (Verna.–**Rai**)–Commonly grown in the terai areas of Nainital, North Bengal and Assam. The young leaves and tender shoots are used as vegetable.

7. *Brassica napus* (Eng.–Rape; Verna.–**Toriya, Kali sarson**)–The seeds are used as vegetable. Cultivated in the Punjab, Bengal and Bihar.

8. *Brassica nigra* (Eng.–Biack mustard; Verna.–**Kali rai**)–The seeds are used as spice and condiment. Cultivated in the Punjab, Uttar Pradesh and Tamil Nadu.

9. *Brassica oleracea* var. *acephala* (Eng.–Kale; Verna.–**Karam-sag**)–The young shoots and leaves are eaten as vegetable. Cultivated in Assam, Kashmir and Maharashtra.

10. *Brassica oleracea* var. *botrytis* (Eng.–Cauliflower; Verna.–**Phulgobhi**)–Grown all over Northern India for its edible inflorescence.

11. *Brassica oleracea* var. *capitata* (Eng.–Cabbage; Verna.–**Bandgobhi**)–A common vegetable of Northern India.

12. *Brassica oleracea* var. *gemmifera* (Eng.–Brussels sprouts; Verna.–**Buttamgobhi**)–Grown mainly in Maharashtra. The young shoots, buds and leaves are edible.

13. *Brassica oleracea* var. *gongylodes*; (Eng.–knol–khol: Verna.–**Ganthgobhi**)–The short swollen stem is edible. Grown in Maharashtra, Uttar Pradesh and the Punjab.

14. *Brassica pekinensis* (Eng.–Chinese cabbage)–Grown in Northern India. The leaves are used as vegetable.

15. *Capsella bursa-pastoris*; (Eng.–Shepherd's purse)–A common weed. Found throughout temperate India particularly in Northern Western Himalayas. The plant is of medicinal value.

16. *Cheiranthus cheiri*; (Eng.–Wall flower; Verna.–**Todrisurkh**)–A herb, native of South Europe. Grown as an ornamental. The seeds are useful in dry bronchitis, fevers and injuries to the eyes. The flowers are used in paralysis and impotency.

17. *Eruca sativa* (Verna.–**Tara**)–A herb, native of South Europe but grown mainly in Northern India. The oil, obtained from the seeds, is used for burning purposes and the plant is used as fodder.

18. *Iberis amara* (Eng.–Rocket candytuft)–A common ornamental herb. The plants are used medicinally in rheumatism and gouts. Seeds are used medicinally in asthma and bronchites.

19. *Mathiola incana* (Eng.–Stock; Verna.–**Todri safed**)–A herb, native of Europe. Grown as an ornamental. The seeds are used for extraction of an oil, which yields methyl. The seeds mixed with wine are given as an antidote to poisonous bites. They are also used as tonic in stomach complaints.

20. *Lepidium sativum*; (Eng.–Garden cress; Verna.–**Halim**)–A common cultivated herb. The young as well as the ripened seeds are used as vegetable and pulse. The leaves and tender shoots are largely eaten for liver complaints. The plant is also used in the treatment of asthma, cough and bleeding piles.

21. *Nasturtium officinale*; Syn. *N. fontanum* (Eng.–Water-cress; Verna.–**Brahmi Sag**)–A small herb cultivated in Bengal, Orissa and the Punjab. It is used as a vegetable.

22. *Raphanus sativus;* (Eng.–Radish; Verna.–**Muli**)–A herb, cultivated in Uttar Pradesh, the Punjab, Maharashtra, etc. The roots, young leaves and the fruits are used as vegetable.

23. *Raphanus sativus* var. *candatus* (Eng.–Rat-tail radish; Verna.–**Sengri**)–The fruits are used as vegetable. The fruits are also used for making pickle.

Systematic Relationships: Bentham and Hooker placed the Cruciferae in the order–Parietales. Engler and Diels included the family within the order—Rhoeadales. Hutchinson placed the single family Cruciferae in the order–Cruciales. On the basis of morphology and anatomy of androecium and gynoecium the family is much allied to the Capparidaceae, but Hutchinson does not agree with the statement.

42

ORDER–CARYOPHYLLALES

According to Hutchinson this is the sixty-second order of the phylum Angiospermae, subphylum Dicotyledones and division Herbaceae. The order comprises of five families, *i.e.,* Elatinaceae, Molluginaceae, Caryophyllaceae, Ficoidaceae and Portulacaceae, of which the Caryophyllaceae has been described in the present text in detail. Bentham and Hooker have placed the Caryophyllaceae along with three others in the fourth order–the Caryophyllinae of class Dicotyledones, division Polypetalae and series Thalamiflorae. Engler and Prantl have placed the Caryophyllaceae along with other eight families including Chenopodiaceae, Amarantaceae, Nyctaginaceae, etc., in the seventeenth order–the Centrospermae of class Dicotyledoneae and sub-class Archichlamydeae.

Hypogynous to perigynous; mostly bisexual; cyclic; syncarpous; axile to free central placentation; stamens mostly definite in number; endosperm copious; embryo curved.

Family—Caryophyllaceae

There are about 80 genera and 2100 species in this family.

Distribution. The members of this family are commonly found in the temperate regions of the Northern hemisphere. Certain genera are found in the Southern hemisphere and few are found in the mountains of tropical regions. In our country the plants of this family are either found in the hilly tracts or they grow in the plains during winter season, *e.g., Stellaria, Spergula, Dianthus,* etc. The family is well represented in the British Flora and many species are cultivated as ornamental garden flowers.

There are about 500 species in the genus *Silene* of this family, which is supposed to be the largest genus; *Dianthus* contains 350 species and *Stellaria* about 100 species.

Habit: Most of them are annual, while some of them are perennial herbs. Certain small shrubs, *e.g.,* some species of *Acanthophyllum* are also found in the warmer parts of the world.

Stem: The stem is erect, branched, green, herbaceous, solid and mostly swollen at the nodes.

Leaves: The leaves are simple, opposite, entire and exstipulate. The leaves sometimes possess shortly connate perfoliate base, *e.g.,* in *Dianthus.* They are linear to lanceolate in shape. The development of the leaves is unusual. At each node one leaf develops earlier than the other. This leaf bears in its axil a more vigorous bud than on the other side and frequently it is only this bud which devlops later.

The chief anatomical feature is the presence of caryophyllaceous type of stomata. Each such stoma possesses two subsidiary cells placed at the right angles to the guard cells.

Inflorescence: The inflorescence is cymose. Usually it is a dichasium which later on becomes a dichasial cyme ending in a monochasial cyme. This is known as **cincinnus** inflorescence which is characteristic of the family. In certain cases the flowers are arranged in racemes. In some of genera, *e.g., Arenaria, Githago,* etc., the flowers are solitary.

Flower: The flowers are pedicellate, actinomorphic, usually hermaphrodite and pentamerous, but rarely unisexual or tetramerous. They are regular, complete and hypogynous. Each typical flower bears five whorls, each consisting of five members, such as, five sepals, five petals, two pentamerous whorls of stamens, five carpels, five styles and five double rows of ovules and sometimes five partition walls in the basal part of the ovary.

Calyx: It is composed of five and very rarely of four sepals. The sepals may be free or united together into a tube. They are usually persistent with membranous margins. The aestivation is imbricate (quincuncial).

Corolla: It is composed of five and rarely of four petals. The petals are always free (polypetalous). The petals are mostly notched, sometimes bifid, *e.g., Stellaria media.* Usually the aestivation is imbricate.

Androecium: Usually it consists of ten stamens. Sometimes the number of stamens reduces to eight, five, four, three or even one. They are polyandrous, obdiplostemonous, *i.e.*, the stamens are arranged in two whorls of five each, the stamens of the outer whorl are seen to be opposite the petals and of inner whorl alternating the petals. The stamens are hypogynous and occasionally perigynous. The anthers are two-lobed and dehisce longitudinally. Sometimes the petaloid staminodes are present. In *Spergula,* the five or ten stamens are found to be arranged on a perigynous disc.

Gynoecium: It consists of two or five carpels; syncarpous. The styles are free. The ovary is superior and unilocular; the ovules are many, campylotropous and arranged on a central column. The placentation is free-central which is most characteristic of the family. The number of carpels corresponds to the number of styles and stigmas. In *Stellaria,* the number of carpels is reduced to three.

Fruit: Generally the fruit is an unilocular capsule, *e.g., Stellaria, Arenaria, Spergula,* etc. In some cases the fruit, may be an achene or a nut, *e.g., Herniaria, Dysphania, Scleranthus,* etc.

Seeds: The seeds are small and endospermic. The embryo is curved in the endosperm. Sometimes the funicle is conspicuous. They are dispersed by censor mechanism.

Pollination: It usually takes place by means of insects (*i.e.,* entomophily).

Floral Formula: ⊕ ⚥ K 5, C 5, A 5 + 5 or 5, G (2 – 5).

Bentham & Hooker (1862)	Engler & Prantl (1931)	Hutchinson (1959)
Phanerogams	Phanerogams	Angiospermae
Dicotyledones	Dicotyledoneae	Dicotyledones
Polypetalae	Archichlamydeae	Herbaceae
Thalamiflorae	Centrospermae	Caryophyllales
Caryophyllinae	Caryophyllaceae	Caryophyllaceae
Caryphyllaceae		

Description of some important plants in semi–technical language:

Saponaria vaccaria L. Eng. Cow–herb; Verna. **Musna. Stem**-aerial, erect, herbaceous, cylindrical branched, solid, green glabrous. **Leaf**-cauline, ramal, opposite decussate, exstipulate, simple, sessile, entire, acute, glabrous,lanceolate, coriaceous, unicostate reticulate. **Inflorescence**–cymose, dichasial cyme. **Flower**–bracteate, bracteolate, pedicellate, complete, actinomorphic, hermaphrodite, pentamerous, hypogynous, cyclic. **Calyx**–5, gamosepalous, valvate, tubular, persistent. **Corolla**–5, polypetalous, twisted, caryophyllaceous, petal distinguished into claw and limb. **Androecium**–5+5, obdiplostemonous, polyandrous, filaments, long dithecous, versatile, introrse. **Gynoecium**–bicarpellory, syncarpous, superior, unilocular free-central, ovules many, styles 2, stigmas 2. **Fruit**–capsule.

Floral Formula– Br, Brl ⊕ ⚥ $K_{(5)}$ C_5 A_{5+5} $G_{(2)}$.

Stellaria media Linn.; Verna. **Morolia;** Eng. Chickweed.

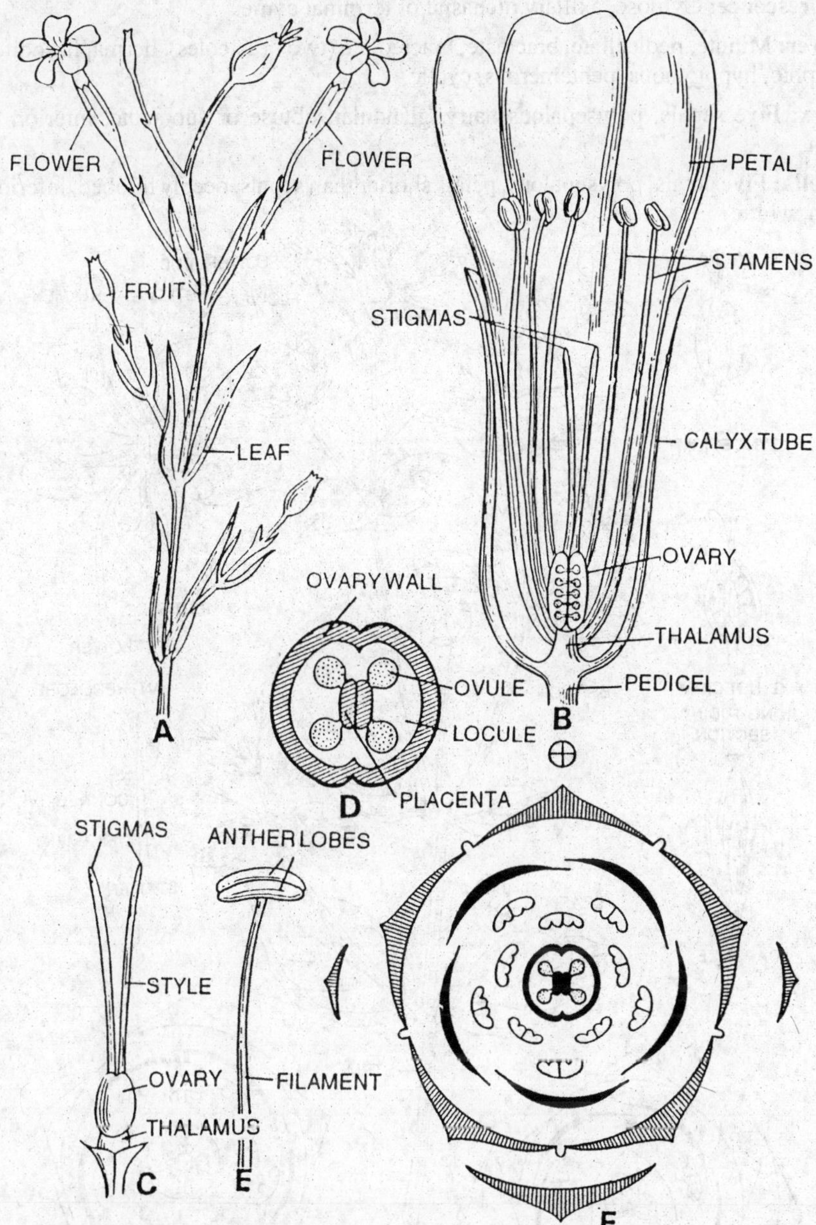

Fig. 42.1. Caryophyllaceae—Saponaria vaccaria L. (Musna). A, Flowering twig; B, L.S. of flower; C, gynoecium; D, T.S. of ovary; E, stamen; F, floral diagram.

Stellaria media Linn. (Eng. Chickweed). A common weed. Distribution: throughout N. India.

Habit: An annual herb, weed, six inches to two feet high.

Root: Tap, branched.

Stem: Erect, aerial, herbaceous, cylinderical, branched, green, glabrous, solid, nodes swollen.

Leaves: Cauline and ramal, simple, opposite decussate, exstipulate, $\frac{1}{2}$ to 1 inch long, lower petiolate, upper sessile, entire, ovate, acuminate, smooth, unicostate reticulate venation.

Inflorescence: Cymose, axillary dichasial or terminal cyme.

Flower: Minute, pedicellate, bracteate, bracteolate (two bracteoles), hermaphorodite, complete, actinomorphic, hypogynous, pentamerous, cyclic.

Calyx: Five sepals, polysepalous hairy, glandular, obtuse or sub-acute, inferior, quincuncial aestivation.

Corolla: Five petals, polysepalous, petals shorter than sepals, deeply bilobed, inferior, imbricate aestivation, white.

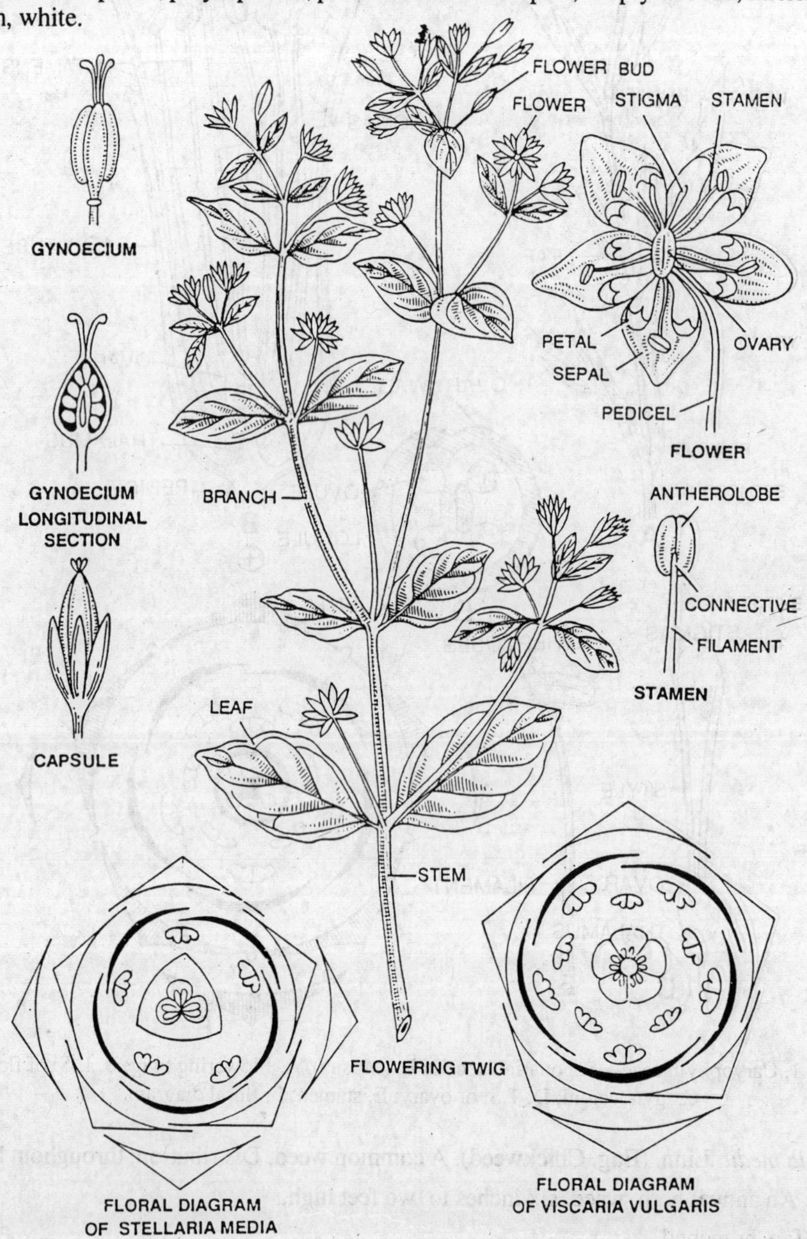

Fig.42.2. Caryophyllaceae. *Stellaria media* Linn. Eng. Chickweed; Verna. **Morolia.**

Economic value. The natives of Assam eat the leaves and tender stalks, boiled in *Khar* water, either by themselves or with fish. It is used also as vegetable by the natives of the Nilgiris. The plant is employed in plaster for broken bones and swellings, as it is supposed to be cooling and binding.

Androecium: Ten stamens, arranged in two whorls of five each, polyandrous, obdiplostemonous (two alternating whorls of stamens, the stamens of outer whorl being arranged opposite the petals), generally outer three or all five stamens reduced to staminodes, filaments long and slender, dithecous, basifixed, introrse.

Gynoecium: Three carpels, syncarpous, ovary superior, unilocular, free-central placentation, styles three, each with its own stigma.

Fruit: An unilocular capsule.

Floral formula: Br. Brl. ⊕ ⚥ K 5, C 5, A 5+5, G (3).

Spergula arvensis Linn.; Eng. Corn spurrey.

Spergula arvensis Linn. (Eng. Corn spurrey), plentiful in cultivated ground, flowering during the cold season. Distribution : throughout N. India.

Habit: Annual herb.

Root: Tap, branched.

Stem: Erect, cylindrical, branched, solid or fistular, green, herbaceous, thickened at the joints, glabrous.

Leaf: Cauling and ranal, simple, opposite, linear, sessile, fleshy, leafy buds in their axils and looking apparently whorled, lanceolate, acute, stipulate (stipules small).

Inflorescence: Cymose, dichasial cyme with suppression of branches on one side, *i.e.* cincinnus.

Flower: Pedicellate, bracteate, bracteolate (two bracteoles), hermaphrodite, complete, actinomorphic, hypogynous, greenish white, pentamerons, cyclic.

Calyx: Five sepals, oikyseoakiysm greeb, inferior, quincuncial aestivation, persistent.

Corolla: Five petals, polypetalous, white, inferior, entire, imbricate aestivation.

Androecium: Ten stamens, arranged in two whorls of five each obdiplostemonous, anthers bicelled, introse, dorsifixed.

Gynoecium: 3 or 5 carpels, syncarpous, ovary superior, unilocular, free-central placentation, styles three or five, stigmas 3 or 5.

Fruit: An ovoid capsule.

Floral formula: Br. Brl ⊕ ⚥ K 5, C 5, A 5, G (3 or 5).

Dianthus caryophyllatus Linn.; Eng. Carnation plant.

Habit: An annual herb, ornamental, cultivated.

Root: Tap, branched.

Stem: Erect, aerial, herbaceous, branched, green smooth, solid.

Leaf: Cauline and ramal, simple, opposite decussate, sessile, exstipulate, sessile, lanceolate, entire, acute, glabrous, unicostate reticulate venation, leaf-base sheathing.

Inflorescence: Cymose, axillary or terminal dichasial cyme.

Flower: Pedicellate, ebracteate, hermaphrodite, actinomorphic, complete, hypogynuous, pentamerous, cyclic.

Epicalyx: 4, in two whorls of 2 each, outer whorl anterio-posterior.

Calyx: Five sepals, polysepalous, inferior, quincuncial aestivation.

Corolla: Five petals, polypetalous, petals fimbriate, inferior, imbricate aestivation, variously coloured, caryophyllaceous, each petal possesses a claw and limb.

Androecium: Ten stamens arranged in two whorls of five each, obdiplostemonous, polyandrous, anthers bicelled, dorsifixed, introrse.

Gynoecium: Two-five carpels, syncarpous, ovary superior, unilocular, free-central placentation, ovules many, styles two, stigmas two, coiled and feathery.

Fruit: A capsule.

Floral formula: ⊕ ⚥ Epi K 2+2, K (5), C 5, A 5+5, G ($\underline{2}$ – $\underline{5}$)

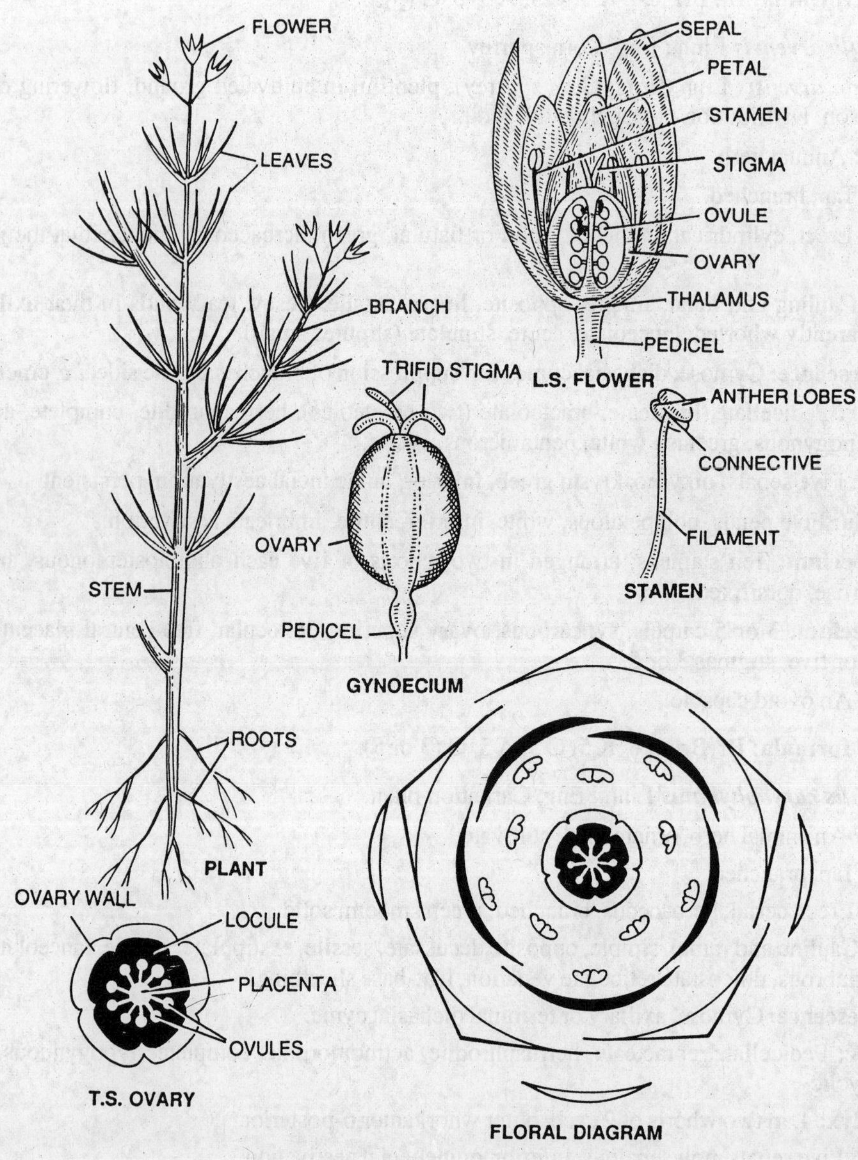

Fig.42.3. Caryophyllaceae. *Spergula arvensis* Linn.; Eng., corn spurrey.

Economic value. Cultivated in Europe as a fodder plant. Used as a diuretic.

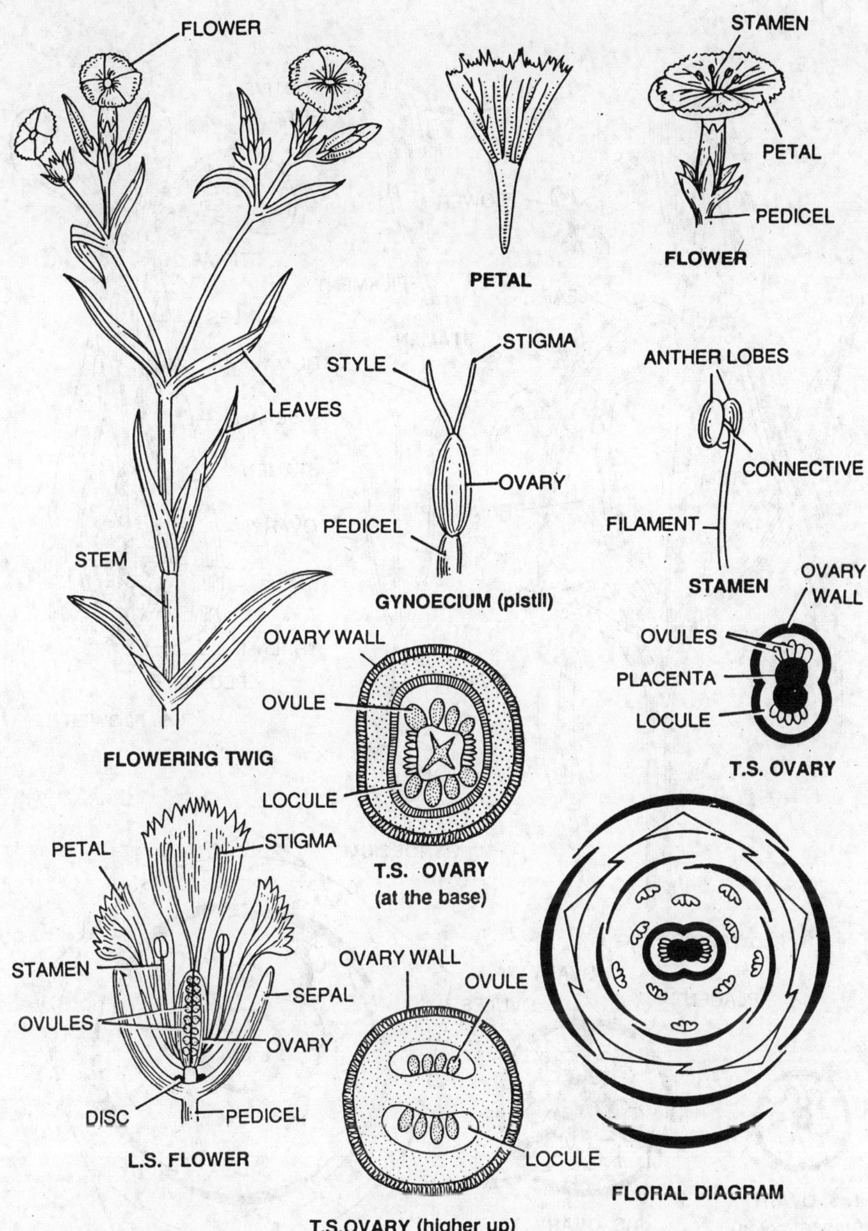

Fig .42.4. Caryophyllaceae. *Dianthus caryophyllatus* Linn.; Eng., carnation plant.

Economic value. Grown as an ornamental in the gardens. The flowers are considered cardiotonic, diaphoretic, nervine and antiseptic.

Silence conoidea Linn

Habit: An annual herb.

Root: Tap, branched.

Stem: Erect, herbaceous, aerial, branched, solid, cylinderic, flat at the nodes, hairy and green.

Leaf: Cauline and ramal, exstipulate, simple, sessile, opposite decussate, leaf base semiamplexicaul, linear, entire, acute, hairy, leathery, unicostate reticulate.

Inflorescence: Cymose, solitary axillary.

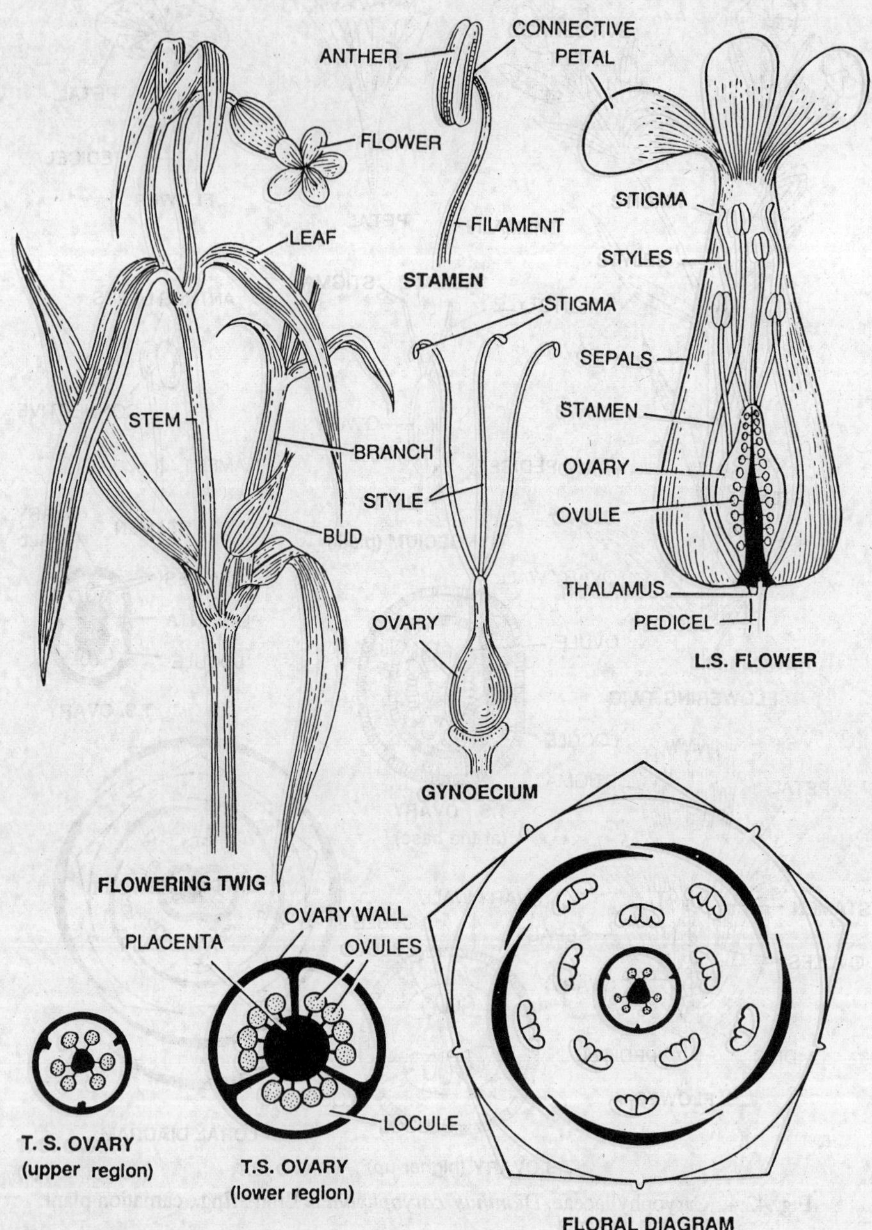

Fig.42.5. Caryophyllaceae. *Silene conoidea* Linn.

Economic value. The plant is considered emollient and used in baths or as a fumigant. The plant juice is used in opthalmia.

Flower: Pedicellate, ebracteate, hermaphrodite, actinomorphic, complete, cyclic, pentamerous, hypogynous.

Calyx: 5 sepals, gamosepalous, forming a flask-like structure, acute, inferior, green.

Corolla: 5 petals, polypetalous, twisted, caryophyllaceous, pink, inferior.

Androecium: 10 stamens in two whorls, 5 in each whorl, polyandrous, obdiplostemonous, dithecous, basifixed, introrse.

Gynoecium: 3 carpels, syncarpous, ovary superior, upper region of the ovary possesses free central placentation while the lower region possesses axile placentation, many ovules, style three, stigmas three and globular.

Fruit: A capsule.

Floral formula: ⊕ ⚥ K (5), C5, A 5+5, G (3).

Economic Importance of the Family

1. *Dianthus barbatus;* Eng.–Sweet William–An annual herb, grown in the gardens as an ornamental.

2. *Dianthus caryophyllatus*; Eng.–Carnation–It is grown as an ornamental.

3. *Dianthus chinensis;* Syn. *D. sinensis*; Eng.–Rainbow pink–This is grown as an ornamental. An annual herb.

4. *Arenaria* sp.–It is grown as an ornamental.

5. *Saponaria vaccaria*; Eng.–Cow herb: Verna.–**Sabuni**–This is an ornamental herb, found throughout India. The plant juice is used as a substitute of soap.

6. *Spergula arvensis*–It is a common winter season weed. In Europe it is used as fodder. It is diuretic.

7. *Stellaria aquatica*; Eng.–Chick weed; Verna.–**Badeola**–The shoots are eaten as vegetable.

8. *Gypsophila* sp.–An extremely ornamental plant.

9. *Lychnis coronaria*–In Spain the decoction of roots is used medicinally for the treatment of lung and liver diseases.

10. *Silene armeria;* Eng.–Sweet William catchfly. This is an annual herb, grown as an ornamental. It is native of South Europe.

11. *Cerastium* sp.–This is an ornamental plant.

Systematic Relationships: The taxonomists have presented many views for the evolution of Caryophyllaceae. Acccording to Eichler (1878) the family has been originated from the Phytolaccaceae by conversion of the outer whorl of stamens to petals and outer whorl of carpels to stamens. Pax, Wettstein and Rendle have also accepted this view in principle. Wernham (1911) proposed another view. According to him the family has been evolved from ranalian ancestors and to have been the source of origin for the Primulaceae and other such reduced families, *e.g.*, Amarantaceae and Chenopodiaceae. This view has been accepted by Bessey, Hutchinson and others. According to Dickson (1936) the family has been derived from the Geraniaceae. Thompson did not agree with this view. The veiw proposed by Wernham, seems more correct in tracing the probable origin and the systematic relationships of the family.

43

ORDER–POLYGONALES

According to Hutchinson this is the sixty third order of the phylum Angiospermae, sub-phylum Dicotyledones and division Herbaceae. This order consists of only two families, *i.e.*, Polygonaceae and Illecebraceae. The Polygonaceae has been described here in detail. Bentham and Hooker have placed the Polygonaceae along with other six families, including Nyctagineae, Amarantaceae, Chenopodiaceae, etc., in the first series–the Curvembryae of class Dicotyledones and division Monochlamydeae or Incompletae. Engler and Prantl have placed single family Polygonaceae in the sixteenth order–the Polygonales of class Dicotyledoneae and sub class Archichlamydeae.

More or less as in Caryophyllales but apetalous; ovary 1–locular; 1–ovuled; embryo straight to curved; endosperm still copious; stipules often sheathing.

Family—Polygonaceae

According to Lawrence there are about 33 genera and 800 species in this family.

Distribution: The members of the family are chiefly found in the north temperate zone of the world. Only a few genera are tropical, *e.g., Coccoloba* is confined to tropical and sub-tropical America. *Oxyria digyna* and *Koenigia islandica* are found in the colder parts of Northern hemisphere. A few genera are widely distributed, *e.g., Polygonum* is world-wide in its distribution and found in all the five continents. Several genera, *e.g., Atraphaxis, Colligonum* and *Pteropyrum* make the Asiatic steppe-plants. In our country the family is represented by several important genera such as, *Polygonum, Antigonon, Rumex, Fagopyrum, Muehlenbeckia, Coccoloba*, etc. *Fagopyrum* is found in Himalayas upto 12,000 feet elevation.

Habit: Mostly they are annual or perennial herbs. Climbing species also occur, *e.g., Polygonum convolvulus, Antigonon leptopus*, etc. *Koenigia islandica* is small low-growing annual, often not an inch in height. Certain species of *Polygonum* are xerophytes whereas some are aquatic, *e.g., Polygonum amphibium* which may also occur on the land.

Stem: The stem is mostly herbaceous, and sometimes woody. It is very often conspicuousty swollen at nodes. In *Muehlenbeckia platyclada,* the stem and its branches are flattened forming flat phylloclades or cladodes which are jointed at the nodes.

Leaves: The leaves are scattered, mostly alternate, rarely opposite, simple and generally entire. Sometimes they are lobed, *e.g.,* in *Rheum* and *Rumex acetosella.* They are mostly smooth but sometimes in mountain regions, the species become woolly and covered with a thick felt. In bud condition the edges of the leaves are rolled back. The most characteristic feature of the family is the constant presence of an ochrea (sheathing stipule). However, in *Eriogonum* and *Chorizanthe,* the ochrea are not found. Calcium oxalate crystals are frequently found in the cells.

Inflorescence: Primarily the inflorescence is of racemose type. Usually the flowers which are small and large in number are found to be arranged in compound inflorescences. Sometimes the flowers are solitary. The flowers of *Eriogonum* are found to be arranged in cymose umbels or heads. The flowers of *Polygonum* species are either found solitary or in panicled racemes. The flowers of *Polygonum plebejum* are solitary axillary whereas the flowers of *P. orientale* are found to be arranged in panicled racemes. The flowers of *Fagopyrum esculentum* are found in capitate cymes.

Flowers: The flowers are generally bisexual, rarely unisexual, actinomorphic (regular), small, trimerous, more rarely dimerous, cyclic and acyclic, usually borne in large numbers in compound inflorescences. The flowers with cyclic arrangement possess regularly alternating whorls of the perianth having the formula P 3+3, here the perianth leaves are similar, A 3+3 and G (3). Very rarely the cyclic flowers are dimerous as in *Oxyria,* the outer staminal whorl is doubled. In *Eriogonum* and

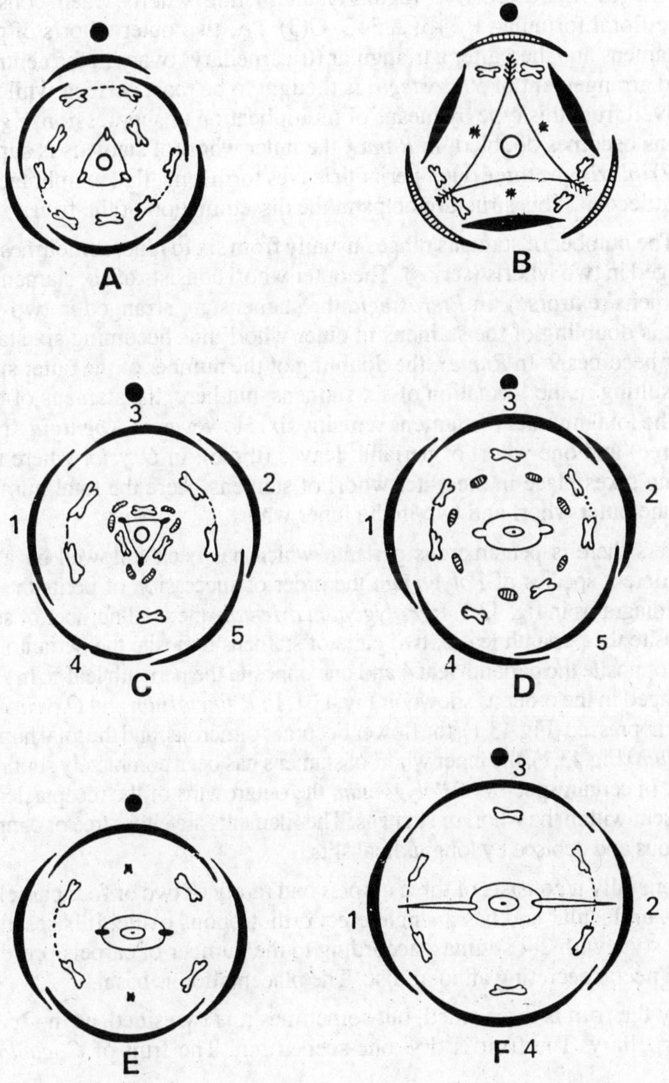

Fig.43.1. Polygonaceae–Floral diagrams: A, *Rheum;* B, *Rumex*; C, *Polygonum bistorta;* D, *Polygonum orientale;* E, *Polygonum diospyrifolium;* F, *Polygonum tinctorium.*

Rheum the outer stamens are branched. The flowers of acyclic arrangement have the formula P 5, A 5-8, G(3). The perianth leaves are arranged according to 2/5 phyllotaxy, *e.g., Polygonum. Polygonum aviculare* possesses cleistogamous flowers under its ochrea, whereas the flowers of the lower portion of the inflorescence have been replaced by bulbils in *P. viviparum.*

Perianth: The flowers with cyclic arrangement possess six similar perianth leaves arranged in two whorls of three each, *i.e.,* P 3+3. *Oxyria digyna* and *Polygonum tinctorium* possess four perianth leaves arranged in two whorls of two each, *i.e.,* P 2+2. However, the flowers of acyclic arrangement possess five perianth leaves arranged according to 2/5 phyllotaxy, *i.e.,* in *Polygonum.* The perianth leaves are generally all alike, green, white or red. They may be free or more or less connate. The perianth leaves persist till the ripening of the fruit and help in its dispersal. In *Rumex,* the three outer perianth leaves remain small but the inner three develop considerably and completely envelope the fruit which form three wing-like membranous structure. In *Pterostegia,* a monotypic Californian genus, the two bracteoles remain persistent developing each in a winged bladder covering the fruit completely. In *Pterostegia,* there are five regularly alternating whorls, each consisting of three members, making the floral formula– P 3+3, A 3+3, G(3), *i.e.,* two outer whorls of perianth leaves, then two whorls of stamens and thereafter a triangular (tricarpellary) ovary in the centre of the flower. The above mentioned arrangement of *Pterostegia* is thought to be most perfect, while other arrangements have been derived from this type by means of multiplication or suppression, *e.g.,* in *Rheum* the outer whorl of stamens becomes doubled; in *Rumex* the inner whorl of stamens is suppressed. In the South African genus *Triplaris,* the three outer perianth leaves form long, flat, membranous, erect wings which look like a shuttlecock. This structure helps in the dissemination of the fruit.

Androecium: The number of stamens ranges usually from six to nine, sometimes fewer. Usually the stamens are arranged in two whorls (series). The outer whorl consists of six stamens (introrse) and the inner of three stamens (extrorse). In *Pterostegia* the stamens are arranged in two whorls of three each. In *Rheum* there is doubling of the stamens in outer whorl thus becoming six stamens, and this way, the total number becomes 9. In *Rumex,* the doubling of the number of the outer stamens of outer whorl takes place, resulting in the formation of six stamens, but here, the stamens of the inner whorl get suppressed, thus the total number of stamens remains six. However, in *Koenigia,* there is only one whorl of stamens (three) and one whorl of perianth leaves (three). In *Oxyria,* where the flowers are dimerous, the doubling takes place in the outer whorl of stamens. Here the total number of stamens becomes six, four in the outer whorl and two in the inner whorl.

In acyclic flowers there is pentamerous perianth which has been followed by a 5 to 8-merous androecium. In the different species of *Polygonum* the order of succession of perianth leaves has been illustrated by separate diagrams in fig. 43.1. In *Polygonum orientale* the total number of stamens is 7 one posterior stamen opposite the perianth leaf 3, two pairs of stamens opposite the perianth leaves 1 and 2, one stamen is situated opposite the perianth leaf 4 and one opposite the perianth leaf 5. In *P. bistorta,* there are eight stamens arranged in the order as shown in Fig.43.1. In *P. tinctorium* and *Oxyria digyna,* the fifth perianth leaf has been suppressed (Fig.43.1), the flower becomes dimerous, and the total number of stamens is six. In *P. diospyrifolium* (Fig.43.1), the inner whorl of stamens has been completely suppressed and there are only four stamens. In certain species of *Polygonum,* the outgrowths of the receptacle, which are like the glands are also present within the whorl of stamens. The filaments are either free or connate at the base. The anthers are dithecous and dehisce by longitudinal slits.

Gynoecium: Generally it consists of three carpels and rarely of two or four carpels; syncarpous. The ovary is superior, unilocular and has a single erect orthotropous ovule. It is sessile, compressed and triangular. Single style with 2-4 stigmas, according to the number of carpels present. At the base of the ovary there is a nectar secreting annular disc. The placentation is basal.

Fruit: Generally the fruit is three sided, but sometimes it is two sided, as in *Oxyria,* where the gynoecium was bicarpellary. The fruit is dry, one-seeded nut. The fruit of *Coccoloba,* possesses fleshy perianth.

Seeds: The seeds are endospermic. The curved or straight embryo remains more or less embedded in the endosperm.

Pollination: The pollination is anemophilous (by wind) or entomophilous (*i.e.* by insect). The flowers of *Rumex* are wind pollinated. Here, the pendulous flowers are found to be situated on slender stalks, they possess large brush-like stigmas in which the pollens are entangled. The genera of *Rheum* and *Polygonum* are entomophilous, nectar being secreted by the glands situated at the base of the stamens. The flowers are produced in large numbers. They are beautifully coloured (red and white) and attract insects. In *Polygonum bistora,* the fragrant flowers are very much crowded.

Floral Formula: ⊕ ⚥ P3 + 3 or 2 + 2 or 5, A6–9 or 5–8, $G(\underline{2-4})$.

Bentham & Hooker (1962)	**Engler & Prantl (1931)**	**Hutchinson (1959)**
Phanerogams	Phanerogams	Angiospermae
Dicotyledones	Dicotyledoneae	Dicotyledones
Monochlamydeae or	Archichlamydeae	Herbaceae
Incompletae	Polygonales	Polygonales
Curvembryae	Polygonaceae	Polygonaceae
Polygonaceae		

1. ***Polygonum barbatum*** Linn. Verna. **Narri, Bekhunjubaz.**

Polygonum is world-wide in distribution and found in all the five continents. They are found in moist places.

Habit: Annual herb.

Root: Tap, branched.

Stem: Erect or prostrate, branched, herbaceous, cylindrical, glabrous, solid, conspicuously swollen at nodes, purple.

Leaf: Simple, alternate, sessile, cauline, stipulate, ochreate stipules (short and fimbriate), linear, lanceolate, unicostate reticulate venation, entire, acute.

Inflorescence: Axillary, 2 or 3 flowers in axil of leaf (*e.g., P. plebejam*); flowers of *P. orientale* are found to be arranged in panicled racemes.

Flower: Pedicellate, short pedicel, pedicel jointed under perianth, bracteate, bract membrane like, complete, hermaphrodite, actinomorphic, pink, hypogynous, small.

Perianth: 5, polyphyllous, perianth leaves short, broad and rounded, rarely two outer ones acute, pink, quincuncial aestivation.

Androecium: 8, stamens arranged in two whorls, 5 in outer whorl and three in inner, anthers of outer five introrse and inner three extrorse, filaments flat at base, basifixed anthers.

Gynoecium: 3, syncarpous, ovary superior, unilocular, single ovule, basal placentation, 3 angled, nectar secreting disc present below ovary.

Fruit: Three sided nut.

Floral formula: Br ⊕ ⚥ P5, A 5 + 3, $G(\underline{3})$.

2. ***Rumex dentatus*** Linn., Verna. **Lalbibi, Jangli palak.**

Habit: Annual herb.

Root: Tap, branched.

Stem: Erect, aerial, herbaceous, branched, angular, solid, glabrous, green.

Leaf: Cauline and ramal, exstipulate, petiolate, alternate, simple, elliptic-lanceolate, undulate, acute, smooth, unicostate reticulate venation.

Inflorescence: Panicled racemose clusters.

Flower: Ebracteate, pedicellate, actinomorphic, hermaphrodite, trimerous, hypogynous, complete, cyclic.

Perianth: 6 tepals in two whorls of three each, polyphyllous, valvate, sepaloid, persistent.

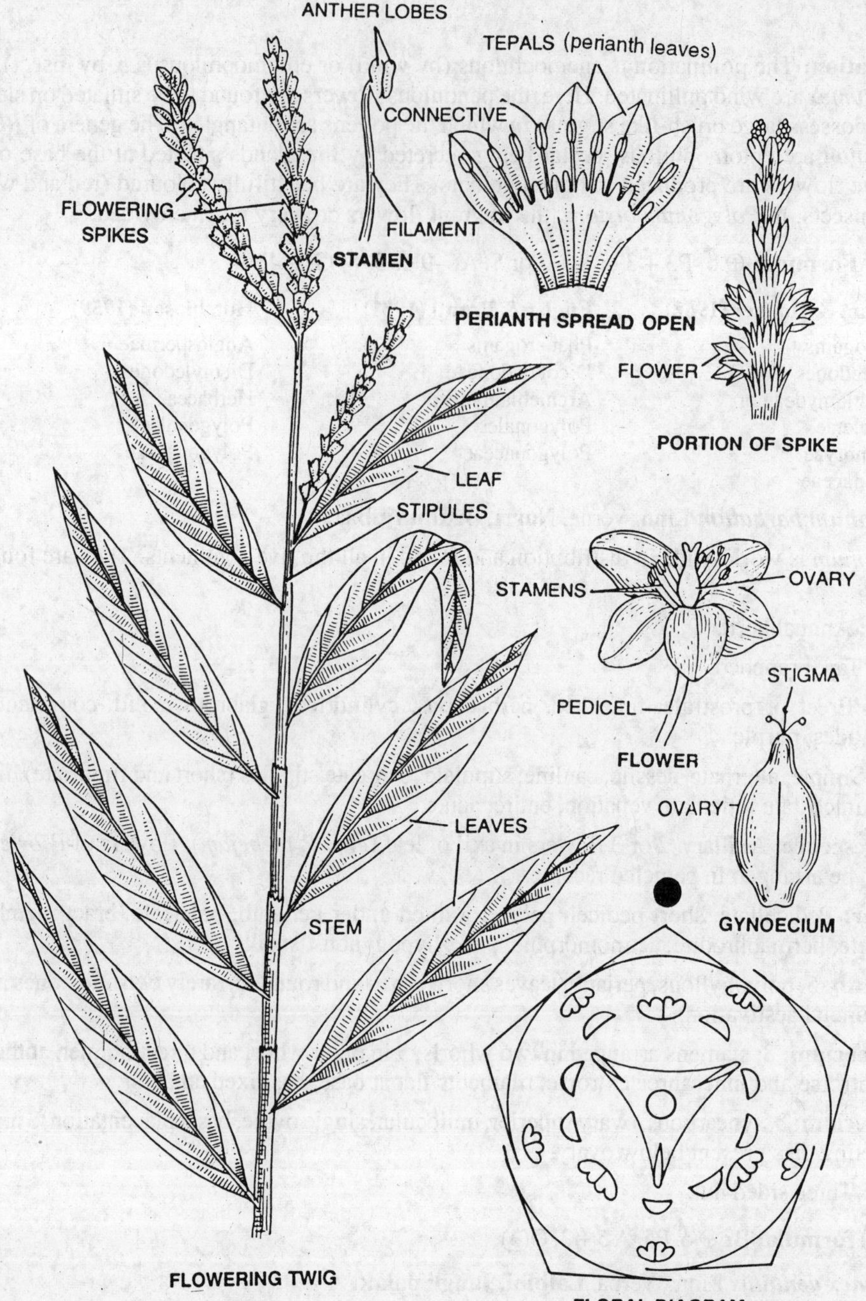

Fig.43.2. Polygonaceae. *Polygonum barbatum* Linn.; Verna. **narri, bekhunjubaz.**

Economic value. The plant is used medicinally. The seeds are used to relieve gripping pains of colic. The root is astringent and cooling.

Androecium: 6 stamens in two whorls of three each, the stamens of outer whorl are opposite to tepals of outer whorl and that of inner whorl are opposite to tepals of inner whorl, polyandrous, dithecous, basifixed, introrse.

Gynoecium: 3 carpels (tricarpellary), syncarpous, ovary superior, unilocular, basal placentation, style short, stigma 3 hanging downward.

Fruit: Nut.

Floral formula: ⊕ ⚥ P3 + 3, A 3 + 3, G(3).

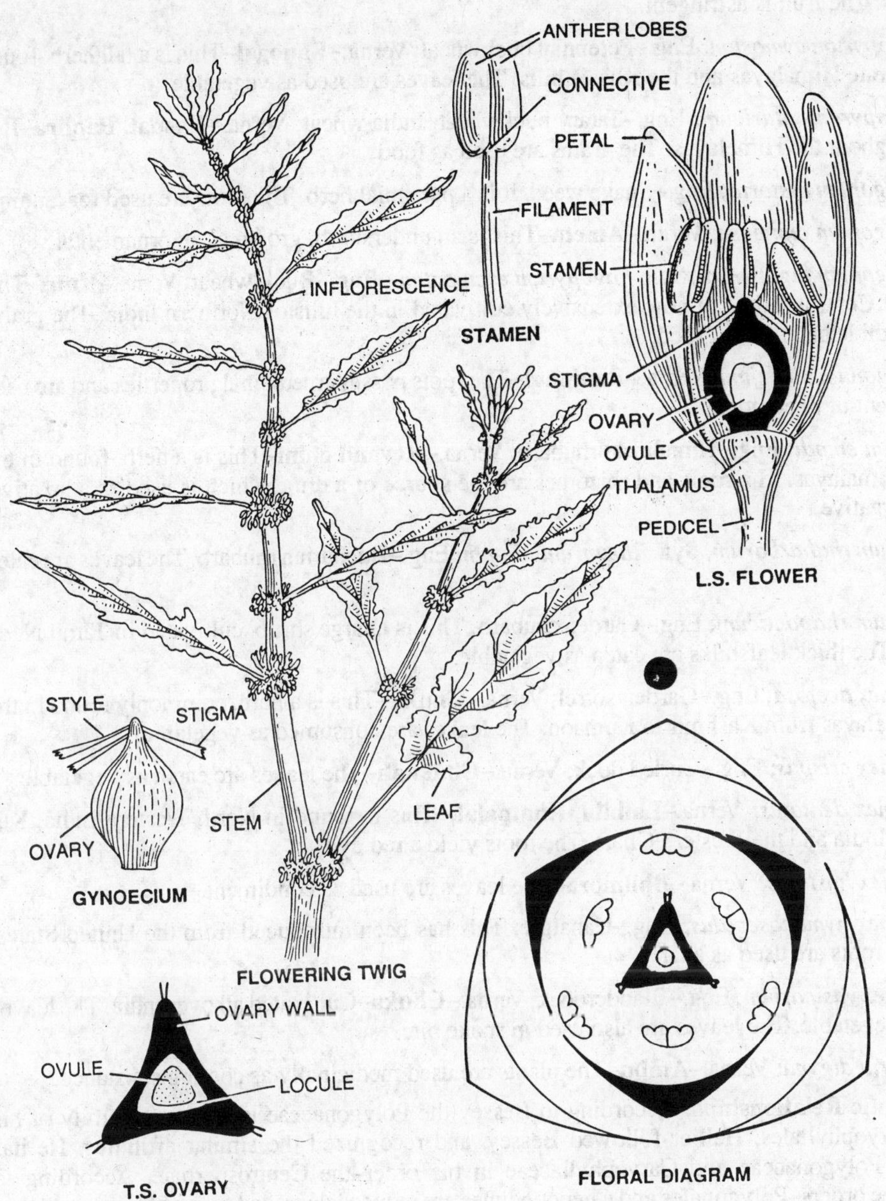

Fig.43.3. Polygonaceae. *Rumex dentatus* Linn.; Verna. **Lalbibi, jangli palak.**

Economic value. The roots yield a red dye. The root is used as astringent application in cutaneous disorders.

Economic Importance of the Family

The family is of less economic value. Some plants are ornamental, some are used as food and some possess medicinal properties. A list of few important plants is given below :

1. *Antigonon leptopus*; Eng.–Coral creeper. This is a large climbing shrub with beautiful pink flowers. It is native of South America. Grown as an ornamental in the gardens.

2. *Muehlenbeckia platycladus*; Syn. *Coccoloba platyclada*; Eng.–Centipede plant. This is a small shrub. It is native of the Solomon islands. It is grown as an ornamental. It possesses flat, green phylloclades. The fruit is astringent.

3. *Fagopyrum cymosum;* Eng.–Perennial buckwheat; Verna.–**Banogal**–This is a tall herb, found in the temperate Himalayas and the Khasia hills. The leaves are used as vegetable.

4. *Fagopyrum tataricum;* Eng.–Tatary buckwheat, India wheat; Verna.–**Kaspat, Paphra**–It is grown throughout the Himalayas. The grains are used as food.

5. *Polygonum bistorta*; Eng.–Snake weed. It is a perennial herb. The roots are used for tanning.

6. *Polygonum chinense*; Verna.–**Ameta**–This is an undershrub, grown as an ornamental.

7. *Polygonum fagopyrum*; Syn. *Fagopyrum esculentum;* Eng.–Buck-wheat; Verna.–**Kutu**–This is a native of Central Asia, but now extensively cultivated in the hills of Northern India. The grains (seeds) are edible and used as food.

8. *Polygonum plebejum*; Verna.–**Raniphul**–The roots possess medicinal properties and are used in the treatment of pneumonia.

9. *Rheum emodi*; Eng.–Himalayan rhubarb; Verna.–**Revand chini**–This is a herb, found in the sub-alpine Himalayas. The roots and rhizomes are the source of a drug, which is used as a laxative, tonic and purgative.

10. *Rheum rhabarbarum;* Syn. *Rheum undulatum;* Eng.–Bucharian rhubarb. The leaves are eaten as vegetable.

11. *Rheum rhaponticum*; Eng.–Garden rhubarb. This is a large shrub, cultivated in Tamil Nadu and Bengal. The thick leafstalks are eaten as vegetable.

12. *Rumex acetosa*; Eng.–Garden sorrel; Verna.–**Chuka**–This is a herb, commonly found in the Western Himalayas from Kashmir to Kumaon. The leaves are consumed as vegetable.

13. *Rumex crispus*; Eng.–Curled dock; Verna.–**Chukkah**–The leaves are eaten as vegetable.

14. *Rumex dentatus;* Verna.–**Lalbibi, Khatpalak**–This is found in North Western India, Kumaon, South India and the Western Ghats. The roots yield a red dye.

15. *Rumex hastatus;* Verna.–**Bhilmora**–The leaves are used as condiment.

16. *Rumex hymenosepalus;* Eng.–Canaigre. This has been introduced from the United States. The tuberous roots are used as a tan.

17. *Rumex vasicarium*; Eng.–Bladderdock; Verna.–**Chuka**–Cultivated all over India. The leaves are eaten as vegetable. The leaves are also used in snake bite.

18. *Oxyria digyna*; Verna.–**Ambu**–The plants are used medicinally as cooling substance.

Systematic Relationships: According to Bessey the Polygonaceae is advanced family of his order–the Caryophyllales. Hallier followed Bessey, and recognized the similar affinities. He has included the Polygonaceae and Caryophyllaceae in his order–the Centrospermae. According to Hutchinson the orders–Polygonales and Chenopodiales are reduced degraded type of Caryophyllales. He has considered the Polygonales together with the Chenopodiales representing the culmination of a line derived from the Ranales via Caryophyllaceous ancestors. This has been very clearly understood that the Polygonales have been derived from the Caryophyllaceous stocks.

44

ORDER-CHENOPODIALES

According to Hutchinson this is the sixty fourth order of the phylum Angiospermae, sub-phylum Dicotyledones, and division Herbaceae. The order comprises of ten families, of which only two (*i.e.*, Chenopodiaceae and Amaranthaceae) have been described here in detail. Bentham and Hooker have placed these families along with five others including Nyctagineae, Polygonaceae, etc., in the first series–the Curvembryae of class Dicotyledones and division Monochlamydeae or Incompletae. Engler and Prantl have placed these families along with seven others including Nyctaginaceae, Caryophyllaceae, etc., among the seventeenth order–the Centrospermae of class Dicotyledoneae and sub-class Archichlamydeae.

As in Polygonales but stipules absent or very small; carpels numerous to solitary, free or connate; embryo curved.

Flowers actinomorphic; usually pentamerous, hermaphrodite, monoecious or dioecious, perianth either single or double; stamens arranged in one or two whorls, five or ten in number when five opposite the petals; ovary superior, unilocular, 1 to 5 carpels, syncarpous; fruit nut-let, achene or capsule; embryo coiled or curved, perisperm present.

Family—Chenopodiaceae

There are 102 genera and about 1400 species in this family.

Distribution: The plants of this family are cosmopolitan in their distribution, *i.e.,* they are found throughout the world. The plants are particularly found in xerophytic and halophytic areas. They are mostly found in the Prairies, North America, Mediterranean coasts, Red sea coasts, Australia and central Asia. In our country the family is represented by several important genera, such as, *Chenopodium, Beta, Spinacia, Kochia, Suaeda, Salsola,* etc.

Habit: Usually the plants are annual or biennial herbs but sometimes the shrubs are also found. The small trees are very rare. *Haloxylon* is a small tree which reaches a height of 20 feet. *Salsola kali* is known as salt wort and grows near the sea shores. The species of *Salicornia* grow in marshy places.

Root: Tap and branched.

Stem: The stem is usually erect, herbaceous, branched, cylindrical solid hairy and green.

Leaves: The leaves are simple and alternate. In *Salicornia, Nitrophila,* etc., the leaves are opposite. In many cases the leaves are minute, fleshy and covered with hairs, *e.g., Salicornia, Suaeda,* etc.

Inflorescence: Cymose, dichasial cyme.

Flowers: The flowers are minute, greenish in colour and arranged in dense cymose inflorescence. Hermaphrodite or unisexual. The plants may be monoecious (*e.g., Sarcobatus*) or dioecious (*e.g., Grayia*). They are bracteate, actinomorphic and hypogynous.

Perianth: Usually consists of 5 perianth leaves (tepals), sepaloid. Sometimes there are 3 or 4 perianth leaves, *e.g., Salicornia* or only 2, *e.g., Atriplex.* The perianth leaves may be free or united. Usually the perianth leaves remain persistent in the fruit. Sometimes the male flowers are devoid of perianth leaves. The aestivation is imbricate.

Androecium: Five stamens, usually the number of stamens corresponds to the number of perianth leaves. The stamens are usually situated opposite the tepals. The number of stamens varies from species to species and sometimes even from plant to plant of the same species, *e.g., Chenopodium.* They are hypogynous or situated on a disc. The anthers are bicelled and dehisce longitudinally.

Gynoecium: 2 or 3 carpels, syncarpous. The ovary is superior. However, in *Beta* the ovary is inferior or half inferior. It is unilocular with a single basal campylotropous ovule.

Fruit: A nut or achene enclosed in the perianth.

Seeds: The seeds are albuminous with a curved or twisted embryo.

Pollination: The pollination takes place through the agency of insects.

Floral formula: ⊕ ⚥ or ♂ or ♀ P 5 or (5), A 5, G $\underline{(2–3)}$ or $\overline{(2–3)}$

Bentham & Hooker (1862)	**Engler & Prantl (1931)**	**Hutchinson (1959)**
Phanerogams	Phanerogams	Angiospermae
Dicotyledones	Dicotyledoneae	Dicotyledones
Monochlamydeae or	Archichlamydeae	Herbaceae
Incompletae	Centrospermae	Chenopodiales
Curvembryae	Chenopodiaceae	Chenopodiaceae
Chenopodiaceae		

1. **Beta vulgaris** Linn.; Verna. **Chukandar;** Eng. Beet root.

Beta vulgaris is cosmopolitan in distribution. The roots and leaves are used as vegetable. In European countries the roots are used as main source for the manufacture of sugar.

Habit: Annual or biennial herbs.

Root: Tap and branched; swollen, possessing dark-red juice.

Stem: Herbaceous, erect, branched, solid, cylindrical, pubescent, purplish green.

Leaves: Cauline and ramal, alternate, exstipulate, simple, petiolate, entire, elliptic lanceolate, unicostate, reticulate, coriaceous.

Inflorescence: Cymose, condensed cyme.

Flower: Bracteate, sessile, complete, actinomorphic, hermaphrodite, pentamerous, peri or epigynous.

Perianth: Tepals 5, polyphyllous, sepaloid; imbricate aestivation.

Androecium: 5 stamens; antetepalous, situated on a disc; anthers dithecous, introrse, dehiscing longitudinally.

Gynoecium: 3 carpels (tricarpellary), syncarpous; ovary half-inferior or inferior; unilocular with a single basal campylotropous ovule.

Fruit: A nut or achene enclosed in the perianth.

Seeds: Albuminous with a curved embryo.

Floral Formula: Br ⊕ ⚥ P5, A5, G$\overline{(3)}$

2. Chenopodium album Linn.; Eng. Pigweed, lambs-quarters; Verna. **Bathua.**

Habit: An annual herb.

Root: Tap and branched.

Stem: Erect, aerial, herbaceous, branched, angular, solid, hairy, reddish green or green.

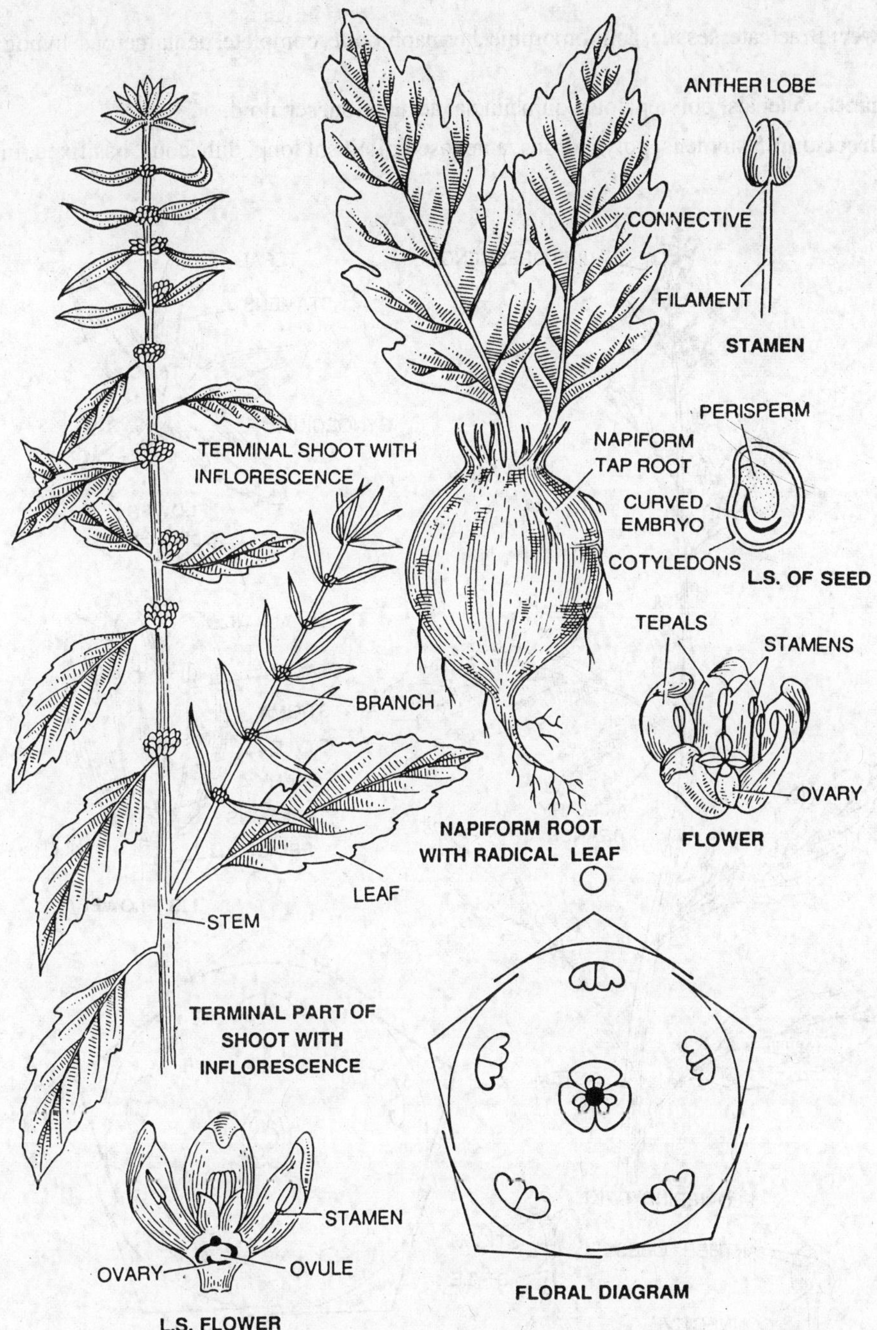

Fig.44.1. Chenopodiaceae. *Beta vulgaris* Linn.; Eng., beet root; Verna. **chukandar.**

Economic value. Beet root is used in vegetable salad. Refreshing juice is extracted. Sugar is manufactured from juice in European countries. The seeds are cooling and diaphoretic. Leaves are applied to burns and bruises. Beet greens contain more iron and are richer in vitamins, particularly vitamin A.

Leaf: Cauline and ramal, exstipulate, petiolate, alternate, simple, elliptic-lanceolate, coriaceous, unicostate reticulate.

Inflorescence: Cymose, condensed cyme.

Flower: Bracteate, sessile, actinomorphic, hermaphrodite, complete, pentamerous, hypogynous, cyclic.

Perianth: 5 tepals, polytepalous, quincuncial aestivation, sepaloid.

Androecium: 5 stemens, polyandrous, anteposed, filament long, dithecous, basifixed, introrse.

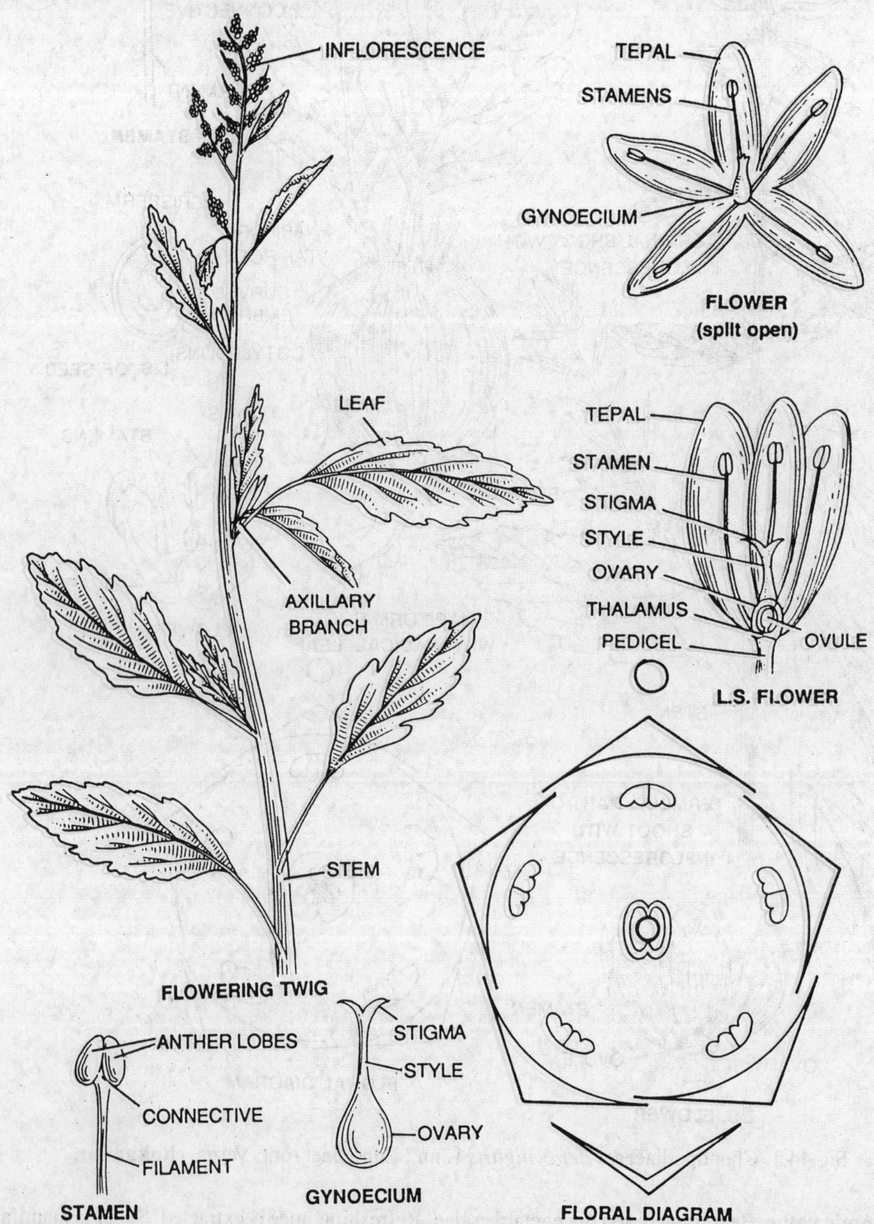

Fig.44.2. Chenopodiaceae. *Chenopodium album* Linn.; Eng., pigweed, lambs-quarters; Verna., **bathua.**

Economic value. The tender twigs are used as vegetable and fodder. The plant is laxative and anthelmintic.

Gynoecium: 2 carpels (bicarpellary), syncarpous, ovary superior, unilocular, single basal ovule, style short, stigma bifid.

Fruit: Utricle.

Floral Formula: Br ⊕ ⚥ P5, A5, G($\underline{2}$)

Economic Importance of the Family

The family consists of a few plants of economic importance. Most of them are used as vegetable.

1. *Atriplex hortensis* (Verna.–**Chakwat**)–The branches and leaves are used as vegetable.

2. *Beta vulgaris* (Eng.–Beet root; Verna.–**Chukandar**)–A biennial herb, the roots and leaves are used as vegetable.

3. *Beta vulgaris* var. *bengalensis* (Eng.–Beet root; Verna–**Chukandar**)–The leaves are eaten as vegetable.

4. *Beta vulgaris* var. *rapa* (Eng.–Sugar beet)–The roots are used for the manufacture of sugar.

5. *Chenopodium album* (Eng.–Pigweed; Verna.–**Bathua**)–A herb. The leaves and tender twigs are used as vegetable and fodder.

6. *Chenopodium ambrosioides* var. *anthelminticum*–It is used as an anthelmintic, particularly effective in the expulsion of hookworms.

7. *Chenopodium murale.* (Verna.–**Khartua**)–It is used as a fodder.

8. *Kochia indica* (Verna.–**Bui**)–A herb. It is used as fodder.

9. *Kochia scoparia* (Eng.–summer–cypress)–Grown as an ornamental.

10. *Salsola baryosma* (Verna.–**Loonuk**)–A herb, used as fodder.

11. *Salsola kali* (Eng.–glass wort; Verna.–**Buti**)–The leaves are used as vegetable.

12. *Suaeda fruticosa;* Syn. *Chenopodium fruticosum* (Verna.–**Lunak**)–An undershrub, used as fodder.

Family—Amaranthaceae

There are about 64 genera and 800 species in this family.

Distribution: The family is well represented in the tropical regions of the world. The plants are very commonly found in the tropics of New World, Africa and India. Some plants are also found in temperate regions. In our country and others, several species are cultivated as ornamental plants *e.g., Celosia cristata* (cockscomb), *Amaranthus* (foxtail), and *Gomphrena globosa.*

The family resembles Chenopodiaceae in several ways. From the point of view of its stem anatomy, the vascular bundles in the stem are arranged in several concentric rings or more or less irregularly like that the stem of the members of Chenopodiaceae. Biallon, the French botanist, however, united this family with Chenopodiaceae, because of these resemblances.

Habit: The plants are herbs or shrubs. Mostly they are annual or perennial herbs. The trees are rare.

Leaves: The leaves are generally opposite or alternate, simple, entire, exstipulate and usually covered with hairs.

Inflorescence: Usually the inflorescence is of racemose type. This may be a simple or branched spike or a raceme. Mostly the small flowers are being arranged in dense fascicles, *e.g., Amaranthus* (foxtail) and *Celosia cristata* (cockscomb). Sometimes the branching of the inflorescence takes place in the axil of the bracteoles of the flowers giving rise to dichasia, which may be borne on a simple or racemosely or cymosely branched main axis. In certain cases the lateral flowers are being aborted and on their place tufts of hair or prickles are developed which persist in the ripe fruit and help in dissemination.

Flower: Usually the flowers are very minute and arranged in dense facicles, *e.g., Celosia cristata* (cockscomb) and *Amaranthus* (foxtail) etc. Each flower bears a pair of large membranous persistent bracteoles; which are sterile. The flower in its structure quite resembles to that of Chenopodiaceae. Usually the flowers are bracteate and bracteolate, hermaphrodite, rarely unisexual by abortion, actinomorphic (regular) and hypogynous.

Perianth: The perianth is dry, membranous and very often white or coloured. It consists of 4 or 5 perianth leaves. The perianth leaves are green and herbaceous. The membranous perianth leaves are free or more or less united. Sometimes they become hairy.

Androecium: It consists of 4 or 5 stamens. They are situated opposite the perianth leaves. Usually the stamens are united at the base forming a membranous tube. Sometimes these stamens bear fringed outgrowths in between them. Mostly the anthers are two–celled but in certain cases they may become one celled. They dehisce longitudinally.

Gynoecium: It consists of 2 or 3 carpels, syncarpous. The ovary is unilocular and superior. In *Celosia* and certain other allied genera there is a single basal ovule. Generally, the ovules are campylotropous. The styles are one, two or three.

Fruit: The fruit is generally dry. This may be a nut, drupe or berry.

Seeds: The seeds are endospermic usually having a rough or polished testa. The embryo is curved and lies close to the seed–coat.

Pollination: Entomophily.

Floral Formula: Br Brl ⊕ ⚥ P 4–5, A 4–5, C($\underline{2–3}$).

Bentham & Hooker (1862)	**Engler & Prantl (1931)**	**Hutchinson (1959)**
Phanerogams	Phanerogams	Angiospermae
Dicotyledones	Dicotyledoneae	Dicotyledones
Monochlamydeae or Incompletae	Archichlamydeae	Lignosae
Curvembryae	Centrospermae	Chenopodiales
Amaranthaceae	Amaranthaceae	Amaranthaceae

1. ***Achyranthes aspera*** Linn.; Verna. **Chirchita, Latzira, Apmarg.**

Habit: Perennial herbs.

Root: Tap and branched.

Stem: Erect, herbaceous, quadrangular, branched, solid, green, pubescent.

Leaf: Cauline and ramal, opposite, exstipulate, simple, sub-sessile, ovate, entire, acute, unicostate reticulate, rough, coriaceous, hairy.

Inflorescence: Racemose, spike.

Flower: Bracteate, bracteolate, sessile, bracts and bracteoles, spinous and persistent, hermaphrodite, actinomorphic, regular, complete, pentamerous, hypogynous, cyclic.

Perianth: 5 tepals, polyphyllous, dry, membranous, white or coloured, hairy.

Androecium: 5 stamens, antetepalous, stamens alternated by fringed outgrowths, anthers dithecous, introrse, dehiscing longitudinally.

Gynoecium: 2, bicarpellary, syncarpous, ovary unilocular, superior, single basal ovule, style short, stigma bifid.

Fruit: Dry, utricle.

Floral Formula: Br Brl ⊕ ⚥ P5, A5, G($\underline{2}$)

2. ***Amaranthus spinosus*** Linn.; Verna. **Jangli chaulai.**

Habit: A spiny herb.

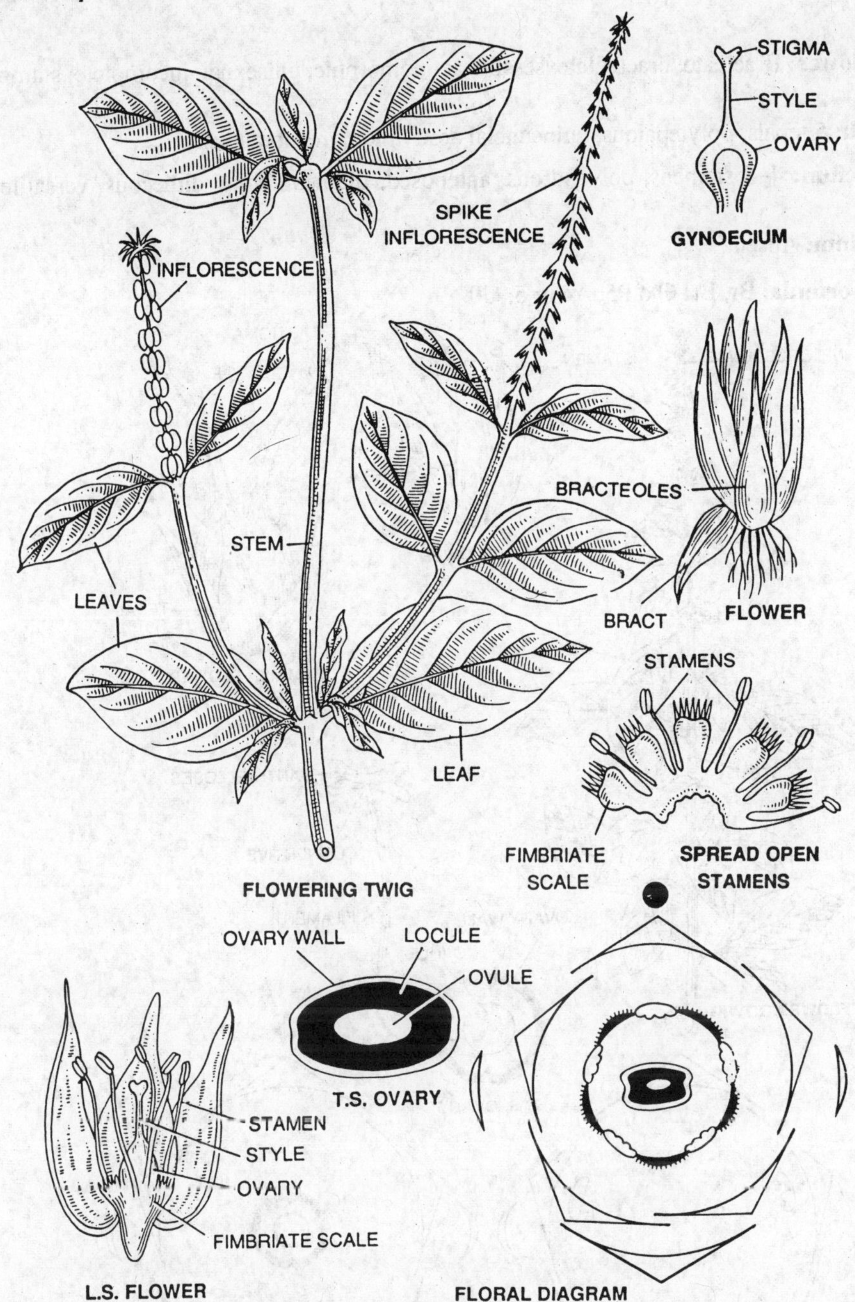

Fig.44.3. Amaranthaceae. *Achyranthes aspera* Linn.; Verna. **Chirchita, latzira, apmarg.**

Economic value. The plants are used medicinally for piles, colic, boils, etc. It is pungent, purgative, diuretic and astringent. Roots are used for pyrrhoea. Also used in cough and fevers.

Root: Tap and branched.

Stem: Erect, aerial, branched, herbaceous, terete, solid, smooth, green.

Leaf: Cauline and ramal, exstipulate, alternate, simple, petiolate, ovate, entire, acute, unicostate reticulate; 2 spines present in the axil which represent the modified axillary branch.

Inflorescence: Spike.

Male Flower: Bracteate, bracteolate, sessile, actinomorphic, unisexual, incomplete, staminate, cyclic.

Perianth: 5 tepals, polytepalous, quincuncial aestivation, membranous.

Androecium: 4–5 stamens, polyandrous, anteposed, filaments long, dithecous, versatile, introrse.

Gynoecium: Absent.

Floral Formula: Br, Brl ⊕ ♂ P5, A4 or 5, G0

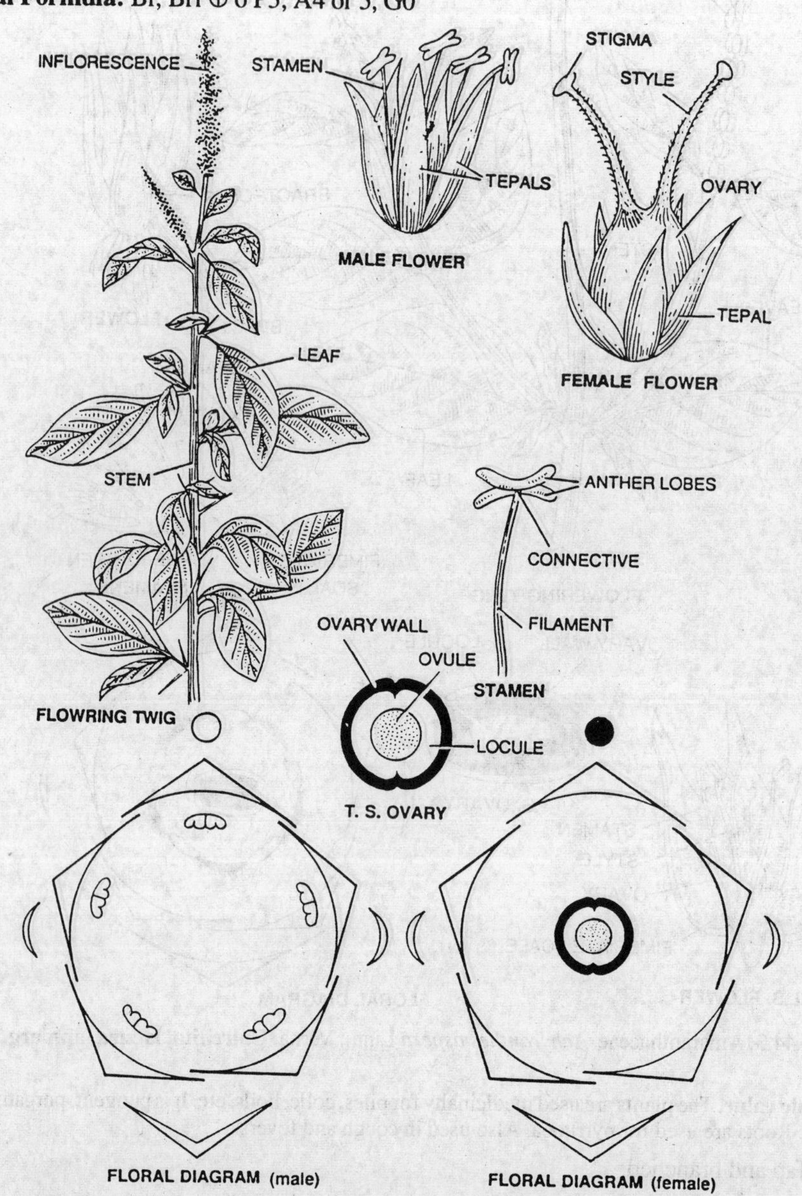

Fig.44.4. Amaranthaceae. *Amaranthus spinosus* Linn.; Verna. **jangli chaulai.**

Economic value. The root is used in eczema and colic. The boiled leaves and roots are given to children as laxative and applied as emollient poultice to abscesses, boils and burns. Plant is also used in snakebite.

Female flower: Bracteate, bracteolate, sessile, unisexual, incomplete, pistillate, actinomorphic, cyclic.

Perianth: 5 tepals, polytepalous, quincuncial, membranous.

Androecium: Absent.

Gynoecium: 2 carpels (bicarpellary), syncarpous, ovary superior, unilocular, single basal ovule, style short, stigma bifid and hairy.

Fruit: Utricle.

Floral formula: Br Brl ⊕ ♀ P5, A0, G($\underline{2}$)

Economic Importance of the Family

The family is of little economic value. A few species are grown as ornamental plants. A list of some plants is given here:

1. *Amaranthus blitum* var. *oleracea*; Eng.–Amaranth; Verna.–**Chaulai**–This is a herb, its leaves are used as vegetable.

2. *Amaranthus spinosus*; Eng.–Prickly Amaranth; Verna.–**Katailchaulai**–This is spiny herb. The ash of the leaves is used in dyeing. The roots possess medicinal properties and are used to cure eczema, colic and gonorrhoea. This is also used as vegetable and fodder.

3. *Amaranthus caudatus*; Syn. *A. paniculatus; A. cruentus*; Eng.–Valvet flower; Verna.–**Ramdana**–This is grown as a vegetable in North India. This is also grown as ornamental. It possesses dense red or golden coloured spikes.

4. *Amaranthus tricolor*; Syn. *A gangeticus*; Eng.–Amaranth; Verna.–**Bari chaulai, lal sag**–The leaves are used as vegetable. Cultivated in the Punjab, Assam, Maharashtra and Tamil Nadu.

5. *Achyranthes aspera*; Verna.–**Chirchita, Latzira**–This is a common weed with spike inflorescence. The plants are used medicinally for several diseases such as piles, colic, boils, etc. This is also used as an antidote for snake bite and scorpion sting.

6. *Alternanthera amoena*; Verna.–**Kanchari**–This is a herb, used as vegetable, cultivated in Bengal, Orissa and Tamil Nadu.

7. *Aerva* sp.–This is a perennial herb with red leaves. Used as edge plant to border the lawns.

8. *Celosia cristata*; Eng.–Cockscomb; Verna.–**Morshikha**–This is grown as an ornamental for its velvety inflorescence.

9. *Digera muricata*; Syn. *D. arvensis; Achyranthes muricata*; Verna.–**Latmohuria**–This is a herb. Its tender twigs and inflorescences are used as vegetable.

10. *Gomphrena globosa*–Grown as an ornamental for its beautiful globose violet inflorescences.

Systematic Relationships: Hutchinson has placed the families Chenopodiaceae and Amarantaceae in his order the Chenopodiales. Bentham and Hooker included seven families in the series Curvembryae. These families are–Nyctagineae Illecebraceae, Amarantaceae, Chenopodiaceae, Phytolaccaceae, Batideae and Polygonaceae. Engler and Prantl have placed the single family Polygonaceae in the order–Polygonales. They have included Chenopodiaceae, Amarantaceae and Nyctaginaceae in the order–Centrospermae. Bessey included the family Polygonaceae as an advanced taxon of his Caryophyllales. Hallier included the family in his Centrospermae. Hutchinson considered his order–Chenopodiales as reduced types of Caryophyllales. This is certain that Polygonales have been derived from Caryophyllaceous ancestors.

45

ORDER–LYTHRALES

According to Hutchinson this is the 65th, order of the phylum Angiospermae, subphylum Dicotyledones and division Herbaceae. This order comprises of five families, *i.e.,* Lythraceae, Onagraceae, Trapaceae, Holoragidaceae and Callitrichaceae, of which Lythraceae and Onagraceae have been described here in detail. Bentham and Hooker have placed the families Lythraceae and Onagraceae in the order Myrtales of Polypetalae and series (c) Calyciflorae. Engler and Prantl have placed these families along with other sixteen families in their 29th, order Myrtiflorae of class Dicotyledoneae and subclass Archichlamydeae.

Flowers actinomorphic, bisexual, cyclic, peri to epigynous, often the receptacle is cupshaped and is adnate with the calyx tube forming hypanthium, 4-5 merous; stamens indefinite to definite; style simple; ovary wall often adnate to the receptacular cup; herbs, shrubs or trees with opposite or alternate leaves.

Family—Lythraceae

Distribution: There are 23 genera and 450 species in this family. The members of this family are mainly distributed in tropical America, while certain genera are Asiatic. Some species, such as *Lythrum salicaria* and *L. hyssopifolia* are found in temperate regions. *Hydrolythrum wallichi* is an aquatic plant and occurs in West Bengal. *Lawsonia inermis* is a common woody shrub, and grown as a hedge plant and for dye (**mehdi**) in hotter parts of India. *Lagerstroemia speciosa* is grown as an ornamental plant in almost all parts of India. *Ammannia baccifera* is a common herb of marshy ground; *Woodfordia fruticosa* is a common shrub of W.Bengal, Bihar and Orissa.

Habit: The plants are herbs, shrubs or trees. *Hydrolythrum wallichi* is an aquatic plant; *Lawsonia inermis* is a common woody armed shrub.

Leaves: The leaves are opposite or sometimes whorled, exstipulate or minutely stipulate, entire and simple; net veined, petiolate and with acute apex.

Inflorescence: Cymose panicles or solitary.

Flower: Actinomorphic, regular, bisexual and perigynous; the flowers are subtended by epicalyx-like united pair of bracteoles (sepal stipule). Flowers are usually actinomorphic but zygomarphic flowers occur in *Cuphea.* The variation in the number of perianth leaves (*i.e.,* sepals and petals) is striking; the perianth is usually 4-6 merous, however, in *Lafoensia* there is 8-16 merous perianth. Petals are absent in *Peplis* and some species of *Ammannia.*

Calyx: Sepals 4, 6 or 8, or rarely more on the margin of the cup-shaped raceptacle (hypanthium), valvate and free.

Corolla: Petals are as many as sepals, *i.e.,* 4, 6 or 8, rarely absent, often clawed, free, aestivation crumpled often, inserted on the margin of the raceptacular cup and alternating with sepals. Petals are absent in *Peplis* and some species of *Ammannia.*

Androecium: Stamens usually twice the number of petals in two whorls or many, such as in *Lagerstroemia;* those of the outer whorl alternate with petals and are inserted on the inside of the receptacle cup or hypanthium much lower down, filaments unequal; sometimes stamens of the inner whorl are absent; anthers dithecous, dehiscence longitudinal introrse. Reduction in the number of androecium is of rather rare occurrence; in some species of *Rotala* stamen may be one.

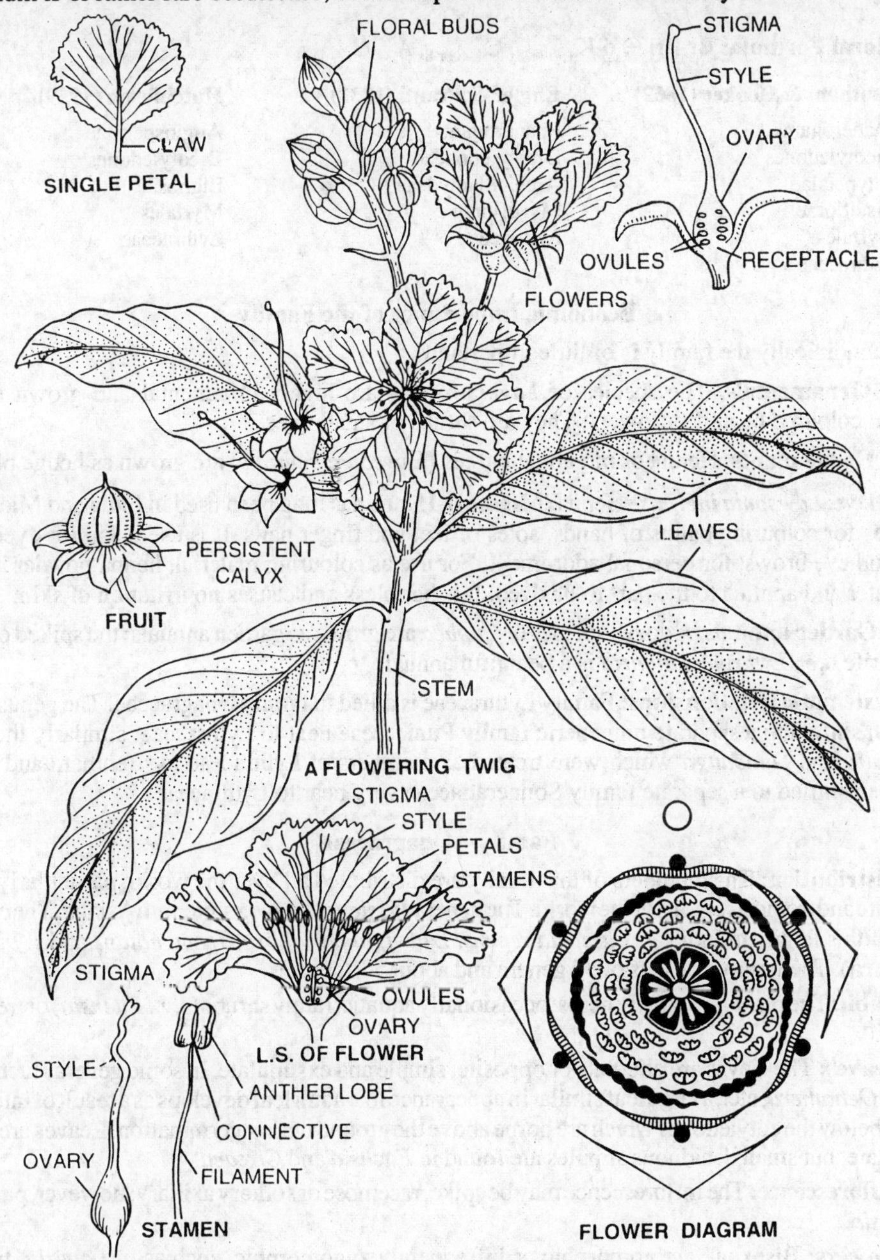

Fig.45.1. *Lagerstroemia speciosa* (L.) Pers.

Gynoecium: Carpels 2-6, united in a superior or semi-inferior 2-6 celled ovary with simple style and capitate stigma; ovules anatropous, usually many an axile placenta; sometimes partition septa disappear in the inner part of the ovary.

Fruit: A capsule; seed with straight embryo, non-endospermic.

The family is distinguished by the presence of the hypanthium, the superior ovary, the crumpled corolla, the often unequal stamens of typically twice the number of sepals, and the seed without endosperm.

Anatomical features: Intraxylary phloem, mucilaginous cells (in leaf) and secretory cavities are present.

Floral Formula: Br Brl $\oplus$ ⚥ $K_{4,\ 6\ or\ 8}$, $C_{4,\ 6\ or\ 8}$, A_{∞} $G_{(\underline{2-6})}$

Bentham & Hooker (1862)	Engler & Prantl (1931)	Hutchinson (1959)
Phanerogams	Phanerogams	Angiospermae
Dicotyledones	Dicotyledoneae	Dicotyledones
Polypetalae	Archichlamydeae	Lignosae
Disciflorae	Myrtiflorae	Myrtales
Myrtales	Lythraceae	Lythraceae
Lythraceae		

Economic Importance of the Family

Economically the family is of little importance.

1. **Ornamentals.** The species of *Lagerstroemia* are highly ornamental and grown for their different coloured beautiful flowers. They are shrubs or small trees.

2. **Hedge plants.** Henna or mignonette trees (*Lawsonia enermis*) are grown as hedge plants.

3. **Dye.** *Lawsonia inermis*–Henna (**Mehndi**). Henna has long been used in India and Middle East countries for colouring palms of hands, soles of feet and finger nails. It is used also for dyeing hair, beard and eye brows, for personal adornment. For use as colouring material, henna powder is pasted with water and applied to the part to be dyed. It is harmless and causes no irritation of skin.

4. **Garden annuals.** A dozen species of *Cuphea* are grown as garden annuals; the spiked or purple loose strife, *i.e., Lythrum salicaria* is a beautiful annual.

Systematic Relationships: Family Lythraceae is allied to family Onagraceae. The genus *Punica* has been shifted to a separate unigeneric family Punicaceae next to Lythraceae; similarly the genera *Sonneratia* and *Duabanga* which were treated as members of Lythraceae by Bentham and Hooker have been shifted to a separate family Sonneratiaceae very near to Lythraceae.

Family—Onagraceae

Distribution: The members of this family are distributed all over the world, particularly in both temperate and tropical regions of the world. The genera *Jussieua, Ludwiga,* etc., are Asiatic. The common plants of this family in India are *Jussieua repens, Ludwigia parviflora, Trapa natans* and *T. bispinosa* (**Singhara**). The family consists of 30 genera and about 500 species.

Habit: The plants are mostly herbs, occasionally aquatic, rarely shrubs (*e.g., Fuchsia*) or trees (*e.g., Hauya*).

Leaves: The leaves are alternate or opposite, simple and exstipulate. In some genera, such as *Epilobium, Oenothera,* etc., a large leaf similar in appearance like adult leaf develops as a result of intercalary growth below the cotyledones which are borne above the ground during germination. Leaves are usually exstipulate, but small deciduous stipules are found in *Fuchsia* and *Circaea.*

Inflorescence: The inflorescence may be spike, racemose or solitary axillary; however, paniculate in *Fuchsia.*

Flowers: Bisexual, actinomorphic or infrequently zygomorphic, such as in *Lopezia,* typically tetramerous; epigynous, complete and regular. The flower of *Trapa* is semi-epigynous with half-inferior ovary of two carpels and single pendulous ovule in each carpellary chamber.

Calyx: Sepals 4, rarely less or more, valvate, forming the calyx tube being basally united and ultimately adnate to the receptacular cup (hypanthium), sometimes persistent as in *Jussieua* and *Ludwigia*; 6 sepals in *Jussieua* spp. The persistent sepals in *Trapa* assume spinous structure above the fruit.

Corolla: Petals 4 usually, sometimes 2 or more or small or none; free and imbricate; mostly clawed and convolute. In *Lopezia,* two posterior petals are smaller than antero-lateral ones and are bent upwards a little from the base.

Androecium: Stamens usually the same number as petals or in 2 whorls and twice as many; 1 fertile stamen and 1 staminode in *Lopezia;* when biseriate, those of outer whorl alternate with petals, distinct, arising from or near hypanthium rim; the anthers typically dithecous (bicelled), dehiscing longitudinally, sometimes each cell divides transversely into two (*e.g.,* in *Circaea*). Pollen grains in *Epilobium* are united in tetrads.

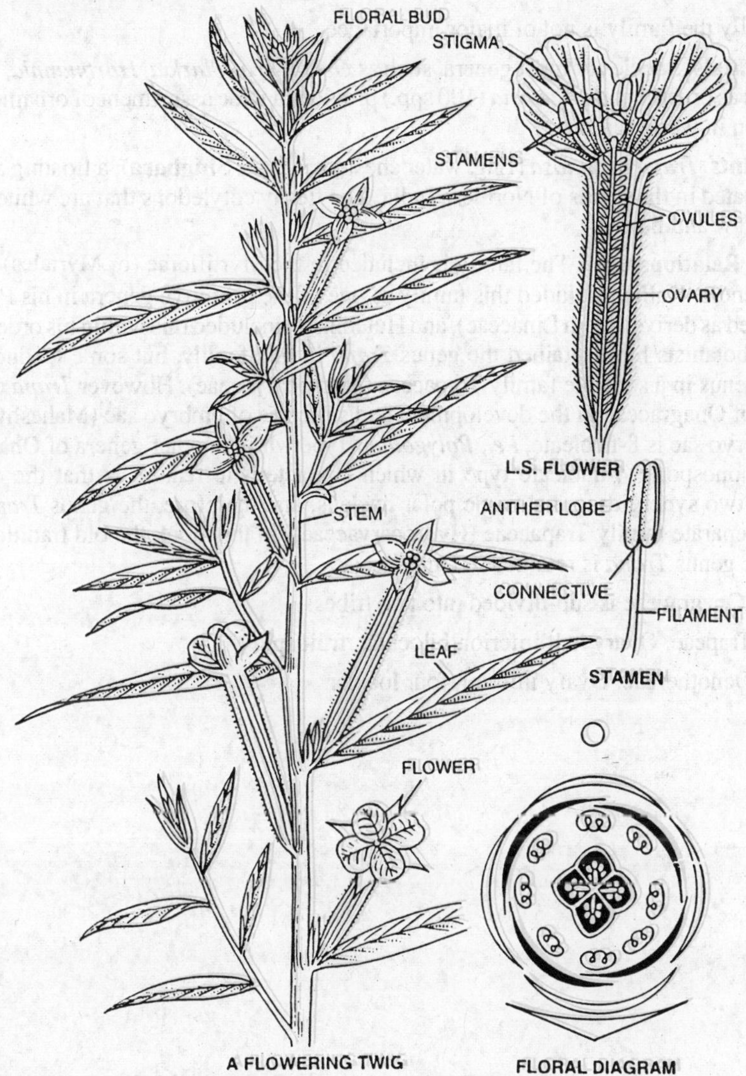

Fig.45.2. Onagraceae. *Jussieua suffruticosa* Linn.

Gynoecium: Carpels usually 4, rarely 2, 3 or 5-6, united into 4-celled inferior ovary, rarely ovary is semi-inferior; style one, simple, slender, the stigma usually capitate, sometimes notched, *e.g.,* in *Circaea* or with radiate branches (*e.g., Oenothera* and *Epilobium*); ovules usually several, sometimes few or one, anatropous, ascending, horizontal or descending on axile placenta.

Fruit: Various, loculicidal capsule, berry, nut or drupe. For example, berry in *Fuchsia,* a bristly 1-2 seeded nutlet in *Circaea* and a 1-seeded smooth nut in *Gaura.*

Seeds: Non-endospermic with small embryo; seeds cymose (*e.g.,* in *Epilobium* and *Zauschneria*) or glabrous. In *Trapa bispinosa* (water chestnut, **Singhara**) the two cotyledons are very unequal in size, one is scaly minute, the other is large and is full of food materials (starch) occupying the entire inner space of the seed; the larger food containing cotyledon remains within the seed, the other scaly is epigeal which is carried up by hypocotyl and turns green.

Floral Formula: $\oplus$ ⚥ K_4, C_4, A_{1-4}, $\overline{G}_{(4)}$.

Economic Importance of the Family

Economically the family is not of major importance.

1. **Ornamentals.** Species of most genera, such as *Oenothera, Clarkia, Hartmannia, Lopezia* are grown in gardens as ornamentals. *Fuchsia* (100 spp.) provides a wide assortment of ornamental shrubs grown in the open in warm regions.

2. **Food plants.** *Trapa bispinosa* (Eng. water chestnut; Verna. **Singhara**), a floating aquatic herb commonly cultivated in the ponds of Northern India. The fleshy cotyledons that are white and rich in starch are eaten raw and boiled.

Systematic Relationships : The family is included in the Myrtiflorae (or Myrtales) by Bessey, Wettstein and Rendle. Hallier included this family as one of the primitive genera in his Polygalineae (an order he treated as derived from Linaceae), and Hutchinson included this within his order Lythrales. Most American botanists have retained the genus *Trapa* in this family, but some taxonomists have transferred this genus in a separate family Trapaceae (Hydrocaryaceae). However *Trapa* differs from all other genera of Onagraceae in the development and structure of embryo sac (Maheshwari, 1945). In *Trapa* the embryo sac is 8-nucleate, *i.e., Polygonum* type while in other genera of Onagraceae the embryo sac is monosporic 4-nucleate type in which characteristic feature is that the embryo sac contains an egg, two synergids and a single polar nucleus, and therefore, the genus *Trapa* has been transferred to a separate family Trapaceae (Hydrocaryaceae). In this book the old tradition has been followed, and the genus *Trapa* is retained in Onagraceae.

The family Onagraceae is sub-divided into two tribes:

Tribe 1. Trapeae. Ovary half inferior, bilocular, fruit spiny.

Tribe 2. Oenothereae. Ovary inferior, four locular.

46

ORDER–SARRACENIALES

According to Hutchinson this is the 70th order of the phylum Angiospermae, subphylum Dicotyledones and division Herbaceae. This order comprises of only two families, *i.e.*, Droseraceae and Sarraceniaceae, of which Droseraceae has been described here in detail. Bentham and Hooker have placed the family Droseraceae along with eight other families, *i.e.*, Connaraceae, Leguminosae, Rosaceae, Saxifrageae, Crassulaceae, Hamamelideae, Brumiaceae and Halorageae in the order-Rosales of Polypetalae and series (c) Calyciflorae. Engler and Prntl have placed the Droseraceae in their 20th. order-Sarraceniales along with two other families, *i.e.*, Sarraceniaceae and Nepenthaceae, of class Dicotyledoneae and subclass Archichlamydeae.

An order of insectivorous plants. Herbs with usually alternate leaves, insectivorous by means of pitchers, or sensitive usually glandular, viscid hairs and actinomorphic flowers with uniseriate to biseriate perianth; flowers bisexual, very rarely unisexual (*e.g.*, in *Nepenthes*) and hypogynous; ovules numerous on axile or parietal placentas.

Family—Droseraceae

This family consists of 4 genera and about 90 species.

Distribution: Three genera are monotypic. *Drosophyllum* is confined to Western Mediterranean; *Dionaea* is found in North and South Corolina, while the species of *Aldrovanda* are found in the old world. *Drosera* with about 85-88 species and most abundant in Australia, is represented by a few species in most regions of all continents. The common Indian plants of this family are *Drosera burmanni, D.indica, Aldrovanda vesiculosa, Dionaea,* etc.

Habit: The members of this family are mostly perennial creeping insectivorous herbs of marshy places with underground stems. Primary root is absent in some species of *Drosera,* the place of the root is taken up by protocorm-like structure developed from hypocotyl and bears filiform attaching organs. On the other hand *Drosophyllum* and *Dionaea* have primary root.

According to different climatic conditions, two types of growth forms, such as photophilous and geophilous have been recorded. In photophilous form, such as in *Drosera indica* the stem is capable of indefinite growth by means of terminal bud. The geophilous forms are adapted to climates where damp cold season alternates with dry hot season, such as in *D. cistiflora* and *D. paucciflora* and they persist by means of swollen root or bulb. In *Drosophyllum* the stem is woody and branched and bears terminal foliage on it.

Generally primary root does not develop, instead numerous adventitious roots are formed.

Leaves: The leaves are alternate, very rarely whorled, lamina covered with tentacular glands, stipules mostly present, membranous. The leaves of *Drosera, Dionea* and other members are metamorphosed for catching insect preys and for photosynthesis. In most cases, the petioles are well developed. In *Dionaea* (Venus, fly-trap), the petiole is winged and acts like leaf. The leaf lamina of *Drosera* has various structures. The upper surface of leaf is provided with tentacles which are glandular hairs.

Morphology of tentacle: Morphologically a tentacle is intermediate between a true leaf part and a trichome. The structure differs according to its situation either on the margin or on the surface. A tentacle is provided with vascular chord, a tracheid ending in a gland, the tracheid end remains covered

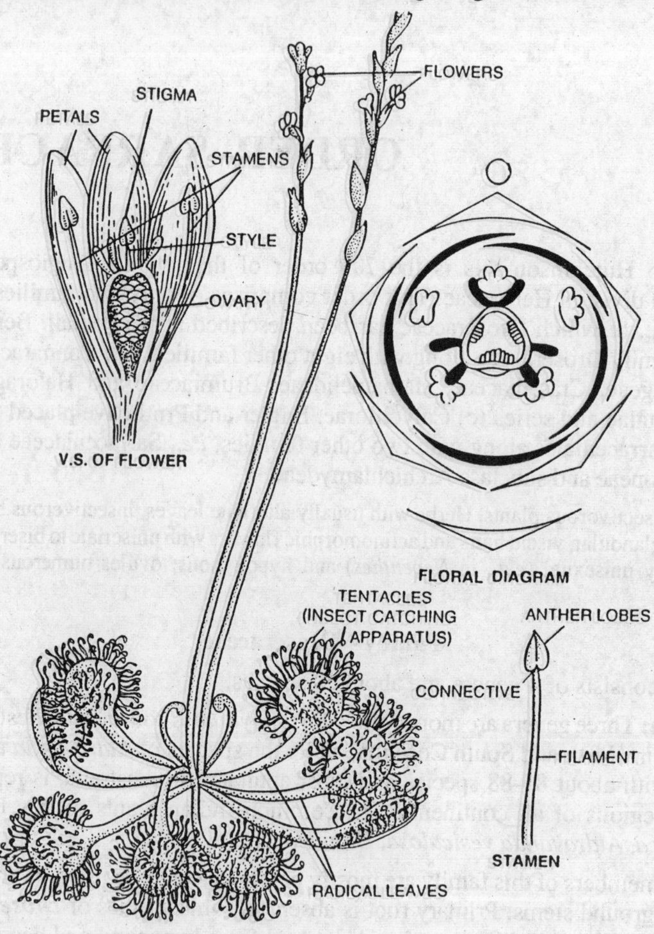

Fig.46.1. Droseraceae. *Drosera burmanii* Vahl.

by two layers of secreting layer made up of secreting epithellium. The tentacles play an important role for catching tiny insects. The tentacles have peculiar power of movement. As soon as an insect gets stuck into the mucilaginous secretions of glands, the tentacles bend down over the body of the entrapped prey and catch them. The insects die, and the proteins are absorbed by the tentacles.

In *Aldrovanda* (an aquatic herb) the leaves are arranged in whorls of 7-9 on the stem, all leaves are united together at the base with one another. Each leaf is modified into an insect catching trap. The leaf possesses wide petiole and several long adnate spinous stipules. The leaf blade has numerous glands, and makes the main trap. The leaf blade is divided into two halves along midrib which are mobile; in touch with insect prey. The leaf halves close, and entrap the victim.

In *Dionaea* (Venus fly-trap) the leaves are radical and arranged in whorls. Each leaf is an insect catching trap. The petiole is winged, and the leaf blade is divided into two halves. The margins of leaf blade are spinous and on touching the insect the prey is entrapped, killed and aborted.

Inflorescence: The flowers may be solitary or in spike or raceme like inflorescence, or scorpioid cyme (cincinnus); racemose to paniculate.

Flowers: Bisexual, actinomorphic, hypogynous, 4-5 merous; perianth biseriate, distinguishing into sepals and petals.

Calyx: Sepals 4-5, mostly briefly and basally connate, persistent, imbricate.

Corolla: Petals 5, distinct, convolute, imbricate, membranous or soft.

Androecium: Stamens 5-20, in one or more pentamerous whorls, distinct or rarely the filaments basally connate; the anthers dithecous (2-celled), the cells sometimes divergent and with broad connective, dehiscing longitudinally, extrorse, the pollen in tetrads.

Gynoecium: Carpels 3-5, syncarpous, united in superior 1-celled (single loculed) ovary; styles 3-5, mostly free, stigma capitate; ovules many, sometimes few, anatropous on parietal placenta, rarely on free central placenta.

Fruit: A loculicidal capsule.

Seed : Seeds numerous, small the embryo straight, endosperm crystalline granular, cotyledons small.

Floral Formula: $\oplus$ ⚥ K_{4-5}, C_{4-5}, A_{5-20}, $G_{(\underline{3-5})}$

Bentham & Hooker (1862)	**Engler & Prantl (1931)**	**Hutchinson (1959)**
Phanerogams	Phanerogams	Angiospermae
Dicotyledones	Dicotyledoneae	Dicotyledones
Polypetalae	Archichlamydeae	Herbaceae
Calyciflorae	Sarraceniales	Sarraceniales
Rosales	Droseraceae	Droseraceae
Droseraceae		

Economic Importance of the Family

The family is little or no economic value.

1. **Ornamentals.** Venus' fly-trap (*Dionaea* spp.) is grown as a novelty in the houses.
2. **Dye.** A kind of violet dye is obtained from *Drosera* spp., but it is not used commercially.

Systematic Relationships: Wettstein and others have placed the family Droseraceae in the order Parietales near Violaceae and Ochnaceae on the basis of the parietal placentation. Lindley and others placed them in the Saxifragaceae because of the determinate inflorescence, the often perigynous condition, and similarities of androecium and perianth as well as of ovule morphology. In many respects, such as the insectivorous and aquatic habit of *Aldrovanda,* and the glandular hairs, the family is similar to Lentibulariaceae, a similarity due to parallel development rather than phylogeny. Hallier, Engler and others have placed the three families, *i.e.,* Sarraceniaceae, Nepenthaceae and Droseraceae into one order–Sarraceniales next to Rhoeadales. Bentham and Hooker have placed the three families in different orders far apart from each other, such as Sarraceniaceae is put under Parietales before Papaveraceae, Droseraceae in Rosales under Calyciflorae. Baillon, states that Sarraceniaceae has affinities with Nymphaeaceae. He also supports its affinities with Aristolochiaceae. Hutchinson (1959) supported Baillon's (1872) view and placed the family Nepenthaceae under Aristolochiales near Berberidales and other two families Sarraceniaceae and Droseraceae in the order Sarraceniales after Saxifragales far apart from Aristolochiales.

47

ORDER-PODOSTEMALES

According to Hutchinson this is the 71st order of the phylum Angiospermae, subphylum Dicotyledones and division Herbaceae. This order comprises of only two families, *i.e.,* Podostemaceae and Hydrostachyaceae, of which Podostemaceae has been described here in detail. Bentham and Hooker have placed the family Podostemaceae in Monochlamydeae or Incompletae and series-2. Multivulatae Aquaticae. Engler and Prantl have placed the Podostemaceae in their 21st order–Rosales along with seventeen other families.

Herbaceous moss-like perennial aquatics; flowers actinomorphic, bisexual or unisexual, haplochlamydous or achlamydous, when haplochlamydous perianth consists of 2-3 membranous tepals; seeds minute, non-endospermic; placentation axile or arie parietal.

Family—Podostemaceae

Distribution: This family consists of 43 genera and about 140 species, distributed pantropically and with a few extensions into north and south temperate regions of all continents. *Dicraea stylosa* and *Podostemon ceratophyllum* are common aquatic plants of India.

Habit and Habitat: The members of this family are identified by their occurrence in streams of running water where they are attached to rocks, by their mosslike or lichenlike appearance. In an Indian species *Podostemon subulatus* the main vegetative body is threadlike thalloid but it gives off exogenously developed branches, it has also root cap. In *Dicraea* the vegetative body is dorsiventral alga-like in form, flat and expanded and developed from the base of the hypocotyl of the embryo, its branches are formed exogenously and have also root like structures. The roots of the plants are often photosynthetic and creeping.

Stem and Leaves: Primary axis small, branching into a dorsiventral thallus-like structure with filiform to laminate segments; secondary leaf-bearing branches usually produced. Milky latex is found in many species.

Leaves alternate, usually simple, linear to lamellate, often basally sheathing. Leaves are small and of various forms ranging from scaly to moss-like. Sometimes the leaves are large, simple and dissected, such as in *Mourera.*

Inflorescence: Solitary and terminal, cymose.

Flowers: Small, bisexual, actinomorphic, sometimes zygomorphic, haplochlamydous, (*i.e.,* perianth of single whorl) or achlamydous (*i.e.,* without perianth), and hypogynous; flowers are usually produced only when plants are exposed by low water; fragrant, pedicellate (with elongated pedicels).

Perianth: The perianth is not differentiated into calyx and corolla. The typical haplochlamydous flowers possess 3-5 perianth members or tepals which may be either free or sometimes somewhat united at the base, imbricate or valvate; sometimes the tepals are very small, scale-like and numerous. In achlamydous flowers the young flower buds remain covered by thin membranes known as 'spathella'.

Androecium: Stamens 1-many, in one to several series, free; sometimes filaments of stamens are united into a tubular sheath round the pistil, such as in *Lacis monadelpha*; anthers typically 4-celled, introrse, one staminode on each side of the stamens; dehiscence longitudinal lateral; pollen grains clustered in twos or fours.

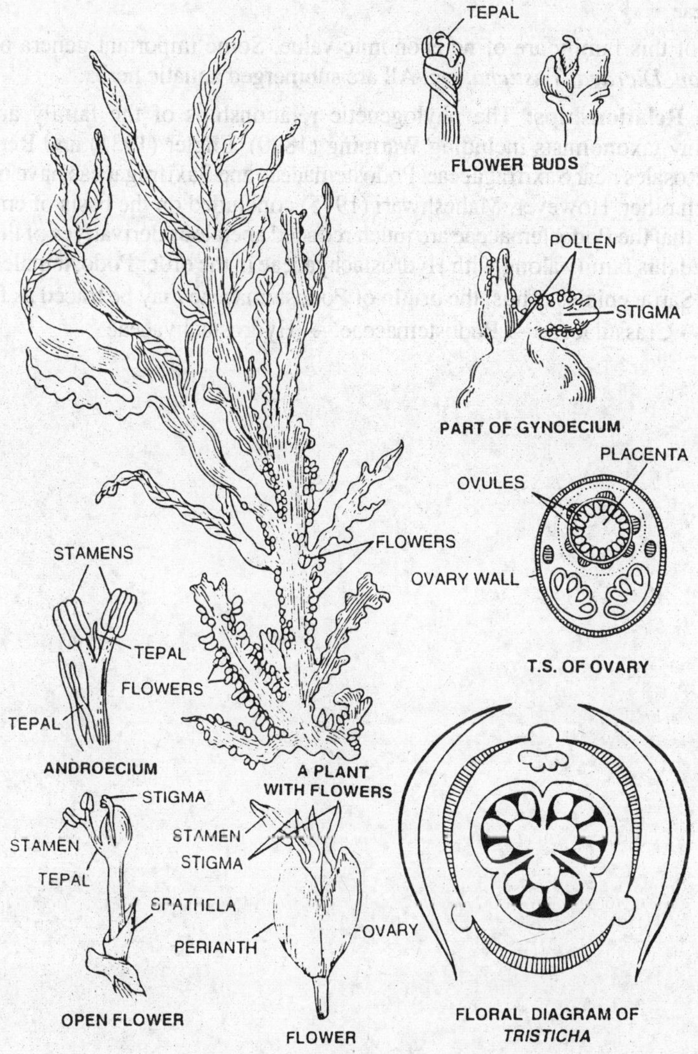

Fig.47.1. Podostemaceae. *Dicraea stylosa* Wight.

Gynoecium: Carpels 2-3, syncarpous, united into a bilocular or trilocular superior ovary; the placentation is axile, sometimes 1-loculed ovary with free-central placentation; the ovules are numerous and anatropous, integuments 2; the styles 2-3 filiform or very short, often strongly papillate, usually distinct, styles as many as carpels; stigmas as many as styles, fimbriate or capitate.

Fruit: Fruit a septicidal capsule.

Seeds: Numerous, minute, non-endospermic.

Floral Formula: ⊕ or ·|· ⚥ P_{3-5}, A_{1-4} or ∝, G(2–3).

Bentham & Hooker (1862)	**Engler & Prantl (1931)**	**Hutchinson (1959)**
Phanerogams	Phanerogams	Angiospermae
Dicotyledones	Dicotyledoneae	Dicotyledones
Monochlamydeae or	Archichlamydeae	Herbaceae
Incompletae	Rosales	Podostemales
Multivulatae Aquaticae	Podostemaceae	Podostemaceae
Podostemaceae		

The plants of this family are of no economic value. Some important genera of the family are *Lacis, Podostemon, Dicraea, Tristicha,* etc. All are submerged aquatic herbs.

Systematic Relationships: The phylogenetic relationships of the family are controversial. According to many taxonomists including Warming (1890), Engler (1931) and Rendle, this family belongs to order Rosales near Saxifragaceae. Podostemaceae and Saxifragaceae have many similarities and resemble each other. However, Maheshwari (1945) concluded on the basis of embryology that it is 'almost certain that the Podostemaceae are much reduced apetalous derivatives of the Crassulaceae'. Hutchinson placed this family along with Hydrostachyaceae in the order Podostemales after the orders Saxifragales and Sarraceniales. Thus, the origin of Podostemaceae may be traced as follows: Ranales → Saxifragales → Crassulaceae → Podostemaceae → Hydrostachyaceae.

48

ORDER–UMBELLALES

According to Hutchinson this is the seventy second order of the phylum Angiospermae, sub-phylum Dicotyledones and division Herbaceae. This order consists of a single family–the Umbelliferae. This family has been described here in detail. Bentham and Hooker have placed the family along with two others in the fifteenth order–the Umbellales of class Dicotyledones, division Polypetalae and series Calyciflorae. Engler and Prantl have put this family along with two others in the 30th order–the Umbelliflorae of class Dicotyledoneae and sub-class Archichlamydeae.

Flowers in umbels; ovary inferior, of two carpels. Herbs, often with hollow stems; leaves without stipules.

Family–Umbelliferae (Apiaceae)

There are about 200 genera and 2900 species in this family.

Distribution: The plants of this family are cosmopolitan in their distribution, however, they are not found in arctic regions. They are very commonly found in northern temperate regions. In the tropical countries they are either found in the hilly tracts or cultivated in the winter season. Majority of the plants are of economic value and many grow as weeds. In our country the family is represented by several important species such as, *Coriandrum sativum, Foeniculum vulgare, Cuminum cyminum, Daucus carota, Carum copticum, Hydrocotyle asiatica,* etc.

Habit: Majority of the plants are annual, biennial or perennial herbs. Sometimes shrubs and undershrubs are also found. *Bupleurum falcatum* is an undershrub and *Pseudocarum* is a climbing plant.

Root: Usually the root is tap and branched.

Stem: Usually, erect, rarely climbing, herbaceous. It becomes fistular at maturity on account of the shrinkage of the pith. The nodes and internodes are quite distinct. The stem is usually green and becomes pinkish as the plants mature. The stems are very often ribbed and angled.

Leaves: The leaves are alternate, exstipulate amplexicaul and much dissected, rarely the leaves are opposite, *e.g., Apiastrum.* Usually the leaves are pinnately compound. However, in some cases the leaves are simple, *e.g., Hydrocotyle, Bupleurum,* etc. The species *Eryngium* and *Aciphylla* possess the leaves with parallel venation and sheathing bases like many monocotyledonous plants. Oil glands are present in all the aerial parts of the plant.

Inflorescence: The inflorescence is umbel. This may be simple or compound. In majority of cases compound umbel is found, *e.g., Coriandrum, Foeniculum,* etc., however, in some cases it is simple umbel, *e.g., Hydrocotyle.* The compound umbels are sometimes cymose in character and a terminal flower may occur, *e.g., Eryngium.* In compound umbels an **involucre** of bracts is present at the top of the main floral axis from where the umbels arise. At the base of each umbel **involucel** of the bracts is found.

Flower: The flowers are usually hermaphrodite but sometimes unisexual flowers are also found. In several genera the outer flowers of the inflorescence are male only. In *Astransia* sp., majority of the flowers of the inflorescence are bisexual and some are only male flowers. In *Echinophora* sp., the monoecious plants bear the unisexual flowers. The plants of *Arctopus* sp. are dioecious and bear unisexual flowers. In majority of the inflorescences the flowers are regular (actinomorphic) but sometimes the outer flowers of the umbles are zygomorphic and irregular, *e.g., Coriandrum sativum.* The flowers are usually pedicellate, bracteate, complete and epigynous.

Calyx: It consists of five sepals. These sepals are very minute in structure. The odd sepal being posterior. In many the calyx is absent, *e.g., Foeniculum vulgare.*

Corolla: It consists of five petals, polypetalous, *i.e.*, the petals are free. They are usually white or yellow in colour. The petals vary in size. In many cases two of the petals being larger than the other three. The tips of the petals are usually reflexed. Sometimes the petals are bifid, *e.g., Coriandrum sativum.* The aestivation is either valvate or imbricate.

Androecium: It consists of five stamens alternate to the petals. The stamens arise from an epigynous disc. The anthers are bilobed, introrse, basi or dorsifixed. The filaments are free. The anthers split by longitudinal slits.

Gynoecium: It consists of two carpels (bicarpellary), syncarpous. The ovary is inferior, bilocular, each loculus contains a single pendulous, anatropous ovule. The placentation is axile. On the top of the ovary a nectar disc is found which surrounds the two capitate stigmas.

Fruit: The fruit is **cremocarp**. The ovary splits into two **mericarps.** The two mericarps remain temporarily attached to a very thin axial prolongation, between them, known as **carpophore**. Each mericarp is usually covered by five longitudinal ridges known as **costae.** Between costae there are furrows known as **valleculae**. Under these furrows there lie special oil ducts known as **vittae.**

Seed: The seed is endospermic containing a minute embryo.

Pollination: The pollination is usually entomophilous. The insects are attracted towards the beautiful condesed umbels having outer zygomorphic flowers. The nectar secreting disc situated at the top of the ovary is the main attraction of the insects. The flowers are protandrous and thus cross pollination is effected.

Floral Formula: ⊕ or ·|· ⚥ K 5, C5, A 5, $G_{\overline{(2)}}$

Bentham and Hooker (1862)	**Engler and Prantl (1931)**	**Hutchinson (1959)**
Phanerogams	Phanerogams	Angiospermae
Dicotyledones	Dicotyledoneae	Dicotyledones
Polypetalae	Archichlamydeae	Herbaceae
Calyciflorae	Umbelliflorae	Umbellales
Umbellales	Umbelliferae	Umbelliferae or
Umbelliferae		Apiaceae

Description of some important plants in semi-technical language:

1. ***Trachyspermum ammi*** (Linn.) Sprague. Verna. **Ajwain;** Eng. Lovage, Ammi.

Habit: Annual, aromatic, cultivated herb.

Root: Tap and branched.

Stem: Erect, branched, glabrous, fistular, ribbed, nodes slightly swollen.

Leaf: Cauline and ramal, compound 2, 3-4 pinnate, ultimate segments linear, alternate, petiolate, petiole sheatihing at the base, exstipulate, aromatic.

Inflorescence: Racemose, compound umbel, primary umbels with an involucre of bracts at the base and the secondary umbel with an involucel of bracteoles.

Flower: Pedicellate, ebracteate, hermaphrodite, actinomorphic, complete, yellow, epigynous.

Calyx: 5 sepals, gamosepalous, calyx tube adnate to the ovary wall, calyx teeth 0.

Corolla: 5 petals, polypetalous, emarginate, yellow, valvate aestivation, superior, yellow.

Androecium: 5 stamens, polyandrous, alternating with petals, filaments long, anthers introrse, bicelled, basifixed.

Gynoecium: 2 carpels, bicarpellary, syncarpous, ovary inferior, bilocular, one ovule in each loculus, axile placentation, style extremely short, stigma bilobed, vittae present in ovary wall.

Fruit: Cremocarp splitting into two meriearps.

Floral formula: ⊕ ⚥ K (5), C5, A 5, G $\overline{(2)}$.

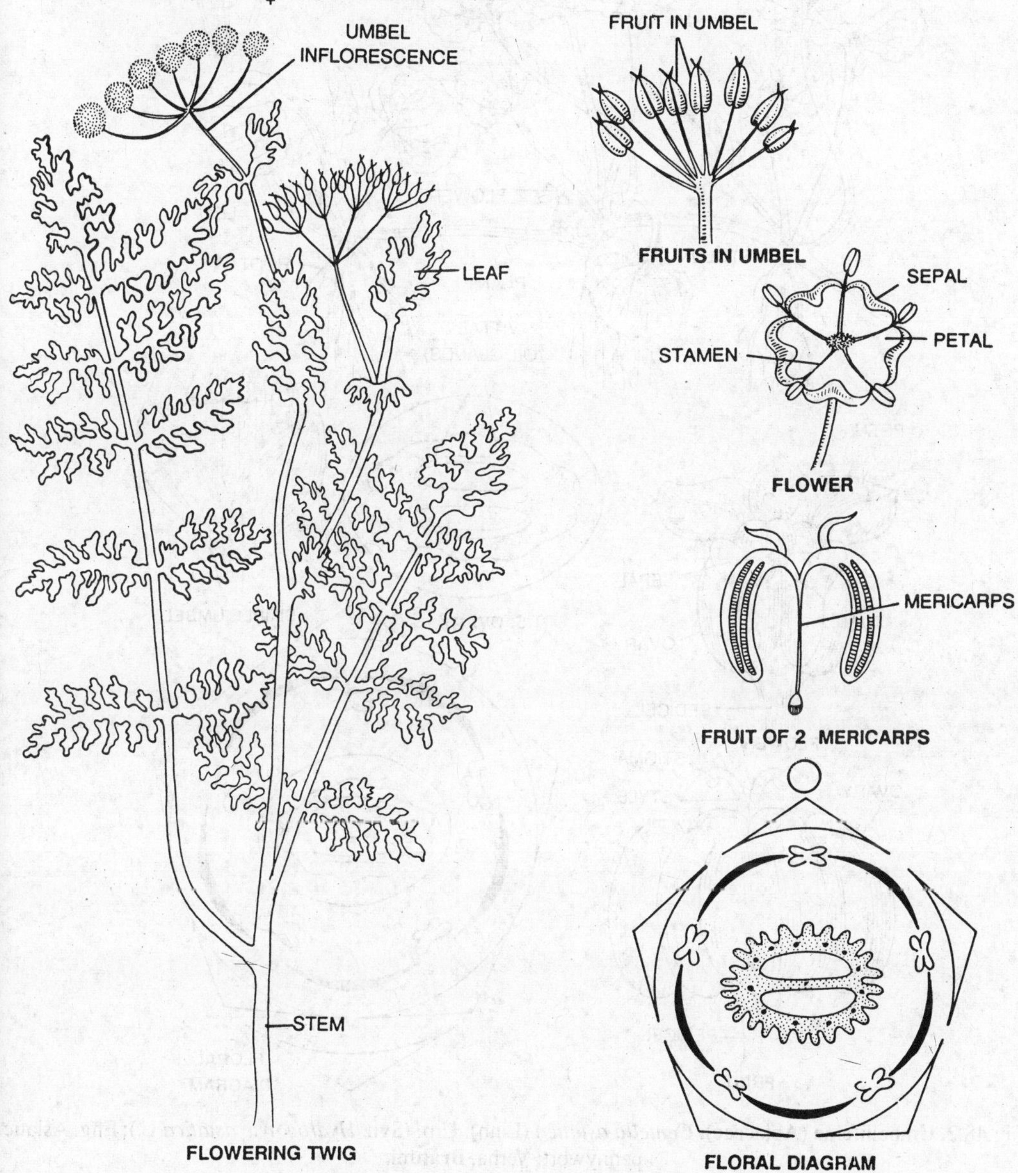

Fig.48.1 : Umbelliferae (Apiaceae). *Trachyspermum ammi* (Linn.) Sprague. (Syn.*Carum copticum*); Eng., lovage ammi, Verna, **Ajwain.**

Economic value : The fruits are used as spice. They are also used medicinally as carminative, stimulant, tonic and in indigestion.

2. ***Centella asiatica*** (Linn.) Urb. (Syn. *Hydrocotyle asiatica* Linn.) Verna. **Brahmi;** Eng. Asiatic pennywort.

Habit: Prostrate herbs, rooting at nodes.

Root: Adventitious roots.

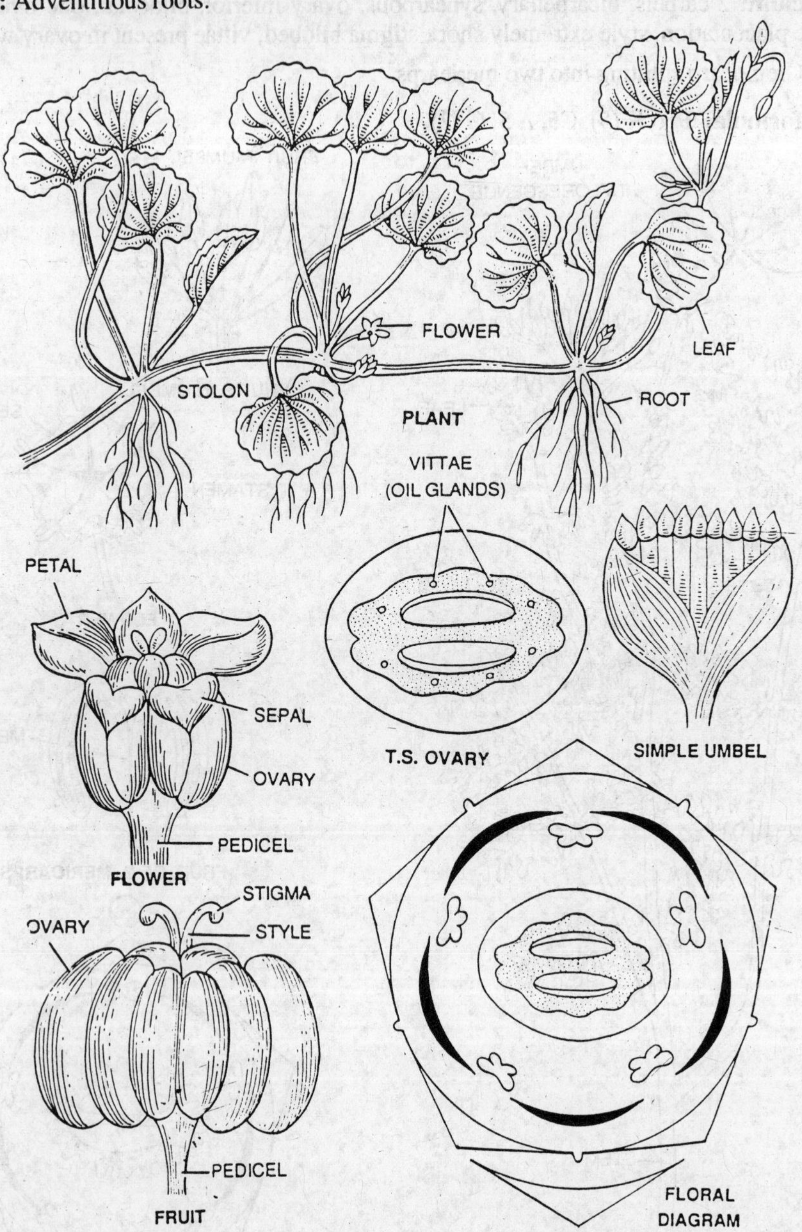

Fig.48.2. Umbelliferae (Apiaceae). *Centella asiatica* (Linn). Urb. (Syn. *Hydrocotyle asiatica* L.); Eng. Asiatic pennywort; Verna. **Brahmi.**

Economic value. The weed is alterative, tonic, diruetic, antiphlogistic, blood purifier and local stimulant. It is a remedy for skin diseases, like chronic eczema, chronic ulcers, etc., enlargement of glands, chronic rheumatism, chronic nervous diseases, madness, cholera, amenorrhoea and piles. An ointmant made of leaf juice and lanoline is of great value in elephantiasis.

Stem: Prostrate, herbaceous, weak, cylindrical, stolon, leaves and roots at nodes.

Leaves: Simple, arising in groups from nodes, orbicular or reniform, crenate and often lobed; glabrous, shining, petiolate, long petioled; stipulate, stipule, adnate to petioles.

Inflorescence: Racemose, simple umbels, 3-6 flowers in each group.

Flower: Pedicellate, small pedicels, bracteate, bracts small, ovate, embracing the flowers, hermaphrodite, actinomorphic, regular, complete, epigynous.

Calyx: 5 sepals, minute, connate, valvate.

Corolla: 5 petals, polypetalous, valvate.

Androecium: 5 stamens, polyandrous, alternate to petals, filament short, anther dithecous, introrse.

Gynoecium: 2 carpels, syncarpous, ovary inferior, 2-celled, disk 2-lobed, vittae present in ovary wall, axile placentation, one ovule in each locule.

Fruit: Cremocarp, laterally compressed.

Floral formula: Br ⊕ ⚥ K (5), C5, A 5, $G\overline{(2)}$.

3. ***Coriandrum sativum;*** Verna. –**Dhaniya.**

Habit: An annual cultivated herb.

Root: Tap and branched.

Stem: Erect, herbaceous, green, solid, branched, cylindrical, aromatic, ribbed, glabrous, nodes, slightly swollen.

Leaf: Decompound, much branched when old, cauline, petiolate, base slightly sheathing, unicostate reticulate venation.

Inflorescence: Compound umbel with involucre of bracts and involucel.

Flower: Pedicellate, bracteate, hermaphrodite, complete, epigynous, the outer flowers of the inflorescence are zygomorphic, *i.e.,*with large and unequal petals, whereas the central flowers are actinomorphic with equal and small petals.

Calyx: 5 sepals, gamosepalous, green, acute, valvate aestivation, superior, calyx tube adnate to the ovary wall.

Corolla: 5 petals, polypetalous, in peripheral flowers two posteior petals bilobed and smallest, one anterior deeply bilobed and largest two lateral large, bilobed, one lobe very small and the other large, zygomorphic, imbricate aestivation, purplish white; the petals of central flowers are small and equal in size, each petal consists of two equal lobes.

Androccium: 5 stamens polyandrous, alternating with petals, filaments long, anthers dorsifixed, introrse and yellow.

Gynoecium: 2 carpels, bicarpellary, syncarpous, ovary inferior, bilocular, one ovule in each loculus, axile placentation, epigynous disc present at the top of ovary prolonged in two short styles, vittae present in ovary wall.

Fruit: Cremocarp.

Floral Formulae:

·|· ⚥ K (5), C 5, A 5, $G\overline{(2)}$ (zygomorphic flowers)

⊕ ⚥ K (5), C5, A 5, $G\overline{(2)}$ (actinomorphic flowers)

4. ***Foeniculum vulgare;*** Verna.–**Sonf.**

Habit : Annual aromatic cultivated herb.

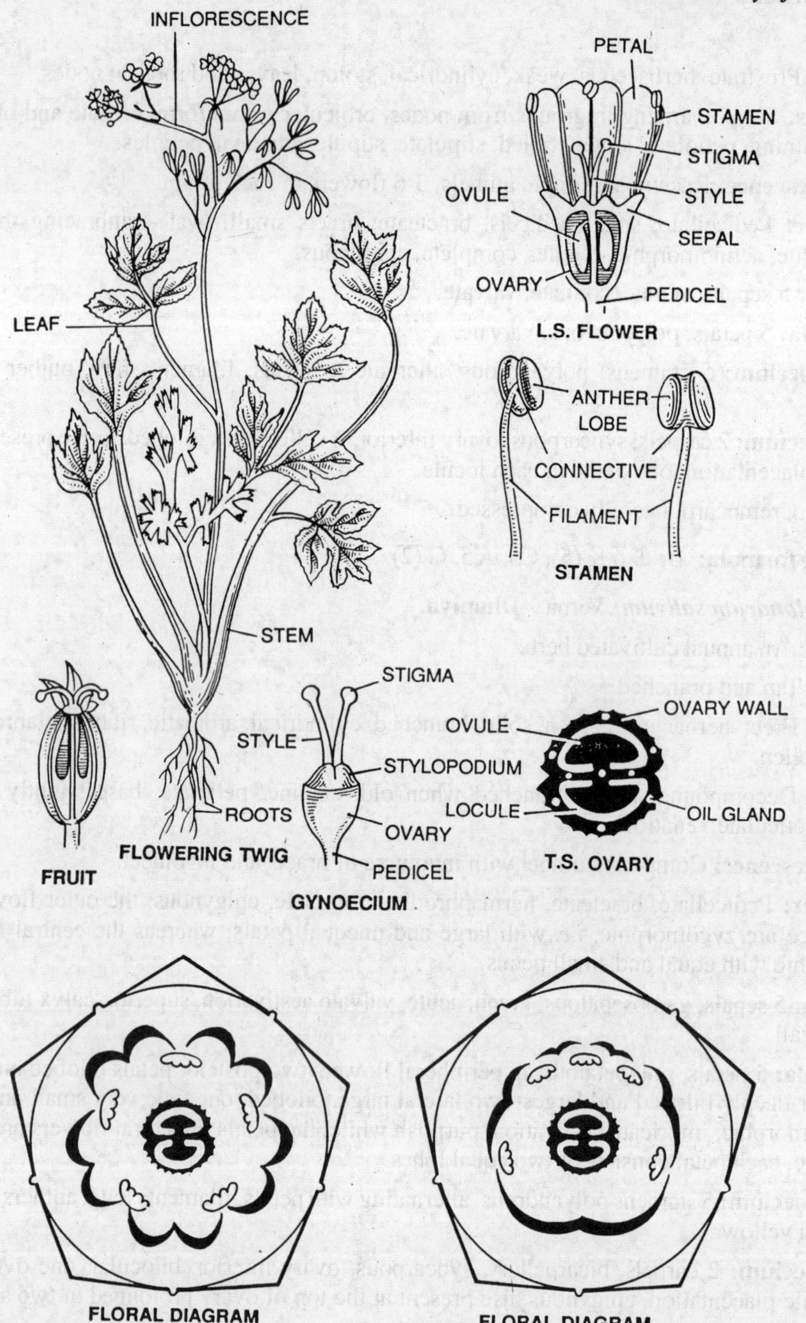

Fig.48.3. Umbelliferae. *Coriandrum sativum* Linn., Eng. Coriander; Verna. **dhaniya.**

Root: Tap and branched.

Stem: Erect, branched, glabrous, fistular, ribbed, nodes slightly swollen.

Leaf: Compound 2,3-4 pinnate, ultimate segments linear, cauline, alternate, petiolate, petiole sheathing at the base, exstipulate.

Inflorescence: Racemose, compound umbel, primary umbels with an involucre of bracts at the base and the secondry umbel with an involucel of bracteoles.

Flower : Pedicellate ebracteate, hermaphrodite, actinomorphic, complete, yellow, epigynous.

Calyx: 5 sepals, gamosepalous, calyx tube adnate to the ovary wall, calyx wall, calyx teeth 0.

Corolla: 5 petals, polypetalous, emerginate, yellow, valvate aestivation, superior, yellow.

Androecium: 5 stamens, polyandrous, alternating with petals filaments long anthers introrse, bicelled, basifixed.

Gynoecium: 2 carpels, bicarpellary, syncarpous, ovary inferior, bilocular, one ovule in each loculus, axile placentation, style extremely short, stylopodium present, stigma bilobed, vittae present in ovary wall.

Fruit: Cremocarp.

Floral formula: $\oplus$ ⚥ K (5), C 5, A 5, $G_{\overline{(2)}}$.

Economic Importance of the Family

The family is very important from the economic point of view. The fruits of some plants are used as condiments whereas the other plants are of medicinal value.

1. *Apium graveolens* var.*dulce* (Eng.–Garden celery; Verna.–**Ajmud,Karas**)–Native of Europe, now cultivated in the North-Western Himalayas and in the hills of Uttar Pradesh, the Punjab and South India. Used as a vegetable. The roots and seeds are used medicinally.

2. *Apium graveolens* var. *rapaceum*; Syn. *A rapaceum* (Eng.–Celeriac; Verna.–**Ajmud, Salari**)-The roots are eaten as vegetable. The fruits are used for extracting an oil, which is much valued, and the seeds as spice. Cultivated in the hills of Uttar Pradesh, the Punjab and South India.

3. *Coriandrum sativum* (Eng.–Coriander; Verna.–**Dhania**)–An aromatic herb, native of the Mediterranean region, now cultivated chiefly in Madhya Pradesh, Maharashtra, Karnataka, Bihar and Uttar Pradesh for its fruits and leaves, which are used as condiment and spice. The fruits are also used as stimulant, carminative, stomachic and tonic.

4. *Cuminum cyminum*; (Verna.-**Jira**)-A herb, native of the Mediterranean region but commonly grown in the punjab and Uttar Pradesh, for the aromatic fruits, which are used for flavouring purposes.

5. *Carum carvi*; (Eng.–Caraway; Verna.–**Shiajira**)–A herb, native of Europe and West Asia, now cultivated in Bihar, Orissa, the Punjab, Bengal and Andhra Pradesh. The fruits are used as condiment, and are also used medicinally as stomachic and carminative.

6. *Bunium persicum;* Syn. *Carum bulbocastanum*; (Verna.–**Kala zira**)–A perennial herb, found in Kashmir. The starchy tubers are eaten as vegetable and the seeds are used as spice.

7. *Trachyspermum ammi*; Syn. *Carum copticum;* (Verna.–**Ajwain**)–A herb, cultivated through out India. The fruits are used as spice, and also used medicinally as carminative, stimulant, tonic and in indigestion.

8. *Trachyspermum roxburghianum*; Syn. *Carum roxburghianum;* (Eng.–Ajmud; Verna.–**Ajmuda, Radhuni**); used as spice and condiment.

9. *Daucus carota* var. *sativa* (Eng.–Carrot, Verna.–**Gajar**)– An annual or biennial herb, grown hiefly in the Punjab, Uttar Pradesh and Madhya Pradesh. The roots are edible.

10. *Centella asiatica*; Syn. *Hydrocotyle asiatica* (Eng.–Asiatic pennywort; Verna.–**Brahmi**)–A creeping herb, used as an antidote against cholera and also to cure madness. The plants are used as remedy of certain forms of leprosy.

11. *Ferula assafoetida;* (Verna.–**Hing**)–A perennial herb, commonly grown in the Punjab and Kashmir. The gum-resin obtained from the roots, is used in perfumery and for flavouring food products. It is also used in medicines in the treatment of asthma, cough and indigestion.

12. *Ferula galbaniflua;* (Eng.–Galbanum; Verna.–**Gandhabiroza**)–Found in North Western India. The oleogum resin, obtained from the stem, used in perfumery, and is also used in the treatment of chronic bronchites and ashthma.

13. *Ferula jaeschkena*-Grows in Kashmir. The roots and seeds produce an essential oil.

14. *Ferula narthex*; (Verna.–**Hing**)–Commonly found in Kashmir. It produces an oleo-gum resin, which is used for flavouring food products.

15. *Anethum graveolens*; Syn. *Peucedanum graveolens*; (Verna.–**Soya**)–Native of Eurasia, now cultivated on a small scale in Jammu and Kashmir. The fruits are used as spice and condiment and are also used as carminative.

16. *Anethum sowa*; (Verna.–**Sowa**)–Cultivated all over India. The leaves are used for flavouring and the fruits are used as carminative, stomachic and stimulant.

17. *Pastinaca sativa*; Syn. *Peucedanum sativum;* (Verna–**Gujur**)–A herb, native of Europe. The roots are eaten as vegetable.

18. *Seseli indicum*; (Verna–**Banajowan**)–A herb, commonly found in Uttar Pradesh, Bengal and Assam. The fruits are used as stimulant, stomachic, and are also used in expelling roundworms.

19. *Foeniculum vulgare*; (Verna-**Saunf**)–An aromatic herb, native of the Mediterranean region, now cultivated mainly in the Punjab, Assam, Uttar Pradesh and Maharashtra. The fruits are used as spice and condiment. Also used medicinally as carminative and stimulant. The residue, left after the distillation of an essential oil, makes a valuable fodder.

Systematic Relationships: Engler has put three families, *i.e.*, Araliaceae, Umbelliferae and Cornaceae in the order Umbelliflorae. Bentham and Hooker also placed these three families in the Umbellales. Rendle has also adopted the same arrangement. Hallier recognized only two families, *i.e.*, Cornaceae and the Umbelliferae. He merged the Araliaceae with Umbelliferae. Hutchinson (1948) has included the single family Umbelliferae in the order Umbellales. He has treated the Araliaceae as a separate order the Araliales, derived from the Cornaceae. According to Rickett (1945) there are two orders the Umbellales and the Cornales. The Umbellales consists of the Araliaceae and Umbelliferae and the Cornales consists of the Cornaceae and the Nyssaceae.

49

ORDER–ASTERALES

According to Hutchinson this is the seventy sixth order of the phylum Angiospermae, subphylum Dicotyledones and divsion Herbaceae. The order consists of a single family, *i.e.*, Compositae, which has been described in the present text in detail. Bentham and Hooker have placed the family Compositae along with three others in their second order–the Asterales of class Dicotyledones, division Gamopetalae and series Inferae. Engler and Prantl, however, have placed this family along with five other families in their 10th order the Campanulatae of class Dicotyledoneae and subclass Sympetalae. They have treated the family to be highest evolved among Angiosperms.

Ovary inferior, 1-locular, 1-ovulate; anthers united into a tube. Herbaceous or woody; flowers collected into heads, surrounded by an involucre of bracts.

Family–Compositae (Asteraceae)

The Compositae is the largest family of the angiosperms, containing about 950 genera and probably 20,000 species.

Distribution: The members belonging to this family are found everywhere on the surface of the earth (cosmopolitan) in each possible type of habitat. Most of them are herbaceous but from tropical regions the trees have also been reported. In our country the family is represented by many genera, *e.g., Helianthus, Tagetes, Xanthium, Zinnia, Ageratum, Chrysanthemum, Cosmos, Vernonia, Carthamus* and many others.

Habit: The plants may be herbs, shrubs and very rarely trees and climbers. Many plants possess a milky juice while in the other members the juice is watery, resinous and bitter.

Root: Usually tap and branched. Sometimes the roots are tuberously thickened (*e.g.*, in *Helianthus tuberosus*).

Leaves: The leaves are usually alternate, sometimes, opposite and rarely whorled; simple or compound; sometimes acicular (needle-like) or reduced to scales in some xerophytes; exstipulate; petiolate; surface glabrous or hairy; margin entire or serrated; apex acute or obtuse; net veined (reticulate venation.)

Inflorescence: Racemose; head or capitulum with an involucre of bracts; rarely in spikes.

Flower: The flowers (florets) of a head may be hermaphrodite (bisexual) or unisexual (monoecious or dioecious) or neuter (asexual); usually pentamerous, actinomorphic or zygomorphic; epigynous.

There are two kinds of florets: (i) Disc florets (tubular flowers), (ii) Ray florets (ligulate flowers).

The arrangement of the florets on the head is as follows:

(a) All the ray and disc florets on a single head may be tubular and actinomorphic e.*g*., *Ageratum, Vernonia* etc.

(b) All the ray and disc florets in the head may be ligulate and zygomorphic, *e.g*., *Launea,Sonchus,* etc.

(c) Disc florets (central flowers) may be actionmorphic and tubular while the ray florets (outer or marginal flowers) may be zygomorphic and ligulate or bilabiate.

When all the flowers in a head are of same type either they may be tubular or ligulate, they are known as homogamous. When the ray and disc florets are unidentical (disc florets tubular and ray florets ligulate), they are described as heterogamous flowers.

DISC FLORET: Sessile, regular, actinomorphic, complete, hermaphrodite, bracteate, epigynous.

Calyx: Either very rudimentary or entirely absent. Sometimes the calyx is modified into a large number of bristles or hair like structures called the *pappus*. However, in most of the cases the calyx is represented by the pappus. The pappus is persistent and acts like a parachute in the dispersal of fruits. When the bristles are few in number, they are modified into barbs or spines which are persistent and may adhere to the body of animals and are disseminated. Sometimes there is an epigynous five-lobed rim-like structure.

Corolla: Petals five, gamopetalous (united), tubular 5-lobed, actinomorphic, swollen near the base of style where nectary is present, valvate aestivation, variously coloured.

Androecium: Stamens five, epipetalous, alternate with corolla lobes. The anthers are introrse, two celled, united laterally into a tube (syngenesious), dehiscing by longitudinal slits. Filaments free, connectives generally prolonged above the anthers.

Gynoecium: Carpels two (bicarpellary), syncarpous; ovary inferior, unilocular, containing a single anatropous basal ovule; basal placentation, style single; stigmas two, curved, the upper surface receptive.

Fruit: Cypsela (achenial single seeded fruit).

Seeds: Exalbuminous.

Floral Fomula: $\oplus$ ⚥ K pappus, $\overbrace{C\ (5), A5}$, $G_{\overline{(2)}}$.

RAY FLORET: Sessile, zygomorphic, irregular, ligulate or bracteate.

Calyx: Either represented by pappus or absent.

Corolla: Petals five, gamopetalous, the corolla may be bilabiate or ligulate.

Androecium: Absent.

Gynoecium: As in disc floret.

Fruit: As in disc floret.

Seed: As in disc floret.

Floral Formula: ·|· ⚥ K Pappus, C (5), A 0, $G_{\overline{(2)}}$

Bentham and Hooker(1862)	**Engler and Prantl(1931)**	**Hutchinson(1959)**
Phanerogams	Phanerogams	Angiospermae
Dicotyledones	Dicotyledoneae	Dicotyledones
Gamopetalae	Sympetalae	Herbaceae
Inferae	Campanulales	Asterales
Asterales	Compositae	Compositae or Asteraceae
Compositae		

Special features of the family : The special features that characterise Compositae (Asteraceae) as an advanced family are as follows:

1. They are predominantly herbs.

2. The flowers are massed together in a head or capitutum inflorence, *i.e.,*a perfect type of inflorescence. The advantages of the head inflorescence are as follows :

(i) There is great conspicuousness to attract insects for cross pollination.

(ii) There is considerable saving of corolla material.

(iii) The cross pollination takes place by a single insect, within a very short time.

3. There is very simple but effective type of floral mechanism to achieve cross pollination without at the same time losing a chance for self-pollination, if the former method fails.

4. There is easy access of insects to the nectary which lies at the base of the style and is protected from rain.

5. The flowers are typically 5-merous, gamopetalous, bisexual, epigynous with definite number of stamens, *i.e.,* 5 and carpels, *i.e.,* (2), an almost similar type of floral construction throughout the family.

6. There is efficient protection of floral buds by the involucre of bracts.

7. There is very effective mechanism of seed (fruit) dispersal by parachute like pappus (calyx) or by hooks or glands developing on the fruit.

A few important plants have been described here in semi-technical language:

1. ***Helianthus annuus*** Linn.; Verna. **Surajmukhi;** Eng. Sunflower.

Habit: An annual large cultivated herb.

Root: Tap and branched.

Stem: Erect, herbaceous, cylindrical, somewhat hairy.

Leaf: Simple, alternate, petiolate, exstipulate, ramal and cauline, persistent, large, ovate, serrate, acute, unicostate reticulate venation.

Inflorescence: Racemose, head or capitulum, consisting of ray and disc florets, involucre 3-seriate.

Ray-floret: Ebracteate, sessile, ligulate, zygomorphic, unisexual, incomplete, pistillate or neuter, epigynous.

Calyx: Either absent or represented by 2 or 3 scales superior.

Corolla: 2-5, gamopetalous, ligulate, superior, yellow.

Androecium: Absent.

Gynoecium: 2, bicarpellay, sycarpous, ovary inferior, unilocular, single ovule, basal placentation, style single, stigma bifid.

Disc floret: Bracteate, sessile, actinomorphic, hermaphrodite, complete, epigynous.

Calyx: As in ray floret.

Corolla: 5, syngenesious, epipetalous, alternate to petals, filaments free, anthers bicelled, introrse, superior.

Androecium: 5, syngenesious, epipetalous, alternate to petals, filaments free, anthers bicelled, introrse, superior.

Gynoecium: 2, bicarpellary, syncarpous, ovary inferior, unilocular, single basal ovule, basal placentation, style single, stigma bifid.

Fruit: Cypsela

Floral formula: **Ray Floret:** ⊣· ♀ K 0(2–5 scales), C (5), A 0, $G\,\overline{(2)}$

Disc Floret: Br ⊕ ⚥ K 0(2–5 scales), $\overset{\frown}{C\,(5),\,A\,(5)}$, $G\,\overline{(2)}$

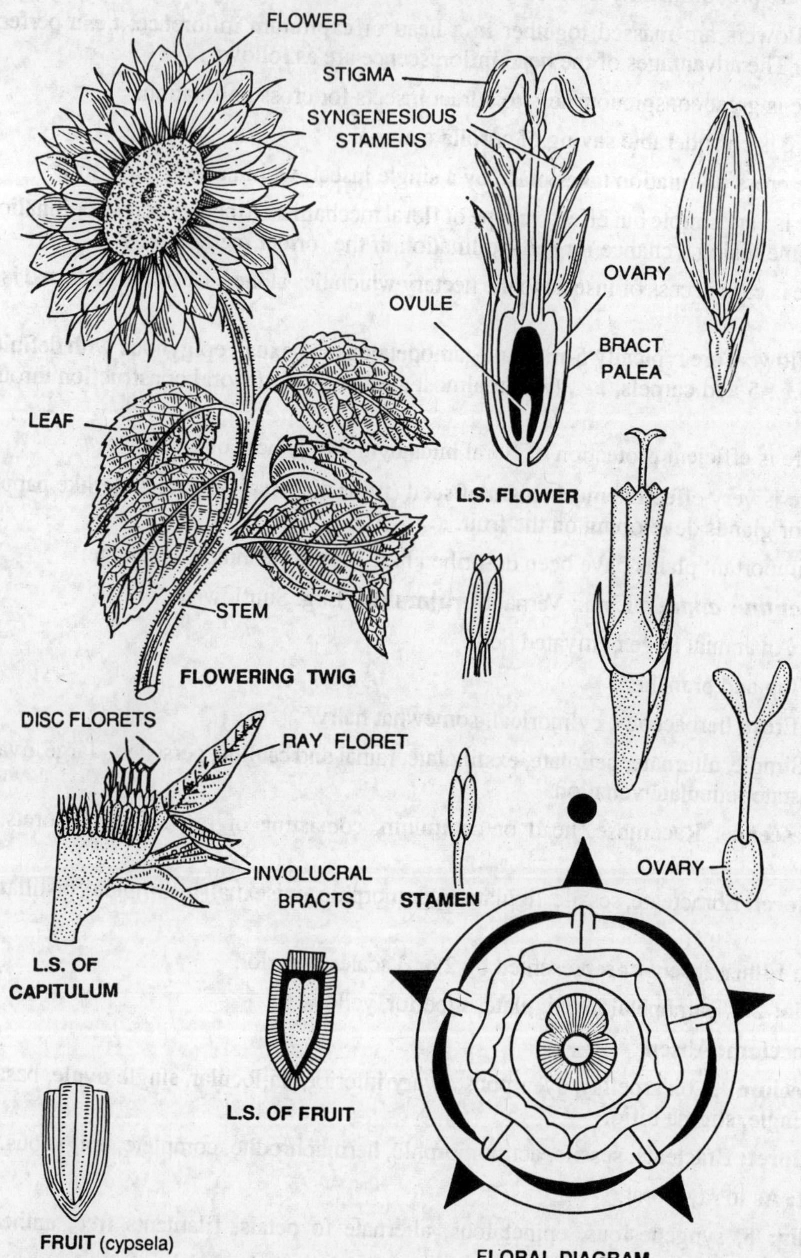

Fig.49.1. Compositae (Asteraceae). *Helianthus annuus* Linn.; Eng., Sunflower; Verna. **Surajmukhi.**

Economic value. The seeds are the source of an oil, which is used in foods, salads and for burning purposes. It can be used in place of groundnut oil in the preparation of *Vanaspati*. It can be safely used by heart patients. The oil does not raise the cholesterol level in the blood. It contains proteins, vitamins A, D and E. The oil is easily digested. Seeds are diuretic, expectorant, used in bronchial, lyrangeal, pulmonary affections, coughs and colds, and in scorpion sting.

2. ***Vernonia*** sp.

Habit: A perennial herb.

Root: Tap and branched.

Stem: Erect, branched, herbaceous, cylindrical, hairy.

Leaf: Simple, alternate, short petioled, ovate-lanceolate, apex acute, unicostate reticulate venation.

Inflorescence: Racemose, capitula arranged in panicled corymbs or simple corymbs.

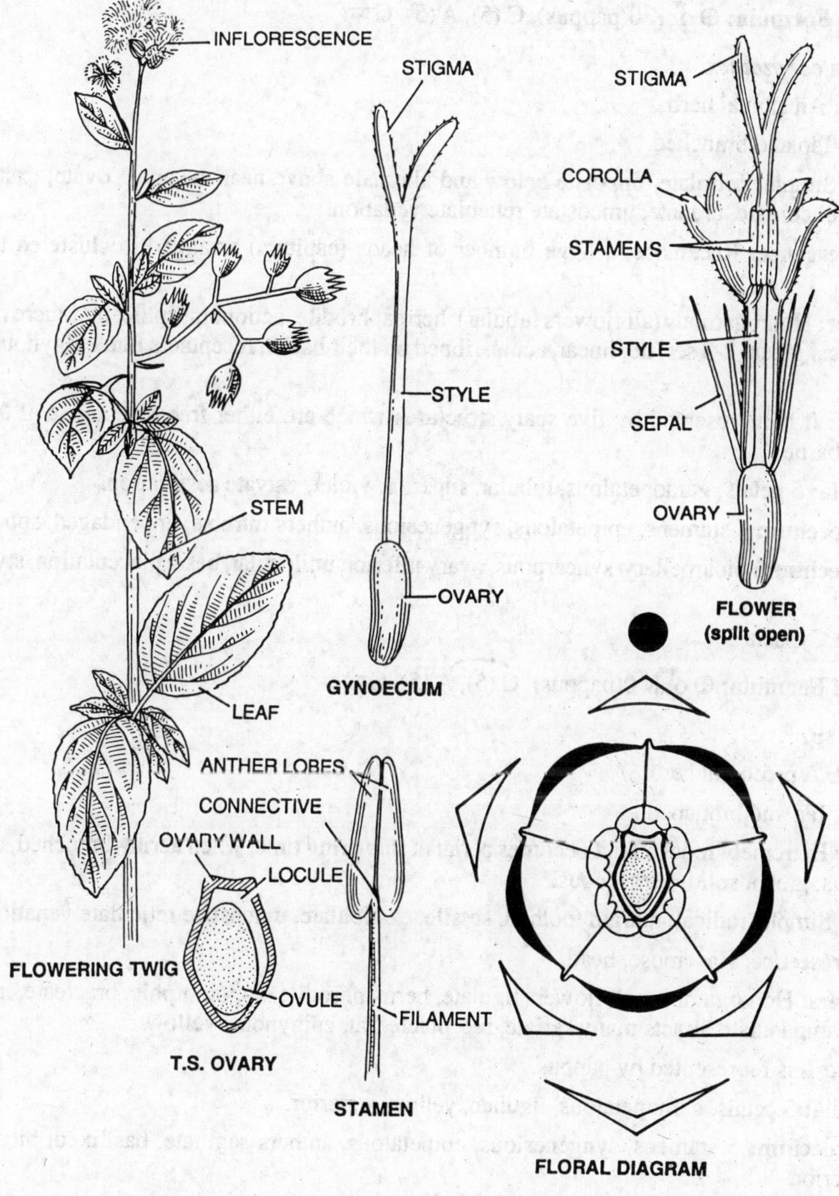

Fig.49.2. Compositae. (Asteraceae). *Ageratum conyzoides* Linn; Verna. **Uchunti.**

Flowers: Homogamous, all flowers tubular, hermaphrodite, actinomorphic, bracts many seriate, involucre of bracts present, receptacle slightly hairy or naked.

Calyx: It is represented by pappus, superior.

Corolla: 5 petals, gamopetalous, tubular, pink or reddish, superior, valvate aestivation.

Androecium: 5 stamens, syngenesious, epipetalous, superior, anther base obtuse, two celled, introrse.

Gynoecium: 2, bicarpellary, syncarpous, ovary inferior, unilocular, basal placentation.

Fruit: Cypsela with persistent parachute like hairy pappus.

Floral Formula: ⊕ ⚥ K 0(pappus), $\overset{\frown}{C\ (5),\ A}$ (5), $G_{\overline{(2)}}$.

3. ***Ageratum conyzoides.***

Habit: An annual herb.

Root: Tap and branched.

Leaf: Simple, petiolate, opposite below and alternate above near the apex, ovate, exstipulate, cuneate to subcordate, crenate, unicostate reticulate venation.

Inflorescence: Racemose, a large number of heads (capitula) arranged in clustered terminal corymbs.

Flower: Homogamous (all flowers tubular), hermaphrodite, actionomorphic; involucre of bracts present, bracts green, 2-3 seriate, linear, acute, ribbed on their backs; receptacle flat; epigynous, violet in colour.

Calyx: It is represented by five scaly structures which are either free or connate at the base, awned and barbed.

Corolla: 5 petals, gamopetalous, tubular, superior, violet, valvate aestivation.

Androecium: 5 stamens, epipetalous, syngenesious, anthers introrse, appendaged, superior.

Gynoecium:2,bicarpellary, syncarpous, ovary inferior, unilocular, basal placentation, style long, stigma bifid.

Fruit: Cypsela

Floral Formula: ⊕ ⚥ K 0(pappus), $\overset{\frown}{C\ (5),\ A}$ (5), $G_{\overline{(2)}}$

4. ***Launea* sp.**

Habit: A perennial herb.

Root: Tap and branched.

Stem: Extremely reduced but becomes aerial at flowering time, when aerial–branched, cylindrical, glabrous, green, solid, herbaceous.

Leaf: Simple, radical, margin toothed, sessile, exstipulate, unicostate reticulate venation.

Inflorescence: Racemose, head.

Flowers: Homogamous, all flowers ligulate, hermaphrodite, zygomorphic, bracteate, involucre of bracts campanulate, bracts many seriate, receptacle flat, epigynous, yellow.

Clayx: It is represented by pappus.

Corolla: 5 petals, gamopetalous, ligulate, yellow, superior.

Androecium: 5 stamens, syngenesious, epipetalous, anthers sagittate, basifixed, bicelled, introrse, superior.

Gynoecium: 2, bicarpellary, syncarpous, ovary inferior, unilocular, basal placentation.

Fruit: Cypsela.

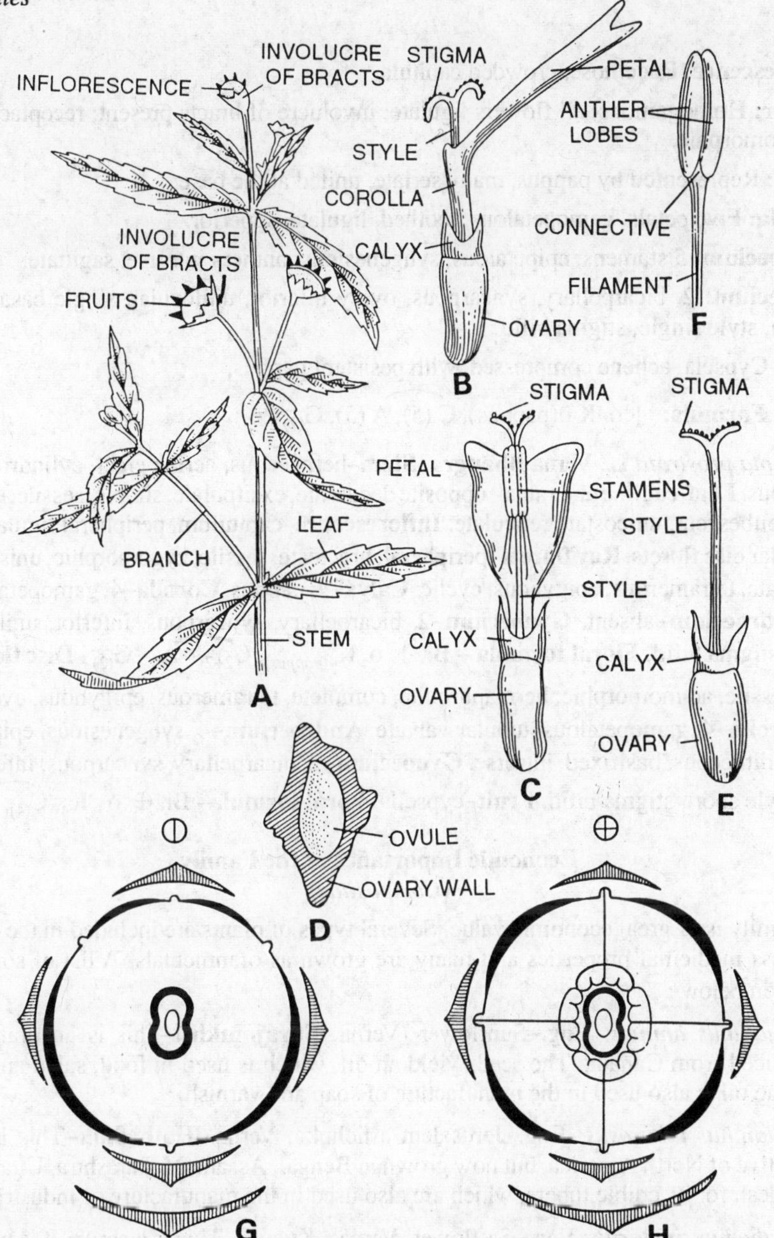

Fig.49.3 : Compositae(Asteraceae). *Eclipta prostrata* L. (**Bhangra**). A, flowering twig; B, ray floret; C, disc floret; D, T.S. of ovary; E, gynoecium; F, stamen; G, floral diagram; of ray florat; H, floral diagram of disc floret.

Floral Formula: ·|· ⚥ K 0(pappus), $\overparen{C\ (5),\ A}$ (5), $G_{\overline{(2)}}$

5. ***Sonchus* sp.**

Habit: An annual herb, 2-3 feet high.

Root: Tap and branched.

Stem: Erect, branched, herbaceous, fistular, angular, latex present.

Leaf: Simple, radical, cauline, entire, semi-amplexicaul, acute, unicostate reticulate venation, smooth, toothed.

Inflorescence: Racemose, crowded capitula.

Flower: Homogamous, all flowers ligulate; involucre of bracts present; receptacle flat, naked; yellow, zygomorphic.

Clayx: Represented by pappus, many seriate, united at the base.

Corolla: Five petals, gamopetalous, toothed, ligulate, superior.

Androecium: 5 stamens, epipetalous, syngenesious; anthers introrse, sagittate.

Gynoecium: 2, bicarpellary, syncarpous, ovary inferior, unilocular, single basal ovule, basal placentation, style single, stigma bifid.

Fruit: Cypsela, achene compressed, with pesistent pappus.

Floral Formula: ·|· ⚥ K 0(pappus), $\overparen{C\,(5), A\,(5)}$, $G_{\overline{(2)}}$

6. *Eclipta prostrata* L., Verna **Bhangra. Stem**–herbaceous, aerial, erect, cylindrical, branched, solid, glabrous. **Leaf**-cauline and ramal, opposite decussate, exsitpulate, simple, sessile, ellipticlanceolate, acute, pubescent, unicostate reticulate. **Inflorescence**–capitulum, peripheral ligulate ray florets, central tubular disc florets. **Ray florets**–peripheral, bracteate, sessile, zygomorphic, unisexual, incomplete, pistillate, tetramerous, epigynous, cyclic. **Calyx**–4, pappus, **Corolla**–4, gamopetalous, ligulate, valvate. **Androecium**–absent. **Gynoecium**–2, bicarpellary, syncarpous, inferior, unilocular, basal, style short, stigma bifid. **Floral formula** – Br. ·|· ♀ $K_{4(pappus)}$, C_{2+2}, A_0 $G_{\overline{(2)}}$. **Disc florets**–central, bracteate, sessile, actinomorphic, hermaphrodite, complete, tetramerous, epigynous, cyclic. **Calyx**–4 pappus. **Corolla**–4, gamopetalous, tubular valvate. **Androecium**–4, syngenesious, epipetalous, filaments long, dithecous, basifixed, introrse. **Gynoecium**–2, bicarpellary, syncarpous, inferior, unilocular, basal, style short, stigma bifid. **Fruit**–cypsella. **Foral formula**– Br. ⊕ ⚥ K_4, $\overparen{C_{(4)}, A_4}$, $G_{\overline{(2)}}$.

Economic Importance of the Family
(*Compositae*)

The family is of great economic value. Several types of plants are included in the family. Many plants possess medicinal properties and many are grown as oranmentals. A list of some important plants is given below :

1. *Helianthus annuus;* Eng.–Sunflower (Verna.–**Surajmukhi**)–This is an oranmental large herb, introduced from Canada. The seeds yield an oil, which is used in food, salad and for burning purposes. The oil is also used in the manufacture of soap and varnish.

2. *Helianthus tuberosus;* Eng.–Jerusalem artichoke; Verna.–**Hathichuk**–This is a perennial herb. It is native of North America, but now grown in Bengal, Assam, Maharashtra, Uttar Pradesh and Andhra Pradesh for its edible tubers, which are also used in the manufacture of industrial alcohol.

3. *Carthamus tinctorius*; Eng.–Safflower; Verna.–**Kusum**–This is a shrub. It is native of India and chiefly cultivated in Madhya Pradesh, Andhra Pradesh, Karnataka and Maharashtra. The flowers are used as dye for dyeing food and cloth. The seeds yield an edibe oil, which is also used for making soaps, varnishes, paints and burning purposes. As a result of hydrogenation and bleaching the refined edible oil is produced.

4. *Carthanus oxycanthus*; Eng.–Wild safflower; Verna.–**Kantiari**–This is a thorny herb, found in Uttar Pradesh and the Punjab. The seed oil is used for burning and edible purposes.

5. *Artemisia absinthium;* Verna.–**Vilaiti afsantin**–This is a herb. The flowers yield a drug santonin, which is used as stimulant and tonic. It is found in Kashmir.

6. *Artemisia dracunculus*– This is used as a condiment. It is found in the Western Himalayas.

7. *Artemisia maritima*; Verna.–**Kirmala.**–The flowering tops are used as an anthelmintic. The flowers yield a drug santonin. It is found in the Western Himalayas.

8 *Arctium lappa*– This is a pereninal herb, found in the Western Himalayas. The roots are used medicinally as diuretic and diaphoretic.

9. *Cosmos sulphurens;* Eng.–Yellow cosmos–This is a herb. It is native of tropical America. It is cultivated in gardens for its beautiful flowers. The flowers yield an orange red dye.

10. *Lactuca sativa,* Eng.–Lettuce; Verna.–**Salad**–This is a herb. It is native of South Europe and West Asia, but now cultivated throughout India for its edible leaves.

11. *Lactuca serriola*; Syn. *L. scariola*; Eng.–Wild lettuce; Verna.–**Kahu**–This is found in the Western Himalayas from Kumaon to Kashmir. The seeds yield a semi–drying oil; which is used in food.

12. *Cichorium endivia;* Eng.–Endive;Verna.–**Kasmi**–This is a herb. It is native of India. It is cultivated mainly in Northern India. The leaves are eaten as vegetable.

13. *Cichorium intybus;* Eng.–Chicory; Verna.–**Kasmi**–It is native of Europe but commonly found in the Punjab and Andhra Pradesh. The leaves are used as vegetable.

14. *Chrysanthemum cinerarifolium*–This is a herb, grown, in Kashmir and Nilgiris. The dried leaves are used as an insecticide.

15. *Chrysanthemum indicum;* Syn. *C. japonicum;* Verna.–**Guldaudi**–This is grown as an ornamental. It is native of China and Japan.

17. *Tagetes erecta;* Eng.–Big marigold; Verna.–**Genda**–This is a herb. It is native of Mexico. It is grown in gardens. The flowers are the source of yellow dye, which is used to dye cotton.

18. *Tagetes paluta*; Eng.–French marigold; Verna.–**Genda**–It is grown in the gardens as an ornamental.

19. *Solidago canadensis*; Eng.–Golden rod–This is a shrub, grown as an ornamental and hedge plants. It is native of North America.

20. *Vernonia anthelmintica*; Verna–**Kaliziri**–This is a common herb. The seeds are used medicinally for leucoderma and other skin diseases, and also used as an anthelmintic.

21. *Callistephus chinensis*; Syn. *Aster chinensis*; Eng.–China Aster–This is an ornamental herb.

22. *Centaurea cyanus;* Eng.–Corn flower–This is an ornamental herb, grown in the gardens.

23. *Centaurea moschata*; Eng.–Sweet sultan–This is an ornamental herb, grown in the gardens.

24. *Calendula officinalis*; Eng.–Pot marigold; Verna.–**Zargul**–This is an ornamental herb, grown for its beautiful flowers in the gardens.

25. *Gaillardia pulchella*; Eng.–Blanket flower–This is grown as an ornamental herb.

26. *Dahlia* sp. They are perennial herbs, grown as an ornamental for their large sized variously coloured beautiful flowers.

27. *Zinnia elegans*– This is a herb. It is grown as an ornamental. It is native of Mexico.

28. *Centipeda minima*; Syn.*C.orbicularis;* Verna.–**Nakch-hikni**–This is a common herb. A snuff obtained from the leaves and seeds, is used for colds.

29. *Blumea balsamifera;* Verna.–**Kakronda**–This is a small plant, found in the subtropical Himalayas, Assam and the Khasia hills. The leaves are aromatic and have been thought of as a possible source of camphor.

30. *Spilanthes acmella;* Verna.–**Pipulka**–This is a herb, found in the temperate Himalayas. The leaves are used as a remedy of toothache. They are also used as mosquito larvicide.

31. *Sphaeranthus indicus*; Verna.–**Mundi**–This possesses medicinal properties. The powder of roots is given in stomachache; the powder of bark is used as a remedy for piles.

32. *Pluchea lanceolata*; Verna.–**Marmandai**–This is a small shrub, found commonly in the Punjab and Uttar Pradesh. It is used as fodder.

33. *Taraxacum officinale;* Verna.–**Kanphul**–The plant possesses medicinal properties. Roots and leaf decoction is used in the disorder of bowels; kidney and liver.

34. *Saussurea lappa;* Eng.–Costus; Verna–**Kuth**–This is a herb. An essential oil, obtained from the roots is used in perfumery. The roots are used medicinally as tonic, carminative, stimulant and in controlling bronchial asthma. The roots are also used as preservative for woollen clothes.

35. *Silybum marianum*– This is an ornamental herb, grown in the gardens.

Systematic Relationships: Bentham and Hooker have placed four families, *i.e.,* Valerianeae, Dipsaceae, Calycereae and Compositae in the order Asterales, Engler and Rendle have placed the Compositae in the Campanulales while the Dipsaceae and Valerianaceae are included in the Rubiales. Many authors have used the name Asteraceae instead of Compositae. Some taxonomists are of the opinion that the family Compositae may be of a polyphyletic origin. All botanists are in agreement that the Compositae is most advanced family among the Dicotyledons.

50

ORDER–SOLANALES

According to Hutchinson this is the seventy-seventh order of the phylum Angiospermae, subphylum Dicotyledones and divison Herbaceae. This order comprises of three families, *i.e,* Solanaceae, Convolvulaceae, and Nolanaceae, of which first two (*i.e.,*Solanaceae and Convolvulaceae) have been described here in detail. Bentham and Hooker have placed these families along with three others in their 8th order–the Polemoniales of class Dicotyledones, division Gamopetalae and series Bicarpellatae. Engler and Prantl have included these families along with eighteen others including Verbenaceae, Labiatae, Scrophulariaceae, Pedaliaceae, Acanthaceae, etc., in the sixth order the Tubiflorae of class Dicotyledoneae and subclass Sympetalae.

Corolla actinomorphic to very slightly zygomorphic; stamens the same number and alternate with the corolla-lobes, epipetalous; ovary superior, 4-1-(mostly 2)-locular with numerous ovules on axile placentas, or rarely basal. Herbaceous to rarely woody often climbing.

Family–Solanaceae

There are about 85 genera and 2200 species in this family.

Distribution: The members of this family are primarily distributed in tropical America and South America, but they are also well represented in temperate regions. The largest genus of the family is *Solanum* which includes about 1500 species. The genera *Lycium, Physalis* and *Nicotiana* each containing 100 species.

Habit: The species belonging to this family are either erect or climbing herbs, shrubs or small trees. *Solanum nigrum* is a small annual herb; species of *Datura* are large annual herbs. *Cestrum nocturnum* is a shrub.

Root: Tap or adventitious.

Leaves: Alternate or becoming opposite at or near the inflorescences, rarely whorled, exstipulate, entire, simple, rarely dissected or pinnate.

Inflorescence: Cymose usually a typical axillary cyme or combination of cymes, rarely helicoid.

Flowers: Usually pedicellate, hermaphrodite, actinomorphic, regular (sometimes slightly zygomorphic), hypogynous complete, bracts and bracteoles absent.

Calyx: 5 sepals, gamosepalous (limb usually 5-lobed or-toothed), persistent and often much enlarged in fruit.

Corolla: 5 petals, gamopetalous, funnel shaped, rotate to tubular, corolla of various shapes but rarely 2-lipped, usually plicate or convolute (rarely valvate).

Androecium: 5 stamens, sometimes 4 and didynamous or not so, or only 2, *e.g., Schizanthus;*

when the stamens are less than five the lost stamen is usually represented by a reduced (staminode). Stamens epipetalous and alternate, the petals, usually not equal in size. Anthers are 2-celled (dithecous), introrse, dehiscence by longitudinal slit or by apical pore, *e.g.*, in *Solanum*, the connective sometimes enlarged, usually hypogynous disc present which is quite apparent.

Gynoecium: 2 carpels (bicarpellary), syncarpous, ovary superior, bilocular (sometimes becomes tetralocular due to false septation), rarely 3-5 celled ovary, placentation axile, ovules usually many on prominently peltate plcentas, anatropous or somewhat amphitropous. The carpels placed obliquely in the flowers. Single style, stigma bi-lobed and capitate.

Fruit: A berry or septicidal capsule.

Seed: Albuminous (endospermic) smooth or pitted, curved embryo embedded in the endosperm.

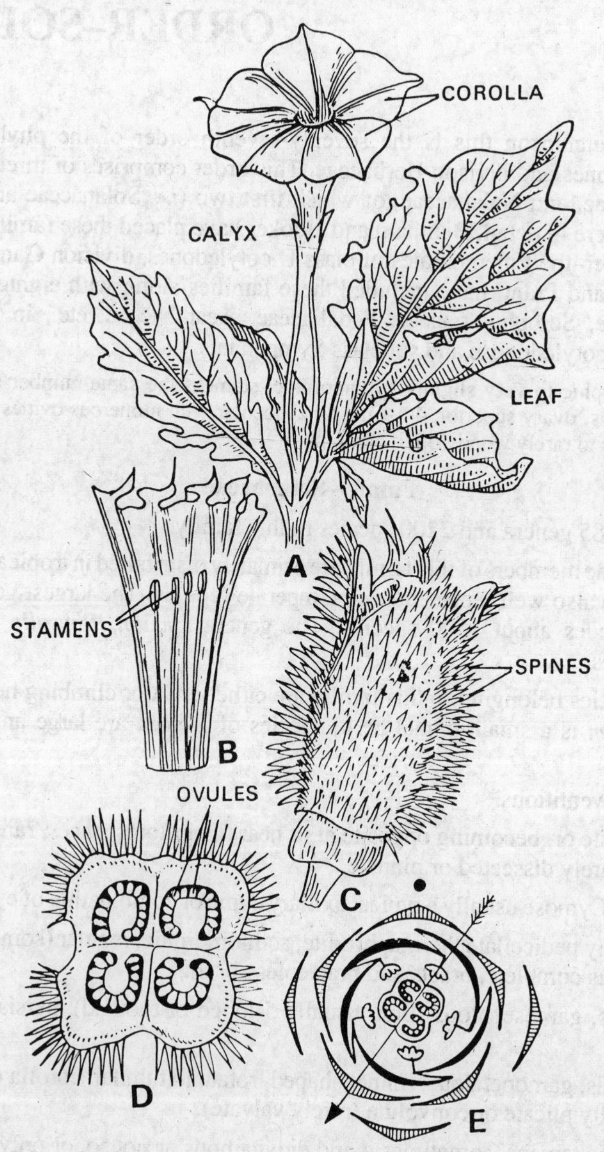

Fig. 50.1 : Solanaceae. *Datura stramonium* (Verna–**Dhatura**); A, a flowering twig; B, corolla spread out with epipetalous stamens; C, opening fruit (capsule); D, cross-section of capsule; E, floral diagram.

Pollination: (Entomophilous) through the agency of insects.

Floral Formula: ⊕ ⚥ K 5, $\widehat{C\ (5), A}$ 5, $\underline{G}_{(2)}$

Bentham & Hooker (1862)	Engler & Prantl (1931)	Hutchinson (1959)
Phanerogams	Phanerogams	Angiospermae
Dicotyledones	Dicotyledoneae	Dicotyledones
Gamopetalae	Sympetalae	Herbaceae
Bicarpellatae	Tubiflorae	Solanales
Polemoniales	Solanaceae	Solanaceae
Solanaceae		

Some important plants have been described here in semi-technical language.

Solanum **sp.**

Habit: Erect or climbing herbs, shrubs or small trees. Most of them are pubescent, some are spiny, and some unarmed.

Leaf: The leaves are alternate, or subopposed in unequal pairs, exstipulate and either lobed or pinnatifid.

Inflorescence: The inflorescence may be cymose, racemose, umbellate or rarely panicle, in most cases extra axillary or opposite the leaves.

Calyx: The calyx is bell-shaped or rotate, 5-lobed or 5-parted, sometimes 10-lobed, rarely 4-lobed or subentire accrescent or not in fruit.

Corolla: The corolla is rotate, rarely campanulate, tube short, limb usually 5-lobed, plicate in bud.

Androecium: The stamens are 5, rarely 4 or 6, inserted on the corolla throat, filaments short, anthers oblong, often narrowed upwards, connivent round the style, dehiscence by apical pores or short slits, introrse.

Gynoecium: The ovary is superior bi-celled and many ovuled (rarely 3-4 celled), style simple, linear, stigma small and terminal.

Fruit: The fruit is globose or elongated berry with persistent calyx.

Seeds: Many, smooth, discoid.

Generic Characters: Corolla rotate or campanulate, lobes plicate in bud; anthers connivent in a cone, dehiscence by apical pores; calyx small, persistent sometimes fleshy.

Nicotiana **sp.**

Habit: Usually herbaceous plants, some species are shrubby or of arborescent type; annual or perennial in habit.

Stem: Usually tall growing, often viscid pubescent, strongly scented with narcotic poisonous properties.

Leaf: Alternate, simple, entire or rarely sinuate, shortly petiolate or sessile, decurrent.

Inflorescence: Either panicle or one sided raceme; sometimes lower flowers solitary axillary.

Flower: Flowers usually large and showy and of different colours (white, yellow, greenish, purple or rosy).

Calyx: Tubular–campanulate and 5–lobed.

Corolla: Funnel or salver-shaped, 5–lobed with a long tube.

Androecium: 5 stamens in number, attached near the base of corolla tube, epipetalous; filaments long.

Gynoecium: Two carpels, syncarpous, superior ovary, bilocular, slender style with capitate stigma.

Fruit: 2-celled capsule splitting at maturity into 2 to 4 valves.

Seeds: Many and small.

Generic Characters: Fruit, a capsule; capsule splitting lengthwise; calyx nearly or quite covering the fruit; fruit not spiny.

Solanum nigrum; Verna.–**Makoi.**

Habit: An annual small wild herb.

Root: Tap and branched.

Stem: Erect, branched, herbaceous, cylindrical, solid, glabrous or somewhat hairy.

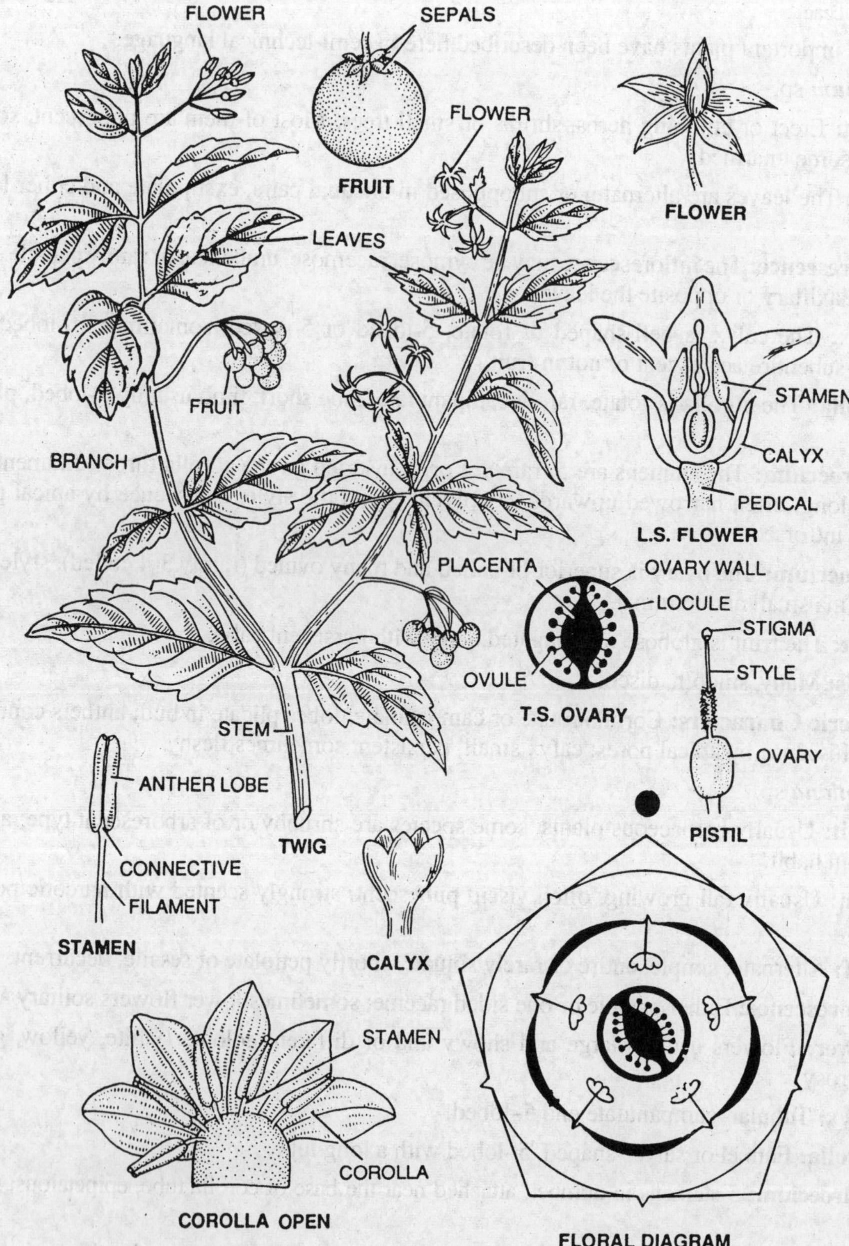

Fig. 50.2 Solanaceae. *Solanum nigrum* Linn. Eng.black nightshade; (Verna. **Makoi**)

Leaf: Simple, alternate, cauline, petiolate or sub-sessile, exstipulate, ovate, dentate, acute, unicostate reticulate venation.

Inflorescence: Cymose, extra-axillary subumbellate cyme.

Flower: Pedicellate, ebracteate, hermaphrodite, actinomorphic, pentamerous except pistil, complete, small, white, hypogynous.

Calyx: 5 sepals, gamosepalous, lobes small, oblong, acute, green, hairy, valvate or imbricate aestivation.

Corolla: 5 petals, gamopetalous, rotate, lobes oblong, acute, white, twisted or valvate aestivation.

Androecium: 5 stamens, alternate to petals, polyandrous, epipetalous, filaments short, anthers oblong, connivent in a cone like structure, bicelled, introrse, yellow, dehiscence by apical pores.

Gynoecium: 2 carpels (bicarpellary), syncarpous, ovary obliquely placed in the flower, ovary superior, globose, bilocular, axile placentation, many shining ovules present on placenta, style simple, hairy at the base, stigma one.

Fruit: Berry: with persistent calyx.

Floral Formula: $\oplus$ ⚥ K 5, $\overset{\frown}{\text{C (5), A}}$ 5, G($\underline{2}$).

Petunia nyctaginifolia Juss

Habit: An annual ornamental herb.

Root: Tap and branched.

Stem: Erect,branched, herbaceous, solid, cylindrical, green and hairy.

Leaf: Simple, cauline, opposite, decussate, sessile, exstipulate, ovate, entire, acute, multicostate reticulate venation.

Inflorescence: Cymose, solitary axillary.

Flower: Pedicellate, pedicel hairy, bracteate, hermaphrodite, actinomorphic, regular, complete, hypogynous.

Calyx: 5 sepals, gamosepalous, free above and connate below, green, hairy, quincuncial or imbricate aestivation.

Corolla: 5 petals, gamopetalous, tubular, twisted or imbricate aestivation.

Androecium: 5 stamens, polyandrous, alternate to petals, epipetalous, filaments long, anthers two celled, basifixed.

Gynoecium: 2 carpels (bicarpellary), syncarpous, ovary superior, obliquely placed in the flower, bilocular, axile placentation, numerous ovules in each loculus, style simple, long, stigma bifid.

Fruit: Capsule

Floral formula: $\oplus$ ⚥ K 5, $\overset{\frown}{\text{C (5), A}}$ 5, G($\underline{2}$).

Withania somnifera Dunal; Verna. **Ashwagandha.**

Habit: A perennial shrub.

Root: Tap and branched.

Stem: Erect, aerial basal part woody, terminal part herbaceous, branched, cylindrical, solid, hairy, green.

Leaf: Cauline and ramal, exstipulate, petiolate, alternate, simple, ovate, entire, acute, glabrous, coriaceous, unicostate reticulate.

Inflorescence: Cymose, axillary umbellate cyme.

Flower: Ebracteate, pedicellate, actinomorphic, regular, hermaphrodite, pentamerous, hypogynous, complete, cyclic.

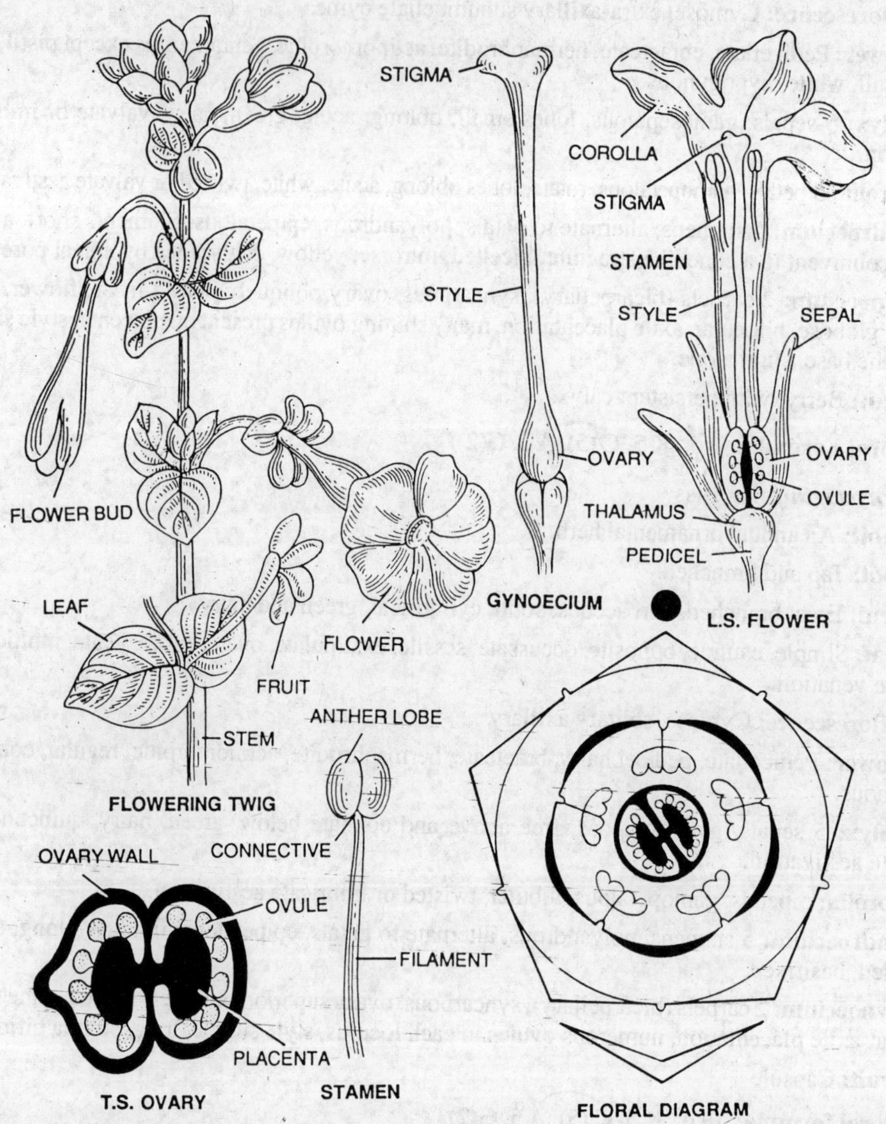

Fig.50.3 : Solanaceae. *Petunia nyctaginifolia* Juss.

Calyx: 5 sepals, gamosepalous, valvate, persistent, hairy, green, inferior.

Corolla: 5 petals, gamopetalous, valvate, campanulate, whitish-yellow, inferior.

Androecium: 5 stamens, polyandrous, epipetalous, anthers connivent, dithecous, basifixed, introrse.

Gynoecium: 2 carpels (bicarpellary), syncarpous, ovary superior, bilocular, many ovules in each locule, axile placentation, ovary obliquely placed in the flower, placenta swollen, style long, stigma capitate.

Fruit: Berry

Floral Formula: ⊕ ⚥ K 5, $\widehat{C(5), A}$ 5, G($\underline{2}$).

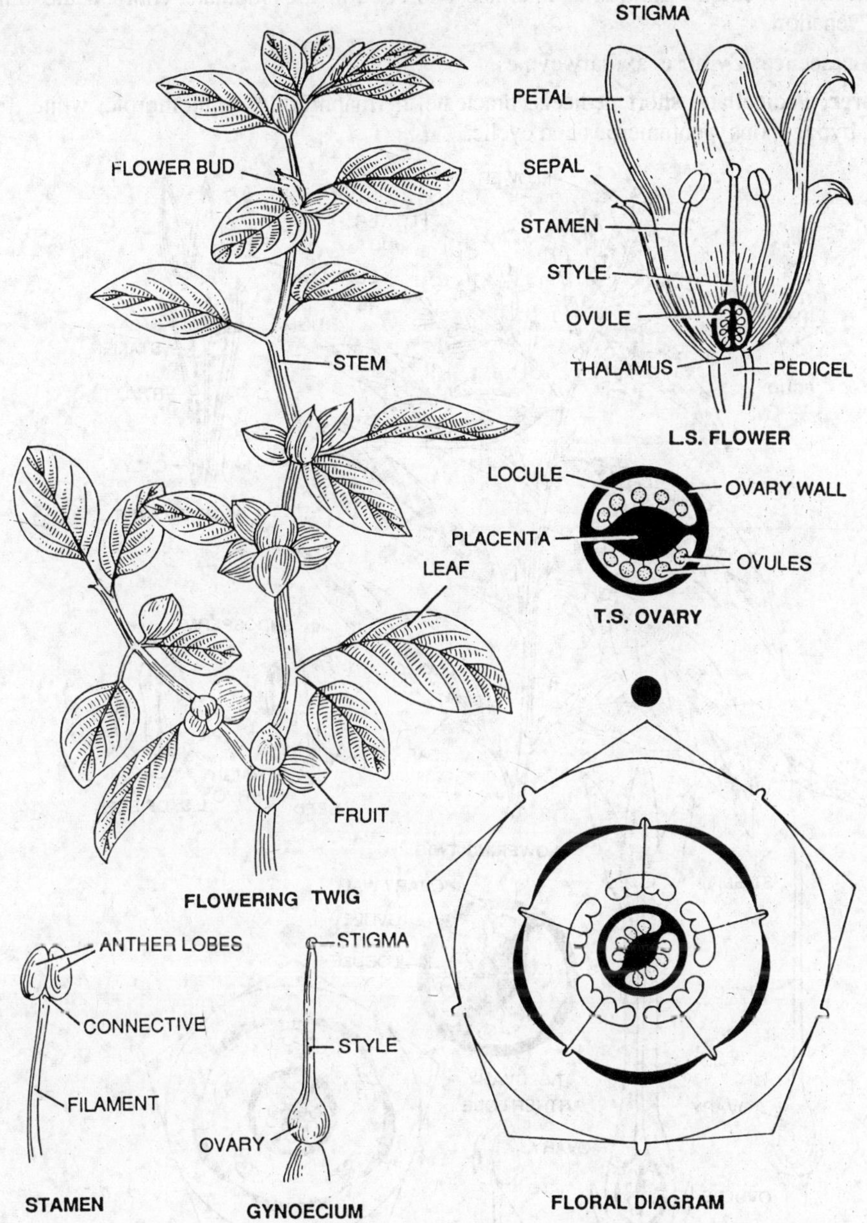

Fig.50.4. Solanaceae. *Withania somnifera* Dunal.; Verna. **Ashwagandha.**

Economic value. The root is considered alterative, aphrodisiac, tonic, deobstruent, diuretic, narcotic, abortifacient; used in rheumatism, consumption, debility from old age, emaciation of children, etc.

Cestrum nocturnum Linn.;Verna.**Rat-ki-Rani;** Eng. Night jessamine.

Habit: An ornamental perennial shrub.

Stem: Erect, branched, woody, solid, glabrous and green.

Leaf: Simple, cauline and ramal, alternate, short petioled, exstipulate, entire, acute, unicostate reticulate venation.

Inflorescence: Cymose, axillary cymes.

Flower: Pedicellate, short pedicels, bracteate, hermaphrodite, actinomorpic, white, fragrant, complete, hypogynous, pentamerous and cyclic.

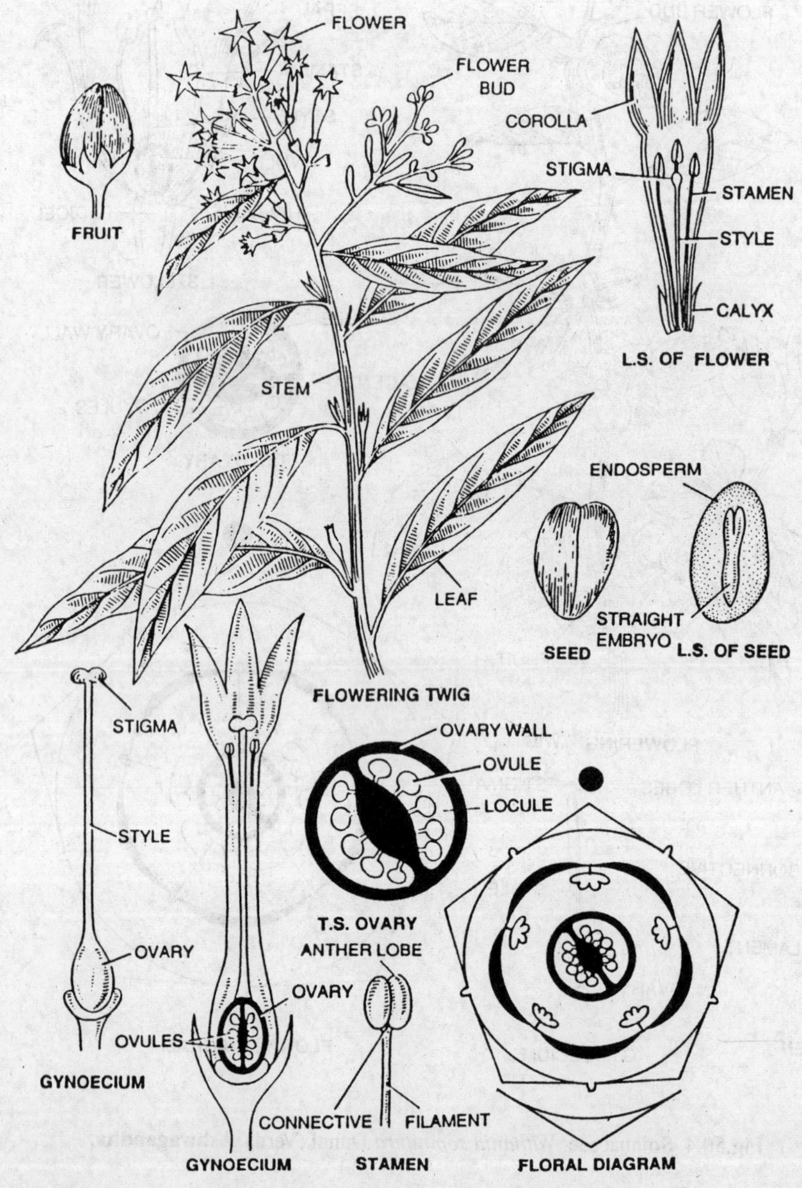

Fig.50.5. Solanaceae. *Cestrum nocturnum* Linn.; Eng., night jessamine; Verna, **Raat-ki-Rani.**

Economic value. Grown for its flowers, which are seented at night.

Calyx: 5 sepals, gamosepalous, green, valvate aestivation, inferior.

Corolla: 5 petals, gamopetalous, tubular, white, fragrant, valvate aestivation, inferior.

Androecium: 5 stamens, polyandrous, epipetalous, alternate to the petals, filaments long, anthers bicelled, introrse.

Gynoecium: 2 carpels (bicarpellary), syncarpous, ovary superior, obliquely placed in the flower, bilocular, axile placentation, many ovules in each loculus.

Floral formula: Br $\oplus$ ⚥ K 5, $\widehat{C\ (5),\ A\ 5}$, G $\underline{(2)}$.

Economic Importance of the Family

Economically the family is fairly important. It comprises of several plants of food value, medicinal, vegetables and ornamentals. A list of few important plants is given here:

1. *Solanum tuberosum*; Eng.–Potato; Verna–**Alu**–The potato is chiefly grown in Assam, Uttar Pradesh, West Bengal, Bihar and the Punjab. The tubers are used as vegetable, and are also the source of starch. It is naive of South America.

2. *Solanum melongena*; Eng.–Brinjal: Verna–**Baigun**–The fruits are eaten as vegetable. It is native of India, and cultivated throughout the country. The seeds are often used as stimulant, and the leaves as a narcotic.

3. *Solanum ferox*; Verna.–**Rambegun**–This is small prickly herb. The fruits are eaten as vegetable. It is found in Assam and South India.

4. *Solanum indicum*; Eng.–Indian night shade; Verna.–**Bari katai**–This is a herb. The fruits are eaten as vegetable. It is found in the sub–Himalayan tract, Assam and the Khasia hills.The roots and the fruits possess several medicinal properties. The fruits are given in colics. The leaves are used in toothache, cough and fevers.

5. *Solanum grandiflorum*; Eng.–Large flowered night shade, potato tree. This is a shrub or small tree. It is native of South America. Grown in gardens as an ornamental for its large bluish, violet flowers.

6. *Solanum jasminoides;* Eng.–Potato vine. This is a climbing shrub. It is native of Brazil. This is a most useful deciduous climber for cool house. It has variegated leaves.

7. *Solanum pseudocapsicum*–This is a shrub grown as an ornamental in the gardens.

8. *Solanum rantonnetii*–This is grown as an ornamental in the gardens for its violet–coloured flowers. It is native of Argentina.

9. *Solanum xanthocarpum*; Verna.–**Kateli**–This is a common prickly herb. The roots are used in cough, asthma and leaves for rheumatism.

10. *Solanum nigrum*; Eng.–Black night shade; Verna.–**Makoi**–It possesses medicinal properties. The fruits are given in fevers; diarrhoea, eye diseases and hydrophobia. The juice of plants is given in chronic enlargements of the liver, in bleeding piles, dysentery, etc. The fruits are edible and very muck liked by the children.

11. *Solanum torvum*; Verna–**Titbaigun**–The fruits are used as vegetable. It is found in Orissa, Bengal, Bihar and Manipur.

12. *Solanum seaforthianum*; Eng.–Potato creeper. It is grown as an ornamental for its blush purple flowers. Native of Brazil.

13. *Cyphomandra betacea*; Syn. *Solanum betaceum*; Eng.–tree tomato–This is a small tree. The fruits are edible. It is native of Peru, but now cultivated in Bengal Assam, Maharashtra, Tamil Nadu, Karnataka and Travancore.

14. *Capsicum frutescens*; Eng.–Red pepper, chillies; Verna.–**Lal Mirch**–This is a herb. The

fruits are extremely pungent and used as a spice and condiment. According to some authors this species is the synonym of *Capsicum annuum*; while others consider the two as separate species. It is native of the West Indies and tropical America, now cultivated throughout our country.

15. *Nicotiana tabacum*; Eng.– Tobacco; Verna–**Tambaku**–The leaves are used for smoking and also contain alkoloids, which are used as insecticides. The oil obtained from the seeds is used for burning purposes, and is also used in the manufacture of paints and varnishes. The oil cake makes a good manure. The snuffs are prepared from tobacco wastes. The tobacco dust is used for making khamira tobacco for smoking and manufacture of nicotine sulphate, an insecticide. It is native of tropical America now cultivated mainly in Andhra Pradesh, Maharashtra, Karnataka, Uttar Pradesh and West Bengal.

16. *Nicotiana rustica*; Eng.– Turkish or East Indian tobacco; Verna.–**Vilaiti–Tambaku**–The leaves are used for smoking and for manufacturing insecticides. This is a herb, native of central America, now cultivated in the Punjab, Bihar and Bengal.

17. *Datura stramonium*; Eng.– Thorn apple; Verna.–**Dhatura**–This is a common weed. The leaves and seeds are narcotic and sometimes used for criminal poisoning. It is used in medicine in the treatment of asthma. It is native of Asia.

18. *Datura innoxia*; Verna.– **Sadahdhatura**–This is a herb. The dried leaves are used in the treatment of asthma. It is native of Mexico, but commonly found in the Western Himalayas and South India.

19. *Datura metel*; Verna.–**Sadahdhatura**–The leaves are used in the treatment of asthma.

20. *Datura suaveolens*; Eng.–Tree datura. This is a small tree. It is grown in gardens for its fragrant creamy–white flowers, This is native of Mexico.

21. *Physalis peruviana*; Syn. *P.edulis*; Eng.–Cape gooseberry; Verna.–**Rasbhari, Tipari, Tankari**–It is cultivated throughout our country for its edible fruits.

22. *Physalis minima*; Eng.–Sunberry; Verna.–**Papotan, Tulatipati**–This is a common herb yielding edible fruits.

23. *Physalis ixocarpa*; Eng.–Tomatillo. The fruits are used as vegetable. They are also used in making chilli sauce and sometimes as a dressing for meat. It is native of Mexico.

24. *Atropa belladona;* Eng.–Belladona; Verna.–**Sag–angur**–The roots possess medicinal properties. They are used as sedative stimulant and antispasmodic. It is native of Europe, but now grown in Kashmir.

25. *Cestrum nocturnum*; Eng.–Night Jessamine; Verna.–**Raat–ki–Rani**–It is grown in the gardens for its fragrant flowers, which are scented at night.

26. *Cestrum diurnum*; Eng.–Day Jessamine–Verna.–**Din–ka–Raja**–This is grown in the garden as an ornamental for its fragrant flowers. It is native of the West–Indies.

27. *Cestrum aurantiacum*–This is a bushy shrub, grown as an ornamental.

28. *Cestrum elegans*–It is grown in the gardens as an ornamental. This is native of Mexico.

29. *Cestrum parqui*; Eng.-Willow—leaved jessamine; Verna.—**Raat—ki—Rani**-This is grown as an ornamental. It is native of Chile.

30. *Petunia axillaris*; Syn. *P. nyctaginflora*; Eng.–Large white Petunia–This is an ornamental herb grown for its beautiful flowers.

31. *Lycopersicon esculentum;* Syn. *Lycopersicum esculentum*; *Solanum lycopersicum*; Eng.– Tomato; Verna.–**Tamatar**–This is cultivated for its edible fruits. This is a common herb. It is native of South America.

32. *Lycopersicon glandulosum*; Eng.–Tomato; The fruits are edible.

33. *Lycopersicon peruvianum;* Eng.–Tomato; Verna–**Tamatar**–It is introduced from Great Britain and grown for its edible fruits.

34. *Lycopersicon pimpinellifolium;* Eng.–Current tomato; Verna.–**Tamatar**–It has been introduced from United Kingdom and grown for the edible fruits.

35. *Lycopersicon pissisi*; Eng.–Tomato–The fruits are edible.

36. *Withania somnifera*; Verna.–**Asgand, punir**–It contains substances with narcotic and sporific properties. This is a undershrub, found in the drier parts of the country.

37. *Withania coagulans*; Eng.–Vegetable rennet; Verna.–**Akri**–The fruits used for coagulating milk. It is found in the Punjab.

Family—Convolvulaceae

There are about 50 genera and more than 1200 species in this family.

Distribution: The members, belonging to this family are, common in the tropical and sub–tropical regions of the world. They are rare in temperate regions. In our country the family is represented by many important genera, *e.g.,Ipomoea, Convolvulus*, *Evolvulus, Cuscuta*, etc. The largest genus *Ipomoea* containing about 400 species.

Habit: Usually twining herbs or shrubs, sometimes erect. rarely trees. The plants may be xerophytes, hydrophytes and sometimes parasites. Usually the species of *Ipomoea* are twining or prostrate herbs, rarely shrubby or erect, the most popular species *Cuscuta reflexa* (**Amar bel**) is leafless twining parasitic herb. The species of *Evolvulus* are small usually silky pubescent, prostrate or erect herbs sometimes undershrubs. *Ipomoea reptans* (**Nari**) is an aquatic herb. The family is very much varied in its habit. The plants usually contain some milky sap. Sometimes when the plants are shrubby they produce thorns.

Root: The root system is quite variable from plant to plant. Sometimes as in *Convolvulus arvensis* the roots are long and slender, where they serve the purpose of vegetative propagation. Sometimes they are thick, fleshy and tuberous as in *Ipomoea batatas* where they store great quantity of food. In *Convolvulus scammonia,* thick tuberous rhizomes are found. In *Cuscuta reflexa* (**Amar bel**), they are represented by small haustoria, through which they absorb their nutrition from the host.

Leaves: Simple, exstipulate, usually petiolate and alternately arranged. On the dodders (*Cuscuta* sp.) they are reduced to scales or altogether absent. Usually entire but sometimes pinnately divided.

Inflorescence: Usually dichasial cyme or solitary axillary. Sometimes it is racemose or panicle as in *Porana*. In *Cuscuta* the flowers are either solitary or in clusters or shortly racemose.

Flower: Bracteate, bracteolate, usually pedicellate with small jointed pedicels, bisexual (rarely by abortion they may be unisexual and the plants then dioecious); actinomorphic, usually showy, hypogynous, usually pentamerous (except gynoecium). The bracts may be in pairs and form an involucre. Sometimes flowers are cleistogamous (*Dichondra*).

Calyx: Five sepals, usually free (polysepalous) but sometimes united at the base. It is persistent and imbricate. The sepals may be considerably variable in shape and size, and may be glabrous or pubescent.

Corolla: The corolla is five–lobed or entire. Usually campanulate, funnel shaped or salver shaped. Rarely the corolla is appendaged, *e.g., Cuscuta*. The aestivation is twisted in bud but in most of the cases it is induplicate valvate, rarely imbricate (*e.g.,* in *Cuscuta*).

Androecium: Stamens 5, free, epipetalous at base of corolla tube, alternate with petals, anthers 2–celled (dithecous), dorsifixed, dehiscence by longitudinal slit, usually introrse, usually intrastaminal disc present (ring or cup–shaped).

Note : Hutchinson, however, has included the genus *Cuscuta* in the family Cuscutaceae of his 80th order the Polemoniales. Here the old concept has been followed and the genus in included in the Convolvulaceae.

Gynoecium: 2 carpels, syncarpous, pistil 1, ovary superior, axile placentation, bilocular, each locule contains 1 to 2 ovules (sometimes 4 locules may form due to false septation), ovules are anatropous, sessile and erect. Usually a single, simple and filiform style (sometimes 2 styles, *e.g.*, in *Cuscuta*). The stigma is terminal and capitate, sometimes 2 distinct stigmas, *e.g.*, in *Convolvulus*.

Fruit: Usually a loculicidal capsule with two or three chambers, or by false septation apparently four or six chambers. Each chamber contains one or two seeds.

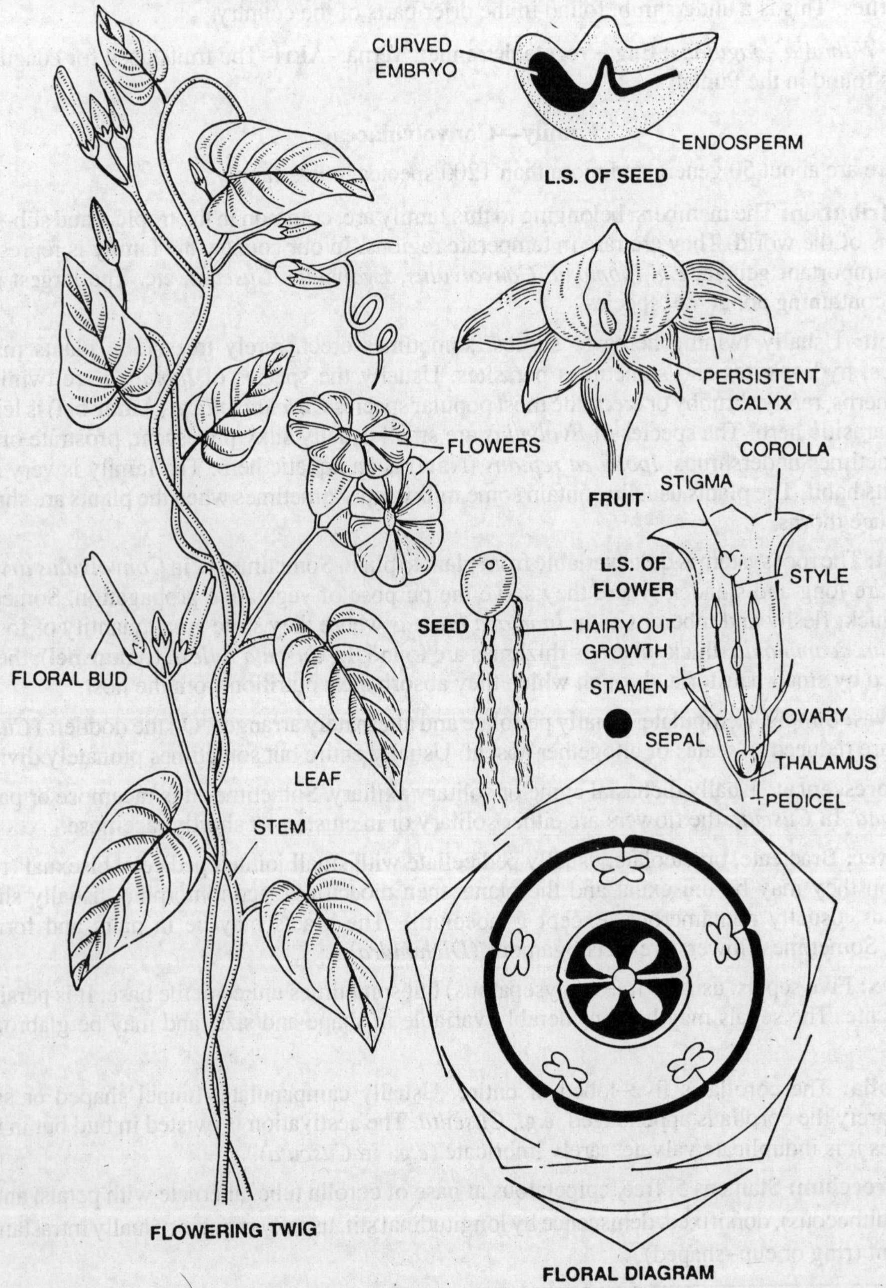

Fig.50.6. Convolvulaceae. *Impomoea tropica* Santapau & Patel; Syn. *I. calycina* Clarke.

Seed: Scanty endospermic, smooth or hairy with large more or less curved embryo. The folded and bilobed cotyledons are present (in *Cuscuta* the embryo is filiform, spirally twisted with no apparent cotyledons), which are surrounded by hard cartilaginous endosperm.

Pollination: (Entomophilous) usually through the agency of insects. The flowers are showy and the honey is secreted in the hypogynous disc. Due to these attractions the insects are naturally attracted and thus pollination takes place.

Floral Formula: Br Brl ⊕ ⚥ K 5, or (5), $\overparen{\text{C (5), A 5}}$, G $(\underline{2})$

Bentham & Hooker(1862)	**Engler & Prantl(1931)**	**Hutchinson(1959)**
Phanerogama	Phanerogams	Angiospermae
Dicotyledones	Dicotyledoneae	Dicotyledones
Gamopetalae	Sympetalae	Herbaceae
Bicarpellatae	Tubiflorae	Solanales
Polemoniales	Convolvulaceae	Convolvulaceae
Convolvulaceae		

Some important plants have been described here in semitechnical language:

1. ***Ipomoea tropica*** Santapau & Patel.

Habit: A perennial twiner.

Stem: Prostrate, twining, branched, herbaceous, solid, glabrous and green.

Leaf: Simple, cauline and ramal, entire alternate, petiolate, stipulate, stipules small in size unicostate reticulate venation.

Inflorescence: Cymose, solitary axillary or in small cymes.

Flower: Pedicellate, pedicel long, campanulate, bracteate, bracteolate, hermaphrodite, actinomorphic, complete, hypogynous, violet coloured, pentamerous and cyclic.

Calyx: 5 sepals, polysepalous, sepaloid, quincuncical aestivation, inferior.

Corolla: 5 lobed, gamopetalous, campanulate, induplicate valvate aestivation, inferior, violet coloured.

Androecium: 5 stamens, polyandrous, epipetalous, filaments long, of different lengths, two long and three short, hair at the base, anther bicelled, introrse, basifixed.

Gynoecium: 2 carpels (bicarpellary), syncarpous, nectar secreting disc present below the ovary, superior, bilocular, axile placentation, two ovules in each loculus, style long, stigma bifid.

Floral formula: Br ⊕ ⚥ K 5, $\overparen{\text{C (5), A 5}}$, G $(\underline{2})$.

2. ***Ipomoea fistulosa*** Mark ex Choisy.

Habit: Perennial shrub.

Stem: Erect, woody, branched, cylindrical, solid, glabrous and green.

Leaf: Cauline and ramal, simple, alternate, petiolate, exstipulate, entire, acute, glabrous, unicostate reticulate venation.

Inflorescence: Cymose, dichasial cyme.

Flower: Pedicellate, bracteate, hermaphrodite, actinomorphic, regular, pentamerous except gynoecium, hypogynous.

Calyx: 5 sepals, polysepalous, imbricate aestivation, inferior.

Corolla: 5 petals, gamopetalous, infundibuliform, corolla tube long, limbs, spreading, inferior.

Androecium: 5 stamens, polyandrous, epipetalous, inferior, filaments long, anthers bicelled, introrse, basifixed.

Gynoecium: 2 carpels (bicarpellary), syncarpous; ovary superior, bilocular, axile placentation, two ovules in each loculus, style dividing near its top into two linear stigmas, honey secreting disc present beneath the ovary.

Fruit: A capsule dehiscing by four valves.

Floral Formula: Br ⊕ ⚥ K 5, C (5), A 5, G (2).

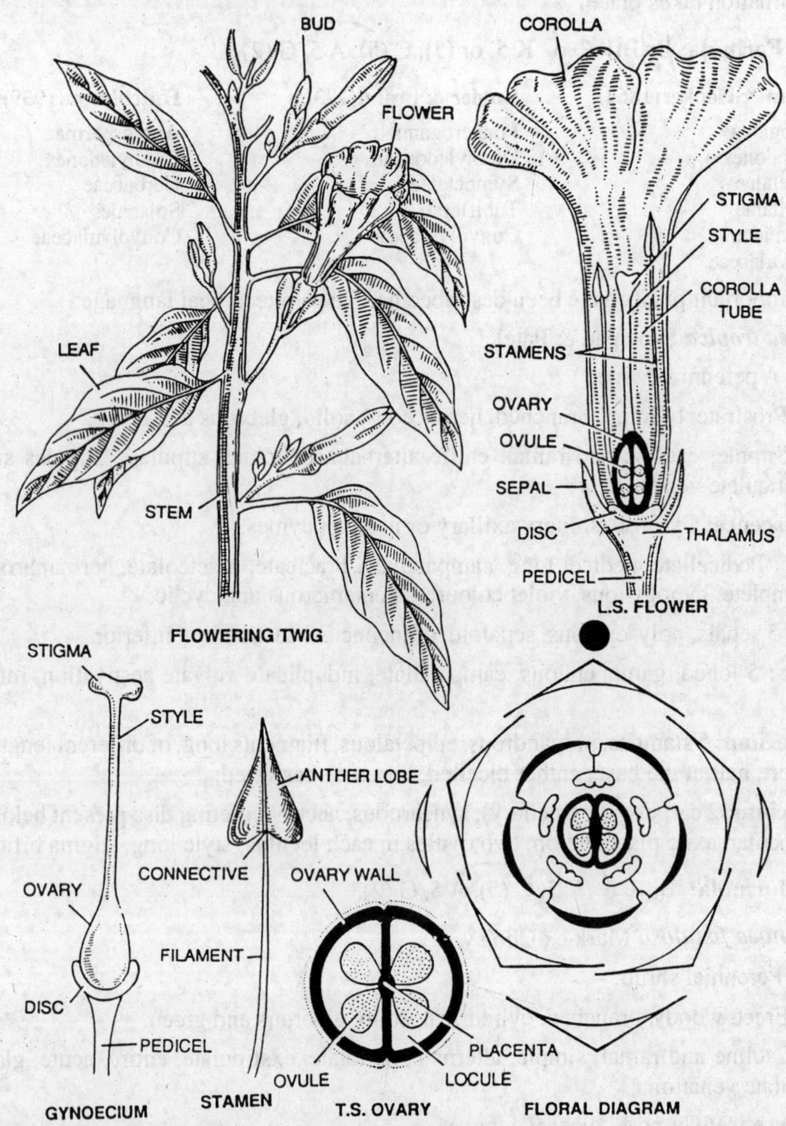

Fig.50.7. Convolvulaceae. *Ipomoea fistulosa* Mark. ex Choisy; Syn.; *I.carnea,* Jacq.

Economic value. A shrub; grown as an avenue plant.

Habit: A perennial twiner.

Stem: Prostrate, twining, branched, herbaceous, solid, glabrous and green.

Leaf: Simple, palmately lobed, alternate, petiolate stipulate, stipules small in size but leaf–like in appearance, multicostate reticulate venation.

Inflorescence: Cymose, solitary axillary or in small cymes.

Flower: Pedicellate, pedicel long, campanulate, bracteate, bracteolate, hermaphrodite, atinomorphic, complete, hypogynous, violet coloured.

Calyx: 5 sepals, polysepalous, sepaloid, quincuncial aestivation, inferior.

Corolla: 5–lobed, gamopetalous, campanulate, induplicate valvate aestivation, inferior, violet coloured.

Androecium: 5 stamens, polyandrous, epipetalous, filaments long, of different lengths, two long and three short, hairy at the base, anther bicelled, introrse, basifixed.

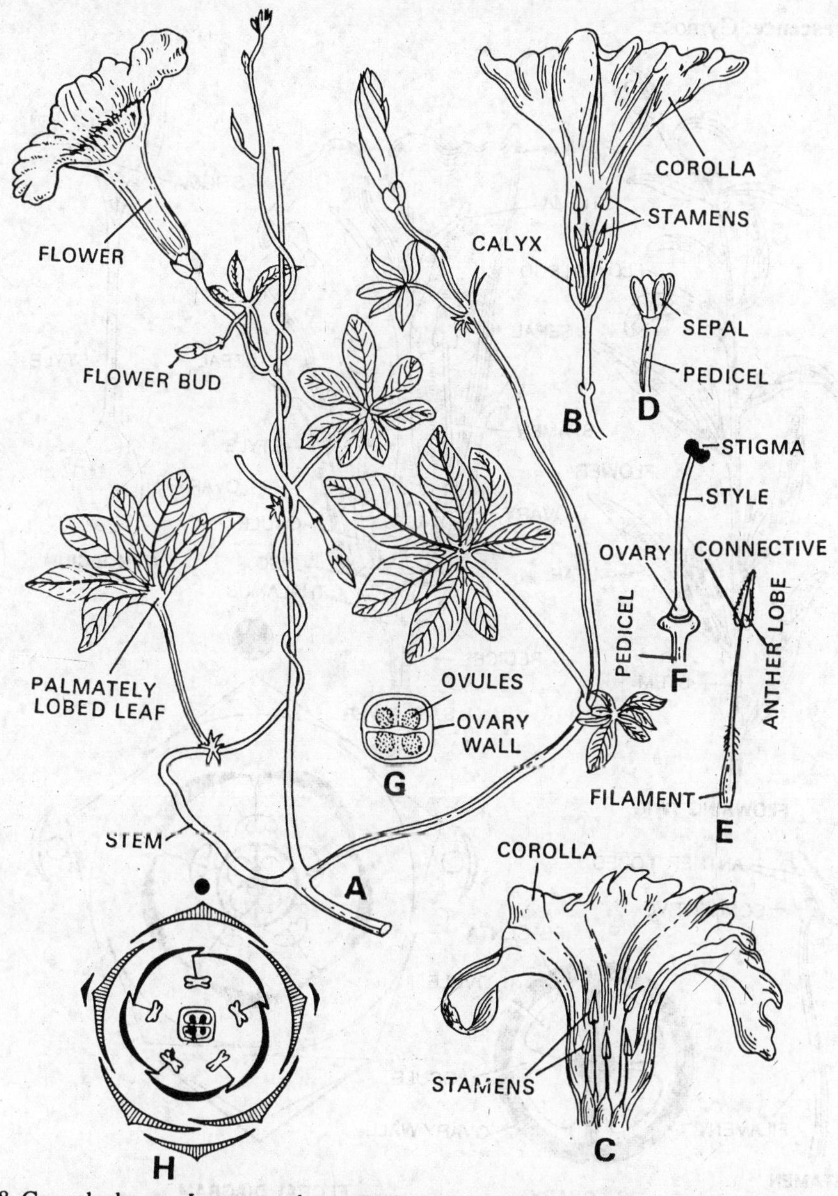

Fig. 50.8. Convolvulaceae. *Ipomoea palmata* (Railway creeper); A, a twig; B, vertical section of flower; C, corolla with epipetalous stamens; D, calyx; E, a stamen; F, the pistil; G, cross–section of ovary; H, floral diagram.

Gynoecium: 2 carpels (bicarpellary), syncarpous, nectar secreting disc present below the ovary, ovary superior, bilocular, axile placentation, two ovules in each loculus, style long, stigma bifid.

Floral Formula: Br ⊕ ⚥ K 5, $\widehat{C (5), A}$ 5, G $(\underline{2})$.

3. ***Convolvulus pluricaulis;*** Verna.–**Sankhapushpi**

Habit: An annual wild herb.

Stem: Prostrate, week, twining, branched, herbaceous, cylindrical, solid, glabrous and green.

Leaf: Simple, alternate, exstipulate, lanceolate, entire, acute, glabrous, unicostate reticulate venation.

Inflorescence: Cymose.

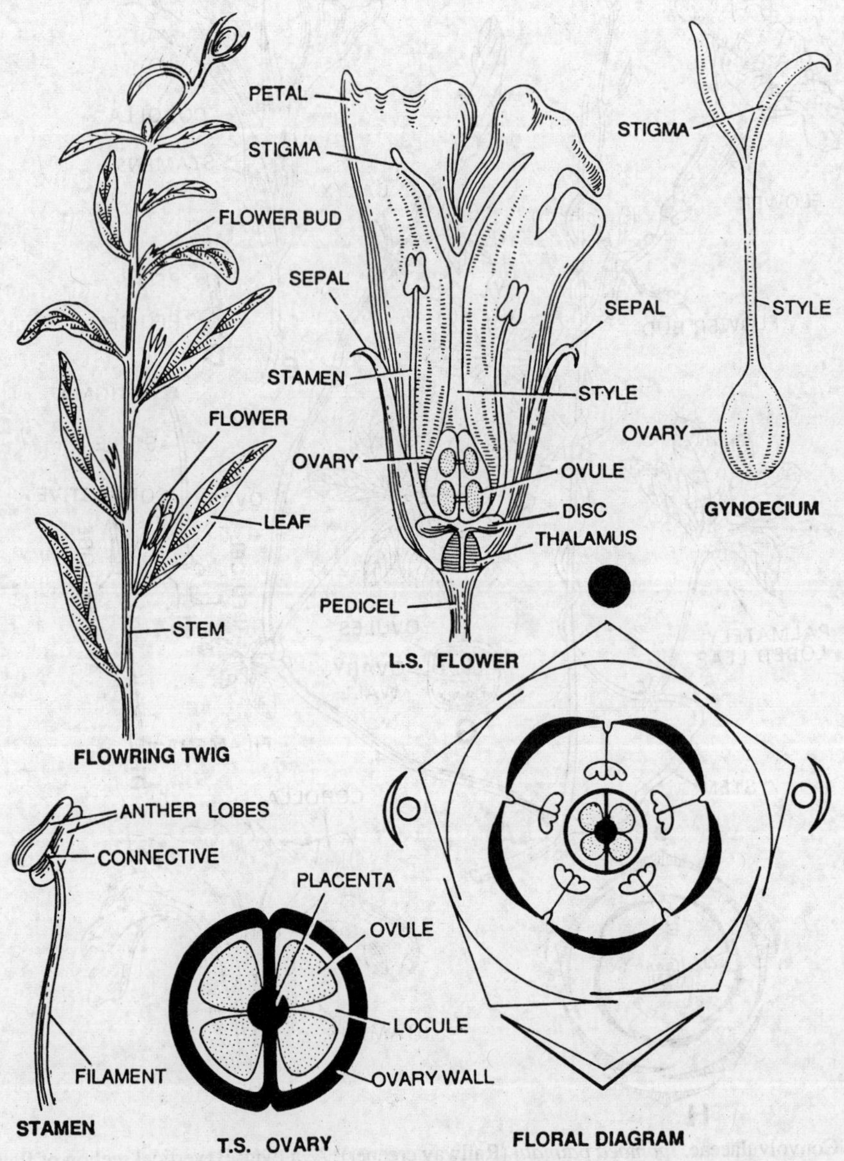

Fig. 50.9 : Convolvulaceae. *Convolvulus pluricaulis* Choisy; Verna. **Sankhapushpi.**

Flower: Pedicellate, bracteate, hermaphrodite, actinomorphic, regular, pentamerous except gynoecium, hypogynous.

Calyx: 5 sepals, polysepalous, imbricate aestivation, inferior.

Corolla: 5 petals, gamopetalous, infundibuliform, corolla tube short, limbs spreading, inferior.

Androecium: 5 stamens, polyandrous, epipetalous, inferior, filaments long, anthers bicelled, introrse, basifixed.

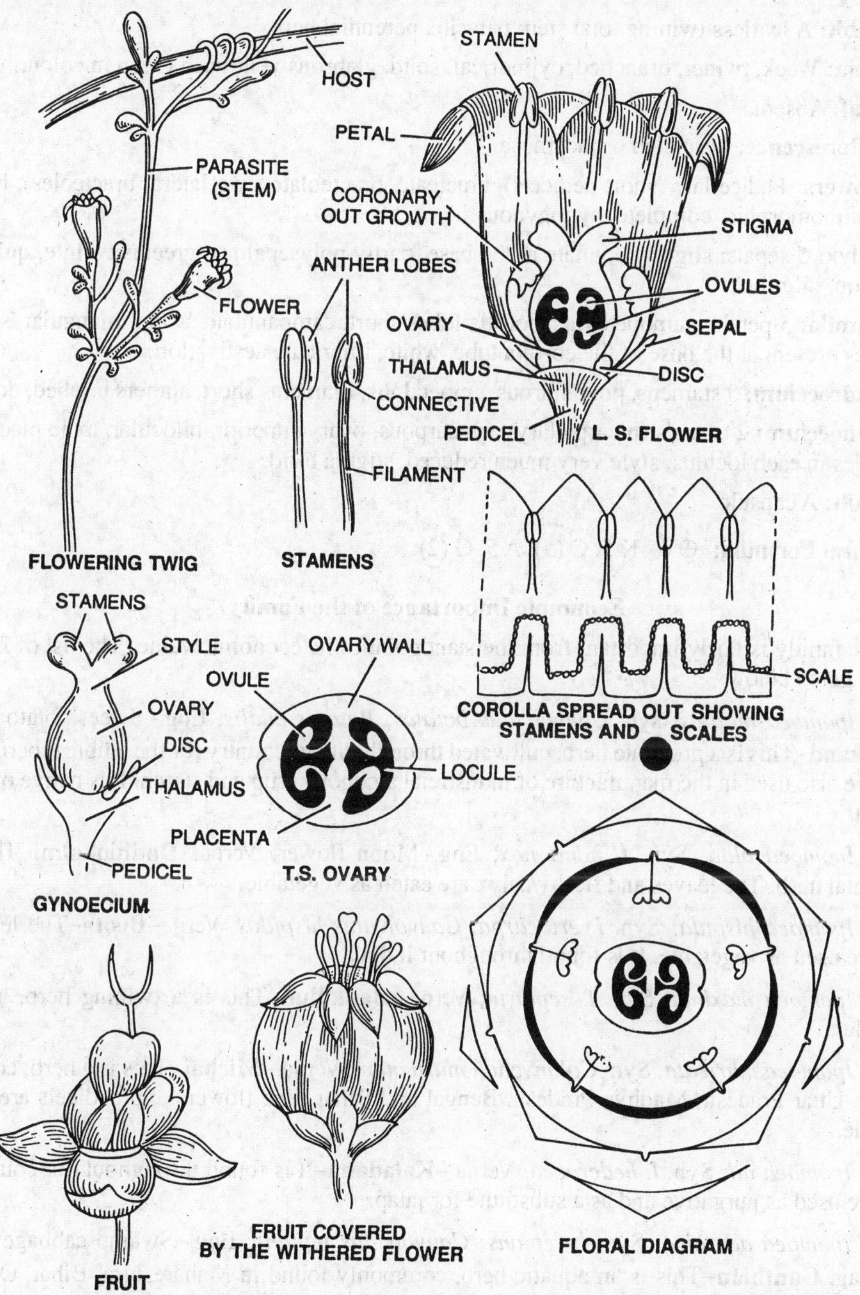

Fig.50.10. Convolvulaceae. *Cuscuta reflexa* Roxb. Eng. dodder; Verna. **Amar bel.**

Gynoecium: 2 carpels (bicarpellary), syncarpous, ovary superior, bilocular, axile placentation, two ovules in each loculus. Style dividing near its top into two linear stigmas, honey secreting disc present beneath the ovary.

Fruit: A capsule dehiscing by four valves.

Floral Formula: $\oplus$ ⚥ K 5, $\overparen{\text{C (5), A 5}}$, G $(\underline{2})$.

4. ***Cuscuta reflexa.*** Verna.–**Amer bel.**

Habit: A leafless twining total stem parasite, perennial herb.

Stem: Week, twiner, branched, cylindrical, solid, glabrous yellowish green in colour.

Leaf: Absent.

Inflorescence: Fascicled or racemose.

Flowers: Pedicellate (short pedicels), bracteate, bracteolate (two lateral bracteoles), hermaphrodite, actinomorphic, complete, hypogynous.

Calyx: 5 sepals, slightly connate at the base, partly polysepalous, greenish white, quincuncial aestivation, inferior.

Corolla: 5 petals, gamopetalous, corolla lobes short, campanulate, acute, triangular, spreading out scales present at the base of the corolla tube, white, imbricate aestivation.

Androecium: 5 stamens, polyandrous, epipetalous, filaments short, anthers bilobed, dorsifixed.

Gynoecium: 2 carpels (bicarpellary), syncarpous, ovary superior, bilocular, axile placentation, two ovules in each loculus, style very much reduced, stigma bifid.

Fruit: A capsule.

Floral Formula: $\oplus$ ⚥ K 5, $\overparen{\text{C (5), A 5}}$, G $(\underline{2})$.

Economic Importance of the Family

The family is fairly important from the standpoint of its economic value. The list of important plants is given below:

1. *Ipomoea batatas*; syn. *Convolvulus batatas*; *Batatas edulis*; Eng.–Sweet potato: Verna.–**Shakarkand**–This is a prostrate herb, cultivated throughout our country for the edible tuberous roots, which are also used in the manufacture of industrcial alcohol, syrup and starch. It is native of tropical America.

2. *Ipomoea alba*; Syn. *I. bona–nox*; Eng.–Moon flower; Verna.–**Dudhiakalmi**–This is an ornamental herb. The leaves and fleshy calyx are eaten as vegetable.

3. *Ipomoea hispida;* Syn. *I. eriocarpa*; *Convolvulus hispidus*; Verna–**Boota**–The leaves and stems are used as vegetable. It is found throughout India.

4. *Ipomoea maxima*; Syn. *I. sepiaria*; Verna.–**Bankalmi**–This is a twining herb, used as a vegetable.

5. *Ipomoea muricata*; Syn. *Colonyction muricatum*; Verna.–**Michai**–This is a herb, commonly found in Uttar Pradesh, Madhya Pradesh, Bengal and Bihar. The flowers and pedicels are eaten as vegetable.

6. *Ipomoea nil*; Syn. *I. hederacea;* Verna.–**Kaladana**–It is found throughout our country. The seeds are used as purgative and as a substitute for jalap.

7. *Ipomoea aquatica*; Syn. *I. reptans; Convolvulus reptans;* Eng.–Swamp cabbage; Verna.–**Kalmisag, Ganthian**–This is an aquatic herb, commonly found in Maharashtra, Bihar, Orissa and South India. The leaves and shoots are eaten as vegetable.

8. *Ipomoea digitata*; Verna.–**Bilaikand**–This is a common climbing herb, used as a fodder.

9. *Ipomoea angulata*; Eng.–Star Ipomoea–This is a climbing shrub, grown as an ornamental in the gardens for its scarlet flower. It is native of North Mexico.

10. *Ipomoea cairica*; Syn. *I. palmata; Convolvulus cairicus;* Eng.–Railway creeper–This is climbing shrub, grown as an ornamental. The flowers are campanulate and violet coloured.

11. *Ipomoea crassicaulis*; Syn. *I. carnea*–This is a stout straggling shrub grown as a hedge plant.

12. *Ipomoea leari*; Eng.–Blue dawn flower–This is an ornamental climbing shrub. It is native of tropical America.

13. *Ipomoea tricolor*; Syn. *I. rubrocaerulea*; Eng.–Morning glory–This is grown as an ornamental in the gardens.

14. *Ipomoea quamoclit*; Syn. *Quamoclit pinnata*; Eng.–Indian pink; Verna.–**Kamalata**–This is a twining shrub; grown as an ornamental. It is native of tropical America.

15. *Exogonium purga*; Syn. *Ipomoea purga*; Eng.–Jalap. This is a twining herb, cultivated in the Nilgiris and Poona. The roots are the source of a resinous drug, which is used as a purgative.

16. *Operculina turpethum*; Syn. *Ipomoea turpethum*; Eng.–Indian Jalap; Verna.–**Nishoth**–This is a common twining herb. A resinous substance known as turpethin, obtained from the root bark, is used as a purgative and acts as a substitute for jalap.

17. *Porana paniculata*; Eng.–Christmas vine; Verna.–**Dela**–This is an ornamental shrub, grown in the gardens.

18. *Argyreia nervosa*; Syn. *Convolvulus nervosus*; Verna.–**Ghabel**–This is a climbing shrub; grown as an ornamental.

19. *Argyreia splendens*; Syn. *Convolvulus splendens;* Eng.–Silver morning glory. This is grown as an ornamental in the gardens.

20. *Cuscuta reflexa*; Eng.–Dodder; Verna–**Amarbel.,Akasbel**–This is total stem parasite. It is twining leafless herb. The seeds are carminative.

21. *Evolvulus alsinoides*; Verna.–**Sankhpushpi**–It is used to cure bronchites and asthma. It is also used in dysentery.

Systematic Relationships: According to Bentham and Hooker the Polemoniales consists of five families, they are–Polemoniaceae. Hydrophyllaceae, Boragineae, Convolvulaceae and Solanaceae; Engler and Prantl have placed all these families in his Tubiflorae along with fifteen other families. Hutchinson has splitted the order Tubiflorae into a number of smaller groups (orders). He has placed the families Solanaceae, Convolvulaceae and Nolanaceae in the order Solanales. Rendle separated the Convolvulaceae into a distinct order the Convolvulales, which he also separated from the Tubiflorae. According to the Wettstein the Convolvulaceae may have been originated from the other groups of the Tubiflorae and considers them closely related to the Malvales or Geraniales. Hutchinson has also shown the close systematic relationship of the Geraniales with the Convolvulaceae.

Bentham and Hooker and Bessey included the family Solanaceae among the Polemoniales. Hallier considered the Solanaceae to be primitive member of the Tubiflorae and supposed it to have been derived from Linaceae, Hutchinson considers the family to be primitive most of the order Solanales.

51

ORDER–PERSONALES

According to Hutchinson this is the 78th order of the phylum Angiospermae, subphylum Dictoyledones and division Herbaceae. This order comprises of six families, *i.e.,* Scrophulariaceae, Acanthaceae, Gesneriaceae, Orobanchaceae, Lentibulariaceae, Columelliaceae, of which the Scrophulariaceae, the Acanthaceae and Lentibulariaceae, have been described in the present text in detail. Bentham and Hooker have placed the families Scrophulariaceae and Acanthaceae along with six others including the Pedalineae and the Acanthaceae in their ninth order the Personales of class Dicotyledones division Gamopetalae, and series Bicarpellatae. Engler and Prantl have included the Scrophulariaceae and the Acanthaceae along with other eighteen families such as convolvulaceae, Verbenaceae, Labiatae, Solanaceae, Pedaliaceae, etc.,in their sixth order the Tubiflorae of class Dicotyledoneae and subclass Sympetalae.

Corolla nearly always zygomorphic with stamens fewer than the corolla lobes. Mostly herbaceous; leaves alternate to opposite.

Family—Scrophulariaceae

There are about 210 genera and 3000 species in this family.

Distribution: The family is world-wide (*i.e.,* cosmopolitan) in distribution and represented on all continents of the world. The largest genera of the family are–*Pedicularis* (about 600 sp.), *Penstemon* (about 250 sp.), *Verbascum* (about 150 sp.), *Linaria* (about 250 sp.), *Mimulus* (about 150 sp.), *Veronica* (about 150 sp.). All these genera are commonly found in the northern hemisphere. The genera *Hebe* (about 140 sp.), and *Calceolaria* (about 50 sp.), are found in Western South America. In our country the family is represented by several important genera, such as– *Verbascum, Linaria, Antirrhinum, Scrophularia, Digitalis, Veronica, Striga* and others. In India, the plants are commonly found on the hills.

Habit: They are either herbs or shrubs, very rarely small trees. Certain plants are lianous, *e.g.,Maurandia, Rhodochiton.* A few plants are chlorophyll-less parasites, *e.g., Hyobanche* and others are chlorophyll containing parasites or saprophytes, *e.g., Melampyrum Pedicularis, Castilleja, Striga.* The species of *Verbascum, Linaria, Antirrhinum* are either annual or perennial herbs. The spicies of *Mimulus* are erect, prostrate or creeping herbs. The species of *Veronica* are either herbs, shrubs or sometimes trees. The species of *Paulownia* found in China and Japan are trees. The species of *Maurandia* are leaf climbers and possess sensitive petioles. *Limnophila racemosa* is an aquatic herb. *Hydrotriche* and *Ambulia* are other water plants.

Root: Tap and branched.

Stem: The stem may be erect, prostrate, creeping or ascending; herbaceous or woody; solid; branched.

Leaves: The leaves may be alternate, opposite or whorled. In the species of *Verbascum* they are alternate throughout; in *Antirrhinum* the lower leaves are opposite and the upper alternate, or all opposite, *e.g.*, in *Mimulus*. In the species of *Veronica* the leaves are whorled. All varieties of leaf arrangement occur in the species of *Veronica.* Usually the leaves are simple, exstipulate, entire or pinnately lobed. In the aquatic genera *Ambulia, Hydrotriche* and others possess heterophylly. In such cases the aerial leaves are entire and submerged are much dissected.

Inflorescence: The inflorescence shows great variations. The flowers may be solitary, *e.g.*, in *Linaria, Mimulus* and others; or they may be arranged in terminal, spikes or recemes, *e.g.*,in species of *Digitalis, Veronica* and others. In certain other species of *Veronica* the flowers are axillary. In many plants the inflorescence is cymose, forming dichasial cymes arranged in terminal or axillary racemes, spikes or panicles. Bracts and bracteoles are generally present. The bracts are sometimes brightly coloured, *e.g.*, in *Castilleja*. In *Verbascum* the arrangement of the flowers is being complicated by the formation below each 3-7 flowered dichasial cyme, of 2-5 accessory dichasial cymes, with gradually decreasing number of flowers.

Flowers: The flowers are bisexual, zygomorphic, sometimes showing actinomorphic tendency, *e.g.*, in *Verbascum*, hypogynous.

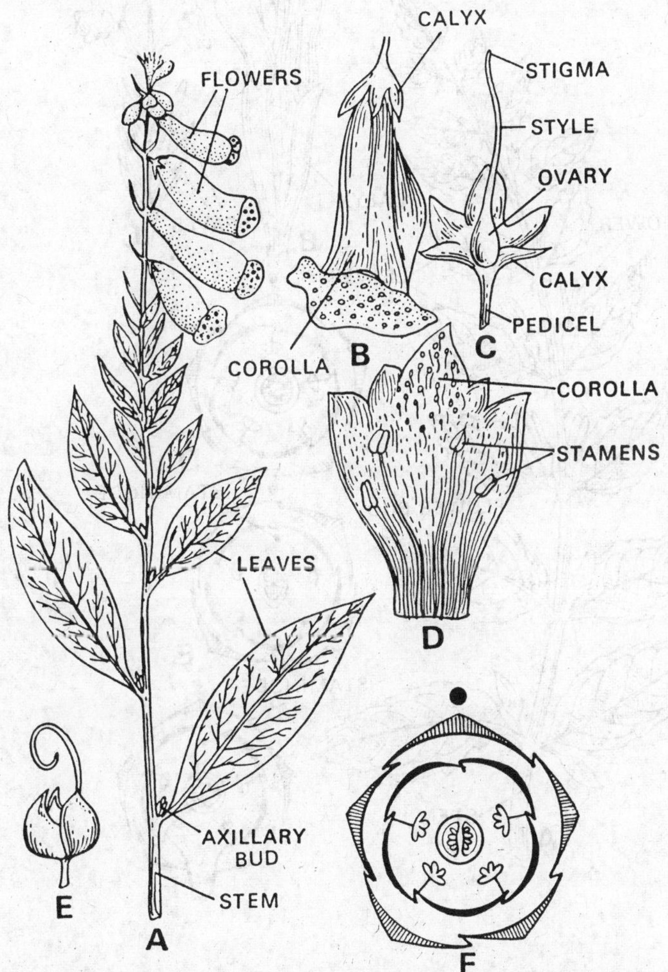

Fig. 51.1. Scrophulariaceae. *Digitalis purpurea* ; A, flowering twig ; B flower ; C, calyx with young fruit; D, corolla expanded; E, dehisced fruit, ; F, floral diagram

Calyx: The calyx consists of typically five sepals, but in *Euphrasia* and many species of *Veronica,* the posterior sepal is suppressed. In *Pedicularis* two to five calyx teeth are present, in *Calceolaria*, the two anterior sepals are completely united.

Corolla: Generally the corolla consists of five lobes, gamopetalous, zygomorphic showing great diversity in form. In *Verbascum,* the corolla is somewhat actinomorphic. In *Veronica officinalis*, the corolla is obsolete (rudimentary); in *Scrophularia nodosa* it is large and bell-shaped; it forms a pair of closed lips, *e.g.*, in *Linaria, Antirrhinum.* In *Linaria,* the anterior petal is represented by spur; in

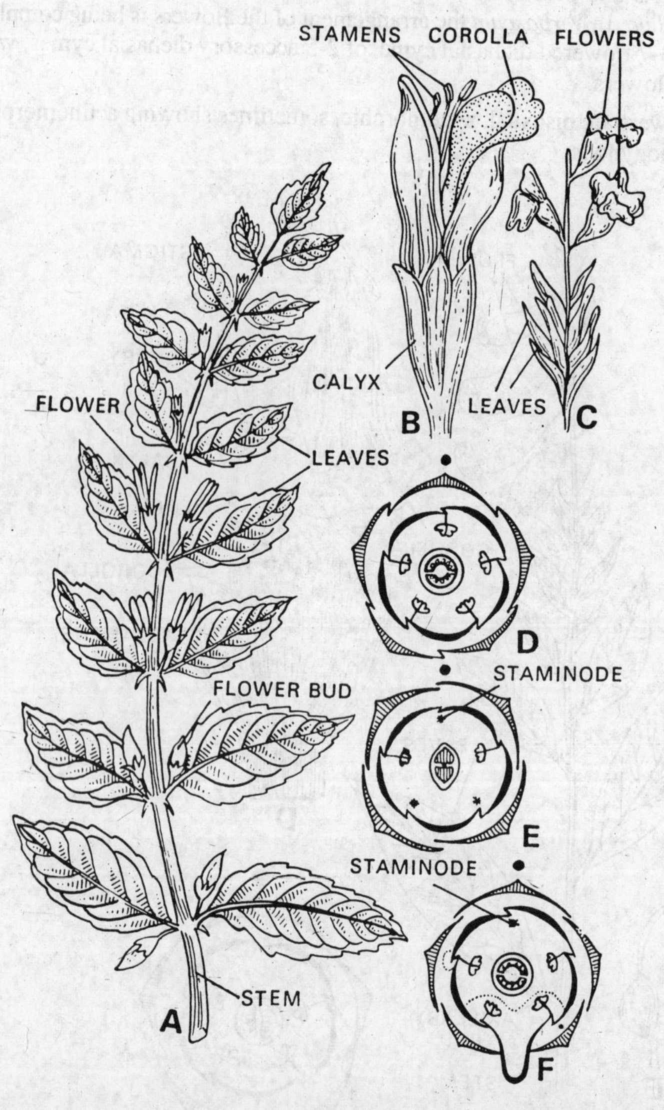

Fig. 51.2. Scrophulariaceae. *Lindenbergia indica* (Linn.) O. Ktze.

Diascia, the spurs are present on each of the lateral anterior petals; in *Calceolaria*, the corolla tube is very short and the limb forms two lips. The upper lip formed by the union of the two petals of posterior side, and the lower larger lip formed by the union of three anterior petals. In the species of *Veronica*, *Gratiola*, etc., the corolla becomes four lobed by a complete union of two postertor petals. In *Antirrhinum* and others the corolla tube possesses gibbous or saccate throat. The aestivation is imbricate.

Androecium: Generally the stamens are 4, didynamous and epipetalous. The number of stamens varies in different genera. In some species of *Verbascum*, the stamens are equal in size and five in number; in other species of *Verbascum*, the three posterior stamens are longer than the anterior ones. In most of the cases the posterior stamen is more or less aborted, and of the remaining 4 fertile stamens the anterior ones are generally longer than the posterior pair, *e.g.*, in *Linaria, Digitalis*, etc. In *Veronica*, *Calceolaria*, etc., only two stamens are present. Anthers bicelled, appendages or hairy structures may be found, introrse, dehisce by longitudinal slit.

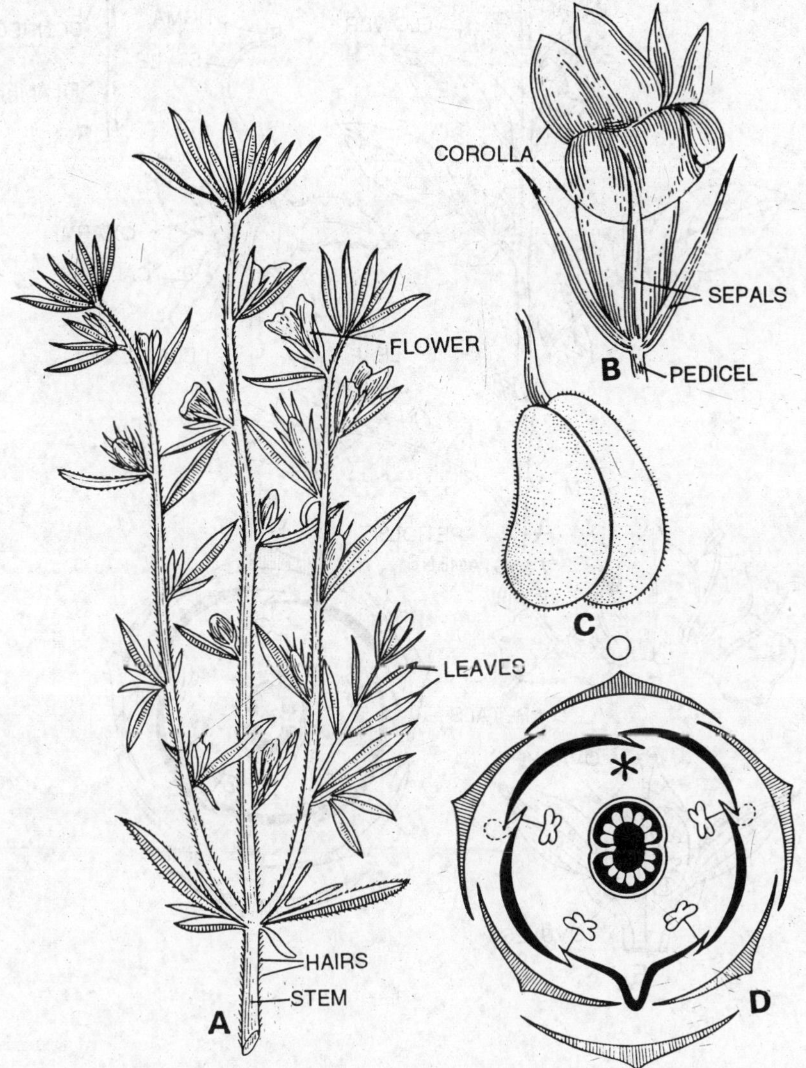

Fig.51.3. Scrophulariaceae. *Antirrhinum orontium*. A, flowering twig; B, flower; C, fruit; D, floral diagram.

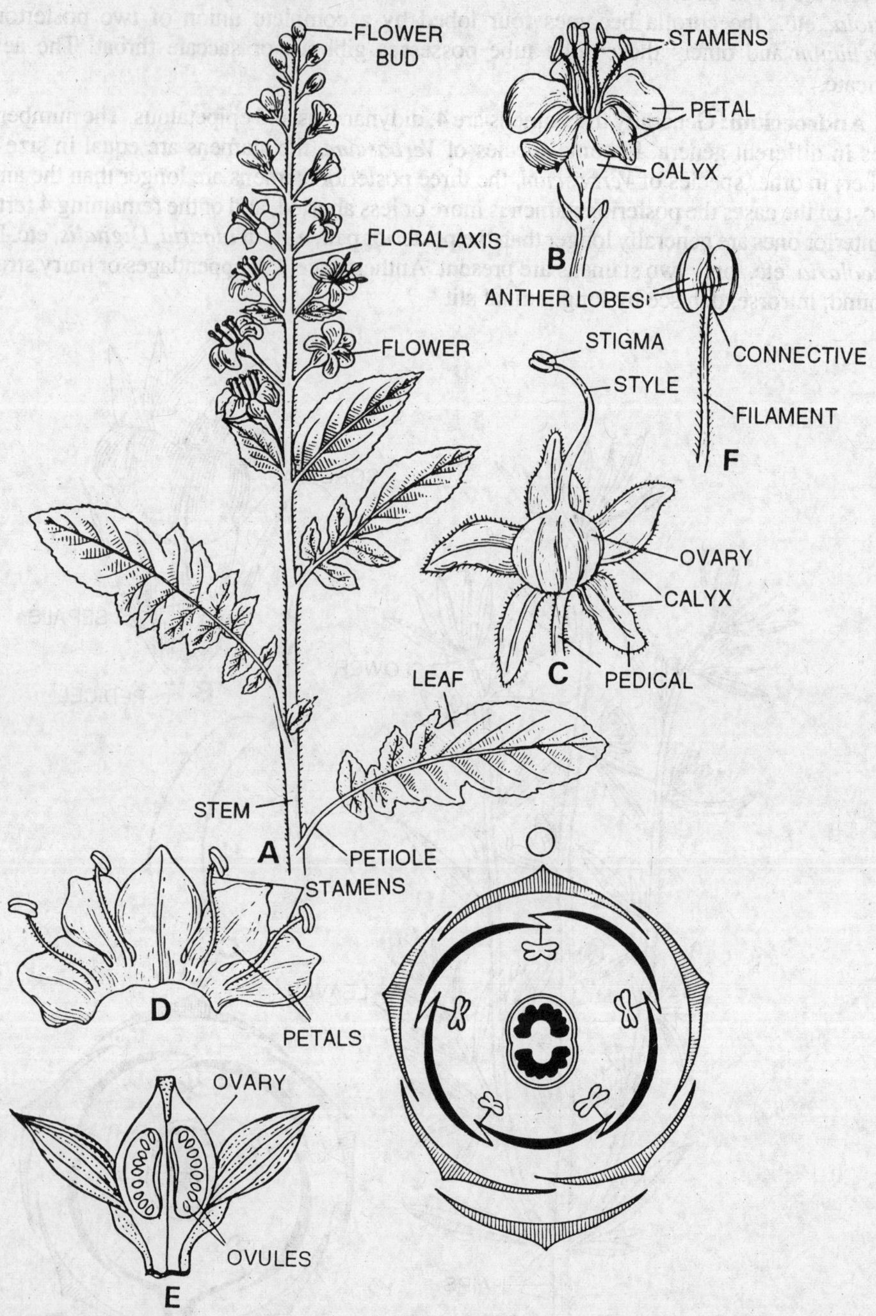

Fig. 51.4. Scrophulariaceae. *Celsia coromandelina*. A, flowering twig; B, flower; C, gynoecium; D, petals with epipetalous stamens; E., L.S. of ovary; F, stamen; G, floral diagram.

Gynoecium: The gynoecium consists of two carpels (syncarpous), situated on a nector secreting disc. Ovary superior, bilocular, rarely unilocular, carpels median, and not obliquely placed as in the Solanaceae. Many anatropous ovules are arranged on a thick axile placenta. In *Veronica,* however, the ovules are few in number. The style is simple and the stigma is bilobed.

Fruit: It is generally a capsule and sometimes a berry. The capsule is surrounded by the persistent calyx at its base.

Seed : Generally small and numerous. However, in *Veronica,* they are few and large. The seeds are endospermic with straight or curved embryo.

Pollination: The pollination takes place through the agency of insects. The flowers are generally adapted for insect pollination. The nectar secreting disc is present below the ovary, which makes the centre of attraction for the bees. Generally the flowers are beautiful and attractive.

Floral Formula: ⊕ ⚥ K 5, $\overparen{\text{C (5), A 2 + 2}}$, G $(\underline{2})$.

Bentham & Hooker(1862)	**Engler & Prantl (1931)**	**Hutchinson(1959)**
Phanerogams	Phanerogams	Angiospermae
Dicotyledones	Dicotyledoneae	Dicotyledones
Gamopetalae	Sympetalae	Herbaceae
Bicarpellatae	Tubiflorae	Personales
Personales	Scrophulariaceae	Scrophulariaceae
Scrophulariaceae		

Fig.51.5. Scrophulariaceae. *Mazus japonicus* (Thunb) Kuntze. A, flowering plant; B, L.S. of flower; C, gynoecium; D, T.S. of ovary; E, stamen; F, floral diagram.

Mazus japonicus (Thunb.) Kuntze; **Habit**– herb. **Root**– tap, branched. **Stem**– reduced. **Leaf**– radical, exstipulate simple, sessile, spathulate, dentate, obtuse, glabrous, unicostate reticulate. **Inflorescence**– Racemose, raceme. **Flower**– bracteate, pedicellate, complete, hermaphrodite, zygomorphic, pentamerous, hypogynous, cyclic. **Calyx**– 5, gamosepalous, valvate, campanulate, persistent. **Corolla**– 5, gamopetalous, bilabiate, personate, upper lip two–lobed, lower lip three–lobed, valvate, gibbous throat, pale blue. **Androecium**– 4, polyandrous, epipetalous, didynamous, filaments long, dithecous, basifixed, introrse. **Gynoecium**– 2, bicarpellary, syncarpous, superior, bilocular, many ovules in each locule, axile, style long, stigma bifid.

Fruit– capsule. **Floral formula** — Br ·|· ⚥ $K_{(5)}$, $\overset{\frown}{C_{(2+3)}, A_{2+2}}$, $G_{(\underline{2})}$.

2. ***Veronica anagalis*** Linn.

Habit: Annual shade and moist loving herb.

Root: Tap and branched.

Stem: Erect, aerial, herbaceous, solid, cylindrical, branched, pubescent, green.

Leaves: Cauline and ramal, semi–amplexicaul, exstipulate, sessile, simple, entire, acute, unicostate reticulate, pubescent.

Inflorescence: Racemose, terminal raceme.

Flower: Bracteate, pedicellate, zygomorphic, hermaphrodite, complete, tetramerous, small, white, hypogynous and cyclic.

Calyx: 4 sepals, polysepalous, imbricate or quincuncial.

Corolla: 4 petals, gamopetalous, imbricate.

Androecium: Stamens 2, epipetalous, free, filaments long, anther dithecous, dorsifixed, introrse.

Gynoecium: Bicarpellary, syncarpous, ovary superior, bilocular, axile placentation, several anatropous ovules in each locule on swollen placenta, style simple, stigma lobed.

Fruit: Capsule.

Floral formula: Br ·|· ⚥ K4, $\overset{\frown}{C(1/3), A2}$, $G(\underline{2})$.

Economic value. The plant is used medicinally. It is alterative, diuretic, antiscorbutic, given in scurvy, impurity of blood, etc. Used as a remedy for scrofulous affections, especially of the skin bruised and applied externally for healing burns and ulcers.

Economic Importance of the Family

The family is of little economic value. Some plants are ornamental and some possess medicinal properties. A list of few important plants is given below.

1. *Linaria moroccana.*– This is a herb. It is native of Morocco but now grown in our country as ornamental. It is used medicinally in diabetes.

2. *Antirrhinum majus*; Eng.–Snapdragon.–This is a popular ornamental herb, largely grown in gardens and public parks for beautiful variously coloured flowers.

3. *Celsia coromandeliana*; Verna.–**Gadar–tambaku**–This is a small herb, with yellow flowers. The plant is of medicinal value. The juice is given in dysentery and diarrhoea.

4. *Calceolaria* sp.–The plants are highly ornamental.

5. *Mimulus luteus*; Eng.–Yellow monkey flower.–This is an ornamental plant.

6. *Veronica beccabunga*; Eng.–**Baccabunga.**–This is a herb found in Western Himalayas and Kashmir. The young stems are used as vegetable.

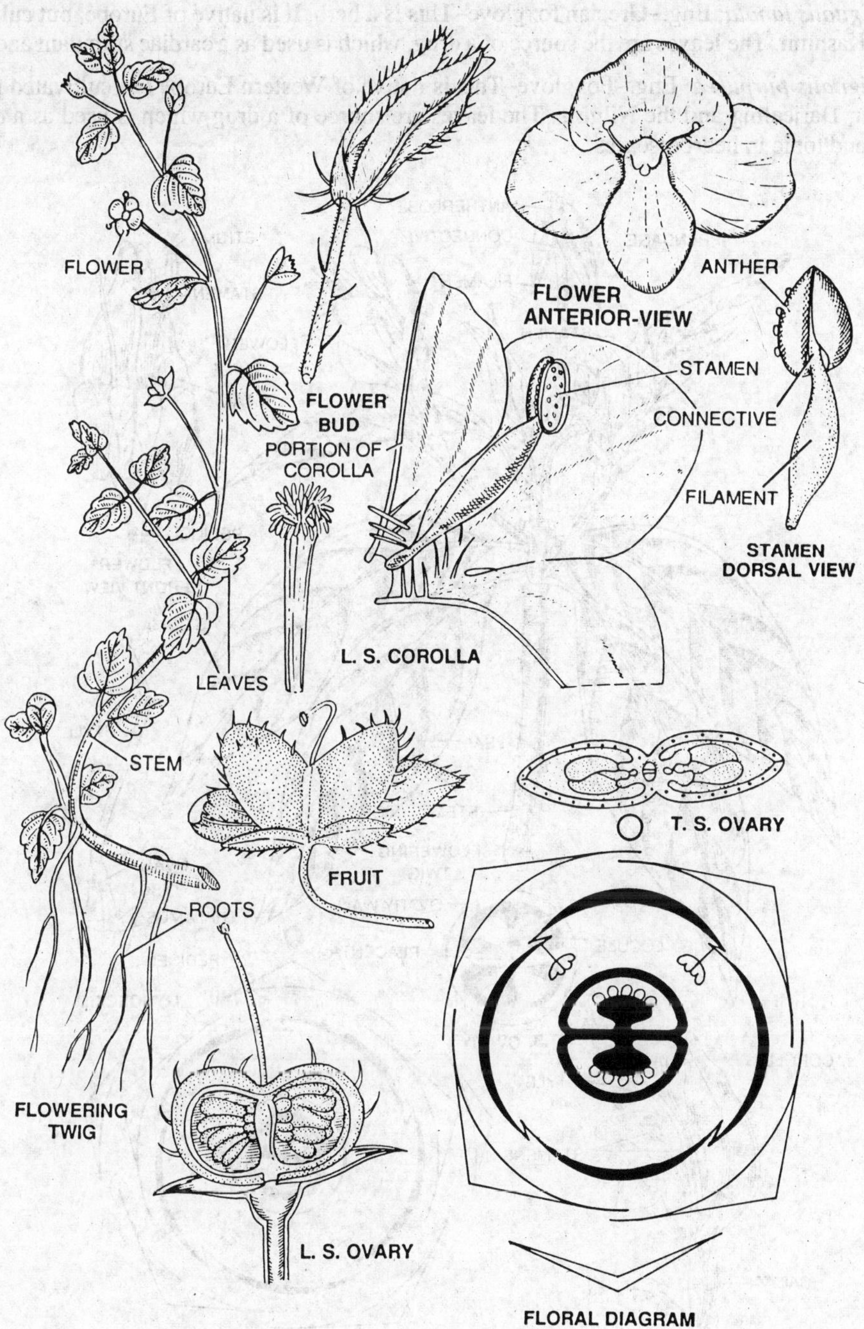

Fig.51.6. Scrophulariaceae. *Veronica anagalis* Linn.

7. *Verbascum thapsus*; Eng.– Jackal's tobacco; Verna.– **Gidar–tambaku.**–The plant possesses narcotic and medicinal properties. It is used to cure asthma and other pulmonary diseases.

8. *Digitalis lanata*; Eng.–Grecian foxglove–This is a herb. It is native of Europe, but cultivated chiefly in Kashmir. The leaves are the source of a drug, which is used as a cardiac stimulant and tonic.

9. *Digitalis purpurea*; Eng.–Foxglove–This is native of Western Europe but cultivated mainly in Kashmir, Darjeeling and the Nilgiris. The leaves are source of a drug which is used as a cardiac stimulant and tonic in heart diseases.

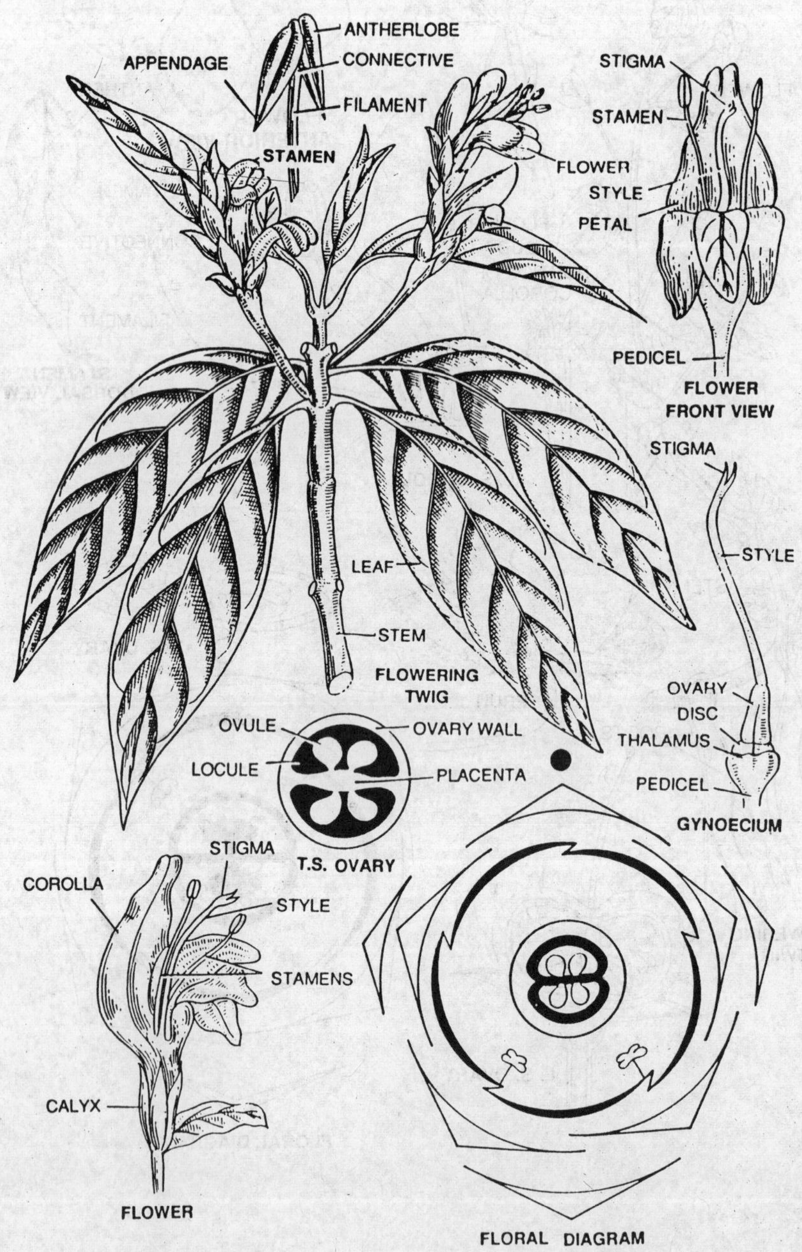

Fig. 51.7 : Acanthaceae. *Adhatoda vasica* Nees. Eng. Malabar nut; Verna. *Arusa*

10. *Russelia coccinea*; Syn.–*R.sarmentosa*;this is a shrub; it is native of Mexico. Grown in gardens for its deep scarlet flowers.

11. *Russelia equisetiformis*; Eng.–Fountain plant, Firecracker plant–This is native of Mexico; grown as an ornamental.

Family–Acanthaceae

There are about 240 genera and 2,200 species in this family.

Distribution: The members of this family are mostly found in tropical to subtropical forests especially in the damp and marshy places.

Habit: Usually they are herbs or shrubs. Trees are rarely found. Some are however xerophytic, such as, *Barleria, Blepharis, Acanthus,* etc. Some plants are climbing in habit, *e.g., Mendoncia, Thunbergia* and other allied plants. *Thunbergia alata* is a herbaceous twiner; *Peristrophe bicalyculata, Andrographis paniculata, Runzia parviflora, Ruellia prostrata, Justicia simplex, Hemigraphis* sp; *Haplanthus* sp., etc., are typical examples of herbs. Several plants are undershrubs, *e.g., Barleria cristata, Ruellia tuberosa, Justicia betonica, Strobilanthes coloratus, Eranthemum pulchellum,* etc. Some good examples of shrubs are, *Barleria alba, Barleria prionitis, Adhatoda vasica,* etc. *Acanthus ilicifolius* is a halophyte and represents mangrove vegetation.

Leaves: The leaves of usual mesophytes are simple, usually opposite decussate, entire, exstipulate, thin and delicate. In xerophytes the leaf blade is more or less spiny.

The characteristic anatomical feature of the family is the presence of cystoliths in the epidermal cells of leaves and stems. Sometimes double cystoliths are found. The cystoliths are calcium carbonate crystals.

Inflorescence: The inflorescence is of cymose type. The dichasial cyme passes into monochasial one in the higher branching. Sometimes the flowers are found to be arranged in the short axillary clusters. Spikes and racemes are also frequently found. Bracts and bracteoles generally well developed and often brightly coloured. These bracts, etc.,add to the beauty of the inflorescence. In *Mendoncia* sp.,and *Thunbergia* sp.,the bracteoles are large and form an involucre around the corolla tube. The flowers of *Dicliptera* sp., are found to be arranged in axillary peduncled cymes. The flowers of *Phialacanthus griffithi* are arranged in typical cymes. The flowers of *Strobilanthes discolor* are found to be arranged in panicled cymose heads whereas in *Strobilanthes anisophyllous* these are arranged in cymose heads. In the species of *Barleria*, the flowers are either solitary or arranged in short spikes, or pairs. In the species of *Thunbergia* the flowers are found to be arranged either in cymose or in racemose types of inflorescence. This way, there is great variation in the types of inflorescence in various members of the family.

Flower: The structure of flower is very uniform. The flowers are bracteate; bracteolate, hermaphrodite, complete, zygomorphic (irregular) and hypogynous. Usually the flowers are tetramerous or pentamerous, but generally there is a reduction to 4 or 2 in the androecium, and there are two carpels in the gynoecium.

Calyx: It consists of five, sometimes four or rarely three sepals. Usually they are gamosepalous. In *Thunbergia* the sepals are reduced to narrow beam. In this case large bracteoles serve the purpose of protection. The aestivation is either contorted or imbricate.

Corolla: Usually it consists of five or sometimes four gamopetalous petals. The corolla is often bilabiate, *e.g.,* in *Ruellia* in such cases the upper lip of the corolla is erect and bifid at the apex and the lower lip is horizontal and three–lobed forming the landing–platform for the insects. Usually the corolla possesses a large or short slender corolla tube which passes above into an equally five lobed limb, *e.g.,* in *Thunbergia.* The inner side of the lip of the corolla possesses dense hairs. These hairs often extend to the mouth of the corolla. In *Acanthus* and other allied genera the upper lip of corolla is altogether absent. Here, the corolla is found to be cut upto the base of the corolla tube. The aestivation is either contorted or imbricate.

Androecium: Usually there are 4 didynamous stamens. In certain cases only two stamens are found. Very rarely the number of fertile stamens reaches to five, *e.g., Pentstemonacanthus*. In the cases where there are only four stamens, the fifth posterior stamen reduces to a staminode or disappears completely. In the **Imbricatae** group of the family, there are only two stamens. In such cases, firstly the posterior stamens reduce to staminodes or disappear completely. However, in *Brillantaisia* the anterior pair of stamens reduces to staminodes.

The filaments are generally quite free and project out from the mouth of the corolla tube. The stamens of *Thunbergia,* are however short and remain inside the corolla tube. The stamens are inserted in the corolla, *i.e.,* epipetalous. The anthers may be two or one celled. In the cases where one celled anthers are found, a rudiment of the second cell may be seen, whereas, in the case of two–celled anthers the cells are equal and more or less separated by a connective. Sometimes the lower cell of the anther becomes spurred. The shape and size of pollen grains are quite different from genus to genus.

Gynoecium: The gynoecium consists of two carpels, syncarpous. The ovary superior, bilocular, two or more ovules are found in each loculus. The Placentation is axile. The style is long and slender and projects out from the mouth of the corolla tube. There are two small stigmas; the posterior stigma is usually reduced.

Fruits: Usually the fruit is bilocular capsule, which dehisces loculicidally. In *Mendoncia* the fruit is drupe.

Seed: In *Nelsonia,* the seeds are small and many. In *Nelsonia* and *Thunbergia,* the funicle of the seeds forms a papilla, the seeds are rounded. In the *Acanthus* and allied genera, *e.g., Ruellia,* the funicle of the seeds forms a hook–like projection, known as **jaculator**, in which the seed rests. These hook–like jaculators make the fruits burst and the seeds become dispersed in different directions. The seeds are exalbuminous.

Pollination: The pollination takes place through the agency of insects.

Floral Formula: ·|· ⚥ K (4–5), $\widehat{C\,(4\text{–}5), A}$ 4 or 2, $G\,\underline{(2)}$.

Bentham and Hooker(1862)	**Engler and Prantl(1931)**	**Hutchinson(1959)**
Phanerogams	Phanerogams	Angiospermae
Dicotyledones	Dicotyledoneae	Dicotyledones
Gamopetalae	Sympetalae	Herbaceae
Bicarpellatae	Tubiflorae	Personales
Personales	Acanthaceae	Acanthaceae
Acanthaceae		

Description of some important plants in semi–technical language:

1. ***Barleria prionitis.***

Habit: A prickly bushy shrub.

Root: Tap and branched.

Stem: Erect, branched, woody, slightly angular, solid, glabrous, green.

Leaf: Simple, opposite, decussate, upper leaves short petioled, lower leaves long petioled, interpetiolar spines, lanceolate, entire acute, unicostate reticulate venation.

Inflorescence: A terminal spike, lower flowers arranged in cymes.

Flower: Sessile, bracteate (leaf bracts), bracteolate (six bracteoles–two large and four small narrow and spiny at the tips), hermaphrodite, zygomorphic, hypogynous.

Calyx: 4 sepals, polysepalous, slightly connate at the base, sepaloid, arranged in two whorls of two each, two outer bigger and inner smaller, inferior, imbricate aestivation.

Corolla: 5 petals, gamopetalous, slightly bilabiate, upper lip consisting of four lobes and the lower of single lobe, inferior, imbricate aestivation, yellowish in colour.

Androecium: 4 stamens, two stamens large and two small, epipetalous, sometimes fifth stamen is represented by staminode, anthers bicelled dorsifixed.

Gynoecium: 2 carpels (bicarpellary), syncarpous, ovary superior, bilocular, axile placentation, one ovule in each loculus style terminal, stigma slightly elongated, ovary situated on a honey secreting disc.

Fruit: An elongated capsule.

Floral Formula: ·|· ⚥ K 2+2, C(4 + 1), A 4, G (2).

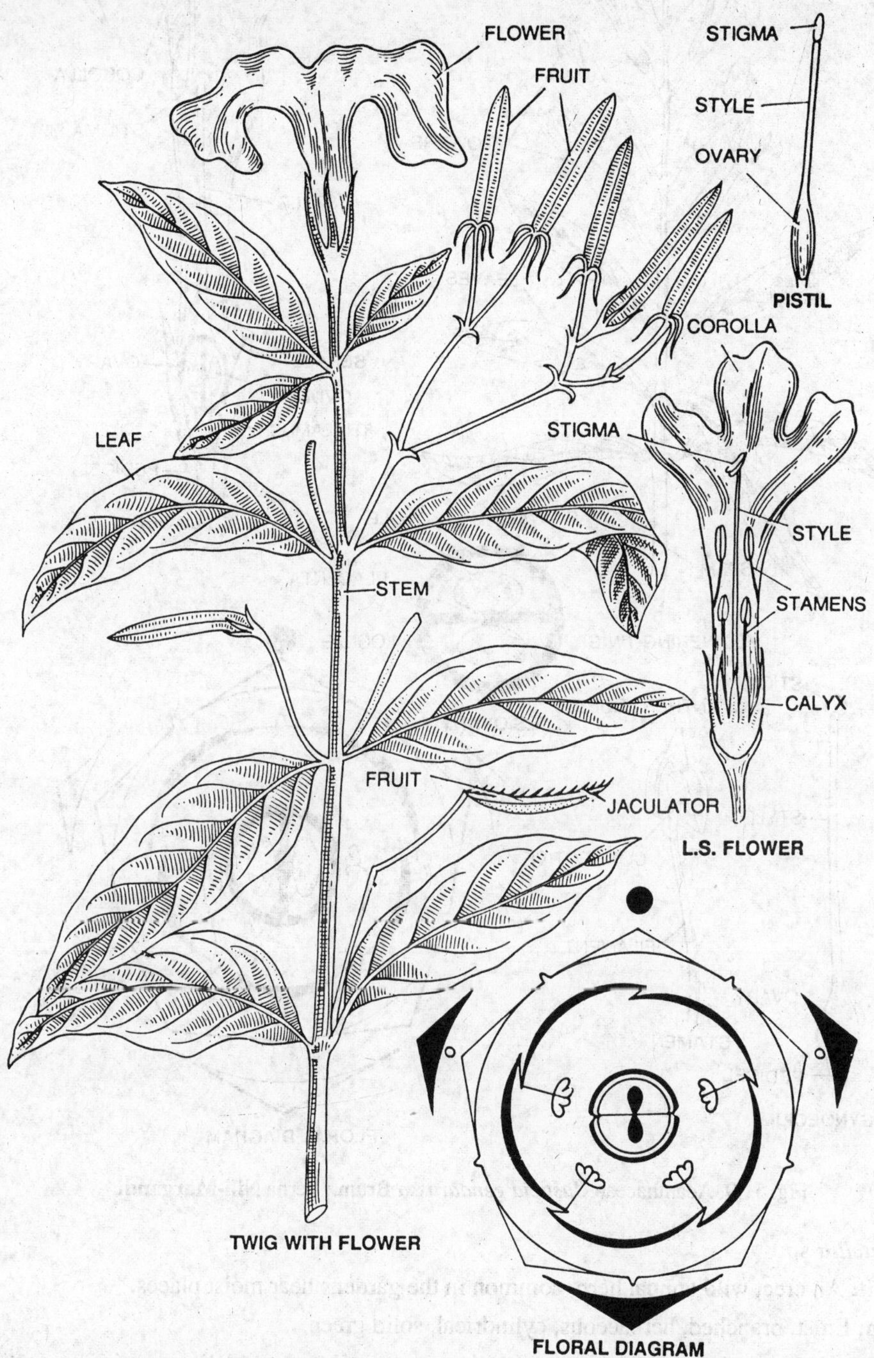

Fig.51.8. Acanthaceae. *Ruellia prostrata* Lamk.; Verna.**Kalighawani.**

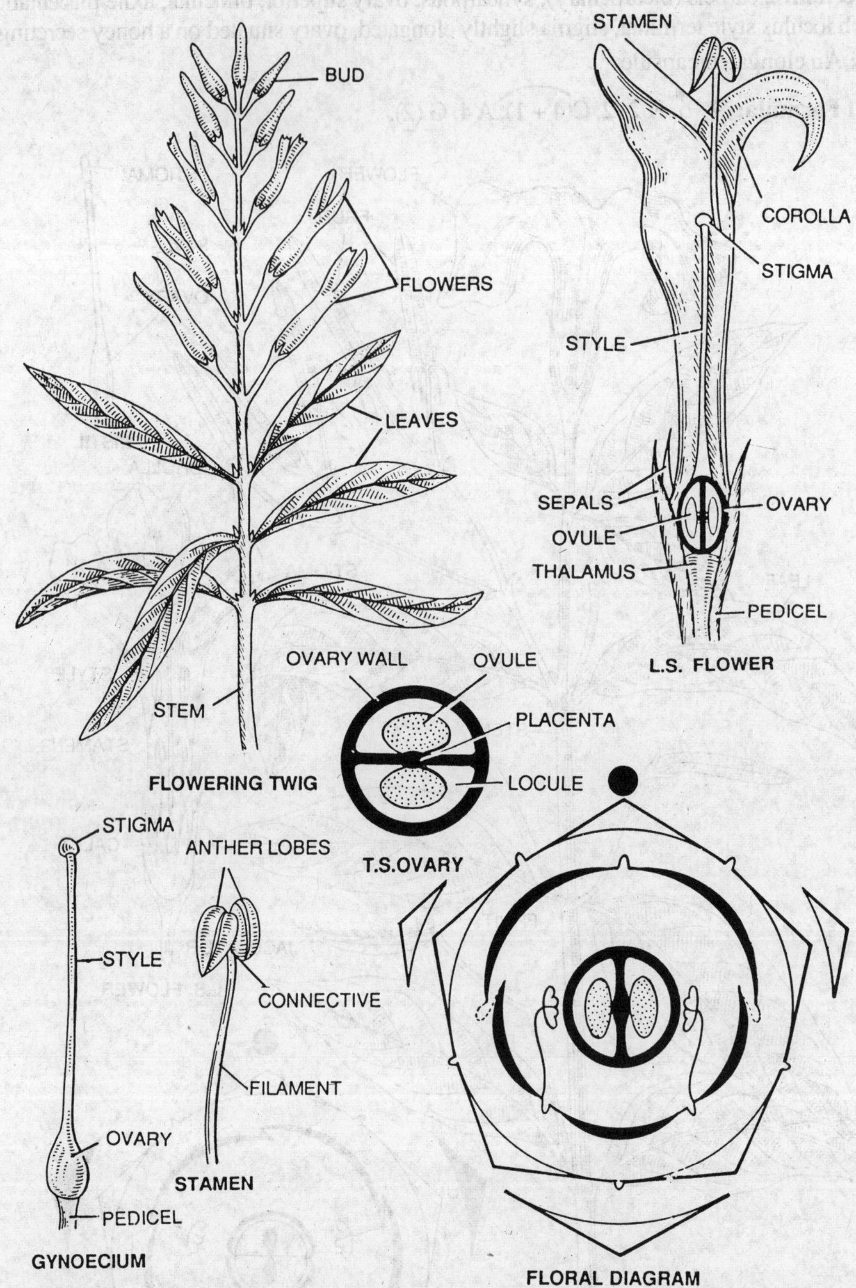

Fig. 51.9. Acanthaceae. *Justicia gendarussa* Brum.; Verna.**Nili-Margandi.**

2. ***Ruellia*** sp.

Habit: An erect wild annual herb, common in the gardens near moist places.

Stem: Erect, branched, herbaceous, cylindrical, solid green.

Leaf: Simple, opposite, decussate, subsessile, ovate, entire, acute, glabrous, unicostate reticulate venation.

Inflorescence: Cymose, dichasial cyme.

Flower: Pedicellate, bracteate, bracteolate, hermaphrodite, zygomorphic, funnel shaped, bluish pink, complete, hypogynous.

Calyx: 5 sepals, gamosepalous (connate at the base), acute, sepaloid, valvate asetivation, inferior.

Corolla: 5 lobes, funnel shaped, gamopetalous, bluish–pink, twisted aestivation, inferior.

Androecium: 4 stamens, polyandrous, epipetalous, anthers bicelled, basifixed, introrse.

Gynoecium: 2 carpels (bicarpellary), syncarpous, ovary superior, bilocular, axile placentation, one ovule in each loculus, style slender, stigma single.

Fruit: An elongated capsule with tapering end, jaculators present.

Floral formula: $\cdot|\cdot$ ⚥ $K(5), \overset{\frown}{C(5), A}\,4, G\underline{(2)}$.

3. ***Justicia*** sp.

Habit: An undershrub, commonly grown as hedge plant.

Stem: Erect, branched, cylindrical, solid, herbaceous or somewhat woody, glabrous.

Leaf: Simple, opposite decussate, cauline, petiolate, lanceolate, entire, acute, glabrous, unicostate reticulate venation.

Inflorescence: Racemose and cymose mixed.

Flower: Sessile, bracteate (bracts lanceolate), hermaphrodite, zygomorphic, complete, hypogynous.

Calyx: 5 sepals, gamosepalous, connate at the base and free above, green, inferior.

Corolla: 5 petals, gamopetalous, bilabiate, upper lip consists of two lobes, lower of three lobes, yellowish white with pink or violet dots, imbricate aestivation, inferior.

Androecium: 2 stamens, free, epipetalous, inserted on the upper part of the corolla, filaments long, anthers bicelled, basifixed.

Gynoecium: 2 carpels (bicarpellary), syncarpous, ovary superior, bilocular, axile placentation, two ovules in each loculus, style filiform, stigma bifid.

Fruit: A capsule.

Floral Formula: $\cdot|\cdot$ ⚥ $K(5), \overset{\frown}{C(2+3)A} + 2\ G\underline{(2)}$.

4. ***Peristrophe bicalyculata*** Nees. **Stem**–aerial, erect, angular, branched, solid, pubescent, green, herbaceous. **Leaf**–cauline and ramal, opposite decussate, simple, exstipulate, petiolate, ovate, acute, hairy, unicostate reticulate. **Inflorescence**–cymose, branched or panicled cyme. **Flower**–bracteate, two bracts (anterior and posterior), bracteolate, 4 bracteoles, bracts and bracteoles persistent, pedicellate, bisexual, complete, zygomorphic, pentamerous, hypogynous, cyclic. **Calyx**–5, polysepalous, valvate. **Corolla**–5, gamopetalous, bilabiate 2/3, valvate, purple. **Androecium**–2, polyandrous, epipetalous, filaments long, dithecous, basifixed, introrse. **Gynoecium**–2, bicarpellary, syncarpous, superior, bilocular, 2 ovules in each locule, axile, style filiform stigma bifid. **Fruit**–capsule, **Floral formula**– Br Brl $\cdot|\cdot$ ⚥ $K_5, C_{(2+3)}, A_2, G_{\underline{(2)}}$.

Economic Importance of the Family

The family is of little economic value. Some plants are used as hedge plants and some are ornamental. Some plants possess medicinal properties. A list of few important plants is given below.

1. *Adhatoda vasica*; Syn.*Justicia adhatoda*; Eng.–Malabar nut; Verna.–**Arusa, Adulasa**–An Ayurvedic drug is obtained from its leaves, which is used in cough, chronic bronchitis and rheumatism. The twigs and leaves are used as green manure. The plant is also used as an antiseptic and insecticide. This is an evergreen shrub commonly found in Northern India.

2. *Barleria alba*; This is a shrub grown as an ornamental.

3. *Barleria cristata*; This is an ornamental small shrub. The juice of the leaves is given in cough.

4. *Barleria prionitis*; Verna.–**Katsareya**–A plant with medicinal properties. The dried bark is given in the cough.

5. *Barleria trigosa*; Verna.–**Nila kusum**–The root is used as a remedy of severe cough.

6. *Andrographis paniculata*; Verna.–**Kiryat.** The leaves possess medicinal properties. The plants are grown as ornamental herbs.

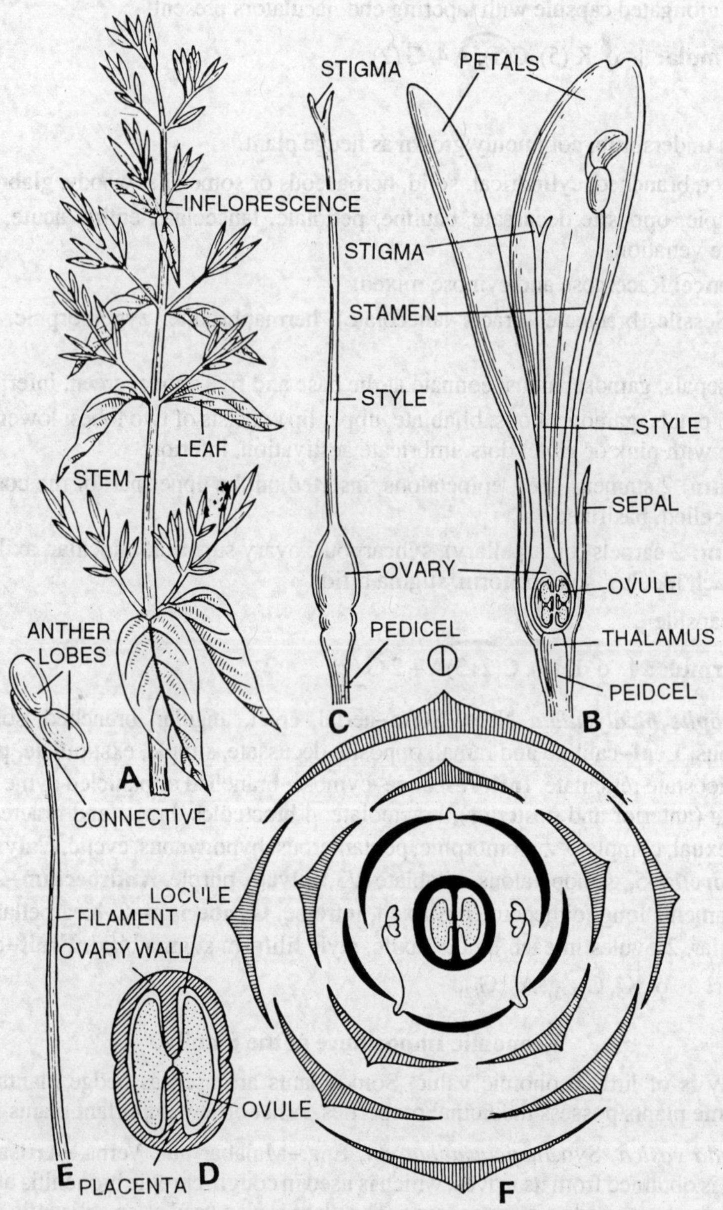

Fig.51.10 : Acanthaceae. *Peristrophe bicalyculata* Nees. A, flowering twig; B, L.S. of flower; C, gynoecium; D, T.S. of ovary; E, stamen; F, floral diagram.

7. *Asteracantha longifolia*; Verna.–**Tal makhana**–This is a medicinal plant. Its leaves, roots and seeds are given in jaundice, rheumatism and dropsy.

8. *Ruellia prostrata*; the plants possess medicinal properties. They are used as a remedy in ear troubles.

9. *Ruellia tuberosa*; This is herb or a small shrub, grown as an ornamental.

10. *Justicia betonica*; An undershrub, grown as hedge plant.

11. *Thunbergia alata*; It is a herbaceous twiner; grown as an ornamental, as it flowers throughout the year.

12. *Thunbergia coccinea*; This is grown as an ornamental.

13. *Thunbergia fragrans*; Grown in the gardens as ornamental.

14. *Thunbergia grandiflora*; Verna.–**Kukua lata**–This is an ornamental; the leaves are edible.

15. *Aechmanthera wallichii*; The lower surface of the leaves is used in making a kind of cloth by natives.

16. *Runzia parviflora*; Verna.–**Pindi**–The juice of its leaves is given to children in small–pox.

17. *Peristrophe bicalyculata*; Verna.–**Atrilal.**–Used as an antidote for snake bite.

18. *Peristrophe bivalvis*; Syn. *P. tinctoria*; A shrub; a yellow orange dye is obtained from its leaves and young stems.

19. *Ecbolum viride*; Syn. *E.linneanum;* Verna.–**Udajat**–An ornamental shrub, grown in the gardens. The roots are given as a remedy for jaundice and rheumatism.

20. *Phlogacanthus thyrsiflorus*; Verna.–**Lal bahuk**.–The flowers are used as vegetable. The fruits and leaves are given in fevers. The leaves of *Tubiflorus* are used as a washing soap.

21. *Rhinacanthus communis;* Verna.–**Palakjuhi.**–The leaves and roots are used as a remedy for skin diseases, *e.g.*, ring–worm.

22. *Jacobinia tinctoria*; This is a diffused shrub and grown as an ornamental in the gardens.

23. *Crossandra infundibuliformis*; Verna.–**Priya darsa**– This is a shrub grown as an ornamental.

24. *Asytasia gangetica*; Syn. *Justicia gangetica*; A herb, grown as an ornamental.

Systematic Relationships: Bentham and Hooker have included eight families in the order Personales.–These families are Scrophulariaceae, Orobanchaceae, Lentibulariaceae, Columelliaceae, Gesneraceae, Bignoniaceae; Pedalineae, and Acanthaceae. Engler and Prantl have included all these families in the Tubiflorae along with other twelve families. Hutchinson has included these families in his Personales. He however, puts the family Pedaliaceae in the order Bignoniales which is not related to the Personales.

Hallier considered the Scrophulariaceae to be one of the primitive families of the Tubiflorae. Bessey considered the family as an advanced member of his Scrophulariales, derived perhaps from the Bignoniaceae.

Many botanists have thought that the Acanthaceae has been derived from the Scrophulariaceae. Hutchinson considered the Acanthaceae to be most advanced among his Personales.

Family–Lentibulariaceae

Distribution: A widely spread family of 5 genera and about 260 species, occurring on all continents of the world. It comprises of insectivorous plants of aquatic and marshy habit. The largest genera are *Utricularia* 200, *Pinguicula* 32 and *Genlisea* 12. *Utricularia* with many species occurs in many parts of India. The common insectivorous aquatic plants (bladder wort) of India are *U.stellaris, U.flexuosa, U.exoleta, U.scandens*, etc.

Habit: The plants are annual and perennial herbs; aquatic and often rootless; or plants usually found in wet or marshy places; predominantly insectivorous. They are insectivorous herbs, such as *Pinguicula, Genlisea, Polypompholyx*, etc., are land plants while many species of *Utricularia, Biovularia, Saccoloria*, etc. are aquatic plants.

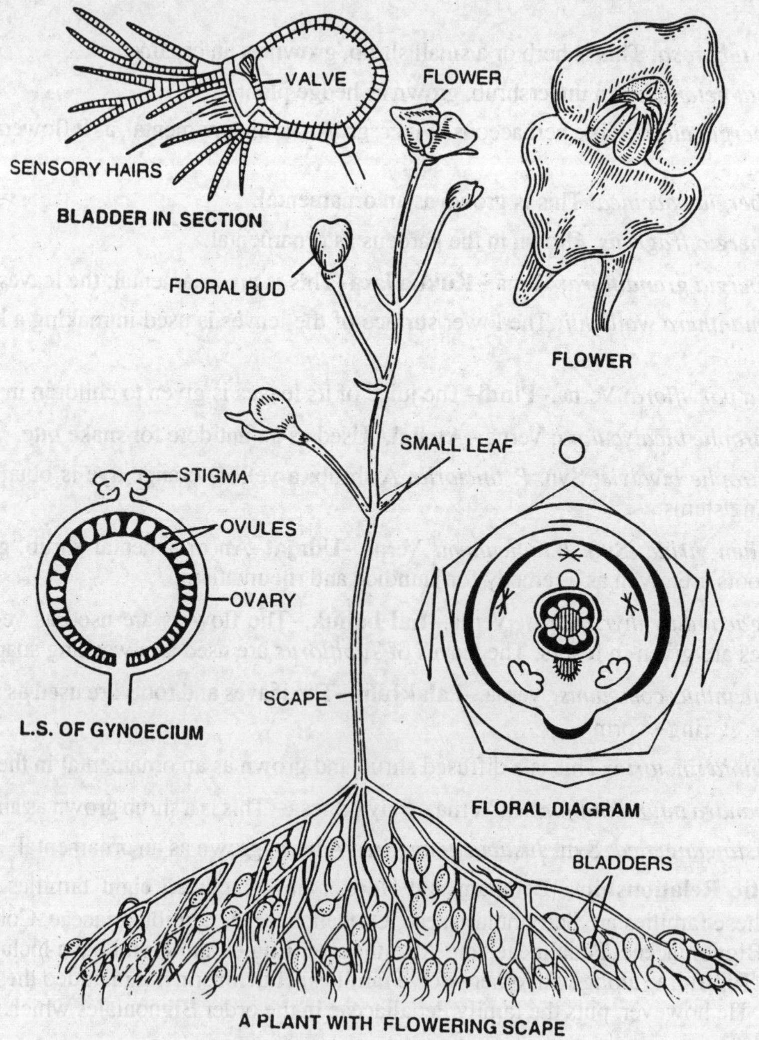

Fig. 5.11. *Utricularia eoleta* R. Br.

***Utricularia* exoleta R. Br.**

Leaves: The leaves are alternate or in basal rosettes; often dimorphic in aquatic members with submerged leaves usually finely divided, and bearing insectivorous bladders. In *Utricularia*, the bladders are thus metamorphosed ultimate leaf segments meant for catching aquatic small insects. The mouth of each bladder is fitted with a valve which opens inwards but closes outwards as a result of the stimulus carried by sensory hairs found at the entrance of the mouth of the bladder. After closing up of the bladder by valve the tiny insects trapped inside cannot come out, they die and their decayed proteinaceous bodies are absorbed by special glandular hairs lining inside the wall of the bladder.

Pinguicula has true roots, rosette of simple entire leaves bearing glandular hairs on the dorsal surface and single–flowered scape. In *Pinguicula* the leaves, like *Drosera*, are the insect catching traps. The glandular hairs of leaves with adhesive secretion infold and capture the insect preys.

Various kinds of glands are found on the surface of leaf. In *Utricularia* these are small capitate gland found while in *Pinguicula*, the glands are stalked and each gland has a basal cell, stalk cell and disc–shaped multicellular secretory cap; some sessile glandular hairs are also present.

Inflorescence: Racemose, racemes of spikes, scapes or solitary on scape (*e.g.*, in *Pinguicula*)

Flowers: Flowers bracteate, pedicels of some each with a pair of bracteoles, bisexual, zygomorphic and hypogynous.

Calyx: Sepals 5, gamosepalous, united in open bilabiate or five lobed, usually persistent, valvate or sometimes imbricate.

Corolla: Petals 5, gamopetalous, united in bilabiate or personate corolla, lower lip spurred or saccate at base, with the palate very variable in form near the throat.

Androecium: Stamens 2, with 2 staminodes sometimes produced, arising from extreme base of corolla tube, anthers monothecous (1–celled), the anther to be sometimes medianly constricted, dehiscing longitudinally.

Gynoecium: Carpels 2, syncarpous united in 1–celled superior ovary, style very short or stigma sessile, bilobed, the posterior lobe of the stigma sometimes aborted; ovules many, anatropous, rarely only two; placentation free central; ovules often sunken in the placental tissue.

Fruit: Many seeded capsule dehiscing along 2–4 valves or along annular or irregular slits; rarely indehiscent one seeded.

Seeds: Usually small, exendospermic with a poorly differentiated embryo.

Pollination: Entomophily; the wide lower lip of the corolla affords landing of the insects that are attracted by nectar stored in the spur. Insects after sucking nectar from one flower already become pollen dusted and when visit another flower touch the stigmatic lobe that is sensitive and closes up on being touched. Thus eross pollination is effected by insects.

Floral Formula: ·|· ⚥ $K_{(5)}, C_{(5)}, A_2, G_{\underline{(2)}}$

Bentham & Hooker(1862)	**Engler & Prantl(1931)**	**Hutchinson(1959)**
Phanerogams	Phanerogams	Angiospermae
Dicotyledones	Dicotyledoneae	Dicotyledones
Gamopetalae	Sympetalae	Herbaceae
Bicarpellatae	Tubiflorae	Personales
Personales	Lentibulariaceae	Lentibulariaceae.
Lentibulariaceae		

Economic Importance of the Family

The members of the family are of little economic value. A few species of *Pinguicula* (butterwort) are grown in the houses as ornamentals, and species of *Utricularia* (bladderwort) are grown in aquaria for ornamental purposes.

Differences: The Lentibulariaceae superficially resemble some of the Scrophulariaceae, but are distinguished from most of the Scrophulariaceae in having 2 stamens, and from all by the presence of free–central placentation. The presence of the small bladders on aquatic members, or other forms of insect traps, is also distinctive.

52

ORDER–GERANIALES

According to Hutchinson the family Geraniaceae belongs to his 79th order–the Geraniales of division Herbaceae, sub-phylum Dicotyledones and Phylum Angiospermae. He has placed four other families, *i.e.*, Limnanthaceae, Oxalidaceae, Trapaeolaceae and Balsaminaceae in this order. Bentham and Hooker have placed this family in the seventh order the Geraniales along with ten other families including Linaceae, Ochnaceae, Meliaceae, Rutaceae, etc., in their series Disciflorae of Dicotyledones. Engler and Prntl also included in their twenty third order Geraniales along with other nineteen families including Linaceae, Rutaceae, Meliaceae, Euphorbiaceae, etc.

Herbs, shrubs or trees; flowers cyclic, actino or zygomorphic, pentamerous, hypogynous; calyx usually persistent; androecium obdiplostemonous; carpels 5-2, syncarpous, usually 2-2 ovuled, sometimes, more ovuled; ovules provided with ventral raphe.

Family–Geraniaceae

There are about 11 genera and 600 species in this family.

Distribution. The members of this family are distributed in both temperate and tropical regions of the world including alpine regions of mountains. *Erodium* is found in Europe. *Dirachma socotrana* is endemic species in Socotora. The genus *Geranium* is largest and found in all parts of the world including India. In India 3 genera and about 25 species occur mostly in Himalayas. The common examples are *Geranium* and *Pelargonium.*

Habit : They are commonly annual herbs or undershrubs and very rarely shrubs (*e.g., Dirachma*).

Stem : Stems are often woody below in herbs. In many genera stems are rhizome or tuber-like at the base. They are thick and fleshy by which plants can face long dry seasons.

Leaves : The leaves are alternate, stipulate rarely opposite. They are incised or palmitifid or incised upto the base, or compound, rarely entire. A xerophytic species of *Sarcocaulon* possesses small leaves metamorphosed into spines. Stipules are small, but in *Geranium wallichianum* and *Pelargonium zonale* they are large in size.

Inflorescence : Cymose or solitary or biclustered. In *Pelargonium* the flowers are arranged in umbels.

Flowers : The flowers are bracteate, bracteolate, bisexual, actinomorphic, rarely zygomorphic, hypogynous, pentamerous, complete. In *Pelargonium* the flowers are zygomorphic.

Calyx : Sepals 5, rarely 4 or 8, polysepalous, rarely gamosepalous, imbricate, rarely twisted, the posterior sepal sometimes spurred (*e.g.,* in *Pelargonium*).

Corolla : Petales 5, rarely 4 or 8, polypetalous, imbricate, rarely twisted. Petals alternate to sepals.

Androecium : Stamens 10 in two whorls, obdiplostemonous, outer stamens shorter than inner, rarely 15 in three series, all fertile or some are sterile, or due to abortion they are 2-9 or 7 in number, anthers dithecous, basifixed with small connectives. In *Erodium* the stamens of outer series are reduced to scale-like staminodes. In *Pelargonium* only 2-4 stamens are fertile. In *Geranium* a nectary is found at the base of each stamen of inner series. Filaments at the base are united, rarely free.

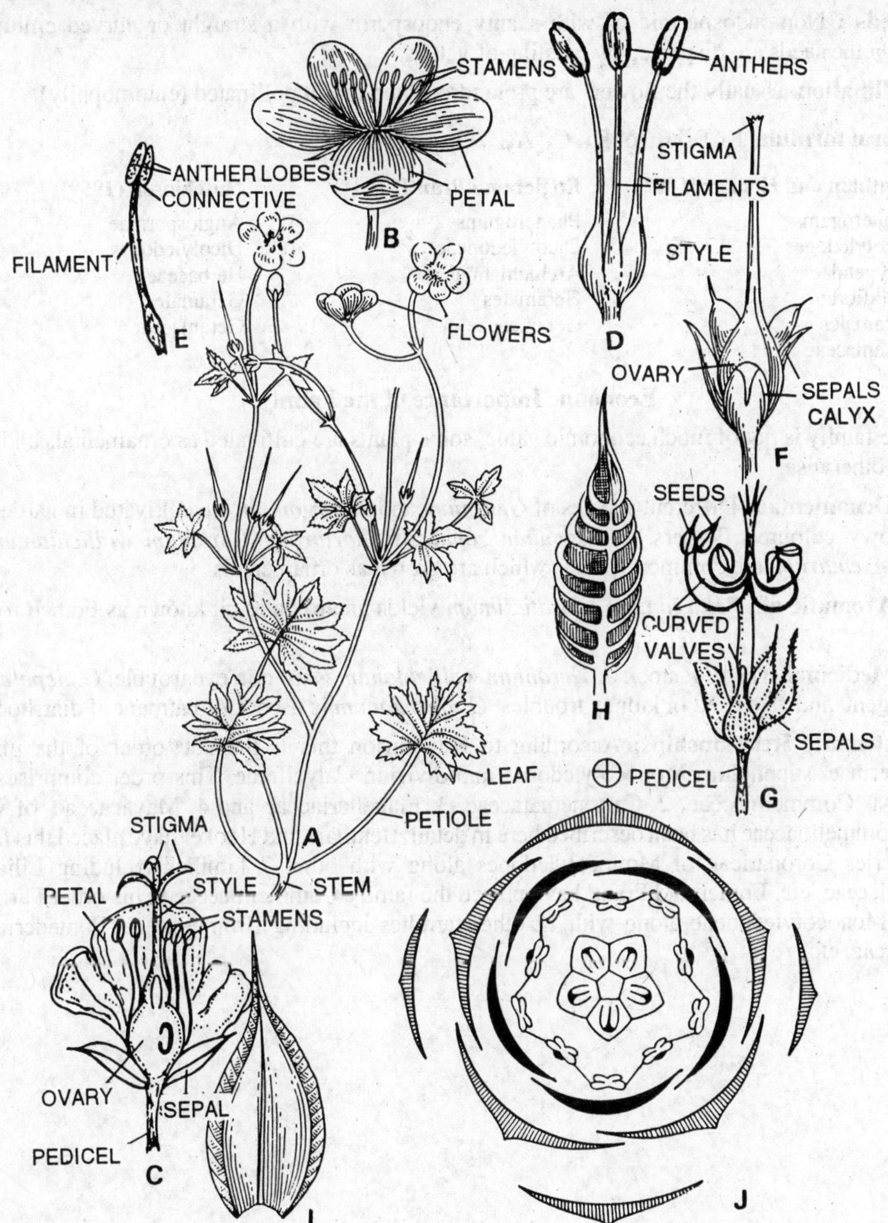

Fig.52.1. Geraniaceae. *Geranium ocellatum*. A, flowering twig; B, flower top view; C, L.S. of flower; D, stamens; E, stamen; F, gynoecium; G, fruit dehiscing showing curved valves with seeds; H, dorsal view of the fruit valves; I, ventral view of the same; J, floral diagram.

Gynoecium: Carpels 3-5, rarely 2-3 or 8, syncarpous, ovary superior, 3-5 locular; usually 2 ovules in each locule, rarely many or solitary in each locule, mostly campylotropous, superposed, pendulous, micropyle directed upwards, raphe ventral; ovary terminates into a beaked structure at the apex; stigma ligulate; style long; placentation axile.

Fruit : Usually a capsule, dehiscing into 3-5 one or more-seeded mericarps. Each part (valve) is twisted or rolled.

Seeds : Non-endospermic or with scanty endosperm with a straight or curved embryo. In *Geranium* the seeds are dispersed by twisting of valves.

Pollination. Usually the flowers are protandrous and insect pollinated (entomophily).

Floral formula. Br Brl ⊕ ⚥ $K_5, C_5, A_{5+5}, \underline{G}_{(5)}$.

Bentham and Hooker (1862)	Engler and Prantl (1931)	Hutchinson (1959)
Phanerogams	Phanerogams	Angiospermae
Dicotyledones	Dicotyledoneae	Dicotyledones
Polypetalae	Archichlamydeae	Herbaceae
Disciflorae	Geraniales	Geraniales
Geraniales	Geraniaceae	Geraniaceae
Geraniaceae		

Economic Importance of the Family

The family is not of much economic value, some plants are cultivated as ornamentals and some are used otherwise.

1. **Ornamentals.** Different species of *Geranium* and *Pelargonium* are cultivated in gardens for their showy coloured flowers. *Pelargonium zonale, P. hortorum, Geranium wallichianum* and *Erodium cicutarium* are common species which are grown as ornamentals.

2. **Aromatic oil.** *Pelargonium odoratissimum* yields an aromatic oil known as French rose oil or 'otto'.

3. **Medicinal.** The root stock of *Geranium wallichianum* is used in eye trouble. *G. nepalense* is an astringent and also used in kidney troubles. *G.robertianum* is used for treatment of diarrhoea.

Systematic Relationships: According to Hutchinson this is the 90t^h order of the phylum Angiospermae, subphylum Monocotyledones and division Calyciferae. This order comprises of 4 families–1. Commelinaceae, 2. Cartonemataceae, 3. Flagellariaceae and 4. Mayacaceae, of which family Commelinaceae has been described here in detail. Bentham and Hooker have placed this family in the series Coronarieae of Monocotyledones along with other 7 families including Liliaceae, Pontederiaceae, etc. Engler and Prantl have placed the family Commelinaceae in his order Farinosae of class Monocotyledoneae along with 12 other families including Bromeliaceae, Pontederiaceae, Mayacaceae, etc.

53

ORDER–BORAGINALES

According to Hutchinson this is the 81st order of the phylum Angiospermae, subphylum Dicotyledones and division Herbaceae. This order has only one family, *i.e.*, Boraginaceae. This family has been described here in detail. Bentham and Hooker have placed this family in Gamopetalae, series Bicarpellatae and order Polemoniales along with four other families, *i.e.*, Polemoniaceae, Hydrophyllaceae, Convolvulaceae and Solanaceae. Engler and Prantl have placed the Boraginaceae in the order Tubiflorae of subclass Sympetalae along with 19 other families, such as Convolvulaceae, Polemoniaceae, Verbenaceae, Labiatae, Solanaceae, Scrophulariaceae, Bignoniaceae, Pedaliaceae, Lentibulariaceae, Acanthaceae, etc.

Herbs, shrubs or trees, rarely lianes; inflorescence determinate, usually composed of 1 or more scorpioid or helicoid cymes that uncoil as the flowers open; flowers bisexual, actinomorphic, hypogynous; sepals 5; corolla 5-lobed, imbricate or contorted in bud; 5 stamens, epipetalous; carpels 2, axile placentation; style 1, gynobasic or terminal; fruit of 4 nutlets; seeds non-endospermic.

Family-Boraginaceae

Distribution: This family consists of 100 genera and about 2,000 species, distributed in temperate and tropical regions of the world. Mediterranean region is the home of many genera. The important genera of this family are—*Cordia, Heliotropium, Myosotis, Mertensia,* etc.

Habit: The members of this family are usually annual or perennial herbs, shrubs or rarely trees, such as *Cordia myxa* (Verna.**Lasora**). One of the main features of the family is the presence of thick-walled stiff hairs in the herbaceous species, usually scabrous or hispid hairy and sometimes glabrous herbs.

Leaves: Generally alternate, the lowermost leaves sometimes opposite, simple, usually entire, exstipulate. In some genera, such as in *Pulmonaria, Cynoglossum,* etc., heterophylly is marked, where the radical leaves differ in structure from cauline leaves; the cystoliths present in the leaves.

Inflorescence: Usually scorpioid cyme (cincinnus) or helicoid cyme, that uncoil as the flowers open; it may be glomerate-racemose or spicate, sometimes loosely cymose or with solitary flowers; determinate.

Flowers: Mostly bisexual, actinomorphic or rarely zygomorphic, such as in *Lycopsis* and *Echium;* hypogynous.

Calyx: Sepals 5, free or basally connate, imbricate or rarely valvate, sometimes irregular.

Corolla: Petals 5, united, imbricate or contorted in bud, rotate, salverform, funnelform, or campanulate; the corolla tube with folds or sometimes terminated or partially closed by scales or faucal appendages.

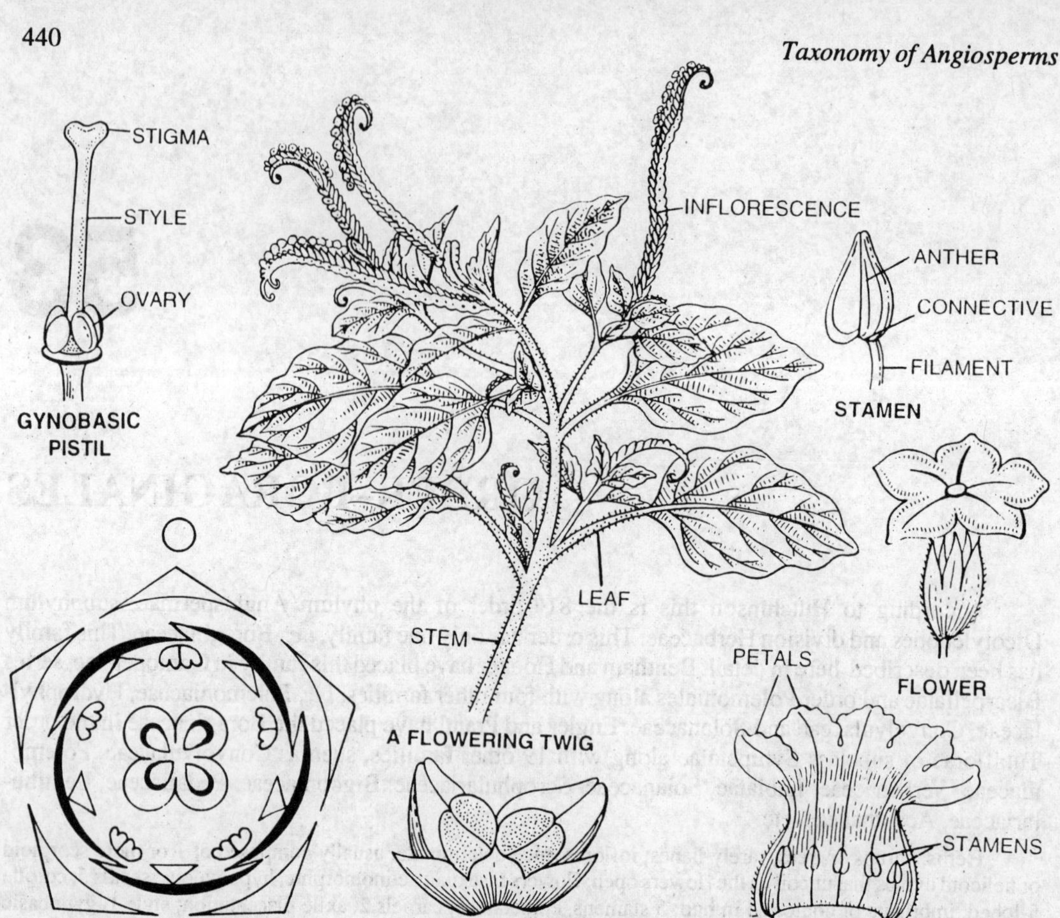

Fig. 53.1. Boraginaceae. *Heliotropium indicum* Linn.

Androecium: Stamens 5, epipetalous, equal or less commonly unequal, dorsally appendaged, such as in *Borago;* alternate with corolla lobes; anthers dithecous (2-celled), dehiscing longitudinally, basifixed or basally dorsifixed, introrse.

Gynoecium: Carpels 2, syncarpous, united in bilocular or sometimes in unilocular superior ovary, placentation axile, or sometimes apparently central or parietal; the style 1, gynobasic or terminal, generally simple; the stigma mostly simple, capitate or bilobed, such as in *Anchusa* or *Pulmonaria,* or 4-lobed, as in *Cordia;* ovules anatropous or amphitropous, many or 2 in each locule of ovary.

Fruit: Fruit usually loculicidal capsule; or fruit of 4 nutlets (2 nutlets in *Cerinthe*) or a 1-4 seeded nut or a drupe as in *Cordia.*

Seed: Seeds minute, usually without endosperm, or the endosperm is scanty and fleshy (abundant in *Heliotropium*).

Floral Formula: $\oplus$ ⚥ $K_5, \overbrace{C_{(5)}, A_5}, G_{(\underline{2})}$.

Bentham & Hooker (1862)	**Engler & Prantl (1931)**	**Hutchinson (1959)**
Phanerogams	Phanerogams	Angiospermae
Dicotyledones	Dicotyledoneae	Dicotyledones
Gamopetalae	Sympetalae	Herbaceae
Bicarpellatae	Tubiflorae	Boraginales
Polemoniales	Boraginaceae	Boraginaceae
Boraginaceae		

Economic Importance of the Family

This family is of slight economic value :

1. **Ornamentals.** Many species of about 30 genera of this family are cultivated as ornamentals, such as *Heliotropium*(helitrope), *Mertensia* (blue bells), *Myosotis* (forget-me-not), *Pulmonaria* (lung-wort), *Borago* (borage), *Anchusa* (alkanet), *Cerinthe* (honeywort), *Cynoglossum* (Hound's tongue), *Symphytum* (comfrey), *Echium* (viper's bugloss), and *Cordia.*

2. **Dye yielding**. *Alkanna tinctoria*, a dye yielding plant. Dye is obtained from the roots of the plants.

3. **Medicinal.** The species of *Cordia* and *Symphytum* are used medicinally. The fruit of *C.myxa* is diuretic and demulcent, used in afffections of urinary passages, diseases of lungs and spleen.

4. **Wood.** The wood of *Eheretia serrata* is very strong and is used for making agricultural implements and sword-hilts.

Systematic Relationships: Most of botanists have accepted the elevation of the subfamilies, such as cordioideae, Ehretioideae, Heliotropioideae, and Boraginoideae to family status, such as Cordiaceae, Ehretiaceae, Heliotropiaceae and Boraginaceae. Lawrence (1937) is of opinion, that the subfamilies were best treated as components of one family Boraginaceae. He concluded, that 'without doubt the Boraginoideae with their deeply cut nutlets, are the most highly evolved.' Hutchinson (1948), was of opinion that the woody subfamilies should be kept distinct, as Ehretiaceae.

Distinguishing Characters: The Boraginaceae are distinguished from Labiatae (Lamiaceae), Verbenaceae and other families by the leaves predominantly alternate, terete stems, the circinate cymose inflorescence, usually actinomorphic corollas, frequently with faucal appendages, and the characteristic fruit usually with an erect embryo.

54

ORDER–LAMIALES

According to Hutchinson this is the eighty-second order of the phylum Angiospermae, subphylum Dicotyledones and division Herbaceae. This order comprises of four families, *e.g.*, Myoporaceae, Selaginaceae, Globulariaceae and Labiatae, of which the Labiatae has been described in the text in detail, Bentham and Hooker have placed the Labiatae along with three other families including Verbenaceae, in the tenth order the Lamiales of class Dicotyledones, division Gamopetalae and series Bicarpellatae. Engler and Prantl have placed the Labiatae along with nineteen other families including Convolvulaceae, Verbenaceae, Solanaceae, Scrophulariaceae, Pedaliaceae, Acanthaceae, etc., in the sixth order the Tubiflorae of class Dicotyledoneae and sub class Sympetalae.

Leaves mostly opposite or whorled; corolla zygomorphic; stamens 4 or 2; ovary becoming lobed with a gynobasic style; ovules mostly paired. Largely herbaceous, with inflorescence tending to capitula or verticillasters.

Family–Labiatae (Lamiaceae)

There are about 200 general and 3200 species in this family.

Distribution: The members of this family are found throughout the world. They are being distributed both in temperate and tropical regions. The chief centre of distribution of the family is the mediterranean region. In our country the family is represented by several important genera, such as, *Mentha, Ocimum, Leucas, Coleus, Pogostemon,* etc.

Habit: Majority of the plants are annual or perennial herbs, inhabiting the temperate regions. In warmer climates, the plants become shrubby in nature. Trees are rare. Certain plants are xerophytes with extremely reduced leaves. A few species are climbers, but the climbing habit is very rare.

Some important herbs are, *Coleus aromaticus, C. forskohlii, Leucus cephalotes, Mentha arvensis, Ocimum sanctum, O. basilicum,* etc., *Salvia aegyptiaca, Coleus rotundifolius* are undershrubs, *Ocimum gratissimum, Meriandra bengalensis* are shrubs. Certain species of Brazilian genus *Hyptis* and Indian genus *Leucosceptrum* are trees. The certain species of *Mentha* and *Lycopus* are found in marshy places. A few American species of *Scutellaria* are climbers. Certain plants are extreme xerophytes, *e.g., Rosmarinus*. In this plant the leaves are rolled back and stomata are found among the hairs in the grooves on the underside of the leaf.

The propagation of many plants of this family takes place by means of new shoots which remain attached to the plant, *e.g.*, in *Saliva, Ballota.* In *Ajuga reptans* the new shoots form aerial runners with somewhat reduced leaves; in *Lycopus*, the subterranean stolons (suckers) are found with short internodes and scaly colourless leaves. The suckers are found in the species of *Mentha.*

Stem: The young shoots are usually four-sided or quadrangular.

Leaves: Usually the leaves are simple, opposite, decussate and exstipulate. They show many variations from an entire blade to toothed, lobed, cut or finely dissected as in many species of *Salvia.*

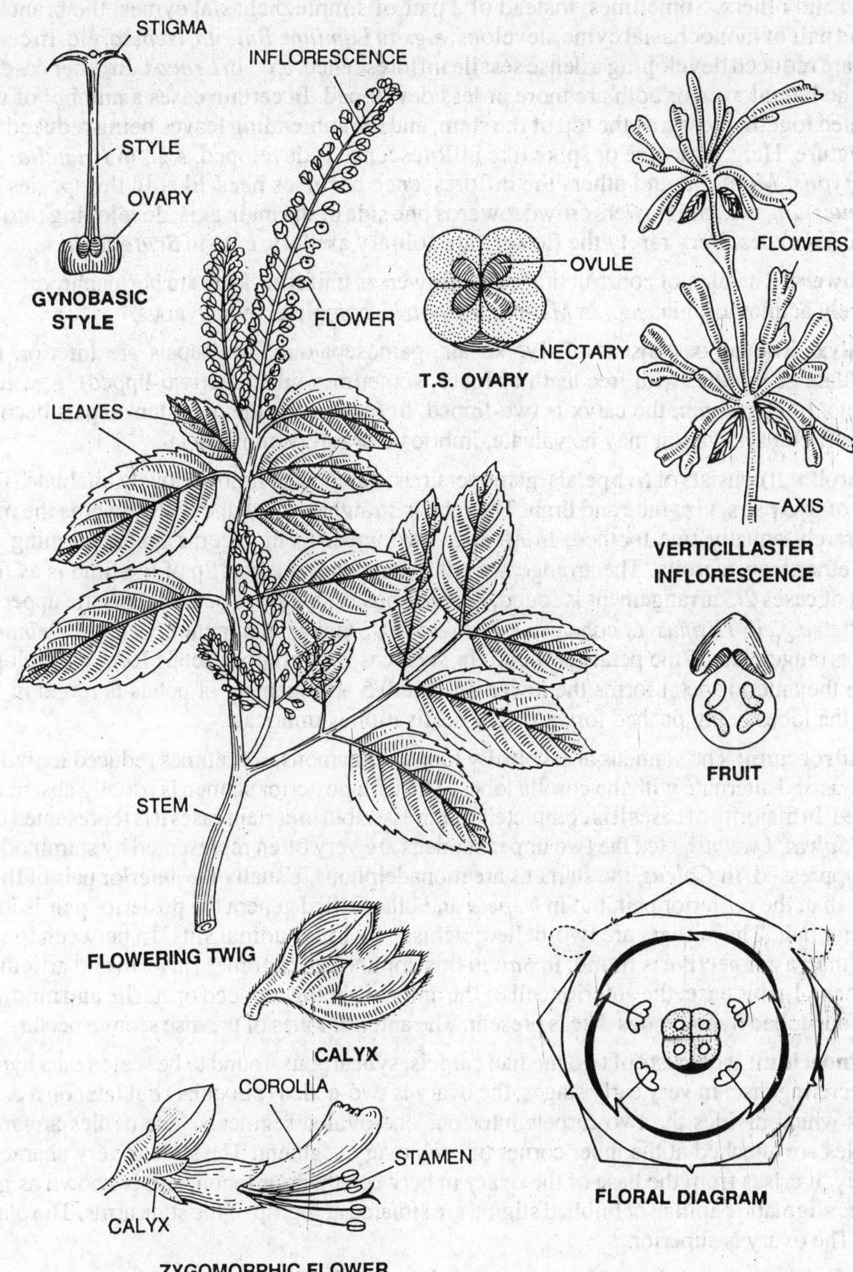

Fig.54.1. Labiatae-*Ocimum sanctum* Linn. (Verna.**tulsi**).

A whorled leaf arrangement of 3 to 8 leaves is found in some genera, *e.g.,Ocimum*. All parts of the plants, such as stem, leaves and inflorescence are more or less hairy and possess glandular hairs, which secrete characteristic scent of the genus or species. Sessile scented oil secreting glands are also found frequently on the epidermis.

Inflorescence: The characteristic inflorescence of the family is **verticillaster.** In this type of inflorescence the whorls of flowers develop at the nodes. This consists of a pair of cymose inflorescences. Each cyme forms a simple three-flowered dichasium, *e.g.*, in the species of *Salvia, Prunella* and others. Sometimes, instead of a pair of simple dichasial cymes, the branching takes place and pair of monochasial cyme, develops, *e.g.*, in *Lamium, Ballota, Nepeta,* etc. In certain cases all axes are reduced developing a dense sessile inflorescence, *e.g.*, in *Leucas*. In other cases the main axis or the lateral axes or both are more or less developed. In certain cases a number of whorls are aggregated together towards the top of the stem, and the subtending leaves being reduced to a bract-like structure. Here, a raceme or spike like inflorescence is developed, *e.g.*, in *Prunella, Teucrium,* etc. In *Hyptis, Monarda* and others the inflorescence becomes head-like. In the species of *Scutellaria, Teucrium,* etc., the flowers crowd towards one side of the main axis, developing into dorsiventral inflorescences. Very rarely the flowers are solitary axillary, *e.g.*, in *Scutellaria.*

Flowers: The plan of construction of the flowers is uniform. They are hermaphrodite, zygomorphic, rarely actinomorphic, *e.g.*, in *Mentha, Elshotzia,* complete, hypogynous.

Calyx: The calyx consists of five sepals, gamosepalous. The sepals are inferior, persistent, campanulate or tubular, with free teeth or lobes, sometimes bilabiate (two-lipped), *e.g.*, in *Thymus, Salvia,* etc. In *Orgianum,* the calyx is two-lipped, in *Hoslundia,* the persistent sepals become fleshy in the fruit. The aestivation may be valvate, imbricate or rarely quincuncial.

Corolla: It consists of five petals, gamopetalous, tubular and limb variously bilabiate. The corolla consists of two pairs, *viz.*, tube and limb. The tube is straight or bent and wide towards the mouth. The limb is rarely equally fine toothed. In *Mentha,* two upper teeth become united forming an almost regular tetramerous corolla. The arrangement of the upper and lower lip of the limb is as follows–In majority of cases 2/3 arrangement is found, here the posterior pair of petals forms the upper lip which may be flat, *e.g.*, in *Thymus,* or concave, *e.g.*, in *Salvia, Stachys,* etc. In the species of *Ocimum, Hyptis* etc., 4/1 arrangement of the petals is found. In such cases four upper petals form a developed upper lip while the anterior petal forms the lower lip. The 0/5 arrangement of petals is found in *Teucrium.* Here all the lobes being pushed forward. The aestivation is imbricate.

Androecium: The stamens are typically four, didynamous, sometimes reduced to two. They are epipetalous and alternate with the corolla lobes. The fifth posterior stamen is usually absent and rarely developed. In majority of cases it is completely suppressed but in certain cases it is represented by a staminode. In *Salvia, Lycopus,* etc., the two upper stamens are very often represented by staminodes or altogether suppressed. In *Coleus,* the stamens are monadelphous. Usually the anterior pair of the stamens is longer than the posterior pair, but in *Nepeta* and other allied genera the posterior pair is longer than the anterior pair. The anthers are two-celled; dehisce by longitudinal slits. In between the two cells of the anther, a connective is found. In *Salvia* this connective becomes filiform and articulated with the filament. In this case, the anterior cell of the anther is being reduced or sterile and modified. Generally a four-lobed hypogynous disc is present. The anterior lobes of the disc secrete nectar.

Gynoecium: It consists of two median carpels, syncarpous, found to be seated on a hypogynous, nectar secreting disc. In very early stages, the ovary is two-celled (bilocular) but later on a constriction develops which divides the two carpels into four one-ovuled segments. The ovules are anatropous. The ovules are attached at the inner corner of each ovary segment. The style is very characteristic of the family. It arises from the base of the ovary in between the four loculi, and is known as **gynobasic style.** The stigmatic papillae of bilobed stigma are situated at the tip of the style arms. The placentation is axile. The ovary is superior.

Fruit: The fruit consists of four one-seeded nutlets, included within the persistent calyx.

Seeds: Exalbuminous or with scanty endosperm. The testa is thin; the embryo possesses flat cotyledons.

Pollination: The pollination takes place through the agency of insects (*i.e.*, entomophilous).

Floral Formula: ·|· ⚥ K (5), $\widehat{C(5), A}$ 4 or 2, G $\underline{(2)}$.

Bentham & Hooker (1862)	**Engler & Prantl (1931)**	**Hutchinson (1959)**
Phanerogams	Panerogams	Angiospermae
Dicotyledones	Dicotyledones	Dicotyledones
Gamopetalae	Sympetalae	Herbaceae
Bicarpellatae	Tubiflorae	Lamiales
Lamiales	Labiatae	Labiatae or
Labiatae		Lamiaceae

Description of some important plants in semitechnical language.

1. ***Ocimum sanctum;*** (Verna.–**Tulsi**).

Habit: A perennial herb with typical aromatic smell.

Stem: Erect, branched, quadrangular, somewhat woody, solid, branches covered with soft hairs.

Leaf: Simple, opposite, short petioled, exstipulate, ovate, serrate, acute, gland dotted, unicostate reticulate venation.

Inflorescence: Verticillaster, 6-10 flowered whorls present, emitting fine smell.

Flower: Pedicellate, bracteate, bracts small and caducous, hermaphrodite, zygomorphic, complete, purple, hypogynous.

Calyx: 5 sepals, gamosepalous, bilabiate, petaloid (purple coloured), posterior lip broad and boat shaped, anterior lip with 4 small lobes possessing mucronate teeth, gland dotted, imbricate aestivation, inferior.

Corolla: 5 petals, gamopetalous, bilabiate, corolla tube short, upper lip four lobed, lower lip large, imbricate aestivation, inferior.

Androecium: 4 stamens, polyandrous, didynamous, epipetalous, fifth posterior stamen completely suppressed, anthers bicelled, introrse, dorsifixed.

Gynoecium: 2 carpels (bicarpellary), syncarpous, ovary bilocular in early stage but becomes tetralocular in later stage, ovary superior, four chambered, axile placentation, single ovule in each loculus, gynobasic style (*i.e.,* it arises from the base of the ovary), stigma bifid.

Fruit: Schizocarpic, carcerulus, 4 nutlets developed.

Floral Formula: ·|· ⚥ **K** (1 + 4), $\overset{\frown}{C(4+1), A}$ 2 + 2, G $\underline{(2)}$.

2. ***Salvia officinale.***

Habit: An annual ornamental herb with typical aromatic smell.

Stem: Erect, branched, herbaceous, quadrangular, solid, hairy and green.

Leaf: Simple, opposite, decussate, petiolate, exstipulate, ovate, serrate, acute, unicostate reticulate venation.

Inflorescence: Verticillaster with usually six-flowered whorls.

Flower: Pedicellate, bracteate, hermaphrodite, zygomorphic, complete, coloured, hypogynous.

Calyx: 5 sepals, gamosepalous, bilabiate, 2 lobes consisting of one lip and three the other lip, petaloid, imbricate aestivation, inferior.

Corolla: 5 petals, gamopetalous bilabiate, 2 lobes in one whorl and 3 in the other, imbricate aestivation, inferior.

Androecium: 2 stamens (anterior), posterior, 3 stamens absent or staminodes present, epipetalous, short filament connected to a long connective, lower end of connective is flat, short and sterile, upper end of it with a large, fertile anther lobe, anthers bicelled, introrse.

Gynoecium: 2 carpels (bicarpellary), syncarpous pistil present on nectar secreting disc, ovary superior, bilocular when young, but becomes tetralocular in the later stage, single ovule in each loculus, axile placentation, gynobasic style, bifid stigma.

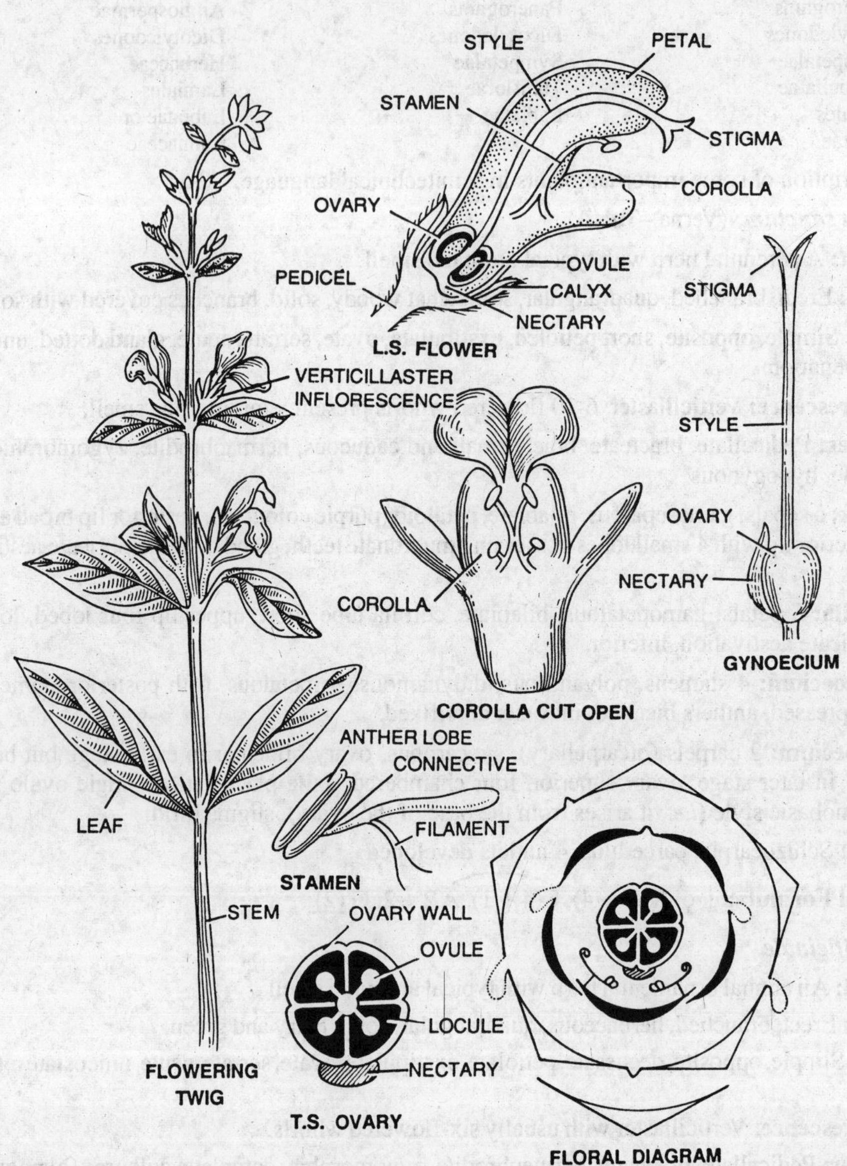

Fig.54.2. Labiatae (Lamiaceae). *Salvia officinalis* Linn; Eng. sage; Verna. **salbia sefakuss.**

Fruit: Four nutlets in a group.

Floral Formula: ·|· ⚥ **K** (2 + 3), **C**(2 + 3), **A** 2, **G** (2).

Economic Importance of the Family

The family is of little economic value. Some plants are grown in the gardens as ornamentals, some plants yield essential oil while certain possess medicinal properties. A list of few important plants is given below.

1. *Coleus* sp.; Several species of *Coleus* are grown as ornamentals in the gardens for their variously coloured beautiful foliage.

2. *Coleus amboinicus;* Syn. *C. aromaticus*; Eng.–Indian borage; Verna.–**Pathorchur**–This is a common herb. Its leaves are used for flavouring food products.

3. *Coleus forskohlii;* Syn. *C.barbatus*: This is a herb, cultivated in South India. The roots are used as spice and condiment.

4. *Coleus rotundifolius;* Eng.–Madagascar potato; Verna–**Koorkan**–This is a herb or shrub commonly grown in South India. The underground tuberous stems are eaten as vegetable.

5. *Colebrookea oppositifolia*; Verna.–**Binda**–The leaves are applied to wounds. The wood is used for making gun powder.

6. *Leucas cephalotes*; Verna.–**Goma**–This is a common herb. The flowers are used medicinally for cough and cold.

7. *Leucas zeylanica;* Verna.–**Gattatumba**–The fresh juice of leaves is used as a remedy for headache and cold. It is also applied to skin diseases. The flowers are used for cough and cold.

8. *Meriandra bengalensis;* Verna.–**Kafur ka pat**– This is a shrub grown as an ornamental. The leaves possess medicinal properties. They emit a camphor–like scent. The leaves are used to preserve clothes against insects.

9. *Mentha arvensis*; Eng.–Field mint; Verna.–**Pudina**–This is a perennial herb. The leaves yield an essential oil which is used in the manufacture of certain kinds of cigarettes and pharmaceuticals. It is also used as a carminative, stimulant and refrigerant. Usually grown in the Punjab, Western Himalayas, Kashmir and Kumaon hills.

10. *Mentha arvensis* var *piperascens;* Eng.–Japanese peppermint. The leaves are the source of an essential oil, from which menthol is prepared. It is used in the treatment of colds. The plant is a herb and commonly cultivated in Jammu and Kashmir.

11. *Mentha longifolia;* Syn. *M. sylvestris*; Eng.–Mint; Verna.–**Pudina**–This is an aromatic herb. The leaves are used for flavouring purposes. The dried leaves are used as stimulant and carminative. Usually cultivated in Uttar Pradesh, Kashmir, Maharashtra and the Punjab.

12. *Mentha piperita*; Eng.–Peppermint; Verna.–**Vilayati pudina**–The leaves are used for flavouring purposes. An essential oil is obtained from the leaves, which is used in perfumery and soap making industries. The oil is also used medicinally as stimulant, carminative and for sickness and vomitting. This is a herb, cultivated in the Punjab, Kashmir and Maharashtra.

13. *Mentha pulegium;* Eng.–Europcan pennyroyal–An essential oil is extracted from the leaves and tops of the plants, which is used in perfumery and cosmetics. It is herb, cultivated in Jammu and Kashmir.

14. *Mentha spicata*; Syn. *M. viridis*; Eng.–Spearmint; Verna.–**Pahari pudina**–The leaves yield an essential oil, known as spearmint oil, which is used for flavouring food products. This herb is cultivated in the Punjab, Uttar Pradesh and Maharashtra.

15. *Lavandula officinalis;* Syn. *L.vera; L. spica*; Eng.–Lavender–This is a small shrub. It is native of Europe but now grown in Jammu and Kashmir. An essential oil is extracted from its flowers which is used in perfumery.

16. *Ocimum sanctum*; Eng.–Holy basil; Verna.–**Tulsi**–This is a small shrub and treated as holy plant by Hindus. Usually grown in the courtyards of Hindus and by the side of temples. This plant possesses several medicinal properties. The juice of the leaves is used as a stimulant and as a remedy for bronchitis. The seeds are used as a remedy for urino–genital troubles. The plant is used as an antidote for snake–bite.

17. *Ocimum kilimandscharicum*; Eng.–Camphor basil. The camphor is obtained from the leaves It is cultivated on a very small scale in West Bengal, Assam, Tamil Nadu and Dehradun.

18. *Ocimum basilicum*; Eng.–Basil; Verna.–**Bantulsi**–This is an aromatic herb cultivated throughout our country. The leaves are used for flavouring purposes. The seeds are used medicinally as a remedy for dysentery and diarrhoea.

19. *Ocimum gratissimum*; Eng.–Shruby basil; Verna.–**Ranitulsi**–This is a common ornamental, aromatic shrub. The plant possesses medicinal properties. The seeds are given in headache. This is also used as a mosquito repellant.

20. *Pogostemon perilloides;* Verna.–**Pacholi**–This is a herb. The leaves yield an essential oil, which is used for making soaps and perfumes. The dried leaves are used for scenting clothes to keep off insects. It is commonly cultivated in the Western Ghats and Nilgiris.

21. *Rosmarinus officinalis;* Verna.–**Rusmori**–This is a herb grown as an ornamental. It yields an essential oil which is used in perfumery.

22. *Roylea calycina*; Verna.–**Patkarru**–The plant possesses medicinal properties and used as a bitter tonic.

23. *Salvia officinalis*; Eng.–Sage; Verna.–**Salbia, sefakuss, Seesti**–This is a small shrub, grown as an ornamental. This is native of Mediterranean region.

24. *Salvia splendens;* Syn. *S.colorans;* Eng.–Scarlet sage–This is grown as an ornamental.

25. *Satureia hortensis*; This is a herb found in Kashmir. The aromatic stem and leaves are used for flavouring food products.

26. *Anisomeles indica*; Syn. *A. ovata;* Verna.–**Gopali, Kalabhangra**–This is a woody herb found throughout our country used as fodder and for flavouring purposes.

Systematic Relationships: Bentham and Hooker have included five families in the order Lamiales. They are Myoporineae, Selagineae, Verbenaceae, Labiatae and Plantagineae. Engler and Prantl have included (excluding Selagineae) in the order Tubiflorae. Hutchinson has included the families Myoporaceae, Selagineae, Globulariaceae and Labiatae in his order the Lamiales and the Verbenaceae in the Verbenales.

The Family Verbenaceae is related phylogenetically to the Labiatae. Bessey has also included these families within the Lamiales. Originally Hutchinson (1926) included the Verbenaceae in the Lamiales but later on (1948) he transferred the family Verbenaceae in the Verbenales. He considered that the Verbenaceae and Labiatae are quite unrelated and the Verbenaceae have been derived from Rubiaceae.

THE MONOCOTYLEDONES

55

ORDER–BUTOMALES

According to Hutchinson this is the eighty-third order of the phylum Angiospermae, subphylum Monocotyledones and division Calyciferae. This order comprises of the families Butomaceae and Hydrocharitaceae. The Hydrocharitaceae has been described here in detail. Bentham and Hooker have placed the Hydrocharideae along with two others including Orchideae in the first series the Microspermae of class Monocotyledones. Engler and Prantl have placed the family Hydrocharitaceae along with six other families in the second order the Helobiae of class Monocotyledoneae. The family has been considered to be highest evolved within the order.

Aquatics; apocarpous or syncarpous; ovary superior or inferior; ovuless numerous, scattered over the walls of the carpels, no endosperm.

Family–Hydrocharitaceae

There are sixteen genera and about ninety species in this family.

Distribution: All plants are aquatic herbs. They are found either in sea or in fresh water. The marine genera are found in tropical or subtropical regions, the fresh water genera are also found in the temperate zones. In our country the family is represented by the genera, *e.g., Hydrilla, Vallisneria, Hydrocharis,* etc. *Hydrilla verticillata,* the only species, is widely distributed throughout the warmer part of the world. *Vallisneria* (two species only) is commonly found in the warmer parts of both hemispheres. *Halophila, Enhalus* and *Thalassia* are found in the sea.

Habit: As mentioned, the plants are small aquatic herbs. Most of them occur in fresh water and some are marine. The plants are either free floating or submerged. *Hydrocharis morsusranae* floats on the surface of still water. *Stratiotes* and *Vallisneria* are quite submerged, the flowers only appear at the surface of water when mature. *Vallisneria* grows in the mud at the bottom of fresh water. The species of *Halophila* remain completely submerged. *Elodea canadensis* is quite submerged, the flowers appear at the surface when mature.

Stem and Leaves: In *Hydrocharis,* the stem is quite reduced, sending out numerous spreading fibrous roots and stolons. At the end of stolon new leaf-rosettes arise. The plants of *Stratiotes aloides* also possess stolons. *Elodea canadensis,* possesses an elongated, slender branched stem bearing whorls of narrow toothed, leaves. In *Vallisneria,* the shortened stem bears a tuft of roots and a crowded cluster of long, linear leaves. It also sends out horizontal runners, at the end of which a new.plant is developed. In *Halophila* a marine genus the elongated stem possesses roots at the nodes. In *Enhalus* (marine-monotypic genus), the stout stem becomes clothed with smooth black threads, the persistent hard bast strands of the leaf. *Thalassia* (marine genus) possesses the creeping rooting system which bears numerous scale leaves and several upright branches with distichous ribbon-like leaves. *Hydrilla*, a fresh water form possesses weak, slender, elongated stem bearing whorls of leaves at nodes.

Hydrocharis sp., possesses rosettes of rounded kidney-shaped leaves. *Stratiotes aloides* has a rosette of stiff tapering leaves with a spiny margin. The young leaf-rosettes sink down in the water during winter season and again come to the surface on the approach of spring season, and thus, the plant overwinters and protects itself from frost, etc., and therefore, the plant is known as water-soldier. *Elodea* possesses whorls of narrow toothed leaves. Generally the leaves are basal and often crowded, or cauline, when cauline they are alternate, opposite or whorled, usually sessile.

Inflorescence: The inflorescence is always lateral. Usually there is one inflorescence in each leaf-axil. However, in *Vallisneria spiralis* three floral shoots as well as a foliage shoot arise from the leaf axil. The young inflorescence remains enclosed within a spathe formed by a two free or somewhat united bracts which very often persist upto the maturation of the fruit. Generally the flowers are unisexual, and very rarely, they become hermaphrodite. The plants may be monoecious or dioecious. In *Halophila*, both male and female flowers are found within the same spathe. The male inflorescence consists of one to many flowers. The female and bisexual inflorescences are generally one-flowered.

Flower: Usually the flowers are actinomorphic, rarely zygomorphic (*e.g.*, in *Vallisneria*), trimerous and epigynous.

Perianth: It is generally white and may easily be distinguished into an outer calyx and inner corolla. The petals remain inserted on the united base of the sepals. The calyx is tough and protective in function, whereas the corolla is fugacious and delicate. In *Hydrocharis,* the flowers are white and of one inch in diameter. The sepals are quite small, and the petals are broad and membranous. The female flowers generally possess staminodes, but the male flowers do not have any trace of pistil. The flowers of *Halophila* do not have petals and the perianth is represented by three sepals.

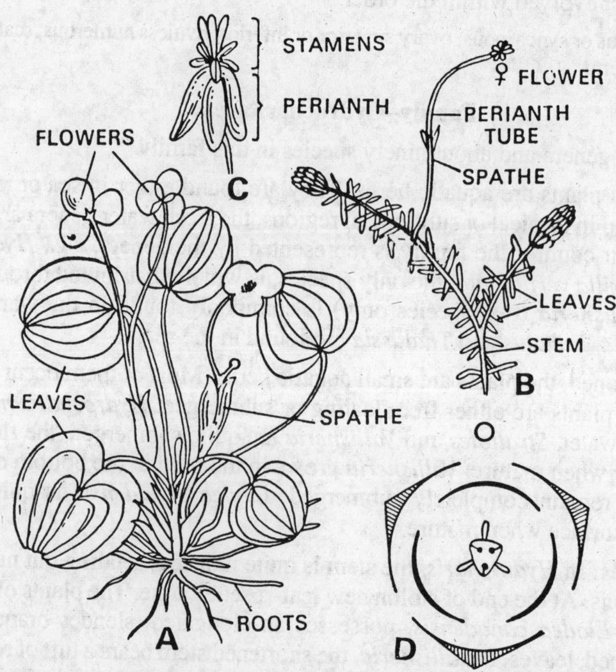

Fig.55.1. Hydrocharitaceae. A, male plant of *Hydrocharis* sp.; B, branch of female plant of *Elodea*; C, staminate flower of *Anacharis canadensis;* D, floral diagram of female flower of *Elodea.*

Androecium: In *Hydrocharis,* the androecium consists of four whorls, the innermost whorl comprises of only staminodes, while in the third whorl there are half-anthers and the two whorls having three stamens in each whorl.

Gynoecium: There are 2-15 carpels which are syncarpous. The ovary is inferior, unilocular with parietal placentation. The placentas are often produced into the centre. Generally the ovules are numerous.

Fruit: The fruits are berry-like, indehiscent, leathery or fleshy, submerged and opening irregularly.

Seeds: The seeds are many, non-endospermic and containing a large embryo.

Pollination: The pollination takes place through the agency of water; (*i.e.*, hydrophily). In several species, the female flowers are raised to the surface of the water by their long stalks to receive the pollen grains from the detached floating male flowers.

Floral Formula : ⊕ ·|· ♂, ♀ or ⚥, K(3), C3, A3—15 $(2\overline{—15})$.

Bentham & Hooker (1862)	Engler & Prantl (1931)	Hutchinson (1959)
Phanerogams	Phanerogams	Angiospermae
Monocotyledones	Monocotyledoneae	Monocotyledones
Microspermae	Helobiae	Calyciferae
Hydrocharideae	Hydrocharitaceae	Butomales
Hydrocharitaceae		

Economic Importance of the Family

The family is of little economic value. The list of some important plants is given below:

1. *Hydrilla verticillata*–The plants are used for refining the sugar. They are also used in experiments of photosynthesis conducted in the laboratory. They can be easily grown in the small ponds of botanical gardens.

2. *Vallisneria spiralis*–The plants are submerged, and found in water ponds.

3. *Hydrocharis* sp.–The plants are floating, and may be grown as ornamental in the garden ponds for their beautiful foliage and flowers.

4. *Elodea canadensis*–They are submerged plants.

Systematic Relationships: According to Bentham and Hooker the Hydrocharideae is placed in the first series the Microspermae of class Monocotyledones, including three families, *i.e.*, Hydrocharideae, Burmanniaceae and Orchideae. Engler and Prantl have placed these families in separate orders. They placed families in Hydrocharitaceae along with other six families in the second order Helobiae, Orchidaceae and Burmanniaceae in the eleventh order the only two families naming Butomaceae and Hydrocharitaceae in the order Butomales. According to Hutchinson the Hydrocharitaceae is closely related to the Butomaceae. Miki (1973), however, placed the Hydrocharitaceae higher within the order and considered it the ancestor of Najadaceae on the basis of an assumed inferior ovary in Najadaceae.

56

ORDER–ALISMATALES

According to Hutchinson this is the 84th order of the phylum Angiospermae, subphylum Monocotyledones and division Calyciferae. According to him this order consists of *three* families–1. Alismataceae, 2. Scheuchzeriaceae and 3. Petrosaviaceae, of which Alismataceae has been described here in detail. Bentham and Hooker have placed family Alismaceae along with *two* other families–Triurideae and Najadaceae in their sixth series, Apocarpae of class Monocotyledones. Engler and Prantl have included the family Alismaceae along with six families in the order Helobiae of class Monocotyledoneae.

Perennial or annual marsh or aquatic herbs; plants cauline or erect or with floating leaves; roots fibrous from stout rhizome; leaves basal, long petioled, sagittate or hastate; raceme or panicle; flowers trimerous; fruit achene or follicle; seed endospermic.

Family–Alismataceae
(Water Plantain Family)

According to Lawrence this family consists of 14 genera and about 55-60 species.

Distribution: The family is of wide distribution, but mainly in temperate and tropical regions of the northern hemisphere. In this country the family is represented by 7 genera and 10 species. The important genera are–*Sagittaria* and *Alisma.*

Habit: Mostly the plants are perennial or annual marsh or aquatic herbs.

Roots and stem: The fibrous roots are developed from stout rhizomes. The plants may be cauline and erect or with leaves floating (*e.g.,Sagittaria* spp).

Leaves: The leaves are basal, long-petioled, sheathing basally; the blades vary from linear to ovate or the bases sagittate or hastate, reticulately veined with few to many primary parallel veins converging apically and numerous close and parallel transverse veins.

Inflorescence: Raceme or panicle.

Flower: Pedicellate, often in whorls, bracteate, bisexual or unisexual (*e.g.,* in *Sagittaria*), complete or incomplete, actinomorphic, trimerous, hypogynous.

Perianth: 4 tepals, polytepalous (free segments), tepals arranged in two series, regular, imbricate, the outer 3 tepals are green and sepal-like, herbaceous and persistent, the inner three larger and petaloid segments are deciduous. In *Butomus* the perianth segments of both series are petaloid.

Androecium: Stamens 6, or more, rarely 3, free, anthers dithecous, extrorse or dehiscing by lateral slits. In *Alisma,* there is a single whorl of six stamens. In *Butomus* and *Echinodorus* the first whorl of 6 stamens is followed by a second whorl of 3 stamens. In *Sagittaria* the outer whorl of six

stamens is followed by numerous stamens. In *Hydrocleis* and *Limnocharis* the fertile stamens remain surrounded by sterile stamens or staminodes. In *Wisneria* there is a whorl of three stamens.

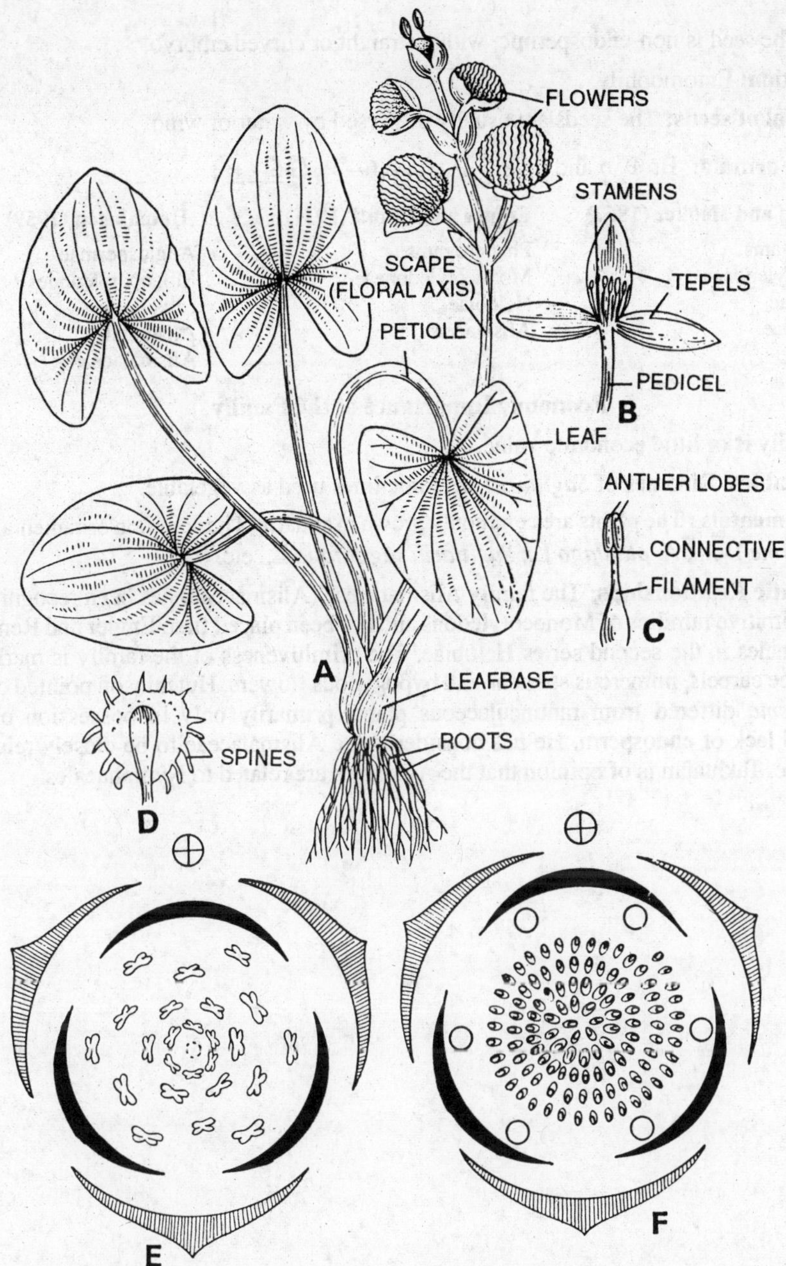

Fig.56.1. Alismataceae. *Sagittaria* spp. A, flowering plant; B, male flower; C, stamen; D, fruit (achene); E, floral diagram of male flower; F, floral diagram of female flower.

Gynoecium: 6 to many free carpels, spirally arranged or in a single whorl. In *Wisneria* there are 3 carpels; in *Butomus*, 6 carpels in two whorls of three each; in *Alisma* about 18 carpels whereas in *Sagittaria* the number of carpels is indefinite (many); the receptacle is flat or convex; ovary superior, unilocular, the ovules/are solitary or many, basal, the style mostly needle-like and persistent, the stigma simple and 1 and hardly may be distinguished from style.

Fruit: An achene (*e.g.*, in *Alisma*), rarely follicular (*e.g.*, in *Butomus*), when follicle it dehisces basally.

Seed: The seed is non-endospermic, with a straight or curved embryo.

Pollination: Entomophily.

Dispersal of seeds: The seeds are usually dispersed by water or wind.

Floral Formula: Br ⊕ ♀ and ♂ or ⚥ P_{3+3}, A 6—∝, G $\underline{6—∝}$.

Bentham and Hooker (1862)	Engler and Prantl (1931)	Hutchinson (1959)
Phanerogams	Phanerogams	Angiospermae
Monocotyledones	Monocotyledoneae	Monocotyledones
Apocarpae	Helobiae	calyciferae
Alismaceae	Alismaceae	Alismatales
		Alismataceae

Economic Importance of the Family

The family is of little economic value.

1. **Vegetables:** Rizomes of *Sagittaria* are sometimes used as vegetable.

2. **Ornamentals:** The plants are commonly grown in aquaria. The common ornamental plants are *Alisma reniforme* Don.; *A. plantago* L.; *Sagittaria sagittifolia* L., etc.

Systematic Relationships: The family Alismataceae (Alismaceae) has been recognized as one of the most primitive families of Monocotyledons and has been placed first. Engler and Rendle placed it after Pandanales in the second series Helobiae. The primitiveness of the family is marked by the presence of free carpels, numerous stamens and hypogynous flowers. Hutchinson pointed out that the genus *Ranalisma* differed from ranunculaceous plants primarily only in possession of a single cotyledon and lack of endosperm. He has considered the Alismataceae to be closely related to the Ranunculaceae. Takhtajan is of opinion that the Alismales are related to Nymphaeales.

57

ORDER–COMMELINALES

According to Hutchinson this is the 90th order of the phylum Angiospermae, subphylum Monocotyledones and division Calyciferae. This order comprises of 4 families-1. Commelinaceae, 2. Cartonemataceae, 3. Flagellariaceae and 4. Mayacaceae, of which family Commelinaceae has been described here in detail. Bentham and Hooker have placed this family in the series Coronarieae of Monocotyledones along with other 7 families including Liliaceae, Pontederiaceae, etc. Engler and Prantl have placed the family Commelinaceae in his order Farinosae of class Monocotyledoneae along with 12 other families including Bromeliaceae, Pontederiaceae, Mayacaceae, etc.

Herbs or rarely undershrubs; bisexual or unisexual, cyclic; outer perianth segments sepaloid or hyaline; inner often petaloid; frequent reduction in androecium; ovules mostly orthotropous; endosperm mealy.

Family—Commelinaceae

According to Lawrence there are 37 genera, of which 9 are monotypic and about 600 species.

Distribution: The members of this family are mostly tropical and subtropical plants, and found to be distributed over warmer parts of the earth. About 10 genera and 76 species are found in Southern and Western parts and in the tropical Eastern Himalayas of this country. The common examples are from the genera–*Tradescantia, Commelina, Zebrina* and *Rhoeo.*

Habit, Stem and Roots: They are succulent perennial or annual herbs having nodose stems. The roots are fibrous or sometimes much thickened and tuberlike (*e.g.*, in *Aneilema*). Climbing plants are rare (*e.g., Streptolirion*) and *Spatholirion*; *Cochliostema* is the only epiphyte.

Leaves : The leaves are alternate, flat or troughlike, entire, with parallel venation, sheathing by a basal membranous and often closed sheath.

Inflorescence : The inflorescence is usually monochasial cyme (cincinnus). It may be terminal, terminal and axillary or less often axillary, a simple or compound helicoid cyme or thyrse, or the flower solitary, sometimes subtended by a spathe or foliaceous bracts.

Flower : The flowers are usually actinomorphic, or in some genera zygomorphic, hermaphrodite, trimerous, hypogynous.

In *Commelina benghalensis* small cleistogamic flowers are found on subterranean branches of rhizome.

Perianth : The perianth consists of 2 series. The outer series consists of 3 free, green sepal-like tepals, imbricate (rarely gamotepalous and rarely the tepals petaloid); the inner series consists of three petaloid (coloured) tepals, free, equal or unequal, rarely united into a tubular structure, the third tepal sometimes much reduced.

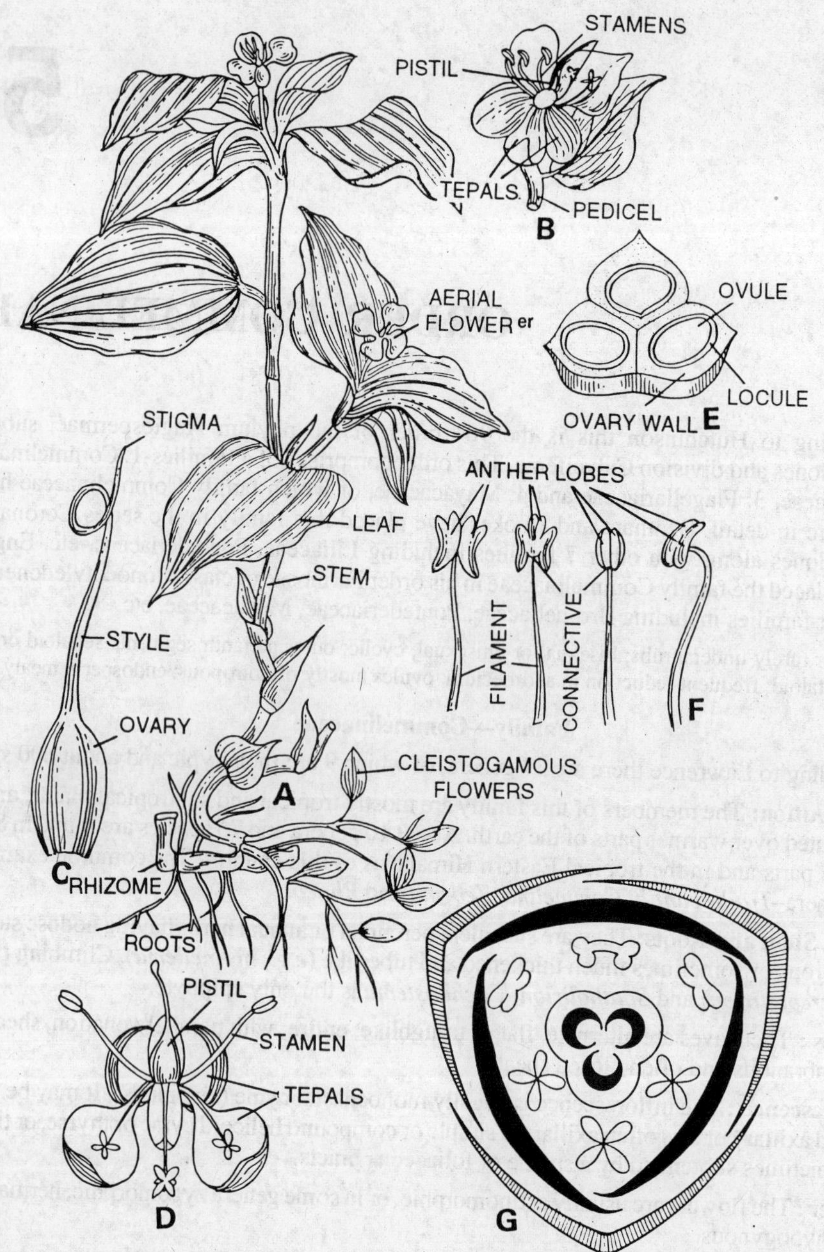

Fig.57.1. Commelinaceae. *Commelina benghalensis*. A, flowering plant also with cleistogamous flowers; B, a cleistogamous flower; C, gynoecium; D, aerial flower; E, T.S. of ovary; F, stamens; G, floral diagram.

Androecium: Stamens 6, but often the fertile stamens reduced to 3 by abortion (*Pollia* and *Commelina*), aborted stamens are generally represented by staminodes or rarely 1 stamen functional, no staminodes in *Callisia,* the anthers are dithecous, 2 parallel or divergent cells, dehisce by longitudinal slits or by an apical pore (*e.g.,* in *Dichorisandra*), the filaments are distinct and rarely connate, often bearded with coloured moniliform hairs.

Gynoecium. Carpels 3, syncarpous, ovary superior, trilocular, the ovules 1-few per locule and orthotropous, placentation axile, style 1, stigma 1 and capitate or trifid.

Fruit. A loculicidal capsule, sometimes enclosed by fleshy outer perianth segments. Rarely the fruit is fleshy and indehiscent (*e.g.*, in *Pollia*).

Seeds. The seeds are endospermic with copious and mealy endosperm. They are provided with a punctiform to linear funicular scar, usually netted, muricate or ridged. In *Dichorisandra* the seed is arillate. The embryo is situated beneath a conspicuous disc-like callosity on seed coat. The fruits and seeds are dispersed by birds and other small animals.

Pollination. Entomophilous. The insects are attracted by attractive colours of the petals and moniliform coloured hairs on the stamens.

Floral Formula. Br ⊕ ⚥ P_{3+3} A_6 $G_{\underline{(2)}}$

Bentham and Hooker (1862)	Engler and Prantl (1931)	Hutchinson (1959)
Phanerogams	Phanerogams	Angiospermae
Monocotyledones	Monocotyledoneae	Monocotyledones
Coronarieae	Farinosae	Calyciflorae
Commelinaceae	Commelinaceae	Commelinales
		Commelinaceae

Economic Importance of the Family

The family is of little economic value.

1. **Ornamentals.** The species of *Rhoeo, Tradescantia, Zebrina* and *Commelina* are grown in the pots as ornamental for their coloured foliage.

2. **Vegetables.** The rhizomes of *Commelina benghalensis* and the leaves of *C. diffusa* are eaten as vegetable.

3. **Fodder.** Some species of *Commelina* are used as cattle fodder.

Systematic Relationships. The family Commelinaceae is one of the four families of Hutchinson's 90th order the Commelinales which forms a link between Helobieae and Liliflorae because it resembles Liliaceae in nuclear endosperm and trimerous flowers. However, it differs from Liliaceae in nature of endosperm. Engler and Rendle placed this family in Farinosae after Spathiflorae. This family has strong relationship with Pontederiaceae. According to Hutchinson this family has its origin from Helobieae (*i.e.*, Butomales and Alismatales).

58

ORDER–ZINGIBERALES

According to Hutchinson this is the ninety fourth order of the phylum Angiospermae, subphylum Monocotyledones and division Calyciferae. The order comprises of six families, *i.e.,* Musaceae, Strelitziaceae, Lowiaceae, Zingiberaceae, Cannaceae and Marantaceae of which Musaceae, Zingiberaceae and Cannaceae have been described in the present text in detail. Bentham and Hooker have given the name Scitamineae to the Musaceae and have placed in their second series the Epigynae of class Monocotyledones. However, they have not recognized the Zingiberaceae and Cannaceae. Engler and Prantl have included these families (*i.e.,* Musaceae, Zingiberaceae and Cannaceae) along with one more, *i.e.,* Marantaceae in the tenth order the Scitamineae of class Monocotyledoneae.

Herbs with rhizomes; leaf sheath usually open and with a ligule; calyx and corolla in separate whorls; stamens 6-5, or often only 1, the remainder staminodes and petaloid; ovary inferior; seeds with endosperm, often arilate.

Family–Musaceae

According to Hutchinson there is only one genus *Musa* including about forty five species in this family. He has placed the genera, *Strelitzia, Ravenala, Heliconia* in a separate family known as Strelitziaceae. Engler and Prantl have included all these genera together with *Musa* and two others in the family Musaceae.

Distribution: The species of the family are confined to tropical regions. Most of them are moisture loving plants. In our country the family is represented by *Musa paradisiaca* (Verna.–**Kela**) and *Musa textilis,* cultivated in South India.

Habit: Mostly perennial herbs of big dimensions. They are tree like in appearance and described as gigantic herbs. Most of them possess an underground branched rhizome and a pseudoaerial stem.

Root: Adventitious.

Stem: Underground rhizome or root stock. The pseudoerect aerial stem of the so called banana 'tree' reaching upto 10-15 feet height is actually formed by the long, stiff and overlapping leaf sheaths. The erect aerial shoots die after they bear fructifications (fruits). The apparent aerial stem is termed as **shaft.**

Leaf: The leaves are simple, sometimes very large upto 6 feet or more, entire, glabrous, with a strong midrib and lateral parallel veins. The leaves possess large and overlapping sheaths; petiolate, the petiole is long and thick. The leaves are rolled up in early stages but later on they spread in large laminas. Due to wind pressure the leaves are usually torn into ribbon like structures, because of the absence of secondary veins.

Inflorescence: The flowers are borne in terminal spikes or panicles which are covered by spathes and the floral clusters often found in the axils or large and showy bracts. In *Musa,* the large red bracts are arranged in three spiral lines round the peduncle, each bract covers several unisexual flowers which arise without any bract from the axis of the spike. At the opening of the flowers, the bracts roll back and ultimately fall off.

Flower: Sessile, hermaphrodite, sometimes unisexual, *e.g.,* in some species of *Musa,* zygomorphic, epigynous. When unisexual, the plants are monoecious. The male and female flowers are arranged on the same inflorescence, the male flowers are found within the upper bracts while the female ones are within the lower bracts towards the base of the inflorescence. The flowers are complete when hermaphrodite and incomplete when unisexual. In *Musa,* sometimes the hermaphrodite flowers occur in the middle of the inflorescence (between male and female clusters).

Perianth: Six perianth leaves are found to be arranged in two whorls of three each; gamophyllous or polyphyllous (united or free). In *Musa,* the three outer and the two inner anterior perianth leaves are united into a tube-like structure. The inner perianth leaf is free. Usually perianth leaves (tepals) are coloured.

Androecium: Six stamens, free, usually arranged in two whorls; five stamens are perfect while the sixth one is either absent or rudimentary (staminode). In *Musa* the sixth stamen is absent. The anthers are bicelled, basifixed, introrse.

Gynoecium: Three carpels, syncarpous; ovary inferior trilocular, axile placentation, many ovules in each loculus; style filiform; three branched stigmas.

Fruit: An elongated fleshy, three chambered, indehiscent berry. In cultivated varieties the ovules are aborted and the fruits are seedless. In wild varieties the seeds are present.

Seeds: Exalbuminous with a straight embryo.

Pollination: Entomophilous, by insects and birds.

Bentham & Hooker(1862)	**Engler & Prantl (1931)**	**Hutchinson (1959)**
Phanerogams	Phanerogams	Angiospermae
Monocotyledones	Monocotyledoneae	Monocotyledones
Epigynae	Scitamineae	Calyciferae
Scitamineae	Musaceae	Zingiberales
Musaceae		Musaceae

According to Engler and Prantl the family Musaceae is included in the order Scitamineae of monocots. Rendle has also included the family in the same order. Bentham and Hooker anyhow placed this family in the order Epigynae. Hutchinson puts this family in his own way in the order Zingiberales.

Floral Formula: ⚥, ♀ or ♂, P 3 + 3 or (3 + 2) + 1, A 3 + 2 + 1 (staminode), $\overline{G_{(3)}}$.

Description of *Musa* in semi-technical language.

Musa sp.

Habit: Perennial gigantic herb, sometimes 15 feet high, cultivated or wild.

Root: Adventitious.

Stem: Underground rhizome, also present a false erect aerial stem formed by the long, stiff and overlapping leaf sheaths, apparent aerial stem known as **shaft**.

Leaf: Simple large sometimes six feet long or more, entire, glabrous, apex obtuse, with a strong midrib and parallel venation, petiolate, with long and thick petiole, exstipulate, lamina tears between the veins and often makes its appearance like pinnately divided leaf.

Inflorescence: A terminal spike covered by red bracts arranged in three spiral lines round the floral axis, each bract covers many unisexual flowers, bracts are caducous and red in colour.

Flower: Sessile, monoecious, unisexual, zygomorphic, epigynous, male flowers occur above the female in the bracteate inflorescence, bisexual flowers also found within the middle bracts.

Perianth : Six perianth leaves (tepals), arranged in two whorls of three each, three outer and two inner anterior perianth leaves unite to form a tube like structure, sixth perianth leaf (tepal) is free, petaloid.

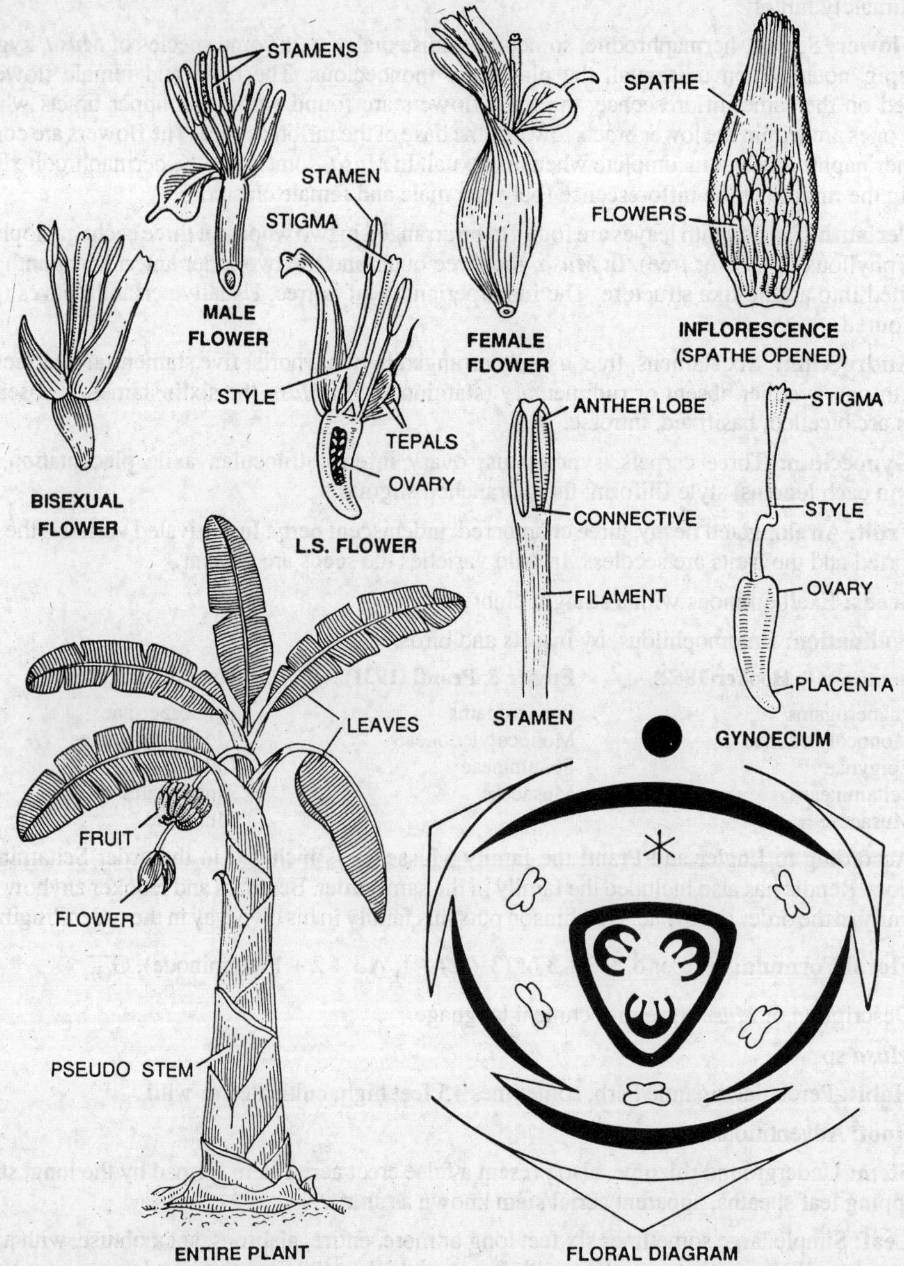

Fig.58.1. Musaceae-*Musa paradisiaca* Linn. Verna.–**Kela;** Eng.–Banana.

Androecium: Six stamens, five stamens perfect and one (posterior) rudimentary (staminode), sometimes the sixth stamen altogether absent, stamens free, anthers bicelled and basifixed.

Gynoecium: Three carpels, syncarpous; ovary inferior, trilocular; axile placentation; each loculus with many ovules, in *Musa paradisiaca* and *Musa sapientum* the ovules are very short and do not develop in seeds, and therefore, the fruits are seedless; style simple, filiform; stigma with three branched lobes.

Fruit: An elongated berry. The mesocarp is edible.

Seeds: Undeveloped, if present exalbuminous.

Generic characters: Apparent aerial stem 'shaft' is present; inflorescence terminal spike with large, leathery and reddish bracts; six stamens, five stamens perfect, one staminode; ovlues do not develop in seeds; fruit an elongated berry.

Floral Formula : ·|· ♀, ♂ or ⚥ , P (2+3) + 1, A5, $G_{(3)}$.

Economic Importance of the Family

1. *Musa paradisiaca;* Eng.–Banana; Verna.–**Kela**–This is most useful among all other species. The fruits are rich in starch and vitamins. Unripe fruits are cooked as vegetable. They are also very rich in sugars and useful for health and digestion troubles. The juice of leaves and pseudo stem is also used in many ways medicinally. In many places the banana leaves are used as dining plates on festive occasions. Among Hindus the banana fruits and plants are known as most sacred. They are used on every festive occasion. It is native of India and Malaya, chiefly cultivated in Assam, Madhya Pradesh, Bihar, Uttar Pradesh, Andhra Pradesh, Maharashtra, Tamil Nadu and Kerala.

2. *Musa chinensis*; (Verna.–**Chinea kela**), yields small edible bananas which are sweet, rich in starch and vitamins.

3. *Musa textilis*; Eng.–Manila hemp, abaca, the fibre yielding plant yields good fibre which is woven and made into abaca cloth. It is native of the Philippines but cultivated in South India. The fibre is also used for twine, bagging and wrapping paper.

Family–Zingiberaceae

According to Lawrence, there are about 47 genera and 1400 species in this family.

Distribution: The plants of this family are found mainly in the tropics of both worlds, but chiefly in the Indo-Malayan region where about 50% of the total genera are found.

Habit: The plants are perennial herbs, perennating for a short or long period by means of a sympodial, fleshy, either elongated or tuberous rhizomes. The rhizomes of *Zingiber officinale* (**Adark**) is creeping, thick jointed, branched and sympodial. These branches grow obliquely upwards and develop leafy aerial shoots from their terminal bud. The plant of *Curcuma longa* (**Haldi**) possesses a thick tuber like rhizome with rounded segments having transverse ring-like leaf scars. The aerial shoots, whenever present are short and covered by leaf-sheaths, with the result a pseudo-stem like that of *Musa* of Musaceae may be formed.

Root: Usually the roots are thick and fleshy. Sometimes they are spindle shaped, *e.g., Globba* and in some plants they are slender with tuber like ends, *e.g., Curcuma.*

Leaves: They are usually two ranked, but sometimes spiral, *e.g.,* in *Costus*. The leaves possess short stalks and sheathing bases. The long sheath grows upwards into a ligule. This ligule varies in shape and size in different genera. It is often of a different colour from rest of the leaf. The leaf blade is lanceolate, ovate or oblong with a strong mid-rib. The venation is parallel possessing numerous ascending, parallel lateral veins. The leaf lamina remains convoluted in the bud.

Inflorescence: The inflorescence may be terminal, *e.g.,* in the species of *Alpinia* or it may develop on a special scale leaf bearing shoot arising from the rhizome, *e.g., Zingiber officinale.* Sometimes it arises from the base of the leafy stem. It is a bracteate spike or raceme. Sometimes the flowers are arranged either in heads of cymes, *i.e.,* monochasial cyme (a cincinnus). The monochasial cyme is found to be developed in each bract-axil. The bracts are spirally arranged. They are stiff and coloured. In the secies of *Amomum* and *Zingiber* the bracts closely overlap, thus giving the inflorescence a cone like appearance. The bracteoles are often sheathing.

Perianth: It consists of six tepals (perianth leaves) arranged in two whorls of three each. It is usually distinguished into distinct calyx and corolla.

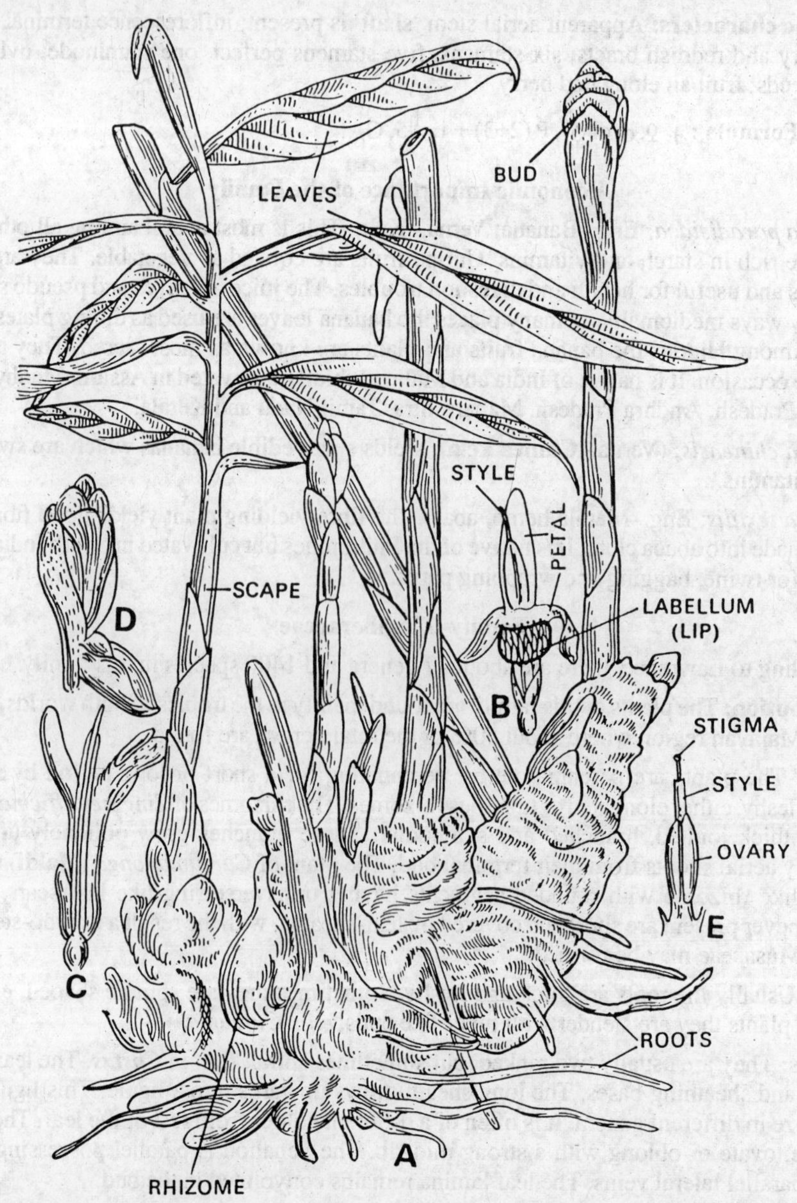

Fig. 58. 2. *Zingiber officinale*. (Hindi—**Adrak**); A, plant of Ginger (**adrak**) showing branching rhizome bearing leaves anf flower shoots: B, flower; C, flower in vertical section; D, flower bud; E, pistil.

Calyx: It consists of three sepals, gamosepalous. The odd sepal is anterior. The calyx is tubular or campanulate, dividing above into three short teeth. Sometimes splits on one side.

Corolla: It consists of three petals, polypetalous. It is tubular below and separates above into three similar or dissimilar limbs. The edges of the posterior petal cover the edges of the lateral ones.

Androecium: Usually it consists of six stamens arranged in two whorls of three each. The stamens are found inserted on the throat of the corolla tube. The outer whorl of the stamens may be entirely suppressed, *e.g.,* in the species *Costus, Renealmia,* but it is usually represented by two lateral leafy staminodes, the anterior one being always completely suppressed. The inner staminal whorl is complete and out of the three stamens the median or the posterior stamen is fertile and epipetalous, while the lateral ones unite to form the labellum, which may be two or three lobed embracing the fertile stamen.

Gynoecium: It consists of three carpels (tricarpellary), syncarpous. The carpels are situated opposite the sepals. The ovary is trilocular and inferior. The placentation is axile, but sometimes the ovary remains unilocular with parietal placentation. In such cases three parietal placentas are present. The style is long, simple and slender, lies in a channel along the fertile stamen. The stigma, however, projects beyond the anther. The shape of stigma varies greatly in the different genera. A pair of nectar secreting epigynous glands is present.

Fruit: The fruit is generally a loculicidal capsule. Sometimes, it opens irregularly, *e.g.*, in *Globba.* The fleshy and indehiscent fruits are found in *Amomum* and other genera.

Seeds: The seeds are endospermic with a straight embryo. They are roundish or angular. The testa is smooth.

Pollination: The pollination is entomophilous, *i.e.,* it takes place through the agency of insects.

Floral formula: ⊕ or ·|· ⚥ K (3), C3, A 1, $G_{\overline{(3)}}$.

Bentham & Hooker (1862)	Engler & Prantl (1931)	Hutchinson (1959)
Phanerogams	Phanerogams	Angiospermae
Monocotyledones	Monocotyledoneae	Monocotyledones
Epigynae	Scitamineae	Calyciferae
Not recognized	Zingiberaceae	Zingiberales
		Zingiberaceae

Economic Importance of the family

1. *Zingiber officinale*; Eng.–Ginger; Verna.–**Adrak**–It is native of South East Asia, but now cultivated mainly in Kerala, Uttar Pradesh, West Bengal, Maharashtra, Sirmur district of Himachal Pradesh and Andhra Pradesh for the rhizomes, which are used as spice and condiment, and are also used in medicines as carminative and a digestive stimulant. The essential oil obtained from the rhizomes is used for flavouring purposes.

2. *Zingiber cassumunar;* Eng. Cassumunar ginger; Verna.–**Jangli-adrak**–A perennial herb found throughout India, the rhizomes are used as condiment.

3. *Zingiber zerumbet;* Eng.–Zerumbet ginger; Verna. **Banadrak**–Cultivated throughout India; the rhizomes are used as spice and condiment.

4. *Curcuma longa;* Syn. *C. domestica*; Eng.–Turmeric; Verna.–**Haldi**. Cultivated commonly in Maharashtra, Bengal, Tamil Nadu, Andhra Pradesh and Orissa. The rhizomes are used as condiment and dyestuff. Turmeric is used as an auspicious article in all religious ceremonies of Hindus. It is used as stomachic, tonic and blood purifier. Juice of fresh rhizomes is used for many skin diseases.

5. *Curcuma amada*; Eng.–Mango-ginger; Verna.–**Amahaldi**–A herb, grown in Bengal, Andhra Pradesh and Tamil Nadu. The rhizomes are used as spice and condiment. The paste made from the powdered rhizomes is remedy for inflamed joints.

6. *Curcuma angustifolia*; Eng.–East Indian arrowroot; Verna.–**Tikora**–Cultivated in South India, Bihar, Bengal and Maharashtra. The rhizomes are the source of starch.

7. *Curcuma aromatica*; Verna.–**Jangali Haldi**–Cultivated chiefly in Bengal and Travancore. The rhizomes are used as tonic and carminative.

8. *Curcuma caesia*; Eng.–Black zedoary; Verna.–**Kali haldi**–Cultivated mainly in Bengal for the tubers which are used in cosmetics.

9. *Curcuma leucorhiza*; Verna.–**Tikar**–Found in Bihar. The rhizomes yield starch.

10. *Curcuma zedoaria*; Syn. *C. zerumbet*; Eng.–Zedoary: Hindi–**Kachura**–Cultivated throughout India for its tuberous rhizomes, which are used in perfumery and cosmetics.

11. *Costus speciosus;* Verna.–**Keu**–An ornamental herb, commonly cultivated in Bengal and South India. The root is used as a purgative and in snake–bite. The tuberous roots are edible.

12. *Alpinia galanga*; Syn. *Languas galanga*; Eng.–Greater galangal; Verna.–**Kulinjan**–A perennial herb; native of Indonesia and Malaya. The rhizomes are used as a condiment and the source of an essential oil. The flowers are eaten raw, and the seeds are used as spice. In India, it is found in Eastern Himalayas and South West India.

13. *Alpinia speciosa*; Syn. *A.nutans*; Eng.–Shell–flower, Verna.–**Punnag Champa**–An erect, leafy ornamental plant.

14. *Amomum aromaticum*; Eng.–Bengal cardamom; Verna.–**Bari elaichi**–A perennial herb, used as a condiment. Grown in North Bengal and the Khasia hills.

15. *Amomum dealbatum*–Found in the Eastern Himalayas, the Khasia hills and Bengal. The seeds are used as a substitute for cardomom.

16. *Amomum subulatum*; Verna.–**Bari elaichi**–The seeds are used in the preparation of sweet–meats and for flavouring beverages. Grown in swampy places in Bengal, Sikkim, Assam and Tamil Nadu.

17. *Hedychium coronarium*; Eng.–Ginger lily–A perennial herb found in Khasia hills and South India. The aerial stems are used for manufacturing paper.

18. *Hedychium spicatum*; Eng.–Spiked ginger lily; Verna–**Kapur kachri**–Found in the sub–tropical Himalayas and Kumaon. The rhizomes are used as flavouring agent and the leaves are woven into mats. An essential oil, obtained from the rhizomes, is used in perfumery and pharmaceutical preparation.

19. *Kaempferia galanga*; Verna.–**Chandramula**–A common ornamental herb. The rhizomes are used as condiment; and are also used medicinally as stimulant, carminative and expectorant.

20. *Kaempferia rotunda*; Verna.–**Bhuichampa**–An aromatic ornamental herb.

21. *Renealmia*–This is an ornamental herb.

Family—Cannaceae

There is only one genus containing about fifty species in this family.

Distribution: The species of *Canna* are chiefly found in the warmer parts of America. Several species are cultivated for their beautiful flowers. *Canna indica* (Indian shot; Verna.–**Keli**) is commonly grown in parks and gardens in our country and cultivated throughout the warmer parts of the world.

Habit: The species of *Canna* are perennial herbs.

Stem: The plants persist by means of a rhizome. The aerial stem is also found.

Leaves: The aerial stem bears large pinnately veined leaves. The leaves resemble to those of Zingiberaceae, but without the ligule. The leaves are large, foliaceous, oblong to broadly elliptical.

Inflorescence: The inflorescence is terminal, forming a spike. Sometimes it is branched, and therefore, each bract found on the main axis subtends a two–flowered cincinnus, instead of a single flower.

Flower: The flowers are hermaphrodite, zygomorphic, bracteate, bracteolate, showy, trimerous and epigynous.

Perianth: The perianth consists of two trimerous series, the outer consisting of sepals and the inner of petals.

The sepals are free, small, herbaceous, greenish purple, showing imbricate aestivation.

The petals are united into a tube below, larger than the sepals, coloured, subequal, imbricate.

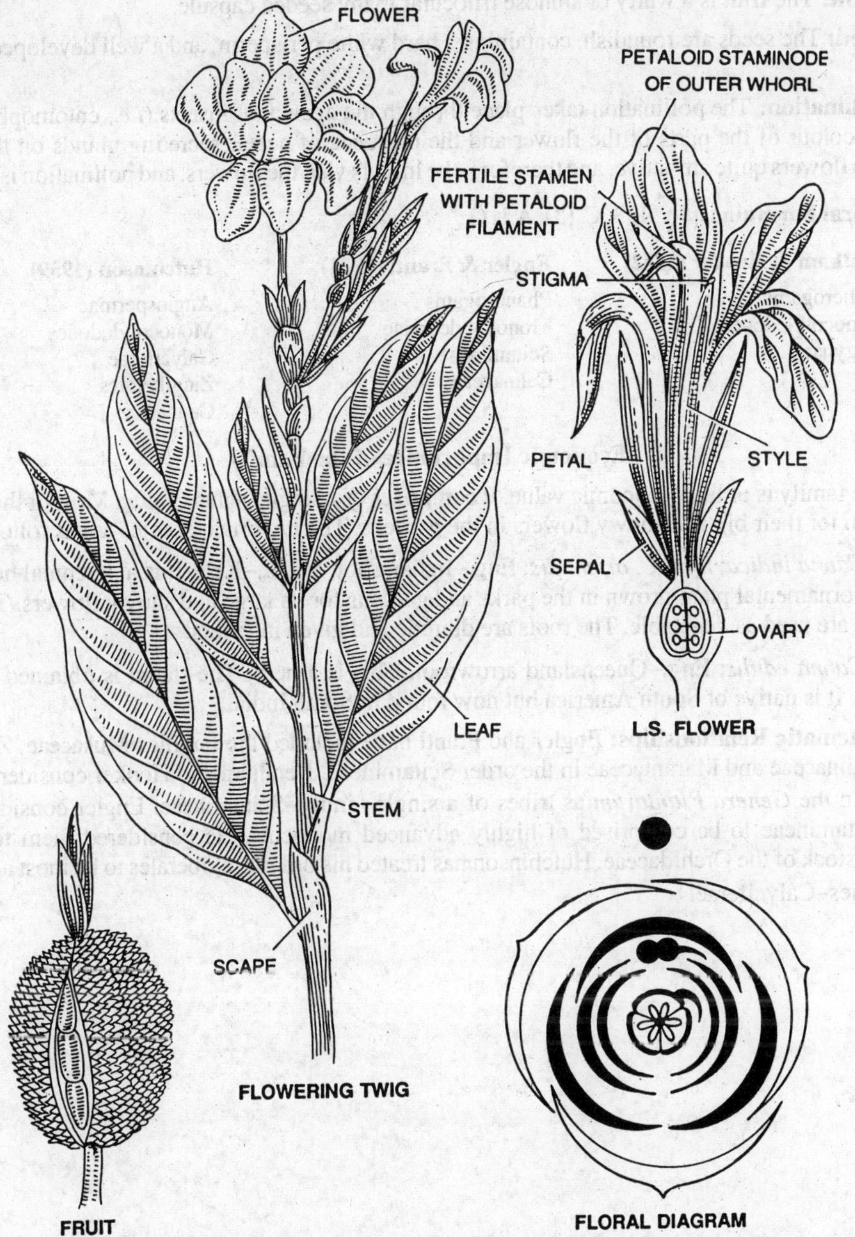

Fig.58.3. Cannaceae. *Canna indica* Linn.; Verna.–**Keli**; Eng.Indian shot.

Androecium: The androecium consists of six stamens, sometimes reduced to four. All stamens are petaloid, showy and slightly connate at the base. The outer three stamens are always sterile, out of which, the **labellum** is rolled back on itself. The two of the inner stamens are also sterile and somewhat connate. The remaining part of the androecium consists of a pataloid stamen bearing a half–anther on one edge.

Gynoecium: 3 carpels, syncarpous, ovary inferior,trilocular , axile placentation. The anatropous ovules are being arranged in two series at the inner angle of each loculus. The style is petaloid; the stigma is terminal and oblique.

Fruit: The fruit is a warty or spinose trilocular many seeded capsule.

Seed: The seeds are roundish, containing a hard white perisperm, and a well developed straight embryo.

Pollination: The pollination takes place through the agency of insects (*i.e.,* entomophily). The brilliant colour of the parts of the flower and the presence of nectar secreting glands on the ovary, make the flowers quite attractive, and therefore, the insects visit the flowers, and pollination is effected.

Floral Formula : ·|· ⚥, K 3, C (3), A 1, $G_{(3)}$.

Bentham & Hooker (1862)	**Engler & Prantl (1931)**	**Hutchinson (1959)**
Phanerogams	Phanerogams	Angiospermae
Monocotyledones	Monocotyledonae	Monocotyledones
Epigynae	Scitamineae	Calyciferae
	Cannaceae	Zingiberales
		Cannaceae

Economic Importance of the Family

The family is of little economic value. It comprises of a single genus *Canna.* Most of the species are grown for their big and showy flowers in the gardens. The important species are as follows:

1. *Canna indica*; Syn. *C. orientalis*; Eng.–Indian shot; Verna.–**Keli**–It is a perennial herb. This is highly ornamental plant grown in the parks and gardens for its large and showy flowers. Tuberous rhizomes are used as vegetable. The roots are diuretic and given in fevers.

2. *Canna edulis*; Eng.–Queensland arrowroot. This is a herb. The starch is obtained from its rhizomes. It is native of South America but now found in South India.

Systematic Relationships: Engler and Prantl have included the families Musaceae, Zingiberaceae, Cannaceae and Marantaceae in the order Scitamineae. Bentham and Hooker considered these families in the *Genera Plantarum* as tribes of a single family–Scitamineae. Engler considered the order Scitamineae to be comprised of highly advanced monocots. He considered them to be the ancestral stock of the Orchidaceae. Hutchinson has treated his order Zingiberales to be most advanced of his series–Calyciferae.

59

ORDER–LILIALES

According to Hutchinson this is the ninety-fifth order of the phylum Angiospermae, subphylum Monocotyledones and division Corolliferae. The order consists of six families, *i.e.*, Liliaceae, Tecophilaeaceae, Trilliaceae, Pontederiaceae, Smilacaceae and Ruscaceae, of which Liliaceae, Smilacaceae and Ruscaceae have been described in this text in detail. Bentham and Hooker have placed the Liliaceae along with seven other families in the third series –Coronarieae of class Monocotyledones. They have merged the Smilacaceae and Ruscaceae (of Hutchinson) within Liliaceae. Engler and Prantl have placed the Liliaceae along with eight others in the ninth order– the Lilliflorae of class Monocotyledoneae. They have also not recognized the Hutchinson's Smilacaceae and Ruscaceae.

Herbs with rhizomes, corms, as bulbs, rarely climbing; corolla-like perianth a dominant feature, the two series similar and often fusing together into one tube; stamens often 6; ovary superior or rarely semi-inferior; capsule or berry; seeds with copious endosperm.

Family—Liliaceae

There are about 240 genera and 4000 species in this family.

Distribution: The plants belonging to this family are cosmopolitan in distribution. They are abundantly found in temperate and sub-tropical regions. The plants are usually annual and perennial herbs. Sometimes, the soft, wooded shrubs are also found. In our country the family is represented by several genera such as, *Lilium, Aloe, Asparagus, Asphodelus, Gloriosa,* etc.

Habit: The plants are annual, biennial and perennial herbs and shrubs. The plants survive under the most varied conditions. Majority of the plants are xerophytic. The species of *Aloe* are xerophytic and may be either shrubs or tree-like with perennial aerial stems. One of the species of *Lilium* bears the bulbils in the axis of the leaves to furnish the vegetative reproduction. The species of *Aloe* are xerophytic bearing succulent and fleshy leaves. The plants of *Gloriosa* and *Asparagus* are climbers. The plants of *Asparagus* bear cladodes.

Roots: The roots are adventitious and fibrous.

Stem: In some species, the stems are under-ground, bulbs or rhizomes. The inflorescences may develop on a special aerial stem known as **scape** (*e.g.*, in *Asphodelus*).

Leaf: Usually the leaves are simple, radical or cauline. They may be alternate or whoried. The leaves of *Aloe* are fleshy and succulent. The plants of *Asparagus* bear scale-like leaves. However, they possess phylloclades for the purpose of carbon assimilation. The plants possess parallel veined leaves.

Inflorescence: It is usually of racemose type, but the flowers are never arranged in umbels. In *Rohdea,* the native of China and Japan the flowers are found to be arranged spirally on a spadix. In the species of *Gloriosa*, the flowers are solitary and terminal. In *Tulipa, Lilium* and *Fritillaria* flowers are solitary and terminal. In *Gonioscypha*, the flowers are arranged in spikes.

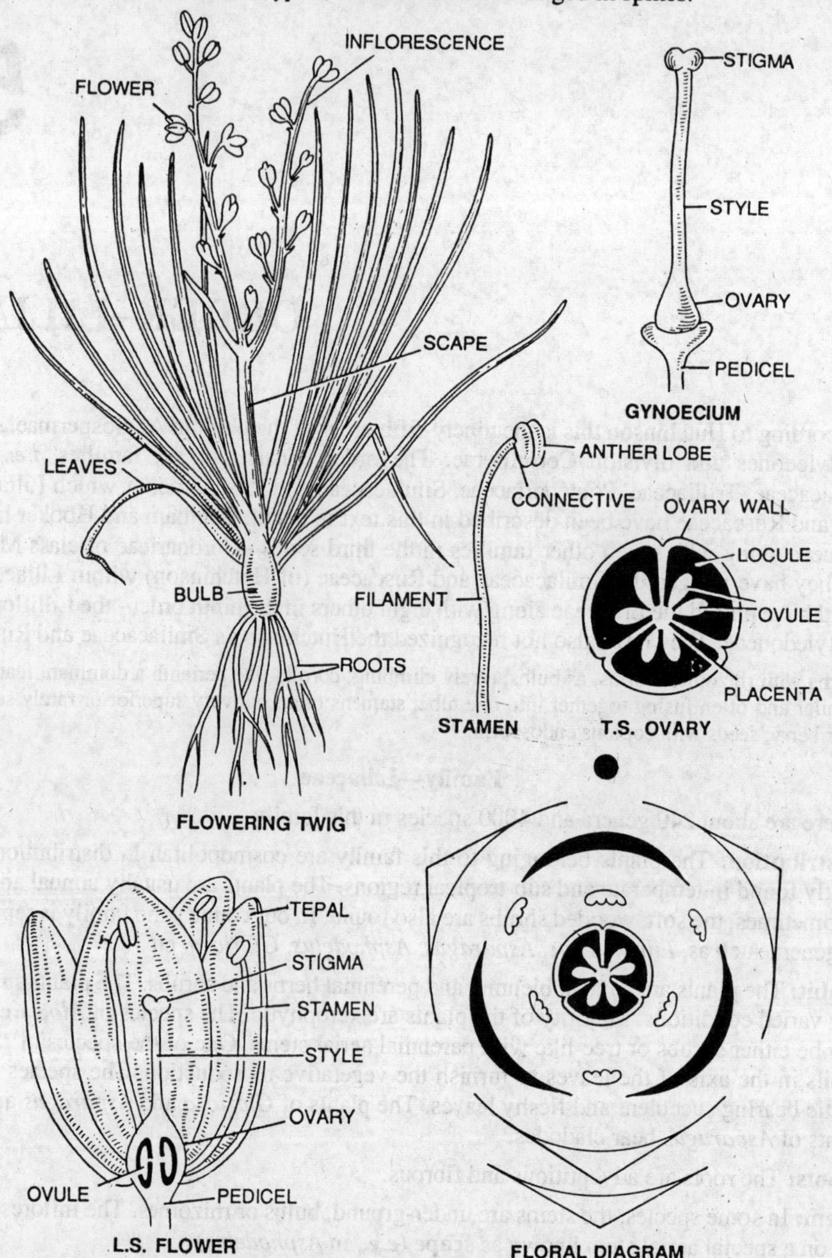

Fig.59.1. Liliaceae. *Asphodelus tenuifolius* Cav. Verna.–**Piazi.**

Flower: The flowers are pedicellate, bracteate (sometimes ebracteate), hermaphrodite, complete, regular, hypogynous and usually trimerous. In *Maianthemum bifolium* (British species) the flowers are tetramerous and in *Paris polyphylla* pentamerous.

Perianth: Six perianth leaves arranged in two whorls each whorl consisting of three segments. The segments may be free or united, valvate or imbricate, petaloid or sepaloid and regular. In *Haworthia*, a native of South Africa, the perianth is bilipped.

Androecium: It consists of six stamens arranged in two whorls; the stamens are opposite to the perianth leaves; the stamens may be free or adnate to the perianth leaves. The filaments may be free or connate; the anthers are dithecous, introrse and basifixed, they open usually by a longitudinal slit and rarely by a terminal pore.

Gynoecium: It is tricarpellary, syncarpous. The ovary is superior and trilocular. Placentation axile; style single; stigma trilobed; usually the ovules are numerous and arranged in two rows in each loculus.

Fruit: Fleshy, berry or a capsule. In *Asparagus* it is berry. In *Gloriosa* and *Asphodelus* they are septicidal and loculicidal capsules respectively.

Seeds: Endospermic or albuminous with straight embryo.

Pollination: It takes place through the agency of insects (*i.e.* entomophilous).

Floral Formula : ⊕ ⚥ P_{3+3}, A_{3+3}, $G_{(3)}$.

Bentham & Hooker (1862)	**Engler & Prantl (1931)**	**Hutchinson (1959)**
Phanerogams	Phanerogams	Angiospermae
Monocotyledones	Monocotyledoneae	Monocotyledones
Coronarieae	Lilliflorae	Corolliferae
Liliaceae	Liliaceae	Liliales
		Liliaceae

Description of *Asphodelus tenuifolius* in semi-technical language:

Asphodelus tenuifolius; Verna.–**Piazi**

Habit: Annual weed.

Root: Adventitious and fibrous.

Stem: Very small bulb.

Leaves: Radical, slender, erect, semiterete, fistular, glabrous and pointed.

Scape: One or more than one; one to two feet long having racemes at their terminal ends.

Inflorescence: Racemose, raceme.

Flower: Pedicellate, bracteate, small, actinomorphic, regular, hermaphrodite and hypogynous.

Perianth: Six perianth leaves in two whorls of three each; white; free or united at the base.

Androecium: Six stamens in two whorls of three each; epiphyllous; filaments broad at base and surround the ovary; anthers dithecous, basifixed or versatile, introrse, brown.

Gynoecium: Three, tricarpellary, syncarpous; ovary superior, trilocular, each loculus bearing two ovules; placentation axile; style short, stigma three-lobed.

Fruit: Small capsule.

Seeds: Albuminous with straight embryo.

Floral Formula: ⊕ ⚥ $P_{(3+3)}$, A_{3+3}, $G_{(3)}$

Economic Importance of the Family

The family is of little economic value. Some are grown as ornamentals while others possess medicinal properties. A list of few important plants is given below:

1. *Lilium candidum*; Eng.–Madonna lily.–This is a perennial herb, grown as an ornamental in the gardens. It is native of South Europe and South West Asia.

2. *Gloriosa superba*; Eng.–Glory lily; Verna.–**Kalihari**–This is an ornamental climbing herb with beautiful flowers, grown in the gardens.

3. *Asparagus adscendens*; Verna.–**Safed Musli**–This is a small herb. The root is used as a

demulcent and also in diarrhoea and dysentery. It is found in Maharashtra and Madhya Pradesh.

4. *Asparagus racemosus*; Verna.–**Satavar**–This is a common herb. It is consumed as vegetable.

5. *Asparagus officinalis*; Eng.–Garden Asparagus; Verna.–**Seetmuli, Merchuba, Halyum.**–The shoots are consumed as vegetable. It is native of Europe and West Asia.

6. *Asparagus plumosus;* Eng.–Asparagus fern.–This is a beautiful climbing herb, grown as an ornamental in the gardens. It is native of South Africa.

7. *Asparagus sarmentosus;* The roots are eaten as vegetable. The roots are also used as demulcent. It is found in Northern India.

8. *Aloe barbadensis*; Syn. *A.vera*: Verna.–**Ghikanvar**–This is a succulent shrub. The leaves yield a resinous drug, which is used mainly as purgative. It is native of West Indies but grows wild in South India.

9. *Colchicum luteum*; Verna.–**Hirantutiya**–This is a perennial herb. The roots are used medicinally in the treatment of rheumatism and gout. It is found in the temperate Himalayas from Kashmir to Chamba valley and the Punjab.

10. *Chlorophytum tuberosum*; Verna.–**Kulai**.–This is an undershrub. The leaves are consumed as vegetable. It is found in Madhya Pradesh, Andhra Pradesh, Tamil Nadu and Travancore.

11. *Fritillaria* sp.; Many species are grown as ornamentals in the gardens. Some possess medicinal properties.

12. *Tulipa* sp. The plant is grown as ornamental in the garden.

13. *Convallaria majalis*; They are highly ornamental plants and grown in the gardens for their beautiful flowers.

14. *Urginea indica*; Syn. *Scilla indica*; Eng.–White squill; Verna.–**Jangli piyaz**–This is a herb. The bulbs yield a raticide, and a drug, which is used as an expectorant and stimulant. It is found in the Western Himalayas, Bihar and along the Coromandal coast.

15. *Asphodelus tenuifolius*; Verna.–**Piazi**–This is a main weed of Rabi crop. The seeds are used medicinally.

Family—Smilacaceae

Hutchinson has recognised four genera (*i.e., Rhipogonum*; *Smilax; Pseudosmilax* and *Heterosmilax*) in this family. There are about 200 species in the genus *Smilax*.

Distribution: The members of the family are widely distributed in the temperate and tropical regions of the World.

Habit: The plants are climbing or straggling shrubs. Usually they possess tendril-like petioles and spiny, branched stems. The spines of the stem are very often found in the form of recurved hooks which help in climbing.

Root: The adventitious roots arise from the rhizome.

Leaves: The leaves are small petioled, alternate or opposite, leathery and net veined, *i.e.,* an unusual character in Monocotyledons. In many species the leaves possess two tendril-like stipules at their base.

Inflorescence: Racemose; generally the flowers are being arranged in axillary umbels. Sometimes they are found to be arranged in racemes of spikes.

Flower: The flowers are small, dioecious, unisexual and rarely hermaphrodite, trimerous and hypogynous.

Perianth: The perianth consists of six perianth leaves arranged in two whorls of three each. Perianth leaves are generally similar and petaloid. They are either free or rarely united to form a dentate tube.

Androecium: In the male flowers it consists of six stamens arranged in two whorls of three each. The filaments are either free or united. The anthers are introrse and basifixed. Rudimentary ovary is not found.

Gynoecium: In female flowers it consists of three carpels which are syncarpous. The ovary is superior, trilocular with axile placentation. One to two ovules are found in each loculus. Styles separate or united. Staminodes are present.

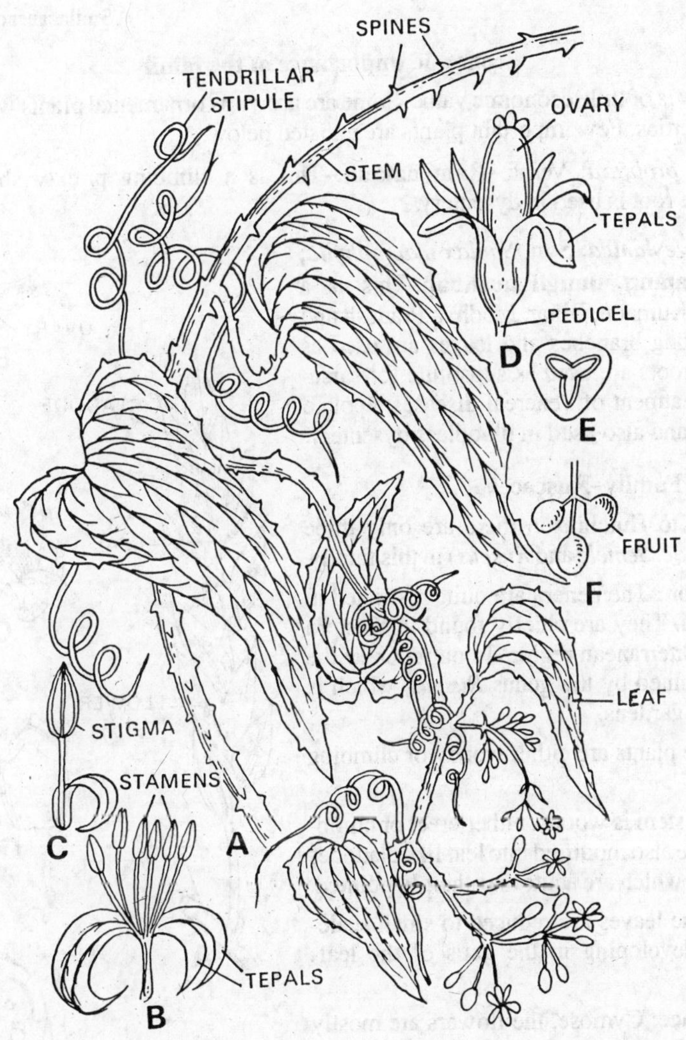

Fig.59.2. Smilacaceae-*Smilax aspera*. A, a flowering twig with tendril-like stipules; B, male flower; C, perianth leaf and stamen; D, female flower; E, cross-section of ovary; F, fruits.

Fruit: A berry.

Seeds: The seeds contain a copious fleshy endosperm (*i.e.*, endospermic), and a small embryo.

Pollination: It takes place through the agency of insects (*i.e.*, entomophilous).

Floral Formulae:

Male: $\oplus$ ♂ P_{3+3}, A_{3+3}

Female: $\oplus$ ♀ P_{3+3}, $G_{(3)}$

Bentham & Hooker (1862)	**Engler & Prantl (1931)**	**Hutchinson (1959)**
Phanerogams	Phanerogams	Angiospermae
Monocotyledones	Monocotyledoneae	Monocotyledones
Coronarieae	Lilliflorae	Corolliferae
Liliaceae	Liliaceae	Liliales
		Smilacaceae

Economic Importance of the family

The family is of little economic value. Some are grown as ornamental plants while others possess medicinal properties. Few important plants are enlisted below:

1. *Smilax prolifera*; Verna.–**Ram dataun.**–This is a climbing prickly shrub, grown as an ornamental. The root is used in dysentery.

2. *Smilax zeylanica*; Syn. *Smilax macrophylla*; Verna.–**Ram–dataun, Jangli aushbah**–This is a shrub found i n Kumaon, Bihar, Madhya Pradesh and Assam. The young branches and leaves are used as vegetable. The roots are used as substitute for sarsaparilla in the treatment of venereal diseases, applied for rheumatism and also used in bloodless dysentery.

Family–Ruscaceae

According to Hutchinson there are only three genera (*i.e., Danae, Semele* and *Ruscus*) in this family.

Distribution: The genera are quite restricted in their distribution. They are chiefly found in Western Europe and Mediterranean region. In our country the family is represented by the genus *Ruscus,* which is cultivated in the gardens.

Habit: The plants are either woody or climbing shrubs.

Stem: The stem is woody, either erect or climbing. The stems are also modified into leaf like cladodes (*e.g.*, in *Ruscus*), which are acute and sharply pointed.

Leaves: The leaves are reduced to small scale-like structures developing in the axils of the leaf-cladodes.

Inflorescence: Cymose, the flowers are mostly solitary axillary. Sometimes terminal racemes are developed which remain free from cladodes.

Flowers: The flowers are small, hermaphrodite or unisexual, when unisexual the plants are dioecious.

Perianth: It consists of six perianth leaves arranged in two whorls of three each. The perianth leaves are either free or somewhat connate. When connate, there is a fleshy corona.

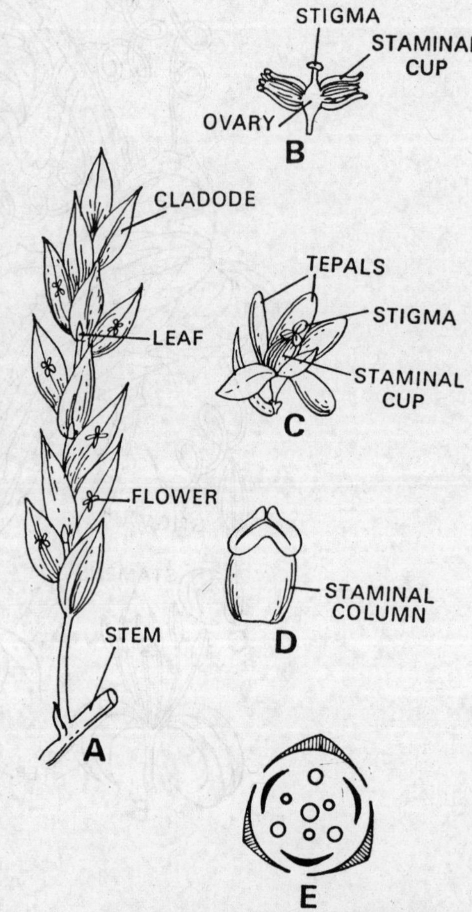

Fig. 59.3. Ruscaceae—*Ruscus sp.*: A. flowering twig; B, pistil with cut away staminal cup; C, female flower with pistil covered by fleshy staminal cup bearing aborted anthers ; D, staminal column, formed by the union of three stamens; E, floral diagram of male flower.

Androecium: In male flowers, the androecium consists of six or three stamens. The filaments form a column. The anthers are extrorse. Sometimes rudimentary ovary is found in the male flower.

Gynoecium: In female flowers, the gynoecium consists of three carpels, which are syncarpous. The ovary is superior uni or trilocular, two ovules in each loculus when three chambered. The staminodes are present.

Fruit: A roundish berry.

Seeds: The seeds are endospermic with a small cylindrical embryo.

Pollination: The pollination takes place through the agency of insects.

Floral Formulae:

Male: ⊕ ♂ P_{3+3}, A_{3+3}, or (3)

Female: ⊕ ♀ P_{3+3}, $G_{(3)}$

Bentham & Hooker (1862)	**Engler & Prantl (1931)**	**Hutchinson (1959)**
Phanerogams	Phanerogams	Angiospermae
Monocotyledones	Monocotyledoneae	Monocotyledones
Coronarieae	Lilliflorae	Corolliferae
Liliaceae	Liliaceae	Liliales
		Ruscaceae

Economic Importance of the Family

The family is of little economic value, having a few ornamental plants grown in the gardens and homes, *e.g., Ruscus.*

Systematic Relationships. Engler and Prantl have placed the Liliaceae in the order–the Lilliflorae. Bentham and Hooker have placed the Liliaceae in the series—Coronarieae, Hutchinson has placed the Liliaceae together with other five families (*i.e.,* Trilliaceae, Smilacaceae, Tecophilaeaceae, Pontederiaceae and Ruscaceae) in his order—the Liliales. He included the genera, *Lilium, Aloe, Asparagus, Asphodelus, Gloriosa,* etc., in the Liliaceae; *Smilax* in the Smilacaceae and *Ruscus* in the Ruscaceae. He included the genus *Allium* in Amaryllidaceae and *Dracaena, Yucca, Agave, Cordyline,* etc., in the Agavaceae of his order—Agavales. Hutchinson, however has distinguished his Liliaceae and Amaryllidaceae on the basis of the inflorescence character which is probably supposed to be more stable and fundamental. Hutchinson, however, transferred the umbellate spathaceous genera to Amaryllidaceae. The genera *Agapanthus* and *Allium* have umbellate inflorescences and are closely related anatomically and, therefore, Hutchinson has placed them in the Amaryllidaceae.According to him the flowers in Liliaceae are never arranged in umbellate inflorescences. The morphologists, such as Cheadle (1942) the Maia (1941), supported the transfer of Agapantheae, Allieae and Gilliesieae to the Amaryllidaceae on the basis of their pollen studies. Lawrence stated, that from the stand-point of its phylogeny, the Liliaceae is now believed to represent basic monocotyledonous stock from which ancestral liliaceous stocks have evolved the majority of present day petaloid monocot families, such as Amaryllidaceae, Iridaceae, Palmae, Araceae and Orchidaceae. It is also thought that its aneestral stocks have been possibly derived from those of Commelinales and Butomales. There is a general belief that within the Liliaceae the groups that possess rhizomes are more primitive than the groups which possess bulbs.

60

ORDER–ARALES

According to Hutchinson this is the ninety-seventh order of the phylum Angiospermae, subphylum Monocotyledones and division Corolliferae. The order comprises of only two families, *i.e.*, Araceae, and Lemnaceae. Both of these families have been described in the present text in detail. Bentham and Hooker have placed the families Aroideae (*i.e.*, Araceae) and Lemnaceae along with three others including Pandaneae and Typhaceae in the fifth series –the Nudiflorae of class Monocotyledones. Engler and Prantl have placed these two families in their seventh order– the Spathiflorae of class Monocotyledoneae.

Flowers much reduced and arranged on a thickened spike (spadix) subtended by a large often coloured bract (spathe); bisexual to unisexual; perianth very small or entirely reduced; ovary superior; fruit a berry; copious endosperm.

Family—Araceae

There are about 110 genera and 1500 species in this family (Rendle).

Distribution: The plants belonging to this family are mostly common in tropical and sub–tropical regions, but many are common in temperate regions also. The family is represented in our country by few important genera, *e.g., Colocasia, Alocasia, Pistia, Arisaema* (in Himalayan regions, about 3000 feet elevation), *Amorphophallus,* etc.

Habit: The plants may be small or large herbs, sometimes shrubs with sympodial branches climbing by aerial roots (*e.g., Pothos*). Some plants are epiphytic and climbing, some are aquatic while some are marshy in habit. The species of *Remusatia* are herbs with tuberous rhizomes. The species of *Cryptocoryne* are found in marshy places. *Pistia* is a free floating aquatic plant. Most of the plants contain watery or milky juice (latex). Sometimes the sap is acrid and with pungent odour. Resin canals and calcium oxalate crystals are also present within the tissues.

Root: The roots are adventitious, sometimes epiphytic plants produce aerial roots. Usually the velamen layer is found on the aerial roots to absorb the moisture from atmosphere. In epiphytic and climbing plants *e.g., Pothos,* the roots serve both the purposes of absorption as well as of clasping. The clasping roots are negatively phototropic and fix the plant to its support.

Stem: Erect or climbing; underground stems are usually developed in the form of roots-stocks, tubers, or corms, which often have a very pungent taste.

Leaves: The leaves are mostly large, simple or compound, usually with parallel venation (*e.g.*, in *Acorus*), sometimes venation is reticulate (*e.g.*, in *Pothos, Arum*, etc.), petiolate (long petioled), petiole with a sheathing base. Cordate, sagittate or hastate, lamina entire, lobed, pinnate or perforate (with holes); the arrangement of the leaves in shrubby plants is alternate, distichous or spiral. In herbaceous species, leaves are few, clustered or solitary, radical.

Inflorescence: It consists of many small, sessile unisexual (rarely bisexual) flowers arranged on a spadix which is more or less completely enclosed in a green (*e.g.,* in *Arum* sp.) or coloured (*e.g.,* in *Richardia, Anthurium*, etc.) leaf like bract (spathe). When flowers are unisexual, they are usually monoecious with the male flowers towards the apex of the spadix and females below. Racemose, spadix with spathe. The entire inflorescence may be covered with flower (*e.g.,* in *Acorus*), or this ends in a sterile conical portion (*e.g.,* in *Arum, Alocasia*).

Flowers: Sessile, small, actinomorphic, regular, bi to trimerous or dimerous (*e.g.,* in *Anthurium*), unisexual, rarely hermaphrodite, monoecious or dioecious. If monoecious, the male flowers are always towards the apex of the spadix and female flowers below. The male and female flower zones are separated from each other by neuter flowers zone.

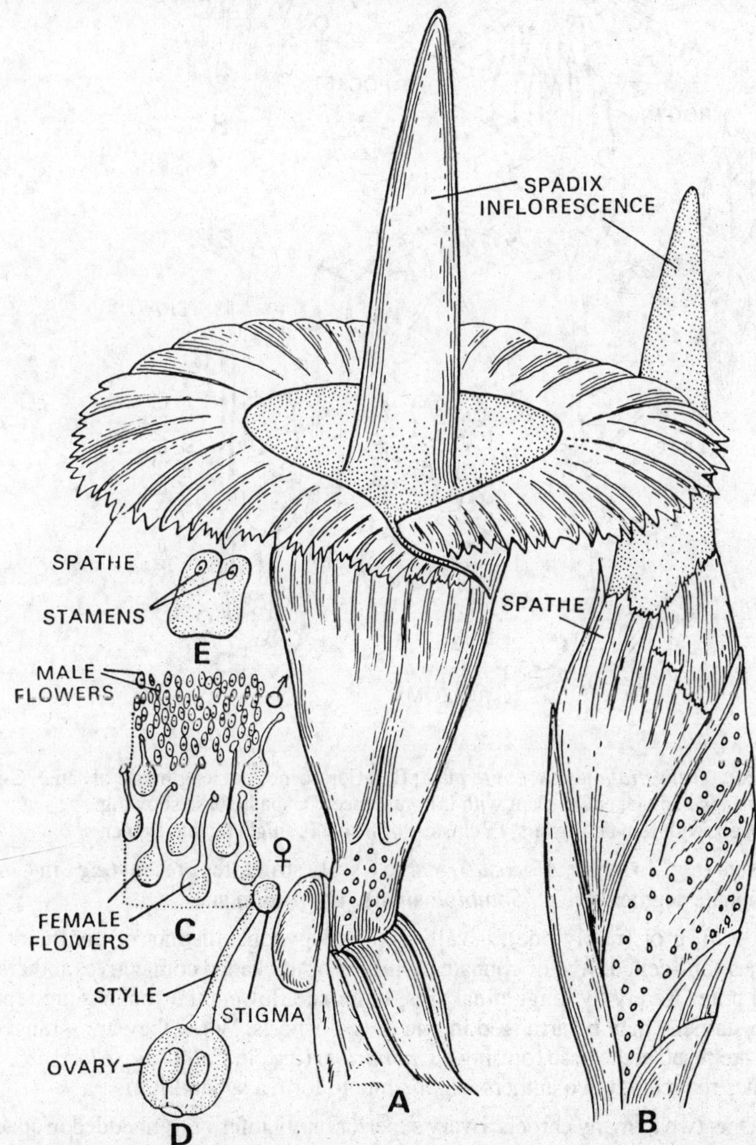

Fig. 60.1. Araceae–*Amorphophallus titanum*: A, spathe and spadix open; B, spathe and spadix in bud; C, part of spadix showing male and female flowers; G, longitudinal section of ovary; E, stamen.

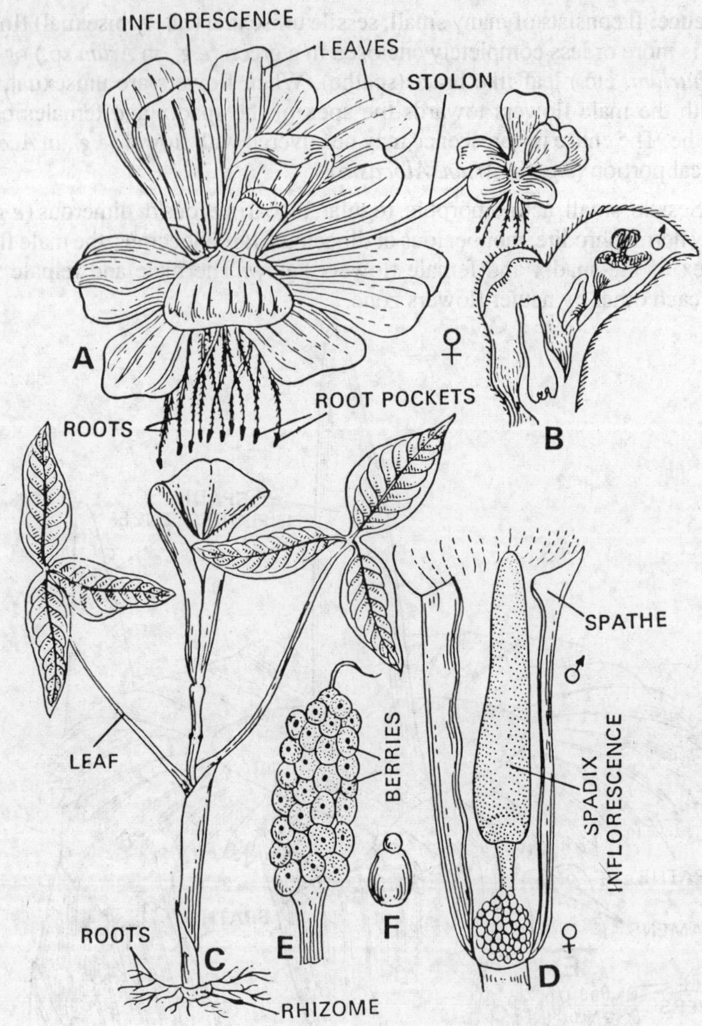

Fig.60.2. Araceae–*Pistia stratiotes*: A, entire plant; B, inflorescence cut lengthwise of same; C-F, *Arisaema* sp.; C, entire plant with inflorescence; D, spathe base showing spadix with female flowers, E, clusters of fruits; F, single female flower.

Perianth: Absent (*e.g.,* in *Arum, Alocasia*), or 4 to 6 scaly structures present (*e.g.,* in *Calla*), free (*e.g., Acorus*), or connate segments (*e.g., Spathiphyllum*), rarely cupular.

Androecium: 2, 4, 6 or (rarely one), usually six, hypogynous, distinct or confluent; anthers two-celled sometimes four-celled, free or connate by means of thickened connective, anthers usually dehisce by terminal pores, rarely by longtudinal slits. In female flowers the stamens are represented by staminodes. The stamens may be arranged in one or two whorls. When they are arranged in one whorl the filaments are fused at the base forming a *synandrium* (*e.g.,* in *Colocasia, Alocasia*). In *Pistia* the male is sometimes reduced to two anthers which unite to form a synandrium.

Gynoecium: One, two to many carpels; ovary superior rarely inferior (embedded in spadix), one to many loculed; the ovules one to many in each loculus. The placentation may be basal, axile or apical; style short; stigma disc-like or lobed.

Fruit: A berry, sometimes many small, free or connate berries or drupes adnate to the spadix.

Seed: Endospermic (albuminous), in each berry or drupe, one to many seeds, sometimes non-endospermic.

Pollination: Usually entomophilous.

Floral Formula : ⊕ ♂ or ♀ or ⚥ P4—6, A2—8, $G_{\overline{1-\infty}}$

Bentham & Hooker (1862)	Engler & Prantl (1931)	Hutchinson (1959)
Phanerogams	Phanerogams	Angiospermae
Monocotyledones	Monocotyledoneae	Monocotyledones
Nudiflorae	Spathiflorae	Corolliferae
Aroideae	Araceae	Arales
		Araceae

Description of *Alocasia* sp., in semitechnical language.

Alocasia sp.

Habit: Annual herb, either grown for its beautiful foliage in the gardens or it is cultivated for its edible rhizomes.

Root: Adventitious, some are aerial.

Stem: Usually rhizome, but sometimes rises a considerable distance above the ground.

Leaves: Usually large, entire, peltately attached or not to the long petiole, more or less ovate and cordate or sagittate, venation parallel, petioles are well developed and swollen at the base.

Inflorescence: Spadix, which is covered by a leafy bract known as spathe, which is tubular at the base but spread above. Spadix free, somewhat shorter than the spathe. Female flowers are arranged towards the top of the spadix while male flowers on the lower side. In the gap between males and females the neuters are present.

Perianth: Absent in *Alocasia.*

Androecium: 3 to 5 stamens arranged in a single whorl. The filaments are united throughout their length forming a **synandrium,** anthers are dithecous (2-celled), dehiscence by pores.

Gynoecium: Two to four carpels, syncarpous, ovary ovoid, one celled (unilocular), with few to many basal orthotropous, ovules; style short; stigma 2-4 lobed.

Fruit: Berry, almost spherical, red, few seeded, enclosed by the accrescent tube of spathe.

Seeds: Albuminous, roundish, with thick smooth testa and axile embryo.

Generic Characters: Fruit a berry, enclosed by the accrescent tube of spathe; perianth absent; inflorescence spadix, monoecious female flowers above, males below; synandrium is formed by the fusion of the filaments of the stamens; stem usually rhizome.

Floral Formula: ⊕ ⚥, ♂ or ♀, PO, A3–5, $G_{\overline{(2-4)}}$

Economic Importance of the Family

The family is of little economic value. A few plants are grown for their edible underground stems. Some are grown as ornamentals for their beautiful foliage, others possess medicinal properties. A list of some important plants is given here:

1. *Alocasia indica*; Eng.–Giant taro: Verna.–**Mankanda**–The stems and root stocks are edible. This is grown in Assam, Bengal, Maharashtra, Tamil Nadu, Malabar and Travancore.

2. *Alocasia cucullata*; Eng.–Giant taro. It is grown for its edible rhizomes. It is cultivated in Bengal and Assam.

3. *Alocasia macrorhiza*; Eng.–Giant taro; Verna.–**Mankanda**–The corms are edible. It is grown chiefly in Assam.

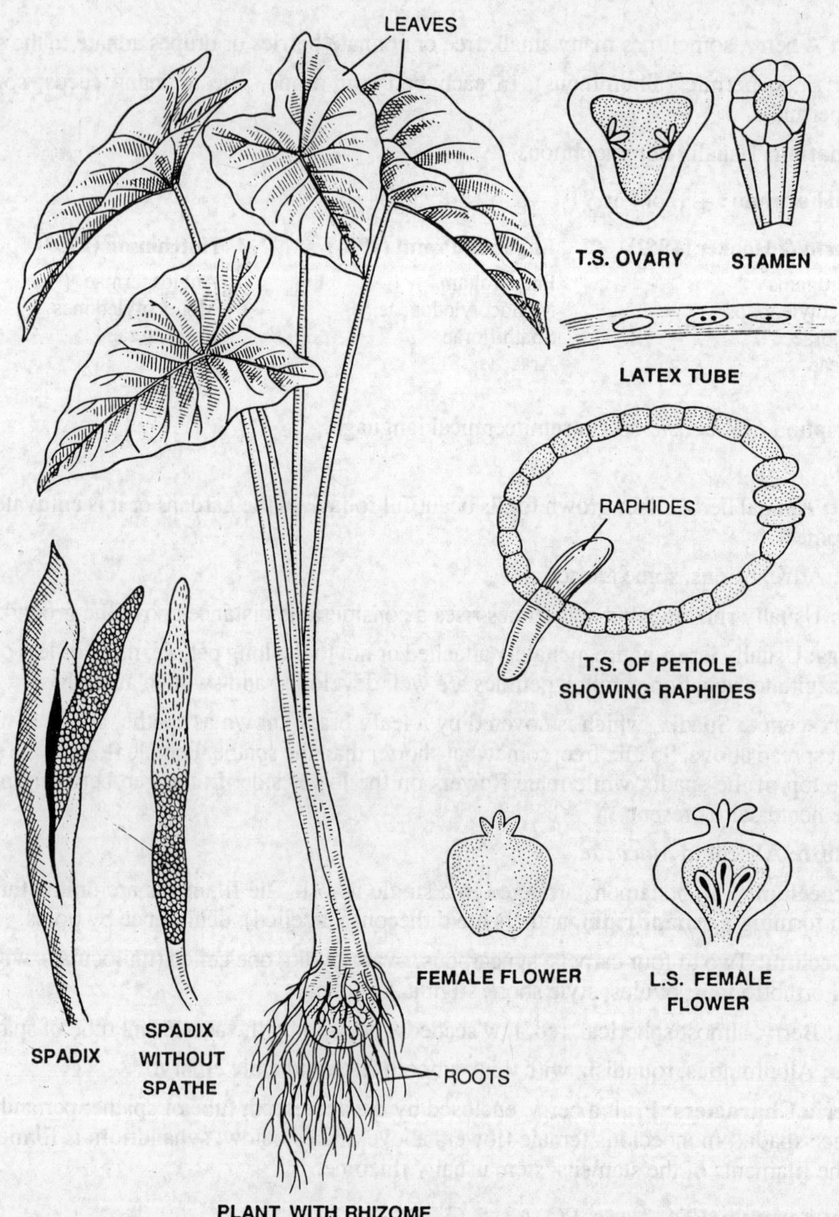

Fig. 60.3. Araceae-*Alocasia indica* Schott.; Eng.giant taro; Verna. **Mankanda.**

4. *Amorphophallus campanulatus*; Syn. *Arum campanulatum*; Verna.–**Zamin–kand**–This is an undershrub, whose tuberous roots are eaten as vegetable. It is chiefly cultivated in Maharashtra, Surat, Bengal, Malabar, Tamil Nadu and Northern India.

5. *Arisaema tariuosum*; Verna.–**Samp-ki-kumb-**This is a tall herb. It is poisonous and its tubers are used as an insecticide. It is found in the Khasia hills, Manipur, Bihar, Tamil Nadu and the temperate Himalayas.

6. *Homalomena aromatica*; Syn. *Calla aromatica*, Verna.–**Kuchugundubi**–This is an aromatic herb. The rhizomes of the plant are used in tobacco and snuff preparations. It is found mainly in Assam.

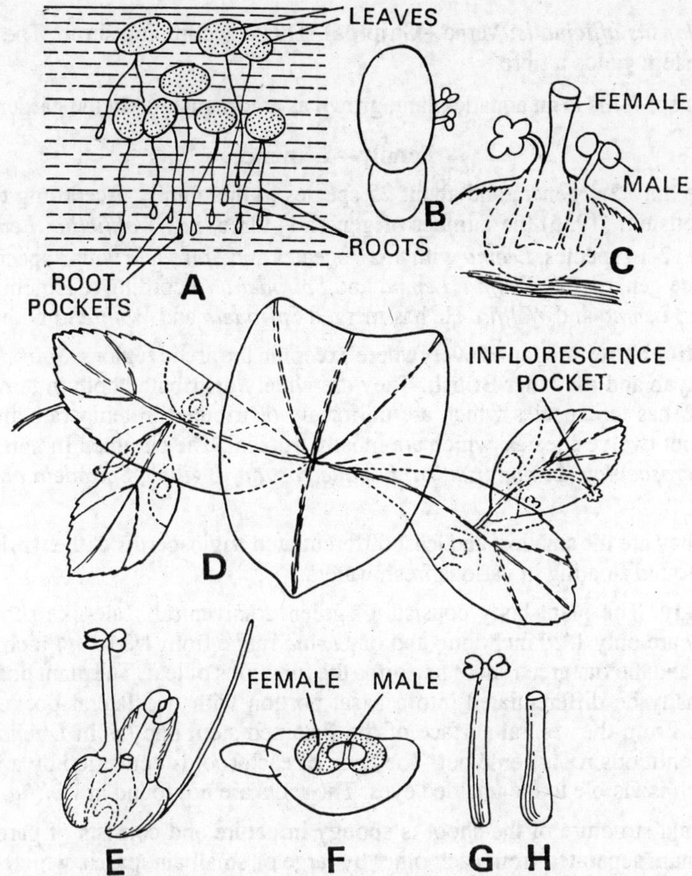

Fig.60.4. Lemnaceae *Lemna minor*: A, plants with rhizoids; B, plant in flower; C, sheath with inflorescence having three flowers; D, flowering plant of *Lemna trisulca;*E, inflorescence of same; F, flowering plant of *Wolffia* sp., with male and female flowers; G, staminate flower of *Lemna minor;* H, pistillate flower of *Lemna minor.*

7. *Homalomena rubescens*; Verna.–**Kuchugundubi**–The rhizomes yield an essential oil, which is used in perfumery. It is found in the Sikkim and Khasia hills.

8. *Zantedeschia aethiopica*; Syn. *Calla aethiopica*; Eng.–Arum lily. This is a herb, grown as an ornamental in the gardens. It is native of South Africa.

9. *Pothos scandens*–This is a climbing or trailing herb, grown as an ornamental in the gardens and homes.

10. *Acorus calamus*; Eng.–Sweet flag; Verna.–**Safed bach**–This is a marshy aromatic herb. The rhizomes are used medicinally as carminative, stimulant and as a tonic. An essential oil, obtained from the rhizomes is used in perfumery. It is commonly found in Kashmir, Manipur and Karnataka.

11. *Acorus gramineus*–The root stock is used as an insecticide. It is found in the Khasia hills and Eastern Himalayas.

12. *Colocasia esculenta*; Syn. *C. antiquorum*; Eng.–Taro, Arum; Verna.–**Arvi, Ghuiyan, Kachalu**–This is a herb. It is grown throughout the country for its edible starchy, tuberous corms. The leaves are also used as vegetable. The sap gives a permanent black stain to the clothes. It is native of South East Asia.

13. *Monstera deliciosa*; Syn. *M. lennea*; Eng.–Ceriman; Verna.–**Amarphal**–This is a common ornamental herb or shrub with large leaves. The fruits are edible. It is native of Central America.

14. *Scindapsus officinalis*; Verna.–**Gajpipal**–This is a climbing shrub. The leaves are eaten as vegetable and stem yields a fibre.

15. *Pistia* sp.–This is an aquatic plant, grown as an ornamental in the garden ponds.

Family—Lemnaceae

There are only 2-4 genera and about 25 species in this family. According to Engler and Diels (1936) and Wettstein (1935) the number of genera is three, *i.e., Wolffiellla, Lemna* and *Spirodela. Wolffiella* with 12-14 species, *Lemna* with 8–10 species and *Spirodela* with 3 species. Rendle has also recognized three genera, *i.e., Wolffia, Lemna* and *Spirodela.* According to Hutchinson there are only two genera, *i.e., Lemna* and *Wolffia.* He has merged *Spirodela* and *Wolffiella* with *Wolffia.*

Distribution: They are found everywhere except in the arctic regions. *Spirodela* has two species one Indo-Malayan and the other British. They are widely distributed both in temperate and tropical regions. *Lemna* has ten species which are uniformly distributed in temperate and tropical regions. *Wolffia* has about twelve species, which are mostly tropical. The common Indian species are *Wolffia arrhiza, W. microscopica, Lemna trisulca, L. paucicostata, L.gibba, Spirodela oligorrhiza, S. polyrrhiza.*

Habit: They are the smallest and least differentiated angiosperms of the world. They are aquatic in nature, and found floating in various fresh waters.

Plant Body: The plant body consists of green dorsiventral scale-like shoots. The shoots of *Wolffia arrhiza* are only 1/20 inch long and of *Lemna* range from 1/8 to 3/4 inch. The plants do not possess leaves and the flat green shoot performs the functions of leaf. The plant body of *Spirodela* and *Lemna* may easily be differentiated into a basal portion with two lateral pockets from which the branches arise. From the ventral surface of the flattened stem one (as in *Lemna*) or several (as in *Spirodela*) adventitious roots come out. The apex of each root is converted by a few layered sheath (root cap) which is visible to the unaided eyes. The roots are not found in *Wolffia.*

The internal structure of the shoot is spongy in nature and consists of parenchymatous cells. These cells remain separated from each other by large or small air spaces, which communicate with the outside by stomata on the upper surface. Vascular tissue is not found in *Wolffia.* In *Lemna* and *Spirodela,* it is represented by a single median vascular bundle of very simple structure.

Inflorescence: In temperate zones, the flowers are rarely developed. The inflorescence is quite simple, and in *Lemna* and *Spirodela* it arises in the pocket.

Flowers: The flowers are unisexual. The plants are monoecious, *i.e.,* both male and female flowers develop in the same inflorescence. The flowers are without perianth (naked). The **male flower** consists of a single stamen. The filament is stout bearing at its apex a pair of dithecous anther-halves. The pollen grains are spherical and covered with small warty outgrowths. The **female flower** consists of a single carpel. The pistil is flask shaped with a short funnel shaped stigma. The ovary is unilocular with one to six basal, erect, orthotropous or more or less completely anatropous ovules.

Seed: The seed possesses a thick fleshy outer and a thin inner coat. The embryo consists of a large cotyledon surrounded by scanty endosperm.

Pollination: The pollination is effected by wind, water or animals.

Floral Formulae: Male flower- ♂ P0, A1

Female flower- ♀ P0, $\underline{G1}$

Bentham & Hooker (1862)	**Engler & Prantl (1931)**	**Hutchinson (1959)**
Phanerogams	Phanerogams	Angiospermeae
Monocotyledones	Monocotyledoneae	Monocotyledones
Nudiflorae	Spathiflorae	Corolliferae
Lemnaceae	Lemnaceae	Arales
		Lemnaceae

Economic Importance: Practically the family is of no economic value.

Systematic Relationships: There is a great controversy about the systematic relationships of the Araceae. According to Engler (1920) the Araceae has been derived from the Palmae via the Cyclanthaceae. Wettstein (1935) considered the aroids to be advanced over the orchids. He also thought that they have been derived directly from Helobiae–Lilliflorae stocks. According to Bessey (1915), the Palmae and the Araceae are parallel or divergent in their evolution from the Liliaceae. Hutchinson (1934) has considered their derivation from stocks ancestral to the Aspidistrae tribe of Liliaceae.

According to many taxonomists the Lemnaceae represents a degenerate offshoot of the Araceae with origins probably from *Pistia* or some other ancestral stocks of close affinity. According to Brooks (1940), the Lemnaceae have been derived from the Araceae. He further stated that the Lemnaceae have a very reduced structure and show a reduction series from *Spirodela* through *Lemna* to *Wolffiella* and *Wolffia*. Lawrence (1945) however, was not in agreement with the above mentioned statement. He considered, that the Lemnaceae was not related phylogenetically with the Araceae. According to him the aroids were derived from the Helobiae.

61

ORDER–TYPHALES

According to Hutchinson this is the 98th order of the phylum Angiospermae, subphylum Monocotyledones and division Corolliferae. This order comprises of only two families, *i.e.,* Sparganiaceae and Typhaceae of which the latter has been described in the present text in detail. Bentham and Hooker have placed the Typhaceae along with four other families including Pandanaceae, Aroideae and Lemnaceae in their fifth series –the Nudiflorae of class Monocotyledones. Engler and Prantl have placed the Typhaceae along with other two families (*i.e.,* Pandanaceae, and Sparganiaceae) in their first order –the Pandanales of class Monocotyledoneae.

Aquatic or marsh herbs with rhizomes; leaves sheathing at the base; flowers unisexual, anemophilous, minute, crowded into clusters or dense spikes; perianth much modified and reduced.

Family—Typhaceae

There is a single genus *Typha* containing about fifteen species.

Distribution: The species are found in the marshy places of both the temperate and tropical regions of northern and southern hemispheres. The plants of *Typha angustifolia* occur in ditches, ponds, lakes, on river banks, etc.

Habit: They are water or marsh-herbs, with a perennial creeping rhizome. The rhizome bears erect simple shoots and distichous, long, narrow entire sheathing leaves. The rhizome bears two lateral rows of scales. The branches may spring out from the axils of the scales. Generally a pair of lateral branches is found at the base of a flowering shoot. These branches become thickened at their tips and form a knee-like band. The lower leaves are reduced to long sheaths and they represent the transitional stage to the scale leaves of the rhizome. The leaves are parallel veined.

Inflorescence: The unisexual flowers are borne on a dense spike or cylindrical spadix, the male flowers in the upper portion while the female flowers in the lower portion (*i.e.,* monoecious). The male flowers arise directly from the flattened axis of the spike. The female flowers are usually borne on crowded, short cylindrical outgrowth. The flowers of each are subtended by one caducous bractlike spathe.

Male Flowers: Each male flower consists of 1-5 stamens, variously monadelphous. The filaments being connate and bear long silky hairs. Commonly 3 stamens are found, often there are only two stamens, sometimes only one stamen present as in *Typha minima.* The hairs or scales developing at the base of the flowers on the axis of the spike surround the flowers. These hairy outgrowths are not arranged in a definite pattern, and probably they do not represent the reduced perianth. In *Typha minima,* such outgrowths are not found. Some botanists regard these outgrowths as a reduced perianth. The function of these hairs or scales is to protect the flower. The connective remains protruded beyond the anther. The pollen grains are either simple or united in tetrads.

Female flowers : Generally the female flowers remain subtended by small, spathulate scale like bracts (*e.g.*, in *Typha angustifolia*). However, such bracts are altogether absent from other species, *e.g.*, *T. latifolia* . The fertile flowers generally occur at the base of short lateral outgrowths, which are situated upon the axis of the spike. The upper flowers situated on the short outgrowths are more or less aborted. The normal flower possesses a long gynophore on which numerous irregular arranged simple hairs have been found. Some botanists consider these hairs as a reduced perianth. The gynophore and hairs, both elongated in the fruit and make a feathery structure which helps in the dispersal of fruit by means of wind. The style is slender which terminates in a unilateral elongated stigma. The ovary is superior, unilocular, with a single anatropous pendulous ovule; parietal placentation.

Fruits: The fruit is an achene, covered with a membranous or leathery pericarp.

Seeds: The seeds are small and endospermic with a straight embryo.

Pollination: It takes place through the agency of wind (*i.e.* anemophilous).

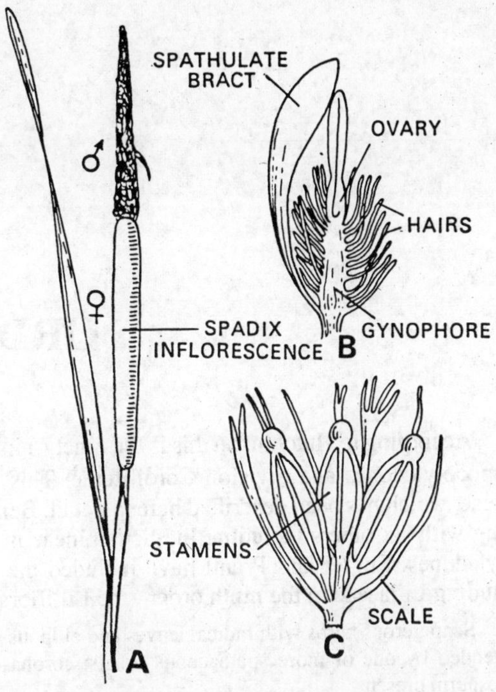

Fig. 61.1. Typhaceae–*Typha* sp: A, flowering shoot showing male and female spikes; B, female flower in vertical section; C, male flower with adjacent scales.

Floral Formula:

Male : ♂ P0, A1–5

Female : ♀ P0, $G\underline{1}$

Bentham & Hooker (1862)	**Engler & Prantl (1931)**	**Hutchinson (1959)**
Phanerogams	Phanerogams	Angiospermae
Monocotyledones	Monocotyledoneae	Monocotyledones
Nudiflorae	Pandanales	Corolliferae
Typhaceae	Typhaceae	Typhales
		Typhaceae

Economic Importance of the Family

The family is of little economic value. A list of few important plants is given below:

1. *Typha angustata*; Syn. *Typha elephantina*; Eng.–Elephant grass; Verna.–**Gond-Pater**–This is a small shrub, found in North-Western India and Assam and used as a soil-binder. It is also used for making mats, ropes and baskets.

2. *Typha angustifolia*; Eng.–Cat's tail, Pithgrass; Verna.–**Pater**–The leaves are made into mats and baskets, and the hairs on the fruits are used for stuffing pillows.

Systematic Relationships: Engler and Prantl have included Typhaceae, Pandanaceae and Sparganiaceae in the order–Pandanales. Hutchinson, however, segregated the Pandanaceae from Typhaceae and Sparganiaceae as a separate order–the Typhales. He has interpreted the fusion of the fruits into a syncarp to be more advanced, and therefore, treated the Typhaceae to be more primitive than the Pandanaceae.

62

ORDER–AMARYLLIDALES

According to Hutchinson this is the ninety-ninth order of the phylum Angiospermae, subphylum Monocotyledones and division Corolliferae. This order consists of a single family–The Amaryllidaceae, which has been described here in detail. Bentham and Hooker have placed their Amaryllideae along with six others including the Scitamineae in the second series –the Epigynae of class Monocotyledones. Engler and Prantl have included the Amaryllidaceae along with other eight families including Liliaceae in the ninth order– the Lilliflorae of class Monocotyledoneae.

Scapigerous herbs with radical leaves and bulbous root stock; flowers showy, in 1-many flowered umbels subtended by one or more spathaceous bracts; corona often present; stamens 6; ovary superior or inferior; endosperm present.

Family—Amaryllidaceae

There are about 75 genera and 800 species in this family.

Distribution: They are widely distributed throughout the world. Majority of the members occur on the more level plains, plateaus and steppes areas of the tropics and sub–tropics. The important genera are: *Allium, Crinum, Amaryllis, Zephyranthes, Narcissus, Galanthus, Leucojum, Gethyllis, Agapanthus* and others.

Habit: The plants are generally scapose, perennial herbs, with a tunicated bulb or a rhizome. Most of the plants perennate by bulbs and very rarely by rhizome. The rhizome occurs in those genera which are considered as more primitive.

Leaves: The leaves are few in number and generally arise from the bulb (*i.e.*, radical). They are more or less linear and have parallel venation with transverse secondary veins.

Inflorescence: The flowers are usually found to be arranged in umbellate inflorescences. Rarely they are solitary. They are borne at the top of a leafless scape. The flowers are subtended by an involucre of one or more membranous bracts.

Flowers: The flowers are pedicellate, bracteate (*i.e.,* an involucre of two, or more rarely one membranous bracts), hermaphrodite, actinomorphic, trimerous, hypo or epigynous.

Perianth: It consists of six petaloid perianth leaves arranged in two whorls of three each. They may not be equal in size. Very often a corona is present, and the perianth leaves are frequently united into a tubular structure.

Androecium: The androecium consists of six stamens. The stamens are found to be inserted opposite the perianth leaves. However, in *Gethyllis* there are 12-18 stamens. They are hypogynous and inserted on the tube or at the base of the perianth leaves. The filaments are either free or connate at the base forming a corona. The anthers are bicelled, introrse, versatile or basifixed, dehiscing by longitudinal slits. In *Leucojum* and *Galanthus,* the anthers open by terminal pores.

Gynoecium: The gynoecium consists of three carpels, which are syncarpous. The ovary is superior or inferior. The placentation is axile. Usually the ovary is trilocular. Mostly many ovules in each loculus. The ovules are anatropous and superimposed, arranged in two series. The style is slender with a capitate or trilobed stigma.

Fruit: A capsule or may be fleshy and indehiscent.

Seeds: The seeds are small, numerous and endospermic. The small embryo remains surrounded by a fleshy endosperm.

Pollination: The pollination takes place through the agency of insects.

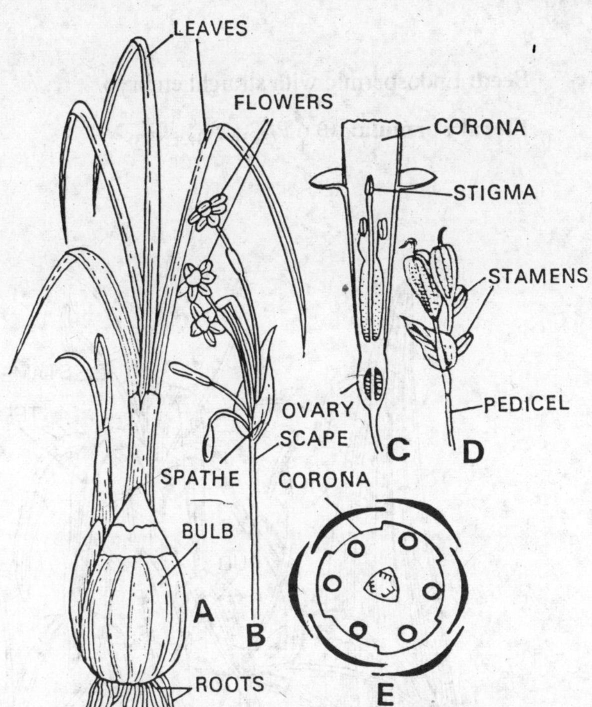

Fig. 62.1 Amaryllidaceae–*Narcissus*; (Verna-**Nargis**); A, entire plant with bulb and leaves; B, inflorescence; C, flower in vertical section; D, fruiting stage; E, floral diagram.

Floral Formula: $\oplus$ ⚥ P_{3+3} or$_{(3+3)}$, A3+3, G$(\underline{3})$ or $(\overline{3})$

Bentham & Hooker (1862)	**Engler & Prantl (1931)**	**Hutchinson (1959)**
Phanerogams	Phanerogams	Angiospermae
Monocotyledones	Monocotyledoneae	Monocotyledones
Epigynae	Lilliflorae	Corolliferae
Amaryllideae	Amaryllidaceae	Amaryllidales
		Amaryllidaceae

Description of *Allium cepa* in semitechnical language:

Allium cepa: Eng.–Onion; Verna.–**Piaj.**

Habit: Annual herb grown for edible bulbs.

Root: Adventitious, fibrous.

Stem: Underground bulb; the bulb bears outer scaly and inner fleshy leaves.

Leaves: Radical, cylindrical, fistular sheathing.

Inflorescence: Racemose, umbellate, actinomorphic, regular, hermaphrodite, small, white and hypogynous.

Perianth: Six perianth segments in two whorls of three each; three outer and three inner; free but united at base; valvate.

Androecium: Six stamens; filaments slender, broad at base, free; anthers bicelled, basifixed and introrse.

Gynoecium: Three, tricarpellary, syncarpous ovary, superior, trilocular; two ovules in each locule; placentation axile; style short, filiform; stigma short.

Fruit: Small capsule.

Seed: Endospermic with straight embryo.

Floral Formula: ⊕ ⚥ $P_{(3+3)}$, A_{3+3}, $G_{(3)}$

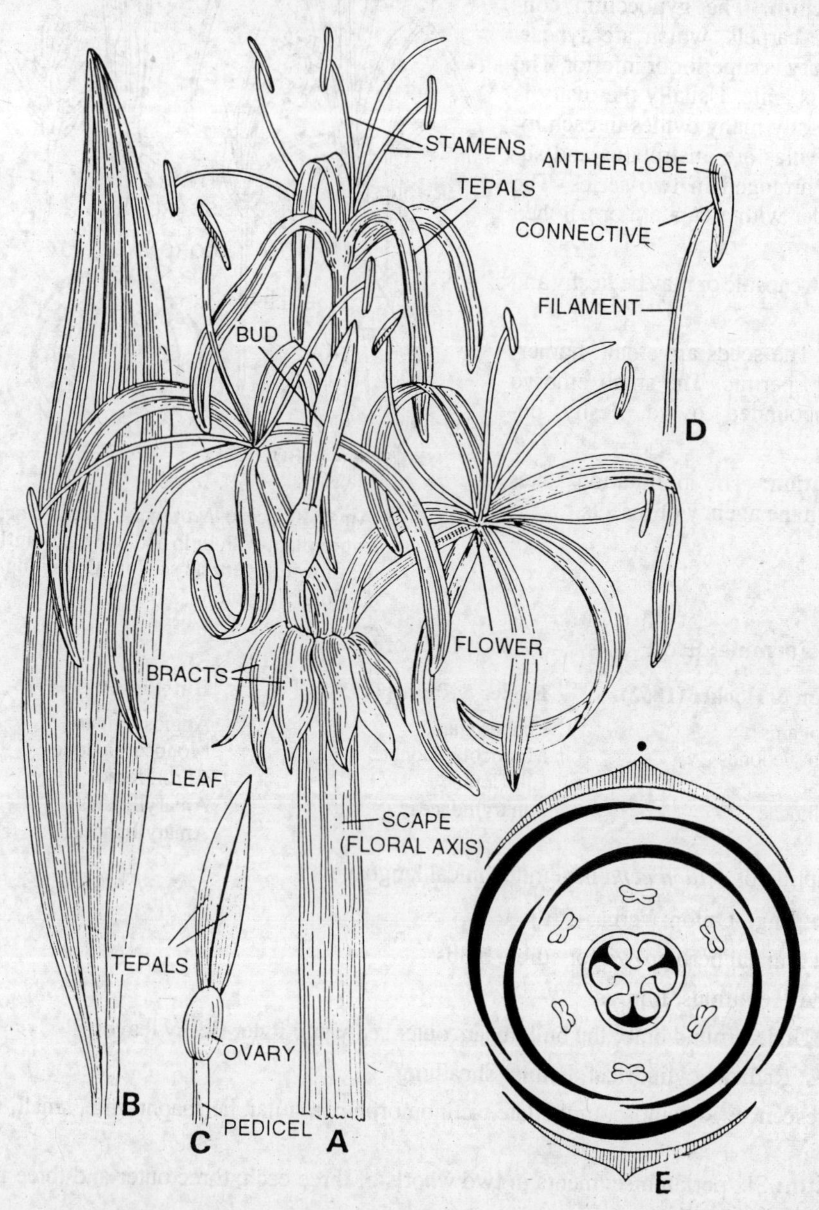

Fig. 62.2. Amaryllidaceae—*Crinum asiaticum*. A, scape with flowers (inflorescence); B. leaf; C, flower bud; D. stamen; E, floral diagram.

Economic Importance of the Family

The family is fairly important from its economic value. It contains several plants, of high ornamental value and many are used as vegetable.

The list of some more important plants is given below:

1. *Allium cepa*; Eng.–Onion; Verna.–**Piaz**–The bulbs are used as vegetable and as flavouring substance. The green leaves are also eaten. The plant possesses medicinal properties. This is stimulant, diuretic and expectorant. It is grown throughout the country.

2. *Allium sativum*; Eng.–Garlic; Verna.–**Lahsun**–The bulbs are used as a condiment and flavouring substance for meat and vegetables. Garlic powder is extensively used as a condiment and also serves as carminative and gastric stimulant in medicinal preparations.

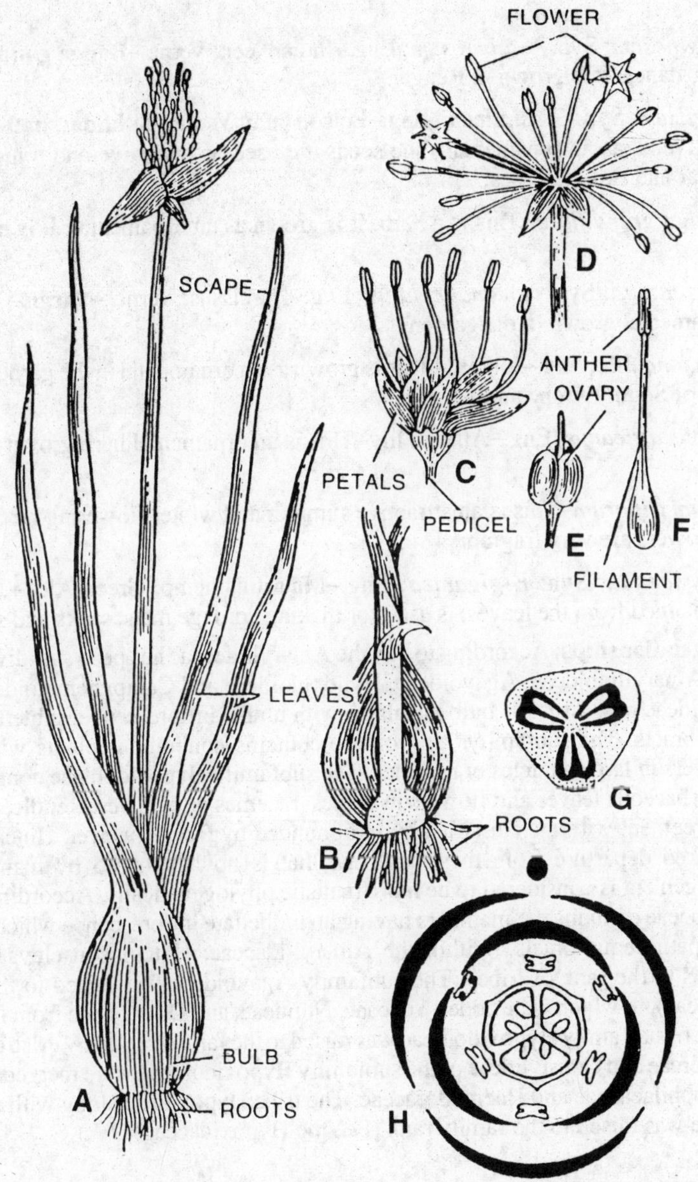

Fig . 62.3. Amaryllidaceae: *Allium cepa*-A, entire plant; B, bulb in vertical section; C, flower; D, inflorescence; E, cross section of ovary; F, pistil; G, single stamen dialated at base; H, floral diagram.

3. *Allium ascalonicum*; Eng.–Shallot; Verna.–**Gandana**–It is chiefly found in Tamil Nadu. The bulbs and green leaves are eaten as vegetable.

4. *Allium cepa* var. *viviparum*; Eng.–Top onion–It is introduced at Coimbatore and Coonoor. The rhizomes are used as condiment.

5. *Allium porrum*; Syn. *A. ampeloprasum*; Eng.–Leek; Verna.–**Vilaiti Lahsun**–It is native of the Mediterranean region. It is eaten as a vegetable.

6. *Allium schoenoprasum*; Verna.–**Piazi**–The leaves and bulbs are used as condiment. It is found in the Western Himalayas from Kashmir to Kumaon and also cultivated on the hill stations in South India.

7. *Allium uliginosum*; Syn. *A. tuberosum*; Eng.–Indian leek; Verna.–**Banga gandina**– It is used as a flavouring substance. It is grown in Bengal.

8. *Crinum defixum*; Syn. *C. asiaticum*; Eng.–Poison bulb; Verna.–**Sukhdarshan**–This is a large bulbous herb, cultivated as an ornamental. The seeds are used as purgative and tonic. The juice of leaves is used in ear and skin deseases.

9. *Zephyranthes grandiflora*–This is a herb. It is grown as an ornamental. It is native of South America.

10. *Narcissus tazetta*; Syn. *N. canaliculatus;* Eng.–Narcissus; Verna.–**Nargis**–This is an aromatic popular ornamental grown in the gardens.

11. *Narcissus jonquilla*; Eng.–Jonquil–This is grown as an ornamental in the gardens and homes. It is a herb; native of South Europe and Algeria.

12. *Agapanthus africanus*; Eng.–African lily–This is an ornamental herb grown in the gardens and homes.

13. *Pancratium triflorum*–This is an attractive summer and winter flowering herb, grown as an ornamental. The flowers are very fragrant.

14. *Furcraea foetida*; Syn. *F. gigantea*; Eng.–Mauritius hemp, Green Aloe–This is a large shrub. The fibre obtained from the leaves, is used for making cordage mats, sacks and soles of shoes.

Systematic Relationships: According to Pax the Amaryllideceae has been sub-divided into four sub-families (*i.e.,* Amaryllidoideae, Agavoideae, Hypoxidoideae and Campynematoideae). The subfamily Amaryllidoideae possesses the bulbous plants with umbel inflorescence, subtended by one or more spathaceous bracts. The subfamily, Agavoideae contains nonbulbous plants with the fibrous leaves and the flowers in large panicles or racemes. The subfamily Hypoxidoideae contains rhizomatous plants with herbaceous leaves and flowers in cymes, racemes or panicles. Rendle, Wettstein and Bessey were in agreement with the classification propounded by Pax. However, Hutchinson (1934) represented a marked departure from these and one that is now favoured by many authorities. Hutchinson's arrangement is considered to be more realistic phylogenetically. According to Hutchinson, the Amaryllidaceae contains the members having an umbellate inflorescence which is subtended by one or more spathaceous bracts. Within the Amaryllidaceae, Hutchinson elevated Pax's and Hoffman's subtribes to the rank of tribes. The subfamily Agavoideae was raised to the rank of the family (*i.e,* Agavaceae), to which were added Yuccaae, Nolineae, and Dracaeneae from Liliaceae. The tribe Alstoemerieae of subfamily Hypoxidoideae was raised to the family rank (*i.e.,* Alstroemeriaceae). The tribes Conanthereae and Constylideae of the subfamily Hypoxidoideae were merged, respectively, in the families Tecophilaeaceae and Haemodoraceae. The tribe Hypoxideae along with the subfamily Campynematoideae was raised to the family rank (*i.e.,* the Hypoxidaceae).

63

ORDER–AGAVALES

According to Hutchinson this is the 102nd order of the phylum Angiospermae, subphylum Monocotyledones and division Corolliferae. This order comprises of only two families, *i.e.,* Xanthorrhoeaceae and Agavaceae, of which the latter has been described in this text in detail. Bentham and Hooker have not recognized this order. They have placed the genera of Agavaceae in the families Liliaceae and Amaryllidaceae. Engler and Prantl have also not recognized the order and the family. They have also included the genera in the Liliaceae and the Amaryllidaceae and have placed them in their order– the Lilliflorae.

Perennials with often woody stock or stem, reaching tree form; flowers becoming dioecious, numerous, and small; perianth becoming dry and often glumaceous.

Family—Agavaceae

Distribution: They are tropical or sub-tropical in distribution. Most of them are Xerophytes. The important genera of the family are–*Yucca, Dracaena, Agave, Dasylirion, Phormium*, etc. The genus *Yucca* contains about thirty species distributed in Mexico and Southern United States. The genus *Dracaena* contains about forty species distributed in the temperate and tropical regions of the world. The genus *Agave* contains about 250 species distributed in tropical America and the Southern United States. The genus *Furcraea* contains about twenty species distributed in tropical America. The genus *Beschorneria* contains ten species distributed in Mexico and *Doryanthes* occurs with three species in Australia. The genus *Dasylirion* contains about ten species distributed in Mexico and Texas. The genus *Phormium* with two species occurs in New Zealand.

Habit: The plants are perennials, with thick woody rhizomes producing stems which assume the form of a tree. The plants of *Yucca* have woody stems. The species of *Dracaena* are mostly trees. The best known genus is *Dracaena draco* (Dragon tree). There was a biggest tree of *D. draco* on Teneriffe, which was blown down in 1868. This was 70 feet in height and 45 feet in girth, and was thought to be 6000 years old. The species of *Agave* have short unbranched stems. The plants of *Phormium* are short statured and arise from a rhizome.

Anatomy: There is peculiar type of secondary thickening of the stem caused by the activity of a cambium which produces fresh vascular bundles. This type of anomalous secondary growth occurs in *Yucca, Dracaena, Cordyline* and a few other genera.

Leaves: Usually the leaves are crowded either at the base or at the apex of the stem. They are thick, fleshy and sometimes prickly at their margins. The leaves of *Yucca* are crowded, long and linear. The leaves of *Agave* are also crowded and fleshy.

Inflorescence: The flowers are usually produced in much branched panicles. In *Yucca,* the flowers are found to be arranged either in racemes or panicles. The flowers of *Agave* are produced in large panicles. The plants of *Agave* are monocarpic and the production of the large inflorescences marks the end of their life. The plants live for a long time and during this time stores up a vast reserve of food, ultimately an inflorescence develops; which is a compound pancile reaching 20 or more feet in height.

Flower: The flowers are hermaphrodite or unisexual, mostly actinomorphic or somewhat zygomorphic, polygamous or dioecious.

Perianth: The perianth consists of six perianth segments arranged in two whorls of three each. The perianth leaves are similar in shape. In *Yucca,* the perianth leaves are free or rarely connate. In *Agave* the perianth is developed into a tube enclosing six stamens. In *Phormium* the perianth segments are connate.

Androecium: It consists of six stamens arranged in two whorls of three each, situated opposite the perianth segments, the anthers are bicelled, introrse, dorsifixed and dehisce by longitudinal slits. In *Yucca,* the stamens are inserted in the base of the perianth In *Agave,* the stamens are enclosed within the perianth tube.

Gynoecium: It consists of three carpels which are syncarpous. The ovary is either superior or inferior, trilocular with axile placentation, one or more anatropous ovules in each loculus.

Fruit: The fruit may be a berry or capsule.

Seeds: The seeds are endospermic with small embryo.

Pollination: The pollination takes place through the agency of insects.

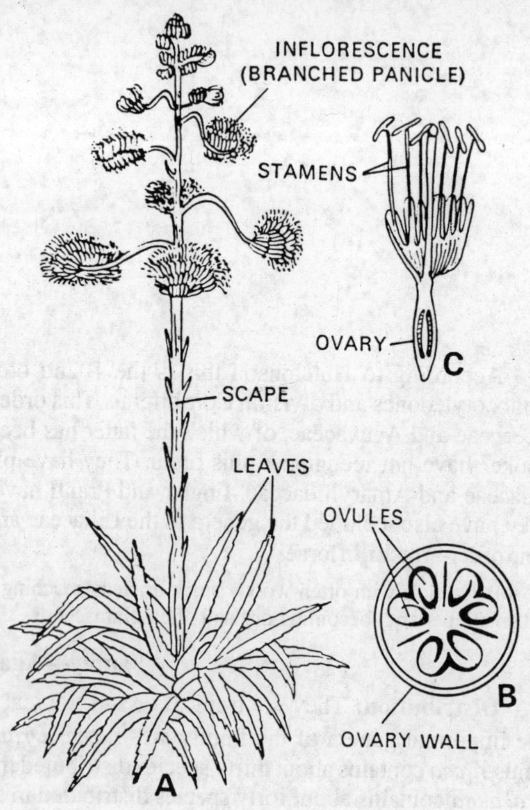

Fig.63.1. Agavaceae–*Agave americana*: A, entire plant; B, transverse section of ovary; C, flowers in vertical section.

Bentham & Hooker (1862)	**Engler & Prantl (1931)**	**Hutchinson (1959)**
Phanerogams	Phanerogams	Angiospermae
Monocotyledones	Monocotyledoneae	Monocotyledones
Epigynae and	Lilliflorae	Corolliferae
Coronarieae	Liliaceae and	Agavales
Liliaceae and	Amaryllidaceae	Agavaceae
Amaryllideae		

Economic Importance of the Family

The family is fairly important from the standpoint of its economic value. Some plants yield fibre while others are grown as ornamentals in the gardens. Some possess medicinal properties. A list of important plants is given below:

1. *Agave americana*; Eng.–Century plant; Verna.–**Kantala.**–This is a stout shrub, grown as an ornamental in the gardens. It is native of South America but now grown throughout our country.

2. *Agave sisalana*; Eng.–Sisal Agave–A fibre, obtained from the leaves is utilized for the

manufacture of ropes. Sisal wax resembles carnauba wax, which is used in the preparation of polisaes and in the manufacture of carbon papers. It is found in Assam, Bihar, Bengal, Maharashtra and Karnataka.

3. *Agave cantala*; Cantala, Manila maguey–This is a short stemmed woody herb. A fibre, obtained from the leaves is utilized for the manufacture of ropes, cordage and twines. It is found in Tamil Nadu, Maharashtra, Bihar, Uttar Pradesh and the Punjab.

4. *Agave-versa–cruz*; Eng.–Blue elephant Aloe; Verna.–**Kuwarbuti**–A fibre, obtained from the leaves is used for manufacturing ropes, cordage and mats. It is found in South India.

5. *Dracaena* sp.–The plants are grown in the gardens as ornamental for their beautiful foliage.

6. *Furcraea foetida*; Syn. *F. gigantea*; Eng.–Mauritius hemp, Green aloe–This is a large shrub, yielding a fibre which is used for making cordage, mats, sacs and soles of shoes. It is native of tropical America, but now found in South India, Bengal and Assam.

7. *Phormium tenax*–This is grown as an ornamental. The leaves of it yield a fibre known as New Zealand flax.

8. *Sansevieria hyacinthoides*; Syn. *S. zeylanica*; Eng.–Ceylon bowstring hemp; Verna.–**Murva**–The fibre obtained from the leaves, is used for manufacturing mats, cordage and paper. It is native of Sri Lanka, now found along the coastal regions of India from Bengal to Tamil Nadu.

9. *Sansevieria roxburghiana*; Eng.–Indian bowstring hemp; Verna.–**Murahri, Murva**–The fibre obtained from the succulent leaves is used for mats and cordage. This is a herb, found along the Coromandel Coast.

10. *Polianthes tuberosa*; Eng.–Tuberose; Verna.–**Gulshaba**–This is a herb, grown as an ornamental. It is native of Mexico.

11. *Yucca gloriosa*; Eng.–Spanish dagger-plant. This is a herb or shrub, grown as a hedge plant. It is native of Central America.

12. *Cordyline terminalis*; Eng.–Palm lily.–This is a shrub, grown in the gardens as an ornamental.

Systematic Relationships: Hutchinson has included two families. *i.e.*, Xanthorrhoeaceae and Agavaceae in the order—Agavales. Engler included the members of Agavaceae in the Amaryllidaceae. Hutchinson based his reclssification on the type of inflorescence and not on ovary position as Engler and others did. Hutchinson isolated the xerophytic fibrous leaved Dracaenoideae as one component of Agavaceae. Hutchinson has elevated the subfamily Agavoideae of Pax and Hoffman to the rank of family–the Agavaceae, to which were added also the Yuccaac, Nolincac and Dracaeneae from Liliaceae. According to him the Agavaceae is placed in between the Liliaceae and the Palmae.

64

ORDER–PALMALES

According to Hutchinson this is the 103rd order of the phylum Angiospermae, subphylum Monocotyledones and division Corolliferae. This order comprises of a single family, *i.e.,* Palmae, which has been described in the text in detail. Bentham and Hooker have placed this family along with two others in their fourth series – the Calycinae of class Monocotyledones. Engler and Prantl have included the single family Palmae in their fifth order – the Princippes of class Monocotyledoneae.

Stem herbaceous to climbing, or tall tree forms and woody; leaves often large, fibrous; flowers small, in panicles, becoming dioecious, subtended by large spathe-like bracts; ovary superior; ovule solitary; berry or drupe; seeds with endosperm.

Family—Palmae (Palmaceae/Arecaceae)

There are about 21 genera and 4,000 or more species in this family.

Distribution: The members of this family are essentially found in the tropical regions. A monotypic genus known as *Nannorhops* is found in the Himalayas, which extends through Afghanistan and to the South-East Persia. The palms are found on the east coast of Chinese-Japanese region. A few small genera are found in the United States and California. In our country the family is represented by several important genera, *e.g., Cocos nucifera* (**Narial**); *Phoenix dactylifera* (**Khajur**), *Areca catechu* (**Supari**) and many others. *Borassus flabellifer* is found along the coastal areas of Bengal, Bihar and the Western and Eastern Peninsulas. *Areca catechu,* the native of Malaya, now grown along the coasts of Karnataka, Tamil Nadu, West Bengal, Assam and Maharashtra. *Cocos nucifera,* cultivated chiefly in Tamil Nadu, Kerala and Karnataka.

Habit: The adult palms have generally a tall, woody unbranched stem bearing a crown of leaves and having its circumference ring marked with the bases of leaves which have perished. Some palms are climbing. They possess slender stems and are known as Rattans or Cane–palms.

Roots: The roots are adventitious. The primary root soon dies off and is replaced by adventitious roots emerging from the base of the stem.

Stem: Usually the plants possess an upright tall and unbranched stem which bears a crown of leaves at its top. The stem of date-palm, *i.e., Phoenix dactylifera,* reaches a height of sixty feet or more than that and remains covered with remnants of persistent leaf bases. *Phoenix acaulis,* a North Indian species possesses a very short bulbiform stem 6-10 cm in diameter. The stem of *Thrinax*, a West Indian species hardly reaches a height of a foot. The stem of *Phytelephus* of tropical America possesses a very short and thick stem; the leaves appear to come out from the ground. Some of the palms may reach a height of 150 feet or more.

The branching is rare in the aerial stems of palms. The branching is found in the Doum-palm, *i.e., Hyphaene thebaica* and few other species of *Hyphaene.*

Some Palms are climbing. They are known as Rattans or Canepalms. They climb over surrounding vegetation and sometimes attain a length of three hundred feet.

In *Rhapis flabelliformis,* the horizontal suckers are present at the base of the stem. Later on they grow erect and attain a bushy appearance.

Leaves: The leaves of this family are very characteristic. Usually the leaves are few in number and often very large. There are two important types of leaves, (1) the palmately divided leaves and (2) the pinnately divided leaves; they are popularly known as fan-palm and feather-palm types respectively. In the fan-palm type the depth of division varies greatly in different genera and species. In the feather-palm type a terminal leaflet may or may not represent, and the structure of pinnae is quite distinctive in different species.

In the fan-leaved palms, *e.g., Borassus,* the leaf blade remains entire and folded while in the bud, and as soon as the leaf expands the folds become incised from the margin inwards. Such foldings and incisions are also found in the feather-leaved palms.

In each case, the leaf possesses the large, stout petiole with broad sheathing base. When the leaf dies, it gradually falls. The sheath persists for sometime, though its fibres form a dense matting around the base of the younger leaves. In rattan palms, just above the petiole, a sheathing stipule or *ochrea* encircles the stem.

Sometimes, the thorns occur on the stem, leaves or even roots, *e.g.,* in the species of *Iriartea.* The stem thorns may also be present within the leafsheath. In rattan-palms, many curved spines occur on the stems and leaves, which help them in climbing.

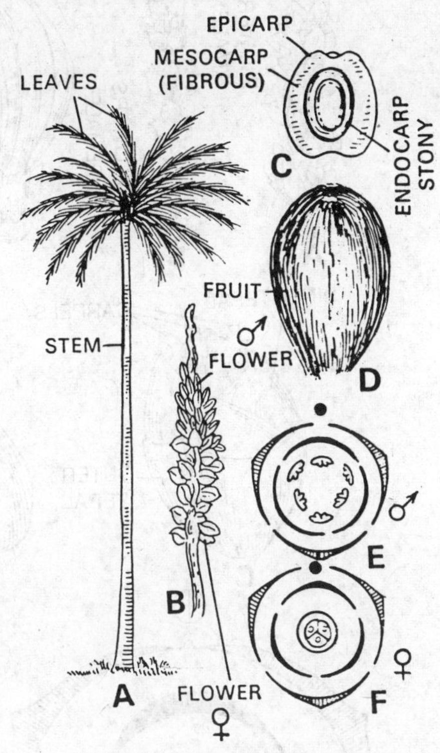

Fig. 64.1. Palmaceae–*Cocos nucifera*; A, a tree; B, a branch of inflorescence showing male and female flowers; C, fruit in long itudinal section; D, fruit (drupe); E, floral diagram of male flower; F, floral diagram of female flower.

Inflorescence: Some of the palms are monocarpic. In such cases the plant grows vegetatively for a number of years producing a stout woody trunk and a crown of leaves. Later on the growing point stops to produce any more leaves and develops a huge inflorescence. This production of the inflorescence exhausts the food reserves of the plant to such extent that it dies after fruiting, *e.g., Corypha* or Talipot palm.

In majority of cases the palms are polycarpic, *i.e.,* they bear axillary inflorescence which wither away after fruiting, *e.g., Borassus flabelliformis.*

The inflorescence may be simple, *e.g., Borassus fiabelliformis*; compound spike, *e.g., Cocos nucifera* and profusely branched panicle, *e.g., Daemonorops adspersus*. The young inflorescence remains enclosed within a spathe; or each branch is separately covered by a small spathe and thus, the inflorescence is usually known as spadix. As the flower shoot grows rapidly, the spathe is torned in definite way, and the whole structure comes out of it. The numerous flowers are sessile and sometimes, as in *Geonoma,* they remain embedded in the fleshy surface of spadix. They are arranged spirally.

Flowers: As a rule the flowers are unisexual. They are actinomorphic, incomplete and hypogynous. They are sessile and rarely possess short pedicels. The flowers are very small in size and present in very large numbers in the inflorescence.

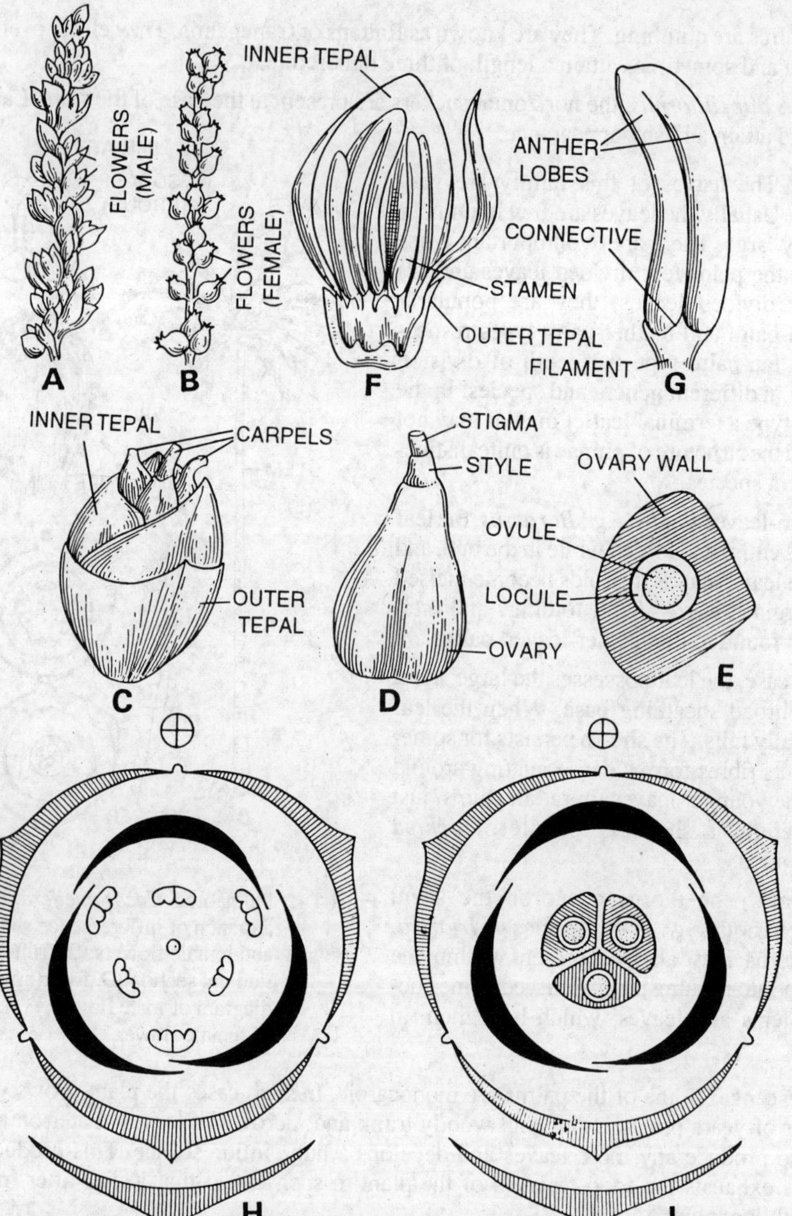

Fig.64.2 Palmae (Palmaceae)–*Phoenix sylvestris* (L.) Roxb. (**Khajur**). A, male inflorescence; B, female inflorescence; C, female flower; D, carpel; E, T.S. of ovary; F, L.S. of male flower; G, stamen; H, floral diagram of male flower; I. floral diagram of female flower.

The majority of plants are monoecious, *e.g.,* the male and the female flowers occur on the same plant, *e.g., Cocos*. In some cases the plants are dioecious, *i.e.,* the male and female flowers occur on two separate plants, *e.g., Phoenix.* In the monoecious plants, the male and female flowers occupy the different positions in the inflorescence in the different genera. In many cases a few female flowers occur at the base of the branches, whereas the male ones are crowded in the upper part. In *Raphia ruffia*, the branches of the spike bear female flowers in the lower half, and the male flowers in the upper half. There may be considerable difference in size between the two kinds of flowers, *e.g.,* in *Borassus,* the female flowers are much larger than the male ones which are very minute in size.

Perianth: It consists of six perianth leaves (tepals) arranged in two whorls of three each. They are polyphyllous and persistent. In *Areca catechu* the perianth leaves of the outer whorl are much smaller than the perianth leaves of the inner whorl. They are green to yellow or white in colour.

Androecium: It consists of six stamens arranged in two whorls of three each. In rare cases the number of stamens may be reduced to three only or increased to indefinite. They are polyandrous. The anthers are bi-lobed and dehisce by longitudinal slits. The pollen grains are produced in the form of powder in huge quantity and escaping like clouds. The filaments are short and distinct.

Gynoecium: It consists of three carpels (tricarpellary), apocarpous or syncarpous. The ovary is superior and trilocular, rarely unilocular. Each carpel contains a single anatropous ovule. The style is short. In *Cocos nucifera*, two of the three carpels may become aborted and the fruit contains one seed only, though the three carpels are marked by the presence of three longitudinal sutures and three round scars on the hard endocarp.

Fruit: Usually the fruit is a berry, either with a fleshy or fibrous covering, ***e.g., Phoenix*** (date palm). In *Cocos nucifera*, the fruit is fibrous drupe. There is great variety in the size of the fruit. The berries are very small and coconut and double coconut (*Lodoicea* sp.) fruits are very large in size.

Seed: The seeds vary in shape and size. The seed of coconut is rounded whereas long and narrow in date. It is endospermic. The embryo is minute cylindrical or conical. In coconut the endosperm contains large amount of oil and proteids.

Pollination: Wind-pollination (anemophily) is probably most general. In certain palms, *e.g., Sabal* and *Chamaedorea*, it takes place through the agency of insects. The flowers are protandrous, and, therefore, self-pollination is checked.

Floral Formulae:

Male Flower: ⊕ ♂ P_{3+3}, A_{3+3}

Female Flower: ⊕ ♀ P_{3+3}, $G_{(3)}$ or $\underline{3}$.

Bentham & Hooker (1862)	**Engler & Prantl (1931)**	**Hutchinson (1959)**
Phanerogams	Phanerogams	Angiospermae
Monocotyldones	Monocotyledoneae	Monocotyledones
Calycinae	Principes	Corolliferae
Palmae	Palmae	Palmales
		Palmae (Arecaceae)

Borassus flabellifer Linn.; Verna. **Tar;** Eng. Palmyra palm.

Habit: Tree.

Stem: Erect, aerial, woody, cylindrical, covered with persistent bases of petioles, unbranched, solid, rough, brown.

Leaf: Cauline, arranged in terminal crown; exstipulate, petiolate, petiole long and stout, glabrous, multicostate parallel venation.

Inflorescence: Spadix.

Male flower: Bracteate, sessile, incomplete, actinomorphic, unisexual, staminate, trimerous, hypogynous, cyclic.

Perianth: 6 tepals, arranged in two whorls of 3 each.

Androecium: 6 stamens, in two whorls of 3 each, polyandrous, filament short, dithecous, dorsifixed, introrse.

Gynoecium: 0.

Floral Formula: Br ⊕ ♂ P_{3+3}, A_{3+3}, G0

Female Flower: Bracteate, sessile, incomplete, pistillate, unisexual, actinomorphic, trimerous, hypogynous, cyclic.

Perianth: 6 tepals, in two whorls of 3 each, imbricate or valvate.

Androecium: 0.

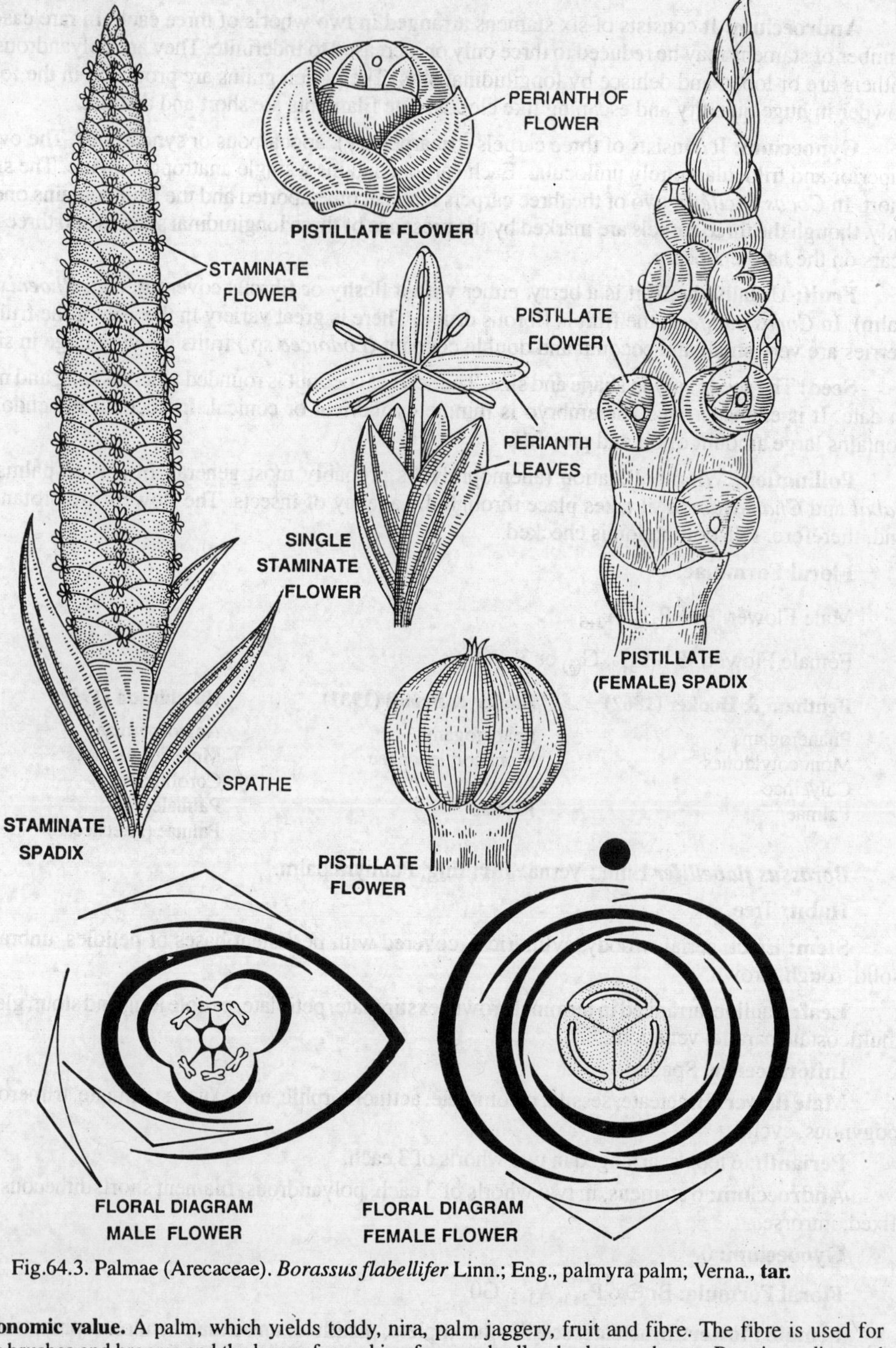

Fig.64.3. Palmae (Arecaceae). *Borassus flabellifer* Linn.; Eng., palmyra palm; Verna., **tar.**

Economic value. A palm, which yields toddy, nira, palm jaggery, fruit and fibre. The fibre is used for preparing brushes and brooms and the leaves for making fans, umbrellas, baskets and mats. Root is cooling and restorative. Juice of plant is diuretic, stimulant, useful in inflammatory affection and dropsy.

Gynoecium: 3 carpels (tricarpellary), syncarpous, ovary superior, 3-locular, single ovule in each locule; style absent, stigma short.

Fruit: Berry.

Floral formula: Br ⊕ ⚥ P_{3+3}, A_0, $G_{(3)}$

Economic Importance of the Family

1. *Areca catechu*; Eng.–Betelnut palm; Verna.–**Supari**. This is a beautiful slender stemmed palm; native of Malaya, now grown along the coasts of Karnataka, Tamil Nadu, West Bengal, Assam and Maharashtra. The nuts are chewed along with betel leaves **(Pan), Katha** (from *Acacia catechu*), lime, spices and tobacco. Besides, they are used as a vermifuge for animals and for the preparation of denitrifices. The husks are used for making hard board and brown wrapping paper. Galls, caused by insects are used as tanning material. The nuts are roasted, powdered and then made into tooth powder by adding some other substances. The beads and other fancy articles are made from the nuts. The trunk is used as water pipe. The leaves are used for thatching and woven into mats.

2. *Borassus flabellifer*; Syn. *B. flabelliformis*; Eng.–Palmyra palm; Verna.–**Tar**–This is a beautiful tall palm which yields several products of commercial importance, such as, toddy, nira, palm jaggery, fruit and fibre. The fibre is used for making brushes and brooms and the leaves for making fans, umbrellas, baskets and mats. The leaf stalks are used for making, writing and printing papers. These palms are found along the coastal areas of Bengal, Bihar and the Western and Eastern Peninsulas. In old times the leaves were used for writing purposes. The wood of the trunk is hard and durable and resists saline water. The trunks of female plants are strongest. They are used in making poles, pillars and posts. The fruit is roasted and eaten. The bows of **Tar** were popular in ancient India.

3. *Cocos nucifera*; Eng.–Coconut; Verna.–**Nariyal**. A tall palm, cultivated chiefly in Kerala, Tamil Nadu and Karnataka. An edible oil, obtained from the nut, is used in the manufacture of soap and candles. **Copra**, obtained by cutting the nut, is used in confectionery, soap and margarine. The copra residue is used as fodder and as manure. The fibrous rind of the nut yields the coir, which is used for making carpels and ropes. The endosperm yields the oil, which is used as hair oil and for several other purposes. The coconut pith and husk are used in the preparation of thermal insulating boards. The inflorescence is tapped when young for toddy, which when fermented gives an alcoholic drink. The root is astringent and is used in uterine diseases.

4. *Caryota urens*; Eng.–Toddy palm; Verna.–**Mari.** It is a palm with large bipinnate leaves. They are found in Orissa, North Bengal and Assam. The plant yields toddy by simply tapping the stem. The fresh toddy is extremely rich in vitamins and recommended for tuberculosis patients. The plant yields juice which is used for making sugar. The trunks are used as water pipes. The fibre obtained from the leaf sheaths, is used for ropes, baskets and soft brushes. The pith of stem yields sago.

5. *Caryota mitis*; This is a tall palm of Andaman islands. The young leaves are used as vegetable. The seed is used as a masticatory.

6. *Calamus acanthospathus*; This is a common climber of North-East India. It is extensively used for making tea-baskets. It is also used in making temporary bridges, walking sticks and chairs.

7. *Calamus latifolius*; Eng.–East India cane; Verna.–**Danagribet**–A climbing palm found in Bengal and Assam. The stems are used for walking sticks and baskets.

8. *Calamus rotang*; Eng.–Rattan; Verna.–**Bet.** They are climbing palms and sometimes reach a length of 500-600 feet. They are commonly found in Bengal, Assam, South India and Madhya Pradesh. The stem and branches are used for making baskets, furniture and canes. The stripped stems are used in making chair seats. The ropes made from this cane are used in dragging heavy loads. It is also used in making suspension bridges over streams and rivers in the hills. The root is given in chronic fever.

9. *Calamus tenuis*; They are found in Assam, sub–Himalayan tract and Dehradun. The walking sticks are made from stems. The mats, screens and baskets are also made.

10. *Copernicia*; It is found in tropical America. The Carnauba palm wax is obtained from it, which is used in making gramophone records, candles, etc. The wood is also used in several ways.

11. *Corypha umbraculifera*; Eng.–Talipot Palm.–This is a palm reaching upto 80 feet height. Found in South India. It flowers once in its life time and thereafter dies. The leaves are used as umbrellas. The mats, fans and baskets are made from the leaves.

12. *Daemonorops jenkinsianus*; Syn. *Calamus jenkinsianus*; Verna.–**Gola bet**–A tall palm found in Assam, the Khasia hills and Sikkim. The baskets are manufactured from its long, soft stems.

13. *Elaeis guineensis*; Eng.–Oil palm; this is native of West Africa but now grown in Travancore. The oil, obtained from the nuts, is used in making soap, margarine and as a fuel for diesel motors. The oil is also used in rheumatism.

14. *Lodoicea maldivica*; Eng.–Double coconut; Verna.–**Daryaka nariyal**–A tall palm. The plant bears one of the largest known fruits which takes ten years to ripen. The fruit is used to check diarrhoea and vomitting. The leaves are used for making thatches and hats. The shells of the large fruits are used as water containers.

15. *Metroxylon sagu*; Syn. *M. rumphii*; Eng.–Sago palm Verna.–**Sago**–The plants flower once in their life time. The fruit matures in three years. The sago (**sabudana**) is obtained from the pith of the old trunk by crushing and washing. This is used as food.

16. *Roystonea regia*; Syn. *Oreodoxa regia*; Eng.–bottle palm–It is a beautiful palm and largely grown in avenues.

17. *Phoenix dactylifera*; Eng.–Datepalm: Verna.–**Pindkhajur**.–A tall tree; native of Arabia or India. It is cultivated in Punjab and Rajasthan for its edible fruits (berries). A fibre is obtained from the leaves which is used for making ropes, baskets and cordage. The walking sticks are prepared from the leafstalks.

18. *Phoenix sylvestris*; Eng.–Wild date; Verna.–**Khajur.**–This is a tall palm. It is native of India. It is cultivated largely in Andhra Pradesh, Madhya Pradesh and Karnataka. The fruits are edible.

19. *Phoenix acaulis*; Verna.–**Khajeria.**–A stemless palm. Found in Bihar, Uttar Pradesh and North Bengal. The young buds are used as vegetable. The fruits are edible.

Systematic Relationships: Bentham and Hooker have placed the Palmae together with the Flagellarieae and the Juncaceae in the series–Calycinae. According to Engler and Prantl the single family Palmae is placed in the order–Principes. Rendle included the Palmae together with Araceae and Lemnaceae in the order–Spadiciflorae. Hutchinson has not followed the above mentioned view. He does not see any systematic relationship in between Araceae and Palmae, and therefore he has placed the Araceae in the order–Arales and the Palmae or the Palmaceae in the order–Palmales which are quite apart from each other and not related phylogenetically.

65

ORDER–PANDANALES

According to Hutchinson this is the 104th order of the phylum Angiospermae, subphylum Monocotyledones and division Corolliferae. This order consists of a single family the Pandanaceae, which has been described in the text in detail. Bentham and Hooker have placed the family Pandaneae along with four other families including Typhaceae, Aroideae and Lemnaceae in the fifth series – the Nudiflorae of class Monocotyledones. Engler and Prantl have placed the family Pandanaceae along with two others including the Typhaceae in the first order – the Pandanales of class Monocotyledoneae.

Stems often with aerial roots; leaves often spirally arranged, crowded, mostly spinulose, flowers dioecious; paniculate or densely crowded into spikes enclosed by large spathe like bracts; perianth vestigeal or absent; ovaries sometimes connate into bundles, in fruit forming a syncarp, the carpels drupaceous or baccate.

Family—Pandanaceae

There are three genera, *i.e., Pandanus, Freycinetia* and *Sararanga* and about 400 species in this family (Rendle). Hutchinson, however, has included one more genus (*i.e., Souleyetia*) in the family.

Distribution: They are commonly distributed in the tropical and subtropical regions of the old world especially in oceanic islands. They occur especially in the islands of the Malaya, Archipelago and of the Indian and Pacific oceans. The genus, *Pandanus* is the best known, which contains about 180 species. The genus *Freycinetia* contains about fifty species occurring from Ceylon to Polynesia. The genus *Sararanga* contains only two species one occuring in the Solomon Islands.

Habit: They are either shrubs or trees of palm like habit. The species of *Pandanus* are shrubs or trees often producing aerial prop roots with multiple membranous root caps. The species of *Freycinetia* are shrubby root climbers.

Root: The main root soon perishes and is replaced by adventitious roots arising from the stem and branches. In *Freycinetia,* the roots are climbing organs while in *Pandanus* they form stout aerial roots which spring out from the stem, obliquely downwards and because of the decay of the lower part of the stem they act as supporting organs.

Leaves: The leaves are mostly spirally arranged. They are cauline and spinulose on the margins. Generally, the leaves are stiff, long, linear, pointed forming a tuft at the ends of the stem and branches. They are arranged in three ranks which form wellmarked spirals. Each leaf possesses a sheathing base and a long blade bearing spines on both the margins and also on the back of the midrib. Buds occur in all the leaf axils, but only a few develop either in vegetable shoots or in inflorescences.

Inflorescence: The inflorescence is terminal. The bracts are found to be arranged in three rows. The lower bracts are leaflike, white the upper ones are smaller, spathe-like and somewhat coloured. The flowers are arranged either in panicles or crowded into spadix-like inflorescence with large spathe-like bracts. In *Pandanus* and *Sararanga* the male inflorescence is generally a branched panicle. In *Pandanus,* the branches of the panicle are either true spikes or look like spikes. In *Freycinetia,* the male inflorescence is a cluster of spikes. In genus *Sararanga,* the inflorescence arises in the axil of a scale like bract. In *Pandanus,* the female inflorescence is simple, forming a spherical or cylindrical head. Sometimes, it is branched, having the heads with or without stalks. In *Freycinetia,* the female inflorescence consists of a few stalked cylindrical clustered spikes. In *Sararanga,* it is a much branched pendulous panicle.

Flowers: The flowers are unisexual. The most perfect flowers of the family occurs in the genus *Sararanga,* where it develops in the axil of a scaly bract. In this genus, the flower is short peduncled and consists of indefinite stamens surrounded at the base by a short lobed perianth. In *Pandanus* and *Freycinetia*, the flowers are ebracteate. In *Freycinetia,* rudimentary ovary is often present, which remains surrounded by indefinite stamens. In *Pandanus* the flower consists only of stamens, which may be solitary or many in number. When many, they are found to be arranged in a spike or umbel on short lateral axes. Each axis is supposed to be a floral axis. The filaments may be long or short, anther filiform, dehiscing by longitudinal slit.

In *Sararanga,* the female flowers are short peduncled; each flower develops in the axis of a bract; a flat undulate-margined perianth surrounds the base of a fleshy pistil; the pistil consists of 70-80, closely cohering carpels; each carpel being indicated by a stigma. Like that of male flowers, the female flowers of the genus *Sararanga* are supposed to be most perfect. The flowers of other genera (*e.g., Pandanus* and *Freycinetia*) are naked. In *Freycinetia*, an indefinite number of staminodes surrounds the base of gynoecium. The staminodes are rarely found in *Pandanus.* In *Pandanus,* the gynoecium consists of several carpels which are usually more or less completely coherent. Each carpel bears a sessile stigma; one ovary chamber corresponding to each carpel. Each ovary chamber contains one to many anatropous ovules.

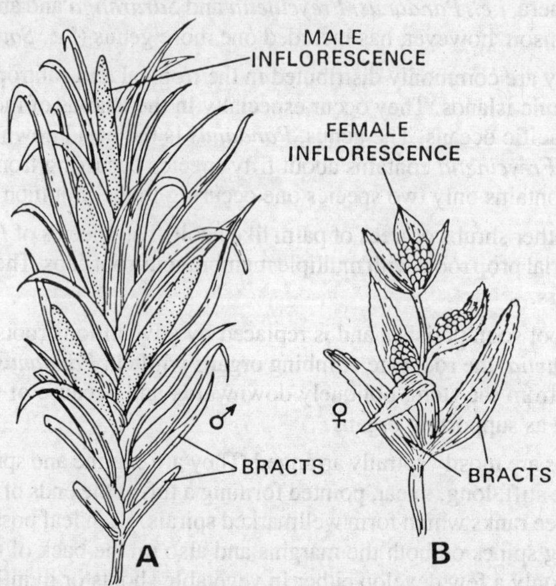

Fig. 65.1. Pandanaceae–*Pandanus;* A, male inflorescence; B, female inflorescence.

Fruit: The fruit is a berry or drupe. In *Pandanus*, the endocarp is hard and the mesocarp is fleshy, fibrous or woody.

Seeds: The seeds are small and endospermic (oily endosperm), with a small basal embryo.

Pollination: The pollination takes place through the agency of insects (*i.e.*, entomophily). The insects are attracted because, of the showy bracts and the strong smell of male inflorescence. The anemophily is also possible because of the production of large quantity of pollen grains.

Floral Formulae: Male flowers- ♂ P0, A∝

Female flowers- ♀ P0, G(∝)

Bentham and Hooker (1862)	**Engler and Prantl (1931)**	**Hutchinson (1959)**
Phanerogams	Phanerogams	Angiospermae
Monocotyledones	Monocotyledoneae	Monocotyledones
Microspermae	Microspermae	Corolliferae
Pandanaceae	Pandanaceae	Pandanales
		Pandanaceae

Economic Importance of the Family

The family is of little economic value. A few important plants are enlisted below:

1. *Pandanus tectorius;* Syn. *P. odoratissimus;* Eng.–Screwpine; Verna.–**Keora**–This is ornamental shrub or small tree. An essential oil, obtained from the male inflorescence, is used in perfumery. A fibre is obtained from the leaves, which is used for making cordage, nets, mats and sacs. The leaves are also used in paper–making. It is found along the sea-coasts of both the Peninsulas, Orissa, Uttar Pradesh and the Andaman Islands.

2. *Pandanus latifolius*–This is a small tree, introduced from Malaya and is now grown on the West Coast. It is used for flavouring cooked rice.

3. *Pandanus lerami*–The fruits are edible and are known as Nicobar Breadfruits.

4. *Pandanus edulis*– The tough long leaves are used for mat making. The fruits are edible.

5. *Freycinetia* sp.–They are climbing shrubs with brightly coloured fleshy bracts, grown as ornamentals.

Systematic Relationships: Engler and Prantl have included the Pandanaceae together with the Typhaceae and the Sparganiaceae in the order–Pandanales. Bentham and Hooker placed the Pandaneae together with the Aroideae, Lemnaceae, Typhaceae in the series–Nudiflorae. Hutchinson, however, segregated the Pandanaceae from other families as a separate order–the Pandanales. He has interpreted the coalescence of fruits into a syncarp, as found in Pandanaceae, to be more advanced than the Typhaceae or Sparganiaceae. Hutchinson considers that the Pandanaceae have been derived from the Palmae.

66

ORDER–ORCHIDALES

According to Hutchinson this is the 108th order of the phylum Angiospermae, subphylum Monocotyledones and division Corolliferae. The order consists of a single family–the Orchidaceae, which has been described here in detail. Bentham and Hooker have placed the family Orchideae along with two others including Hydrocharideae in their first series–the Microspermae of class Monocotyledones. Engler and Prantl have placed the Orchidaceae along with one more (*i.e.,* Burmanniaceae) in their eleventh order– the Microspermae of class Monocotyledoneae.

Terrestrial or epiphytic or saprophytic; flowers strongly zygomorphic; perianth segments in 2 whorls, usually petaloid, variously modified stamens 2 or 1; pollen from granular to waxy and in masses; ovary inferior; seeds very numerous and minute, without endosperm.

Family—Orchidaceae

There are 450 genera and about 10,000–15,000 species in this family (Lawrence).

Distribution: They are distributed throughout the world in the temperate and tropical regions. They are found most abundantly in the tropical regions, where the majority of genera are epiphytes. The genera found in the temperate and arctic regions are mostly terrestrial. The largest genera are *Spiranthes*, *Habenaria, Cypripedium, Listera* and *Epidendrum.* In our country about 1600 species are found in the Himalayas and other hills.

Habit: They vary very much in their habit. The plants are always herbaceous with sympodial stems. Some plants are typical terrestrial herbs, many are epiphytes while several others are saprophytes. Many epiphytes and terrestrial plants are xerophytes and they store water in thickened leaves, aerial roots, swollen internodes and pseudo-bulbs. As already mantioned the epiphytic orchids are common in tropics while terrestrial orchids are found in temperate regions. Epiphytic orchids develop long aerial roots for the absorption of water. These roots are characterized by a special development of the epidermis to form the **velamen,** a tissue consisting of several layers of short tracheids. The sponge-like tissue absorbs water from the atmosphere, and passes it on to the internal tissues. The other roots of epiphytes serve as anchorage organs and for the intake of mineral nutrients.

Burgeff (1909) has demonstrated that all orchids possess the **mycorrhiza,** and in the species where the chlorophyll is poorly developed, it is especially well provided. The species of *Rhizoctonia* (fungus), usually enters the root by the root hairs and occupies a more or less definite zone in the cortical region. Mycorrhiza, however, does not occur in the chlorophyll containing roots.

There are many saprophytic orchids. The important ones are: *Neottia nidus-avis* (Bird's nest Orchid) and *Corallorhiza trifida* (Coral Root). The Malayan saprophyte, *Galeola altissima* possess long thin stems which climb to the tree tops, where they fix themselves by aerial roots arising from nodes.

The other important Orchids are: *Orchis mascula* (Purple Orchid), *Orchis fuchsii* (Spotted Orchid), *Orchis morio* (Green-Winged Orchid), *Anacamptis pyramidalis* (Pyramidal Orchid), *Ophrys apifera* (Bee Orchid), *Platanthera bifolia* (Butterfly Orchid), *Coeloglossum viride* (Frog Orchid), *Listera ovara* (Tway-blade), *Spiranthes spiralis* (Lady's Tresses Orchid), *Cephalanthera demasonium* (White Helleborine), *Cypripedium calceolus* (Lady's Slipper) and several others.

The Orchid plant may be constructed in several ways. The construction may be sympodial or monopodial. In sympodial construction, the main axis is composed of annual portions of successive axes each of which begins with scale leaves and terminates into an inflorescence. In monopodial construction the main axis continues to grow year by year and produces flowers only on lateral branches.

All terrestrial orchids are sympodial in construction and possess a short rhizome. Each annual shoot bends up into the leafy branch of the current year. In many such orchids, the storage organs develop to face the dormant period of the year. In most of the species, the internode of the stem becomes thickened and is known as a **pseudo-bulb**. In many others a thick, fleshy, adventitious root develop into a tuber filled up with food reserve. Such structures develop from the bud for the next year's growth situated at the base of the stem.

The epiphytic orchids are either sympodial or more rarely monopodial. The roots of these orchids are of three types: (1) The clinging roots which are not affected by gravity but are negatively phototropic. (2) The absorptive roots develop between the plants and the support. Usually the humus is accumulated around such roots and they take the nutrition from the humus. (3) The true aerial roots, with velamen tissue, hanging down in the moist air to absorb water. Majority of orchids of tropical regions drop their leaves in the dry season, and then the aerial roots form the only remaining assimilating tissue.

Leaves: The leaves are simple, alternate, rarely opposite or whorled, which may be fleshy. In *Listera* and some others the leaves stand close together in pairs. In most of saprophytic orchids they are often reduced to scales which may be membranous or succulent. They are usually linear, often distichous and sometimes radical. Very often leaf sheath develops which surrounds the stem. When the shoot is reduced to a scaly bulb, the foliage leaves may be entirely absent. In *Pleurothallis* and allied genera the numerous shoots each bear a single leaf, which spreads at right angles to the axis. Sometimes the axis is very short (*e.g.*, in *Masdevallia*) and the leaves appear to come out from the substratum. The margin is entire, the apex often denticulate and parallel venation.

Inflorescence: Usually it is of racemose type, and may be a spike, raceme or a panicle. Sometimes, the flowers are solitary (*e.g.*, in *Cypripedium*). In *Renanthera lowii* the raceme becomes about thirteen feet long. In *Oberonia*, the flowers are crowded, shortly stalked and resemble a spike of Plantain. In many species of *Bulbophyllum*, the flowers remain almost sunk in a fleshy axis. A panicle is found in many species of *Oncidium*. Generally the inflorescence is short lived; but sometimes it becomes perennial producing fresh flowers round the year.

Flower: Generally the flowers are hermaphrodite, sometimes unisexual, medianly zygomorphic, bracteate, sessile or pedicellate and epigynous. The flowers are often of very colourful and attractive form, but sometimes they are colourless or brown and inconspicuous. The flowers are specially designed in relation to special insect visitors.

Perianth: It consists of six perianth leaves arranged in two whorls of three each. The outer whorl is calyx-like while the inner whorl forms a corolla. Sometimes, the outer whorl takes the form of a corolla and the inner whorl may be reduced to a very small size. The perianth-leaves of each whorl may remain free or somewhat united. The segments of the outer whorl are generally imbricate. The median perianth leaves of each whorl are different in size, shape and often in colour from the lateral segments. One of the segments of the inner whorl becomes often much enlarged and modified and is known as a lip or **labellum.** Generally the whole perianth becomes twisted through 180° with the result of a torsion produced by the ovary, and then the labellum in the anterior abaxial position. This phenomenon is known as **resupination.** Sometimes the labellum or very rarely the odd sepal is prolonged backwards into a spur or a sac, which collects nectar or acts as nectar secreting tissue. In *Disa,* the odd sepal is spurred. In *Haemaria* the odd sepal unites with the lateral petals are often absent. In *Epicranthes,* the lateral petals are filiform. In *Cypripedium caudatum,* the petals to form a hood. In *Corysanthes* the lateral petals are often absent. In *Epicranthes*, the lateral petals are filiform. In *Cypripedium caudatum,* the petals are narrow ribbon-like and often of one yard in length. In *Disa,* the lip is small and narrow. In *Odontoglossum,* the labellum is large and spreading. In *Cattleya* the labellum is upwardly curved. In *Cypripedium* the labellum is slipper shaped.

Androecium: The androecium consists of two trimerous whorls of stamens, which are never all present. Generally there are one or two stamens, which unite together and to an extension of the gynoecium forming a **column.** The anthers are bicelled, introrse and dehisce by longitudinal slits. The pollen grains are granular and usually bound up by elastic threads of viscin and agglutinated in two waxy masses called the **pollinia.** The viscin extends into a columnar, organ known as the **caudicle** at its lower end.

Gynoecium: It consists of three carpels, which are syncarpous. The ovary is inferior and unilocular very, rarely trilocular. When unilocular the placentation is parietal, when trilocular the placentation is axile. The stigma is modified. In *Cypripedium* all the three lobes of the stigma are functional, but more often the anterior lobe of the stigma becomes sterile and develops into a small pocket known as the **rostellum,** the remaining two stigmatic lobes are functional. The rostellum is situated between anther and the functional stigmatic lobes. A part of the rostellum develops into a viscid disc which is known as the **viscidium.** The caudicles remain attached to this viscidium. There are numerous anatropous ovules which are extremely minute in size.

Fruit: It is a capsule, opening by three to six longitudinal slits.

Seeds: The seeds are numerous and very minute in size, which may easily be blown away by wind from one place to another.

Pollination: It takes place through the agency of insects (*i.e.,* entomophilous). The insects are attracted to the flowers, because of their attractive colours and sweet smell. The nectar is also secreted either in the spur or in the labellum.

Floral Formula: ·|· ⚥ P3+3, A1–2, $G(\overline{3})$

Bentham & Hooker (1862)	**Engler & Prantl (1931)**	**Hutchinson (1959)**
Phanerogams	Phanerogams	Angiospermae
Monocotyledones	Monocotyledoneae	Monocotyledones
Microspermae	Microspermae	Corolliferae
Orchideae	Orchidaceae	Orchidales
		Orchidaceae

1. *Vanda roxburghi;* R. Br; Verna. **Rasna.**

Habit: Epiphytic herbs.

Root: Adventitious aerial roots, the aerial roots are characterized by a special development of the epidermis to form the **velamen,** other roots serve as anchorage organs.

Stem: Leafy or scapose, monopodially branched, the main axis grows year after year and bears flowers at the axil of leaves, the axis continuously produces new leaves apically.

Leaves: Simple, alternate, exstipulate, fleshy, linear, distichous, leaf sheath present which surrounds the stem, the axis is very short and the leaves appear to come out from the substratum; margin entire, apex acute to obtuse, parallel venation.

Inflorescence: Racemose, raceme.

Flower: Pedicellate, bracteate, hermaphrodite, medianly zygomorphic, irregular, epigynous, trimerous, colourful.

Perianth: 6 tepals arranged in two whorls of three each, outer whorl calyx-like, inner whorl makes corolla, polyphyllous, imbricate, tepals of each whorl are different in size, shape and even in colour from the lateral segments; one of the tepals of inner whorl becomes much enlarged, the **labellum**; labellum prolonged backward into a spur which collects nectar.

Androecium: 6 in two whorls of 3 each, never all stamens present; in *Vanda* one stamen present which unites to an extension of the gynoecium forming a **column**; anthers bicelled, introrse, dehiscing by longitudinal slits; pollen grains granular and bound up by elastic threads of viscin forming two **pollinia;** the viscin extends into a columnar organ, the **caudicle,** at its lower end.

Fruit: Capsule.

Seeds: Numerous, very minute.

Floral formula: Br ·|· ⚥ P3+3, A1, $G(\overline{3})$

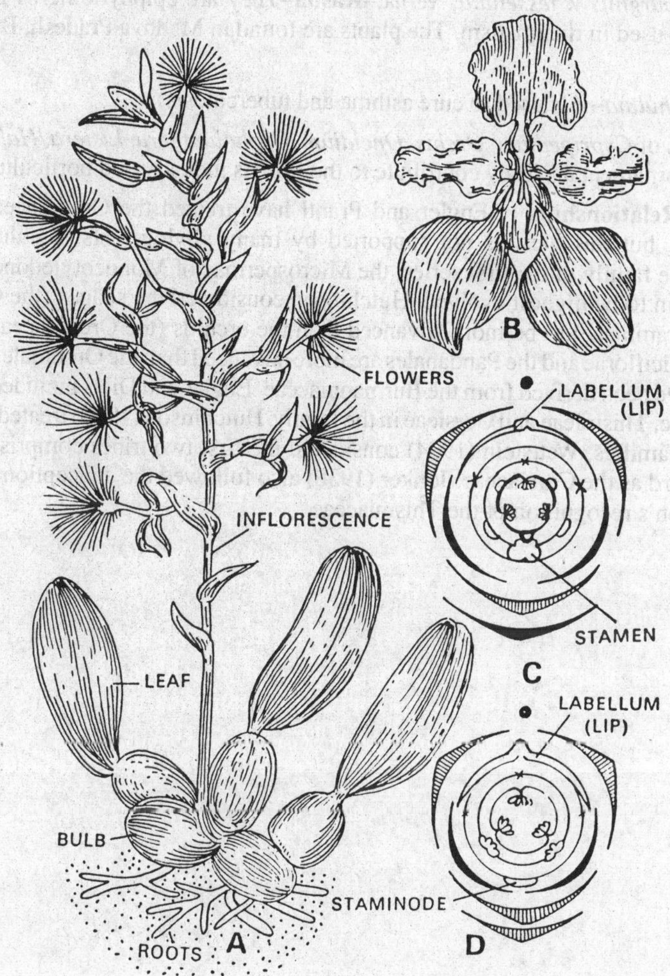

Fig. 66.1. Orchidaceae—A, sympodial orchild, *Bulbophyllum* with inflorescence; B, flower of *Oncidium* (from view ; C, floral diagram of a typical monandrous orchid; D, floral diagram of *Cypripedium*

Economic value. An epiphytic herb with beautiful purple flowers. Grown as an ornamental. The root is useful in rheumatism. The paste of leaves is applied to the body during fever.

Other important members of the family. *Orchis latifolia* Linn. **Salap**–tubers edible; *Vanilla planifolia* Andr., pods are used for flavouring purposes ; *Vanda spathulata* Spreng. flowers given in asthma and mania.

Economic Importance of the Family

The family is of little economic value. Most of the plants are grown for their beautiful flowers. Some possess medicinal properties and some yield scent. A list of few important plants is given below

1. *Orchis latifolia*: Eng.–Orchis; Verna.—**Salad**–This is perennial herb found in Ḳashmir. The tuberous roots are edible.

2. *Vanilla planifolia;* Syn. *V. fragrans;* Eng.–**Vanilla**–This is a climbing shrub. The fruits are used for flavouring confectionery and in perfumery. It is native of tropical America but now grown in the Nilgiris, Coorg and Mysore. The natural Vanilla is extracted from its capsules (fruits).

3. *Vanda roxburghii*; *V. tessellata*; Verna.–**Rasna**–They are epiphytic herbs found on mango trees. The roots are used in rheumatism. The plants are found in Madhya Pradesh, Bihar, Bengal and Travancore.

4. *Vanda spathulata*–It is used to cure asthma and tuberculosis.

5. The species of *Cypripedium, Orchis, Oncidium, Cephalanthera, Listera, Habenaria, Bletilla*, etc., are grown as ornamentals. They contribute to the florists industry and horticulture.

Systematic Relationships : Engler and Prantl have treated the Orchidaceae to be highest evolved monocots, but this view is not supported by many phylogenists. Bentham and Hooker, however, placed the family in the first series, the Microspermae of Monocotyledones and they have not considered them to be highest evolved. Hutchinson considered the sedges (the Cyperaceae) and the grasses (the Gramineae) to be more advanced than the orchids (the Orchidaceae). According to Wettstein, the Spadiciflorae and the Pandanales are more advanced than the Orchidales. Engler thought that the orchids have been derived from the Burmanniaceae. Engler and Diels included the three tribes, *i.e.*, Euburmannieae, Thismieae and Corsieae in the family. Hutchinson (1930) treated each of the three tribes as separate families. Wettstein (1934) considered the first two tribes comprising the Burmanniaceae and the third as the Corsiaceae. Jonker (1938) also followed the conception of Wettstein and rejected Hutchinson's recognition of the Thismiaceae.

67

ORDER–CYPERALES

According to Hutchinson this is the 109th order of the phylum Angiospermae, sub-phylum Monocotyledones and division Glumiflorae. The order comprises of a single family-the Cyperaceae, which has been described here in detail. Bentham and Hooker have placed the Cyperaceae along with four other families including Gramineae in their seventh series–the Glumaceae of class Monocotyledones. Engler and Prantl have included the Cyperaceae along with Gramineae in their fourth order–the Glumiflorae of class Monocotyledoneae.

Herbs rarely sub-shrubby; leaves linear, grass-like, sheathing at the base, blade often quite reduced; flowers anemophilous, becoming unisexual; perianth reduced to scales or setae or absent, stamens 6-1; fruit an indehiscent 'nutlet'; seeds with endosperm; stems rarely hollow.

Family—Cyperaceae

There are about 85 genera and 2500 species in this family (Lawrence).

Distribution: The members of this family are cosmopolitan in their distribution. They are mostly found in the arctic and temperate regions of the northern and southern hemispheres. The plants are usually found in marshy places. The plants grow well in humus soil. The plants constitute about 10% of the flora of arctic regions. The plants of this family are popular as **sedges.** *Carex arenaria* forms an important part of sand-dune vegetation. In our country *Scirpus* and *Cyperus* plants are very common on the margins of the ponds and banks of the river and marshy places.

Habit: The sedges are usually perennial herbs and rush-like in appearance. They perennate by means of creeping sympodial rhizomes. Some sedges are annual herbs. The aerial shoots are given out from their underground rhizomes either solitary or in clusters.

Stem: The stem is usually solid and triangular. The elongated runners are found in many sedges.

Leaves: The leaves are found in three rows. They possess narrow blade and closed sheath.

Anatomy: Big air spaces are found in the stem and leaves. The rhizome also possesses big air spaces. The bundles of rhizome are concentric, *i.e.,* either amphicribral or amphivasal.

Inflorescence: The inflorescence is either spike of spikelets or panicle. The unit of inflorescence is known as spikelet as that of allied family Gramineae. A single flower develops in the axil of the bract, *i.e.,* glume. However, the bracteoles are rarely present. In certain cases the spike like cymes are so aggregated that they look like spikes or heads. The terminal and solitary flowers are rarely found, *e.g.,* in *Oreobolus.*

Flower: The flowers are sessile, bracteate, hermaphrodite or unisexual. The flower is borne in the axil of glume. The female flower of *Carex* sp., is borne in the axil of second glume known as **utricle.** The

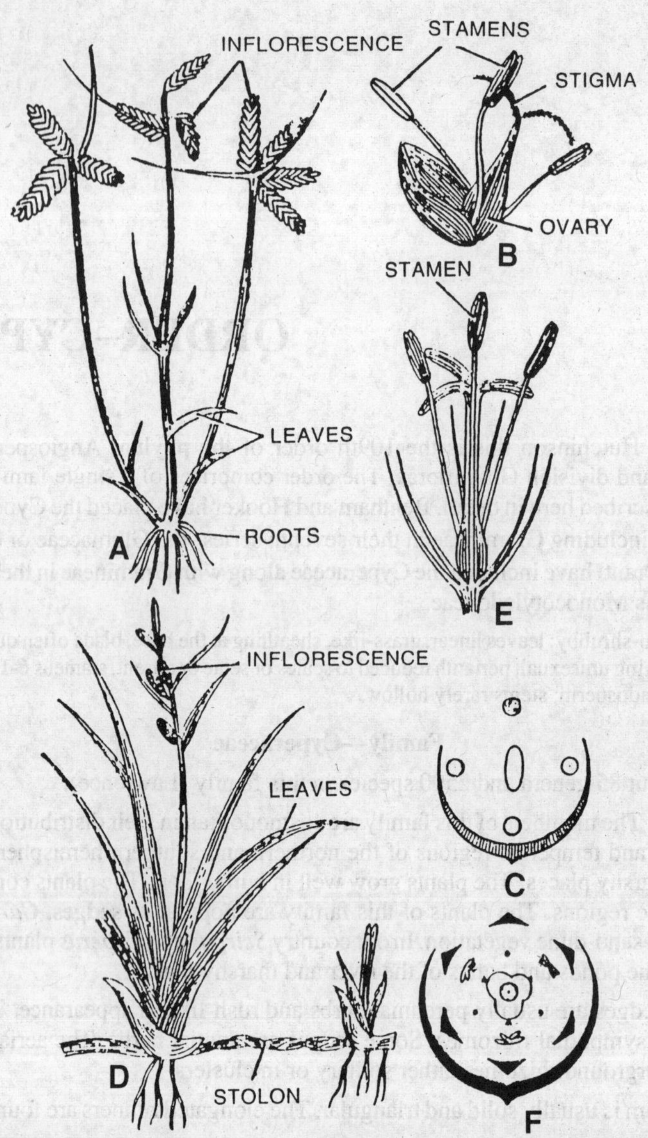

Fig. 67.1. Cyperaceae–A, *Cyperus* sp., habit; B, flower with bract of same; C, floral diagram of same; D, a horizontal rhizome of a sedge with sympodial growth; E, flower of *Scirpus* sp; F, floral diagram of same.

utricle forms a sac-like structure around the ovary. The flowers may be naked as in *Cyperus, Carex,* etc.

Perianth: In *Eriophorum gracile* the perianth is represented by hairs. In *Scirpus lacustris* the perianth is represented by bristles. Sometimes the perianth is represented by scales. The perianth of *Oreobolus* consists of six perianth leaves (tepals) arranged in two whorls of three each.

Androecium: Rarely the androecium consists of six stamens arranged in two whorls of three each. Usually it consists of only three stamens of outer whorl. In these common cases the stamens of inner whorl are altogether absent. Sometimes the number of stamens is reduced to one only, *e.g.,*

Hemicarpha and some others. The anthers are bilobed and basifixed. The filaments are free.

Gynoecium: It consists of three carpels (tricarpellary) or two carpels (bicarpellary), syncarpous. As many feathery stigmas are found as many carpels are present. The base of style is thick and beak–like. The ovary is superior, unilocular. It bears single basal anatropous ovule. The basal placentation is believed to be derived from free–central placentation.

Fruit: Usually the fruit is achene.

Seeds: The seeds are albuminous with small embryo.

Pollination: The pollination is anemophilous *i.e.,* takes place through the agency of wind.

Floral Formula: ·|· ⚥ or ♂ or ♀ P3+3, or O, A3, G$(\underline{2})$ or $(\underline{3})$

Bentham & Hooker (1862)	**Engler & Prantl (1931)**	**Hutchinson (1959)**
Phanerogams	Phanerogams	Angiospermae
Monocotyledones	Monocotyledoneae	Monocotyledones
Glumaceae	Glumiflorae	Glumiflorae
Cyperaceae	Cyperaceae	Cyperales
		Cyperaceae

Economic Importance of Family

1. *Cyperus papyrus*–In ancient times as early as 2400 B.C. Egyptians used the plants for paper making. The word paper is derived from the word papyrus.

2. *Cyperus esculenta*–Verna. **Kaseru**–The rhizomes and the tuberous roots of many species are eaten raw and after cooking.

3. *Cyperus* sp.–The baskets are made from the stems of the species.

4. *Scirpus lacustris*–The matting and chair seats are made from the plants.

5. *Eriophorum* sp.–The perianth leaves convert into long hair-like bristles. They are used for filling purpose of pillows and cushions.

6. *Kyllinga triceps*–The plants possess aromatic roots. The decoction prepared from these roots is given to diabetic patients. An aromatic oil is also extracted from the roots which is used in the liver disorders.

Systematic Relationships: Bentham and Hooker included the families–Eriocauleae, Centrolepideae, Restiaceae, Cyperaceae and Gramineae in the series, Glumaceae. They have treated this series to be highest evolved among Monocots. Engler and Prantl included only two families–the Gramineae and the Cyperaceae in the order–Glumiflorae. However, they have not considered the order to be highest evolved among Monocots. Hutchinson rejected the concept of the older workers of placing the Gramineae and the Cyperaceae into one order, he segregated them into separate orders, the Graminales and the Cyperales. He placed the other 4 families into a third related order–the Juncales. The morphological studies of Belk (1939) and Blaser (1944) supported the segregation of the grasses (the Gramineae) and the sedges (the Cyperaceae) into two separate orders–the Graminales and the Cyperales. Most of the phylogenists are in agreement that the grasses and sedges are very advanced groups. It is also generally agreed that they have been evolved from primitive liliaceous ancestral stocks.

68

ORDER–GRAMINALES

According to Hutchinson this is the 111th order (the last one) of the phylum Angiospermae, subphylum Monocotyledones and division Glumiflorae. He has considered this order to be highest evolved among the phylum Angiospermae. This order comprises of a single family the Gramineae. The family has been described here in detail. Bentham and Hooker have also considered the Gramineae to be highest evolved among Phanerogams. They have placed the Gramineae along with four other families including Cyperaceae in their seventh series – the Glumaceae of class Monocotyledones. Engler and Prantl have placed the Gramineae along with the Cyperaceae in the fourth order– the Glumiflorae of Monocotyledoneae.

Similar to Cyperales but flower subtended by 2 bracts (lemma and palea); perianth usually represented by lodicules; fruit a caryopsis, rarely a nut, berry, or utricle; stems mostly hollow between the nodes.

Characteristic Features: World-wide in distribution; annual or perennial herbs, rarely shrubs or trees, stems erect, ascending or prostrate creeping, usually branched at the base in perennials forming sterile shoots (innovations) and flowering stems (culms), in annuals only the culms present; culms cylindrical rarely flat jointed. Usually hollow in the internodes, solid at the nodes; leaves solitary at nodes, sometimes crowded at the base of the stems, alternate, two rowed, consisting of sheath, ligule and blade; sheath encircles the culm, margins of sheath free and overlapping and more or less connate, swollen at the base; ligule situated at the junction of sheath and blade, membranous or reduced to fringe of hairs, rarely absent; blades long, narrow, rarely broad, passing gradually into the sheath, rarely with a petiole like base, flat, convolute or terete, parallel venation; flowers generally hermaphrodite, rarely unisexual, small and inconspicuous, perianth represented by 2 or 3 minute hyaline or fleshy scales known as lodicules, subsessile between two bracts, *i.e.*, lemma and palea, the whole structure known as floret or a false flower; florets one to many, distichous, sessile on a short slender axis, the rachilla and bearing at the base two empty bracts, the upper and lower glumes, the florets and glumes forming a spikelet; spikelets pedicelled on open or contracted panicles or racemes, or sessile in spikes; stamens 1 to 6, rarely more, usually three, hypogynous, filaments delicate, anthers bicelled versatile, dehiscing by longitudinal slit; ovary unilocular, with one anatropous ovule, often adnate to the adaxial side of the carpel; styles usually two, rarely 1 or 3; stigmas 2, feathery (plumose); fruit caryopsis, with a thin pericarp adnate to the seed, rarely a nut or berry; seeds endospermic, embryo small.

Family—Gramineae (Poaceae)

There are about 500 genera and 4,500 species in this family (Willis).

Distribution: The members (grasses, cereals and millets, etc.) belonging to this family are cosmopolitan in distribution. The plants are commonly distributed in temperate regions, but they are also not uncommon in tropical and sub-tropical areas. In our country the family is represented by many food crops, *e.g.*, *Triticum aestivum* (Wheat), *Oryza sativa* (Rice), *Hordeum vulgare* (Barley), *Zea mays* (Maize), *Sorghum vulgare* (Millet **Jowar**), *Avena sativa* (Oat), *Saccharum officinarum* (Sugarcane), etc. The grasses, *e.g., Poa, Festuca, Andropogon, Cynodon, Cymbopogon,* etc., and bamboos are very common.

The great adaptability of the different species has enabled them to survive under the most varied conditions, some are aquatic while some are xerophytic and found on sea-shores and in arid regions.

Habit: Mostly the plants are annual, biennial or perennial herbs or shrubs. The largest woody species are the bamboos, which may grow to more than 100 feet in height and several inches in diameter. Most of the plants, *e.g.,* grasses, are wild while some are cultivated for valuable food.

Root: Generally adventitious, fibrous, fascicled, stilt (*e.g., Zea mays, Sorghum vulgare,* etc.).

Stem: The stem may be erect, prostrate or even creeping. Usually fistular (hollow), rarely solid (*e.g., Zea, Saccharum,* etc.). The internodes are long and nodes are conspicuous. In many grasses the runners and suckers are also developed. Among the perennial grasses, usually the rhizomes and root stocks are formed, which help in vegetative propagation. Small tubers and corms are also found in so many species.

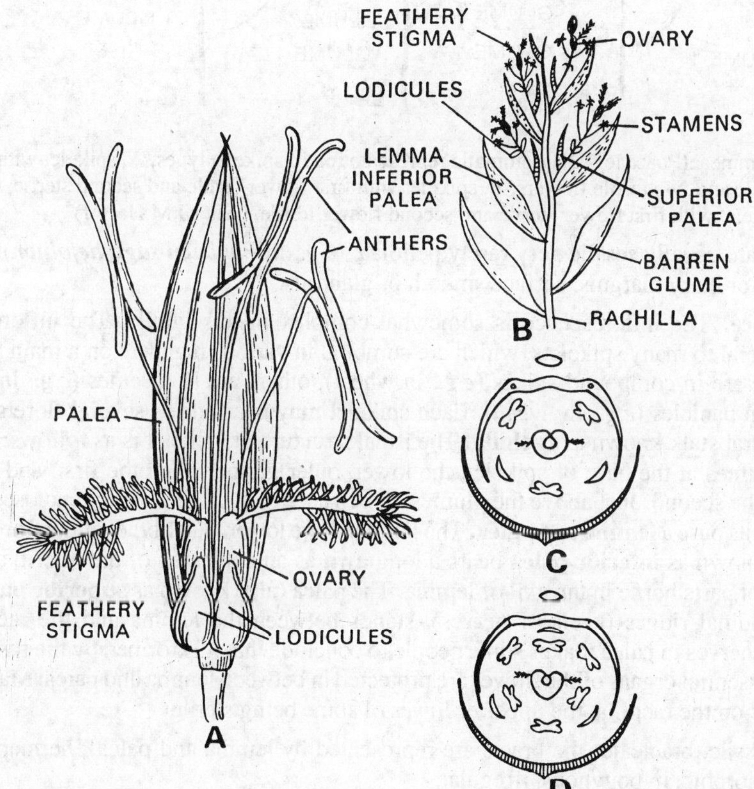

Fig.68.1. Gramineae (Poaceae)–A, a spikelet of *Festuca* sp., after the removal of the glumes showing lodicules, ovary, anthers, stigma, etc; B, spikelet of a grass; C, floral diagram of a grass; D, floral diagram of *Bambusa.*

In this family the aerial stem is commonly known as **'culm'** which is usually terete (cylindrical) but in few cases flattened.

In most grasses the main axis only develops lateral branches from the basal buds. These branches are **'tillers'**.

Leaves: In most of the members of this family, the foliage leaf can easily be divided into two parts, the sheath and the blade (lamina).

The leaf sheath usually covers the internode completely or partially. The inner surface of leaf sheath is usually glabrous (smooth) while the upper surface is variously grooved and either glabrous or hairy.

At the junction of the leaf sheath and leaf blade there occurs a thin membranous outgrowth, the **ligule.** In some species the ligule is represented by only few hairs.

In many species there are two lateral pointed outgrowths above the ligule are known as **auricles.** They are well developed in *Hordeum* and *Oryza.*

The leaf blade is long, narrow, acuminate and with parallel venation.

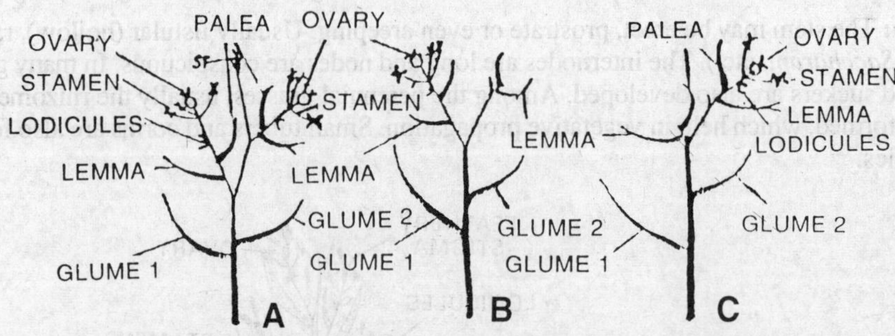

Fig.68.2. Gramineae(Poaceae)–Diagrammatic representation of spikelet types. A, spikelet with one sterile and two fertile flowers; B, spikelet with first flower fertile and second sterile; C, spikelet with first flower sterile and second flower fertile. (After J.M.Hector)

The leaves are usually sessile very rarely petiolate (*e.g.*, in *Panicum sagittaefolium*), alternate, lanceolate, entire or hairy margins, surface smooth or glabrous.

Inflorescence: The inflorescence is somewhat complex in this family. The inflorescence is composed of several to many spikelets, which are combind in various manners on a main axis called the **rachis.** Some are in compound spikes (*e.g.*, in wheat), others are in racemes (*e.g.*, in *Festuca*), while some are in panicles (*e.g.*, in *Avena*). Each spikelet may bear one to several florets (flowers) attached to a central stalk known as **rachilla.** The usual structure of spikelet is as follows: There is a pair of sterile **glumes** at the base of spikelet, the lower, outer glume called the first, and the upper, inner one called the second. Just above the glumes, there is series of florets, partly enclosed by them. Each floret has at its base a **lemma** and **palea.** The lemma is the lower, outer bract of the floret. Usually the lemma also known as **inferior palea** bears a long **awn** as an extension of the midrib at the apex or back. The floral parts borne in the axil of lemma. The palea (also known as **superior palea**), often with two longitudinal ridges (keels or nerves), stands between the lemma and the rachilla. The pressence of two nerves in palea makes some people to conclude that it is formed by the fusion of two bracteoles. The essential organs of the flower are protected in between lemma and palea. Many species have other florets on the rachilla, the upper or lower of some being sterile.

Flower: Sessile, bracteate (the bracts are represented by lemma and palea), hermaphrodite or unisexual, zygomorphic, hypogynous, irregular.

Perianth: Sometimes absent or represented by two or rarely three (*Bambusa*) minute, membranous scale like **lodicules** (here the perianth leaves are known as lodicules).

Androecium: Usually three stamens, sometimes six, *e.g.*, in *Oryza, Bambusa,* etc., and sometimes reduced to two or one, the stamens are with free, long slender filaments and versatile anthers. Anthers two-celled, dehisce longitudinally.

Gynoecium: One carpel (monocarpellary) (though the pistil is tricarpellary but only one carpel is functional, syncarpous); unilocular ovary, superior, single ovuled (anatropous ovule); style short,

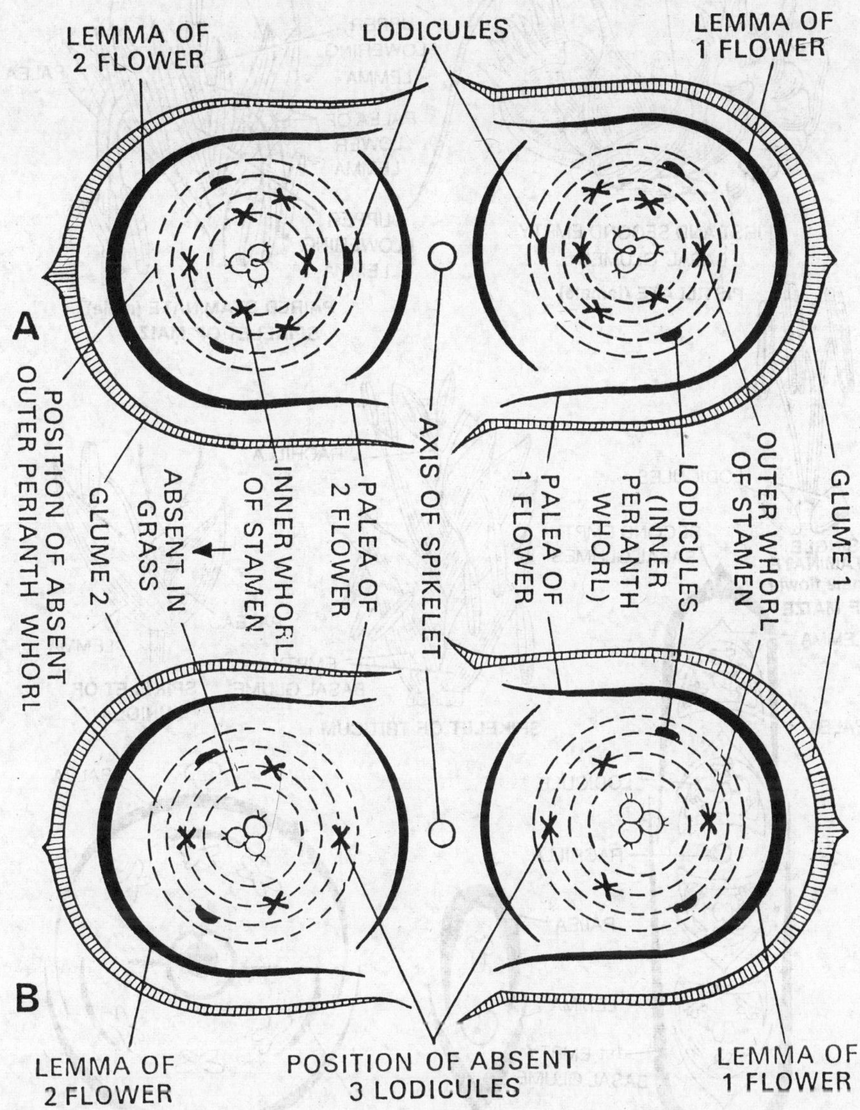

Fig.68.3. Gramineae (Poaceae)–Diagrammatic representation of the Gramineous spikelets; A, the Bamboo type: (3 lodicules; 2 whorls of 6 stamens; 6 fused carpels each with styles); B, the grass type: (2 lodicules; 1 whorl of 3 stamens inner whorl absent); 3 fused carpels with two styles, front style absent. (After J.M.Hector)

usually two, stigma usually two arise from the carpellary wall, feathery. In maize the style is long and silky; placentation basal.

Fruit: Usually **caryopsis** (pericarp completely united with the seed coat), rarely a nut or berry (*e.g.*, in *Bambusa*). In *Dendrocalamus* it is a nut.

Seeds: Endospermic (albuminous).

Pollination: Usually anemophilous (wind pollination), cross pollination, may also take place.

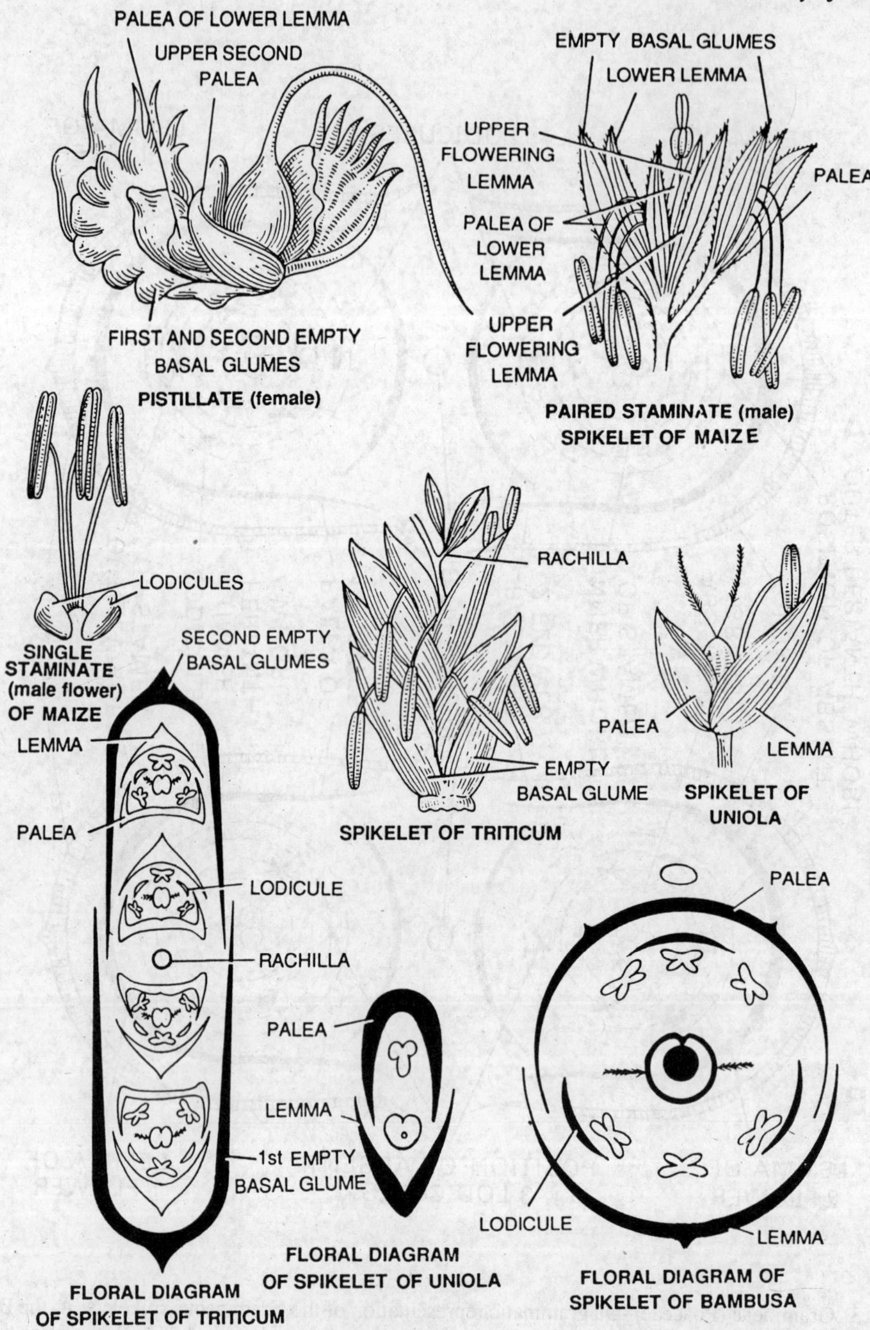

Fig.68.4. Gramineae (Poaceae). *Zea mays; Triticum; Uniola; Bambusa,* etc.

Floral formula: ·|· ⚥, ♂ or ♀, PO or 2 lodicules, A3 or 6, G(1).

Bentham & Hooker (1862)	**Engler & Prantl (1931)**	**Hutchinson (1959)**
Phanerogams	Phanerogams	Angiospermae
Monocotyledones	Monocotyledoneae	Dicotyledones
Glumaceae	Glumiflorae	Glumiflorae
Gramineae	Gramineae	Graminales
		Gramineae/Poaceae

Triticum aestivum (Verna.–**Gehun**):

Habit: The wheat plant is an annual cereal. In India all wheats are winter grown, except in Nilgiri and Palini hills where two crops (winter and summer) are taken.

Root: Adventitious, fibrous.

Stem: The stem is 3 to 4 feet in height or more. The stem (culm) is erect, cylindrical sometimes furrowed, and either glabrous or scabrous. The nodes are 5 to 7 in number, the average number is six. Each node is swollen and solid. The internodes are generally hollow. The lower and upper internodes are covered by the leaf-sheaths. Branching by tillers.

Leaves: The normal leaf of wheat is divisible into two parts, the leaf-sheath and the leaf blade (lamina). The leaf also possesses two accessory organs, the ligule and the auricle.

The sheath is inserted on the node and envelops the stem. The leaf sheath is usually thicker than the leaf blade with thin transparent margins. The surface is either glabrous or hairy.

The leaf blade is long, narrow, lanceolate, acuminate with parallel venation, sometimes hair also present along the veins.

At the junction of leaf blade and leaf sheath thin membranous **ligule** encircles the stem. It is colourless and hairy.

Two claw-like appendages near the ligule region are prominent, these are known as **auricles.** They are usually hairy and pale green.

Inflorescence: Spike of spikelets. This is a compound spike bearing two rows of lateral spikelets on its axis and a single terminal spikelet. There are many short internodes on the main axis. Each internode is narrow at the base and broader at the apex. One side of internode is convex while the other side is concave or flattened. The position of concave side alternates with convex side and the whole axis has a *zig zag* appearance. The spikelets are sessile and arranged alternately, the spikelet is inserted on the apex of the internode.

Spikelets: The solitary and sessile flower (florets) are arranged on a short, jointed axis, the rachilla. The flowers are alternately placed on the spikelet. At the base of each spikelet, two glumes occur. They appear to be opposite of each other. Actually one glume overlaps the other. The normal glume is somewhat boat shaped, with a thick main nerve, dividing it into two unequal halves. The base of the glume is rounded while sometimes the nerve extends in an awn 2 to 5 cms. long. The surface is either glabrous or hairy. In each spikelet the number of flowers varies. Usually two grains, sometimes three, and very rarely four mature in a single spikelet.

Flower: Sessile, bracteate two bracts -a lemma (inferior palea) and a palea (superior palea). They are situated opposite to each other. The **lemma** is somewhat boat-shaped with many nerves. The upper margin is notched and ends in a pointed awn. The colour of lemma is greenish white, sometimes pink. The **palea** is a thin membranous bract just opposite the lemma. It is slipper-shaped with two prominent nerves. The flower is small, zygomorphic, hermaphrodite, hypogynuous, incomplete, irregular and not showy.

Perianth: The perianth is represented by two thin membranous structure, the lodicules. They are colourless, narrow and scale-like, and sometimes hairy.

Androecium: Three stamens. The filaments are long, slender and free; the anthers are bi–celled and versatile.

Gynoecium: Single median carpel (monocarpellary). The ovary is superior, unilocular, hairy and triangular. From the tip of the ovary two styles arise, stigma feathery. Single ovule, basal placentation.

Fruit: A caryopsis, the seed coat firmly united to the ovary wall.

Seed: Albuminous.

Generic characters: Ligule and auricles present, inflorescence spike of spikelets, two rows of lateral spikelets on the axis and a single terminal spikelet, lemma boat shaped, palea slipper shaped.

Floral Formula: ·|· ⚥ P 2 (lodicules), A 3, G (1).

Hordeum vulgare; Verna.–**Jau.**

Habit: Annual herb, annual winter cereal crop (in India).

Root: Adventitious, fibrous, branched. The whole root system may be divided into two types of roots, the seminal roots and the adventitious roots. The seminal roots are branched throughout their length and penetrate to a depth of 5 to 6 feet. The true adventitious roots arise from nodes near ground level. They are unbranched and comparatively thicker. They also penetrate to a depth of 3 to 5 feet. Their colour is white and covered throughout their length by root hairs.

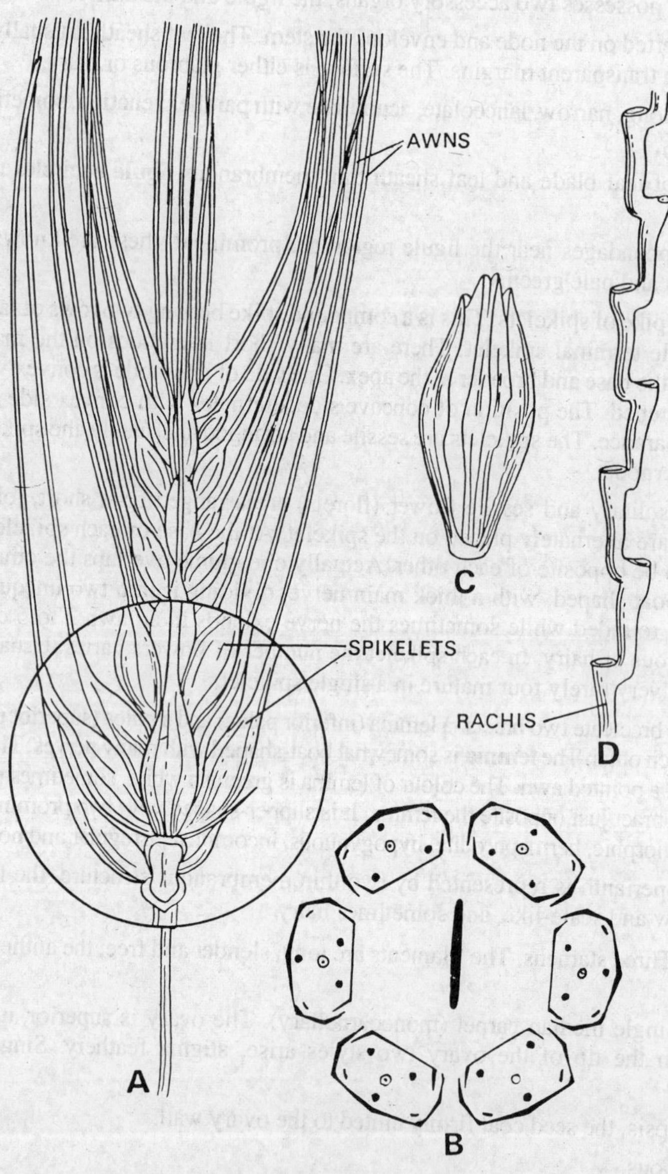

Fig. 68.5. Gramineae (Poaceae)–*Hordeum vulgare*–A, the ear; B, floral diagram; C, husked grain; D, a part of barren rachis.

Stem: Erect, cylindrical, fistular, solid at nodes, 5 to 8 conspicuous nodes. Branching by tillers (tillers develop fewer than wheat). Usually glabrous and covered by leaf sheaths.

Leaves: The leaf can easily be divided into two parts, the leaf sheath and the leaf blade. The leaf sheath covers the stem and inserted at the node, glabrous and parallel veined. The leaf blade is long, linear lanceolate, acuminate, somewhat hairy, margins also hairy, multicostate parallel venation. At the junction of the leaf sheath and the leaf blade a prominent ligule and well developed auricles are present.

Inflorescence: Spike of spikelets. The inflorescence is a cylindrical spike carrying an alternating series of spikelets. The main rachis is compressed and truncate. The rachis consists of a series of straight joints whose ends are more or less flat and upon these flat cushions the spikelets are inserted. The rachis at the base of the ear is surrounded by a ridge of tissue which has been termed the **'collar'.**

Spikelet: There are two long narrow bracts, almost parallel to each other at the base of the spikelet. These bracts may be considered as glumes. Behind these glumes the solitary flower is inserted, enveloped in lemma (inferior palea) and palea (superior palea). The lemma is broad, rounded on the back, 5-nerved and with a barbed awn. The innermost bract known as palea (superior palea) is less obvious, narrow and mostly covered by enveloping lemma. It is awnless and two ridged.

Flower: Sessile, zygomorphic, irregular, hermaphrodite, incomplete, hypogynous.

Perianth: The perianth is represented by two small membranous lodicules, arranged opposite the palea.

Androecium: Three stamens; filaments long, slender, free; anthers 2-celled, versatile.

Gynoecium: Carpel one (monocarpellary); ovary superior, unilocular, single basal ovule, more or less ovate; two feathery stigmas.

Fruit: A caryopsis with pointed ends, usually invested by lemma and palea.

Seed: Albuminous.

Generic characters: Prominent ligules and well developed auricles present; inflorescence spike of spikelets, cylindrical spike carrying alternate series of spikelets; the 'collar' present at the base of rachis, each spikelet represented by a single flower, two long narrow bracts (glumes) with each spikelet; caryopsis usually invested by lemma and palea.

Floral Formula: ·|· ⚥, P2 (lodicules), A3, G$\underline{(1)}$.

Oryza sativa (Rice–**Chawal**)

Habit: Annual herb, cultivated as annual cereal in both tropical and temperate regions. Also sown upto 6,000 feet in Himalayan region (but in extremely humid conditions). This cereal crop is always sown under water and in the rain faded areas. In India, Bengal and Bihar are under rice cultivation. Usually the rice is cultivated in lowland areas.

Root: The whole root system can be divided into two parts, the seminal or primary roots and the adventitious or secondary roots.

The adventitious roots arise from the lower nodes of the plant and its tillers. They are fibrous and branched.

The rice grown in the deep waters, possess adventitious roots at the nodes above ground level.

Stem: Erect, cylindrical, hollow except at nodes, 6 to 8 mm. thick, nodes and internodes conspicuous, the number of internodes may vary from 10 to 20, branching by tillers.

Leaf: They always emerge from the nodes, so the number of leaves on an axis is equal to the number of nodes. A normal leaf possesses the leaf sheath, ligule, auricles and the leaf blade (lamina). The leaf sheath always encircles the internode from the node (pulvinus) upwards. The leaf sheath is split at the base, finely ribbed, glabrous. The ligule is well developed, membranous, colourless or pink coloured. The auricles are well developed, sickle shaped 3 mm. in length, present at the junction of the sheath and the blade (lamina). The leaf blade is long, narrow, lanceolate, acuminate, usually pubescent, parallel venation with a distinct midrib. The uppermost leaf, the flag is always shorter and broader than the lower leaves.

Inflorescence: A terminal panicle. The inflorescence borne on the peduncle or on the uppermost internode. Usually the inflorescence is partly covered by leaf sheath, in extreme cases it is fully covered. The main axis or the rachis of the panicle is slightly angular, more or less glabrous, somewhat hairy at the nodes. It may be erect or curved. One to many branches are given out from each node of the rachis. The panicle may be compact or loose. On the end of each branch the spikelets are borne.

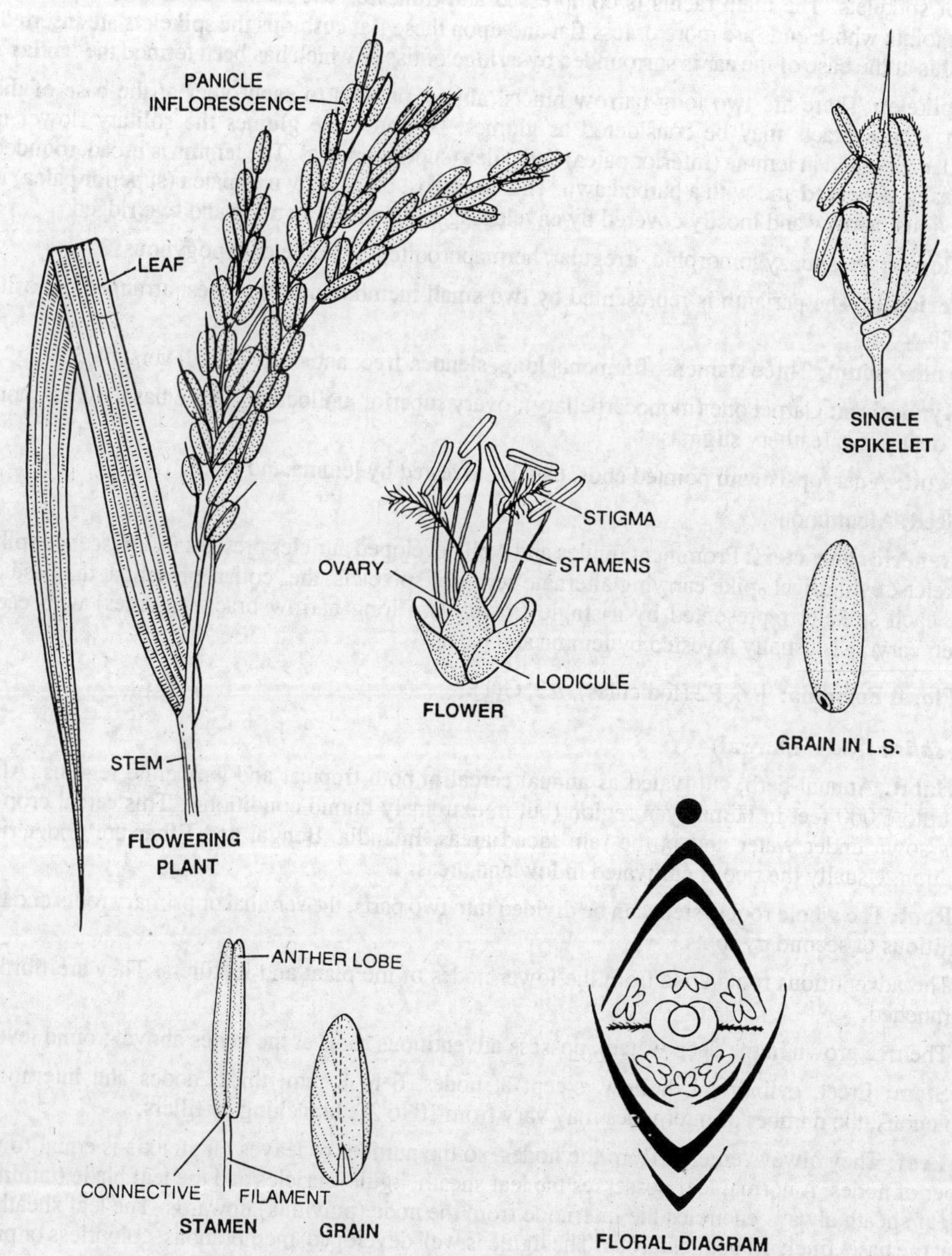

Fig.68.6. Gramineae (Poaceae)–*Oryza sativa* Linn. Verna. **Dhan;** Eng. Rice.

Spikelets: The spikelets are borne either singly on the ends of the branches or in clusters of 2 to 7. The number of spikelets on a panicle varies from 50 to 500. Each spikelet is short pedicelled, generally single flowered, very rarely two flowerd. The apex of the pedicel is enlarged and oblique and is termed as **facet.** On the facet two glumes occur one outer and the other inner. The glumes are small, lanceolate, shiny, coriaceous approximately equal in size, coloured or colourless (white, yellow or pink).

Flower: Bracteate, a large conspicuous bract, the **lemma.** The lemma is boat shaped, 5-nerved and hairy on the nerves. The apex is pointed and usually prolonged into an awn. On each side of the

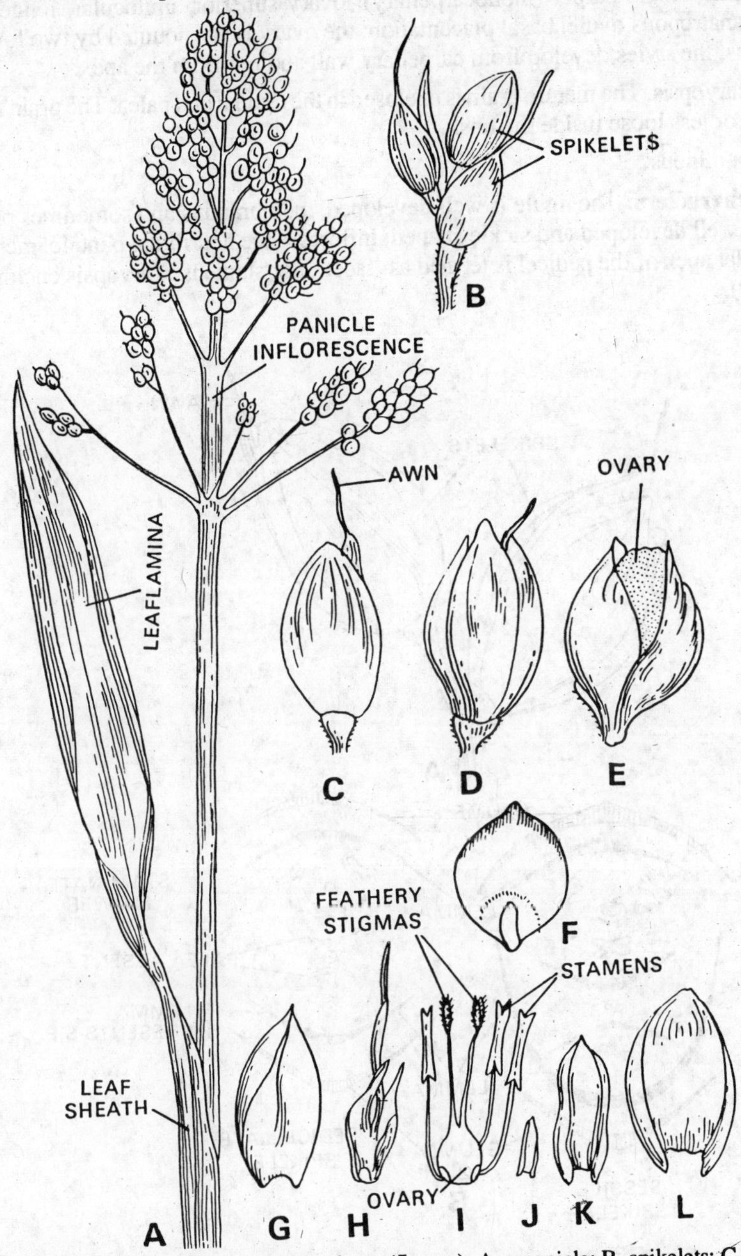

Fig.68.7. Gramineae (Poaceae)–*Sorghum vulgare* (**Jowar**): A, a panicle; B, spikelets; C, dorsal view of a spikelet; D, ventral view; E, opened caryopsis; F, a caryopsis; G, glume; H, lemma; I, ovary and stamens; J, pale; K, pale of fertile spikelet; L, glume.

apex or awn two tooth like projections. Lemma is usually pigmented (coloured). The **palea** is situated just opposite the lemma, it is also boat shaped and 3-nerved. The end point of the palea develops into **apiculus.** The apiculus is usually pigmented (purple coloured).

The flower is zygomorphic, hypogynous, irregular, incomplete and sessile.

Perianth: The perianth is represented by two broad, thick and fleshy lodicules.

Androecium: Six stamens, arranged in two regular whorls. The filaments are long, slender and free; anthers bi-celled, versatile.

Gynoecium: Single carpel (monocarpellary); ovary superior, unilocular, longer than broad, single ovuled, anatropous ovule; basal placentation; the ovary is surmounted by two long styles with feathery stigmas; the styles develop from carpellary wall and not from the apex.

Fruit: A caryopsis. The mature grain is enclosed in the lemma and palea. The grain may be firmly placed or more or less loose inside the husk.

Seed: Albuminous.

Generic characters: The ligule is well developed, membranous and sometimes pink coloured, the auricles are well developed and sickle shaped; inflorescence a terminal pancile, rachis of panicle is angular, the flat apex of the pedicel is termed as 'facet'. Six stamens. Caryopsis enclosed in lemma and palea (husk).

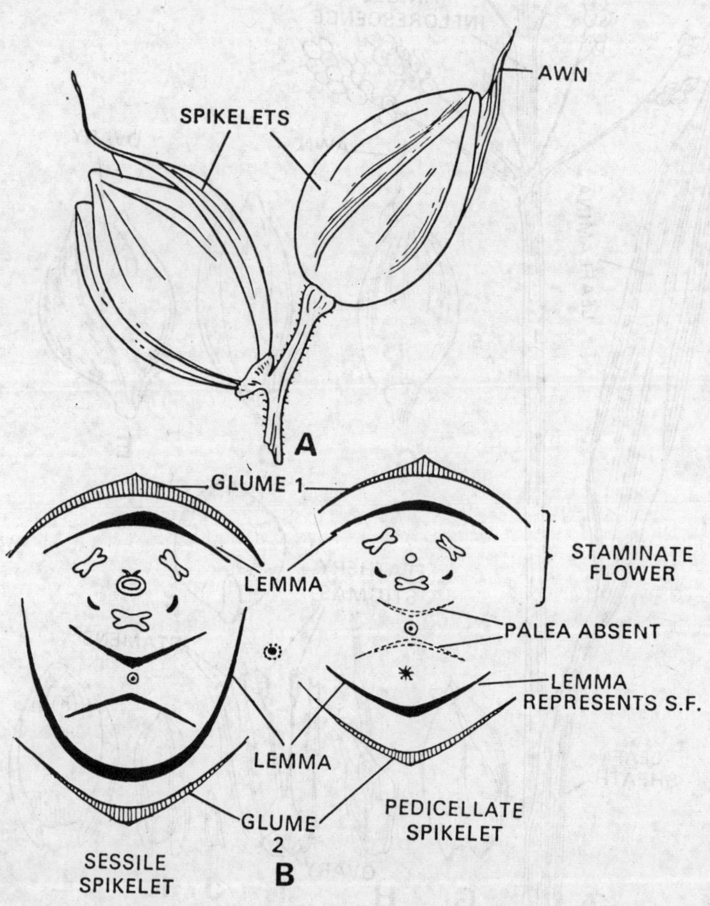

Fig. 68.8. Gramineae (Poaceae)–*Sorghum vulgare*; Verna.–**Jowar**–A, paired spikelets, one sessile another pedicelled; B, floral diagrams of respective spikelets.

Floral formula: ·|· ⚥ P 2 (lodicules), A3+3, <u>G1</u>.

Sorghum vulgare; Verna.–**Jowar.**

Habit: All grain producing Sorghums are annual. They are herbaceous and sown in rainy season.

Root: The root system may be divided into two parts, the seminal roots (primary roots) and the adventitious roots (secondary roots). The seminal roots develop first and penetrate vertically downwards in the soil. The adventitious roots develop from the basal nodes of the stem and extend both vertically downwards and horizontally. The adventitious roots are fibrous and very well developed. They are profusely branched. Stilt roots are also present usually above the ground level.

Stem: Erect, solid, several feet high, slightly furrowed on alternate sides. The nodes and internodes are short at the base and longer above. The terminal internode which bears head is longest of all, those of middle region are equal in length. Branching by **suckers,** each sucker may develop into a head. At each node there is bud except the terminal node.

Leaves: Simple, alternate, sessile; long, lanceolate; acuminate, surface glabrous, flat, distinct sheathing base (leaf sheath), the margins of leaf sheath are membranous and overlapping, amplexicaul, multicostate parallel venation with a very well developed mid-rib. Ligule, the ligule is short, membranous and fringed, present at the junction of the leaf sheath and the leaf blade. The surface of leaf blade is glabrous.

Inflorescence: Usually compact panicle, sometimes loose and spreading panicle. The main axis of rachis narrow towards the apex and it is not identified in the terminal and sub-terminal branches. The rachis is somewhat angular and furrowed, hairy at the nodes. The internodes may be hairy or may not. At the nodes, lateral branches may develop into whorls or in spirals. Each lateral branch may rebranch and on the ultimate branchlets one to many paired spikelets are produced. The lateral branches are somewhat angular and hairy at the base and at the axils, sometimes the heads (panicles) become inverted.

Spikelets: The spikelets usually occur in pairs on the lateral branches of panicles, but sometimes they occur in three towards the tip. When the spikelets are in pairs, one is sessile and perfect while the other one is pedicelled and staminate (male). When they are in threes, one is sessile and perfect while the rest two are pedicellate and staminate (male).

Fertile (perfect) spikelet: The fertile or perfect spikelets are comparatively larger than the staminate spikelets. They may be ovate, oval, elliptical or obovate. Each fertile spikelet possesses two glumes approximately equal in size. Both glumes are nerved. The nerves are more conspicuous on the outer glume than those of the inner one. The outer glume partially encloses the inner one which is narrow and pointed towards the apex. They are thick, hairy and with membranous margins.

Flower: On each spikelet there are two florets (flowers) found within the two glumes. The lower floret is sterile while the upper one is perfect. There is a lemma (inferior palea) of lower floret which also partially envelops the perfect floret. The lemma is usually broad, hairy and membranous. The second and perfect floret has a lemma which is membranous and hairy and two cleft at the apex. In the protruding cleft an awn arises. The palea is usually small and thin.

The flower (perfect) is sessile, bracteate, zygomorphic, irregular and hypogynous.

Perianth: The perianth is represented by two short and broad, truncate and fleshy, hairy lodicules.

Androecium: Three stamens; the filaments are free; anthers bi-celled, versatile.

Gynoecium: Single carpel (monocarpellary); ovary superior, unilocular, with single basal anatropous ovule; basal placentation; two long styles ending in short bushy stigma.

Staminate and Pedicellate Spikelet: The spikelet is narrower than the sessile one; the pedicel is hairy on the margins; two glumes enclose the two florets (flowers). The lower floret is represented by the lemma only; the upper floret is staminate, with short awned lemma; palea absent; two lodicules; three stamens; pistil absent.

Fruit: A caryopsis, vary in shape and colour usually white or pale.

Seed: Albuminous.

Generic characters: Fringed membranous ligule, no auricles; inflorescence loose or compact panicle, rachis angular and furrowed and hairy at nodes, spikelets in pairs on lateral branches, one spikelet is sessile and perfect while the other is pedicelled and staminate, two florets in each spikelet, one fertile other sterile, usually a white oval caryopsis.

Floral Formula: ·|· ⚥ P2 (lodicules), A3, G(1)

Zea mays; Eng.–Maize; Verna.–**Makka.**

Habit: An annual cereal, usually cultivated for its foodgrains.

Root: The roots are fibrous and may be divided into three types, (1) the seminal roots, (2) the adventitious roots, and (3) the prop or stilt roots. The seminal roots penetrate deep in the soil and are branched profusely. The adventitious roots develop from the nodes near ground level, they run horizontally in the ground and are also profusely branched. The colour is either brown or white. The stilt or prop roots arise from the nodes sufficiently above the ground level. The nodes may give out the whorls of prop roots. These roots are thicker and stronger than the normal roots, and somewhat pigmented.

Stem: Erect, several feet long, well defined nodes and internodes, the stem is thickened at the base of nodes. Both nodes and internodes are solid. The stem is furrowed on the other side of the leaf, position of furrow alternates at each node. The last node ends in the **tassel.** Branching sometimes by

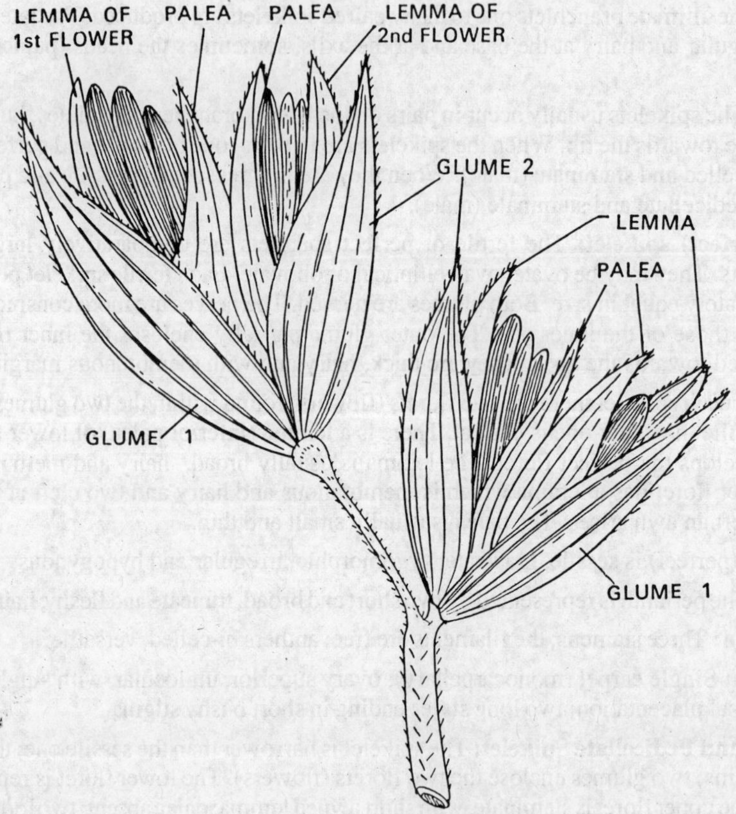

Fig.68.9. Gramineae (Poaceae) –*Zea mays* (Maize: Verna.–**Makka**) paired male spikelets, one sessile other pedicelled.

tillers. A bud is formed in the axil of each leaf, the buds near ground level develop into tillers while at higher levels they form the **'pistillate branches'**.

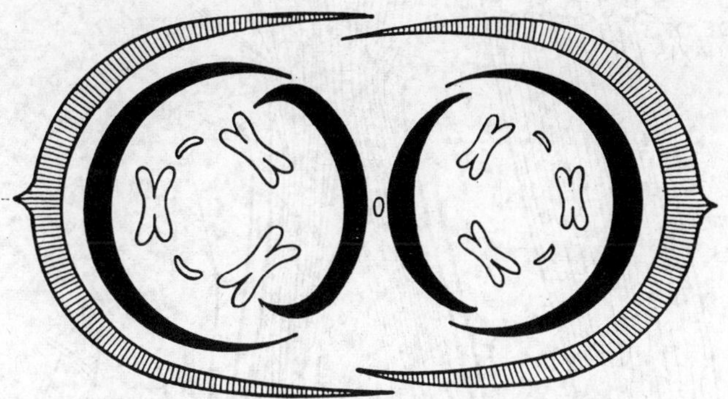

Fig. 68.10 Gramineae (Poaceae)–Floraliagram—Maize(male)—one spikelet

Leaves: Sessile, simple alternate. The leaf may be divided into four parts, the leaf sheath, the leaf blade, the ligule and the auricles. The leaf sheath encloses the internode (amplexicaul). It is entire below and split above. The surface may be glabrous or hairy. The margins glabrous below and hairy above. The leaf blade is thin, flat, ribbon like, acuminate, lanceolate, parallel venation with a thick and prominent mid-rib. At the junction of the leaf sheath and the leaf blade is membranous, colourless, quarter of an inch long **ligule** is present. At the junction there are two lateral outgrgowths, the **auricles.**

Inflorescence: Maize is monoecious, diclinous, *i.e.,* the stamens and pistils are borne on separate inflorescences, but on the same individual plant. The male flowers are carried on terminal panicles known as "**tassel**", while the female flowers are on the **"ear"**.

Staminate or Male Inflorescence: The male or staminate inflorescence is a panicle at the top of the plant; the panicle consists of a central axis and lateral branches of it; the central axis bears several rows (4 to 11) of paired spikelets, while the lateral branches have only two rows of paired spikelets, one sessile and the other pedicelled.

Pistillate or Female Inflorescence: The bud found in the axil of foliage leaf develops in the pistillate inflorescence. Sometimes two and very rarely three buds may develop. In structure the pistillate inflorescence is somewhat similar to the main shoot. All its internodes became very short. The uppermost internode is shortest while the lower ones are progressively bigger in comparison. The leaf at the basal node is the **prophyllum.** The outer leaves constitute the **'husk'** and protect the 'ear'. These protective leaves may be termed as bracts or spathes. Each spathe or bract is a modified leaf possessing leaf sheath, blade and ligule. The ear is a modified spike with thickened central axis, the 'cob'. The 'cob' carries a series of a paired spikelets in longitudinal rows. Each row of paired spikelets develops two grains, and so the rows of grains are always in even number.

Staminate or Male Spikelets: The paired male (staminate) spikelets are quite similar to each other except that one is sessile and the other is pedicelled. At the base of each spikelet there is a pair of glumes. The glumes are somewhat longer than the spikelets and nearabout equal in size. They are glabrous but hairy at magins, nerved (7 to 12 nerves). The outer or the lower glume overlaps the inner or upper one. The apex is acuminate. Within each two glumes there are two staminate florets (flowers) borne on a short common rachilla.

Staminate or Male Flower: Bracteate, the outer bract is the lemma, while the inner one is the palea. The lemma is oval, concave 3-nerved, glabrous, hairy at margins with somewhat rounded apex. The palea is inserted on the opposite side of the lemma. It is thin, membranous, 2-nerved flat and with incurved margins. In the upper flower the palea is larger than lemma while in the lower it is vice versa. The flower is sessile, zygomorphic, unisexual, male.

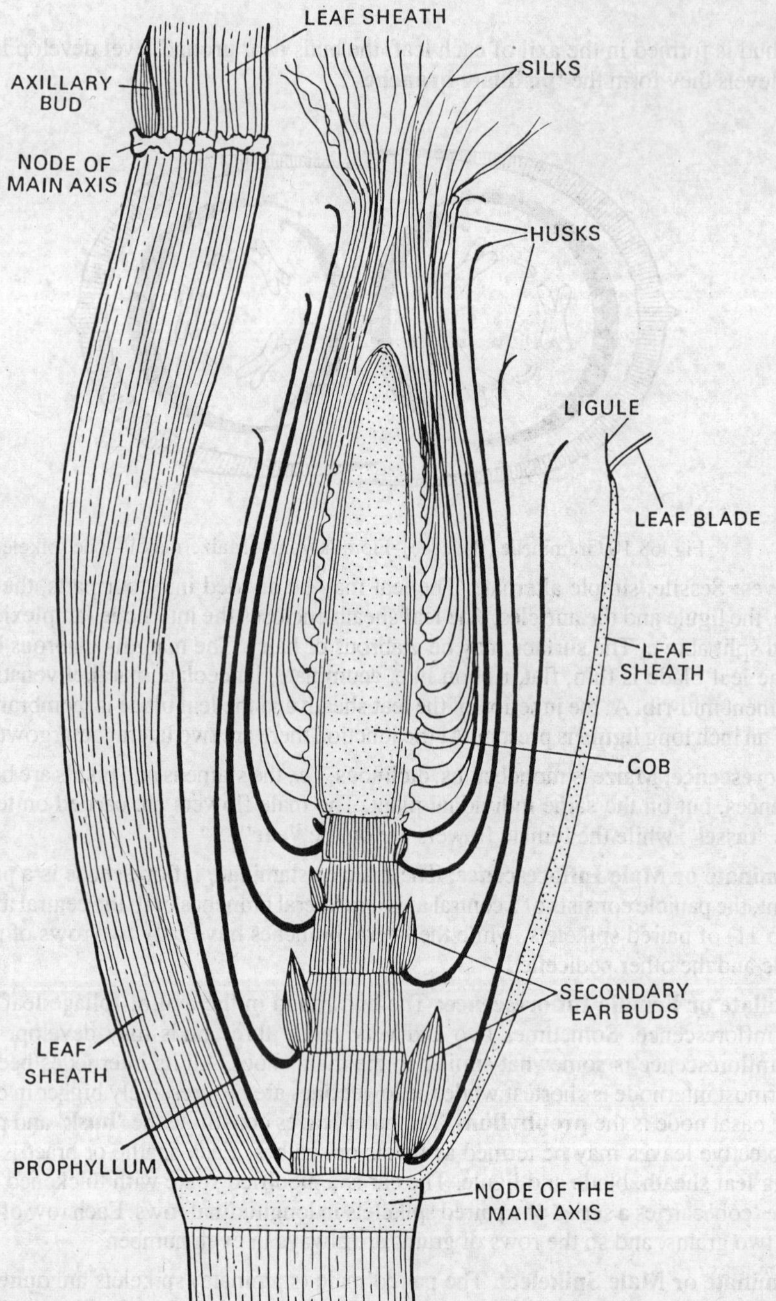

Fig.68.11. Gramineae (Poaceae)–*Zea mays;* (Maize; Verna–**Makka**) L.S. ear bearing branch; A, leaf sheath; B, axillary bud; C, silks; D, node of main axis; E, husks; F, ligule; G, leaf blade; H, leaf sheath; I, cob; J, secondary ear buds; K, node of the main axis; L, prophyllum; M, leaf sheath.

Perianth: The perianth is represented by two fleshy lodicules. They are inserted opposite the lemma and alternate with the stamens.

Androecium: Three stamens (one opposite the palea, remaining two opposite the lemma); the filaments free and short, anthers bilobed, versatile, purple in colour, dehiscence, by a slit at the terminal end.

Gynoecium: Absent, represented by a rudiment.

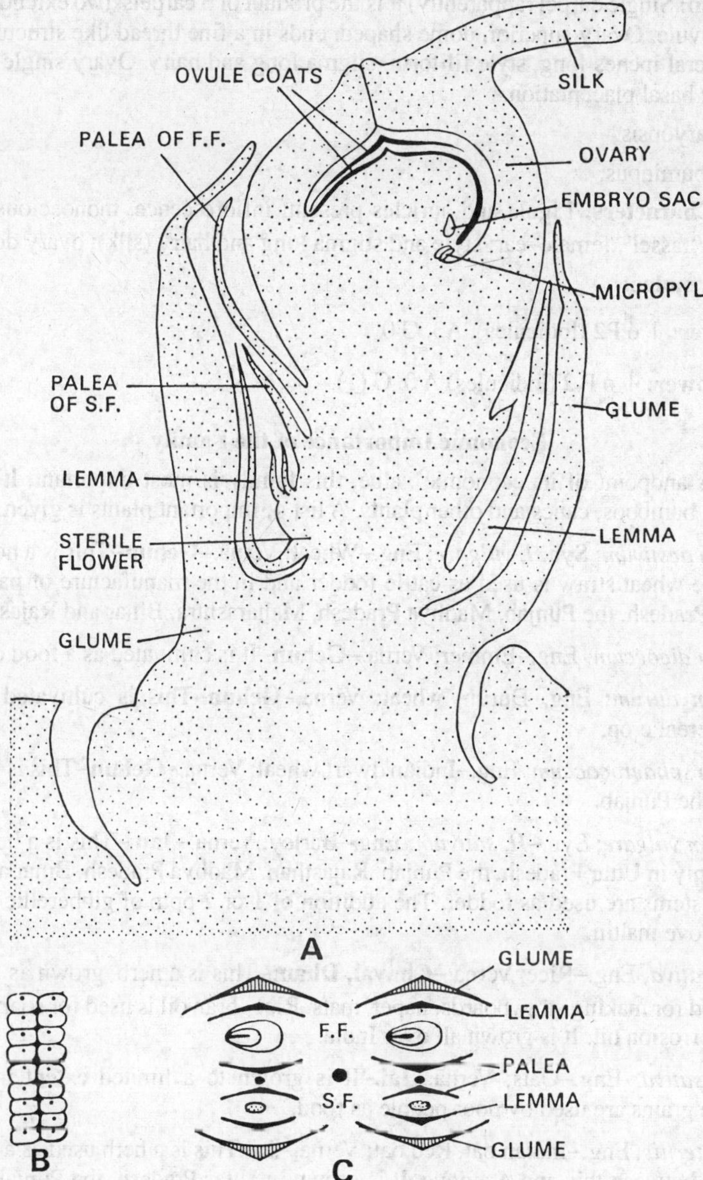

Fig.68.12. Gramineae (Poaceae)–*Zea mays;* (Maize; Verna.–**Makka**): A, a female spikelet with one sterile and other fertile flower; B, paired female spikelets; C, floral diagrams of a female spikelet.

Pistillate or Female Spikelets: The pistillate paired spikelets are arranged completely in rows on the axis or cob. Both the spikelets are sessile and quite similar to each other. There are two glumes at the base of a spikelet. The glumes are thin and membranous and somewhat thicker near the base of the ovary. Each spikelet consists of two flowers, the lower is sterile and the upper one is fertile.

Flower: Bracteate, the bracts are represented by lemma and palea. The lemma is short, broad and membranous. The palea resembles with lemma, equal or sometimes bigger than lemma.

Perianth: Usually absent, sometimes represented by two small, scaly lodicules.

Androecium: Absent, sometimes rudiments of three stamens may be visible.

Gynoecium: Single carpel (apparently) it is the product of 3 carpels, two extend to form silk and the third bears ovule. Ovary superior, dome shaped, ends in a fine thread like structure, the silk. The silk may be several inches long, style filiform, stigma long and hairy. Ovary single ovuled, campylotropous ovule, basal placentation.

Fruit: A caryopsis.

Grain: Albuminous.

Generic Characters: Ligule and auricles present; inflorescence, monoecious, male inflorescence panicles– 'tassel', female–ear; style and stigma long and hairy (silk); ovary dome shaped.

Floral Formulae:

Male flower: ·|· ♂ P2 (lodicules), A3, G 0

Female flower: ·|· ♀ P 2 (lodicules) A 0, G $\underline{(1)}$.

Economic Importance of the Family

From the standpoint of its economic value, this family is most important. It contains all the cereals, grasses, bamboos, canes and other plants. A list of important plants is given below:

1. *Triticum aestivum*; Syn. *T. vulgare*; Eng.–Wheat; Verna.–**Gehun**–This is a herb, cultivated as a food crop. The wheat straw is used as cattle fodder and in the manufacture of paper. It is mainly grown in Uttar Pradesh, the Punjab, Madhya Pradesh, Maharashtra, Bihar and Rajasthan.

2. *Triticum dicoccum*; Eng.–Emmer; Verna.–**Gehun**–It is cultivated as a food crop.

3. *Triticum durum*; Eng.–Durum wheat; Verna.–**Gehun**–This is cultivated mainly in Maharashtra as a cereal crop.

4. *Triticum sphaerococcum*; Eng.–Indian dwarf wheat; Verna.–**Gehun**–This is grown as a grain crop mainly in the Punjab.

5. *Hordeum vulgare*; Syn.–*H. sativum*; Eng.–Barley; Verna.–**Jau.**–This is a herb, cultivated as a food crop mainly in Uttar Pradesh, the Punjab, Rajasthan, Madhya Pradesh, Bihar and West Bengal. The leaves and stems are used as fodder. The addition of 1 or 2 ppm of gibberellic acid to barley is claimed to improve maltin.

6. *Oryza sativa*; Eng.–Rice; Verna.–**Chaval, Dhan**:–This is a herb, grown as a food crop. The rice straw is used for making strawboards, paper, mats. Rice–bran oil is used for soaps and cosmetics, and as an anticorrosion oil. It is grown all over India.

7. *Avena sativa;* Eng.–Oats; Verna.–**Jai.**–It is grown to a limited extent in North-Western Himalayas. The grains are used by poor people as food.

8. *Avena sterilis*; Eng.–Indian oat, Red oat; Verna.–**Jai.**This is a herb used as a fooder. A hybrid has been raised between this and *A. sativa.* It is grown in Uttar Pradesh, the Punjab, Bihar, Bengal, Orissa and in the hills of Tamil Nadu.

9. *Sorghum vulgare*; Syn. *Andropogon sorghum; Holcus sorghum;* Eng.–Sorghum; Verna.–**Jowar.**–The grains are used as food and the stem and leaves are consumed as a cattle fodder. A spirit is distilled from the grain. It is cultivated mainly in Uttar Pradesh, Punjab, Madhya Pradesh, Andhra Pradesh, Maharashtra and Rajasthan.

10. *Sorghum sudanense*; Eng.–Sudan grass.–This is an annual grass found in Maharashtra, Assam and the Punjab. It is used as a fodder.

11. *Sorghum halepense*; Syn. *Holcus halepensis*: Eng.–Johnson grass; Verna.–**Baru Kala mucha.**–This is a perennial grass, used as fodder.

12. *Pennisetum typhoides*; Syn. *P. typhoideum*; Eng.–Pearl millet; Verna.–**Bajra.** It is grown for

its edible grains. The straw is used as a fodder. It is cultivated in Uttar Pradesh, the Punjab, Andhra Pradesh, Rajasthan and Maharashtra.

13. *Pennisetum purpureum*; Eng.–Elephant grass.–It is grown in Assam, Bengal, Bihar, Uttar Pradesh and the Punjab. It is used as a fodder.

14. *Panicum miliaceum*; Verna.–**Chin, Morha, Anu.**–The grains are edible and the straw is used as a fodder. It is grown mainly in Uttar Pradesh.

15. *Panicum miliare;* Eng.–Little millet; Verna.–**Sawa, Kutki.**–The grains are edible. The stems and leaves are used as fodder. It is cultivated in Uttar Pradesh and Madhya Pradesh.

16. *Panicum antidotale*; Eng.–Blue panicum; Verna.–**Bansi, Gunara**–This is a tall grass, used as a fodder and also for the fixation and reclamation of sand dunes. It is found in the Punjab, Uttar Pradesh and the Western Peninsula.

17. *Zea mays;* Eng.–Maize, Corn, Indian corn; Verna.–**Makai, Makka, Bhutta.**–This is a tall annual herb. It is grown as a food crop mainly in Uttar Pradesh, the Punjab, Madhya Pradesh, Bihar, Andhra Pradesh, Jammu and Kashmir. The immature cobs are largely eaten after roasting. The grains are also used in making corn starch and industrial alcohol. It is native of South America.

18. *Setaria italica*; Syn. *Panicum italicum*; *Chaetochloa italica*; Eng.–Italian millet; Verna.–**Kangni, Kangu, Kakun.**–It is chiefly cultivated in Andhra Pradesh and Tamil Nadu for fodder and food. It is used as diuretic and astringent also used externally for rheumatism.

19. *Setaria glauca*; Syn. *Panicum glaucum*; Eng.–Cat tailmillet; Verna.–**Bandra.**–This is a herb, used as a fodder. The grains are also eaten as food.

20. *Eleusine coracana*; Eng.–Finger millet, Ragi; Verna.–**Mandal, Mandua.**–This is a herb, cultivated as a food crop. This is the main food of the farmers in South India; Maharashtra and of hill-men in Northern India. This is made into cakes, porridge and sweetmeats. A beer is brewed from the grain by the hill tribes. The enzymic extract of the germinated grain can be used for unhairing hides and skins. It is chiefly cultivated in Andhra Pradesh, Tamil Nadu, Karnataka, Orissa, Bihar, Uttar Pradesh and Maharashtra.

21. *Saccharum officinarum*; Eng.–Sugarcane; Verna.–**Ganna, Ikh**–The stems yield the cane syrup. The by-products of sugar industry are commercially used for various purposes, *e.g.*, molasses is used in cooking, also for rum, industrial alcohol, synthetic rubber. A mixture of refuge canes and molasses called molas-cuit is used as fodder. Cane refuge is also utilized for manufacturing bagpaper, wrapping-writing and printing papers, card-board and as fuel. It is chiefly grown in Uttar Pradesh, Bihar and the Punjab.

22. *Saccharum spontaneum*; Eng.–Thatch grass; Verna.–**Kans.**–It is used as a sand binder and also for paper pulp. It is found throughout India. The stems and leaves are used for making thatches.

23. *Saccharum fuscum*–This grass is used for making cheaper grades of papers, wrapping–boards, and other cellulose products.

24. *Agropyron repens*; Syn. *Triticum repens.* The decoction of the rhizome is a demulcent and diuretic and used in the treatment of catarrhal diseases of the urino-genital tract.

25. *Aristida depressa*–The panicles are made into brooms.

26. *Aristida setacea*–The panicles are used for making brooms and brushes.

27. *Arundinaria falcata*; Syn. *Bambusa falcata*; Eng.–Himalayan bamboo–The stems are used for making baskets, arrows, fishing rods and the lining of roofs of houses. It is found in Western Himalayas and Sikkim.

28. *Arundinaria spathiflora*–This is a small bamboo growing the undergrowth of the fir, oak and deodar forests of the Western Himalayas at an elevation of 6,500 feet to 9,500 feet. The culms are used for pipe stems and baskets.

29. *Arundinaria racemosa*–It is found in Bhutan and Western Himalayas. The culms are used for making mats and roofs of houses.

30. *Arundo donax*; Eng.–Giant reed; Verna.–**Baranal, Narhal.**–This is a tall perennial grass. The stems are used in the manufacture of baskets, mats, pipes, writing and printing papers. It is chiefly found in Kashmir, Assam and the Nilgiris.

31. *Bambusa arundinacea;* Syn. *B. spinosa*; Eng.–Thorny bamboo; Verna.–**Bans.**–This is a tall woody grass found throughout our country, particularly along river valleys and in moist situations. The culms are used for making paper, thatches, etc. The young buds and grains are consumed as food.

32. *Bambusa vulgaris*; Eng.–Feathery bamboo; Verna.–**Basini bans**–The split culms are made into mats and baskets and the young buds are consumed as vegetable. It is cultivated in the hotter parts of our country.

33. *Bambusa tulda*; Eng.–Bamboo; Verna.–**Peka**–The mature culms are used for manufacturing paper, baskets and fans. The young culms are eaten as vegetable. It is found in Assam and Bengal.

34. *Bambusa balcooa;* Eng.–Plains bamboo; Verna.–**Bhaluka**–It is used for building purposes. It is cultivated in Assam, Bengal, Bihar and Uttar Pradesh.

35. *Bambusa polymorpha*–It is used for making paper, in Bengal and Assam.

36. *Cephalostachyum pergracile.*–This is a tufted bamboo and frequently found in association with teak. The culms are used for building and mat-making, and cooked rice is often carried in the internodes on a journey.

37. *Cymbopogon citratus;* Syn. *Andropogon citratus*; Eng.–West Indian lemon grass; Verna.–**Sugundh rohisha**–The leaves yield an aromatic oil which is used in perfumery, cosmetics and as a flavouring substance. It is chiefly grown in Travancore and Malabar.

38. *Cymbopogon martini*; Syn. *Andropogon martini*; Eng.–Ginger grass; Verna.–**Gandhejghas**–The leaves are the source of an aromatic oil, which is used in perfumery and cosmetics. This occurs in two varieties known as the **Motia** and **Sofia.** The essential oil, obtained from the Motia variety, is of a superior quality and is referred to as palmarosa oil, while the oil, from the Sofia variety is inferior in quality and is known as Gingergrass oil. Next to Sandalwood and lemon-grass oils, palmrosa oil is the important of essential oils produced in our country.

39. *Cymbopogon nardus*; Syn. *Andropogon nardus*; Eng.–Citronella grass; Verna.–**Ganjni**–The oil known as Ceylon type, obtained from the leaves; is used in perfumery and cosmetics. It is also used in mosquito-repellent preparations. Also used in an infusion, as a stomachic and carminative.

40. *Cymbopogon schoenanthus*; Syn. *Andropogon schoenanthus* Eng.–Camel hay; Verna.–**Rousa ghas.**–The leaves yield an essential oil which is used in perfumery.

41. *Cymbopogon winterianus*; Eng.–Citronella grass. The oil commercially known as Java type contains high proportions of geraniol, citronella and citronollol and hence used in the synthesis of organic compounds such as synthetic menthol.

42. *Cymbopogon flexuosus*; Syn. *Andropogon flexuosus*; Eng.–East Indian lemon grass.–The leaves yield an oil which is used in perfumery and cosmetics.

43. *Cymbopogon jwarancusa*–The oil smells like peppermint but the yield is low. The roots are used in the treatment of fevers.

44. *Cymbopogon caesius*; Syn. *Andropogon caesius* Eng. Ginger grass–This is a perennial grass. The leaves yield an essential oil, which is used in soap manufacture. The grass is grown in Tamil Nadu and Travancore.

45. *Cynodon dactylon*; Eng.–Bermuda–grass; Verna.–**Doob.**This is a perennial grass used as fodder. This is the commonest used for tennis lawns.

46. *Dendrocalamus giganteus*–This is the largest of the Indian bamboos. It produces culms upto twenty five cm. in diameter. They are used for water-buckets and boxes. It is cultivated in Assam, Bengal and Malabar.

47. *Dendrocalamus hamiltonii;* Verna.–**Kagzi bans**–This is a tall tufted grass, used in the manufacture of paper, baskets and mats. It is mainly cultivated in Dehradun.

48. *Dendrocalamus strictus*; Eng.–Solid bamboo; Verna.–**Bans kaban**–It is used in the manufacture of paper, baskets and brushes. It is found in Northern India, Madhya Pradesh and South India.

49. *Echinochloa colonum*; Syn. *Panicum colonum*; Eng.–Shama millet; Verna.–**Samak**–The grains are eaten in times of scarcity.

50. *Echinochloa crus-galli*; Syn. *Panicum crus-galli*; Eng.–Baryard millet; Verna.–**Samak**–It is used as a fodder and food crop. It is chiefly cultivated in Maharashtra and Andhra Pradesh.

51. *Echinochloa frumentacea;* Syn. *Panicum crus galli* var. *frumentacea*; Eng.–Japanese millet; Verna.–**Sanwa, Sawa**–It is used as a fodder and food crop. It is used as a basis of a very potent beer. It is mainly grown in Madhya Pradesh and Uttar Pradesh.

52. *Echinochloa stagnina*; Verna.–**Banti**–It is used as a fodder. The grains are edible. It is found throughout India.

53. *Eragrostis curvula*–This is an introduced grass from United States. It is grown for erosion control and fodder.

54. *Erianthus arundinaceus*; Syn. *Saccharam arundinaceum*; Eng.–Pin reed grass; Verna.–**Ramsar, Sarkanda**–This is a perennial grass. The leaf sheaths yield a fibre which is used for making ropes, twines and paper, and the stems are used for making chairs, stools, tables, baskets and screen.

55. *Erianthus munja*; Syn. *Saccharum munja*; Verna.–**Munj sentha, Sarkanda, Munja**–The stem fibre is used for making baskets, mats and cordage. The leaves are used for thatching, and are also the source of paper. It is also used as soil-binder. It is commonly found in the Punjab and Uttar Pradesh.

56. *Erianthus ravennae*; Syn. *Andropogon ravennae*; *Saccharum ravennae*; Verna.–**Moonj**–The stem fibre is used for making chairs, muddas, chhappars and ropes.

57. *Heteropogon contortus*, Syn. *Andropogon contortus*; Verna.–**Kumeria**–This is a common perennial grass used as a fodder and the fibre is made into mats. It is also used to check soil erosion and in the manufacture of paper and board.

58. *Imperata cylindrica*; Syn. *I. arundinacea*; Verna.–**Ulu.**This is a perennial grass, the stem is used for thatching and a raw material for paper making. The pillows and cushions are stuffed with its fruits. It is also used in soil conservation.

59. *Indocalamus wightianus*; Syn. *Arundinaria wightiana*–It is used for making baskets, mats and walking sticks.

60. *Phragmites karka*; Syn. *P. maxima*; Verna.–**Narkul.**This is a perennial grass, the stems are made into pipes, and are also used for making writing and printing paper. They are also used for making chairs, baskets and mats.

61. *Pseudostachyum polymorphum*–This is a shrubby bamboo found in Sikkim, Assam. It is used by tea planters for making baskets. It is also used for making umbrella handles and walking sticks.

62. *Urochloa panicoides*–A herb used as fodder. The grains are also edible.

63. *Urochloa reptans*–A herb used as a fodder. The grains are also eaten.

Systematic Relationships : Bentham and Hooker have included five families in the series Glumaceae, *i.e.*, the Eriocauleae, Centrolepideae, Restiaceae, Cyperaceae and Gramineae. Engler and Prantl have placed the only two families, *i.e.*, Gramineae and Cyperaceae in the order–Glumiflorae.

Hutchinson has placed the family Cyperaceae in the order–Cyperales; and Gramineae in the order–Graminales. Hutchinson has treated the Graminales to be more highly advanced than the Cyperales. According to him both the Cyperales and Graminales have been derived from liliaceous ancestral stocks in a Juncaceae complex. According to Wettstein the Cyperaceae and Gramineae have been derived on parallel lines from Juncaceae. It is generally agreed that the grasses and sedges are very advanced groups whose simple inflorescence and floral structures represent drastic reductions from unknown ancestral types.

The Gramineae (grasses) and the Cyperaceae (sedges) are not as closely related as was formerly believed to be the case. It has been shown that the Gramineae differ in having terminal flowers whose ovaries evolved from ancestral types having parietal placentation, from that of the Cyperaceae which possess axillary flowers whose ovaries evolved from types having free-central placentation. The spikelet, present in both the families is now not considered to be homologous and it does not indicate any phylogenetic relationship.

69

HOW TO DESCRIBE AN ANGIOSPERMIC PLANT

1. Habitat. The natural abode or locality of plant, *i.e.,* whether cultivated as an ornamental plant, a food crop or occurs in a wild state.

2. Habit. (a) Herb, undershrub, shrub or tree.

(b) annual, biennial or perennial.

(c) Any special feature, *e.g.,* parasite, epiphyte, mesophyte, xerophyte or hydrophyte.

3. Root. (a) Tap or adventitious.

(b) Branched or unbranched.

(c) Any special modification, *i.e.,* fibrous, prop, stilt, aerial, climbing, respiratory, tuberous, parasitic, conical, fusiform, napiform, fasciculated, beeded, nodulated, etc.

4. Stem. (a) Nature of stem, *i.e.,* erect, prostrate, twining or climbing; if climbing, write mode of climbing whether by leaf tendrils, leaflet tendrils, stem tendrils or stipular tendrils or by hooks or some other means.

(b) Any special modification, *i.e.,* rhizome, bulb, corm, tuber, offset, stolon, runner, sucker, phylloclade, etc.

(c) Branched or unbranched. If branched, write the mode of branching, *i.e.,* whether racemose or cymose. If cymose, whether uniparous or multiparous.

(d) Texture, *i.e.,* herbaceous.

(e) Solid or fistular.

(f) Cylindrical (terete), angular flattened or reduced.

(g) Hairy, glabrous, waxy or spiny.

(h) Colour. Whether green, grey, etc.

5. Leaf. (a) Insertion. Radical, cauline or ramal.

(b) Arrangement. Alternate, opposite or whorled (verticillate). If opposite, whether superposed or decussate.

(c) Petiolate, sessile or subsessile.

(d) Stipulate or exstipulate. If stipulate, describe the nature of stipules, *i.e.,* leafy (foliaceous), spiny, tendrillar, petiolar, etc.

(e) Leaf base. Sheathing connate or ligulate or amplexicaul.

(f) Kind. Simple or compound.

(g) If simple, write about form of lamina, *i.e.,* acicular, linear, lanceolate, oval, ovate, obovate, cordate, obcordate, spathulate, orbicular, rotund, reniform, sagittate, hastate, lyrate, etc.

(h) If compound, whether pinnate or palmate, paripinnate, imparipinnate, bifoliate, trifoliate, etc.

(i) Incision of lamina. Pinnatifid or palmatified, pinnatisect or palmatisect.

(j) Margin of lamina. Entire, denate, serrate, hairy, spiny, convolute, etc.

(k) Apex. Acute, obtuse, acuminate, emarginate, truncate, mucronate, cuspidate, etc.

(l) Surface. Hairy, glaucous, spiny.

(m) Venation. Reticulate or parallel, unicostate or multicostate.

(n) Texture. Coriaceous, fleshy, succulent, etc.

(o) Colour. Green or pigmented.

Note. In a compound leaf, the leaflets should be described in the manner as a simple leaf.

6. Inflorescence. Simple mixed, compound and special. If simple, racemose or cymose. If racemose, whether raceme, corymb, spike, catkin, spadix, umbel, hypanthodium, capitulum, etc. If cymose, uniparous, biparous or multiparous cyme, helicoid or scorpioid, or solitary axillary. Compound and mixed, compound spike, compound umbel, compound corymb, spike of spikelets, panicle and panicle of spikelets, etc. Special verticillaster, cyathium, etc.

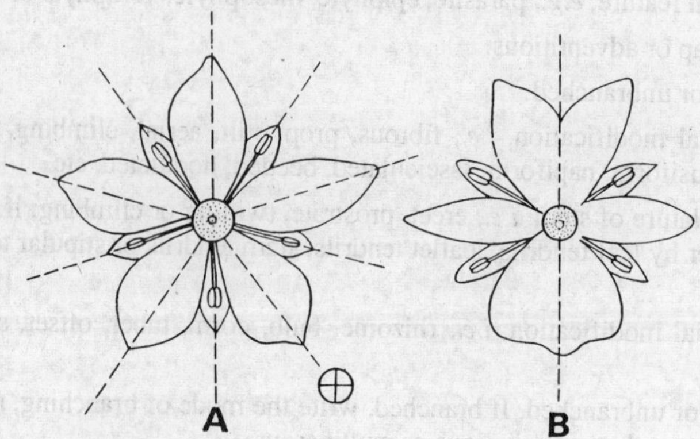

Fig. 69.1. Symmetry of flower. A, regular or actinomorphic; B, Zygomorphic or irregular.

7. Flower. (a) Pedicellate or sessile.

(b) Bracteate or ebracteate; if bracteate, the nature of the bract should be described.

(c) Actinomorphic or zygomorphic.

(d) Unisexual or hermaphrodite (bisexual).

(e) Complete or incomplete.

(f) Regular or irregular.

(g) Hypogynous, perigynous or epigynous.

(h) Colour of the flowers.

8. Calyx. (a) Number of sepals.

(b) Polysepalous or gamosepalous. If polysepalous, describe the nature and form of sepals, *i.e.,* deciduous, caducous, persistent, etc. If gamosepalous, give special forms, *e.g.,* tubular, campanulate, bilabiate, etc.

(c) Sepaloid (green) or petaloid (coloured).

(d) Inferior or superior.

(e) Aestivation of the calyx, *e.g.*, imbricate, valvate, quincuntial, etc.

9. Corolla. (a) Number of petals.

(b) Poly or gamopetalous, if polypetalous, one should describe the form and nature of petals, *i.e.*, rosaceous, papilionaceous, etc. If gamopetalous, give special form, *e.g.*, campanulate, tubular, bilabiate, spurred, ligulate, etc.

(c) Colour of the petals.

(d) Inferior or superior.

(e) Aestivation of the corolla.

10. Androecium. (a) Number of stamens; 1 to 10 or indefinite.

(b) Free or fused; if fused, adelphous or syngenesious; if adelphous, mono, di or polyadelphous. If free, any special form, *e.g.*, didynamous, tetradynamous, etc.

Fig. 69.2. A, relation of a flower to the mother axis or the peduncle; B, it shows antero-posterior, lateral and diagonal planes of the flower.

(c) Epipetalous, epiphyllous or free from petals or tepals.

(d) Filament long, short or flattened.

(e) Fixation of anthers. Adnate, basifixed, dorsifixed or versatile.

(f) Anthers. Monothecous or dithecous (single-celled or bicelled).

(g) Dehiscence of anthers. Longitudinal, transverse, porous, extrorse, introrse.

(h) Colour of the anthers.

(i) Sometimes stamens represented by staminodes.

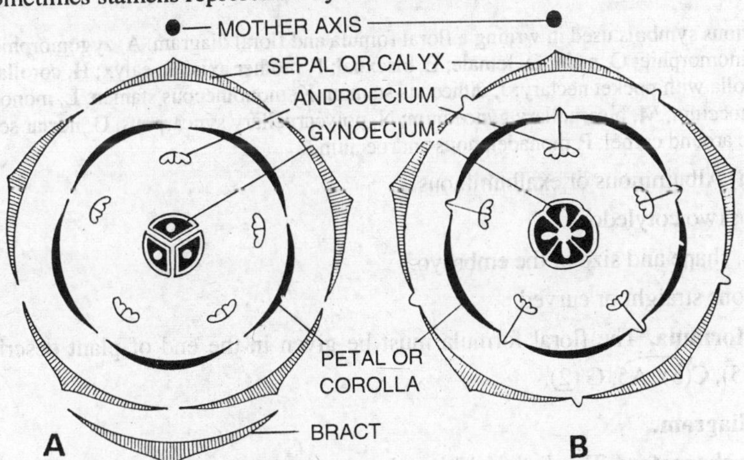

Fig. 69.3. Floral diagrams of polypetalous (A) and gamopetalous (B) flowers, with apocarpous and syncarpous gynoecia respectively.

11. Gynoecium. (a) Number of carpels.

(b) Apocarpous or syncarpous.

(c) Ovary superior or infferior.

(d) Number of loculi.

(e) Number of ovules in each loculus.

(f) Placentation. Marginal, parietal, axile, free central, basal, superficial, etc.

(g) Nature of style, *i.e.*, short, slender, terminal, divided, etc.

(h) Nature of stigma, *i.e.*, simple or lobed, feathery, bearded, hairy, capitate, curved, etc.

(i) Any disc present below the ovary should be described.

12. Fruit. Whether it is simple, aggregate, multiple; pod, legume, follicle, lomentum, capsule, siliqua, silicula, achene, cypsela, cremocarp, caryopsis, samara, nut, pome, drupe, berry, carcerule, hesperidium, etaerio of achenes, drupes, follicles or berries, sorosis, syconus.

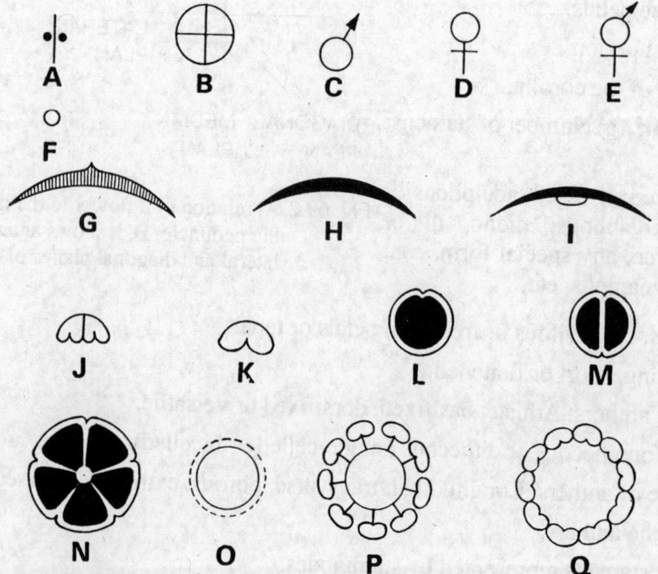

Fig. 69.4. Various symbols used in writing a floral fomula and floral diagram. A, zygomorphic; B, actinomorphic; C, male; D, female; E, bisexual; F, mother axis; G, calyx; H, corolla; I, corolla with pocket nectary; J, dithecous stamen; K, monothecous stamen; L, monocarpellary gynoecium; M, bicarpellary gynoecium; N, polycarpellary syncarpous; O, nectar secreting disc around carpel; P, monadelphous androecium.

13. Seed. (a) Albuminous or exalbuminous.

(b) Single or two cotyledons.

(c) Position, shape and size of the embroyo.

(d) Cotyledons straight or curved.

14. Floral formula. The floral formula must be given in the end of plant description, *e.g.*, *Solanum*: ⊕ ⚥ K(5), $\widehat{\text{C(5), A5}}$, G $\underline{(2)}$.

15. Floral diagram.

16. Generic characters. The distinguishing characters of the genus should be underlined in the description.

17. Systematic position. In the end of the description the systematic position of the genus should be given with the help of key.

70

GLOSSARY WITH ILLUSTRATIONS

HABITAT

The natural abode or locality of plant, *i.e.*, whether cultivated as an ornamental plant, a food crop or occurs in a wild state.

HABIT

Herb. Plant with no persistent parts above ground, as distinct from shrubs and trees, *e.g.*, *Ranunculus* or Ranunculaceae.

Shrub. Perennial woody plant, typically with several stems arising from or near the ground, *e.g.*, *Capparis* of Capparidaceae.

Tree. A perennial woody plant with a single trunk, *e.g.*, *Melia* of Meliaceae.

Annual. Plant that completes its life-cycle, from seed germination to seed production, followed by death within a single season, *e.g.*, *Brassica* of Cruciferae.

Biennial. Plant that continues its growth from year to year. In *herbaceous* perennials serial parts die away in autumn, replaced by new shoots in the following year from underground structures, *e.g.*, *Delphinium*; in *woody* perennials, permanent woody stems above ground from starting point for each new year's growth, a characteristic that enables some of them to reach a large size, *e.g.*, shrubs and trees.

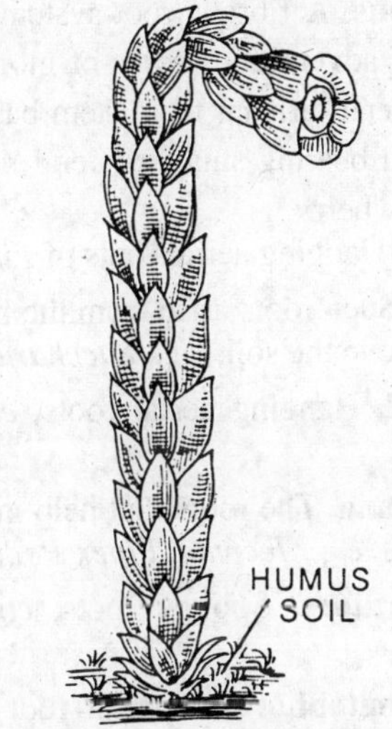

Fig.70.1. Saprophyte. *Monotropa.*

Parasite. Plant living in or on another plant (its host) from which it obtains food, *e.g.*, *Cuscuta* of Convolvulaceae.

Epiphyte. Plant attached to another plant, not growing parasitically upon it but merely using it for support, *e.g.*, *Vanda* (an orchid) of Orchidaceae.

Mesophyte. Plant growing under average conditions of water supply.

Hydrophyte. Plant whose habitat is water or very wet places, *e.g.*, *Ranunculus aquatilis* of Ranunculaceae; *Neptunia oleracea* of Mimosoideae.

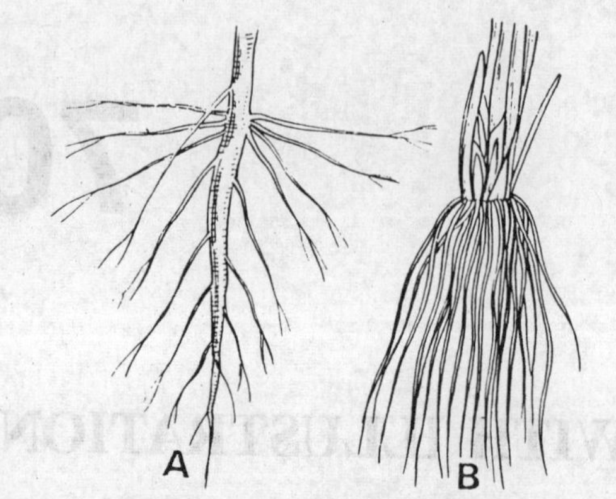

Fig.70.2. Roots. A, tap and lateral roots of a dicotyledon; B, fibrous roots of a monocotyledon.

Xerophyte. Plant of dry habitat able to endure conditions of prolonged drought, *e.g.*, *Capparis decidua* of Capparidaceae.

Saprophyte. Plant which obtains organic matter in sloution from dead and decaying tissues of plants (or animals), *e.g.*, *Monotropa.*

Root. That part of vascular plants that usually grows downwards into the soil, anchoring plant and absorbing water and nutrient salts.

Tap. Root system with a prominent main root, directed vertically downwards and bearing smaller lateral roots, *e.g.*, most of dicots; *Cajanus* of Papilionatae.

Adventitious. Root developing from part of plant other than roots, *e.g.*, from stem or leaf cutting; of buds, developing from part of plant other than in axil of leaf, *e.g.*, from root, most of monocots.

Branched. Roots possessing branches.

Unbranched. Roots without branches.

Fibrous. A fibrous root system consists of a tuft of adventitious roots of more or less equal diameter, arising from stem base or hypocotyl and bearing smaller lateral roots *e.g.*, wheat, strawberry.

Prop. Hanging aerial roots of *Ficus.*

Stilt. Such roots arise from the nodes of the stem above the soil, *e.g.*, *Saccharum.*

Aerial. Hanging aerial roots, *e.g.*, Orchids.

Climbing. The roots that help in climbing of plants, *e.g.*, *Tecoma, Piper, Pothos.*

Respiratory. Spongy roots, *e.g.*, *Jussiaea.*

Pneumatophores. Special root branch produced in large numbers by some vascular plants growing in water or in tidal swamps, *e.g.*, Mangrove; grows erect, projecting into the air above and contains well developed intercellular system of air spaces in communication with atmosphere through pores on aerial portion.

Tuberous. Irregularly swollen roots laden with starch, *e.g.*, *Ipomoea batatas* of Convolvulaceae.

Parasitic. The sucking roots of parasites, *e.g.*, *Cuscuta, Dendrophthoe, Viscum, Orobanche.*

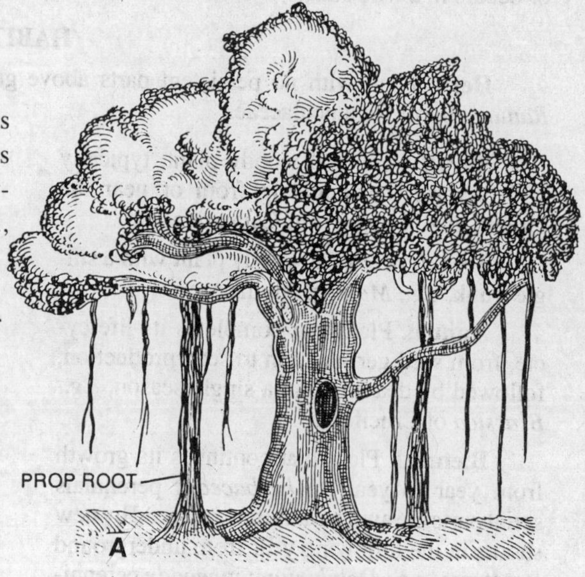

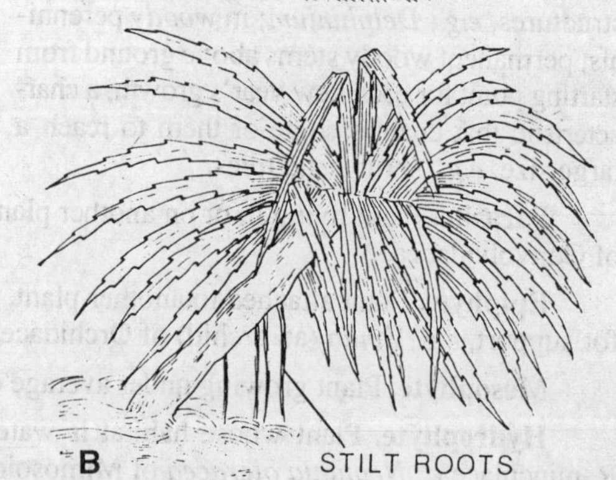

Fig.70.3 . Adventitious modified roots. A, prop roots of banyan; B, stilt roots of **Keora.**

Epiphytic. Hanging aerial roots of Orchids.

Assimilatory. The aerial roots with chlorophyll, *e.g., Tinospora.*

Aquatic. The roots found in water plant. They do not possess root caps and root hairs, *e.g., Pistia.*

Conical. Cone like fleshy roots, *e.g.,* carrot.

Fusiform. *e.g.,* fleshy root of radish (*Raphanus sativus* of Cruciferae).

Napiform. *e.g.,* fleshy root of turnip.

Fasciculated. Clusters of fleshy roots, *e.g., Asparagus.*

Nodulated. The nodules of the roots contain bacteria, *e.g.,* Leguminous roots; *Cicer, Arachis, Trifolium.*

Beaded or Moniliform. The roots possessing beaded structures, *e.g., Vitis.*

Annulated. When the root has a series of ring-like swellings on its body, *e.g.,* ipecacuanha.

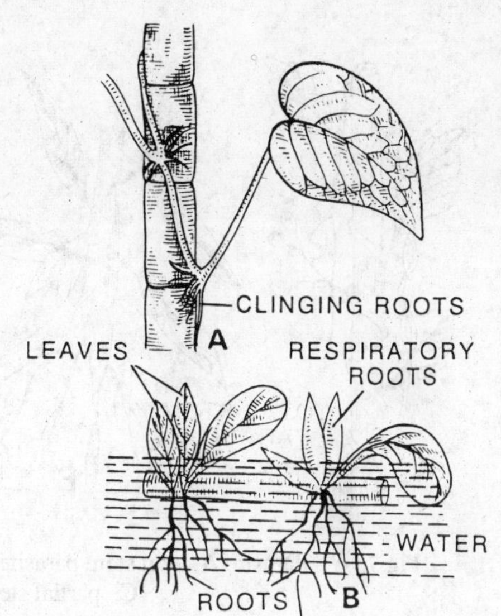

Fig.70.4. Adventitious root modified roots. A, climbing roots of betel, B, respiratory roots of *Jussiaea.*

Nodulose. When the slender root becomes suddenly swollen near the apex, *e.g., Curcuma amada* (**Am–haldi**).

STEM

Stem. Normally aerial part of axis of vascular plants, bearing leaves and buds at definite positions (nodes) and reproductive structures, *e.g.,* flowers.

Erect. Rigid, strong and upright stem.

Prostrate. Trailing stem lying flat on the ground, *e.g., Portulaca.*

Twiner. Long, slender and branched stem climbing by twining its body round the support, *e.g., Cuscuta.*

Climbers. A weak stem attaching itself to any neighbouring support by means of special structures such as rootlets, hooks, leaf tendrils, stem tendrils, stipular tendrils, etc.

BREATHING PORES
PNEUMATOPHORES

Fig.70.5. Roots. Pneumatophores arising vertically upwards from an underground root.

Rhizome. Underground stem, bearing buds in axils of reduced scale-like leaves; saving as a means of perennation vegetative propagation, *e.g., Zinger, Mentha.*

Bulb. Modified shoot consisting of very much shortened stem enclosed by fleshy, scale leaves, *e.g., Allium cepa.*

Corm. Swollen stem base containing food material and bearing buds in the axils of scale-like remains of leaves of previous season's growth, *e.g. Colocasia, Amorphophallus, Gladiolus.*

Tuber. Swollen end of underground stem bearing buds in axils of scale-like rudimentary leaves

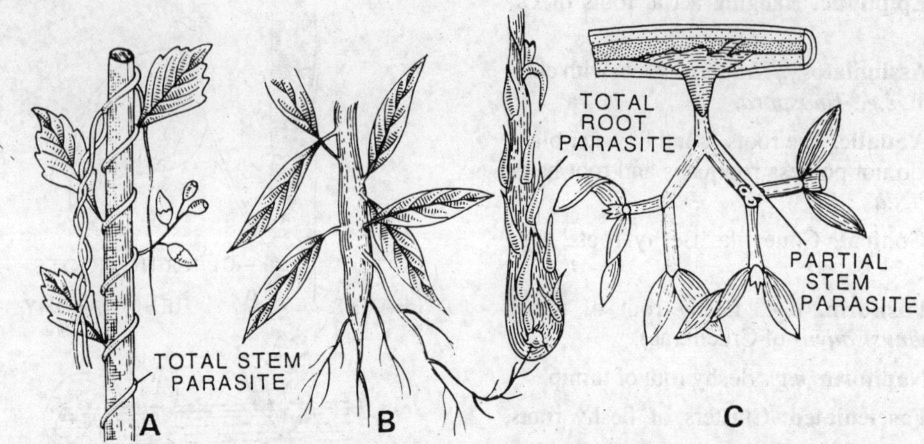

Fig.70.6 . Parasite.A, total stem parasite of *Cuscuta*; B, total root parasite of *Orobanche*; C, partial stem parasite of *Viscum*.

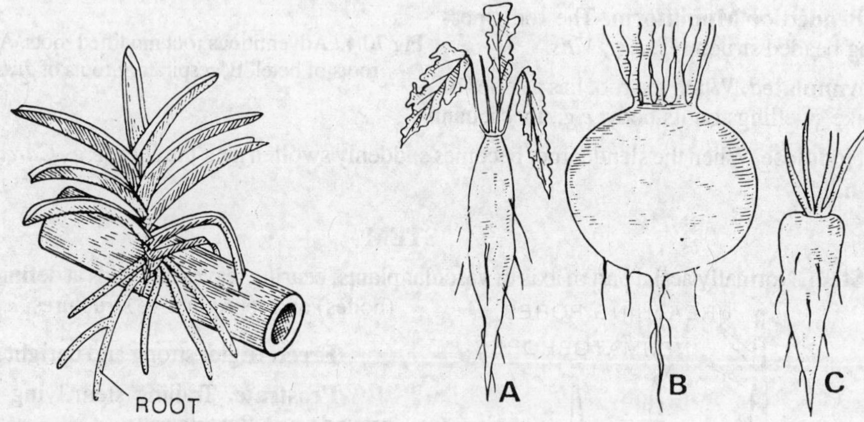

Fig.70.7. Epiphytic roots of an orchid (*Vanda*).

Fig.70.8. Modified roots. A, fusiform roots of radish; B, napiform root of turnip; C, conical root of carrot.

(stem tuber), *e.g.*, potato.

Offset. A horizontal, short, more or less apex a tuft of leaves above and a cluster of thickened, prostrate branch producing at the small roots beneath, *e.g., Pistia.*

Stolon. Horizontally growing stem that roots at nodes, *e.g.*, strawberry runner, *Colocasia.*

Runner. Stolon that roots at tip forming new plant that eventually is freed from connection with parent by decay of runner, *e.g., Oxalis.*

Sucker. A creeping stem but growing obliquely upwards directly giving rise to a leafy shoot, *e.g., Chrysanthemum.*

Phylloclade. Modified stem having appearance and function of a leaf, *e.g., Ruscus,* Cocoloba.

Cladode. A phylloclade of single internode, *e.g., Asparagus.*

Branched. Stem possessing branches, *e.g.*, Nim tree.

Unbranched. Stem having no branches, *e.g.*, Palm.

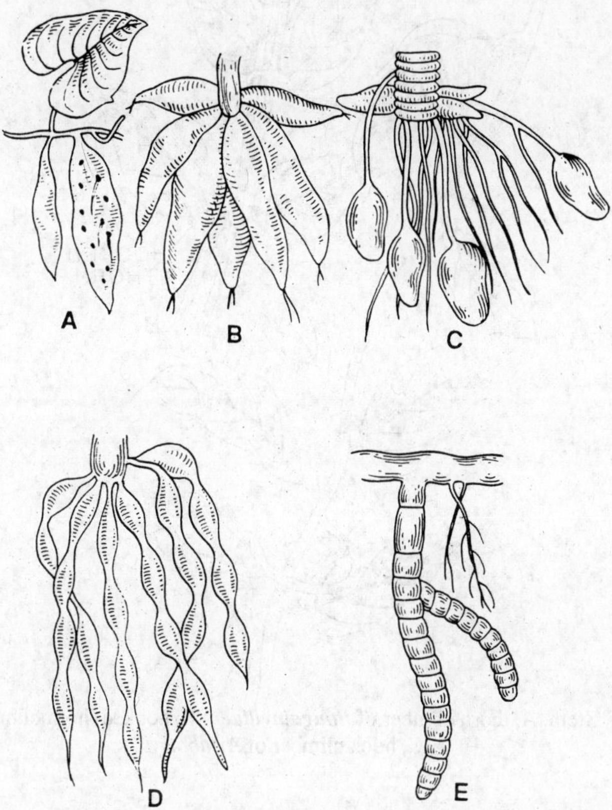

Fig.70.9. Adventitious modified roots. A, tuberous roots of sweet potato; B, fasciculated tuberous roots of *Dahlia;* C, nodulated roots of mango ginger; D, moniliform roots of *Momordica;* E, annulated roots.

Fig.70.10. Rootlet climbers of *Ficus pumila.*

If branched whether racemose or cymose type of branching. If cymose whether uniparous, biparous or multiparous.

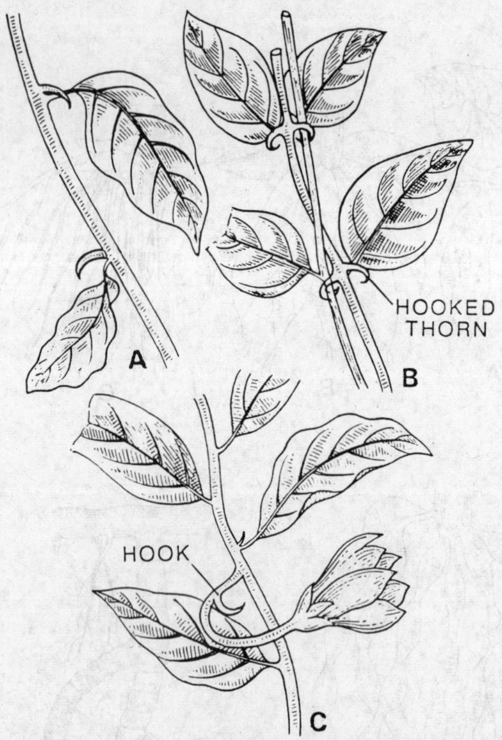

Fig.70.11. Stem. A, thorn climber of *Bougainvillaea*; B, hooked thorn climber of *Uncaria;* C, hook climber of *Artabotrys.*

Fig.70.12 . Stem. A, prickle climber of cane; B, prickle climber of rose.

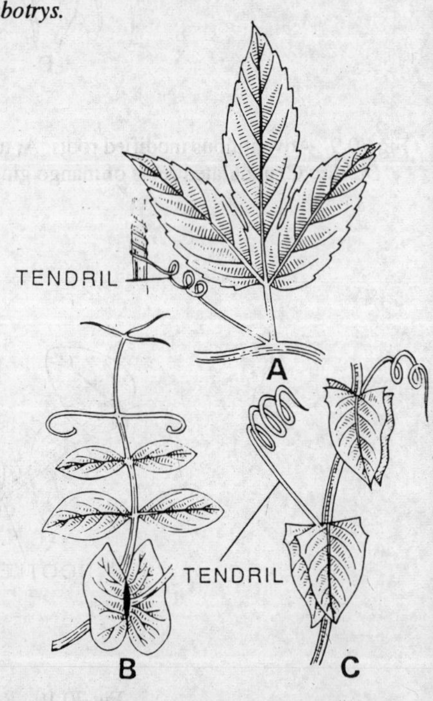

Fig.70.13 . Tendril climbers. A, stem tendril climber of passion flower; B, leaf tendril climber of pea; C, leaf tendril climber of *Lathyrus.*

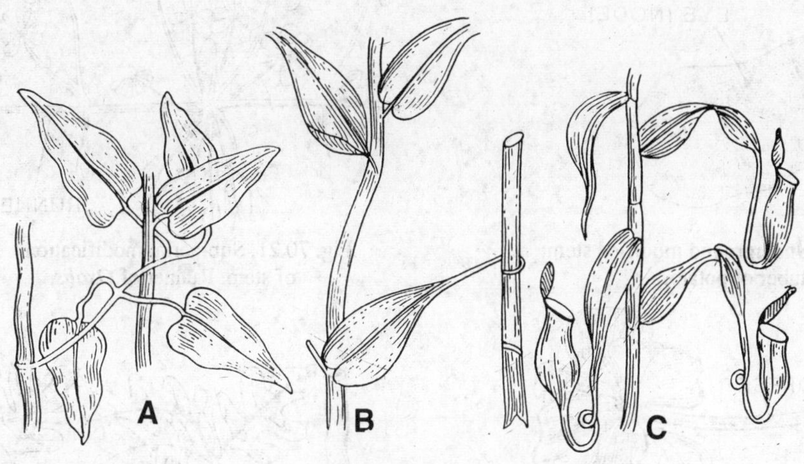

Fig.70.14. Leaf climbers. A, leaf climber of *Clematis*; B, leaf climber of *Gloriosa* ; C, leaf climber of pitcher plant.

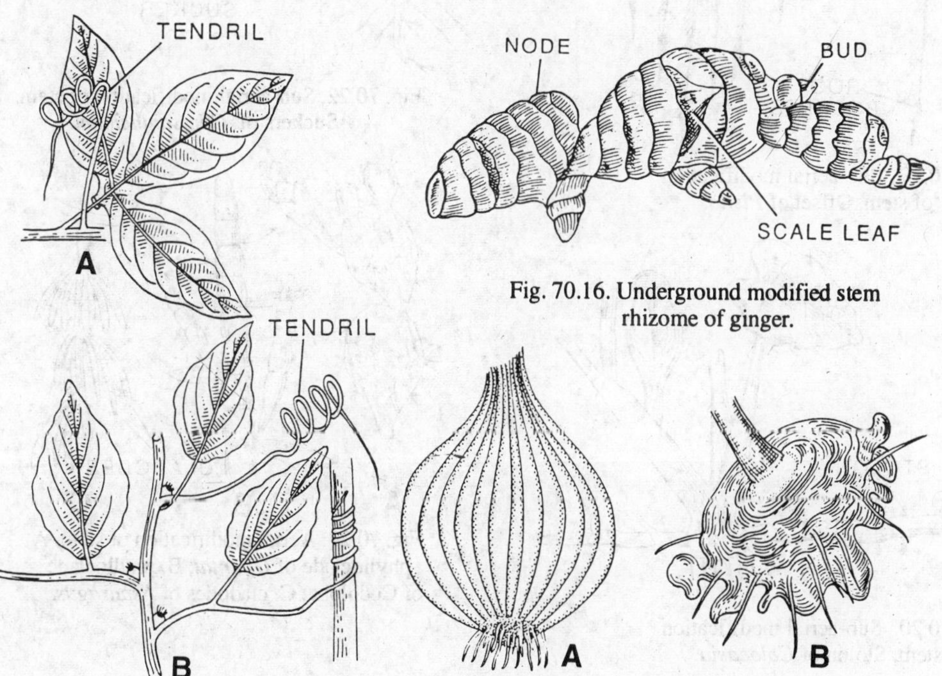

Fig. 70.16. Underground modified stem rhizome of ginger.

Fig.70.15. Stem tendrils. A, *Passiflora*; B, *Gouania*.

Fig. 70.17. Underground modified stem. A, bulb of onion; B, com of *Amorphophallus*.

Herbaceous. Having the characters of a herb, *e.g., Ranunculus.*

Woody. Having the characters of a shrub and tree, *e.g., Capparis, Melia.*

Solid. Interior portion of the stem is filled up with matter.

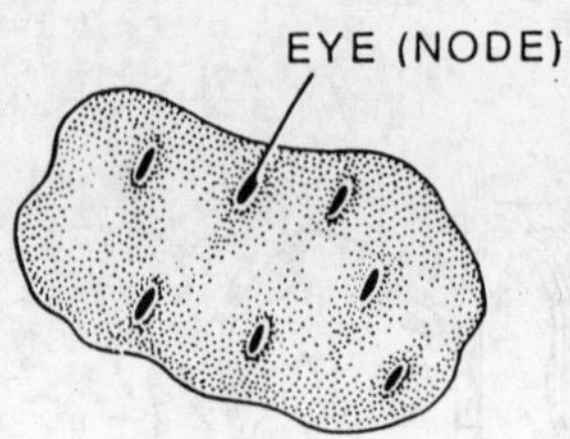

Fig.70.18. Underground modified stem; tuber of potato.

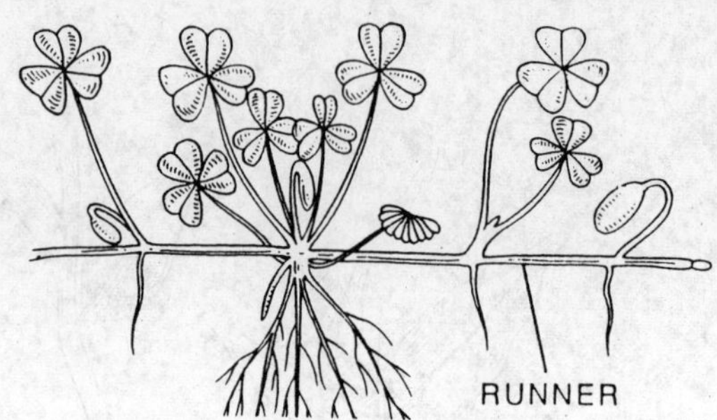

Fig. 70.21. Sub-aerial modification of stem. Runner of *Oxalis*.

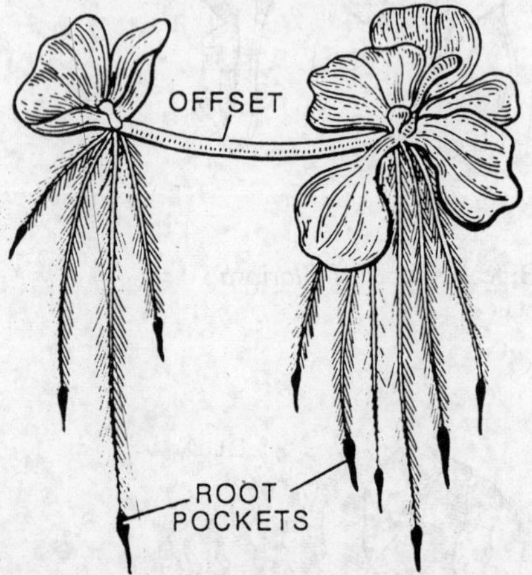

Fig.70.19. Sub-aerial modification of stem. Offset of *Pistia*.

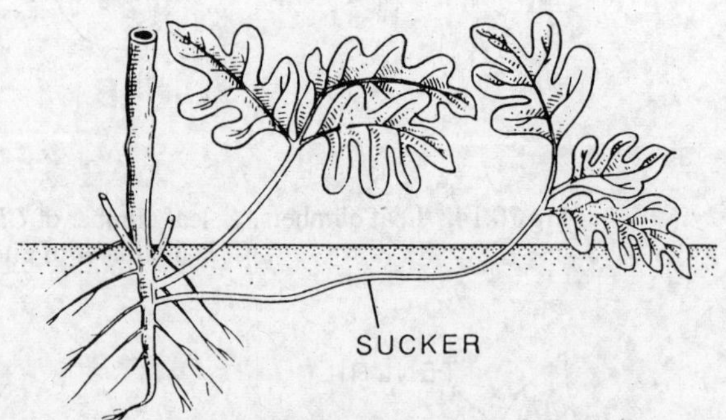

Fig. 70.22. Sub-aerial modification of stem. Suckers of *Chrysanthemum*.

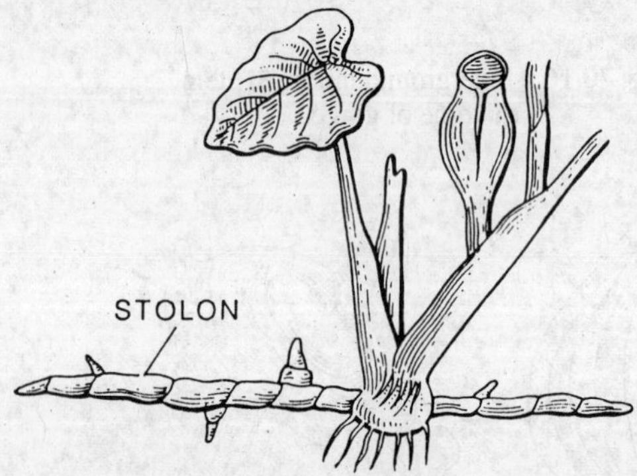

Fig.70.20. Sub-aerial modification of stem. Stolon of *Colocasia*.

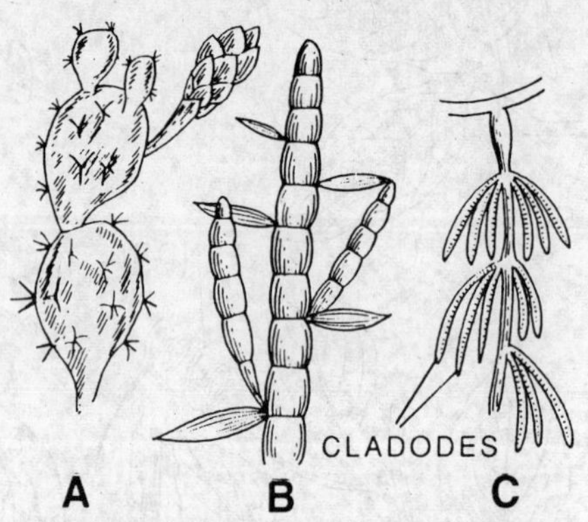

Fig. 70.23. Aerial modification of stem. A, phylloclade of *Opuntia*; B, phylloclade of Cocoloba; C, cladodes of *Asparagus*.

Fistular. A stem having hollow interior, *e.g.*, wheat, bamboo.

Cylinderical (terete). Circular stem as seen in T.S.

Angular. A stem shows many angles in T.S., *e.g., Cucurbita.*

Flattened. Flat stem in T.S.

Hairy. Stem possessing hairs on its surface.

Glabrous. Smooth stem.

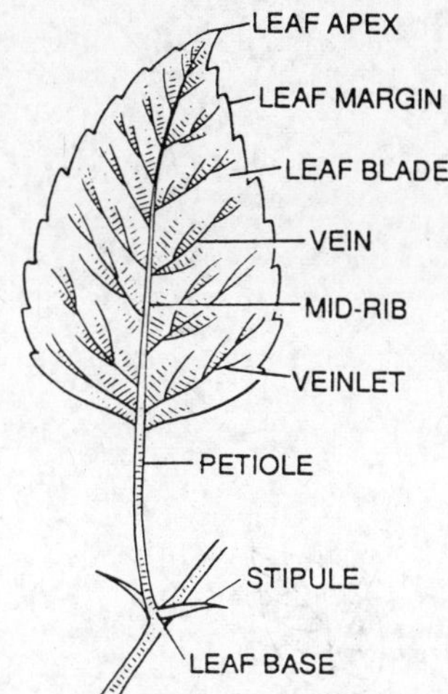

Fig.70.24. Leaf. Parts of a leaf.

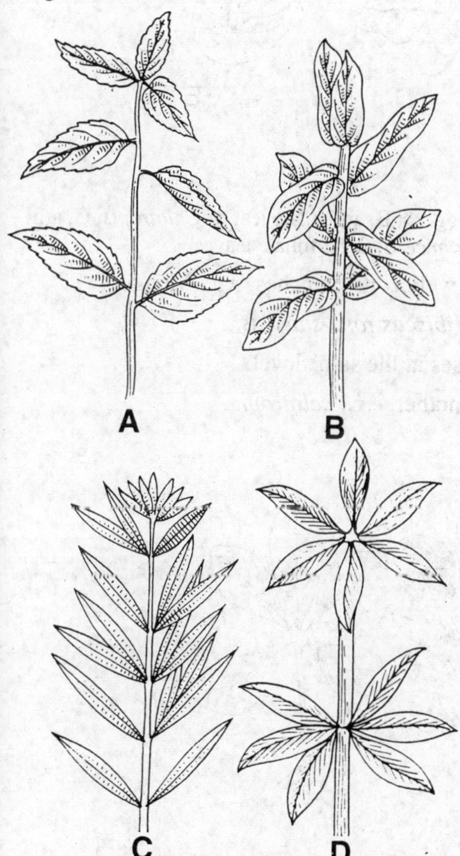

Fig.70.25. Phyllotaxy of leaves. A, alternate; B, opposite decussate; C, whorled leaves of *Nerium;* D, whorled leaves of *Alstonia*.

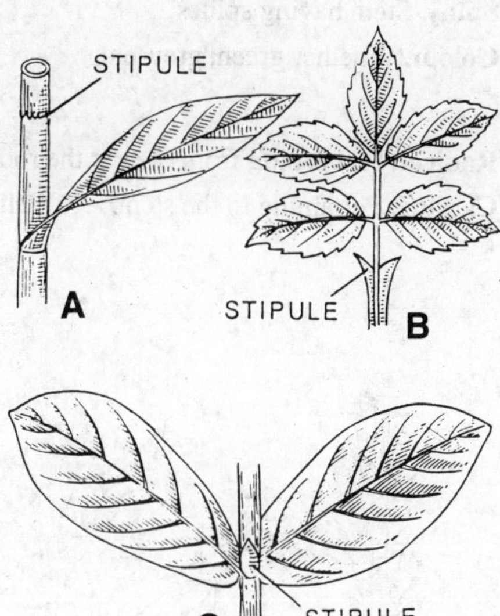

Fig.70.26. Stipules. A, ochreate stipule of *Polygonum*; B, interpetiolar stipule of *Ixora*; C, adnate stipule of rose.

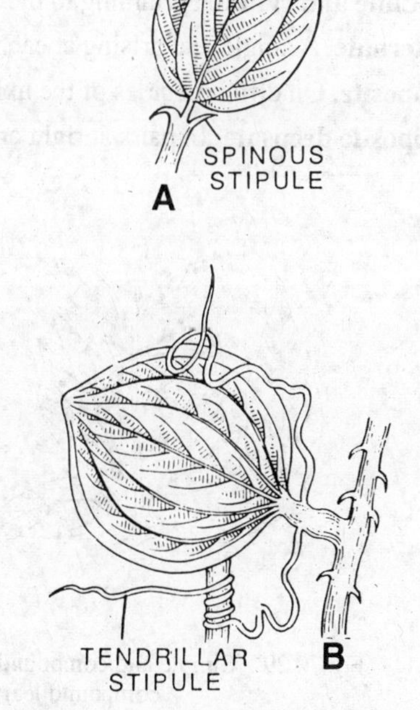

Fig.70.27. Stipules, A, spinous stipules of *Zizyphus*; B, tendrilar stipules of *Smilax*.

Waxy. Stem having wax coating, *e.g., Calotropis*.

Spiny. Stem having spines.

Colour. Whether green, grey, etc.

LEAF

Radical. Proceeding from or near the root, *e.g.,* Onion.

Cauline. Pertaining to the stem, *e.g.,* Palms.

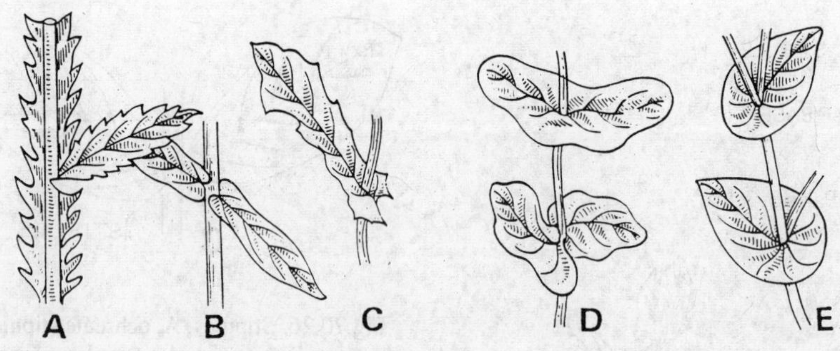

Fig.70.28. Leaf base of sessile leaves. A, decurrent leaf of *Laggera*; B, auriculate leaf of *Calotropis*; C, amplexicaul leaf of *Emilia*; D, connate leaf of *Lomicera*; E, perfoliate leaves.

Cauline and ramal. Pertaining to the main stem as well as its branches, *e.g.,* Mango.

Alternate. A single leaf arising at each node, *e.g., Hibiscus rosa-sinensis.*

Opposite. On different sides of the axis with the bases at the same level.

Opposite decusate. In pairs at right angles to one another, *e.g., calotropis.*

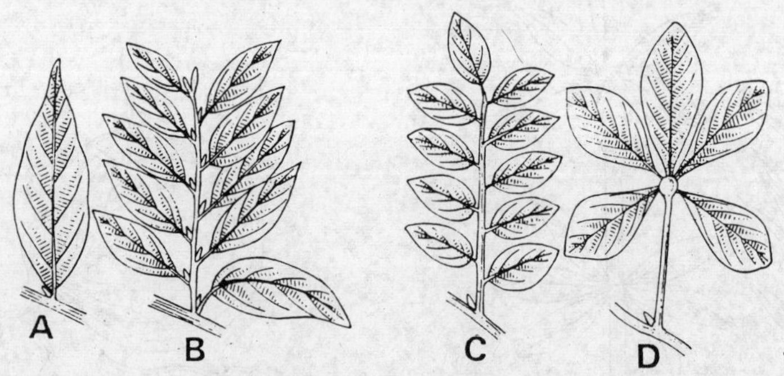

Fig.70.29 . Simple and compound leaves. A, simple leaf; B, a branch; C, pinnately compound leaf; D, palmately compound leaf.

Opposite superposed. A pair of leaves that stands directly over the lower pair in the same plant, *e.g.,* Guava.

Whorled. More than two leaves arranged in a circle round an axis, *e.g., Spergula.*

Petiolate. The leaf blade is situated on the petiole.

Sessile. Without a petiole or stalk.

Sub-sessile. Having short petiole.

Stipulate. With stipules, *e.g.*, rose.

Extipulate. Having no stipules, *e.g., Ipomoea.*

Stipules. An appendage of the leaf. Normally two stipules are developed at base of a leaf petiole, they may be leafy (foliaeous), *e.g., Lathyrus*; free lateral, *e.g.*, China rose; adnate, *e.g.*, rose; interpetiolar, *e.g., Ixora; Spergula*; spiny, *e.g., Acacia*; tendrilar, *e.g., smilax.*

Leaf Base

Connate. Two sessile opposite leaves meeting each other across the stem and fusing together, *e.g., Lomicera flava.*

Amplexicaul. Clasping or surrounding the stem, as base of leaf, *e.g., Sonchus.*

Auriculate. Leaf with expanded bases surrounding stem, *e.g., Calotropis.*

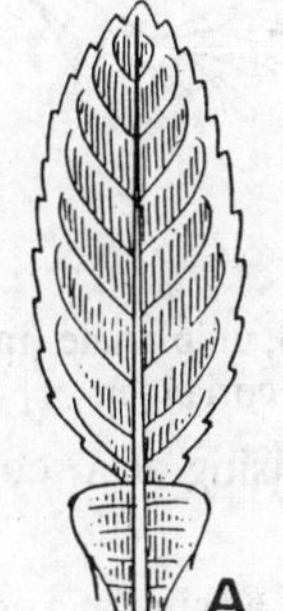

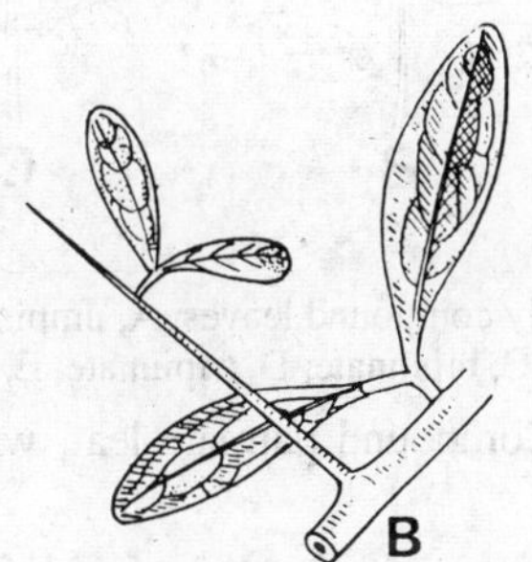

Fig.70.30. Compound leaf. A, unifoliate leaf (*Citrus*); B, bifoliate leaf (*Prinsepia*).

Decurrent. Having leaf base prolonged down stem as a winged expansion or rib, *e.g., Laggera pterodonta.*

Perfoliate. A leaf with basal lobes so united as to appear as if stem ran through it, *e.g., Aloe perfoliata.*

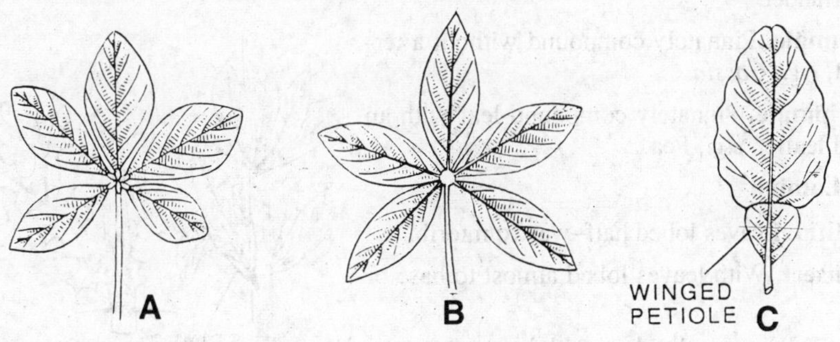

Fig.70.31. Palmate compound leaves. A, digitate leaf of *Gynandropsis*; B, digitate leaf of *Bombax*; C, unifoliate compound leaf of *Citrus* with winged petiole.

Simple. A leaf which may be entire or incised to any depth, but not down to the midrib or petiole.

Compound. A leaf made up of two or more leaflets, *e.g.*, pea and several other members of Leguminosae.

Unifoliate. Having one leaflet only, *e.g.*, *Citrus.*

Bifoliate. Palmate compound leaf with two leaflets, *e.g.*, *Prinsepia, Balanites.*

Trifoliate palmate. Compound leaf having three leaflets growing from same point, *e.g.*, *Oxalis.*

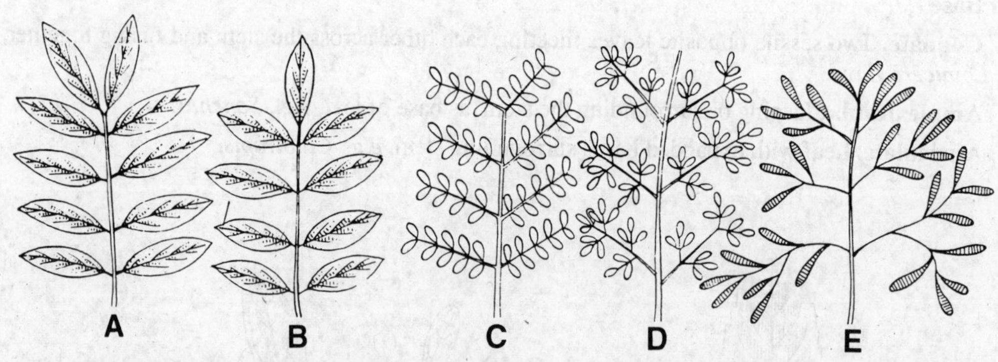

Fig.70.32. Pinnately compound leaves. A, unipinnate (paripinnate); B, unipinnate (imparipinnate); C, bipinnate; D, tripinnate; E, decompound leaf of coriander.

Quadrifoliate. Compound palmate leaf, with four leaflets arising at a common point, *e.g.*, *Marsilea.*

Multifoliate. Compound palmate leaf with five or more leaflets arising at a common point, *e.g.*, *Gynandropsis pentaphylla.*

Pinnate. A compound leaf having leaflets on each side on an axis or mid-rib.

Unipinnate. Having leaflets on each side of an axis, *e.g.*, *Cassia.*

Bipinnate. The central axis produces secondary axis which bears the leaflets, *e.g.*, *Acacia.*

The secondary axes produce the tertiary axes which bears the leaflets, *e.g.*, *Acacia.*

The secondary axes produce the tertiary axes which bear the leaflets, *e.g.*, *Moringa.*

Decompound. More than thrice pinnate, *e.g.*, old leaves of coriander.

Paripinnate. Pinnately compound without a terminal leaflet, *e.g.*, *Cassia.*

Imparipinate. Pinnately compound leaf with an odd terminal leaflet, *e.g.*, Pea.

Incision of Lamina

Pinnatifid. Leaves lobed half–way to midrib.

Pinnatisect. With leaves lobed almost to base or midrib.

Palmatifid. Leaves divided into lobes to about the middle, at acute angles to each other, *e.g.*, Castor.

Palmatisect. Palmate with division nearly to base.

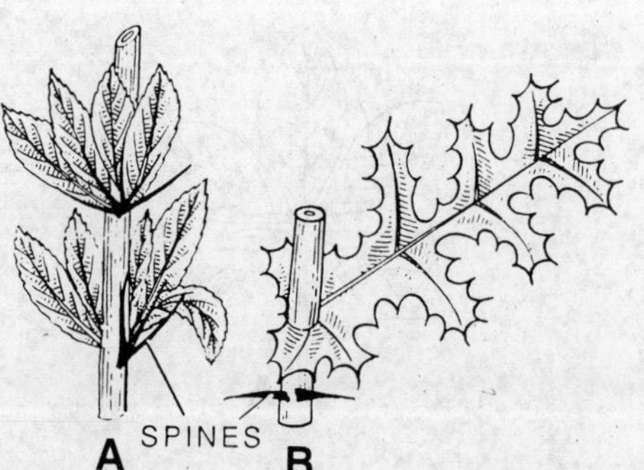

Fig.70.33. Modified leaves. A, leaves modified into spines; B, leaf of *Argemone* with spines.

Margin of Lamina

Entire. With continuous margin, *e.g., Psidium.*

Dentate. With large saw-like teeth on the margin.

Serrate-dentate. With serrate edges themselves toothed.

Undulate. Wavy, *e.g., Polyalthia.*

Convolute. Rolled together.

Crenate. With scalloped margin, *e.g., Bryophyllum.*

Lacerate. Having margin or apex deeply cut into irregular lobes, *e.g.*, many members of Ranunculaceae.

Laciniate. Irregularly incised; fringed.

Laciniolate. Minutely incised or fringed.

Ciliate. Bearing fine hairs on the margin.

Crispate. Curled or extremely undulate margin.

Spinous. Bearing many spines, *e.g., Argemone.*

Pectinate. Comb-like.

Lobed. Leaf margin divided into many lobes, *e.g., Ranunculus.*

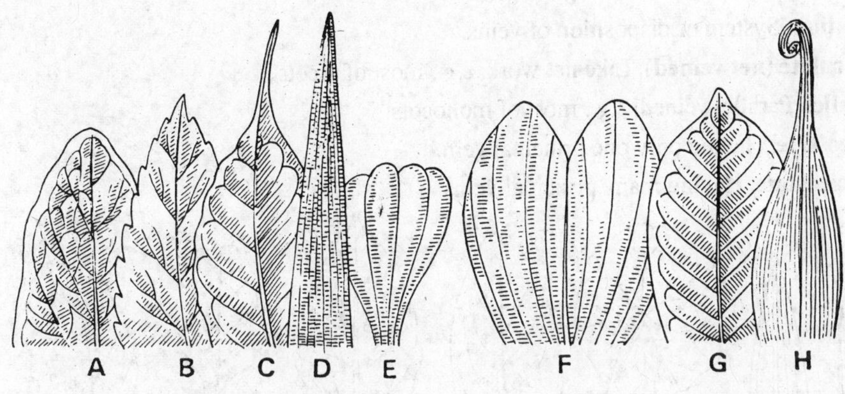

Fig.70.34. Leaf apex. A, obtuse; B, acute; C, acuminate; D, cuspidate; E, retuse; F, emarginate; G, mucronate; H, cirrhose.

Leaf Apex

Acute. Ending in a sharp point forming an acute angle, *e.g.*, mango.

Acuminate. Drawn out into long point; tapering; pointed, *e.g., Ficus religiosa.*

Obtuse. With blunt or rounded end, *e.g.*, Banyan.

Emarginate. Having a notch at apex, *e.g., Bauhinia.*

Truncate. Terminating abruptly, as if tapering end were cut off, *e.g., caryota urens.*

Mucronate. Abruptly terminated by a sharp spine, *e.g.*, apex of leaflet of *Cassia obtusifolia.*

Cuspidate. Terminating in a point.

Aristate. Provided with awns, or with a well developed bristle.

Retuse. Obtuse with a broad shallow notch in middle, *e.g., Oxalis.*

Cirrhose. Leaf with prolongation of midrib forming a tendril, *e.g., Gloriosa.*

Apiculate. Forming abruptly to a small tip, *e.g., Dalbergia.*

Leaf Surface

Hairy. Leaf surface covered with fine hairs. The hairs are of various types.

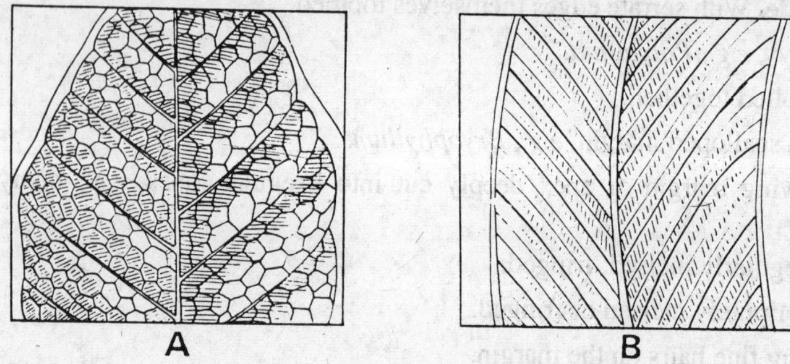

Fig.70.35. Leaf venation. A, reticulate in a dicot leaf; b, parallel in a monocot leaf.

Glabrous. With a smooth even surface, without hairs, *e.g.,* China rose.

Glaucous. Shiny green, *e.g., Citrus.*

Spiny. Covered with spines, *e.g., Argemone.*

Venation. System or disposition of veins.

Reticulate (net veined). Like net work, *e.g.,* most of dicots.

Parallel. Parallel veined, *e.g.,* most of monocots.

Unicostate. Having only one principal vein.

Multicostate. Having many principal veins.

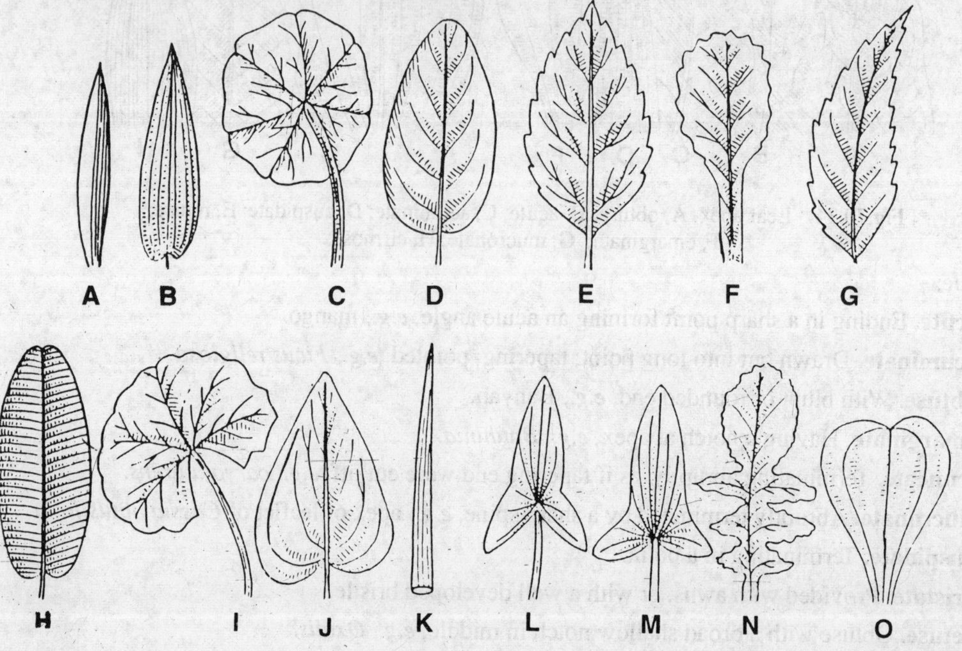

Fig.70.36. Shape of leaf. A, linear; B, lanceolate; C, rotund; D, oval or elliptical; E, ovate; F, spathulate; G, oblique; H, oblong; I, reniform; J, cordate; K, sagittate; L, hastate; M, lyrate; N, acicular; O, cuneate.

Texture

Coriaceous. Leathery, *e.g., Calotropis.*

Fleshy. Soft and thick, *e.g., Spergula.*

Succulent. Full of juice or sap, *e.g., Aloe.*

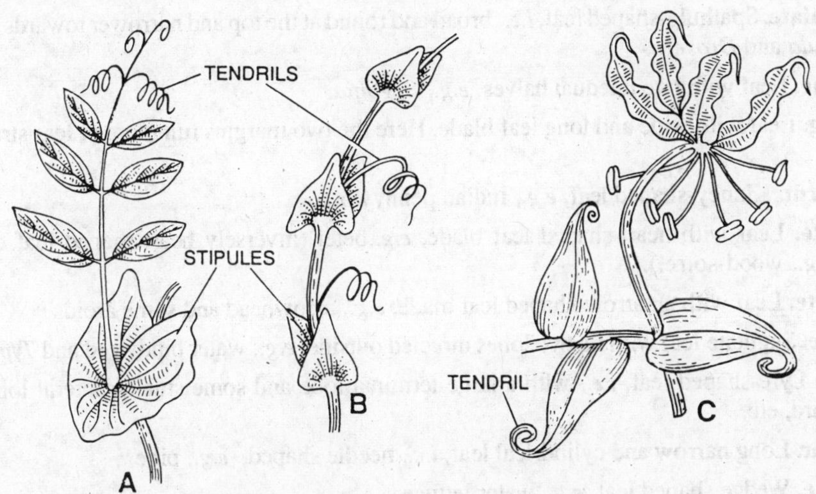

Fig.70.37. Modified leaves A, leaf tendrils of pea; B, leaf tendrils of *Lathyrus*; C, leaf tendrils (cirrhose) of *Gloriosa.*

Colour

Green. Usually the leaves are green in colour.

Pigmented. In certain leaves the pigments are developed, *e.g., Aerva.*

Note. In a compound leaf the leaflet should be described in the manner as a simple leaf.

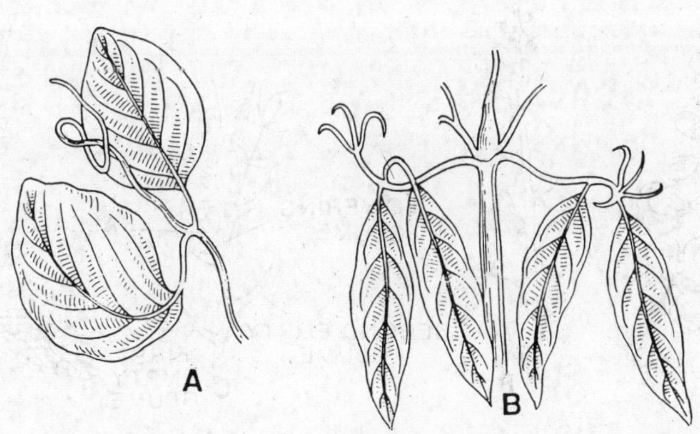

Fig. 70.38. Modified leaves. A, leaf of *Naravelia* with the terminal leaflet modified into a tendril; B, leaflet modified into hooks (*Bignonia*).

Shape of the Leaf

Linear. Long and narrow leaf, *e.g.,* many grasses.

Lanceolate. Lance-shaped leaf, *e.g.,* bamboo, *Nerium,* etc.

Round or orbicular. Leaf with a circular leaf blade, *e.g.,* lotus, garden nasturtium, etc.

Elliptical. An ellipse-shaped leaf, *e.g.,* guava, jack, etc.

Ovate. Leaf with an egg-shaped leaf blade, *i.e.,* slightly broader at the base than at the apex, *e.g.,* banyan, China rose, etc.

Spathulate. Spathula-shaped leaf, *i.e.,* broad and round at the top and narrower towards the base, *e.g., Calendula* and *Drosera.*

Oblique. Leaf with two unequal halves, *e.g., Begonia.*

Oblong. Leaf with wide and long leaf blade. Here the two margins run more or less straight up, *e.g.,* banana.

Reniform. Kidney-shaped leaf, *e.g.,* Indian pennywort.

Cordate. Leaf with heart-shaped leaf blade, *e.g.,*betel (inversely heart-shaped leaf called as obcordate, *e.g.,* wood-sorrel).

Sagittate. Leaf with an arrow-shaped leaf blade, *e.g.,* arrowhead and some aroids.

Hastate. Sagittate leaf with its two lobes directed outside, *e.g.,* water bindweed and *Typhonium.*

Lyrate. Lyre-shaped leaf, *i.e.,* with a large terminal lobe and some smaller lateral lobes, *e.g.,* radish, mustard, etc.

Acicular. Long narrow and cylindrical leaf, *i.e.,* needle shaped, *e.g.,* pine.

Cuneate. Wedge shaped leaf, *e.g.,* water lettuce.

Modified Leaves

Leaves of many plants are often modified into some specified structures to do various specialized functions. Leaf tendril is one of these structures.

Leaf-tendrils. Leaf or part of the leaf modified as a slender, branched or unbranched, thread like structure, used by many climbing plants for attachment to a support in various ways.

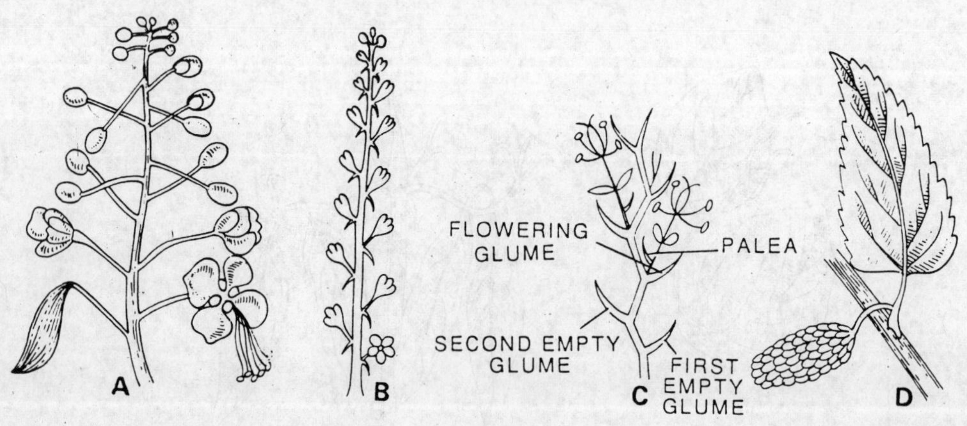

Fig.70.39. Inflorescence-racemose. A, raceme of **gul-mohur**; B, spike; C, spikelet of a grass; D, female catkin of mulbery.

In the leaf of pea upper leaflets modify into tendrils while in *Gloriosa* leaf-apex modifies into a tendril. In *Naravelia* the terminal leaflet modifies into a tendril. Likely in *Bignonia unguis-cati* these leaflets are modified into the hooks. *Lathyrus aphaca* is an example of whole leaf modification into a single tendril.

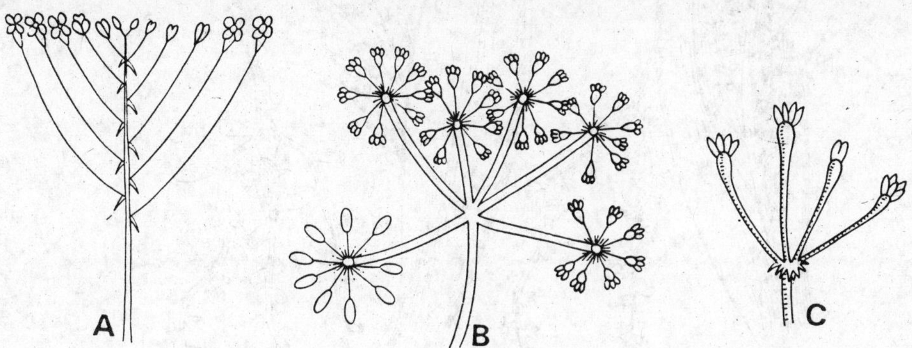

Fig. 70.40. Inflorescence. A, corymb; B, compound umbel (umbel of umbels); C, simple umbel.

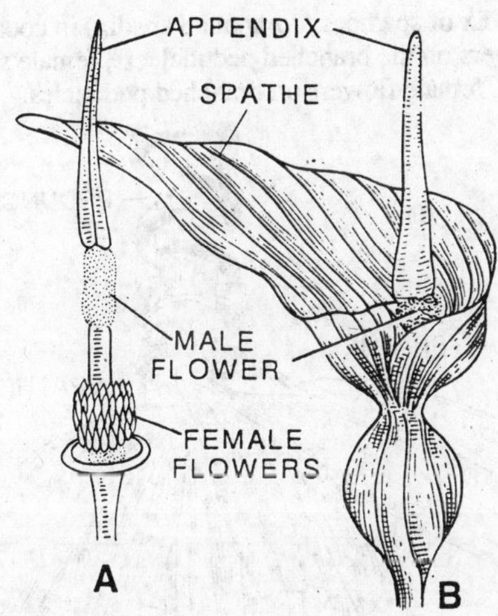

Fig.70.41. Inflorescence-spadix. A, spadix of an aroid without spathe; B, same with spathe.

INFLORESCENCE

Inflorescence. Method in which flowers are arranged on an axis.

Racemose. Inflorescence with monopodial branching.

Raceme. Inflorescence having a common axis and stalked flowers in acropetal succession, *e.g., Brassica.*

Corymb. A raceme with lower pedicels elongated so that the top is nearly flat, *e.g.,* Candytuft.

Spike. Inflorescence with sessile flowers along axis, *e.g., Achyranthes.*

Catkin. A spike with unisexual flowers and pendulous rachis, *e.g., Morus.*

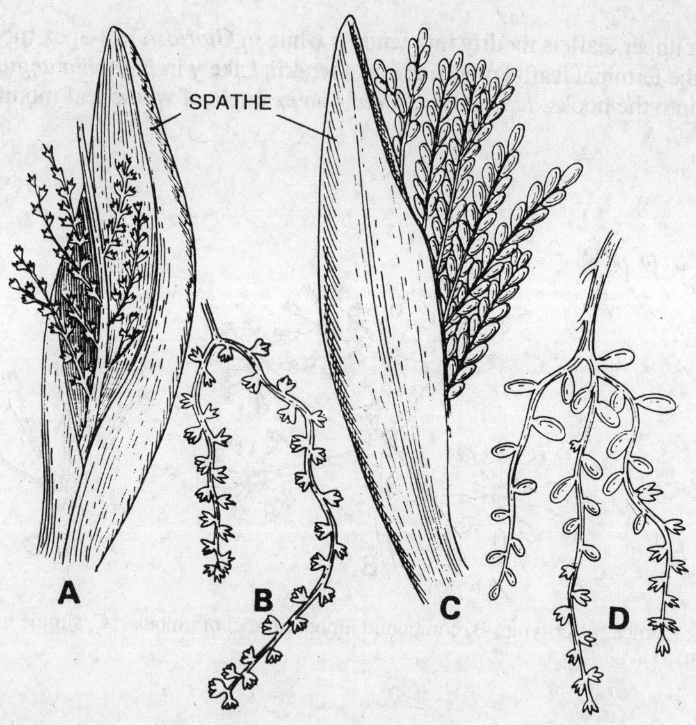

Fig.70.41(a). Inflorescence. Spadix of spadices (compound spadix) in coconut palm; A, male spadix of spadices; B, male flowers on the branched peduncle; C, female spadix of spadices; D, female flowers on branched peduncles.

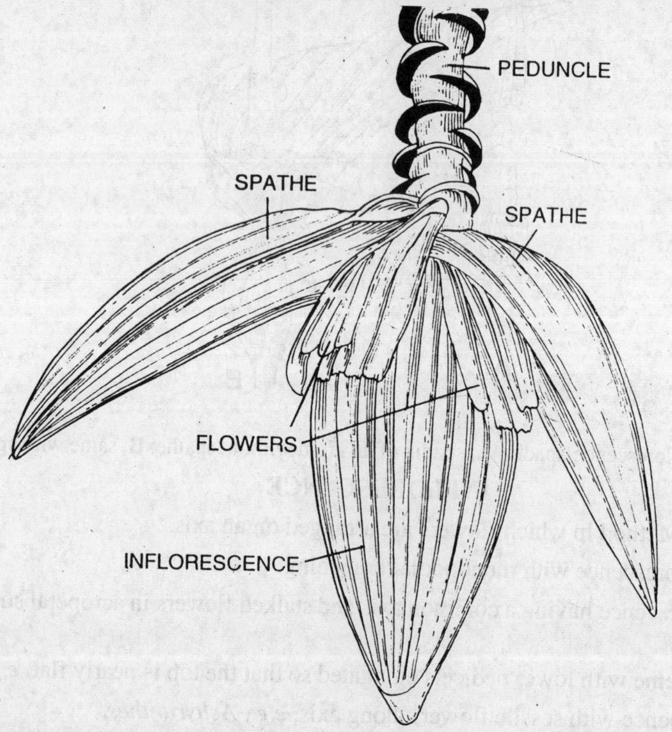

Fig.70.41(b). Inflorescence-spadix. Spadix of spadices (compound spadix) in *Musa*.

Umbel. An arrangement of flowers springing from a common centre and forming a flat or rounded cluster, *e.g., Coriandrum, Foeniculum,* etc.

Spadix. A racemose inflorescence with elongated axis, sessile flowers, and an enveloping spathe, *e.g., Colocasia, Alocasia,* Palm, *Musa,* etc.

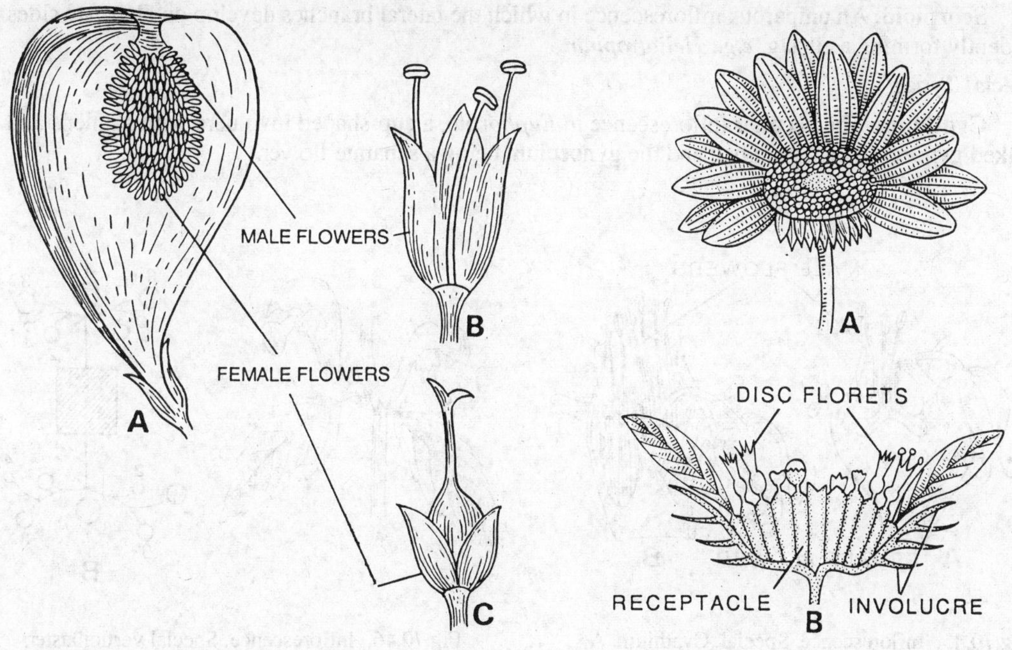

Fig.70.42. Inflorescence. A, hypanthodium of *Ficus;* B, male flower; C, female flower.

Fig.70.43. Inflorescence. Head (capitulum). A, a head; B, a head in L.S.

Hypanthodium. An inflorescence with concave capitulum on whose walls the flowers are arranged, *e.g., Ficus.*

Capitulum. An inflorescence of sessile flowers or florets crowded together on a receptacle and usually surrounded by an involucre.

Cymose. Sympodially branched.

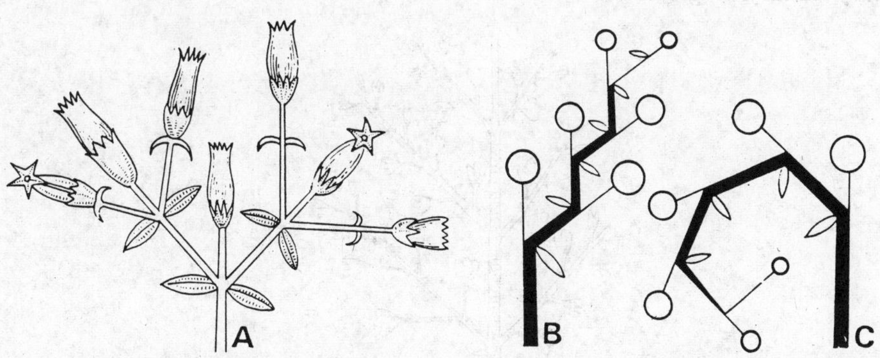

Fig.70.44. Inflorescence. Cymose. A, biparous (dichasial) cyme; B, scorpioid cyme; C, helicoid cyme.

Uniparous (monochasial). Having a cymose inflorescence with one axis at each branching.

Biparous (dichasial). Dichotomously branched cymose inflorescence, *e.g., Ixora, Saponaria.*

Helicoid cyme. An uniparous inflorescence produced by suppression of successive axes on same side, thus causing the sympodium to be spirally twisted, *e.g., Begonia, Juncus.*

Scorpioid. An uniparous inflorescence in which the lateral branches develop on alternate sides evidently forming a zigzag, *e.g., Heliotropium.*

Special Types

Cyathium. The peculiar inflorescence in *Euphorbia,* a cup-shaped involucre with stamens and stalked gynoecium, each stamen and the gynoecium being a separate flower.

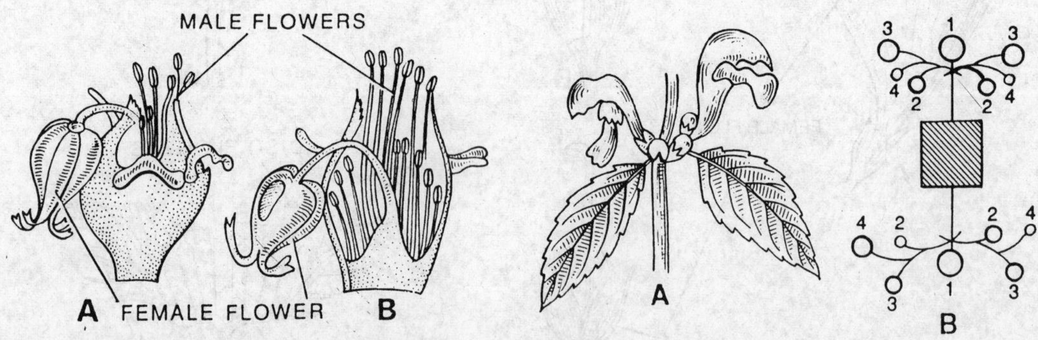

Fig.70.45. Inflorescence. Special. Cyathium. A, cyathium of *Euphorbia*; B, L.S. of same.

Fig.70.46. Inflorescence. Special verticillaster. A, verticillaster of Labiatae; B, diagram of same.

Verticillaster. A much condensed cyme with appearance of whorl, but in reality arising in axils of opposite leaves, *e.g.,* characteristic of Labiatae.

Panicle. A compound branched raceme, *e.g.,* Mango, *Melia.*

Spikelet. A secondary spike of grasses, bearing few flowers.

Compound umbel. Umbel of umbels.

Compound corymb. Corymb of corymbs.

Solitary flower. Single flower.

Solitary axillary. Single flower in the axil of leaf, *e.g., Cucurbita.*

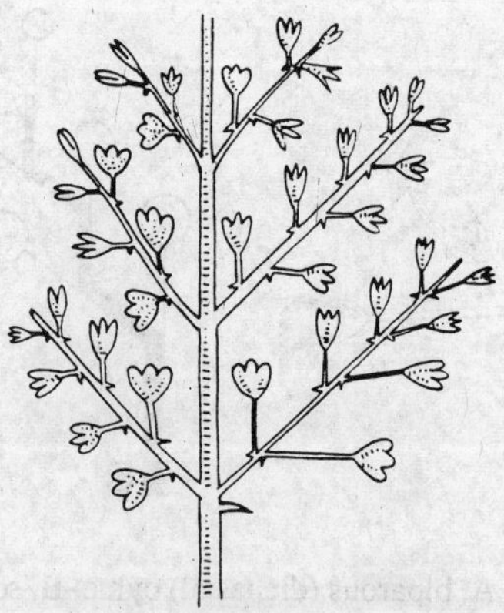

Fig.70.47. Inflorescence. Panicle.

Solitary terminal. Single flower at the apex, *e.g.,* Poppy.

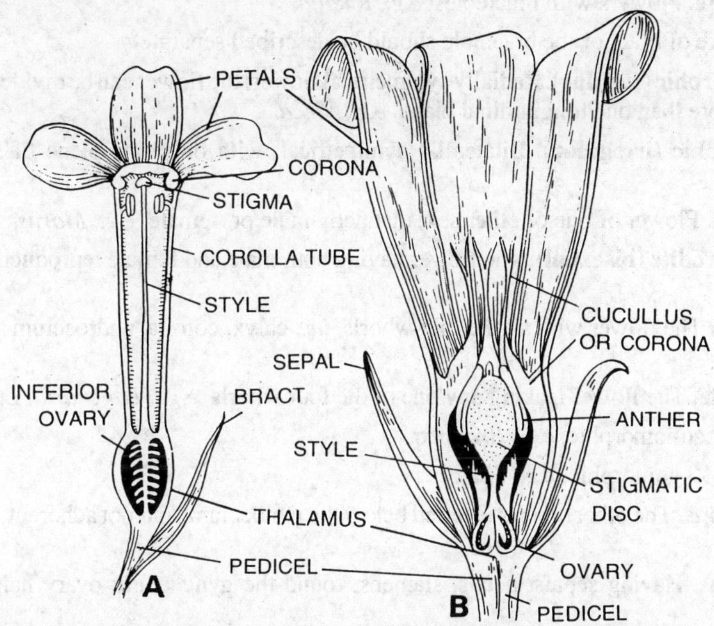

Fig.70.48. Flower. A, Parts of a flower with inferior ovary; B, L.S. of flower showing position of whorls on the thalamus with superior ovary.

FLOWER

Flower. The blossom of a plant, comprising generally sepals, petals, stamens and pistil a leafy shoot adapted for reproductive purposes.

Pedicel. Stalk of a flower.

Pedicellate. Supported by a pedicel, *e.g.,* Mustard.

Sessile. Without pedicel, *e.g., Achyranthes.*

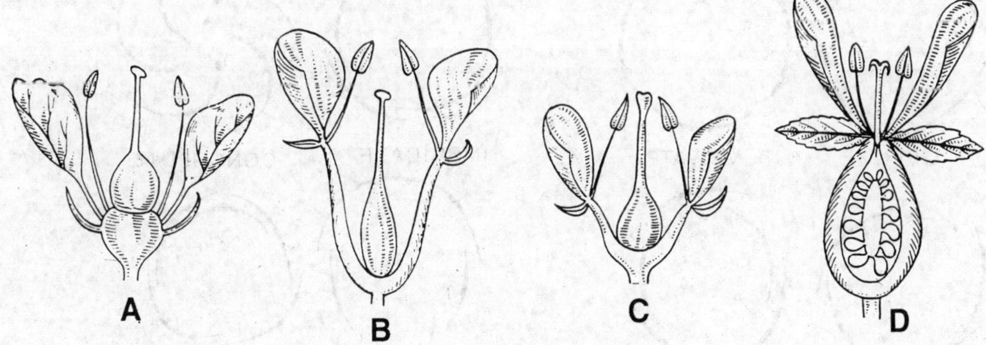

Fig.70.49. Position of floral leaves on the thalamus. A, hypogyny; B and C, perigyny, D, epigyny.

Bract. A modified leaf in whose axil a flower arises.

Bracteole. Secondary bract at the base of an individual flower.

Bracteate. Flowers having bracts, *e.g., Adhatoda.*

Ebracteate. Flowers without bracts, ***e.g., Solanum.***

Bracteolate. Flowers with bracteoles, ***e.g., Ruellia.***

N.B. Nature of bract or the bracteole should be described separately.

Actinomorphic (regular). Radially symmetrical flower; the flower can be divided into two equal halves along more than one longitudinal plane, ***e.g., Vinca.***

Zygomorphic (irregular). Bilaterally symmetrical, with only one plane of symmetry, ***e.g., Pisum.***

Unisexual. Flower of one or other sex; distinctly male or female, ***e.g., Morus.***

Hermaphrodite (bisexual). The flower having both male and female reproductive organs, ***e.g., Brassica.***

Complete. The flower with all the four whorls, ***i.e.,*** calyx, corolla, androecium and gynoecium, *e.g., Hibiscus.*

Incomplete. The flower lacking any one of the four whorls, ***e.g., Lagenaria, Luffa, Cucurbita.***

Regular. Actinomorphic, ***e.g., Brassica.***

Irregular. Zygomorphic, ***e.g.,*** Pea.

Hypogynous. The other whorls inserted below the gynoecium, and not adherent; ovary superior, *e.g., Solanum.*

Perigynous. Having sepals, petals, stamens, round the gynoecium; ovary half superior, ***e.g., Prunus.***

Epigynous. Having the various whorls adnate to ovary, thus apparently inserted in ovary; inferior ovary, *e.g., Hamelia; Coriandrum.*

Acyclic. Flowers with floral leaves arranged in a spiral, ***e.g., Ranunculus.***

Cyclic. Having parts of flower arranged in whorls, ***e.g., Solanum.***

Colour. Variously coloured.

CALYX

Calyx. The outer whorl of floral leaves (sepals).

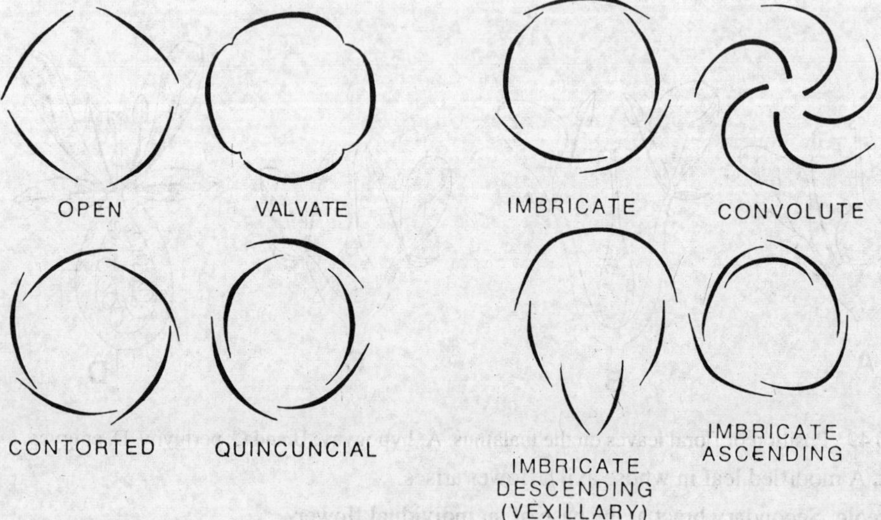

Fig.70.50. Aestivation of corolla.

Sepal. A leaf-like division of calyx.

Sepaloid. Like a sepal; green in colour.

Petaloid. Like a petal; coloured.

Number of Sepals

Polysepalous. Having free or distinct sepals, *e.g.,* Mustard.

Gamosepalous. With coherent sepals, *e.g.,* China rose.

Aestivation. The mode in which sepals are disposed in flower bud.

Valvate. Meeting at edges, *e.g., Solanum.*

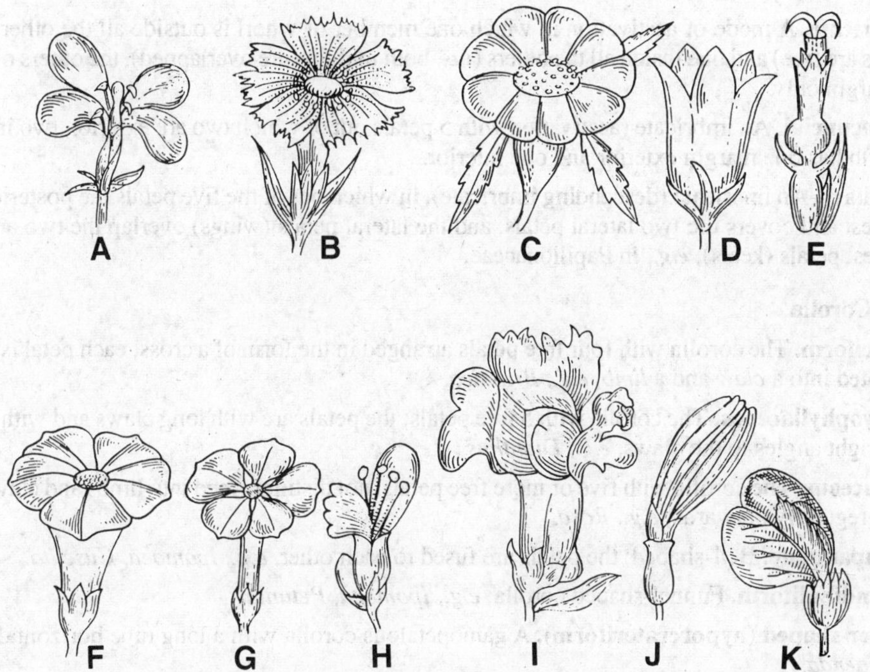

Fig.70.51. Forms of corolla. A, cruciform; B, caryophyllaceous; C, rosaceous; D, campanulate; E, tubular; F, funnel-shaped; G, rotate; H, bilabiate; I, personate; J, ligulate K, papilionaceous.

Induplicate. In aestivation having bud sepals folded inwards at points of contact, *e.g., Ipomoea.*

Contorted (twisted). Aestivation in which one sepal overlaps the next with one margin, and is overlapped by the previous on the other.

Imbricate. Having sepals overlapping each other, one sepal completely internal, one completely external and the remaining three partly internal and partly external.

Quincuncial (quincunx). Arrangement of five sepals, of which two are exterior, two interior and the fifth partly interior, partly exterior.

Caducous. The sepals that fall off early, *e.g., Argemone, Papaver.*

Deciduous. Falling at maturity.

Persistent. Remaining attached till maturation, *e.g., Solanum* and other members of Solanaceae.

COROLLA

Corolla. The whorl of petals.

Petal. One of the floral leaves of corolla.

Polypetalous. With free petals, *e.g., Brassica, Rosa.*

Gamopetalous. United petals, *e.g, Ipomoea, Petunia.*

Aestivation. The arrangement of the parts of the floral envelopes in bud.

Valvate. An aestivation when the segments of corolla are so placed that their edges touch each other, not overlap, *e.g., Solanum.*

Valvate induplicate. Valvate aestivation with the margins of the petals folded inwards, *e.g., Ipomoea.*

Twisted. Also known as contorted; one margin of the petal overlaps that of the next one, and the next margin overlaps the third one.

Imbricate. A mode of aestivation in which one member of whorl is outside all the others (*i.e.*, its margins are free) and one inside all the others (*i.e.*, both margins are overlapped), the others overlap by one margin only.

Quincuncial. An imbricate (aestivation with 5 petals out of which two are exterior, two interior and the fifth has one margin exterior and one interior.

Vexillary. An imbricate (descending imbricate), in which out of the five petals the posterior one is the largest and covers the two lateral petals, and the lateral petals (wings) overlap the two anterior and smallest petals (keels), *e.g.*, in Papilionaceae.

Shape of Corolla

Cruciform. The corolla with four free petals arranged in the form of a cross; each petal is being differentiated into a *claw* and a *limb, e.g., Brassica.*

Caryophyllaceous. The corolla with 5 free petals; the petals are with long claws and with limbs placed at right angles to the claws, *e.g., Dianthus.*

Rosaceous. The corolla with five or more free petals, not distinguished into limbs and claws and spreading regularly outwards, *e.g., Rosa.*

Campanulate. Bell-shaped; the petals are fused to each other, *e.g., Ipomoea, Cuscuta.*

Infundibuliform. Funnel-shaped corolla; *e.g., Ipomoea, Petunia.*

Salver-shaped (hyporcrateriform). A gamopetalous corolla with a long tube horizontal limb, *e.g., Mussaenda.*

Rotate. Wheel-shaped gamopetalous corolla, *e.g., Solanum*; the gamopetalous corolla with a flat and circular limb at right angles to the short tube.

Papilionaceous. Butterfly-like, as corolla of pea family; the corolla possesses one large posterior *standard,* two lateral *wings* and two innermost and smallest *keels.*

Bilabiate. Two lipped; zygomorphic gamopetalous corolla, *e.g., Justicea, Ocimum Salvia.*

Personate. Zygomorphic, gamopetalous corolla with two lips, as in *Antirrhinum.*

Ligulate. Zygomorphic, gamopetalous corolla forming a short, narrow tube below and ligule–like flat structure above, *e.g., Sonchus.*

Tubular. Tube-like, gamopetalous corolla, *e.g., Helianthus, Ageratum.*

Spur. A sac-like or tubular projection of a petal, *e.g., Viola, Tropaeolum.*

Nectary. A nectar-secreting gland in petals, *e.g., Salvia, Ranunculus,*

Corona. The appendages found in between corolla and stamens, or on the corolla, *e.g., Calotropis, Asclepias.*

PERIANTH

Perianth. The floral envelopes including both calyx and corolla, no differentiation, into calyx and corolla.

Tepal. A segment of perianth.

Polytepalous. With free tepals.

Gamotepalous. With united tepals.

Sepaloid. Green resembling sepals.

Petaloid. Cloured resembling petals.

Aestivation. The arrangement of the tepal in bud.

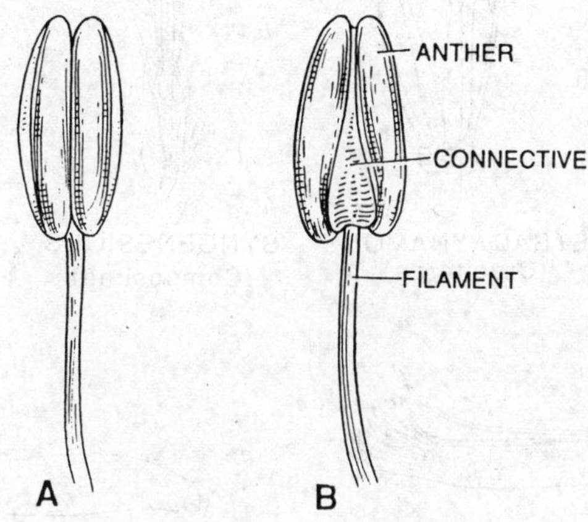

Fig.70.52. Stamens. A, face; B, back.

ANDROECIUM

Androecium. Collective term for all the stamens and staminodes.

Androphore. A stalk bearing the androecium.

Stamen. A modified leaf in flowering plants bearing pollen sacs. It consists of filaments and anthers.

Staminode. Rudimentary stamens which do not bear fertile pollen, *e.g., Stellaria.*

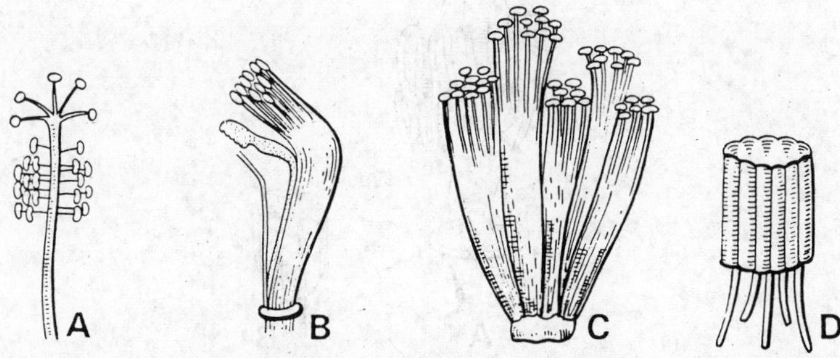

Fig.70.53. Cohesion of stamens. A, monadelphous; B, diadelphous; C, polyadelphous; D, syngenesious.

Polyandrous. Androecium that consists of free stamens, *e.g.*, Poppy.

Monadelphous. United into one bundle by the filaments, *e.g.*, *Hibiscus*.

Diadelphous. Stamens arranged in two bundles, *e.g.*, many members of Papilionaceae.

Polyadelphous. Stamens united in many bundles, *e.g.*, *Citrus, Bombax*.

Syngenesious. With anther cohering in a ring, *e.g.*, many members of Compositae, (*e.g.*, *Helianthus*).

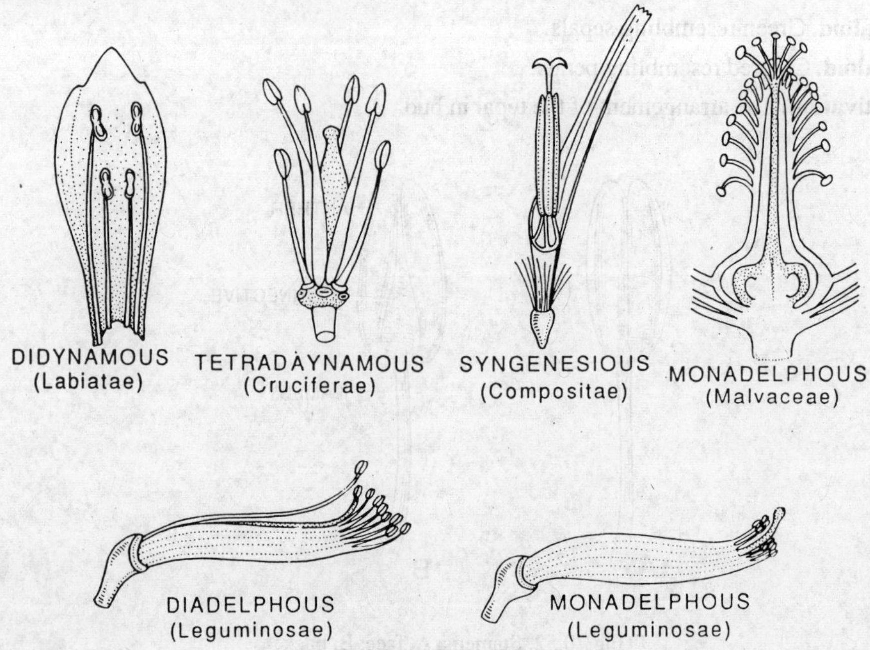

Fig. 70.53 (a). Cohesion of stamens.

Epipetalous. Stamens born on the petals or corolla tube, *e.g.*, *Convolvulus, Ipomoea, Justicea, Solanum,* etc.

Epitepalous. Stamens born on the tepals, *e.g.*, *Asphodelus*.

Gynandrous. Stamens adhering to the carpels, *e.g.*, *Calotropis*.

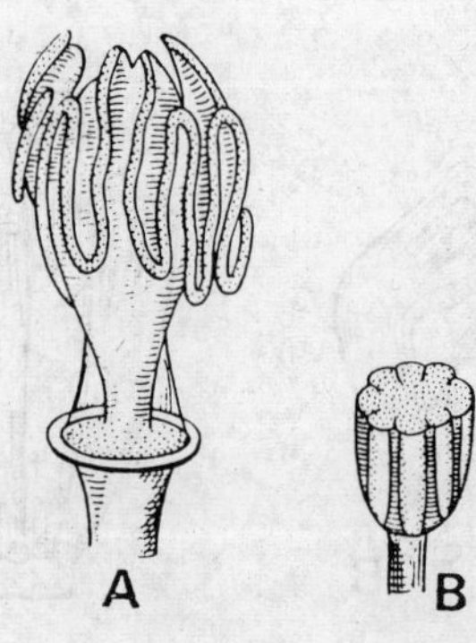

Fig.70.54. Stamens. Synandrous. A, *Cucurbita;* B, *Colocasia*.

Obdiplostemonous. Stamens found in two alternating whorls and that of the outer whorl opposite the petals, *e.g., Murraya, Geranium.*

Tetradynamous. With four long and two short stamens, *e.g., Brassica, Raphanus, Eruca,* etc.

Didynamous. In an androecium four stamens in two pairs one pair shorter than the other, *e.g., Ocimum.*

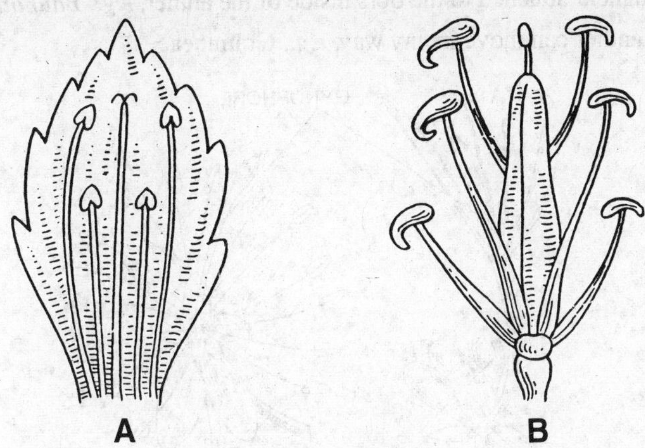

Fig.70.55. Length of stamens. A, didynamous; B, tetradynamous.

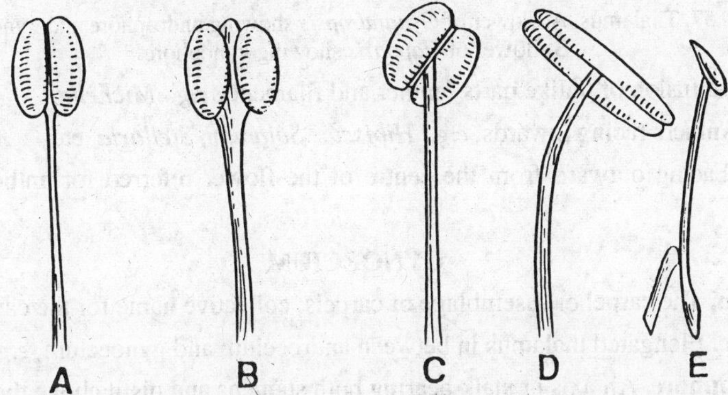

Fig.70.56. Attachment of the filament to the anther. A, basifixed; B, adnate; C, dorsifixed; D, versatile; E, elongated connective of *Salvia,* separating the two anther lobes (upper fertile, lower sterile).

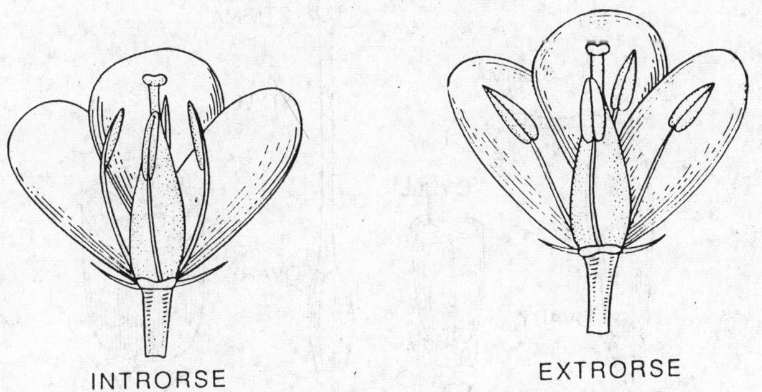

Fig.70.56 (a). Facing of stamens.

Inserted. Stamens shorter than the corolla tube, *e.g., Ixora, Mussaenda.*

Exerted. The stamens projecting beyond the petals, *e.g.*, passion flower.

Monothecous. Single-celled anther, *e.g., Hibiscus.*

Dithecous. Two-celled anther, *e.g., Papaver.*

Basifixed. Fixed to the filament (stalk) at the base, *e.g., Brassica, Papaver.*

Dorsifixed. Filament attached to the dorsal side of the anther, *e.g., Bauhinia.*

Versatile. The anther can move in any way, *e.g.*, Gramineae.

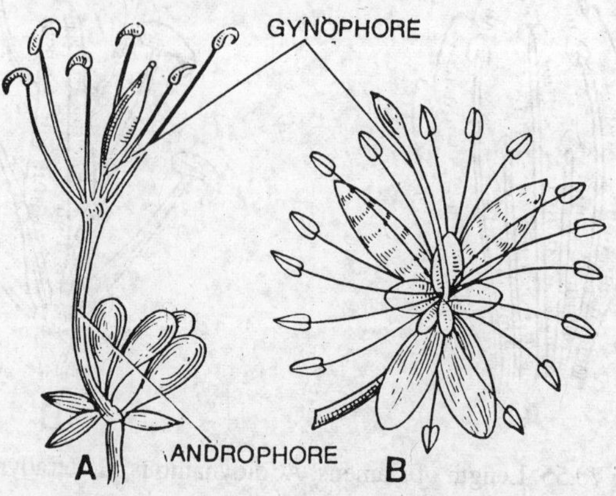

Fig.70.57. Thalamus. A, flower of *Gynandropsis* showing androphore and gynophore; B, flower of *Capparis* showing gynophore.

Adnate. The fusion of unlike parts (anther and filament), *e.g., Michelia.*

Introrse. Anthers facing inwards, *e.g., Hibiscus, Solanum, Stellaria,* etc.

Extrorse. Facing outward from the centre of the flower referred for anthers, *e.g., Papaver, Argemone.*

GYNOECIUM

Gynoecium. The carpel or assemblage of carpels; collective name for the carpels of a flower.

Gynophore. Elongated thalamus in between androecium and gynoecium, *e.g., Cleome.*

Androgynophore. An axis or stalk bearing both stamens and pistil above the point of perianth attachment, *e.g., Gynandropsis.*

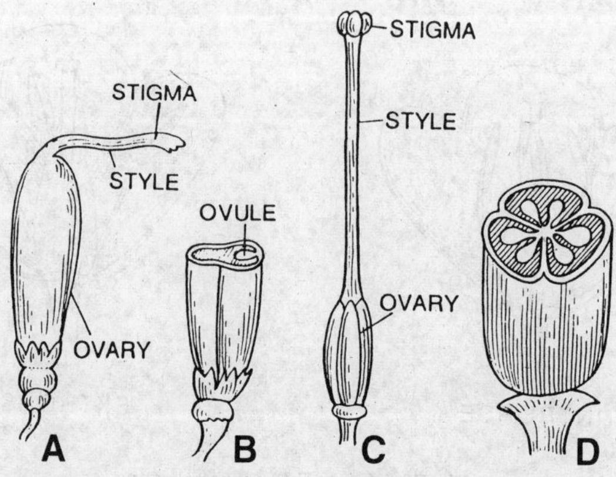

Fig.70.58. Pistil. A, simple pistil; B, unilocular ovary of simple pistil; C, syncarpous pistil; D, trilocular ovary of syncarpous pistil.

Carpel. The modified leaves which bear the ovules, a simple pistil or one of the segments of a compound pistil.

Pistil. A collective term for ovary, style and stigma.

Compound pistil. Where many pistils are aggregated together; the condition many be bicarpellary; *e.g., Brassica;* Tricarpellary, *e.g.,* Lillaceae; tetracarpellary, *e.g., Datura;* pentacarpellary, *e.g., Melia;* polycarpellary, *e.g., Sida, Papaver,* etc.

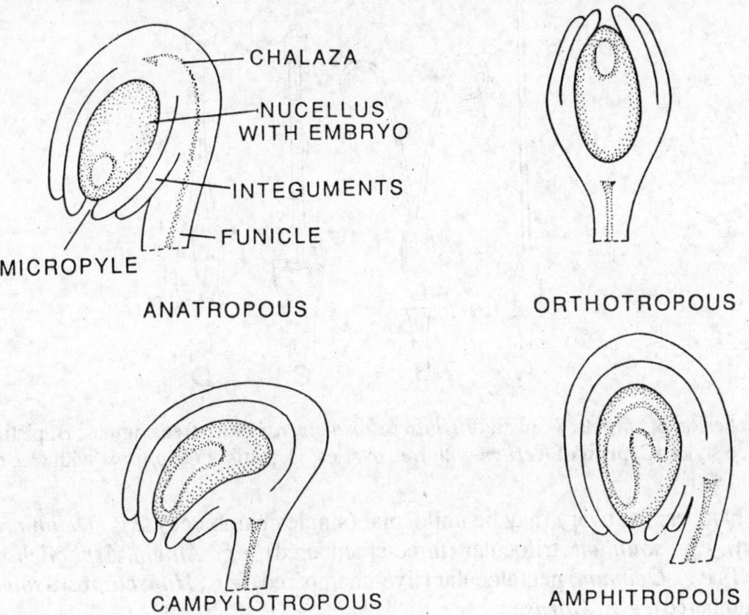

Fig. 70.58 (a). Ovules.

Apocarpous. Carpels in a flower free from each other, *e.g., Ranunculus, Nigella, clematis*, etc.

Syncarpous. United carpels, compound ovary, *e.g., Citrus.*

Superior. Situated above another member, a superior ovary has its base above the insertion of calyx; a superior calyx or corolla is inserted above the ovary, *e.g., Brassica.*

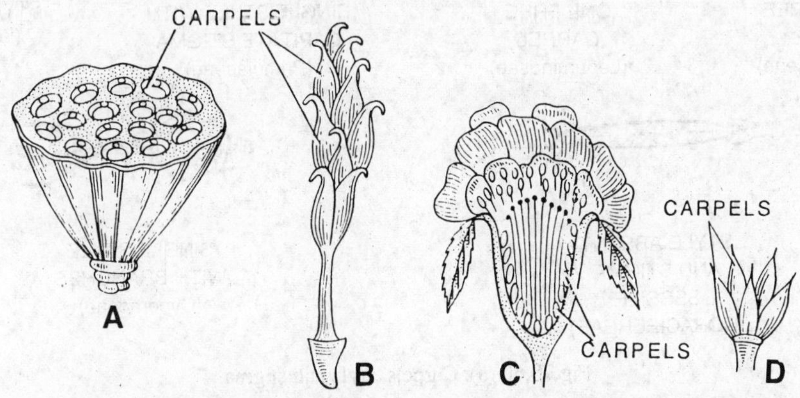

Fig.70.59. Apocarpous pistil. A, Lotus; B, *Michelia;* C, Rose; D, Sedum.

Inferior. Epigynous condition; the ovary is borne below attachment of all other floral envelopes and adnate to them, *e.g., Coriandrum, Ixora, Pyrus,* etc.

Semi-inferior. Intermediate condition between superior and inferior, *e.g.,* Pea.

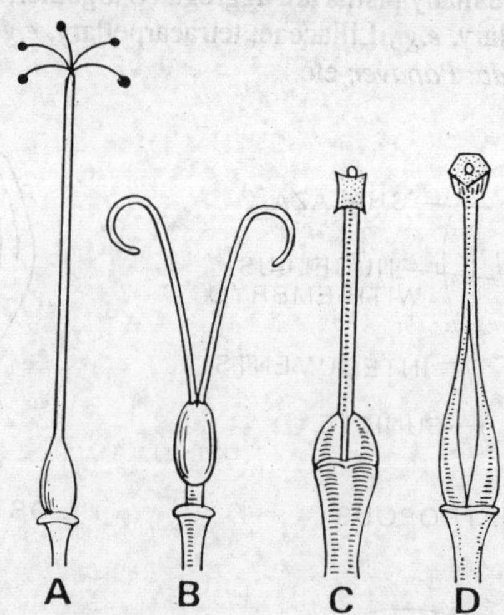

Fig.70.60. Cohesion of carpels. A, pistil of *Hibiscus rosa-sinensis* with free stigmas; B, pistil with free styles; C, pistil of *Nerium* with free ovaries; D, pistil of *Calotropis* with free ovaries and styles.

Locule. Chamber; the ovary may be unilocular (single chambered), *e.g., Delphinium;* bilocular (two-chambered), *e.g., solanum;* trilocular (three-chambered), *e.g., Allium, Asphodelus;* tetralocular (four-chambered), *e.g., Ocimum;* pentalocular (five-chambered), *e.g., Hibiscus rosa-sinensis*; multilocular (many chambered), *e.g., Citrus.*

CARPELS FREE (APOCARPOUS)

CARPELS UNITED (SYNCARPOUS)

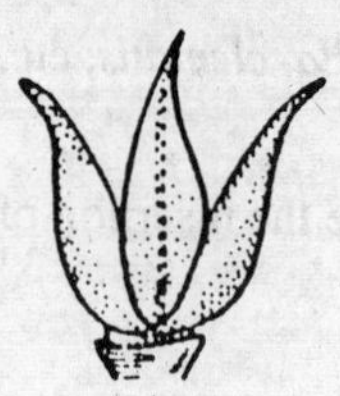

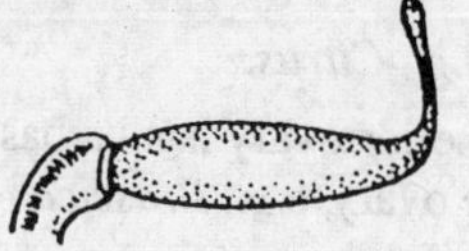

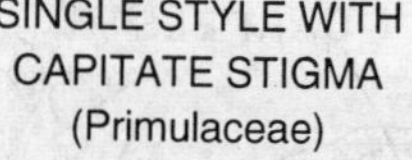

Fig.70.60 (a). Carpels, style and stigma.

Ovule. Minute bodies found inside the ovary, which after fertilization develop into seeds.

Placenta. The organs to which ovules are attached.

Placentation. Type of arrangement of placentas in a syncarpous ovary; (a) **parietal;** carpels are fused only by their margins, placentas then appearing as internal ridges on ovary wall, *e.g., Argemone, Viola, Brassica,* etc. (b) **axile;** margins of carpels fold inwards, fusing together in centre of ovary to form a single, central placenta; ovary is divided into as many compartments (loculi) as there are carpels, *e.g., Asphodelus, Hibiscus;* (c) **free-central**; placenta arising as a central upgrowth from ovary base, *e.g., Stellaria*; (d) **marginal**; placenta develops along the junction of two carpels, in a unilocular ovary, *e.g.,* pea; (e) **basal;** the ovules are few or reduced to one and are borne at the base of ovary, the ovules when solitary often filling the cavity, the ovary is unilocular, *e.g.,* Compositae; (*f*) **superficial;** carpels numerous, the placentae develop all round the inner surface of the partition wall, ovary multilocular, *e.g., Nymphaea*; (g) **lamellate;** this is modified partietal placentation, *e.g., Papaver* (Fig.70.61).

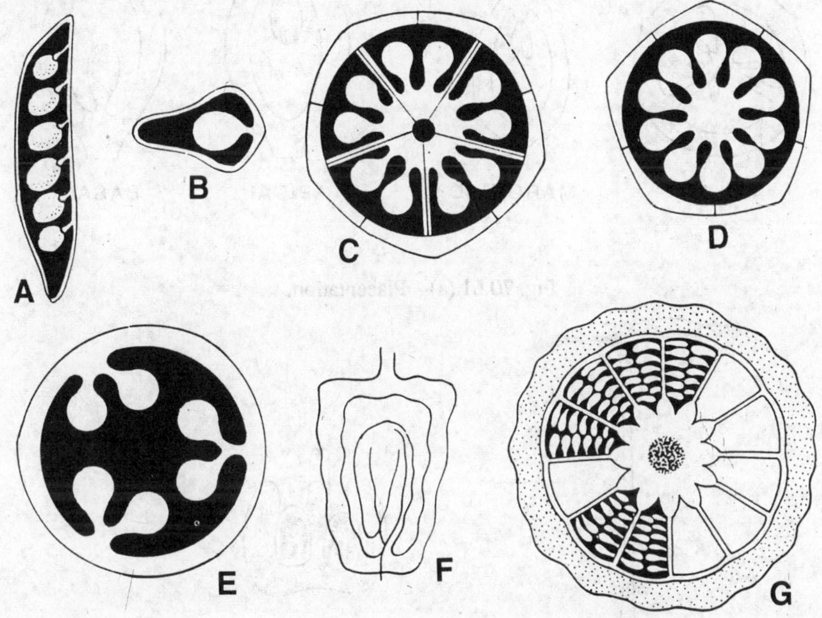

Fig.70.61. Types of placentation. A, marginal in L.S.; B, marginal in T.S.; C, axile; D, free-central; E, parietal; F, basal; G, superfical.

Style. Prolongation of carpel supporting stigma: (a) terminal-style lying in the same straight line with the ovary, *e.g.,* China rose; (b) lateral-style that arises from the side of the ovary, *e.g., Potentilla*; (c) Gynobasic style arises from the depression in the centre of ovary or from the thalamus, *e.g., Ocimum, salvia.*

Stigma. Terminal expansion of style, surface of the carpel which receives pollen; the stigma may be capitate, feathery (*e.g.,* Gramineae), discoid, dumb-bell shaped, bifid to multified, sticky.

FRUIT

Fruit. Ripened ovary of the flower, enclosing seeds.

Simple fruit. When a single fruit develops from the ovary of a flower with or without accessory parts, it is said to be a simple fruit.

Legume or *Pod.* A dry monocarpellary fruit developing from a superior, one chambered ovary and dehiscing by both the sutures, *e.g.,* pea, bean, pulses.

Follicle. A dry, monocarpellary, superior one-chambered fruit like the legume, but it dehisces by one sutre only *e.g., Calotropis, Asclepias,* Vinca, Michelia.

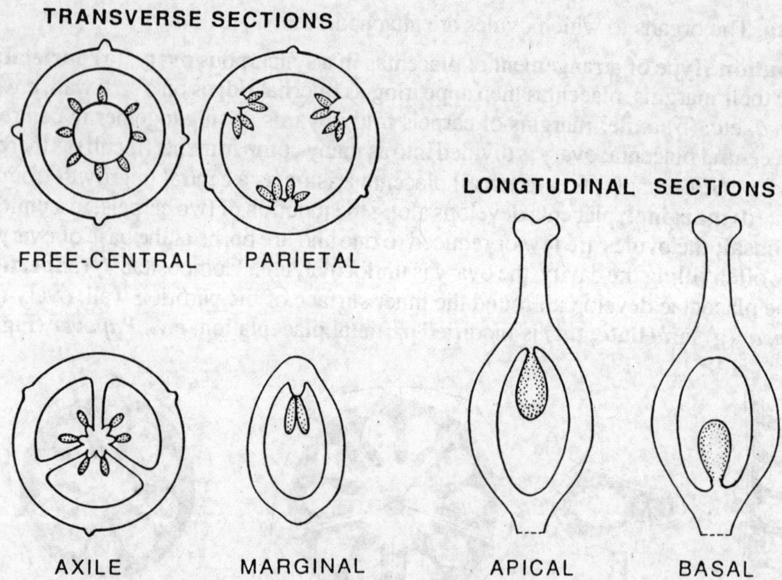

Fig. 70.61 (a) . Placentation.

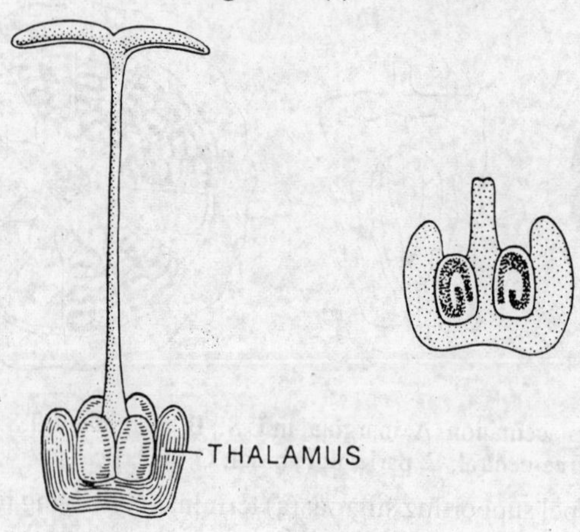

Fig.70.62. Gynobasic style of *Ocimum.* A, entire gynoecium on thalamus; B, the same in L.S.

Siliqua. A long, narrow many seeded fruit developing from a superior, bicarpellary ovary with two parietal placentae, dehiscing from below upwards by both the sutures, *e.g., Brassica, Raphanus.*

Silicula. When a siliqua is much shorter and contains only a few seeds, *e.g., Capsella,* candytuft.

Capsule. A many-seeded, uni or multiocular fruit developing from a superior, bi or polycarpellary ovary, and dehiscing in various ways, a capsule may dehisce by pores, as in poppy; or transversely; as in cock's comb; or loculicidally, as in cotton, lady's finger and *Habiscus mutabilis.*

Caryopsis. A very small, dry one-seeded fruit developing from a superior, monocarpellary ovary, with the pericarp fused with the seedcoat, *e.g.,* Gramineae.

Achene. A small, dry, one-chambered and one-seeded fruit developing from a superior, monocarpellary ovary; but unlike the previous one, the pericarp is free from the seed coat; commonly developing an apocarpous pistil and, therefore, a single flower produces an aggregate of achenes, *e.g., Clematis, Naravelia.*

Cypsela. A dry, one-chambered and oneseeded fruit developing from an inferior, bicarpellary ovary with the pericarp and the seed-coat free, *e.g.*, Sunflower, marigold, *Cosmos.*

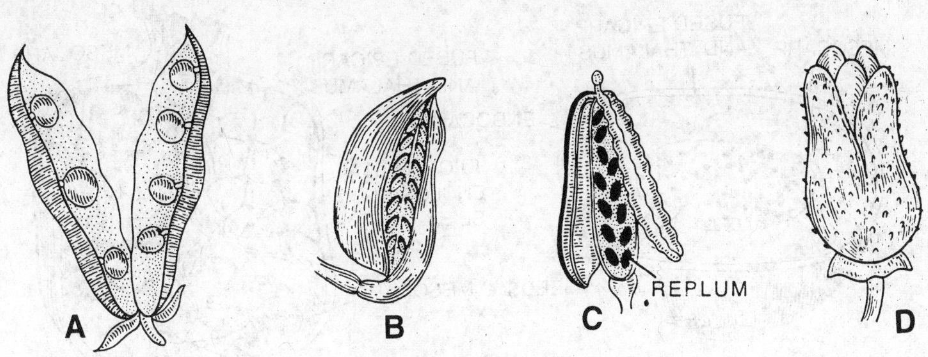

Fig.70.63. Fruits. A, legume or pod of pea; B, follicle of *Calotropis*; C, siliqua of mustard; D, capsule of *Datura.*

Nut. A dry, one chambered and one-seeded fruit developing from a superior, bi or polycarpellary ovary, with the pericarp hard and woody, *e.g.*, chestnut, oak, etc.

Samara. A dry, indehiscent, one or two-seeded fruit developing from a superior, bi-or tricarpellary ovary, with flattened wing-like outgrowths, *e.g., Hiptage.*

Lomentum. The legume is constricted or partitioned between the seeds into a number of one-seeded parts, *e.g., Acacia.*

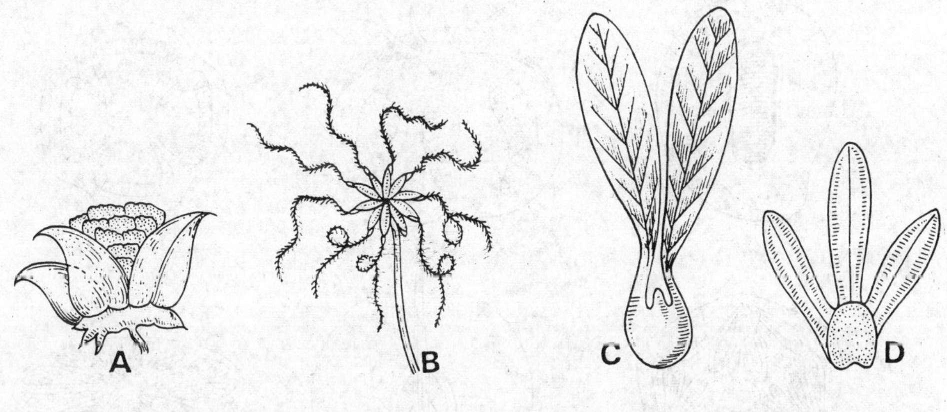

Fig.70.64. Fruits. A, capsule of cotton; B, achenes of *Naravelia* with persistent feathery styles; C, samara of *Dipterocarpus*; D, samara of *Hiptage.*

Cremocarp. A dry, indehiscent, two-chambered fruit developing from an inferior, bicarpellary ovary; when ripe splits into two, indehiscent, one-seeded mericarps which remain attached to the prolonged end (carpophore) or the axis, *e.g.*, Umbelliferae.

Regma. A dry schizocarpic fruit developing from a tricarpellary, syncarpous, superior ovary and splitting at maturity into three cocci, *e.g. Ricinus.*

Carcerulus. The fruit develops from a bicarpellary pistil with a superior ovary which becomes quadrilocular owing to the formation of two false septa; at maturity the four mericarps separate from each other towards the middle, *e.g.*, Labiatae (*Ocimum*) and Boraginaceae. In the carcerulus of *Malva* the superior ovary of the polycarpellary pistil splits into a large number of mericarps.

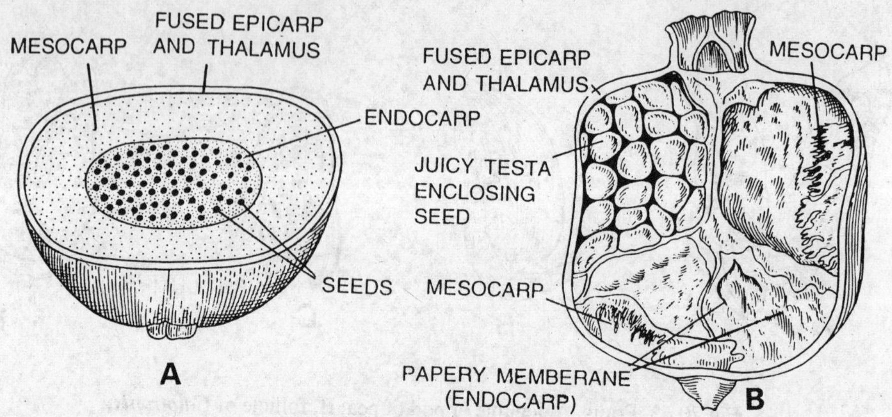

Fig.70.65. Fruits (inferior berries), A, T.S. of inferior berry of guava; B, L.S. of inferior berry of balusta of pomegranate (*Punica granatum*).

Double samara. It consists of two, three or four samaras, *e.g., Acer* and Sapindancae.

Drupe. A fleshy, one or more-chambered and one or more-seeded fruit developing from a monocarpellary or syncarpous pistil, with the pericarp differentiated into the *epicarp* which forms the skin of the fruit, the *mesocarp* which is often fleshy, and the *endocarp* which is hard and stone, *e.g.*, mango, peach, plum, coconut-palm, plamyra-palm, almond, etc.

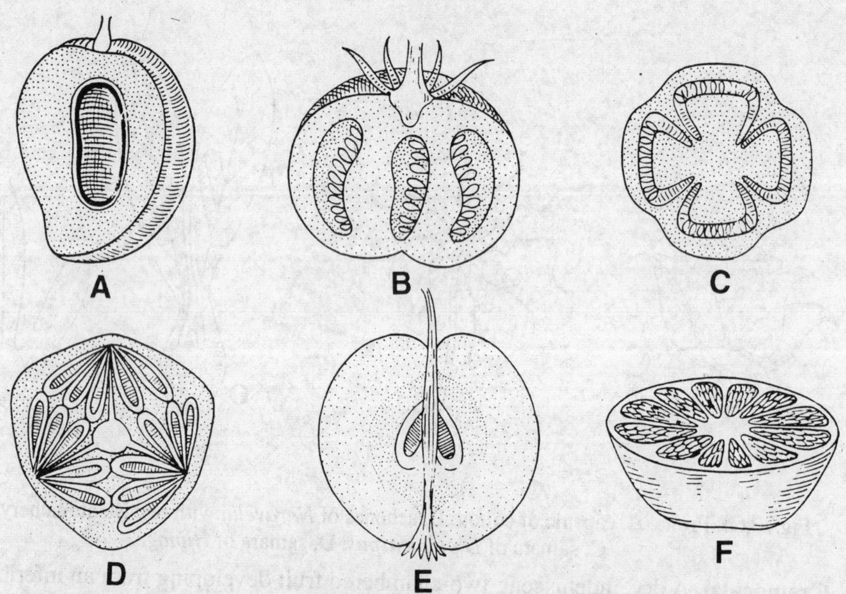

Fig.70.66. Fruits. A, drupe of mango, B, berry of tomato, C, same in T.S., D, pepo of cucumber in T.S., E, pome of apple, F, hesperidium of orange.

Berry. A superior (rarely inferior) indehiscent, usually many-seeded, fleshy or pulpy fruit developing from a single carpel or more commonly from a syncarpous pistil, with axile or parietal placentation, *e.g.*, tomato, grapes, brinjal, guava.

Pepo. A fleshy, many-seeded fruit developing from an inferior, one-celled or three-celled, syncarpous pistil with parietal placentation, *e.g.*, Cucurbitaceae.

Pome. An inferior, two or more-celled fleshy syncarpous fruit surrounded by the thalamus, *e.g.*, Apple, Pear.

Hesperidium. A superior, many-celled, fleshy fruit developing from a syncarpous pistil with axile placentation; the endocarp projecting inwards forming distinct chambers and the epicarp and mesocarp, fused together, forming the separable rind of the fruit, *e.g., Citrus.*

Aggregate fruit. It develops from a single flower with an apocarpous pistil, carpels being free, each of them develops into a simple fruitlet; an aggregate fruit, therefore, consists of a collection of simple fruitlets as may as the number of free carpels, borne by a single flower; an aggregate of simple fruitlets born by a single flower is known as an *etaerio.* Each fruitlet of an etaerio may be a follicle, an achene, a drupe or a berry.

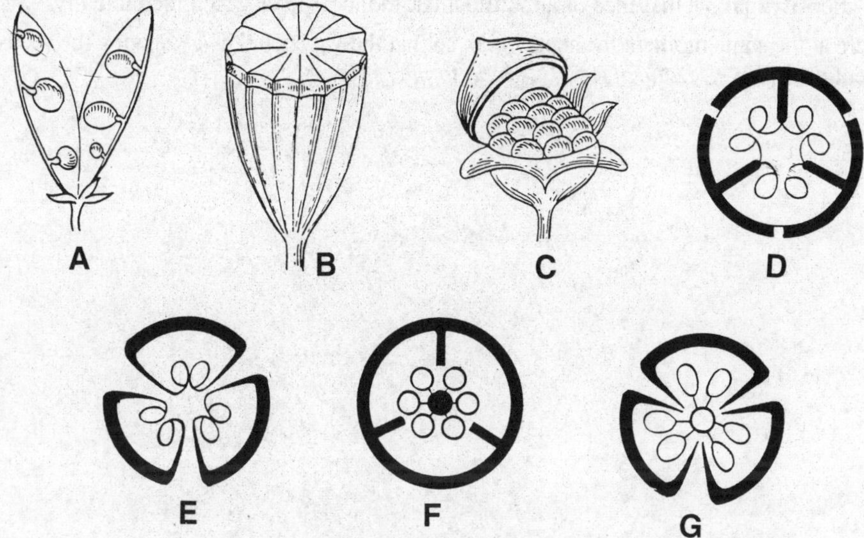

Fig.70.67. Dehiscence of fruits. A, sutural, B, porous, C, transverse, D, loculicidal, E, septicidal, F, and G, septifragal.

Etaerio of follicles. Each etaerio consists of two or more follicles, *e.g., Calotropis, Asclepias, Aconitum, Michelia.*

Etaerio of achenes. Aggregate of achenes, *e.g., Clematis, Naravelia,* strawberry, etc.

Etaerio of drupes. A number of small drupelets, developing from separate carpels of a flower, are aggregated together on a fleshy thalamus, *e.g.*, raspberry (*Rubus*).

Etaerio of berries. Developing from an apocarpous pistil and lying embedded in the fleshy thalamus, *e.g.*, custard apple; in *Artabotrys* a group of distinct and separate berries is produced.

Multiple of composite fruit. A fruit which develops from a number of flowers juxtaposed together, from a inflorescence.

Sorosis. A multiple fruit developing from a spike or spadix; flowers fuse together by their succulent sepals and the axis bearing them grows and becomes fleshy or woody, and the whole inflorescence becomes a compact mass, *e.g.*, pineapple, jack-fruit, mulberry.

Syconus. Develops from a hollow, pearshaped, fleshy receptacle which encloses a number of minute, male and female flowers; the recepatacle becomes fleshy and forms the so called fruit; it really encloses a number of true fruits or achenes, *e.g.*, banyan, peepul.

Dehiscence of Fruits:

(A) Sutural. Opening of monocarpellary fruit by sutures, either ventral, *e.g.,* madar or dorsal, *e.g., Magnolia* or by both sides, *e.g.,* in Leguminosae, pea, bean, etc.

(B) Porous. Rupturing of fruit by numerous little pores, *e.g.,* Poppy, bath sponge, etc.

(C) Transvese. Bursting of fruit transversely in the way of unhinged lid of a box, *e.g.,* in *Celosia, Portulaca*, etc.

(D) Loculicidal. Splitting of monocarpellary fruit through the back of the loculus (or chamber). Here valves separate from the axis, *e.g., Hibiscus mutabilis* and Acanthaceae, *Adhatoda, Ruellia,* etc.

(E) Septicidal. Dehiscence of monocarpellary fruit through septa, *e.g.,* in mustard family or Cruciferae, radish, mustard, etc.

(F) Septifragal. Dehiscence of a multilocular, monocarpellary fruit loculicidally.

Here at the same moment the septa break so that the valves fall away leaving the seeds attached to the central axis, *e.g., Datura,* toon (*Cedrela toona*), etc.

71

GLOSSARY OF BOTANICAL TERMS

Abaxial, the side of an organ away from the axis or centre of the axis; dorsal.

Accrescent, continuing to grow.

Achene, a small, dry, hard, one-chambered, one-seeded fruit.

Achlamydeous, flowers without perianth.

Acicular, needle like.

Actinomorphic, a flower having radial symmetry, regular.

Acuminate, ending in a pointed tapering apex.

Acute, tapering more broadly than acuminate.

Adaxial, the side toward the axis; ventral.

Adelphous, combined in groups.

Adnate, the fusion of unlike parts.

Adventitious, appearing in other than the usual place subsequently and irregularly.

Aestivation, the arrangememnt of the parts of the floral envelopes in bud.

Albumen, endosperm, nutritive tissue in the seed outside the embryo.

Albuminous, containing albumen.

Alternate, arranged in a *zigzag* manner, neither opposite nor whorled.

Amphitropous, curved ovule, one end comes nearer to the other.

Amplexicaul, the petiolar base surrounds the stem half way.

Anatropous, an ovule inverted on the stalk in such a way that the nucellus remains straight and the micropyle is directed towards the point of origin of the stalk or funicle.

Androecium, collective term for all the stamens and staminodes.

Androgynophore, an axis or stalk bearing both stamens and pistil above the point of perianth attachment.

Androphore, a stalk bearing the androecium.

Anemophilous, wind pollinated.

Angiosperm, plants having their seed enclosed inside the ovary.

Annual, of one year's growth.

Annular, in a ring or arranged in circle.

Anterior, the lower one (sepal or petal).

Anther, pollen bearing part of the stamen.

Apetalous, a flower without petals.

Apex, the tip or distal end.

Apocarpous, carpels in a flower free from each other.

Appendage, an attached subsidiary or secondary part, as a projecting part, or a hanging part or supplement.

Arborescent, of tree like habit.

Areole, a small pit or raised spot, often bearing a tuft of hairs, glochids or spines.

Aril, a small appendage around the hilum of a seed.

Arista, an awn.

Aristate, awned.

Ascending, becoming erect from a prostrate base.

Attenuate, drawn out into a long slender point.

Auricle, a claw-like appendage.

Awn, a hair or bristle-like terminal appendage.

Axile, the angle formed by an axis and a lateral member (*e.g.,* stem and leaf).

Axil, situated round an axis.

Axillary, situated in an axil.

Axis, any member which bears lateral members may be called the axis of the lateral members.

Baccate, berry like.

Basifixed, fixed to the filament (stalk) at the base.

Beaked, provided with a solid, narrow, tubular, beak-like prolongation.

Berry, a fleshy indehiscent fruit with many seeds. Sometimes seeds are few. The fruit contains thin epicarp, fleshy mesocarp and hard endocarp.

Biennial, the plants which complete their life cycle in two years.

Bifid, divided into two parts, bifurcated.

Bifurcate, forked, divided into two.

Bilabiate, two lipped.

Bilocular, two celled.

Bipinnate, twice pinnate (compound leaf).

Bisexual, containing both stamens and carpels.

Blade, the expanded part (lamina) of a leaf.

Bract, a leaf-like structure at the base of a flower.

Bracteate, with bracts.

Bracteole, small bracts occurring on the pedicel of flower.

Bracteolate, possessing bracteoles.

Bristly, bearing stiff strong hairs or bristles.

Bulb, a thick part in a resting state and made up of scales on a much shortened axis.

Bulbil, a small bulb, usually in leaf axils.

Caducous, falling off very quickly.

Calyx, the outermost green floral leaves.

Calyx-tube, the tube formed by the cohesion of the calyx leaves (sepals).

Campanulate, bell–shaped.

Campylotropous, the ovule bent so as to bring the hilum micropyle and chalaza close together.

Capitate, knob like (swollen).

Capitulum, a head of flowers.

Capsule, dry dehiscent fruit, when ripe opens by two or more valves.

Carpel, the modified leaves which bear the ovules, a simple pistil or one of segments of a compound pistil.

Caryopsis, the grain (fruit) of cereals and grasses one-seeded, indehiscent.

Catkin, a scaly bracted, usually flexuous spike or spike-like inflorescence.

Cauline, belonging to the stem.

Ciliate, possessing hairs.

Claw, the narrow stalk of some petals.

Coccus, a part of a schizocarpic fruit, each of which is usually derived from a single carpel of the ovary.

Compound, composed of two or more similar parts, generally referred to a compound leaf, compound inflorescence and compound ovary.

Connate, united similar part (*e.g.*, sepal to sepal).

Connivent, the bending towards each other of two or more similar organs.

Convolute, rolled up from one or both margins.

Cordate, heart shaped usually referred for leaves.

Coriaceous, leathery.

Corm, an underground stem with sufficient starch.

Corolla, the whorl of petals.

Corymb, a form of racemose inflorescence in which the several pedicels arising at different levels reach more or less the same level at the top.

Corymbose, corymb like arrangement.

Cristate, ribbed.

Culm, the erect stem in Gramineae.

Cuneate, wedge shaped.

Cuspidate, ending in a sharp rigid point.

Cyclic, parts arranged in whorls.

Cyme, the main axis terminates in a flower.

Deciduous, falling off.

Decompound, branched repeatedly.

Decumbent, having the lower parts prostrate.

Decurrent, prolonged downwards from the base.

Dentate, with sharp teeth.

Diandrous, flowers having two stamens.

Diadelphous, in two bundles.

Diclinous, bearing the androecium and gynoecium in separate flowers.

Didynamous, in an androecium, for stamens in two pairs, one pair shorter than the other.

Digitate, finger like division.

Dimorphic, occuring in two different forms.

Dioecious, male and female flowers on different plants.

Disc, a glandular swelling below the gynoecium.

Distichous, arranged alternately in two opposite rows.

Dissected, divided into many segments.

Dorsifixed, attached to the dorsal side.

Drupaceous, more or less drupe like.

Drupe, a stone fruit, the pericarp is usually differentiated into three layers, the epicarp thin and membranous, the mesocarp thick and fleshy, and the endocarp stony, inside which is the seed, *e.g.*, plum.

Ebracteate, without bracts.
Echinate, with long spines.
Emarginate, noted at the apex.
Embryo, dormant plant inside the apex.
Endocarp, the innermost layer of the pericarp.
Endosperm, the nutritive tissue with albumen of the seed.
Entire, without any teeth at the margin.
Entomophilous, insect pollinated.
Epicalyx, a whorl of bracts just below the calyx.
Epicarp, the outermost layer of the pericarp of the fruit.
Epigynous, borne upon, or on top of the gynoecium, with inferior ovary.
Epipetalous, borne-upon on the petals or corolla tube, usually referred for stamens.
Exalbuminous, without endosperm or albumen.
Exerted, the stamens projecting beyond the petals.
Exstipulate, without stipules.
Extra-axillary, situated away from the axil of the leaf.
Extrorse, facing outward from the centre of the flower referred for anthers.
Falcate, somewhat curved.
False septum, not true septum.
Fascicled, closely aggregated.
Female, pistil bearing (pistillate).
Fid, divided, lobed.
Filament, the stalk of an anther or the lower part of the stamen.
Filiform, thread or filament like.
Fimbriate, fringed along the edge.
Fistular, hollow throughout its length.
Fleshy, thick and of leathery or succulent texture.
Flower, a collection of sporophylls or spore bearing leaves with receptacle and perianth.
Foliaceous, leaf like.
Foliate, referring leaflets.
Follicle, the fruit produced from a single carpel and opening by ventral suture.
Free, not united.
Fruit, the ovary and its contents after the fertilization of the ovules.
Frutescent, becoming shrubby.
Fruiticose, shrubby.
Fugacious, falling off rapidly.
Funicle, stalk of ovule.
Fusiform, spindle shaped.
Gamopetalous, united petals (corolla).
Gamophyllous, united tepals (perianth).
Gamosepalous, united sepals (calyx).
Gibbous, swollen on one side.
Glabrous, not hairy, smooth.
Glandular, supplied with glands.
Glaucous, of a blue green colour.

Globose, nearly spherical.

Glumaceous, resembling the glumes of Gramineae, chaffy.

Glume, the bracts and bracteoles of the spikelets of Gramineae

Gynoecium, the carpel or assemblage of carpels.

Habit, general appearance of a plant.

Habitat, the kind of locality in which plant grows.

Hairy, clothed with hairs.

Head, a compact (capitulum) inflorescence.

Herbaceous, not woody.

Hermaphrodite, bisexual, flower having both androecium and gynoecium.

Heterogamous, bearing two kinds of flowers. In Compositae when ray and disc florets are situated on the same head.

Hispid, with short scattered stiff bristles.

Homogamous, a head of Compositae bearing only one type of flowers.

Homologous, of similar morphological significance.

Hypanthium, tubular or flask-shaped zone of the floral axis which grows above the level of the ovary and bears on its margin the floral leaves and androecium.

Hypogynous, situated below the gynoecium or ovary referring to stamens, petals and sepals.

Imbricate, a mode of aestivation in which one member of the whorl is outside all the others (*i.e.,* its margins are free) and one inside all the others (*i.e.,* both margins are overlapped), the others overlap by one margin only.

Incised, deeply cut.

Incurved, with the ends curved inward.

Indefinite, numerous.

Indehiscent, not opening by valves or pores.

Induplicate, both sides rolled inwards.

Inferior, epigynous condition.

Inflorescence, arrangement of the flowers on the floral axis.

Infundibular, funnel shaped.

Internode, space between two successive nodes.

Introrse, facing inwards usually referred for anthers.

Involucre, whorl or bracts situated below the flower or inflorescence.

Irregular, unsymmetrical, zygomorphic.

Keel, the anterior petal in the Papilionaceae.

Labiate, lipped.

Lanceolate, lance-head shaped, a leaf broad below the middle and tapering towards apex.

Lamina, leaf blade.

Latex, milky juice.

Leaflet, one of the blades of compound leaf.

Legume, the pod of Pea Family, dehiscing along both sutures.

Leguminous, resembling legume.

Ligule, a membranous outgrowth at the junction of leaf blade and leaf sheath in grasses.

Ligulate, with ligule.

Limb, the broader or expanded part of a gamopetalous corolla.

Linear, long and narrow.

Lip, one of the two divisions of a bilabiate corolla.

Lobed, cut less than half way.
Loculicidal, fruits which open by the dorsal suture of the carpels.
Locular, chambered.
Lodicule, rudimentary membranous perianth at the base of the ovary in Grass Family.
Lomentum, a pod or legume constricted between the seeds.
Male flower, androecium bearing (staminate) flower.
Monadelphous, united into one bundle by the filaments.
Monoecious, male and female flowers on the same plant.
Mucronate, tipped with an abrupt point.
Nectary, a nectar secreting gland.
Node, the place of insertion of a leaf on the stem.
Nut, a hard dry one seeded, indehiscent fruit.
Oblong, longer than broad with parallel sides.
Obovate, inversely ovate.
Obtuse, rounded and blunt.
Opposite, arranged on different side of the axis but on the same level.
Orthrotropous, an straight ovule with the micropyle opposite to the chalaza.
Oval, broadly elliptical.
Ovary, part of gynoecium containing the ovules.
Ovate, egg shaped, broader end towards the base.
Ovule, minute bodies found inside the ovary, which after fertilization develop into seeds.
Pale, Palea, the scaly bracts within the glumes in the flowers of the Grass family.
Palmate, with palm-like spreading segments.
Panicle, a compound racemose inflorescence.
Paniculate, bearing panicle type inflorescence.
Papilionaceous, butterfly like, as corolla of Pea family.
Pappus, modified calyx into bristles, *e.g.,* in many Composites.
Parietal, borne or peripheral region reference for placenta.
Pedicel, stalk of flower.
Pedicellate, the flower with pedical.
Peduncle, stalk of an inflorescence.
Pentamerous, with five members in each whorl.
Perianth, the floral envelopes including both calyx and corolla no differentiation into calyx and corolla.
Perennial, can live for more than two years.
Perigynous, sepals, petals, and stamens inserted above the hypanthium.
Persistent, not falling off.
Petal, of the floral leaves of the corolla.
Petaloid, white or coloured.
Petiole, the stalk of the leaf.
Petiolate, the leaf with petiole.
Pinna, the leaflets of a bipinnate leaf.
Pinnate, a compound leaf in which the leaflets are arranged on the two sides of the rachis.
Pistil, a collective term for ovary, style and stigma.
Placenta, the organs to which ovules are attached.
Plumose, feathered.

Pod, a dry fruit developed from a monocarpellary ovary.

Pollination, transference of pollen grains from anthers to stigma.

Polypetalous, bearing uni and bisexual flowers on the same plan.

Polypetalous, with free petals.

Polysepalous, with free sepals.

Polymorphic, with many forms.

Pome, most of the edible part develops from the thalamus, *e.g.,* apple.

Prostrate, lying horizontally on the ground.

Pubescent, densely clothed with hairs.

Pulvinus, the swollen base of the petiole.

Quincuncial, an imbricate aestivation, with 5 petals out of which two are exterior, two interior and the fifth has one margin exterior and one interior.

Raceme, an inflorescence with stalked flowers, in which the main axis continues to grow and the lowest flowers are the oldest and open first (arranged in basipetal succession).

Racemose, an inflorescence in which the main axis continues to grow and remains stronger than the laterals which arise from it, the youngest lateral come out from nearest the apex.

Rachis, the axis of an inflorescence or of a compound leaf.

Rachilla, the axis of the spikelet of the grasses.

Ray florets, the outer flowers of a composite head.

Receptacle, the portion of the axis (convex, concave or flattened) of a head on which the florets are arranged.

Regular, actinomorphic, with radial symmetry.

Reniform, kidney shaped.

Reticulate venation, net veined.

Rhizome, an elongated, horizontal underground stem with much quantity of starch.

Root stock, resembling rhizome.

Rotate, wheel shaped usually referred to a gamopetalous corolla, *e.g., Solanum.*

Runner, an elongated prostrate weak stem with adventitious roots at nodes.

Salver-shaped, with a long tube and horizontal limb usually referred to corolla.

Scabrid, clothed with small and stiff hairs.

Scale, a thin membranous outgrowth.

Scaly, with scale.

Schizocarp, a fruit opening in two or more distinct portions each with its separate wall.

Seed, the ovule after fertilization and the development of the embryo.

Sepal, a segment of calyx.

Sepaloid, green.

Septicidal, capsule dehiscing by ventral sutures.

Serrate, toothed with teeth inclined upwards.

Sagittate, arrow shaped with the basal lobes directed backwards.

Sessile, without a stalk.

Simple, not compound.

Sheath, the tubular basal portion of a leaf, usually which encircles the stem.

Sheathing, the leaf having expanded base of the petiole encircling partially or completely the stem.

Sigmoid, curved, in zigzag fashion.

Spathe, a large showy bract usually covering the whole inflorescence.

Spadix, a type of inflorescence with succulent axis possessing male and female flowers, *e.g.*, the Aroid family.

Spike, a racemose elongated inflorescence bearing sessile flowers.

Spikelet, the laterals of the inflorescence of grasses, the small spikes.

Spine, a sharp pointed hard body.

Stamen, a modified leaf in flowering plants bearing pollen sacs. It consists of filaments and anthers.

Staminate, having stamens, male.

Staminode, rudimentary stamens which do not bear fertile pollen.

Standard, the broad upper petal of Papilionaceae.

Stigma, the pollen receiving part of the carpel, usually it is lobed and sticky.

Stipule, a leaf-life outgrowth at the base of the petiole, usually in parts.

Stipulate, having stipules.

Style, a slender outgrowth of a carpel which bears stigma.

Sucker, a shoot of subterranean origin.

Sulcate, grooved.

Superior, situated above another member. A superior ovary has its base above the insertion of calyx; a superior calyx or corolla is inserted above the ovary.

Suture, the line of dehiscence of dry fruits.

Syncarpous, united carpels, compound ovary.

Syngenesious, with anthers cohering in a ring.

Tap root, primary root, deeply penetrating the soil.

Tassel, the male inflorescence of maize.

Tendril, they are either modified branches or leaves. They are filament like usually coiling round the support.

Tepal, a segment of perianth.

Terete, cylindrical.

Tetradynamous, with four long and two short stamens.

Trifoliate, with three leaflets.

Truncate, cut off at the end, square or so.

Tuber, a short underground stem with starch.

Turgid, swollen.

Twiner, a climber, which climbs by spirally coiling stems.

Umbel, an inflorescence having aerial branches of equal length from a common point.

Umbellate, umbel like.

Unilocular, one chambered ovary.

Valvate, an aestivation with the segments of calyx or corolla are so placed that their edges touch each other not overlap.

Venation, the arrangement of veins in a leaf.

Ventricose, suddenly bulged.

Versatile, can move in any way.

Verticillate, whorled.

Viscid, with sticky secretion.

Whorled, verticillate.

Xerophyte, a desert plant.

Zygomorphic, asymmetrical, irregular.

SELECTED REFERENCES

Anonymous 1948-1976. *The Wealth of India.* Raw Materials, 11 vols. CSIR, New Delhi.

Babu, C.R. 1977. *Herbaceous Flora of Dehradun.* CSIR, New Delhi.

Bhandari, M.M. 1978. *Flora of the Indian Desert.* Scientific Publishers, Jodhpur.

Blatter, E. and C.McCann, 1935. *The Bombay Grasses.* Govt. of India Press, Delhi.

Blatter, E., C.McCann and T.S.Sabnis, 1929. *The Flora of Indus Delta.* Madras.

Blatter, E. and W.S.Millard, 1929. *Some Beautiful Indian Trees.* J.Bom. Nat.Hist.Soc. **33**:624–635.

Bor, N.L. 1941. *Common Grasses of the United Provinces.* Ind. For. Rec. 2:1-220.

Bor, N.L. 1943. *Manual of Indian Forest Botany.* London.

Bor, N.L. 1960. *The Grasses of Burma, Ceylon, India and Pakistan.* Oxford.

Bor, N.L. and M.B.Raizada, 1954. *Some Beautiful Indian climbers and shrubs.* Bom. Nat. Hist. Soc., Bombay.

Bourdillon. T.F. 1908. *The Forest trees of Travancore.*

Chopra, R.N., Nayer, S.L. and Chopra, I.C. 1956. *Glossary of Indian Medicinal Plants.* C.S.I.R., N.D.

Cliford, H.T. and W. Stephenson, 1975. *An Introduction to Numerical Taxonomy.* Academic Press, N.Y.

Cole, A. J. (Ed.) 1969. *Numerical Taxonomy.* Academic Press, N.Y.

Collett, H. 1902. *Flora Simlensis.*

Cook, C.D.K. 1974. *The Water Plants of the World.* The Hague.

Cooke. T. 1901–1908. *The Flora of the Presidency of Bombay.*

Corven, D.V. 1952. *Flowering Trees and Shrubs in India.* Bombay.

Cronquist, A.1968. *The Evolution and Classification of Flowering Plants.* Thomas Nelson and Sons, Ltd., London.

Dastur, J. F. 1951. *Useful Plants of India and Pakistan,* Bombay.

Davis, P. H. and V. H. Heywood, 1963. *Principles of Angiosperm Taxonomy.* Oliver & Boyd, London.

Duthie, J.F. 1903–1922. *Flora of the Upper Gangetic Plain and of the Adjacent Siwalik and Sub-Himalayan Tracts,* 3 vols.

Duthie, J.F. 1906. *The Orchids of the North-Western Himalayas.* Ann. Roy. Bot. Gdn., Calcutta, 9:81–211.

Elliot, W.1859. *Flora Andhrica.*

Fyson, P.F. *The Flora of the Nilgiri and Pulney Hill-Tops.* 3 Vols. Govt. Press, Madras.

Fyson, P.F. *The Flora of the South Indian Hill Stations: Ootacamund, Coonoor, Kotagiri, Kodaikanal, Yercaund and the Country Round,* 2 vols. Govt. Press, Madras.

Gamble, J.S. 1878. List of *Trees, Shrubs and Large Climbers Found in the Darjeeling District, Bengal,* Calcutta.

Gamble, J.S. and C.E.C. Fischer 1915–1936, *Flora of the Presidency of Madras.* 11 parts.

Gamble, J.S. 1922. *A Manual of Indian Timbers,* London.

Graham, J. 1839. *A Catalogue of the Plants Growing in Bombay and its Vicinity,* Bombay.

Haines, H.H. 1910. *A Forest Flora of Chota Nagpur.*

Haines, H.H. 1921–1925. *The Botany of Bihar and Orissa,* 6 parts

Hawkes, J.G. (Ed.) 1968. *Chemotaxonomy and Serotaxonomy.* Acad. Press, London.

Heywood, V.H. 1967. *Plant Taxonomy,* London.

Heywood, V.H. (Ed.) 1968. *Modern Methods in Plant Taxonomy.* Acad. Press, London.

Hill, A.F. 1952. *Economic Botany.* N.Y.

Hooker, J.D. 1875–1897. *The Flora of British India.* 7 Vols.

Hutchinson, J.1959. *The Families of Flowering Plants.* 2 Vols. Oxford.

Hutchinson, J. *Evolution and Phylogeny of Flowering Plants.* Acad. Press, London.

Kanjilal, P.C. 1933. *A Forest Flora of Pilibhit, Oudh, Gorakhpur and Bundelkhand.*

Kanjilal, U. 1928. *Forest Flora of the Chakrata, Dehradun and Saharanpur Forest Divisions, United Provinces.*

Kanjilal, U., A. Das, P. C. Kanjilal and R.N. De 1934–1940. *Flora of Assam.* 5 Vols.

Kirtikar, K.R. and Basu, B.D. 1933. Indian Medicinal Plants.

Lawrence, G.H.M. 1951. *Taxonomy of Vascular Plants.* N.Y.

Lawrence, G.H.M. 1955. *An Introduction to Plant Taxonomy.* N.Y.

Maheshwari, J.K. 1963. *The Flora of Delhi.*

Maheshwari, P. and R.N. Kapil, 1963. *Fifty Years of Science in India: Progress of Botany.* Ind. Sc. Cong. Assn., Calcutta.

Maheshwari, P. and U. Singh, 1965. *Dictionary of Economic Plants in India.* ICAR, New Delhi.

Mayurnanthan, P.V. 1929. *The Flowering Plants of Madras City and its Immediate Neighbourhood.*

McCann, C. 1959. *100 Beautiful Trees of India.* Bombay.

Mooney, H.F. 1950. *A Supplement to the Botany of Bihar and Orissa.* Govt. Press, Ranchi.

Nadkarni, K.M. 1927. *Indian Materia Medica.* 2 Vols. Bombay.

Osmaston, A.E. 1927. *A Forest Flora for Kumaon.*

Pal, B.P. 1960. *Beautiful Climbers of India,* ICAR, New Delhi.

Pandey, B.P. 1980. *Economic Botany.* S. Chand & Co., Ltd., New Delhi.

Pandey, B.P. 1978. *Vargikaran Vanaspati Vigyan.* S. Chand & Co., Ltd., New Delhi.

Parker, R.N. 1918. *A Forest Flora for the Punjab with Hazara and Delhi.*

Parkinson, C.E. 1923. *A Forest Flora of the Andaman Islands.*

Poster, C.L. 1959. *Taxonomy of Flowering Plants.*

Prain, D. 1903. *Bengal Plants.* 2 Vols.

Puri, V. 1952. *Floral Anatomy in Relation to Taxonomy,* Agra Univ. J. Res. (Sci.) **1:** 15–35.

Puri, V. 1958. *Floral Anatomy and Taxonomy.* Ind. Bot. Soc. Mem. **1:** 15–18.

Puri, V. 1962. *Floral Anatomy in Relation to Taxonomy.* Bull. Bot. Sur. India, **4:** 161–165.

Raizada, M.B. 1958. *Name Changes in Common Indian Plants* Ind. For. **85:** 467–538.

Raizada, M.B. 1975. *Supplement to the Duthie's Flora of the Upper Gangetic Plain.*

Raizada, M.B., R.C. Bhardwaja and S.K. Jain, 1961, 1966. *The Grasses of Upper Gangetic Plain.*

Raizada, M.B. and H.O. Saxena, 1978. *Flora of Mussoorie.*

Randhawa, M.S. 1957. *Flowering Trees in India.* ICAR, New Delhi.

Ranga Achariyar, K. 1921. *A Hand-book of Some South Indian Grasses.*

Rau. M.A. 1975. *High Altitude Flowering Plants of Western Himalayas.* BSI, Calcutta.

Rendle, A.B. 1925. *The Classification of Flowering Plants.* 2 Vols. London.

Royle, J.F. 1833. *Illustrations of the Botany and other Branches of the Natural History of the Himalayan Mountains and of the Flora of Kashmir.*

Santapau, H. 1953. *The Flora of Khandala on the Western Ghats of India.*

Santapau, H. 1957. *The Flora of Purandhar.*

Santapau, H. 1958. *History of Botanical Research in India, Burma and Ceylon Part II Systematic Botany of Angiosperms.*

Santapau, H. 1962. *The Flora of Saurashtra.*

Santapau, H. and Z. Kapadia, 1966. *The Orchids of Bombay.*

Singh, V. and D.K. Jain, 1981. *Taxonomy of Angiosperms.* Rastogi Publications, Meerut.

Smith. P.M. 1976. *The Chemotaxonomy of Plants,* London.

Sneath, P.H. A. and R.R. Sokal, 1973. *Numerical Taxonomy.*

Sporne, K.R. 1956. *The Phylogenetic Classification of the Angiosperms.* Biol. Rev. **31:** 1–29.

Stewart, J.L. 1869. *Punjab Plants.*

Subramaniam, K. 1962. *Aquatic Angiosperms.* CSIR. New Delhi.

Sundararaj, D.D. and Balasubramanyam, G. 1959. *Guide to the Economic Plants of South India.*

Swain, T. 1963. *Chemical Plant Taxonomy*. Acad. Press, London.

Swingle, D.B. 1946. *A Text-book of Systematic Botany.* Mc Graw Hill Book Co., New York.

Takhtajan, A. 1969. *Flowering Plants: Origin and Dispersal.*

Talbot, W. A. 1909–1911. *Forest Flora of the Bombay Presidency and Sind.* 2 Vols.

Turrill, W.B. 1942. *Taxonomy and Phylogeny.* Bot. Rev. **8:** 247–270.

Watson, J.F. 1824. *Plants of Kumaon.*

Watt, G. 1889. *A Dictionary of the Economic Products of India.* 6 Vols.

Willis, J. C. 1973. *A Dictionary of the Flowering Plants and Ferns.* Cambridge Univ. Press, London.

Wodehouse, R.P. 1936. *Pollengrains in the Identification and Classification of Plants.* Am. J. Bot. **23:** 535.

Question Bank

TAXONOMY OF ANGIOSPERMS

Long Answer Questions

1. Write a short essay on the 'Origin of Angiosperms' (15 minutes:15 marks).

2. Enumerate the important criteria used for classifying plants. (10 mts.; 10 mks.).

3. Give a broad outline of Hutchinson's system of classification of angiosperms and mention its merits and demerits. (15 mts.; 15 mks.).

4. What is a natural system of classification? Is Bentham and Hooker's system a natural system? Explain. (10 mts.; 10 mks.).

5. Explain the principles on which Bentham and Hooker based system of classification. Why is it considered to be a natural system of classification? (10 mts.; 10 mks.).

6. Summarise the staminal variations observed in the different families of sub-class Gamopetalae according to Bentham and Hooker. (30 mts. ; 10 mks.).

7. What is Disciflorae? Give the distinguishing characters of any *three* families included in the group and mention two examples of economic to morphological interest from each of the family you select. (15 mts. ; 15 mks.).

8. Name any *three* similarities you find between the families Papilionaceae, Caesalpiniaceae and Mimosaceae to include them in one main family. (5 mts. ; 3 mks.)

9. Compare the sub-families under Leguminosae with reference to the floral characters. (20 mts. ; 15 mks.).

10. Give the salient features of family Cucurbitaceae and discuss its economic importance. (15 mts. ; 5+2 mks.).

11. Compare the floral structures of Apiaceae (Umbelliferae) with that of Asteraceae (Compositae). Which of the *two* families is highly evolved and why? Give suitable reasons. (15 mts. ; 15 mks.).

12. Show how the floral structures are integrated with special mechanism of pollination in Asclepiadaceae. (8 mts. ; 5 mks.).

13. Give a comparative account of Acanthaceae and Solanaceae with reference to their floral characters. (20 mts. ; 8 mks.).

14. Describe the range of floral structure in Orchidaceae. Give floral diagram and floral formula. Mention botanical name of any plant belonging to this family. (15 mts. ; 15 mks.).

15. (*a*) What are the different types of roots of Orchidaceae?
   (*b*) Relate their morphological structure in the function.
   (c) Give an example of each type. (15 mts. ; 8 mks.)

16. What is meant by classification of plants? State the difference between 'artificial'. 'natural' and 'phylogenetic' systems of classification of plants. Discuss the principles of any one phylogenetic system of classification which you have studied. (15 mts. ; 15 mks.)

Short Answer Questions

1. Give scientific name of the following plant and assign it to its respective family giving reasons : Castor. (5 minutes : 5 marks).

2. To the generic name Aster, if you add the endings-aceae and -ales, what will the two derived names signify? What will be the importance of the generic name itself? (3 mts. ; 3 mks.).

3. In plant classification features of the reproductive structures are considered more reliable than those of the vegetative structures. Why? (3 mts., 3 mks.).

4. Give the botanical names of any three plants that are now in flower on your college campus. (2 mts. ; 2 mks.).

5. Write critical notes on the following : Main concepts in Engler and Prantl's system of classification of angiosperms. (5 mts. ; 5 mks.).

6. What is the relationship between phylogeny and classification of flowering plants. (5 mts.; 5 mks.).

7. Write short note on : A.P. de Candolle. (5 mts.; 5 mks.),

8. Write short note on : Linnaeus. (5 mts. ; 5 mks.).

9. State any *two* demerits of Bentham and Hooker's classification. (5 mts. : 2 mks.).

10. State the basis of Linnaeus's classification of plants. (3 mts.; 3 mks.).

11. Mention any five differences between the classification of Bentham and Hooker and Engler and Prantl. (8 mts. ; 5 mks.).

12. What is chemotaxonomy? (3 mts. ; 2 mks.).

13. Write critical notes on the following : Rules of nomenclature. (5 mts. : 5 mks.).

14. Write critical notes on the following : Merits of Engler's system of classification. (6 mts. ; 6 mks.).

15. Why is it not advisable to separate two species on the basis of gross morphology alone? What other criteria should be taken into consideration? (5 mts. ; 5 mks.).

16. What do you understand by a type specimen? (3 mts. ; 3 mks.).

17. Give the important floral characters which distinguish the two families in any *one* of the pairs in the following :

(*a*) Myrtaceae and Apiaceae (Umbelliferae)

(*b*) Solanaceae and Acanthaceae. (6 mts. ; 6 mks.)

18. Mention the significant features you come across in any *one* of the following genera and assign it to its respective family : (*a*) *Eucalyptus,* (*b*) *Salvia.* (5 mts. ; 5 mks.).

19. Comment upon any *one* of the following :

(*a*) Corolla in Leguminosae.

(*b*) Flower of Cruciferae.

20. Assign any *one* of the following plants to the respective family:

(*a*) *Leucas aspera*

(*b*) *Carthamus tinctorius* (1mt. ; 1 mk.).

21. Given only the androecium and gynoecium, how would you distinguish any one pair of the following families from each other:

(*a*) Rosaceae from Malvaceae

(*b*) Apocynaceae from Asclepiadaceae. (5 mts. ; 5 mks.)

22. Give *four* important floral characters of Malvaceae. Mention the name of *two* economically important plants belonging to the family. (4 mts. ; 4 mks.).

23. Refer the following genus to its respective family giving reasons : *Cocos.* (5 mts.; 5 mks.).

24. How will you differentiate the following from each other : Cruciferae (Brassicaceae) and Capparidaceae. (5 mts. ; 5 mks.).

25. Give floral diagrams and floral formulae with examples to distinguish between following pair of families.

Papaveraceae and Capparidaceae. (5 mts. ; 5 mks.).

26. Mention one plant each under Cruciferae (Brassicaceae) where :

(*a*) There is no underground storage organ.

(*b*) There are less number of stamens than the usual number. (2 mts. ; 2mks.).

27. Describe the structure of fruit in Umbelliferae (Apiaceae). (3 mts. ; 2 mks.).

28. Comment upon the following :
Thalamus of Rosaceae. (5 mts. ; 5 mks.).

29. Give the distinguishing features of any *one* of the following families. Give botanical names of any *three* economically important plants from each family.
(*a*) Umbelliferae (Apiaceae)
(*b*) Rutaceae. (5 mts. ; 5 mks.).

30. Explain how Compositae (Asteraceae) plants are well adapted for rapid spread. (5 mts.; 3 mks.).

31. Mention any *three* distinguishing features of Rubiaceae. (5 mts. ; 3 mks.).

32. When only the fruits of Compositae and Rubiaceae are given, how will you differentiate between the two? (3 mts. ; 2 mks.).

33. List out *five* economically important plants of Solanaceae. (3 mts. ; 2 1/2 mks.).

34. Construct the floral diagram and give the formula of a typical member of Asclepiadaceae. (5 mts. ; 5 mks.)

35. Give floral diagrams and floral formulae with examples to distinguish between the following pair of families :
Asclepiadaceae and Apocynaceae. (5 mts. ; 5 mks.).

36. Give the distinguishing features of the following family. Give botanical names of any *three* economically important plants from : Solanaceae. (5 mts. ; 5 mks.)

37. How will you differentiate the following from each other :
Solanaceae and Convolvulaceae. (5 mts. ; 5 mks.).

38. Taking into consideration only the vegetative characters how would you differentiate Acanthaceae from Labiatae (Lamiaceae). (5 mts. ; 5 mks.).

39. Write notes on the following :
Inflorescence in Euphorbiaceae. (5 mts. ; 5 mks.)

40. Give a general account and economic importance of the following family
Euphorbiaceae. (6 mts. ; 6 mks.)

41. Mention any *three* types of perennating organs in Liliaceae family. Give an example for each type. (8 mts. ; 6 mks.).

42. Comment upon the following :
Roots of Orchids. (5 mts. ; 5 mks.).

43. Write critical notes on the following :
Flowers of *Musa*. (6 mts. ; 6 mks.)

44. Comment upon the following :
Economic importance of Palmae (Palmaceae). (5 mts. ; 5 mks.).

45. List *three* diagnostic features of family Gramineae (Poaceae). (2 mts. ; 4 mks.).

46. Write notes on the following : spikelet in Gramineae (Poaceae). (5 mts. ; 5 mks.).

47. Give a general account and economic importance of the following family : Gramineae (Poaceae). (6 mts. : 6 mks.).

Simple Questions

1. Which is the most famous botanical garden in the world? (1 minute ; 1 mark.).

2. What is the technical term for the following description : Stamens are in two whorls wherein the outer is opposite to the petals? (5 mts. ; 5 mks.).

3. Who is known as father of Botany? What is his contribution to Botany ? (2 mts. ; 1 mk.).

4. Name the systematic classification mainly based on evolutionary trends. (2 mts. ; 1 mk.).

5. Assign any one of the following plants to the respective family:
(*a*) *Tectona grandis* ; (*b*) *Momordica charantia.* (1 mt. ; 1 mk.).

6. Describe the nature of fruits in the family Papilionaceae. (5 mts. ; 3 mks.).

7. Give the binomials of *two* economically important plants each from :

   (*a*) Papilionaceae ; (*b*) Caesalpiniaceae. (5 mts. ; 3 mks.)

8. Name any one of the plants in Papilionaceae having monadelphous androecium. (1 mt. ; 1 mk.).

9. Name an oil-yielding plant in Myrtaceae. (1 mt. ; 1 mk.).

10. Describe the nature of fruits and seeds of family Cucurbitaceae. (5 mts. ; 2+1 mks.).

11. Give the binomials of two economically important plants each from : (*a*) Umbelliferae (Apiaceae) ; (*b*) Cucurbitaceae, and say how it is useful in everyday life. (2 mts. ; 2 mks.).

12. What is the special modification seen in *Mussaenda* flowers? (2 mts. ; 1 mk.).

13. Name a Compositae (Asteraceae) plant which is used as an insecticide? (1 mt. ; 1 mk.).

14. Describe the nature of fruits and seed of family Umbelliferae (Apiaceae). (5 mts., 2 mks.).

15. Name *four* herbs (give botanical names) of your locality belonging to the family Rubiaceae and mention approximate flowering season for them. (3 mts ; 3 mks.).

16. Describe gynosteigium with a suitable example. (3 mts. ; 3 mks.).

17. Name the plant and the family which yields Quinine. (3 mts. ; 2 mks.).

18. How are the stamens modified in Asclepiadaceae. (2 mts. ; 2 mks.).

19. Draw a sketch of a translator and label its parts. (3 mts. ; 2 mks.).

20. Name the economically important plants of Asclepiadaceae. (3 mts. 3 mks.).

21. Distinguish the families of the following pair : Acanthaceae and Labiatae (Lamiaceae). (5 mts. ; 5 mks.).

22. What is the distinguishing feature of Labiatae (Lamiaceae) gynoecium? (2 mts. ; 1 mk.).

23. Relate the structure of stamens and type of pollination mechanism seen in Labiatae (Lamiaceae). (3 mts. ; 2 mks.).

24. Describe the type of fruit in the Acanthaceae and dispersal mechanism of seeds. (5 mts.; 3 mks.).

25. Name the type of pollination mechanism seen in Labiatae (Lamiaceae). (1 mt. ; 1 mk.).

26. Name an orchid that yields aromatic essence. (1 mt. ; 1 [illegible]

27. What is the botanical name of Lady's slipper—an orchi[illegible] ; 1 mk.).

28. Mention *two* climbers from Liliaceae family. (1 mt. ; 2 [illegible]

29. Write botanical names of Ragi, rye, paddy and wheat. ([illegible] mks.).

Multiple Choice Questions

(1 minute ; 1 mark, each question)

1. Which characters among the Angiosperms of the following are most important taxonomically to differentiate it from the Gymnosperms:

   (*a*) Triploid endosperm ☐
   (*b*) Seeds enclosed in fruit ☐
   (*c*) Attractive petals ☐
   (*d*) Presence of vessels in xylem ☐

2. *Genera Plantarum* is written by :

   (*a*) de Candolle ☐
   (*b*) A. L. de Jussieu ☐
   (*c*) Engler and Prantl ☐
   (*d*) Bessey ☐

3. Which of the following is a phylogenetic system of classification :

   (*a*) Lindlay ☐

(*b*) Bentham and Hooker ☐
(*c*) Hutchinson ☐
(*d*) Linnaeus ☐

4. The classification formed by the Swedish Botanist Linnaeus is mainly based on :
(*a*) Morphology ☐
(*b*) Anatomy ☐
(*c*) Embryology ☐
(*d*) Serum diagnosis ☐

5. Botanical name of carrot weed is :
(*a*) *Argemone mexicana* ☐
(*b*) *Daucus carota* ☐
(*c*) *Parthenium* ☐
(*d*) None of the above ☐

6. 'Families of the Flowering Plants' book is written by:
(*a*) Bentham and Hooker ☐
(*b*) John Hutchinson ☐
(*c*) Engler and Prantl ☐
(*d*) None of the above ☐

7. Botanical name of Larkspur is:
(*a*) *Ranunculus scleratus* ☐
(*b*) *Abelmoschus esculentus* ☐
(*c*) *Delphinium ajacis* ☐
(*d*) None of the above ☐

8. Out of the four widely known systems of classification one remains less phylogenetic and more natural. That is of:
(*a*) Engler and Prantl ☐
(*b*) Bentham and Hooker ☐
(*c*) Bessey ☐
(*d*) Hutchinson ☐

9. The condition of androecium in Cucurbitaceae is :
(*a*) Syngenesious ☐
(*b*) Synandrous ☐
(*c*) Diadelphous ☐
(*d*) Free stamens ☐

10. Obliquely zygomorphic nature of the flower in Solanaceae is due to one of the following whorls:
(*a*) calyx ☐
(*b*) corolla ☐
(*c*) gynoecium ☐
(*d*) androecium ☐

11. Find out the false statement with regards to Compositae (Asteraceae):
(*a*) Head inflorescence ☐
(*b*) Hypogynous flowers ☐
(*c*) Syngenesious anthers ☐
(*d*) Fruit a cypsela ☐

12. The largest family among the dicotyledons given below is:

(*a*) Rubiaceae
(*b*) Rosaceae
(*c*) Compositae (Asteraceae)
(*d*) Cucurbitaceae

13. Epigynous, gamopetalous family is :
(*a*) Umbelliferae (Apiaceae)
(*b*) Rosaceae
(*c*) Compositae (Asteraceae)
(*d*) Myrtaceae

14. Tetramerous flowers are seen in the gamopetalous family of :
(*a*) Cruciferae
(*b*) Mimosaceae
(*c*) Rubiaceae
(*d*) Rutaceae

15. Obdiplostamenous condition of androecium is common in the family of :
(*a*) Anacardiaceae
(*b*) Rubiaceae
(*c*) Rosaceae
(*d*) Rutaceae

16. *Sida*, a member of the family Malvaceae does not show this prominent trait :
(*a*) axile placentation
(*b*) monadelphous stamens
(*c*) alternate stipulate leaves
(*d*) presence of epicalyx

17. Usual fruit in the family of Cucurbitaceae is :
(*a*) Pome
(*b*) Pepo
(*c*) Samara
(*d*) Hesperidium

18. Synandrous and sinuous anthers are common in the family of:
(*a*) Umbelliferae (Apiaceae)
(*b*) Rosaceae
(*c*) Malvaceae
(*d*) Cucurbitaceae

19. Spot out the statement applicable to the family Myrtaceae:
(*a*) zygomorphic flowers
(*b*) hypogynous flowers
(*c*) stamens limited
(*d*) inferior ovary

20. *Rubus* a member of Rosaceae produces:
(*a*) an etaerio of achenes
(*b*) an etaerio of drupes
(*c*) an etaerio of follicles
(*d*) an etaerio of berries

21. Polypetalous, irregular corolla is seen in :
(*a*) Rosaceae

(*b*) Malvaceae ☐
(*c*) Papilionaceae ☐
(*d*) Cruciferae (Brassicaceae) ☐

22. Out of the four statements one doesn't fit into the characters of the family of Anacardiaceae:
(*a*) Resin passages in the parenchyma ☐
(*b*) Well developed intrastaminal disc ☐
(*c*) Presence of lateral style ☐
(*d*) Hesperidium type of fruit ☐

23. Gynobasic style arises from the base of the ovary. This is characteristic of the family:
(*a*) Labiatae (lamiaceae) ☐
(*b*) Solanaceae ☐
(*c*) Compositae (Asteraceae) ☐
(*d*) Ranunculaceae ☐

24. In *Leucas aspera* the type of inflorescence is:
(*a*) Panicle ☐
(*b*) Verticillaster ☐
(*c*) Thyrsus ☐
(*d*) Raceme ☐

25. Gymnosperm has been placed in between Dicots and Monocots in the classification of plants by:
(*a*) Charles E. Bessey ☐
(*b*) Bentham and Hooker ☐
(*c*) Alfred Barton Rendle ☐
(*d*) Hutchinson ☐

26. Cleistogamous flower is seen in:
(*a*) Annonaceae ☐
(*b*) Tiliaceae ☐
(*c*) Rutaceae ☐
(*d*) Cucurbitaceae ☐
(*e*) None of the above ☐

27. Apocynaceae differs from Asclepiadaceae in having :
(*a*) latex ☐
(*b*) translators ☐
(*c*) corona ☐
(*d*) bicarpellary ovary ☐

28. Bicarpellary ovary with parietal placentation is found in :
(*a*) Cruciferae (Brassicaceae) ☐
(*b*) Convolvulaceae ☐
(*c*) Labiatae (Lamiaceae) ☐
(*d*) Solanaceae ☐

29. With reference to the habit the members of Nymphyaceae are :
(*a*) Terrestrial plants ☐
(*b*) Aquatic plants ☐
(*c*) Epiphytic plants ☐
(*d*) Xerophytic plants ☐

30. Numerous stamens are generally present in:
    (*a*) Nymphyaceae ☐
    (*b*) Papilionaceae ☐
    (*c*) Rutaceae ☐
    (*d*) Cucurbitaceae ☐
31. In *Brassica,* the placentation is :
    (*a*) axile ☐
    (*b*) basal ☐
    (*c*) parietal ☐
32. Tetradynamous condition of androecium is one of the striking features of:
    (*a*) Cruciferae (Brassicaceae) ☐
    (*b*) Caesalpiniaceae ☐
    (*c*) Papilionaceae ☐
    (*d*) Rubiaceae ☐
33. The condition of stamens in the cotton flower is:
    (*a*) free ☐
    (*b*) monadelphous ☐
    (*c*) diadelphous ☐
    (*d*) polyadelphous ☐
34. In *Mangifera indica* the fruit is a :
    (*a*) Berry ☐
    (*b*) Drupe ☐
    (*c*) Capsule ☐
    (*d*) Pepo ☐
    (*e*) Hesperidium ☐
35. Tetradynamous androecium characteristic of Cruciferae (Brassicaceae)consists of:
    (*a*) Outer 2 long and inner 2 short stamens ☐
    (*b*) Outer 2 short and inner 4 long stamens ☐
    (*c*) Outer 4 long and inner 2 short stamens ☐
    (*d*) Outer 4 short and inner 2 long stamens ☐
36. Monocarpellary condition is one of the striking features of:
    (*a*) Cucurbitaceae ☐
    (*b*) Rubiaceae ☐
    (*c*) Annonaceae ☐
    (*d*) Papilionaceae ☐
37. Pulses are group of plants belonging to the family :
    (*a*) Leguminosae (Papilionaceae) ☐
    (*b*) Cruciferae (Brassicaceae) ☐
    (*c*) Compositae (Asteraceae) ☐
    (*d*) Solanaceae ☐
38. In Papilionaceae the placentation
    (*a*) Parietal ☐
    (*b*) Axile ☐
    (*c*) Marginal ☐
    (*d*) Free central ☐
39. In *Daucus carota* the edible part is a :
    (*a*) corm ☐

(*b*) rhizome ☐
(*c*) tap root ☐
(*d*) lateral root ☐
(*e*) stem tuber ☐

40. In Mustard flowers are:
(*a*) perigynous ☐
(*b*) pentamerous ☐
(*c*) ebracteate ☐
(*d*) slightly zygomorphic ☐

41. In *Launea* the flowers are arranged in the form of head or capitulum. These florets are:
(*a*) Bisexual, tubular and homogamous ☐
(*b*) Bisexual, ligulate and homogamous ☐
(*c*) Outer ligulate and pistillate and inner tubular and bisexual ☐
(*d*) Unisexual on the same head ☐

42. Seeds of sunflower plant contain :
(*a*) Alkaloids ☐
(*b*) Oils ☐
(*c*) Dyes ☐
(*d*) Crystals ☐

43. Which of the following represents the condition seen in the family Compositae (Asteraceae)?
(*a*) superior ovary, axile placentation ☐
(*b*) inferior ovary, axile placentation ☐
(*c*) superior ovary, single basal ovule ☐
(*d*) inferior ovary, single basal ovule ☐

44. In Asclepiadaceae the stamens are modified and known as :
(*a*) Syngenesious ☐
(*b*) Sinuous ☐
(*c*) Translators ☐
(*d*) Androphore ☐

45. The end of the translator that sticks to the body of the insect is:
(*a*) caudicle ☐
(*b*) pollinium ☐
(*c*) retinaculum ☐
(*d*) corpusculum ☐

46. The botanical name of Chilly is:
(*a*) *Solanum tuberosum* ☐
(*b*) *Lycopersicon esculentum* ☐
(*c*) *Capsicum frutescens* ☐
(*d*) *Nicandra physaloides* ☐

47. In *Coleus* the stamens are:
(*a*) Monadelphous ☐
(*b*) Diadelphous ☐
(*c*) Epipetalous ☐
(*d*) Polyadelphous ☐

48. The botanical name of sage plant is:
(*a*) *Mentha piperita* ☐

(*b*) *Salvia splendens* ☐
(*c*) *Lavandula vera* ☐
(*d*) *Thymus vulgaris* ☐

49. In Labiatae (Lamiaceae) the fruits are usually:
(*a*) Follicle ☐
(*b*) Capsule ☐
(*c*) Nutlets ☐
(*d*) Achene ☐

50. *Lantana* is an example of a :
(*a*) huge tree ☐
(*b*) prickly shrub ☐
(*c*) twiner ☐
(*d*) soft herb. ☐

51. *Avicennia* is a :
(*a*) Mesophyte ☐
(*b*) Epiphyte ☐
(*c*) Hydrophyte ☐
(*d*) Mangrove ☐

52. Explosive fruits are seen in :
(*a*) *Ruellia tuberosa* ☐
(*b*) *Solanum nigrum* ☐
(*c*) *Nicotiana tabacum* ☐
(*d*) *Asclepias curassavica* ☐

53. A plant that belongs to Liliaceae in which stipules are modified into tendrils is:
(*a*) *Gloriosa* ☐
(*b*) *Yucca* ☐
(*c*) *Lilium* ☐
(*d*) *Smilax* ☐

54. The botanical name of the 'bird's nest orchid' is :
(*a*) *Dendrobium* ☐
(*b*) *Neottia* ☐
(*c*) *Cypripedium* ☐
(*d*) *Orchis* ☐

55. Sugarcane inflorescence is a :
(*a*) Spike ☐
(*b*) Catkin ☐
(*c*) panicle ☐
(*d*) Verticillaster ☐

56. The inflorescence of paddy is:
(*a*) racemose ☐
(*b*) catkin ☐
(*c*) panicle ☐
(*d*) verticillaster ☐

U.P.S.C. Civil Services Examination (C.S.E.)and Indian Forest Service (I.F.S.)Examination, Question Papers.

Long Answer Questions

1. Describe any *two* systems of classification of Angiosperms. Which of them is considered better by you? Give reasons in support of your answer. (I.A.S. 1968)

2. Discuss the value of cytological characters in improving the existing systems of classification. (I.A.S. 1969)

3. Write an essay on : classification of flowering plants. (I.A.S. 1972)

4. Outline briefly the main characters of Amentiferae and Scitaminae and add a note on their distribution in India. (I.A.S. 1973)

5. What are the principal functions of different parts of an angiospermic plant? Mention *two* angiosperms which do not develop (*a*) stems; (*b*) roots; (*c*) green leaves. (I.A.S. 1973)

6. Discuss the evolutionary trends in Tubiflorae or Centrospermae. (I.A.S. 1976)

7. Compare the floral characters exhibited by Compositae and Umbelliferae and explain their evolutionary trends.

8. Discuss 'modern trends in taxonomy' giving suitable examples. (I.A.S. 1980)

9. What are the main features of the system of classification proposed by Bentham and Hooker? Discuss its merits and demerits. (I.A.S. 1988)

10. Discuss the differences between the systems of classification of Engler and Hutchinston. Which of the systems do you consider more phylogenetic? Give reasons. (C.S.E. (main)

Short Answer Questions

1. Give briefly the main characters and affinities of the following and indicate their distribution in India: (*a*) Rutaceae; (*b*) Podostemonaceae; (*c*) Umbelliferae. (C.S.E 1974)

2. Exclusively on the basis of the gynoecium and androecium how would you distinguish the following families? (*i*) Rosaceae from Malvaceae; (*ii*) Apocynaceae from Asclepiadaceae; (*iii*) Scrophulariaceae from Acanthaceae; (*iv*) Cruciferae from Capparidaceae. (C.S.E. I.A.S. 1974)

3. Enumerate the salient features of any two families listed below and discuss their systematic position : (*i*) Nymphaeaceae; (*ii*) Scrophulariaceae; (*iii*) Orchidaceae. (C.S.E., I.A.S. 1975.)

4. Write a critical commentary on : Merits and drawbacks of Hutchinson's system of classification. (C.S.E., I.A.S. 1976)

5. Write what is unique in the following : (*a*) Pollination in *Vallisneria;* (*b*) Flower of an orchid. (C.S.E. 1976)

6. Answer the following, using not more than two hundred words for each part :

(*a*) What is the concept of species in plant taxonomy.

(*b*) What is the scientific basis of botanical nomenclature? What are the most important rules in this system?

(*c*) What are the comparative adaptive advantages of woody and herbaceous conditions? (C.S.E., I.A.S. 1982)

7. Discuss the special features of the following :

(*a*) Reproductive structures in Asclepiadaceae.

(*b*) Flower of Magnoliaceae.

(*c*) Pollination in Hydrocharitaceae.

(*d*) Floral morphology of Orchidaceae. (C.S.E., I.A.S. 1982)

8. What are the unique features of the following?
   (*a*) Inflorescence of Gramineae.
   (*b*) Gynoecium of Rosaceae
   (*c*) Androecium of Cucurbitaceae
   (*d*) Fruit of Umbelliferae
   (*e*) Flowers of Orchidaceae (C.S.E., I.A.S. 1983)
9. Answer the following using not more than two hundred words for each part :
   (*a*) Mention the most essential features of the International Code of Botanical Nomenclature and their significance.
   (*b*) Discuss the role of Herbaria in modern taxonomical research.
   (*c*) Highlight the importance of chemotaxonomy in the aid of systematics of Angiosperms. (C.S.E., I.A.S. 1983)
10. Answer in not more than 200 words: What is biosystematics? Explain the role of this approach in modern plant taxonomy. (C.S.E., I.A.S. 1984)
11. Describe the uniqueness of the following :
   (*a*) Androecium in Asclepiadaceae.
   (*b*) Corolla in Scrophulariaceae.
   (*c*) Pollination in Labiatae.
   (*d*) Gynoecium in Cyperaceae
   (*e*) Fruit and seed in Umbelliferae. (C.S.E., I.A.S. 1985)
12. Answer the following, using not more than 200 words for each part :
   (*a*) Discuss the merits and demerits of the system of clasification proposed by Hutchinson.
   (*b*) Give the most important features of the family Palmae and discuss its phylogenetic relationships
   (*c*) Bring out the importance of palynology in the phylogenetic systems of classification. (C.S.E., I.A.S. 1985)
13. Answer the following, using not more than 200 words for each part.
   (*a*) What are the salient features of the system of classification proposed by Takhtajan? How does it compare with that of Engler.
   (*b*) Explain the unique features of the family Liliaceae. Discuss its phylogeny.
   (*c*) Discuss the role of phytochemistry in the classification and the determination of interrelationships. (C.S.E., I.A.S. 1986)
14. Describe the uniqueness of the following : (*a*) Flower in Asteraceae; (*b*) Inflorescence in Euphorbiaceae; (*c*) Gynoecium in Rosaceae; (*d*) Androecium in Cucurbitaceae; (*d*) Spikelet in Gramineae. (C.S.E., I.A.S. 1986)
15. Write the salient features of the following:
   (*a*) Inflorescence of Euphorbiaceae.
   (*b*) Androecium in the Asclepiadaceae.
   (*c*) The primitive carpel.
   (*d*) Hypanthodium. (C.S.E., I.A.S. 1986)
16. Describe the characteristics of the following:
   (*a*) Flower of Magnoliaceae;
   (*b*) Gynoecium of Cruciferae;
   (*c*) Androccium of Cucurbitaceae;

(*d*) Fruit of Umbelliferae;

(*e*) Inflorescence of Gramineae. (C.S.E., I.A.S. 1988)

17. Place the following genera to their respective families and indicate their economic importance:

(*i*) *Helianthus;*

(*ii*) *Digitalis*

(*iii*) *Atropa*

(*iv*) *Vanilla*

18. What is numerical taxonomy? Briefly state the methods employed in and importance of their taxonomy. (C.S.E., I.A.S. 1990)

19. Answer any *two* of the following:

(*i*) Outline the salient features of the classification by Bentham and Hooker. What are the merits and demerits of this system?

(*ii*) Briefly comment on the origin of angiosperms.

(*iii*) Importance of the study of phytochemistry in taxonomy. (C.S.E., I.A.S. 1990)

20. State the distingusihging characters of Rosaceae and the affinities of the taxon with Ranunculaceae and Leguminosae. Mention *five* Indian fruit plants of the family.

(C.S.E., I.A.S. 1990)

21. (*a*) Place the following genera to their respective families and indicate their economic importance:

(*i*) *Elaeis;*

(*ii*) *Manihot;*

(*ii*) *Arachis;*

(*iii*) *Nigella.*

(*b*) Distinguish between :

(*i*) Cereals and millets;

(*ii*) *Utricularia* and *Aldrovanda;*

(*iii*) Sorosis and syconus. (C.S.E., I.A.S. 1990)

22. Answer the following in not more than 200 words:

Which characteristics of flowers are considered to be evolutionarily advanced and which are considered to be primitive. (C.S.E., I.A.S. 1991)

23. Explain very briefly the economic importance of

(*i*) Euphorbiaceae;

(*ii*) Compositae;

(*iii*) Coconut palm. (C.S.E., I.A.S. 1991)

24. Answer the following, in not more than 200 words each : (*a*) How does serology help in taxonomy of plants; (*b*) Explain the significance of palynological characters in systematics; (*c*) Outline the scope, usefulness and limitations of Chemotaxonomy and Numerical Taxonomy.

(C.S.E., I.A.S. 1992)

25. Why International rules of botanical nomenclature are important and when were they propounded? What are the main principles governing these rules. (C.S.E., I.A.S. 1993)

26. Answer in not-more than 200 words : Discuss the role of biosystematics in the understanding of the taxonomy of angiosperms. (C.S.E., I.A.S. 1993)

27 Why the Compositae are regarded as the most highly evolved taxon?

(C.S.E., I.A.S. 1993)

28. Answer the following in not more than 200 words each : (*a*) "Magnoliales possess evolutionarily primitive flower"— substantiate; (*b*) Indicate the role of anatomy in angiosperm systematics. (C.S.E., I.A.S. 1994)

29. Give the floral formula and draw the floral diagram of a diaudrous orchid. (C.S.E., I.A.S. 1994)

30. Compare the following:

(*a*) Systems of Classification of plants by Bentham and Hooker, and Engler and Prantl. (*b*) Inflorescence of *Euphorbia* and *Helianthus.* (C.S.E., I.A.S. 1995)

31. Answer in not more than 200 words : Describe the floral structure of *Musa* along with floral diagram and floral formula. (C.S.E., I.A.S. 1995)

32. Enumerae with brief notes the cytogenetic criteria in biosystematics. (C.S.E., I.A.S. 1995)

33. Describe the characteristic features of the family Cucurbitaceae. Give the floral formula and floral diagram. Write botanical names of *five* plants of the family having economic importance. (C.S.E., I.A.S. 1995)

34. Write short notes on : (*i*) Orchid flower: (*ii*) Flower of a grass; (*iii*) Flower of Umbelliferae. (C.S.E., I.A.S. 1995)

Indian Forest Service (I.F.S.) Examination Question Papers

Q.1. (*a*) Which family is the most primitive one among the dicots? Give reasons in support of your answer.

(*b*) Which family is the most highly evolved among the monocots? Give reasons in support of your answer. (I.F.S. 1988)

2.(*a*) (*i*) Justify the inclusion of the families Papilionaceae, Caesalpiniaceae, and Mimosaceae in the order Leguminales.

(*ii*) Why Cucurbitaceae is placed in Polypetalae though they are having gamopetalous condition?

(*iii*) Distinguish between : Apocynaceae and Asclepiadaceae : Cucurbitaceae and Rutaceae.

(*b*) (*i*) State any *two* merits of Bentham and Hooker's classification.

(*ii*) Draw illustrated diagrams to show the following : Inflorescence of Labiatae; Androecium of Orchidaceae: Gynoecium of *Triticum*: Corolla of Papilionaceae.(I.F.S. 1988)

3. Enumerate important differences between the symptoms of classification proposed by Bentham and Hooker, Engler and Prantl and Hutchinson. Which of these systems do you consider to be the most satisfactory? Give reasons in support of your answer. (I.F.S. 1988)

4. (*a*) Adduce evidences to prove that Asteraceae is an advanced family.

(*b*) Compare the flower and inflorescence of Umbelliferae with that of Gramineae.

(*c*) Outline the scope, usefulness and limitations of Chemotaxonomy and Numerical taxonomy. (I.F.S. 1989)

5. (*a*) Give the distinguishing features of the following families : (*i*) Loranthaceae; (*ii*) Magnoliaceae; (*iii*) Orchidaceae.

(*b*) What do you understand by Principles of classification? Even though Hutchinson's system of classification is natural, it has not been accepted by most of the taxonomists, comment. (I.F.S. 1989)

6. (*a*) Describe the floral and diagnostic characters of Euphorbiaceae. Mention *three* plants of economic importance. Mention *three* plants of economic importance and name the products obtained from them.

(*b*) Describe the merits of Takhtajan's system of classification. (I.F.S. 1989)

7. Write briefly : (*a*) International code of Botanical Nomenclature; (*b*) Hutchinson's contributions to Taxonomy; (*c*) Palynological criteria in systematics; (*d*) α-Taxonomy; (*e*) Flower of orchid. (I.F.S. 1990)

8. Describe the floral and diagnostic features of Asteraceae and Palmaceae. Mention the

botanical names of *two* plants of economic importance and name the products obtained from them. (I.F.S. 1990)

9. Explain the special features of the following: (*a*) Androecium in Leguminosae; (*b*) Gynoecium in Papaveraceae; (*c*) Mangrove vegetation. (I.F.S. 1990)

10. Write critical notes on the following:

(*a*) Merits and demerits of Bentham and Hooker's system of classification; (*b*) Importance of cytology in taxonomy; (*c*) Spikelet of Poaceae; (*d*) Phylogenetic system of classification; (*e*) Numerical taxonomy. (I.F.S. 1991)

11. Describe the distinguishing features of Apocynaceae and Euphorbiaceae. Mention at least *three* important genera of each family and give their economic significance. (I.F.S. 1991)

12. Give a brief account of the following:

(*a*) Flower in Asteraceae; (*b*) Pollination mechanism in Orchidaceae; (*c*) Floral diagram and floral formula of Cruciferae and Malvaceae. (I.F.S. 1991)

13. Give brief account of the following:

(*a*) Evolutionary trends in Angiosperm flower:

(*b*) International botanical nomenclature;

(*c*) Natural system of classification of plants;

(*d*) Discuss how chemical parameters help in the classification of plants. (I.F.S. 1992)

14. (*a*) Describe the distinguishing characters of Leguminosae. Discuss the criteria based on which this family has been divided into *three* sub-families.

(*b*) Describe the distinguishing features of Labiatae. Mention economic importance of the family. Add a note on pollination mechanism. (I.F.S. 1992)

15. Give a brief description of the following:

(*a*) Inflorescence, flower and floral diagram of Euphorbiaceae. Mention the economic importance of the family.

(*b*) Give the distinguishing features and affinities of Magnoliaceae.

(*c*) State the distinguishing features, affinities and economic importance of Scitamineae.

(*d*) Distinguish between Rutaceae and Meliaceae. (I.F.S. 1992)

16. (*a*) Describe the range of floral structure in Rosaceae.

(*b*) Make floral diagrams and write floral formulae of representative species (one each) of Labatae and Rutaceae. (I.F.S. 1993)

17. (*a*) Compare the floral characters of Rubiaceae with those of Myrtaceae.

(*b*) Describe the various rules of botanical nomenclature. (I.F.S. 1993)

18. Write short notes on the following:

Illustrate your answer with diagrams.

(*a*) Androecium in Cucurbitaceae.

(*b*) Hypanthodium

(*c*) Translator mechanism

(*d*) Pollination in *Salvia*. (I.F.S. 1993)

19. Compare the floral characters of Rutaceae with that of Meliaceae. Draw floral diagrams and write floral formulae in each case. (I.F.S. 1994)

20. Write critical notes on any *four* of the following: (*i*) Inflorescence in *Euphorbia;* (*ii*) Botanical Survey of India; (*iii*) Spikelet of *Poaceae*; (*iv*) *Loranthus*; (*v*) Coconut and its importance; (*vi*) Principle of priority in nomenclature. (I.F.S. 1994)

21. Describe a system of classification of angiosperms used in practical classes. Give reasons for its use. (I.F.S. 1995)

22. (*a*) Describe the distinguishing characters of the families Gramineae and Cyperaceae.

(*b*) How does family Acanthaceae differ from Labiatae? Give two economic uses of each. (I.F.S. 1995)

23. Describe the following:

(*a*) Placentation

(*b*) Primitive flower

(*c*) Distinguishing features of sub-families of Leguminosae.

(*d*) Fruits in Rutaceae and Rosaceae. (I.F.S. 1995)

24. What is a natural system of classification? Describe a natural system of classification. How is it useful in field studies. (I.F.S. 1996)

25. (*a*) Describe the floral characters of the family Magnoliaceae.

(*b*) How does family Rutaceae differ from Meliaceae? (I.F.S. 1996)

26. Describe the following:

(*a*) Spike structure in Gramineae.

(*b*) Pollinia and Gynostegium.

(*c*) Botanical gardens.

(*d*) Fruits in Cucurbitaceae. (I.F.S. 1996)

INDEX

(*Orders, Families, Genera And General Topics Described In The Text*)